Phenolic Antioxidants in FoodsChemistry, Biochemistry and Analysis

Alam Zeb

Phenolic Antioxidants in Foods: Chemistry, Biochemistry and Analysis

 Springer

Alam Zeb
Department of Biochemistry
University of Makaland
Chakdara, Pakistan

ISBN 978-3-030-74770-1   ISBN 978-3-030-74768-8   (eBook)
https://doi.org/10.1007/978-3-030-74768-8

This Springer imprint is published by the registered company Springer Nature Switzerland AG
The registered company address is: Gewerbestrasse 11, 6330 Cham, Switzerland

To
My elder Brother (Alamgir Khan):
A friend, teacher, the most decent and
humble person.
Your love, sincerity and support can never
be repaid.

Preface

Plant foods consist of a large number of phenolic compounds, the majority of which are antioxidants. These phenolic compounds play a significant role in maintaining our health. This comprehensive book describes the chemistry, biochemistry and analysis of phenolic compounds present in a variety of foods. The book will serve as the latest source of knowledge in the field of food science and technology. Each chapter has specific learning outcomes, which are supported by study questions. The book covers all those important chemical, biochemical and analytical chemistry aspects of phenolic antioxidants in foods that are needed for a beginner to the expert in the field. This is a textbook based on the up-to-date scientific knowledge. This book was therefore aimed to cover them in three main sections. Part I covers the chemistry of phenolic antioxidants. This section consists of ten chapters covering the basic concept of antioxidants, their chemistry and chemical composition in foods. Chapters 3–10 discuss the chemical composition of phenolic compounds in different types of foods. This section is highly significant and is very helpful for the beginners in food science or interested readers and will equip them with the basic concept of phenolic antioxidants in foods.

Part II covers the biochemical aspects of phenolic antioxidants and consists of four chapters. This section covers the biosynthetic pathways of phenolic antioxidants in foods and biological effects of phenolic antioxidants in foods. The molecular mechanism of antioxidant effects in the biological system has been presented. The section has clarified our understanding of the fundamental biochemical reactions taking place in foods and after digestion and absorption.

Part III covers the analytical chemistry for the analysis of phenolic antioxidants in foods. This section consists of three chapters, which cover the basic analytical procedures and methods for analysis. Chapters 16 and 17 cover chromatographic and spectroscopic analyses of phenolic antioxidants in foods. This section is significant for food chemist and food manufacturers to evaluate the nature and chemistry of phenolic antioxidants in foods.

The book is organised in such a manner as to provide a smooth flow of scientific knowledge from chemistry, biochemistry to analysis, while each chapter holds its objectivity from each other. The book would also serve as the latest reference resource for food scientists, technologists, food chemists, biochemists, nutritionists and health professionals in academia, scientific labs and industries. The book provides fundamental and applied information to benefit those with different backgrounds in science. It will serve as a textbook for undergraduate, graduate and postgraduate students in the relevant disciplines.

Chakdara, Pakistan   Alam Zeb

Acknowledgements

This work would not have been possible without the help of some important people. The chemical structures were drawn based on the IUPAC names obtained from PubChem website of National Library of Medicine, US, which is highly acknowledged. I am also thankful to Huaxin Song, Zhejiang Key Laboratory for Agro-Food Processing, Zhejiang University, China for providing rough data for Figure 3.2 and 4.2, which are redrawn with permission from publisher. I am grateful to my family, my brothers (Alamgir Khan and Raham Zeb) and their loving children especially *Muhammad Umair Khan* and *Uzair Ahmad Khan* with whom I always enjoyed time in hometown. I am also grateful to my wife, sons (*Umar Alam Khan, Hamzah Alam Khan*) and daughter (*Haya Bibi*) for giving me tremendous support at home, while I was writing the manuscript.

Alam Zeb, PhD

Contents

About the Author

Alam Zeb is a Professor of Biochemistry at the Department of Biochemistry, University of Malakand, Pakistan. He is serving for 18 years at the same university teaching chemistry and biochemistry courses to undergraduate, graduate and postgraduate students of biotechnology and biochemistry. Dr. Zeb has received his PhD *with distinction* from the Institute of Biochemistry, Technical University of Graz, Austria, in 2010 funded by a scholarship from the Higher Education Commission (HEC) of Pakistan. During his studies at the Technical University of Graz, he also teaches practical food chemistry courses to postgraduate students. Dr. Zeb published more than 100 research articles regarding the subject in various international peer-reviewed journals. Dr. Zeb had authored two books, the recent one is a textbook *Food Frying: Chemistry, Biochemistry and Safety* published by John Wiley and Sons, UK, in 2019. This book has been ranked 4th among the best top 10 books of all times in the field of chemistry and safety by *Bookauthority.org*. He had supervised several PhD and MPhil research students during his service at the University of Malakand. He is a member of the editorial board of several international journals including *Frontiers in Chemistry* and *Current Functional Foods*. Dr. Zeb is a productive scientist of Pakistan since 2009 and had received research productivity awards from Pakistan Council for Science and Technology, Pakistan, for several consecutive years, representing his contribution in the field of food biochemistry.

Part I
Chemistry of Phenolic Antioxidants

Chapter 1
Concept of Antioxidants in Foods

Learning Outcomes

After reading this chapter, the reader will be able to:

- Know the basic concept of an antioxidant and its different types.
- Discuss the mechanism of hydrogen atom transfer in phenolic compounds.
- Explain the single electron transfer mechanism of antioxidant action.
- Understand the sequential proton loss electron transfer mechanism of phenolic compounds.
- Discuss the different methods of estimation of phenolic contents.
- Name phenolic compounds present in foods.

1.1 Introduction

Food spoilage was one of the main problems the human facing since antiquity. In countries with relatively warm climates, food spoilage was a big issue. Historically peoples in the Subcontinent and most of the Asian and Middle Eastern countries, the extract of garlic, onion and some spices were used to extend the shelf life of foods. However, during the last century, it was found that there are some chemical compounds in foods, which were responsible for preventing or controlling food deterioration and subsequent spoilage. These substances were called antioxidants. Until 1950, very little was known in the scientific literature about the mechanism of antioxidants. The tremendous growth in scientific knowledge has been observed since 1990. Figure 1.1 showed the number of scientific publications (research articles and reviews only) indexed in Scopus from 1990 to 2020 on the keyword "phenolic compounds" (Zeb 2020). The figure showed that the amount of literature available has been doubled since the last decade. The growth of knowledge represented the

© Springer Nature Switzerland AG 2021

A. Zeb, *Phenolic Antioxidants in Foods: Chemistry, Biochemistry and Analysis*,

https://doi.org/10.1007/978-3-030-74768-8_1

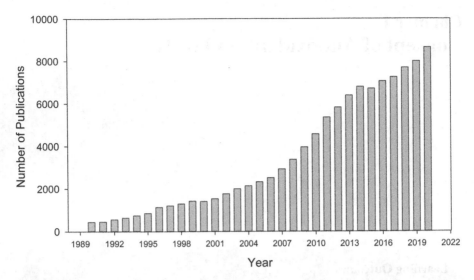

Fig. 1.1 Number of publications with word "phenolic compounds" in title in Scopus from 1990 to 2020. (Redrawn with kind permission of John Wiley and Sons, Zeb 2020)

significance of the subject. Similarly, the library of congress catalogue showed 534 books are available on the subject, with 81 books focused on antioxidants in foods. Fourteen books had been published focusing on phenolic antioxidants in plants. The notable among these books are by Shahidi (2015); Yamane and Kato (2012); Munné-Bosch (2011) and Fraga (2010). All these books have been focused either on a wide range of antioxidants or an only specific class of antioxidants in foods.

Several reviews had been published on the role of phenolic antioxidants in foods. For example, Rice-Evans et al. (1997) and Foti (2007) reviewed the antioxidant properties of polyphenols in *in-vitro* and homogenous solutions, respectively. Similarly, Soobrattee et al. (2005) reviewed the mechanism of action of phenolics. Perron and Brumaghim (2009) discussed the mechanism of scavenging of free radicals by polyphenols. The authors showed that scavenging was not associated with the iron-binding, their copper interactions, and the pro-oxidant role of complexes of iron-polyphenol. Dangles (2012) reviewed the antioxidant role of catechol derivatives of polyphenols. However, these reviews were not focused on the antioxidant role in foods. Heim et al. (2002) reviewed the chemistry, metabolism and structure-activity relationship of flavonoids. Choe and Min (2009) and Brewer (2011) reviewed mechanisms of action of a wide variety of natural antioxidants present foods during oxidation. Leopoldini et al. (2011) reviewed the mechanism of action of stilbenes, flavonoids and phenolic acids. The authors discussed the transfers H-atom and single electron and the metals chelation properties of the selected phenolic compounds (PCs). Recently Shahidi and Ambigaipalan (2015) reviewed antioxidant activity and health benefits of phenolic compounds in spices, beverages and foods. However, these reviews focused inadequately on the concept, mechanism and applications of phenolic antioxidants in foods only. The present review was therefore aimed to comprehensively review the updated knowledge on the concept, mechanism and applications of phenolic antioxidants present in foods.

1.2 Concept of Antioxidant

An antioxidant is a chemical compound that inhibits or stops or controls oxidation of a substrate. It may be an organic or inorganic compound. The substrate is known as an oxidizable substrate. However, the definition of antioxidant varied with its applications. In food science, the term antioxidant is implicitly limited to free lipid radical chain inhibitors. The wide distribution of free radicals, however, demanded the definition to be more precise, broader and widely accepted. Thus antioxidants may be defined as "*any substance present at low concentrations compared with those of an oxidizable substrate, significantly delays or prevent oxidation of that substrate*" (Halliwell 1995, 1990). The oxidizable substrate may be from *in-vitro* or *in-vivo* systems. However, *in-vitro* studies were the first to describe the concept and mechanism of antioxidants. Thus, based on the nature of the substrate and antioxidant itself, the antioxidant in food science has been redefined. *Those substances present in a relatively low amount in foods, which inhibit or delay or control oxidation of food components consequently prevented spoilage and extended shelf life of foods.* Inhibition of oxidation resulted in the delaying of oxidation. The delay in the oxidation process thus extends the shelf life of foods.

1.3 Types of Antioxidants

It is important to know that there are some atoms or groups of atoms that are responsible for the oxidation of the compound. They are called free radicals. A free radical is defined as an atom or group of atoms with one or more unpaired electrons that exist in free form. Free radical is generated by several processes including homolytic, heterolytic cleavage, or redox reactions. The main oxidants are singlet oxygen, ozone, hydrogen peroxide, nitrous acid, hypochlorite, nitric or nitrous oxides etc. These oxidant species are called reactive oxygen species. The free radical formed may be lipid free radical, superoxide anion radical, hydroxyl radical, hydroperoxy radical, peroxy radicals, nitric oxide and nitrosyl cation (Fig. 1.2).

Free radical chain reactions start with the formation of free radical called initiation reaction. The free radical is then propagated by attacking other molecules. The last reactions are termed as termination with the formation of neutral species or

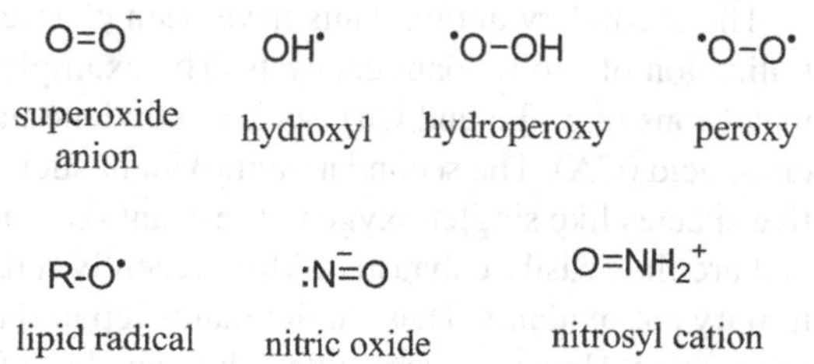

Fig. 1.2 Chemical structures of some important free radicals

$$O=O^-$$ $$OH^•$$ $$^•O-OH$$ $$^•O-O^•$$

superoxide anion   hydroxyl   hydroperoxy   peroxy

$$R-O^•$$ $$:N=O$$ $$O=NH_2^+$$

lipid radical   nitric oxide   nitrosyl cation

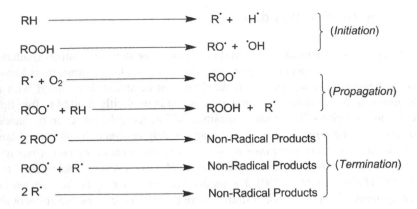

Fig. 1.3 Free radical chain reactions

non-radical products as shown in Fig. 1.3. The substrate (R) for free radicals may be a lipid, protein, carbohydrates, organic, inorganic compound or a metal atom. Free radical formation occurs as a normal process of cellular biochemistry. However, an increase in the free radicals level is causing an imbalance in biochemical pathways, causing oxidative stress. Oxidative stress is responsible for aging or deterioration of foods. The antioxidant is thus needed to overcome oxidative stress.

Based on the nature of these free radicals antioxidants are either organic or inorganic compounds. There are several types of antioxidants based on mechanism of action, nature of origin and natural chemistry.

1.3.1 Primary and Secondary Antioxidants

They are categorized as primary and secondary antioxidants. This classification is based on the mechanism of action of antioxidants. For example, primary antioxidants neutralize free radicals either by donating an H-atom (hydrogen atom transfer, abbreviated as HAT) or by a single electron transfer (SET) mechanism. These antioxidants are very efficient and normally required in very limited amount to neutralize a large number of free radicals (Fig. 1.4). The high catalytic properties of these antioxidants are one main reason for their diversity in nature. Phenolic antioxidants fall under this category. These antioxidants are easily regenerated.

The secondary antioxidants have been characterized by the mechanism of neutralization of pro-oxidant catalysts. The examples include chelators of pro-oxidant metal ions (e.g., Fe and Cu), such as ethylenediaminetetraacetic acid (EDTA) and citric acid (CA). The secondary antioxidant such as β-carotene may neutralize reactive species like singlet oxygen. These antioxidants usually quench one free radical and are thus easily exhausted. More recently a third class has been included called tertiary antioxidants. These antioxidants repair damaged biomolecules such as DNA or proteins. However, very little is known about their role in foods.

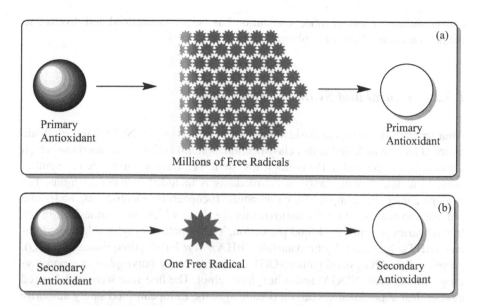

Fig. 1.4 Representative reactions of primary and secondary antioxidants in foods. (**a**) Reaction of primary antioxidant with large number of free radicals, and (**b**) reaction of secondary antioxidant with one free radical. (Reproduced with kind permission of John Wiley and Sons, Zeb 2020)

Fig. 1.5 Chemical structures of synthetic antioxidants used in foods. (**1.1**) Butylated hydroxyanisole (BHA), (**1.2**) butylated hydroxytoluene (BHT), (**1.3**) tertbutylhydroquinone (TBHQ), and (**1.4**) propyl gallate (PG)

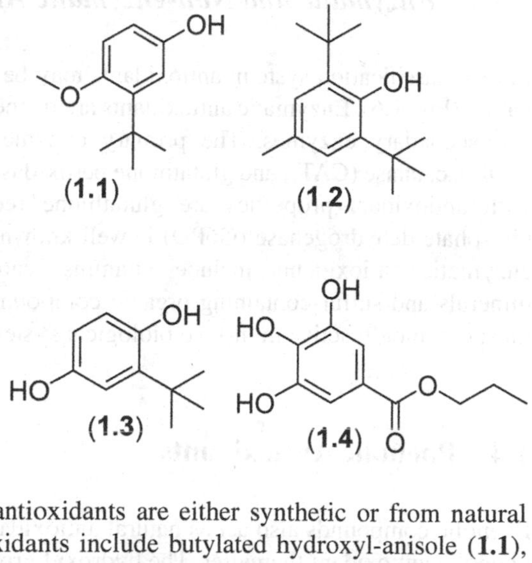

Both primary and secondary antioxidants are either synthetic or from natural sources. Synthetic primary antioxidants include butylated hydroxyl-anisole (**1.1**), butylated hydroxyl-toluene (**1.2**), tertbutyl-hydroquinone (**1.3**), and propyl gallate (**1.4**) (Fig. 1.5). They have been used to control oxidation and off-flavor development by quench free radicals in foods. These antioxidants contain one or more hydroxyl groups or phenol and are very effective (Shahidi 2015). However, owing to the harmfulness and carcinogenic properties of synthetic antioxidants at high concentrations in animal models, attention has been concentrated and diverted to the natural antioxidants from plants and foods.

1.3.2   Natural and Synthetic Antioxidants

Another classification of antioxidants is based on their origin. Natural and synthetic antioxidants are included in this classification. Natural antioxidants are those which are obtained or present in the biological system, i.e., they are originated or synthesized in nature. A wide range of antioxidants is included in natural origins. For example, phenolic compounds, carotenoids, tocopherols, tocotrienols, and some organic substances. Synthetic antioxidants are those which are synthesized in the lab/industries for enhanced food protection. These include butylated hydroxytoluene (BHT), butylated hydroxyanisole (BHA), *tert*-butylhydroquinone (TBHQ), propyl gallate (PG), octyl gallate (OG), 2,4,5-trihydroxy butyrophenone, nordihydroguaiaretic acid (NDGA) and 4-hexylresorcinol. The first four were widely used as chemical antioxidants. Nearly a decade ago, the European food safety authority (EFSA) has reevaluated the usage of these synthetic antioxidants in foods. The EFSA stressed on the uses of natural antioxidants in foods rather than synthetic ones.

1.3.3   Enzymatic and Non-enzymatic Antioxidants

In this classification system, antioxidants may be either enzyme or non-enzyme in nature (Fig. 1.6). Enzymatic antioxidants are further classified into primary enzymes and secondary enzymes. The primary enzymes include superoxide dismutase (SOD), catalase (CAT), and glutathione peroxidase (GPO). The secondary enzymes with antioxidant properties are glutathione reductase (GHR), and glucose-6-phosphate dehydrogenase (G6PD) is well known (Carocho et al. 2014). The non-enzymatic antioxidants include vitamins, carotenoids, phenolic compounds, minerals and sulfur-containing organic compounds. The antioxidant enzymes are most commonly active in in-vivo biological systems.

1.4   Phenolic Antioxidants

Phenolic compounds also act as natural antioxidants, and hence will be referred as phenolic antioxidant thereafter. The hydroxyl group on the benzene ring is responsible for the antioxidant properties of the phenolic compound. When there is more than one phenol moiety in a compound, it is termed as a polyphenolic compound. Recently several computational studies have shown that electron-donating groups (EDGs) increase the antioxidant activity, while electron-withdrawing groups

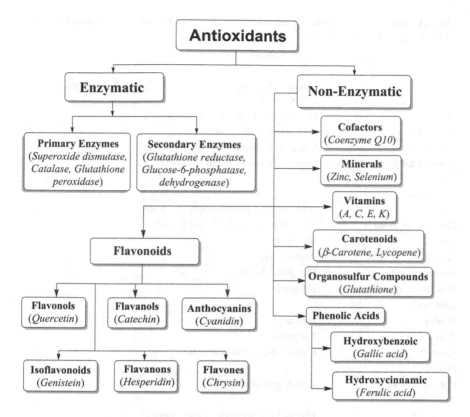

Fig. 1.6 Classification of antioxidants

(EWGs) decrease it (Lee et al. 2020a; Wang et al. 2019). In addition to the ED or EW nature of the substituent, the relative position (*ortho-*, *meta-*, or *para-*) to the hydroxyl group on the benzene ring also plays a significant role. For example, an EDG in the *ortho-* or *para-*position increases the donating potential of H-atom compared to an EDG in the *meta-*position. On the contrary, an EWG in the *ortho-* or *para-*position shows worse activity than an EWG in the *meta-*position. In other words, EDGs increase and EWGs decrease antioxidant activity and the effect is more pronounced if the substituent is in the *ortho-* or *para-*position (Lee et al. 2020a).

H-bonding is another factor that influences the antioxidant activity (Najafi et al. 2013). Any substituent that can form a hydrogen bond (H-bond) with PhOH can alter the H atom donating potential of PhOH (De Heer et al. 2000). Several other factors such as solvent and nature of the free radical also influence the antioxidant activity of the phenolic compound. Phenolic antioxidants have been widely used in cosmetic applications (Cherubim et al. 2020), and foods (Brewer 2011).

Phenolic compounds are present in plant foods. Phenolic compounds are usually biosynthesized from phenylalanine or tyrosine through the shikimic acid pathway. Phenolic compounds ranged from simple to conjugated or complex compounds. The hydroxyl group on the benzene ring is responsible for the antioxidant properties of the phenolic compound. More than two hydroxyl-containing compounds are thus

Table 1.1 Major sources of phenolic antioxidants plant foods. Reproduced with kind permission of John Wiley and Sons (Zeb 2020)

Food	Class of phenolic antioxidants
Fruits	
Berries	Flavanols hydroxycinammic acids, hydroxybenzoic acids, anthocyanins
Cherries	Hydroxycinnamic acids, anthocyanins
Black grapes	Anthocyanins, flavonols
Citrus fruits	Flavanones, flavonols, phenolic acids
Plums, prunes, apples, pears, kiwi	Hydroxycinnamic acids, catechins
Vegetables	
Aubergin	Anthocyanins, hydroxycinnamic acids
Chicory, artichoke	Hydroxycinnamic acids, gallic acid
Parsley	Flavones
Rhubarb	Anthocyanins
Sweet potato leaves	Flavonols, flavones,
Yellow onion, curly	Flavonols
Kale, leek	
Parsley	Flavones
Beans	Flavanols
Spinach	Flavonoids, *p*-coumaric acid
Flours	
Oats, wheat, rice	Caffeic and ferulic acids
Alcoholic drinks	
Red wine	Flavan-3-ols, flavonols, anthocyanins
Cider	Hydroxycinnamic acids
Herbs and spices	
Rosemary	Carnosic acid, carnosol, Rosmarinic acid rosmanol
Sage	Carnosol, Carnosic acid, luteolin, rosmanol, rosmarinic acid, Rosmarinic acid
Oregano	Rosmarinic acid, phenolic acids, flavonoids
Thyme	Thymol, carvacrol, Flavonoids, lubeolin
Summer savory	Rosmarinic, carnosol, carvacrol, flavonoids
Ginger	Gingerols, flavonoids

termed as polyhydroxy phenolic compounds. There is more than one phenol group present in a compound is thus termed as polyphenolic. Phenolic antioxidants are extensively studied in different plant foods such as vegetables, fruits, cereals, seeds, berries, tea, onion bulbs, wine, and vegetable oils (Dimitrios 2006). Table 1.1 showed the phenolic composition of some common foods.

1.5 Mechanism of Antioxidant Action

The mechanism of the antioxidant activity of a phenolic compound may be based either on a HAT mechanism or single electron transfer by proton transfer (SET-PT) (Zhang and Tsao 2016), sequential proton loss electron transfer (Chen et al. 2020) or transition metal chelation (TMC). Each of these mechanisms provides an explanation of the role of phenolic antioxidants as given below.

1.5.1 Hydrogen Atom Transfer

In HAT, phenolic antioxidant containing H-atom represented by AH reacts with free radicals. The free radical is stabilized to form a neutral species, while the antioxidant is converted to antioxidant free radical (A*). Phenolic antioxidant (AH) can provide a H-atom to the free radical substrate and produce a non-radical substrate (RH, ROH, or ROOH) species and antioxidant free radical (A*) as shown in Fig. 1.7. The donation of a hydrogen atom by an antioxidant is explained by the reduction potential. Shahidi and Ambigaipalan (2015) stated that "Any compound that has a reduction potential lower than the reduction potential of a free radical (or oxidized species) is capable of donating its hydrogen atom to that of the free radical unless the reaction is kinetically unfeasible".

The ability or potency of phenolic compounds to have the antioxidant function is dictated by its structure especially the benzene ring and the number and position of the hydroxyl group. The benzene ring is responsible for the stabilization of antioxidant molecules upon reaction with free radicals. Gallic acid (**1.1**) is a phenolic acid containing three hydroxyls and one carboxylic acid group. However, the hydroxyl group is responsible for antioxidant function by forming gallic acid-free radicals. The radical is stabilized by the resonance effects of the benzene ring (Fig. 1.8). Flavonol such as quercetin containing three hydroxyl groups has comparatively greater antioxidant potential than those lacking neutralizing free radical (Tsao 2010).

The antioxidant activity of phenolic acid is enhanced if the COOH group has longer space such as in terms of the CH_2 group to the benzene ring. The longer space affects the electron-withdrawing properties of the COOH group. For example in the case of cinnamic acid or its derivatives have higher antioxidant activity than

Fig. 1.7 Reactions of antioxidant (AH) with free radicals. (Reproduced with kind permission of John Wiley and Sons, Zeb 2020)

$$R^• + AH \longrightarrow RH + A^•$$

$$RO^• + AH \longrightarrow ROH + A^•$$

$$ROO^• + AH \longrightarrow ROOH + A^•$$

$$RO^• + A^• \longrightarrow ROA$$

$$ROO^• + A^• \longrightarrow ROOA$$

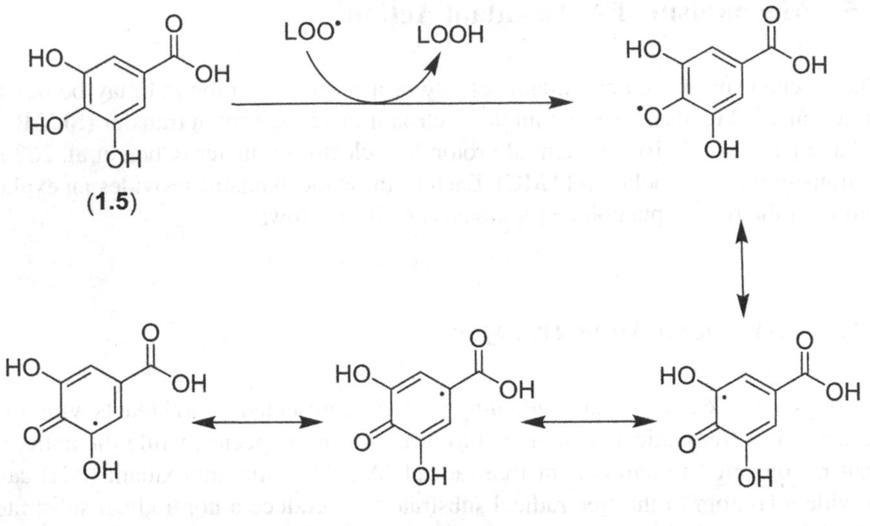

Fig. 1.8 Reaction of gallic acid (**1.5**) with free radical and its stabilization of gallic acid free radical. (Reproduced with kind permission of John Wiley and Sons, Zeb 2020)

benzoic acid or its derivatives. Similarly, the position of hydroxyl groups on the benzene ring also produces effects on antioxidant potential. For example, the hydroxyl group at the *ortho*-position and or *para*-position have enhanced antioxidant activity as compared to other positions and unsubstituted phenol (Göçer and Gülçin 2011). Upon substitution with ethyl or n-butyl group rather than a methyl group at the *para*-position improves the activity of the antioxidant (Shahidi and Ambigaipalan 2015). However, a decrease in the antioxidant activity was due to the presence of chain or branched alkyl groups *para*-position. When bulky groups are present at 2nd and 6th positions, the stability of the phenoxyl radical is further increased as shown in 2,6-di-*t*-butyl-4-methyl phenol (BHT) (Miller and Quackenbush 1957). The bulky groups contribute to the steric hindrance in the radical region, thus reducing the rate of propagation reactions of free radicals. The number of the hydroxyl group on the benzene ring also contributes to the antioxidant potential. For example, flavonoids have shown to possess higher antioxidant activity as compared to phenolic acids (Zhang and Tsao 2016).

It worth to mention, that the bond dissociation enthalpy (BDE) of the hydrogen atom attached to the oxygen atom of phenolic is also a key factor for assessing the role of an antioxidant (Leopoldini et al. 2011). The lower BDE value reflects the easiness of O–H bond dissociation and consequent reaction with a free radical (Table 1.2). The OH group on the *ortho* position of the benzene ring showed a reduction in the BDE value by about 9–10 kcal/mol with reference to a phenol (Wright et al. 2001). The reason may be because the radical formed from H-atom elimination is stabilized by the creation of the intramolecular H-bond with the vicinal hydroxyl (Himo et al. 2000; Thavasi et al. 2006). In addition to BDE, The

Table 1.2 The calculated thermodynamic parameters of 18 phenolic acids ethanol and water. Adopted from Chen et al. (2020)

Compounds	Bond	BDE (kcal/mol)		IP (kcal/mol)		PDE (kcal/mol)		PA (kcal/mol)		ETE (kcal/mol)	
		Ethanol	Water	Ethanol	Water	Ethanol	Water	Ethanol	Water	Ethanol	Water
3-Hydroxybenzoic acid	3-OH	87.4	85.5	126.9	119.3	7.0	13.3	42.3	45.5	91.6	87.2
3-Hydroxycinnamic acid	3-OH	86.8	82.9	123.2	114.5	10.1	15.6	43.4	45.4	89.9	84.6
3-Hydroxyphenylacetic acid	3-OH	85.2	82.4	121.1	113.0	10.6	16.6	44.6	46.6	87.1	83.0
4-Hydroxybenzoic acid	4-OH	88.9	86.7	129.2	120.6	6.2	13.3	38.9	41.5	96.6	92.4
p-Coumaric acid	4-OH	83.8	81.4	118.1	110.1	12.2	18.4	39.1	41.9	91.2	86.7
4-Hydroxyphenylacetic acid	4-OH	83.9	81.5	118.9	111.0	11.5	17.6	44.6	46.7	85.8	82.0
Protocatechuic acid	3-OH	83.3	80.9	120.5	112.6	9.3	15.5	42.9	44.6	87.0	83.5
	4-OH	81.3	79.2			7.3	13.7	35.2	38.1	92.6	88.3
Caffeic acid	3-OH	82.2	80.1	114.4	107.0	14.4	20.3	41.3	48.0	87.4	79.3
	4-OH	77.3	75.7			9.4	15.9	33.9	40.1	89.8	82.7
3,4-Dihydroxyphenylacetic acid	3-OH	80.5	77.9	113.2	105.4	13.8	19.7	44.7	46.4	82.3	78.6
	4-OH	77.2	75.0			10.5	16.8	40.6	43.1	83.1	79.1
Isovanillic acid	3-OH	83.3	80.8	118.8	110.9	11.0	17.0	43.7	45.3	86.1	82.7
Isoferulic acid	3-OH	81.6	78.9	112.5	104.7	15.5	21.4	44.0	45.6	84.1	80.4
Homoisovanillic acid	3-OH	80.1	77.4	112.5	103.5	14.1	21.0	45.4	47.0	81.2	77.6
Vanillic acid	4-OH	83.7	79.8	118.9	110.8	11.3	19.8	39.6	41.3	90.6	89.4
Ferulic acid	4-OH	79.5	76.8	112.2	104.4	13.7	19.5	39.7	41.8	86.3	82.1
Homovanillic acid	4-OH	79.6	76.8	112.6	103.5	13.7	20.5	45.5	47.0	80.6	77.0
Syringic acid	4-OH	78.9	77.3	115.8	108.3	9.7	16.1	39.4	41.7	86.0	82.7
Sinapic acid	4-OH	75.7	71.3	110.2	102.7	12.0	17.5	39.2	41.0	83.0	79.2
4-Hydroxy-3,5-dimethoxyphenylacetic acid	4-OH	75.5	72.8	108.3	100.9	13.5	19.0	45.6	46.6	76.4	73.3

IP Ionization potential, PDE proton dissociation enthalpy, PA proton affinity, ETE electron transfer enthalpy ionization potentials provide diverse trends of reactivity for phenolic compounds (Leopoldini et al. 2011, 2004).

In phenolic acids, when the methoxy and phenolic hydroxyl groups substitution positions on the benzene ring are the same, the BDE values in hydroxyphenyl acetic acid and hydroxycinnamic acid had been found 1.9–13.3 kcal/mol and 0.9–9.2 kcal/mol lower than the corresponding BDE values of hydroxybenzoic acid, respectively. This illustrates that an ethanoic acid moiety can decrease the dissociation energy of the phenolic hydroxyl bond. Thus it enhanced the ability of free radical scavenging as reported by Chen et al. (2020). The authors reported that substituted hydroxybenzoic acid and hydroxycinnamic acid have stronger antioxidant activity than the corresponding hydroxybenzoic acid in the DPPH and FRAP assays.

1.5.2   Single Electron Transfer

In the single electron transfer (SET) mechanism, the anion R formed is an energetically stable species with an even number of electrons, while the cation radical ArOH is a less reactive radical species, and thus stable (Leopoldini et al. 2011). The odd electron formed by the reactions with the free radical in ArOH^{+*} is an aromatic structure that has the chance to be distributed over the whole molecule. This results in a radical stabilization of the antioxidant molecule as shown in Fig. 1.9. The ionization potential (IP) of different atoms is the most important parameter for the evaluation of scavenging activity in the SET mechanism. The easier electron abstraction and consequent reaction with free radicals will occur if the IP value is lower.

1.5.3   Sequential Proton Loss Electron Transfer

In sequential proton loss electron transfer (SPLET) mechanism, a phenolic compound gives up a proton to a free radical and forming an anion, which further donates an electron forming a stable molecule as shown in Fig. 1.10. The proton

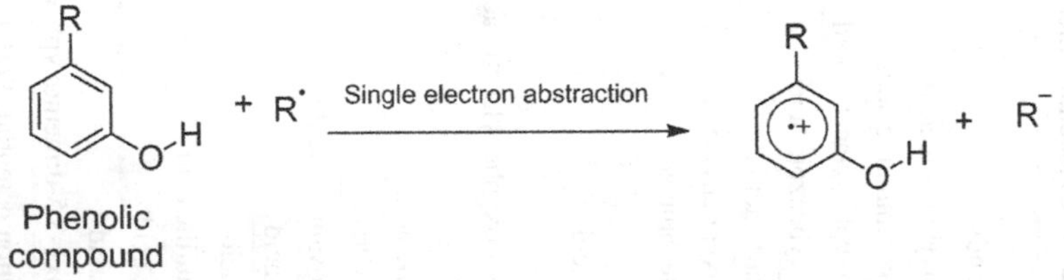

Fig. 1.9 Mechanism of single electron abstraction reaction (SET). (Reproduced with kind permission of John Wiley and Sons, Zeb 2020)

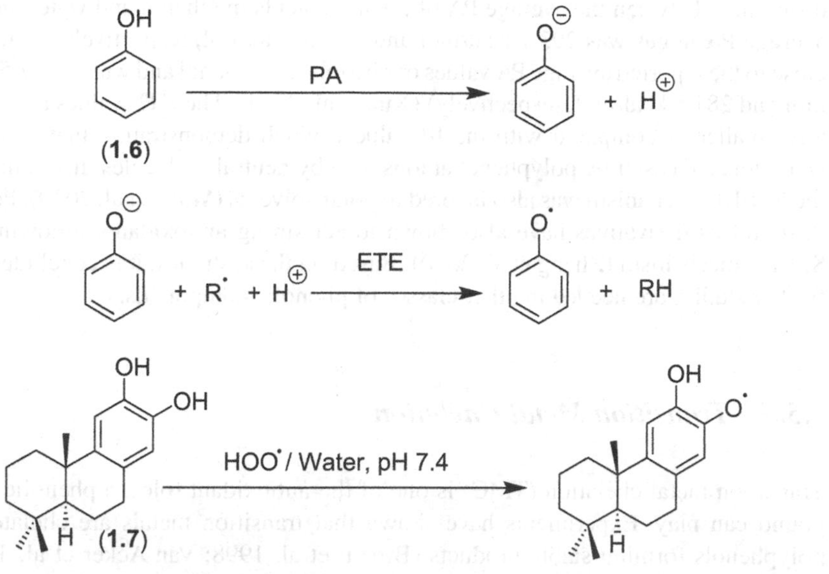

Fig. 1.10 Mechanism of sequential proton loss electron transfer (SPLET). Proton ability (PA) and electron transfer enthalpy (ETE) during antioxidant action of phenol (**1.6**). Third reaction represents SPLET mechanism in 7-deoxynimbidiol (**1.7**) showed 1300 times higher antioxidant activity than Trolox in polar solvents. (Reproduced with kind permission of John Wiley and Sons, Zeb 2020)

ability (PA) and electron transfer enthalpy (ETE) is calculated to determine the antioxidant activity. The PA value represents the degree of difficulty of phenolic hydroxyl dephosphorization, while the ETE value represents the electron donation ability of corresponding phenol ions. The SPLET mechanism is more thermodynamically favorable in water and thus is the most suited for phenolic compounds. The methoxy group (electron-donating group) at the ortho position has been found to slightly increase the values of PA, whereas a nitrogen dioxide (NO_2) group slightly decreases the PA values (Lee et al. 2020b). The methoxy group at the ortho position decrease the ETE values and increased by the NO_2 group. For diterpenes, the polar environment has been reported to favor the SPLET mechanism (Vo et al. 2020). Catechin and epicatechin were studied using density functional theory for antioxidant activity by Anitha et al. (2020). The hydroxyl group present at 4-site from B-ring was the most preferential site for hydrogen donation in both the compounds, whereas the SPLET mechanism was favored in polar solvents. An antioxidant molecule like 1,3,4-oxadiazole (**1.3**) the preferred site for HAT was the same as that of SPLET for each molecule (Alisi et al. 2020).

In the case of phenolic acids like hydroxybenzoic acid and cinnamic acids, Chen et al. (2020) showed that similar to the PDE values, molecule dephosphorization in ethanol was easier than in water. Due to the direct influence of proton solvation enthalpy, the order of PA values was gas > water > ethanol. The order of ETE values was ethanol > water > gas, based on the effect of electron solvation enthalpy. The differences between the average PA of phenolic acids in ethanol and water and the average PA in gas was 293.7 kcal/mol and 291.6 kcal/mol, respectively, which are close to the reported average PA values of phenols in methanol and water (287.5 kcal/mol and 281.5 kcal/mol, respectively) (Xue et al. 2014). The ETE values are generally smaller as compared with the IP values, which demonstrating that electrons were donated easier by polyphenol anions than by neutral molecules. In coumarins, the SPELT mechanism was also favored in polar solvents (Wang et al. 2020). Ferulic acid and its derivatives have also shown to act strong antioxidants following the SPLET mechanism (Zheng et al. 2020). Based on these studies, it is concluded that further studies are needed in other classes of phenolic compounds.

1.5.4   Transition Metal Chelation

Transition metal chelation (TMC) is one of the antioxidant roles, a phenolic compound can play. Experiments have shown that transition metals are chelated by polyphenols forming stable products (Brown et al. 1998; Van Acker et al. 1996). Transition metals such as copper (Cu), manganese (Mg), cobalt (Co) can catalyze such reactions. However, chelation is subject to certain conditions such as these metal ions are not bound to proteins or other chelators.

Phenolic compounds from foods are very good metal chelators. Metal chelation can directly inhibit Fe^{3+} reduction, consequently reducing the formation of reactive OH-free radicals of Fenton reaction (Perron and Brumaghim 2009). The metal chelation depends on the reduction potential of the phenolic compounds. The reducing potential of the compounds was found to be in the order: Propyl gallate > Gallic acid > Caffeic acid > Pyrogallol > Protocatechuic acid > Sinapic acid > BHA > Vanillyl alcohol > Ferulic acid > Guaiacol > Vanillic acid > Coumaric acid (Mathew et al. 2015). Phenolic acids and flavonoids have strong free radical scavenging potential.

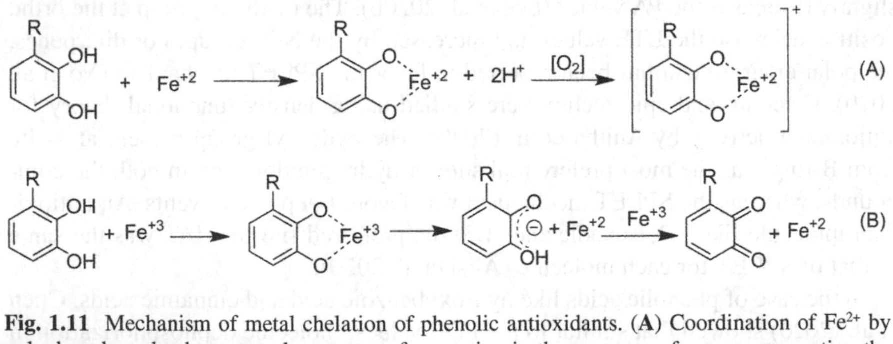

Fig. 1.11 Mechanism of metal chelation of phenolic antioxidants. (**A**) Coordination of Fe^{2+} by polyphenols and subsequent electron transfer reaction in the presence of oxygen generating the Fe^{3+}-polyphenol complex; (**B**) Coordination of Fe^{3+} by polyphenols, subsequent iron reduction and semiquinone formation, and reduction of Fe^{3+} to form a quinone species and Fe^{2+}. R=H, or OH (Perron and Brumaghim 2009)

Nevertheless, the chelating potential with metals and reducing power may differ based on their structural characteristics. For example, in flavonoids, the higher metal chelating and radical scavenging activity depends on the catechol moiety (ring B), 2,3-double bound and conjugated 4-oxo group (ring C) and 3 and 5-hydroxyl moieties (Van Acker et al. 1996).

The ligands of polyphenol strongly stabilize Fe^{3+} over Fe^{2+}. For example, the Fe^{2+} complexes of catecholate and gallate rapidly oxidize in the presence of O_2 to give Fe^{3+}-polyphenol complexes (Chvátalová et al. 2008; Yoshino and Murakami 1998; Hajji et al. 2006; Perron and Brumaghim 2009). This process is commonly called auto-oxidation (Fig. 1.11a). Usually, the oxidation of Fe^{2+} in the presence of O_2 occurs slowly, but binding of polyphenol ligands to Fe^{2+} drops its reduction potential (Cooper et al. 1978) and enhances the rate of Fe oxidation (Powell and Taylor 1982; McBryde 1964; Perron et al. 2010). The Fe oxidation rate differs for polyphenol complexes, i.e., faster oxidation rates for gallate complexes than catecholate complexes (Perron et al. 2010). The oxidation of iron upon attachment to phenol ligands is aided by the higher stability of Fe^{3+} interactions with the hard oxygen ligands of the phenol moieties. The strong electron giving characteristics of the oxygen ligands that stabilize the Fe^{3+} state is also contributing (Fig. 1.11b).

Table 1.3 Important *in-vitro* antioxidant activity measurement methods used for foods. Reproduced with kind permission of John Wiley and Sons (Zeb 2020)

Methods	Foods	Reference
Oxygen radical absorbance capacity (ORAC)	Guava fruit	Thaipong et al. (2006)
Total radical-trapping antioxidant potential (TRAP)	Fruits	Ruiz-Torralba et al. (2018)
Trolox equivalent antioxidant capacity (TEAC)	Legumes, fruits	Thaipong et al. (2006), Koley et al. (2019)
2,2-Diphenyl-1-picrylhydrazyl (DPPH)	Guava fruit, Leafy vegetables, fruits	Ruiz-Torralba et al. (2018), Thaipong et al. (2006), Zeb (2015)
Ferric reducing antioxidant power (FRAP)	Guava fruit, fruits	Ruiz-Torralba et al. (2018), Thaipong et al. (2006), Koley et al. (2019)
ABTS	Guava fruit, grape, bakery products	Ruiz-Torralba et al. (2018), Koley et al. (2019), Sridhar and Charles (2019)
Total antioxidant activity (TAA)	Red beans	Thaipong et al. (2006)
Cupric Reducing Antioxidant Capacity (CUPRAC)	Legumes	Koley et al. (2019)
Peroxyl radical scavenging activity (PRSA)	Beans	Koley et al. (2019), Amarowicz et al. (2017)

1.6   Estimation of Antioxidant Activity

There are several antioxidant activity assays method reported. These methods have their specific target within the food matrix and with advantages and disadvantages. None of these methods can offer unequivocal results. Thus the uses of various methods for comparison are common as compared to a single-dimensional approach. Some of these procedures use synthetic antioxidants or free radicals. They may be specific for lipid peroxidation and aimed for animal or plant systems. Some of these methods have a wider scope, some need least preparation, expertise, and few reagents. Table 1.3 showed some of the most important and widely used antioxidant assays available for phenolic antioxidants in foods. The details of some these methods are discussed in Chap. 15.

- The oxygen radical scavenging capacity assay is optimized for enzymatic antioxidants and relies on the competition kinetics of O radical reduction of cytochrome C (probe) and O radical scavenger (sample) (Fraga 2010; Huang et al. 2005). This assay has been used to study the antioxidant capacity in several fruits including guava fruit (Thaipong et al. 2006).
- The total radical-trapping antioxidant parameter (TRAP) assay is used to evaluate the total antioxidant activity. This assay is based on measuring oxygen consumption during a controlled lipid oxidation reaction induced by thermal decomposition of AAPH (2,2-azobis(2-amidinopropane)hydrochloride) (Ruiz-Torralba et al. 2018).
- The Trolox equivalent antioxidant capacity (TEAC) assay uses the ABTS radical, which undergoes treatment to obtain a dark blue color. A radical solution is mixed with the food extract and the absorbances are measured at 734 nm after specific intervals. The changes in the absorbance are plotted versus the antioxidant amounts to obtain a straight line. The concentration of antioxidants giving the same percentage change in absorbance of the ABTS radical as that of 1 mM Trolox is regarded as TEAC (Huang et al. 2005).
- DPPH (2,2-diphenyl-1-picrylhydrazyl) assay is based on the principle that a hydrogen donation is provided by the phenolic antioxidants. This assay uses the DPPH radical, which changes from purple to yellow in the presence of phenolic antioxidants, and is widely used as a preliminary study (Moon and Shibamoto 2009). This method is widely used for the estimation of antioxidant activity in fruits, leafy vegetables (Thaipong et al. 2006; Zeb et al. 2018; Koley et al. 2019; Ruiz-Torralba et al. 2018).
- The FRAP (ferric reducing antioxidant power) assay is commonly used in a vast number of food matrixes. It is characterized by the reduction of Fe^{3+} to Fe^{2+} depending on the available reducing species followed by the alteration of color from yellow to blue and analyzed through a spectrophotometer (Thaipong et al. 2006; Koley et al. 2019; Ruiz-Torralba et al. 2018).
- The ABTS (2,2-azinobis(3-ethylbenzothiazoline-6-sulfonic acid)) assay is a colorimetric assay in which the ABTS radical decolorizes in the presence of antioxidants (Thaipong et al. 2006) in food like Guava fruit.

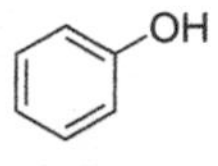

Fig. 1.12 Structure of a phenol (**1.8**)

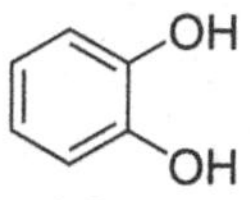

(**1.8**)

Fig. 1.13 Nomenclature of methyl and hydroxyl substituted phenols. 1,2-Benzenediol (**1.9**), 1,3-benzenediol (**1.10**), 2-methylphenol (**1.11**) and 4-methylphenol (**1.12**)

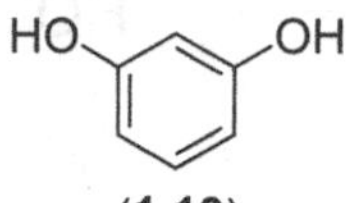

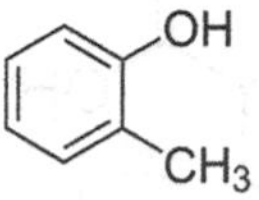

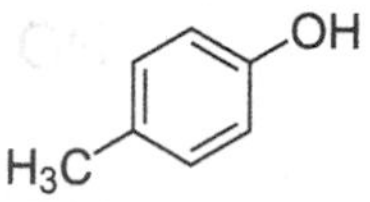

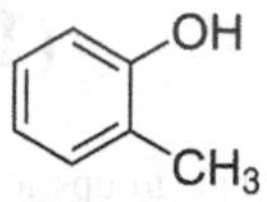

(**1.9**)

(**1.10**)

(**1.11**)

(**1.12**)

Fig. 1.14 Nomenclature of substitution at different position on phenols. o-Cresol (**1.13**), m-cresol (**1.14**), and p-cresol (**1.15**)

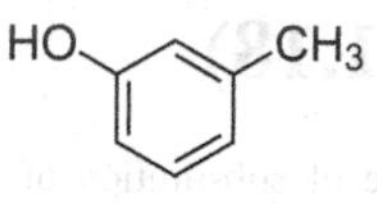

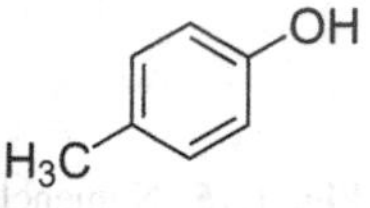

(**1.13**)

(**1.14**)

(**1.15**)

- Total antioxidant activity (TAA) is also used to study antioxidant activity in different beans (Amarowicz et al. 2017)
- Cupric Reducing Antioxidant Capacity (CUPRAC) assay has been used to determine antioxidant activity in legumes (Koley et al. 2019).
- Peroxyl radical scavenging activity (PRSA) assay has been used to determine antioxidant activity in beans (Amarowicz et al. 2017).

1.7 Nomenclature of Phenolic Compounds

Phenolic compounds are also called phenolics or carbolic acids. Chemically they are made of the benzene ring(s) and hydroxyl group(s) and are thus difunctional compounds. The benzene ring and the hydroxyl groups strongly interacted and thus affecting each other polarity. These difunctional properties play a significant role in antioxidant activity in foods. The benzene was formally known as *"phene"*. It hydroxyl derivatives was named as phenols. The first compound is a common phenol (**1.8**) having a single hydroxyl group directly attached to the benzene ring (Fig. 1.12)

The word phenol is an acceptable name in the IUPAC nomenclature. Following are some of the important rules to be remembered while naming a phenolic compound.

- Locate the position of a hydroxyl group attached to the benzene ring.

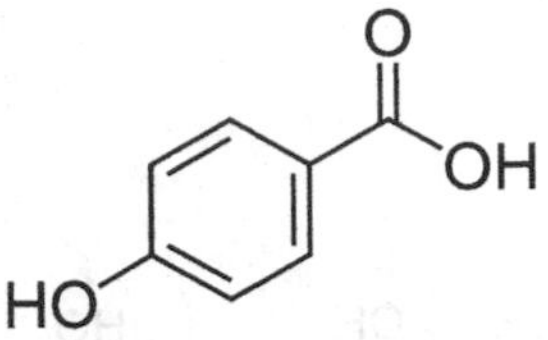

(1.16)

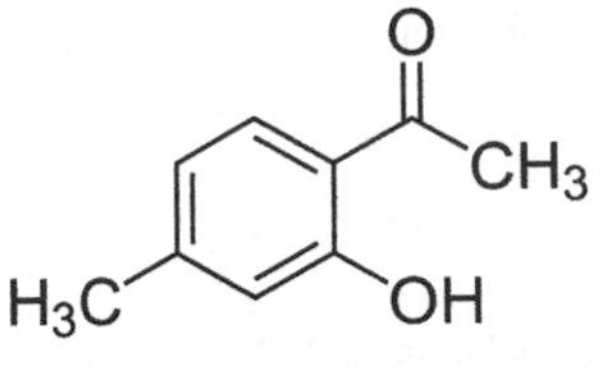

(1.17)

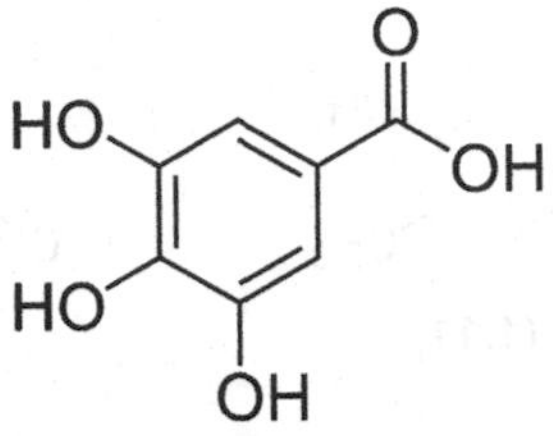

(1.18)

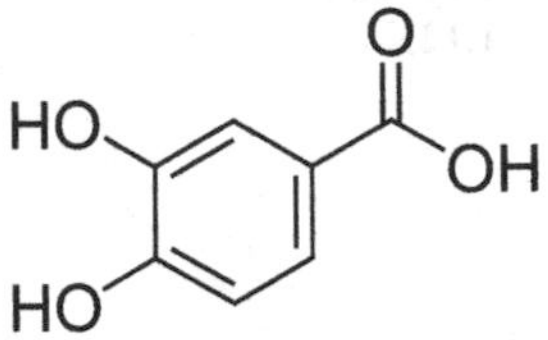

(1.19)

Fig. 1.15 Nomenclature of substitution of functional groups at different position on phenols. 4-Hydroxybenzoic acid (**1.16**), 2-hydroxy-4-methylacetophenone (**1.17**), 3,4,5-trihydroxybenzoic acid (**1.18**), and 3,4-dihydroxybenzoic acid (**1.19**)

- Benzene rings attached to more than one hydroxyl group are labeled with the Greek numerical prefixes such as di, tri, tetra to denote the number of similar hydroxyl groups attached to the benzene ring. If two hydroxyl groups are attached to the adjacent carbon atoms of the benzene ring, it is named as benzene 1,2-diol (**1.9**). Another example is 1,3-benzenediol (**1.10**).
- In the case of substituted phenols, we start locating the positions of the other functional groups with respect to the position where the hydroxyl group is attached. For example, if a methyl group is attached at the second or fourth carbon atom with respect to the hydroxy group, the compound is named as 2-methylphenol (**1.11**) and 4-methylphenol (**1.12**) (Fig. 1.13).
- Depending on the position of substituted functional group with respect to the hydroxyl group, words like *ortho* (when the functional group is attached to the adjacent carbon atom), *para* (when the functional group is attached to the third carbon atom from the hydroxyl group), *meta* (when the functional group is attached to the second carbon atom from the hydroxyl group) are also used for the nomenclature of phenols. See for example cresol (Fig. 1.14).
- Carboxyl and acyl groups take preference over the hydroxyl group in determining the base name. The hydroxyl is treated as a substituent. See the examples of 4-hydroxybenzoic acid (**1.16**), 2-hydroxy-4-methyl acetophenone (**1.17**), 3,4,5-trihydroxybenzoic acid (**1.18**), and 3,4-dihydroxybenzoic acid (**1.19**) (Fig. 1.15). There are several other functional group suffixes, for various substituted phenols.

1.8 Study Questions

- What is an antioxidant? What are the different types of antioxidants?
- Discuss the mechanism of hydrogen atom transfer in phenolic compounds.
- Briefly explain the single electron transfer mechanism.
- What do you know about the sequential proton loss electron transfer of phenolic compounds?
- What are the different methods of estimation of phenolic contents?
- How to name phenolic compounds in foods?

References

Alisi IO, Uzairu A, Abechi SE (2020) Free radical scavenging mechanism of 1,3,4-oxadiazole derivatives: thermodynamics of O–H and N–H bond cleavage. Heliyon 6(3):e03683

Amarowicz R, Karamać M, Dueñas M, Pegg RB (2017) Antioxidant activity and phenolic composition of a red bean (*Phasoelus vulgaris*) extract and its Fractions. Nat Prod Commun 12(4):541–544

Anitha S, Krishnan S, Senthilkumar K, Sasirekha V (2020) Theoretical investigation on the structure and antioxidant activity of (+) catechin and (−) epicatechin – a comparative study. Mol Phys 118:1–12

Brewer MS (2011) Natural antioxidants: sources, compounds, mechanisms of action, and potential applications. Compr Rev Food Sci Food Saf 10(4):221–247

Brown EJ, Khodr H, Hider CR, Rice-evans CA (1998) Structural dependence of flavonoid interactions with Cu2+ ions: implications for their antioxidant properties. Biochem J 330(3):1173–1178

Carocho M, Barreiro MF, Morales P, Ferreira ICFR (2014) Adding molecules to food, pros and cons: a review on synthetic and natural food additives. Compr Rev Food Sci Food Saf 13(4):377–399

Chen J, Yang J, Ma L, Li J, Shahzad N, Kim CK (2020) Structure-antioxidant activity relationship of methoxy, phenolic hydroxyl, and carboxylic acid groups of phenolic acids. Sci Rep 10(1):2611

Cherubim DJL, Martins CVB, Fariña LO, de Lucca RAS (2020) Polyphenols as natural antioxidants in cosmetics applications. J Cosmet Dermatol Us 19(1):33–37

Choe E, Min DB (2009) Mechanisms of antioxidants in the oxidation of foods. Compr Rev Food Sci Food Saf 8(4):345–358

Chvátalová K, Slaninová I, Březinová L, Slanina J (2008) Influence of dietary phenolic acids on redox status of iron: ferrous iron autoxidation and ferric iron reduction. Food Chem 106(2):650–660

Cooper SR, McArdle JV, Raymond KN (1978) Siderophore electrochemistry: relation to intracellular iron release mechanism. Proc Natl Acad Sci 75(8):3551–3554

Dangles O (2012) Antioxidant activity of plant phenols: chemical mechanisms and biological significance. Curr Org Chem 16(6):692–714

De Heer MI, Mulder P, Korth HG, Ingold KU, Lusztyk J (2000) Hydrogen atom abstraction kinetics from intramolecularly hydrogen bonded ubiquinol-0 and other (poly)methoxy phenols. J Am Chem Soc 122(10):2355–2360

Dimitrios B (2006) Sources of natural phenolic antioxidants. Trends Food Sci Technol 17(9):505–512

Foti MC (2007) Antioxidant properties of phenols. J Pharm Pharmacol 59(12):1673–1685

Fraga CG (2010) Plant phenolics and human health: biochemistry, nutrition, and pharmacology. The Wiley-IUBMB series on biochemistry and molecular biology. Wiley, Hoboken, NJ

Göçer H, Gülçin İ (2011) Caffeic acid phenethyl ester (CAPE): correlation of structure and antioxidant properties. Int J Food Sci Nutr 62(8):821–825

Hajji HE, Nkhili E, Tomao V, Dangles O (2006) Interactions of quercetin with iron and copper ions: complexation and autoxidation. Free Radic Res 40(3):303–320

Halliwell B (1990) How to characterize a biological antioxidant. Free Radic Res Commun 9(1):1–32

Halliwell B (1995) Antioxidant characterization. Methodology and mechanism. Biochem Pharmacol 49(10):1341–1348

Heim KE, Tagliaferro AR, Bobilya DJ (2002) Flavonoid antioxidants: chemistry, metabolism and structure-activity relationships. J Nutr Biochem 13(10):572–584

Himo F, Eriksson LA, Blomberg MRA, Siegbahn PEM (2000) Substituent effects on OH bond strength and hyperfine properties of phenol, as model for modified tyrosyl radicals in proteins. Int J Quantum Chem 76(6):714–723

Huang D, Ou B, Prior RL (2005) The chemistry behind antioxidant capacity assays. J Agric Food Chem 53(6):1841–1856

Koley TK, Maurya A, Tripathi A, Singh BK, Singh M, Bhutia TL, Tripathi PC, Singh B (2019) Antioxidant potential of commonly consumed underutilized leguminous vegetables. Int J Veg Sci 25(4):362–372

Lee CY, Anamoah C, Semenya J, Chapman KN, Knoll AN, Brinkman HF, Malone JI, Sharma A (2020a) Electronic (donating or withdrawing) effects of ortho-phenolic substituents in dendritic antioxidants. Tetrahedron Lett 61:151607

Lee CY, Sharma A, Semenya J, Anamoah C, Chapman KN, Barone V (2020b) Computational study of ortho-substituent effects on antioxidant activities of phenolic dendritic antioxidants. Antioxidants 9(3):189

Leopoldini M, Marino T, Russo N, Toscano M (2004) Antioxidant properties of phenolic compounds: H-atom versus electron transfer mechanism. J Phys Chem A 108(22):4916–4922

Leopoldini M, Russo N, Toscano M (2011) The molecular basis of working mechanism of natural polyphenolic antioxidants. Food Chem 125(2):288–306

Mathew S, Abraham TE, Zakaria ZA (2015) Reactivity of phenolic compounds towards free radicals under in vitro conditions. J Food Sci Technol 52(9):5790–5798

McBryde W (1964) A spectrophotometric reexamination of the spectra and stabilities of the iron (III)–tiron complexes. Can J Chem 42(8):1917–1927

Miller GJ, Quackenbush FW (1957) A comparison of alkylated phenols as antioxidants for lard. J Am Oil Chem Soc 34(5):249–250

Moon J-K, Shibamoto T (2009) Antioxidant assays for plant and food components. J Agric Food Chem 57(5):1655–1666

Munné-Bosch S (2011) Phenolic acids: composition, applications, and health benefits. Nova Science Publishers, Inc., New York, NY

Najafi M, Najafi M, Najafi H (2013) Theoretical study of the substituent and solvent effects on the reaction enthalpies of the antioxidant mechanisms of tyrosol derivatives. Bull Chem Soc Jpn 86(4):497–509

Perron NR, Brumaghim JL (2009) A review of the antioxidant mechanisms of polyphenol compounds related to iron binding. Cell Biochem Biophys 53(2):75–100

Perron NR, Wang HC, DeGuire SN, Jenkins M, Lawson M, Brumaghim JL (2010) Kinetics of iron oxidation upon polyphenol binding. Dalton Trans 39(41):9982–9987

Powell HKJ, Taylor MC (1982) Interactions of iron (II) and iron (III) with gallic acid and its homologues: a potentiometric and spectrophotometric study. Aust J Chem 35(4):739–756

Rice-Evans C, Miller N, Paganga G (1997) Antioxidant properties of phenolic compounds. Trends Plant Sci 2(4):152–159

Ruiz-Torralba A, Guerra-Hernández EJ, García-Villanova B (2018) Antioxidant capacity, polyphenol content and contribution to dietary intake of 52 fruits sold in Spain. CyTA J Food 16(1):1131–1138

Shahidi F (2015) Handbook of antioxidants for food preservation. Woodhead Publishing, Cambridge

Shahidi F, Ambigaipalan P (2015) Phenolics and polyphenolics in foods, beverages and spices: antioxidant activity and health effects–a review. J Funct Foods 18:820–897

Soobrattee MA, Neergheen VS, Luximon-Ramma A, Aruoma OI, Bahorun T (2005) Phenolics as potential antioxidant therapeutic agents: mechanism and actions. Mut Res 579(1):200–213

Sridhar K, Charles AL (2019) In vitro antioxidant activity of Kyoho grape extracts in DPPH and ABTS assays: estimation methods for EC50 using advanced statistical programs. Food Chem 275:41–49

Thaipong K, Boonprakob U, Crosby K, Cisneros-Zevallos L, Hawkins Byrne D (2006) Comparison of ABTS, DPPH, FRAP, and ORAC assays for estimating antioxidant activity from guava fruit extracts. J Food Compos Anal 19(6):669–675

Thavasi V, Leong LP, Bettens RPA (2006) Investigation of the Influence of Hydroxy Groups on the Radical Scavenging Ability of Polyphenols. J Phys Chem A 110(14):4918–4923

Tsao R (2010) Chemistry and biochemistry of dietary polyphenols. Nutrients 2(12):1231–1246

Van Acker SABE, Van Den Berg D-J, Tromp MNJL, Griffioen DH, Van Bennekom WP, Van Der Vijgh WJF, Bast A (1996) Structural aspects of antioxidant activity of flavonoids. Free Radic Biol Med 20(3):331–342

Vo QV, Tam NM, Hieu LT, Van Bay M, Thong NM, Le Huyen T, Hoa NT, Mechler A (2020) The antioxidant activity of natural diterpenes: theoretical insights. RSC Adv 10(25):14937–14943

Wang L, Yang F, Zhao X, Li Y (2019) Effects of nitro- and amino-group on the antioxidant activity of genistein: a theoretical study. Food Chem 275:339–345

Wang G, Liu Y, Zhang L, An L, Chen R, Liu Y, Luo Q, Li Y, Wang H, Xue Y (2020) Computational study on the antioxidant property of coumarin-fused coumarins. Food Chem 304:125446

Wright JS, Johnson ER, DiLabio GA (2001) Predicting the activity of phenolic antioxidants: theoretical method, analysis of substituent effects, and application to major families of antioxidants. J Am Chem Soc 123(6):1173–1183

Xue Y, Zheng Y, An L, Dou Y, Liu Y (2014) Density functional theory study of the structure–antioxidant activity of polyphenolic deoxybenzoins. Food Chem 151:198–206

Yamane K, Kato Y (2012) Handbook on flavonoids: dietary sources, properties, and health benefits. Nova Science Publishers, Hauppauge, NY

Yoshino M, Murakami K (1998) Interaction of iron with polyphenolic compounds: application to antioxidant characterization. Anal Biochem 257(1):40–44

Zeb A (2015) Phenolic profile and antioxidant potential of wild watercress (Nasturtium officinale L.). Springerplus 4:714

Zeb A (2020) Concept, mechanism, and applications of phenolic antioxidants in foods. J Food Biochem 44(9):e13394

Zeb A, Haq A, Murkovic M (2018) Effects of microwave cooking on carotenoids, phenolic compounds and antioxidant activity of Cichorium intybus L. (chicory) leaves. Eur Food Res Technol 245(2):365–374

Zhang H, Tsao R (2016) Dietary polyphenols, oxidative stress and antioxidant and anti-inflammatory effects. Curr Opin Food Sci 8:33–42

Zheng Y-Z, Zhou Y, Guo R, Fu Z-M, Chen D-F (2020) Structure-antioxidant activity relationship of ferulic acid derivatives: effect of ester groups at the end of the carbon side chain. LWT Food Sci Technol 120:108932

Chapter 2
Chemistry of Phenolic Antioxidants

Learning Outcomes

After reading this chapter, the reader will be able to:

- Understand the classification of phenolic compounds.
- Describe the classification of flavonoids.
- Learn the structural difference between isoflavones and neoflavones.
- Understand the chemistry of glycosides, its different classes present in foods.
- Discuss the structure of chalcone and their glycosides.
- Describe the physical properties and chemical properties of phenolic compounds.
- Discuss the importance of acidity and hydrogen bonding occurred in phenolic compounds.

2.1 Introduction

Phenolic compounds are made of one or more hydroxyl groups directly attached to one or more aromatic rings. The aromatic ring is responsible for making a phenol a weak acid by donating a H-atom. The carbon-oxygen-hydrogen atoms have an angle of 109° making it a planar molecule similar to methanol with the carbon-oxygen-hydrogen angle of 108.5°. The distance of the carbon-oxygen bond in phenol is slightly less than that of methanol. The detailed chemistry of the phenols can be found in *"The chemistry of Phenols"* by Rappoport (2004). The two volumes book focused extensively on the theoretical aspects, spectroscopy, and synthesis of phenols in the laboratory. Similarly, the edited volume of Baruah (2011) on *"Chemistry of phenolic compounds: State of the art"* focused on the different aspects

© Springer Nature Switzerland AG 2021
A. Zeb, *Phenolic Antioxidants in Foods: Chemistry, Biochemistry and Analysis*,
https://doi.org/10.1007/978-3-030-74768-8_2

of phenolic compounds rather than in foods. In this chapter, the focus is on the chemistry-based classification, physical and chemical properties of phenolic compounds in foods.

2.2 Classification

In nature, phenolic compounds are biosynthesized by the plants. More than 8000 different phenolic compounds have been identified until now (Cheynier 2012; Cheynier et al. 2013). Phenolic compounds are classified based on several properties. However, there is no clear consensus on the classification of phenolic compounds. Phenolic compounds can be classified into four different means, starting from the most general to the most specific (Nollet and Gutierrez-Uribe 2018):

1. Flavonoids and non-flavonoids
2. Number of aromatic rings
3. Carbon skeleton, which portrays in a very basic way how the carbon atoms of the molecule are organized (e.g., C6, C6–C1, C6–C3–C6, etc.)
4. Basic chemical structure, which is a description or image that specifies common atoms (e.g., C, H, O), functional groups (e.g., aromatic rings, hydroxyl groups, keto groups), saturations (e.g., single bonds, double bonds), types of bonds (e.g., covalent, ionic), and how are they all linked to one another.

The number 3 classification was reported by Harborne and Simmonds (1964) and is still widely used as shown in Table 2.1. However, some authors combined classification 3 and 4 to form one single classification (Santana-Gálvez and Jacobo-Velázquez 2018).

Another classification reported by Swain and Bate-Smith (1962) is based on grouping of phenols in to "common" and "less common" categories. Another classification grouped the phenols into three families (Ribéreau-Gayon 1972):

1. Widely distributed phenols—ubiquitous to all plants, or of importance in a specific plant
2. Phenols that are less widely distributed—limited number of compounds known
3. Phenolic constituents present as polymers.

Phenolic compounds are also classified based on the number of hydroxyl groups (Kabera et al. 2014). The solubility based classification includes soluble and insoluble phenolic compounds (Acosta-Estrada et al. 2014; Shahidi and Yeo 2016).

2.2.1 Simple Phenols

The first example of simple phenols is hydroxyl benzene commonly known as phenol. Several substituted phenol is present in nature. Simple phenols should not be confused with those phenols where the substituents are usually of high priority such

Table 2.1 Classification of phenolic compounds in foods

Class	Structure	Example(s)
Simple phenolics	C6	Phenol
Hydroxybenzoic acids	C6–C1	Hydroxybenzoic acid, Gallic acid
Phenolic aldehydes	C6–C1	Vanillin
Acetophenones and phenylacetic acids	C6–C2	Phenylacetic acid
Phenylethanoids	C6–C2	Tyrosol, Hydroxytyrosol
Cinnamic acids, cinnamyl aldehydes, cinnamyl alcohols, esters	C6–C3	Cinnamic acid, Phenylacetone
Coumarins, isocoumarins, and chromones	C6–C3	1-Benzopyran-2-one
Benzophenones, xanthones,	C6–C1–C6	Mangostin
Stilbenes	C6–C2–C6	3,4′,5-Stilbenetriol
Anthraquinones	C6–C2–C6	Emodin
Chalcones, dihydrochalcones	C6–C3–C6	Butein
Aurones	C6–C3–C6	Bracteatin
Flavones	C6–C3–C6	Apigenin
Flavanones	C6–C3–C6	Naringin
Flavonols	C6–C3–C6	2,3-Dihydroflavonol
Flavanols	C6–C3–C6	Catechin, Epicatechin
Isoflavons	C6–C3–C6	Diadzein
Neoflavones	C6–C3–C6	Dalbergin
Anthocyanidins	C15	Cyanidin, malvidin
Biflavonyls	C30	Hinokiflavone
Betacyanins	C18	Betanin
Lignans, neolignans dimers or oligomers	Dimers	Sesamin; Syringaresinol
Lignins	Polymers	Polymer of sinapyl alcohol
Tannins	Oligomers or polymers	Tannic acid, Procyanidin B2, Punicalagin
Phlobaphenes	Polymers	Polymer of tannins
Glycosides	As above	Scopolin, Quercetin-3-glucoside

as carboxylic acid, ketone, or aldehydes. The nomenclature *ortho*, *meta*, and *para* refer to a 1,2-, 1,3- and 1,4-substitution pattern on the aromatic ring, respectively. One of the functional groups, in this case, is the hydroxyl group. When three functional groups are present, the substitution pattern is 1,3,5 (Nollet and Gutierrez-Uribe 2018). When all three substituents are identical, it is designated as a *meta-tri*-substitution pattern. The 1,2,6-substitution pattern is indicated by the prefix '*vic*' (Fig. 2.1).

The most common examples of simple phenols present in foods are *o*-cresol, *m*-cresol, *p*-cresol, *o*-xylenol, *m*-xylenol, *p*-xylenol, guaiacol, 4-ethyl guaiacol, methyl cresol, eugenol, isoeugenol, 2-ethylphenol, 2,3,5-trimethyphenol, syringol, 4-vinylsyringol, and 4-propyl phenol. Simple phenols had been reported to be present in a variety of foods. For example, Cocoa contains different cresol, xylenol and

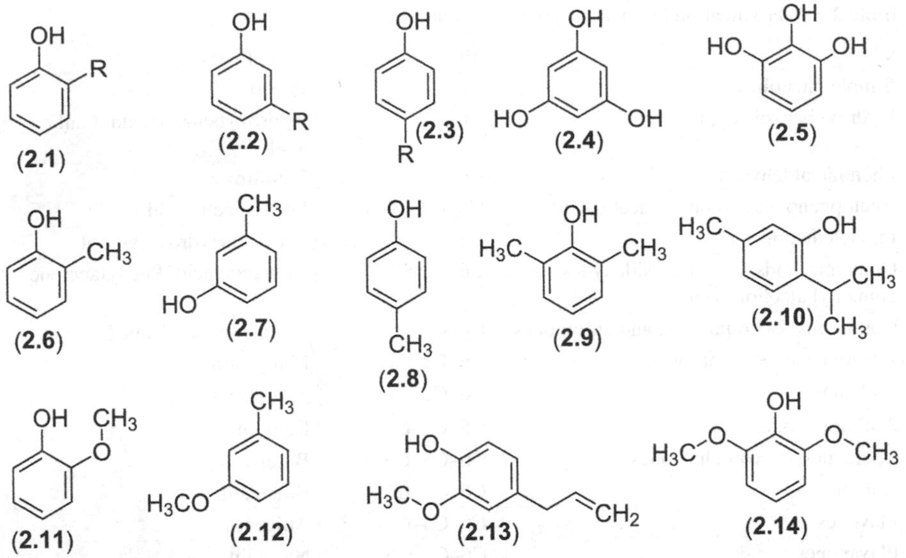

Fig. 2.1 Structures of simple phenolic compounds. (**2.1**) *ortho*-position, (**2.2**) *meta*-position, (**2.3**) *para*-position, (**2.4**) *meta*-tri, (**2.5**) *Vic*-tri, (**2.6**) *o*-cresol, (**2.7**) *m*-cresol, (**2.8**) *p*-cresol, (**2.9**) *o*-xylenol, (**2.10**) thymol, (**2.11**) guaiacol, (**2.12**) 1-methoxy-3-methylbenzene, (**2.13**) 2-methoxy-4-allylphenol, and (**2.14**) syringol

guaiacol derivatives. Coffee, tea, fruit juices, beers, and vine contains simple phenols (Maga 1978). Olive oil obtained from different countries around the globe is also rich in simple phenols like hydroxybenzoic acids and their derivatives (Ocakoglu et al. 2009). Several foods such as smoked foods, tea, and natural vanilla are a rich source of simple phenols. The contents of simple phenols had been reported in some smoked foods (Clifford 2000). The given amount may vary with increased smoking duration. It is important to mention that catechol, and 4-methoxy phenol are carcinogenic at low to high doses (Hirose et al. 1998).

2.2.2   Hydroxybenzoic Acids

The term phenolic acid is usually described to all phenolic compounds having a carboxylic acid group. Thus phenolic acids are classified into derivatives of hydroxybenzoic acid and cinnamic acid. The latter is now classed separately due to relatively different occurrence and properties. In the first case, the presence of a carboxylic acid group on a phenyl ring at a specific position classified it into phenolic acids. The first example of phenolic acid is hydroxyl benzoic acid. In this case, the hydroxyl group is at *para*-position and thus normally called *p*-hydroxybenzoic acid. The hydroxyl group may be one as in *p*-hydroxybenzoic acid and salicylic acid, two in protocatechuic acid, and three in gallic acid. Replacement of a

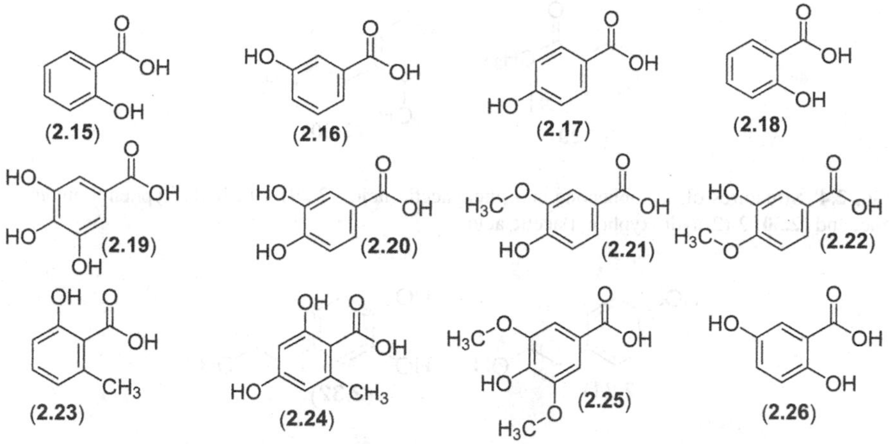

Fig. 2.2 Structures of hydroxybenzoic acids. (**2.15**) 2-Hydroxybenzoic acid, (**2.16**) 3-hydroxybenzoic acid, (**2.17**) 4-hydroxybenzoic acid, (**2.18**) salicylic acid, (**2.19**) 3,4,5-trihydroxybenzoic acid, (**2.20**) protocatechuic acid, (**2.21**) vanillic acid, (**2.22**) isovanillic acid, (**2.23**) 6-methylsalicylic acid, (**2.24**) orsellinic acid, (**2.25**) syringic acid, and (**2.26**) gentisic acid

Fig. 2.3 Structures of phenolic aldehydes. (**2.27**) vanillin, and (**2.28**) ethylvanillin

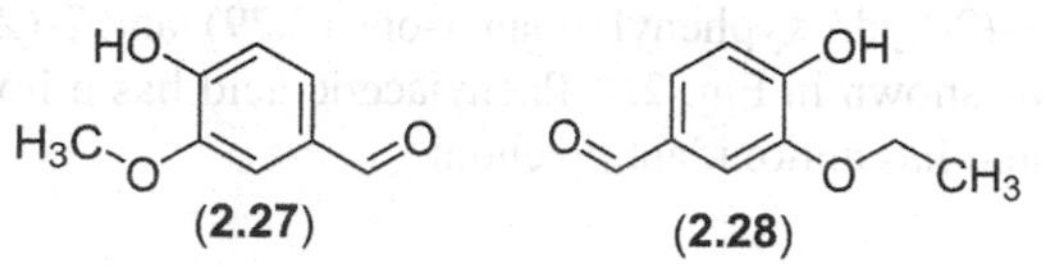

hydrogen atom by a methyl group(s) on the benzene ring may also occur. For example, a methoxy group at *meta*-position as in the case of vanillic acid (**2.21**) as shown in Fig. 2.2. Phenolic acids are present in a variety of foods of plant origin (Khadem and Marles 2010; Heleno et al. 2015).

2.2.3 Phenolic Aldehydes

Phenolic aldehyde is distinguished from hydroxybenzoic acid based on the replacement of the carboxylic acid group by an aldehyde group. The most important example is vanillin (**2.27**). It was first isolated from vanilla in 1858. Another derivative of vanillin is ethylvanillin (**2.28**). It differs from vanillin in having an ethoxy group instead of methoxy as shown in Fig. 2.3. Synthetic vanillin synthesized from guaiacol or lignin is widely used as a flavouring agent in foods for vanilla flavor.

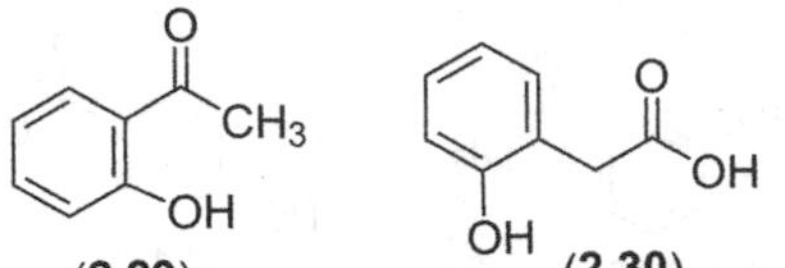

(2.29)   **(2.30)**

Fig. 2.4 Structures of acetophenone and phenyacetic acid. (**2.29**) 1-(2-hydroxyphenyl)ethan-1-one, and (**2.30**) 2-(2-hydroxyphenyl)acetic acid

(2.31)   **(2.32)**

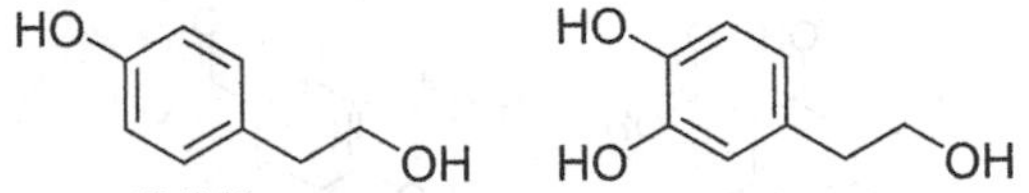

Fig. 2.5 Structures of phenylethanoids. (**2.31**) 4-(2-Hydroxyethyl)phenol (*Tyrosol*), and (**2.32**) 4-(2-Hydroxyethyl)-1,2-benzenediol (*Hydroxytyrosol*)

2.2.4   Acetophenone and Phenylacetic Acids

Acetophenone and phenylacetic acids are classified as the C6–C2 group of phenolic compounds. Phenones are found rarely in foods. Examples of this class are 1-(2-hydroxyphenyl)ethan-1-one (**2.29**), and 2-(2-hydroxyphenyl)acetic acid (**2.30**) as shown in Fig. 2.4. Phenylacetic acid has a honey-like odor and is a white solid and has antioxidant potential.

2.2.5   Phenylethanoids

Phenolic compounds with characteristic phenyl and ethanol group are called phenylethanoid. These compounds are represented by C6–C2. The common examples are tyrosol (4-(2-Hydroxyethyl)phenol), and hydroxytyrosol (4-(2-Hydroxyethyl)-1,2-benzenediol) as shown in Fig. 2.5. Both tyrosol (**2.31**) and hydroxytyrosol (**2.32**) possess in-vitro antioxidant activities (Covas et al. 2003). They are present in wine and olive oils.

2.2.6   Cinnamic Acids, Aldehydes and Alcohols

Cinnamic acids are phenolic compounds with the backbone of C6–C3 and are also known as phenylpropanoids. The first important member of the class is cinnamic acid (**2.33**). The propyl substituent has a double bond, which may be either *cis* or *trans* (represented as a crossed bond). The trans-isomers are common in nature as

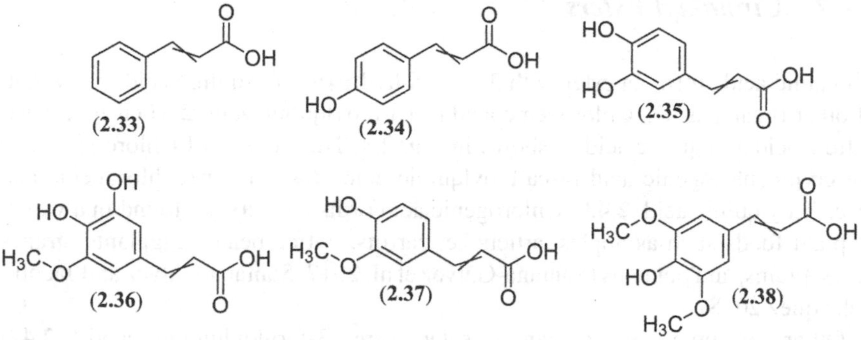

Fig. 2.6 Structures of important cinnamic acids present in foods. The crossed double bond represents both *E* and *Z* isomers. (**2.33**) Cinnamic acid, (**2.34**) p-coumaric acid, (**2.35**) caffeic acid, (**2.36**) 5-hydroxyferulic acid, (**2.37**) ferulic acid, and (**2.38**) sinapic acid

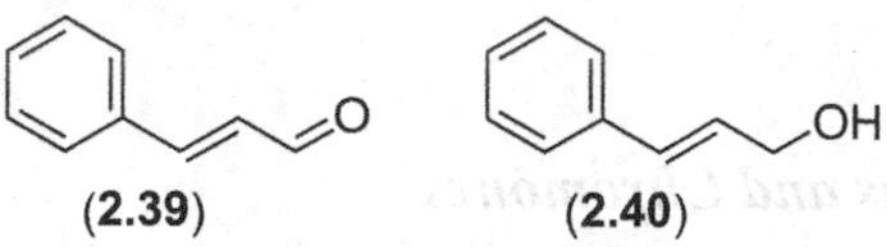

Fig. 2.7 Structures of cinnamic aldehydes and alcohols. (**2.39**) (2*E*)-3-phenylprop-2-enal (*Cinnamaldehyde*), and (**2.40**) (2*E*)-3-phenylprop-2-en-1-ol (*Cinnamyl alcohol*)

compared to their cis-isomers. (**2.33**) Cinnamic acid, (**2.34**) p-coumaric acid, (**2.35**) caffeic acid, (**2.36**) 5-hydroxyferulic acid, (**2.37**) ferulic acid, and (**2.38**) sinapic acid are members of cinnamic acids as shown in Fig. 2.6. Cinnamic acid has a honey-like odor and is a white powder. It was originally obtained from the Cinnamomum family of plants. Cinnamic acid derivatives are more potent antioxidants than its counterpart hydroxybenzoic acid derivatives (Natella et al. 1999). The substitution on the aromatic ring was found to be responsible for varied antioxidant activities. The order of antioxidant activity was *p*-hydroxydimethoxy > dihydroxy > *p*-hydroxymethoxy > *p*-hydroxy.

Cinnamaldehyde (**2.39**) has a similar backbone of C6–C3 having aldehyde functional group (Fig. 2.7). The trans (*E*-) isomer is predominately present in nature. It is a viscous yellowish liquid present in cinnamon. The characteristic odor and flavor of the cinnamon are imparted by cinnamaldehyde. It is widely used as a flavoring agent in ice cream, candies, chewing gum and beverages.

Cinnamyl alcohol (**2.40**) has an alcoholic functional moiety in place of carboxylic or aldehyde groups (Fig. 2.7). Cinnamyl alcohol is also present in cinnamon leaves. It is a white crystalline solid. It has a distinctively sweet, spicy odor, and thus commonly used as an odorant.

2.2.7   Cinnamyl Esters

Cinnamic acids form an ester with 3, 4 or 5-hydroxyl of L-quinic acid or a variety of other organic acids. Chlorogenic acid (3-caffeoylquinic acid, **2.41**) is an ester of caffeic acid and quinic acid as shown in Fig. 2.8. The isomers of Chlorogenic acid are cryptochlorogenic acid (4-caffeoylquinic acid, **2.42**) and neochlorogenic acid (5-caffeoylquinic acid, **2.43**). Chlorogenic acid or its isomers are found in a variety of plant foods such as apples, artichoke, carrots, coffee beans, eggplants, grapes, pears, plums, and potatoes (Santana-Gálvez et al. 2017; Santana-Gálvez and Jacobo-Velázquez 2018).

Other examples of cinnamyl esters are 3-feruloylquinic acid (**2.44**), 4-feruloylquinic acid, 5-feruloylquinic acid (**2.45**), ethyl cinnamate (**2.46**), 3-*p*-coumaroylquinic acid (**2.47**), 4-*p*-coumaroylquinic acid (**2.48**), and 5-*p*-coumaroylquinic acid (**2.49**). These esters are present in stone and pome fruits (Möller and Herrmann 1983). In foods, the second moiety in cinnamic ester may be monosaccharides, which are separately classed as glycosides.

2.2.8   Coumarins and Chromones

Coumarins have the structural backbone of C6–C3 consists of oxygen heterocycle as part of the molecule. They are also known as 1,2-benzopyrones consisting of fused benzene and pyrone rings. Coumarins have been classified as Simple coumarins, Furanocoumarins, Pyranocoumarins, Bis-, and Triscoumarins and Coumarinolignans (Borges et al. 2005). Coumarin (**2.50**), 7-hydroxycoumarin

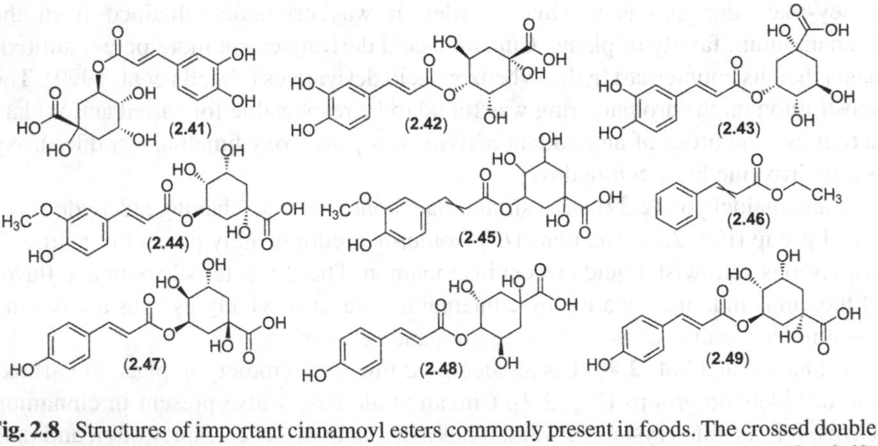

Fig. 2.8 Structures of important cinnamoyl esters commonly present in foods. The crossed double bond represents both *E* and *Z* isomers. (**2.41**) 3-Caffeoylquinic acid (*chlorogenic acid*), (**2.42**) 4-caffeoylquinic acid (*cryptochlorogenic acid*), (**2.43**) 5-caffeoylquinic acid (*neochlorogenic acid*), (**2.44**) 3-feruloylquinic acid, (**2.45**) 5-feruloylquinic acid, (**2.46**) ethyl cinnamate, (**2.47**) 3-coumaroylquinic acid, (**2.48**) 4-coumaroylquinic acid, and (**2.49**) 5-coumaroylquinic acid

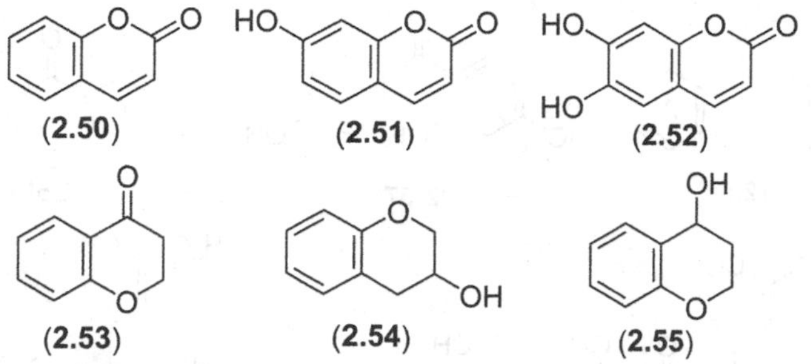

Fig. 2.9 Structures of coumarins and chromones. (**2.50**) Coumarin backbone, (**2.51**) 7-hydroxycoumarin, (**2.52**) esculetin, (**2.53**) chroman-4-one, (**2.54**) chroman-3-ol, and (**2.55**) chroman-4-ol

(**2.51**), bergenin, and esculetin (**2.52**) are common Coumarins present in foods (Fig. 2.9). Coumarins are widely distributed in plants, and also in plant foods such as citrus fruits (Stanley and Jurd 1971). Coumarins and their derivatives are strong antioxidants in in-vitro model systems (Kostova et al. 2011).

Chromones and its derivatives such as isocoumarins, alcohols, and ketones have also a structural backbone of C6–C3. They are distinguished from Coumarins in the absence of a double bond at the oxygen-containing heterocyclic ring (Saengchantara and Wallace 1986). The examples of common chromones are chroman-4-one (**2.53**), chroman-3-ol (**2.54**), and chroman-4-ol (**2.55**). They are present in foods such as Aloe (Sun et al. 2016) and its core contributed to the antioxidant potential (Dias et al. 2011).

2.2.9 Benzophenones and Xanthones

Benzophenones and xanthones are classed with the skeleton of C6–C1–C6 with two aromatic rings. In the case of benzophenones (**2.56**), the C1 is a ketone functional moiety, whereas, in xanthones, an oxygen heterocycle ring is present as shown in Fig. 2.10. The two rings are labeled as A and B. Benzophenones and its derivatives are present in various fruits (Baggett et al. 2005). Several benzophenone derivatives i.e., 13,14-didehydoxyisogarcinol, hydroxygarcimultiflorone B, and garcimultiflorone C were isolated from *Garcinia multiflora* fruits (Chen et al. 2009). In xanthones, the isoprene, methoxy, and hydroxy groups may be present at various locations on the A and B rings forming a diverse array of compounds (Santana-Gálvez and Jacobo-Velázquez 2018). More than 500 xanthones are known (Andrés-Lacueva et al. 2010). Xanthones were therefore classified into five classes; (a) Simple oxygenated xanthones, (b) xanthone glycosides, (c) prenylated and related xanthones, (d) xanthonolignoids, and (e) miscellaneous xanthones (Peres et al. 2000). The prenylated xanthones are commonly present in a variety of fruits including Mangosteen (Kaur et al. 2012). Important examples of xanthones are mangostin and 8-desoxygartanin present in edible fruits (Lyles et al. 2014).

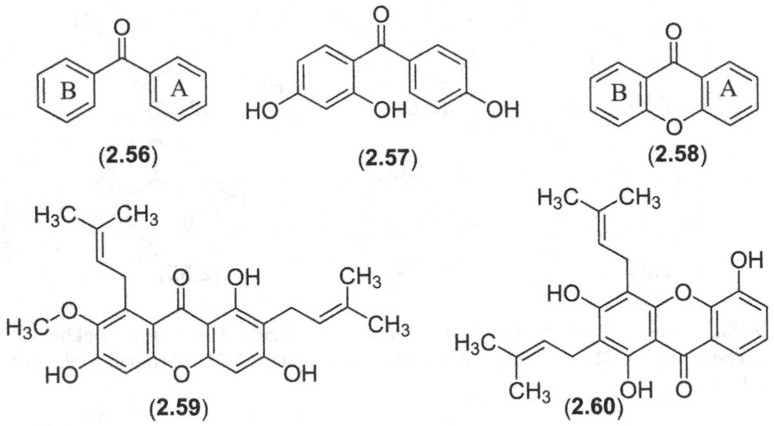

Fig. 2.10 Structures of benzophenones and xanthones. (**2.56**) Benzophenone backbone, (**2.57**) 2,4,4′-trihydroxybenzophenone, (**2.58**) xanthone, (**2.59**) 3,6,8-trihydroxy-2-methoxy-1,7-bis(3-methylbut-2-enyl)xanthen-9-one (*mangostin*), and (**2.60**) 1,3,5-trihydroxy-2,4-bis(3-methylbut-2-enyl)xanthen-9-one (*8-desoxygartanin*)

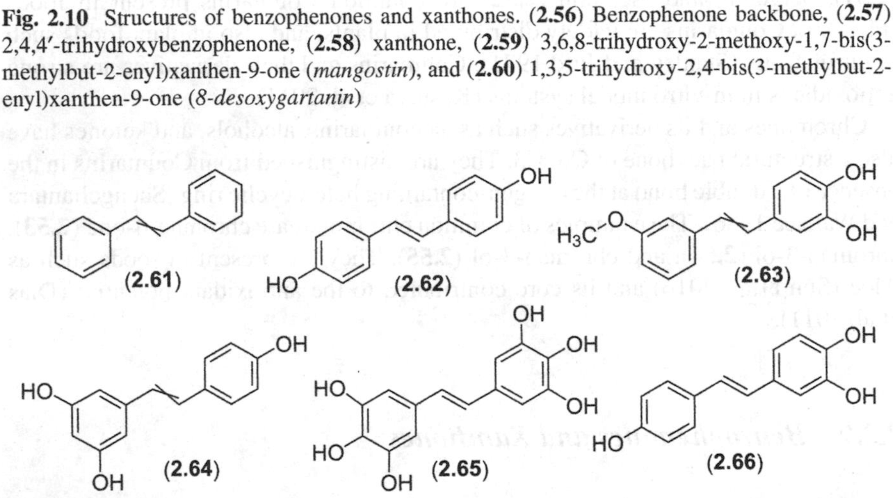

Fig. 2.11 Structures of stilbenes. The crossed double bond represents both *E* and *Z* isomers. (**2.61**) Stilbene, (**2.62**) 4,4′-Dihydroxy-*E*-stilbene, (**2.63**) 3,4-dihydroxy-3′-methoxystilbene, (**2.64**) 3,4′,5-trihydroxystilbene (*resveratrol*), (**2.65**) hexahydroxy-*E*-stilbene, and (**2.66**) 3,4,4′-trihydroxy-*E*-stilbene

2.2.10 Stilbenes

Stilbenes are class of phenolic compounds characterized by the skeleton of C6–C2–C6. They have 1,2-diphenylethylene (**2.61**) nucleus and can be divided into two classes, i.e., monomeric and oligomeric stilbenes (Shen et al. 2009). The ethylene bond may be either cis or trans. Some common examples are 4,4′-dihydroxy-*E*-stilbene (**2.62**), 3,4-dihydroxy-3′-methoxystilbene (**2.63**), resveratrol (**2.64**), hexahydroxy-*E*-stilbene (**2.65**), and 3,4,4′-trihydroxy-*E*-stilbene (**2.66**) (Fig. 2.11).

There are more than 300 derivatives of stilbenes reported, many of these are present in foods. Stilbenes range from simple to oligomers, polymers, and glycosides (Likhtenshtein 2009). Monomeric stilbenes may vary in the number and position of hydroxyl groups; substitution with sugars, methyl, methoxy, and other groups; steric configuration (*E*, *Z*-isomerism). The oligomers such as dimers, trimers, tetramers, etc. are formed from the condensation of different stilbene monomers.

The most important stilbene is resveratrol and its derivatives (Kasiotis et al. 2013). Almonds, for example, contain *trans*-resveratrol, d4-resveratrol, dienestrol, hexestrol, oxyresveratrol, piceatannol, pterostilbene, and resveratrol-3-β-glucoside (Xie and Bolling 2014). Several stilbenes have been found in different wines and their products, legumes, berries, and fruits (Vitrac et al. 2005; Santana-Gálvez and Jacobo-Velázquez 2018).

2.2.11   Anthraquinones

Anthraquinones are those phenolic compounds with the basic skeleton of C6–C2–C6. However, the C2 is different from the stilbenes in respect of two ketonic bridges forming oxygen heterocycle. Anthraquinones are one of the largest group of natural pigments, comprising more than 700 compounds (Diaz-Muñoz et al. 2018; Dufossé 2014). The diversity of the anthraquinones is based on the nature and the position of the substituents. For example, replacing the H atoms on the basic structure (R1–R8), as diverse as –OH, –CH$_3$, –OCH$_3$, –CH$_2$OH, –CHO, –COOH, or more complex groups. When all R groups are replaced by hydroxyl groups, the molecule is known as hydroxyanthraquinone (HAQN). From their structure, HAQN derivatives absorb visible light and are therefore colored (Fouillaud et al. 2016).

They are present in all parts of the plants of families such as Fabaceae, Rhamnaceae, Liliaceae, and Scrophulariaceae. The most common anthraquinones widely reported are anthraquinone (**2.67**), 1,3,8-trihydroxy-6-methylanthracene-9,10-dione (*Emodin*, **2.68**), 1,8-dihydroxy-3-methoxy-6-methylanthracene-9,10-dione (*Physcion*, **2.69**), 1,4,5,7-tetrahydroxy-2-methylanthracene-9,10-dione (*Catenarin*, **2.70**), 4,5-dihydroxy-9,10-dioxoanthracene-2-carboxylic acid (*Rhein*, **2.71**), and 3, 1-hydroxy-3-methylanthraquinone (**2.72**) as shown in Fig. 2.12.

2.2.12   Chalcones

Chalcones and its derivatives are phenolic compounds made of the basic skeleton of C6–C3–C6 (**2.82**). The C3 Bridge is connecting the two aromatic rings, which may contain a double bond, functional group or any substituents as shown in Fig. 2.13. Chalcones such as butein ((E)-1-(2,4-dihydroxy phenyl)-3-(3,4-dihydroxy phenyl) prop-2-en-1-one) is responsible for yellowish color of flowers. Chalcones are converted to food sweetener called dihydrochalcone (Janeczko et al. 2013). Chalcones

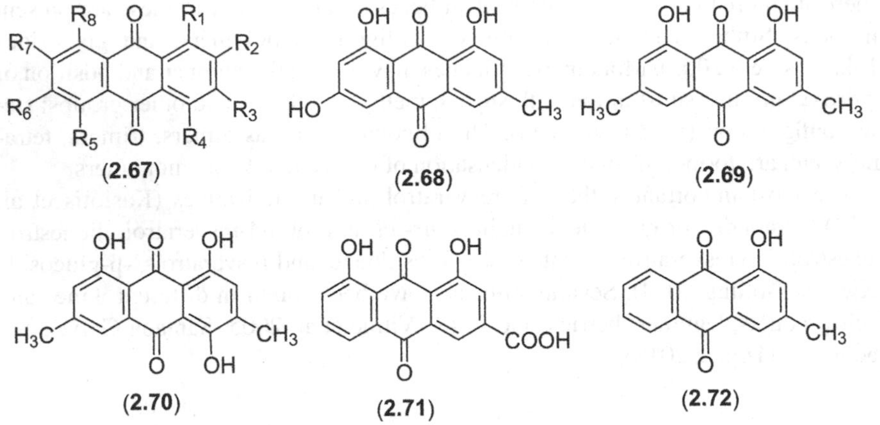

Fig. 2.12 Structures of anthraquinones. (**2.67**) Anthraquinone, (**2.68**) 1,3,8-Trihydroxy-6-methylanthracene-9,10-dione (*Emodin*), (**2.69**) 1,8-dihydroxy-3-methoxy-6-methylanthracene-9,10-dione (*Physcion*), (**2.70**) 1,4,5,7-tetrahydroxy-2-methylanthracene-9,10-dione (*Catenarin*), (**2.71**) 4,5-dihydroxy-9,10-dioxoanthracene-2-carboxylic acid (*Rhein*), and (**2.72**) 3, 1-hydroxy-3-methylanthraquinone

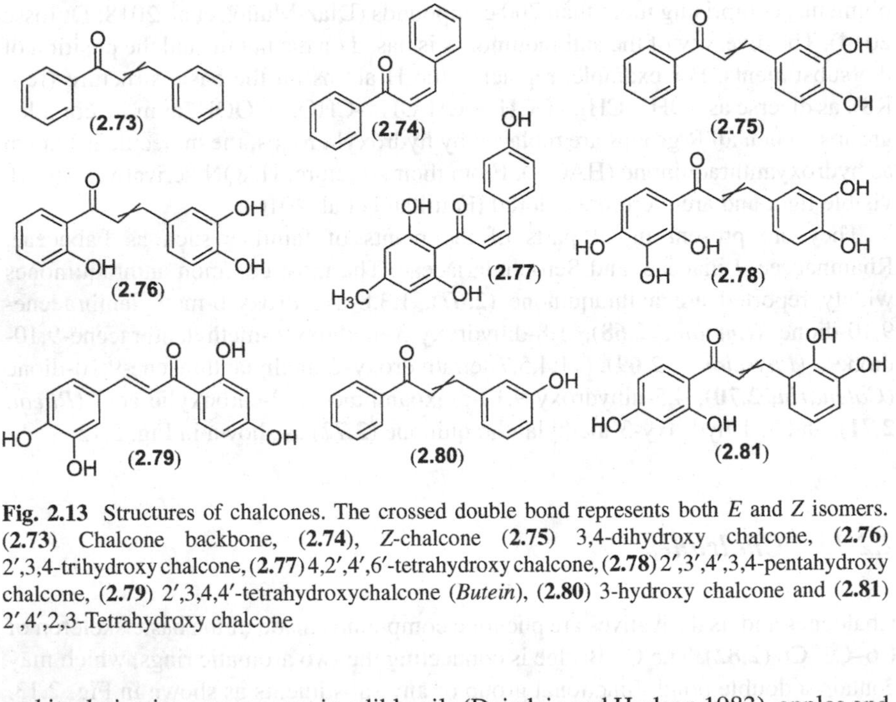

Fig. 2.13 Structures of chalcones. The crossed double bond represents both *E* and *Z* isomers. (**2.73**) Chalcone backbone, (**2.74**), *Z*-chalcone (**2.75**) 3,4-dihydroxy chalcone, (**2.76**) 2′,3,4-trihydroxy chalcone, (**2.77**) 4,2′,4′,6′-tetrahydroxy chalcone, (**2.78**) 2′,3′,4′,3,4-pentahydroxy chalcone, (**2.79**) 2′,3,4,4′-tetrahydroxychalcone (*Butein*), (**2.80**) 3-hydroxy chalcone and (**2.81**) 2′,4′,2,3-Tetrahydroxy chalcone

and its derivatives are present in edible oils (Dziedzic and Hudson 1983), apples and citrus fruits (Tomás-Barberán and Clifford 2000a, b; Mah 2019). Chalcones also form an ester with monosaccharides to form glycosides.

2.2.13 Aurones

Aurones (from Latin *aurum* mean gold), are those phenolic compounds which are formed from cyclization of chalcones. The third oxygen heterocycle is a five-member ring formed from the hydroxyl group on the meta-position of ring A with α-carbon of the C3 mid-chain as shown in Fig. 2.14 (Popova et al. 2019). There are more than a hundred aurones structures identified in a variety of flowers. The flowers of the eudicots and monocots possess aurones such as bracteatin (**2.83**), aureusidin (**2.84**), trihydroxyaurone (**2.85**), rugaurone A, B, and C, etc. (Boucherle et al. 2017). Aurones also form an ester with monosaccharides to form glycosides.

2.2.14 Flavonoids

Flavonoids are phenolic compounds having 15 carbons with a C6–C3–C6 skeleton. The chemical structure consists of three rings i.e. A, B and C. Ring A and B are benzene ring, whereas ring B is a pyran. Thus the backbone is 2-phenyl-3,4-dihydro-2H-chromene as shown in Fig. 2.15. Flavonoids are a highly diverse group of phenolic compounds, which are an essential component of the plant cells. They can be classified on several factors such as structure, functional groups, properties and entropy of information theory (Castellano et al. 2013).

Based on structure, flavonoids can be divided into different subgroups depending on the carbon of the C ring on which the B ring is attached and the degree of

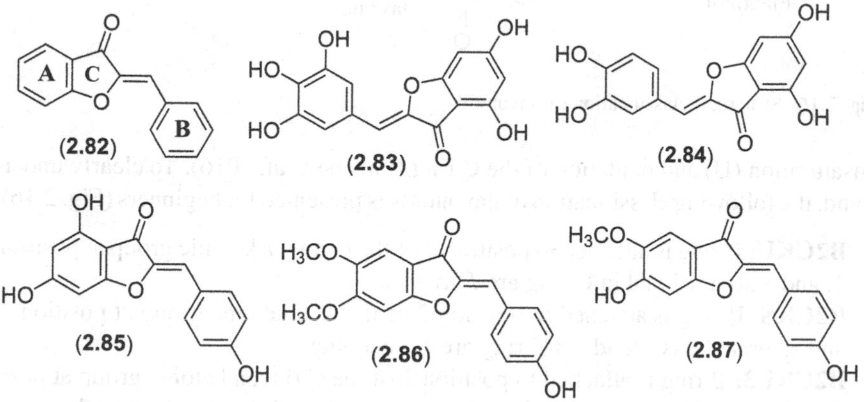

Fig. 2.14 Structures of aurones. (**2.82**) Aurone backbone, (**2.83**), (2Z)-4,6-dihydroxy-2-[(3,4,5-trihydroxyphenyl)methylidene]-1-benzofuran-3-one (*Bracteatin*) (**2.84**) (2Z)-2-[(3,4-dihydroxyphenyl)methylidene]-4,6-dihydroxy-1-benzofuran-3-one (*Aureusidin*), (**2.85**) 4,6,4′-trihydroxyaurone, (**2.86**) (2Z)-2-[(4-hydroxyphenyl)methylidene]-5,6-dimethoxy-1-benzofuran-3-one (*Rugaurone A*), and (**2.87**) (2Z)-5-hydroxy-2-[(4-hydroxyphenyl)methylidene]-6-methoxy-1-benzofuran-3-one (*Rugaurone B*)

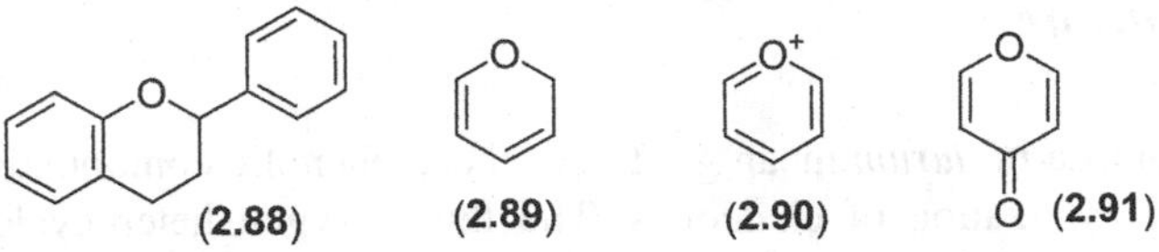

Fig. 2.15 Structural backbones of flavonoids. (**2.88**) Flavan, (**2.89**) pyran, (**2.90**) pyrylium, and (**2.91**) pyrone

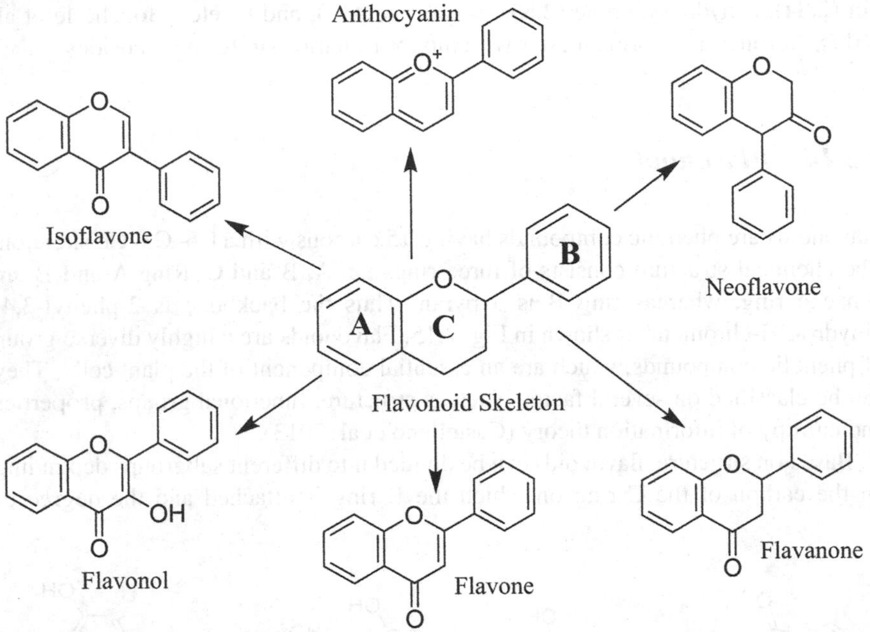

Fig. 2.16 Structural classification of flavonoids

unsaturation (U) and oxidation of the C ring (Panche et al. 2016). To clearly understand, the following classification of flavonoids is presented for beginners (Fig. 2.16):

- **B2CKU:** B ring is attached to position 2 of the C ring, a ketonic group at position 4, and a double bond in C ring are *Flavones*.
- **B2CKS:** B ring is attached to position 2 of the C, a ketonic group at position 4, and saturated C–C bonds in C ring are *Flavanones*
- **B2CKU3:** B ring is attached to position 2 of the C ring, a ketonic group at position 4, and a double bond, hydroxyl group at C-3 in C ring are *Flavonols*.
- **B2CS3:** B ring is attached to position 2 of the C ring, saturation, and hydroxyl group at C-3 in C ring are *Flavanols*.
- **B3C:** B ring is attached to position 3 of the C ring and are *Isoflavones*.
- **B4C:** B ring is attached to position 4 of the C ring are *Neoflavones*.

2.2.14.1 Flavones

Flavones are a diverse subgroup of flavonoids present in nearly all plant foods. They can be denoted as B2CKU, which represents attachment of B ring to C, a ketone group (K) and unsaturation (U). The flavones have a backbone of 2-phenyl-1-benzopyran-4-one (**2.92**) and an unsaturation at C2 and C3 in C-ring as shown in Fig. 2.17. Plant foods are rich in flavones such as 6-hydroxy flavone (**2.93**), apigenin (**2.94**), luteolin, acacetin (**2.95**), diosmetin (**2.96**), chrysoeriol (**2.97**), hispidulin (**2.98**), oroxylin A (**2.99**), chrysin (**2.100**), baicalein (**2.101**), wogonin (**2.102**) and tangeritin (**2.103**) (Hostetler et al. 2017; Jiang et al. 2016).

2.2.14.2 Flavanones

Flavanones are distinguished from the flavone in absence of a double bond in ring-C. The rest of the molecule may have different substituents attached as shown in Fig. 2.18. The absence of a double bond (at C2–C3), the presence of a chiral carbon atom at the C2 position and the absence of substitution at the C3 position of the C ring marks the structural differences characterizing flavanones as a separate class of flavonoids (Barreca et al. 2017; Yamane and Kato 2012). There are more than 300 flavanones present in nature and more than a hundred as glycosides. The most recurrent examples of flavanones are those present in citrus fruits. These include

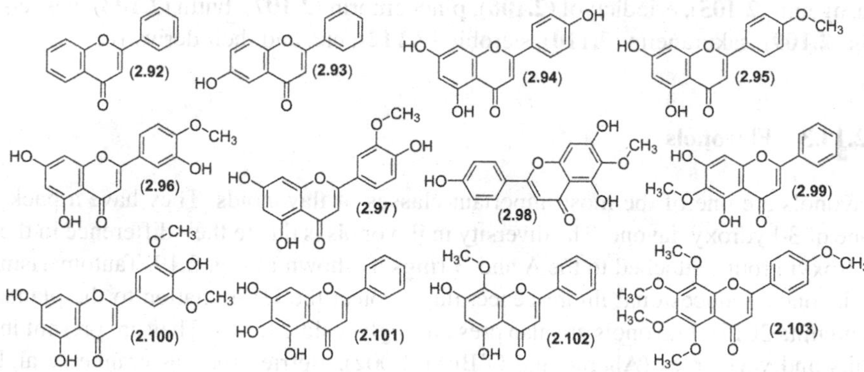

Fig. 2.17 Structures of important flavones present in foods. (**2.92**) Flavone, (**2.93**) 6-Hydroxy flavone, (**2.94**) 5,7-dihydroxy-2-(4-hydroxyphenyl)chromen-4-one (*Apigenin*), (**2.95**) 5,7-dihydroxy-2-(4-methoxyphenyl)chromen-4-one (*Acacetin*), (**2.96**) 5,7-dihydroxy-2-(3-hydroxy-4-methoxyphenyl)chromen-4-one (*Diosmetin*), (**2.97**) 5,7-dihydroxy-2-(4-hydroxy-3-methoxyphenyl)chromen-4-one (*Chrysoeriol*), (**2.98**) 5,7-dihydroxy-2-(4-hydroxyphenyl)-6-methoxychromen-4-one (*Hispidulin*), (**2.99**) 5,7-dihydroxy-6-methoxy-2-phenylchromen-4-one (*Oroxylin A*), (**2.100**) 5,7-dihydroxy-2-phenylchromen-4-one (*Chrysin*), (**2.101**) 5,6,7-trihydroxy-2-phenylchromen-4-one (*Baicalein*), (**2.102**) 5,7-dihydroxy-8-methoxy-2-phenylchromen-4-one (*Wogonin*), and (**2.103**) 5,6,7,8-tetramethoxy-2-(4-methoxyphenyl)chromen-4-one (*Tangeritin*)

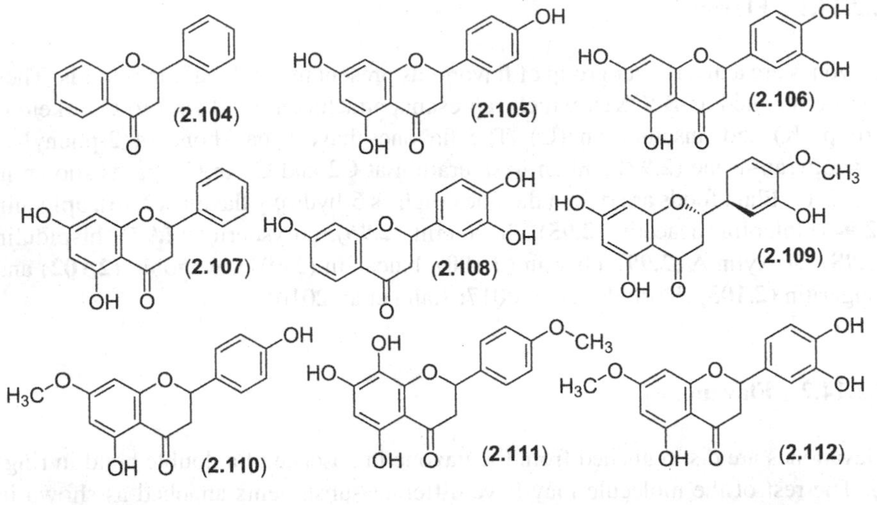

Fig. 2.18 Structures of flavanones present in foods. (**2.104**) Flavanone backbone, (**2.105**) 5,7-dihydroxy-2-(4-hydroxyphenyl)-2,3-dihydrochromen-4-one (*Naringenin*), (**2.106**) (2S)-2-(3,4-dihydroxyphenyl)-5,7-dihydroxy-2,3-dihydrochromen-4-one (*Eriodictyol*), (**2.107**) (2S)-5,7-dihydroxy-2-phenyl-2,3-dihydrochromen-4-one (*Pinocembrin*), (**2.108**) (2S)-2-(3,4-dihydroxyphenyl)-7-hydroxy-2,3-dihydrochromen-4-one (*Butin*), (**2.109**) (2S)-5,7-dihydroxy-2-(3-hydroxy-4-methoxyphenyl)-2,3-dihydrochromen-4-one (*Hesperetin*), (**2.110**) (2S)-5-hydroxy-2-(4-hydroxyphenyl)-7-methoxy-2,3-dihydrochromen-4-one (*Sakuranetin*), (**2.111**) 8-Hydroxynaringenin-4′-methylether, and (**2.112**) 2-(3,4-dihydroxyphenyl)-5-hydroxy-7-methoxychroman-4-one (*Sterobin*)

naringenin (**2.105**), eriodictyol (**2.106**), pinocembrin (**2.107**), butin (**2.108**), hesperetin (**2.109**), sakuranetin (**2.110**), sterobin (**2.112**), etc. and their derivatives.

2.2.14.3 Flavonols

Flavonols are one of the most important classes of flavonoids. They have a backbone of 3-hydroxy flavone. The diversity in flavonols is due to their difference in the hydroxyl groups attached to the A and B rings as shown in Fig. 2.19. Tautomerism in flavonols induces dual fluorescence, thus protect the UV-radiation to the plants (Sengupta 2017). Flavonols are also present in glycosidic forms. They are present in fruits and vegetables (Aherne and O'Brien 2002). Berries such as cranberry and blueberry are especially rich in flavonols (Wang et al. 2019b; Kamiloglu et al. 2019).

2.2.14.4 Flavanols

Flavanols are distinguished from flavonols with the absence of a ketonic group at 4-position and a double bond in C-ring. Catechin (**2.126**), epicatechin (**2.127**), and epigallocatechin (**2.128**) (Fig. 2.20) are the most important Flavanols present in

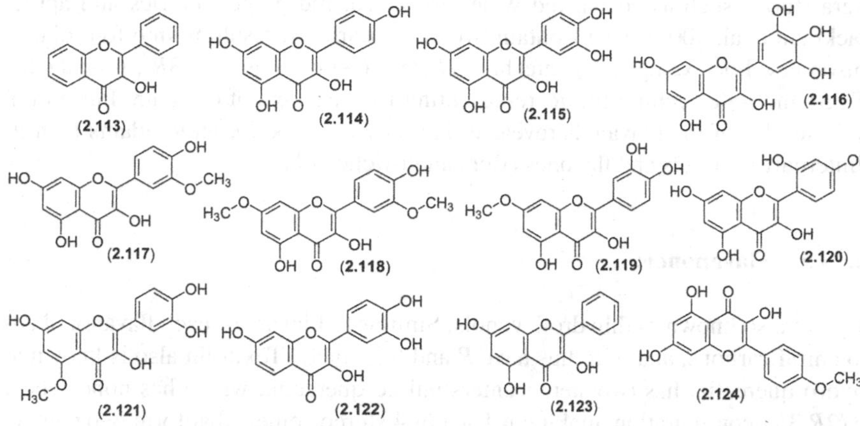

Fig. 2.19 Structures of important flavonols present in foods. (**2.113**) Flavonol, (**2.114**) 3,5,7-trihy droxy-2-(4-hydroxyphenyl)chromen-4-one (Kaempferol), (**2.115**) 2-(3,4-dihydroxyphenyl)-3,5,7-trihydroxychromen-4-one (*Quercetin*), (**2.116**) 3,5,7-trihydroxy-2-(3,4,5-trihydroxyphenyl) chromen-4-one (*Myricetin*), (**2.117**) 3,5,7-trihydroxy-2-(4-hydroxy-3-methoxyphenyl)chromen-4-one(*Isorhamnetin*), (**2.118**)3,5-dihydroxy-2-(4-hydroxy-3-methoxyphenyl)-7-methoxychromen-4-one (*Rhamnazin*), (**2.119**) 2-(3,4-dihydroxyphenyl)-3,5-dihydroxy-7-methoxychromen-4-one (*Rhamnetin*), (**2.120**) 2-(2,4-dihydroxyphenyl)-3,5,7-trihydroxychromen-4-one (*Morin*), (**2.121**) 2-(3,4-dihydroxyphenyl)-3,7-dihydroxy-5-methoxychromen-4-one (*Azaleatin*), (**2.122**) 2-(3,4-dihydroxyphenyl)-3,7-dihydroxychromen-4-one (*Fisetin*), (**2.123**) 3,5,7-trihydroxy-2-phenylchromen-4-one (*Galangin*), and (**2.124**) 3,5,7-trihydroxy-2-(4-methoxyphenyl)chromen-4-one (*kaempferide*)

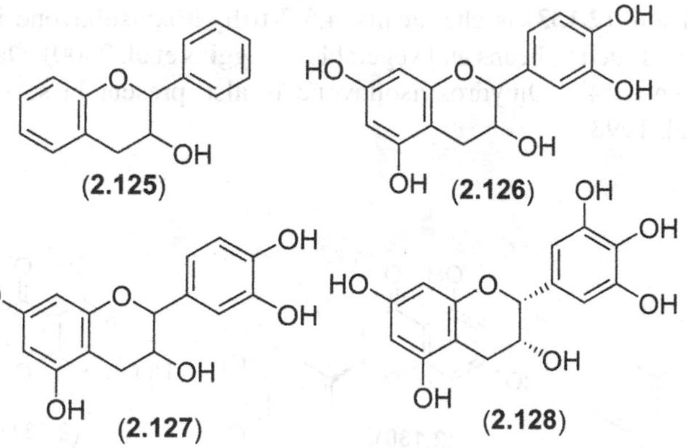

Fig. 2.20 Structures of flavonols present in foods. (**2.125**) Flavan-3-ol, (**2.126**) (2*R*,3*S*)-2-(3,4-dihydroxyphenyl)-3,4-dihydro-2*H*-chromene-3,5,7-triol (*catechin*), (**2.127**) (2*R*,3*R*)-2-(3,4-dihydroxyphenyl)-3,4-dihydro-2*H*-chromene-3,5,7-triol(*epicatechin*),and(**2.128**)(2*R*,3*R*)-2-(3,4,5-trihydroxyphenyl)-3,4-dihydro-2*H*-chromene-3,5,7-triol (*epigallocatechin*)

several foods such as cocoa, red wine, green tea, red grapes, berries and apples (Hackman et al. 2007). They contain two chiral carbons, resulting into four diastereoisomers. For example (+)-catechin (2R,3S), (−)-catechin (2S,3R), (-)-catechin (2S,3S) and (-)-catechin (2R,3R) representing four isomers of catechin. Pulse radiolysis studies of the Flavanols revealed that they are superior antioxidants than its counterpart flavonols and flavones (Bors and Michel 1999).

2.2.14.5 Flavanonols

They are also known as dihydroflavonols. Similar to Flavanols, each flavanonol has two chiral carbons, and will thus have R and S isomers. Toxifolin also is known as dihydro quercetin, has two stereocenters unlike quercetin, which has none. These are (2R,3R)-configuration, making it 1 out of 4 stereoisomers that form two pairs of enantiomers. However, keto-enol tautomerism is absent in taxifolin (**2.130**) and present in quercetin. The absence of a double bond, however, reduces the importance of the hydroxyl group at carbon-3 of C-ring (Trouillas et al. 2006). Similar is the case of Aromadendrin (**2.131**) also known as dihydrokaempferol as shown in Fig. 2.21. These flavanonols also form glycosides in plants.

2.2.14.6 Isoflavones

Isoflavones are those phenolic compounds in which phenyl ring-B is attached to carbon-3 of the ring-C. They have a similar backbone of the C6–C3–C6 skeleton (Fig. 2.22). Many of the isoflavones act as phytoestrogens in animals (Fotsis et al. 1993). Genistein (**2.133**) is chemically 4,5,7-trihydroxyisoflavone is present in a variety of foods such as beans and vegetables (Liggins et al. 2000). Daidzein (**2.135**) is also known as 4′,7-Dihydroxyisoflavone is also present in several vegetables (Coward et al. 1998).

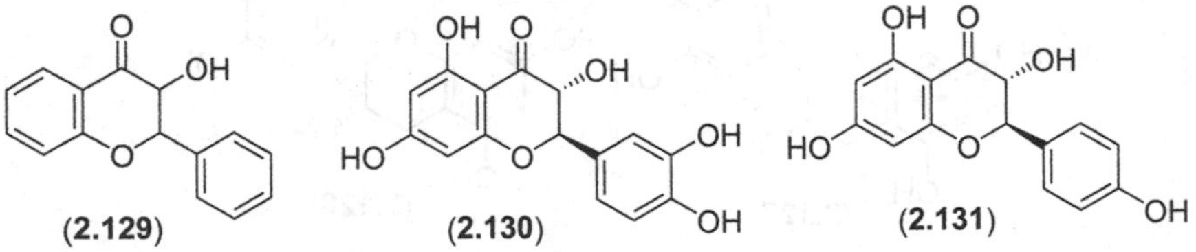

Fig. 2.21 Structures of flavanonols. (**2.129**) Flavanonol, (**2.130**) (2R,3R)-2-(3,4-dihydroxyphenyl)-3,5,7-trihydroxy-2,3-dihydrochromen-4-one (*Taxifolin*), and (**2.131**) (2R,3R)-3,5,7-trihydroxy-2-(4-hydroxyphenyl)-2,3-dihydrochromen-4-one (*Aromadendrin*)

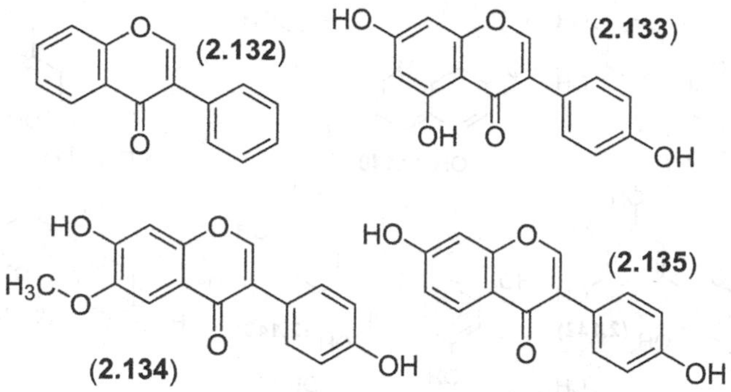

Fig. 2.22 Structures of isoflavones. (**2.132**) Isoflavone, (**2.133**) 5,7-dihydroxy-3-(4-hydroxyphenyl)chromen-4-one (*genistein*), (**2.134**) 7-hydroxy-3-(4-hydroxyphenyl)-6-methoxychromen-4-one (*glycitein*), and (**2.135**) 7-hydroxy-3-(4-hydroxyphenyl)chromen-4-one (*daidzein*)

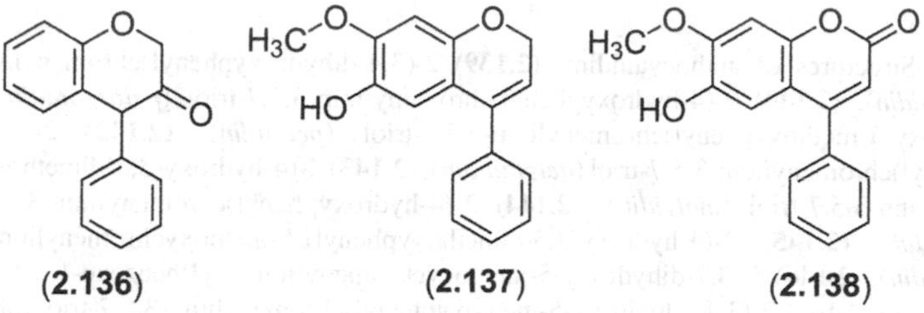

Fig. 2.23 Structures of neoflavones. (**2.136**) Neoflavone, (**2.137**) 7-methoxy-4-phenyl-2*H*--chromen-6-ol (*Dalbergichromene*), and (**2.138**) 6-hydroxy-7-methoxy-4-phenylchromen-2-one (*Dalbergin*)

2.2.14.7 Neoflavones

Neoflavones are those phenolic compounds which have the same skeleton of C6–C3–C6, but the phenyl ring-B is attached to the carbon-4 of the B-ring. The main backbone is known as 2-phenylchromen-4-one. Dalbergichromene (**2.137**), and dalbergin (**2.138**) are important neoflavones (Luqman et al. 2012) as shown in Fig. 2.23.

2.2.15 Anthocyanidins

Anthocyanidins are one of the most important classes of phenolic compounds present in foods. The basic skeleton of the Anthocyanidins is also C6–C3–C6. The C-ring is a pyrilium cation with a hydroxyl group at position-3. The positive charge on the heterocyclic ring can move around the molecule, making it more stable and also contribute to the specific colors. The specific color of some foods may be due

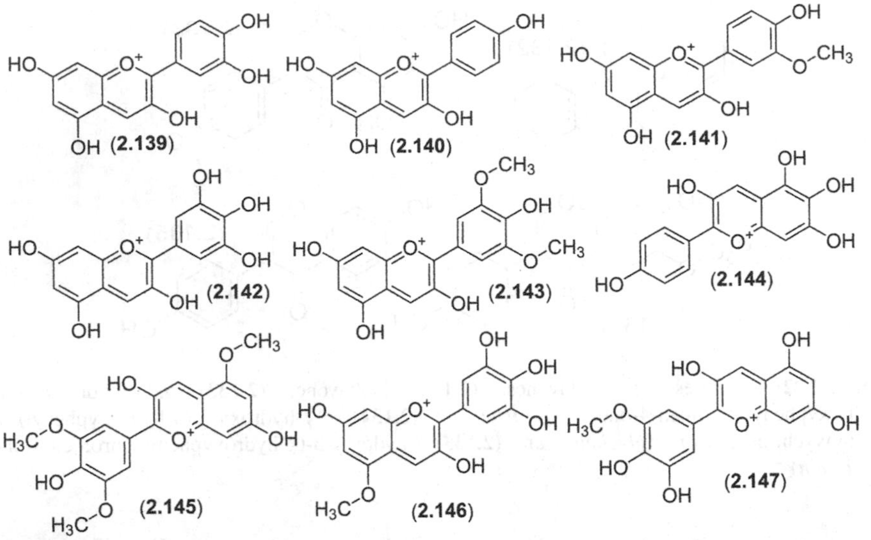

Fig. 2.24 Structures of anthocyanidins. (**2.139**) 2-(3,4-dihydroxyphenyl)chromenylium-3,5,7-triol (*cyanidin*), (**2.140**) 2-(4-hydroxyphenyl)chromenylium-3,5,7-triol (*pelargonidin*), (**2.141**) 2-(4-hydroxy-3-methoxyphenyl)chromenylium-3,5,7-triol (*peonidin*), (**2.142**) 2-(3,4,5-trihydroxyphenyl)chromenylium-3,5,7-triol (*delphinidin*), (**2.143**) 2-(4-hydroxy-3,5-dimethoxyphenyl) chromenylium-3,5,7-triol (*malvidin*), (**2.144**) 2-(4-hydroxyphenyl)chromenylium-3,5,6,7-tetrol (*aurantinidin*), (**2.145**) 2-(4-hydroxy-3,5-dimethoxyphenyl)-5-methoxychromenylium-3,7-diol (*capensinidin*), (**2.146**) 5-(3,7-dihydroxy-5-methoxychromenylium-2-yl)benzene-1,2,3-triol (*pulchellidin*), and (**2.147**) 2-(3,4-dihydroxy-5-methoxyphenyl)chromenylium-3,5,7-triol (*petunidin*)

to the anthocyanidins such as pelargonidin (orange-red), cyanidin (**2.139**, red), peonidin (**2.141**, rose-red), delphinidin (**2.142**, blue-violet), petunidin (**2.147**, blue-purple), and malvidin (purple) (Fig. 2.24). Anthocyanidins rarely occurred in free, and usually form glycosides. There are two other classes of anthocyanidins which are usually derivatives of the parent structures as given below.

2.2.15.1 Leucoanthocyanidins

Leucoanthocyanidins are those anthocyanidins which have two hydroxyl group at position 3 and 4 of the C-ring. Chemically they are flavan-3,4-diols. The hydroxyl groups attached to third and fourth positions and carbon-2 attachment produce stereoisomers such as R and S. For example, (2R,3S,4S)-leucocyanidin. They are usually colorless as compared to their parent anthocyanidins. Examples are leucocyanidin (**2.149**), leucodelphinidin (**2.150**), leucopelargonidin (**2.151**), leucopeonidin (**2.152**), leucofisetinidin (**2.153**), leucomalvidin, leucorobinetinidin and Melacacidin (Fig. 2.25). They are present in nuts, cherry, barley, coca, locust, grape, corn, and legumes (Skadhauge et al. 1997; Zhang et al. 2018).

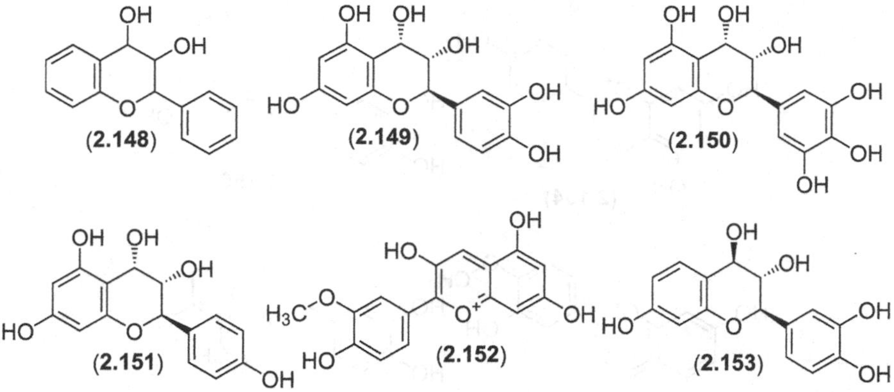

Fig. 2.25 Structures of Leucoanthocyanidins. (**2.148**) Leucoanthocyanidin, (**2.149**) 2-(3,4-dihydroxyphenyl)-3,4-dihydro-2*H*-chromene-3,4,5,7-tetrol (*leucocyanidin*), (**2.150**) 2-(3,4,5-trihydroxyphenyl)-3,4-dihydro-2*H*-chromene-3,4,5,7-tetrol (*leucodelphinidin*), (**2.151**) 2-(4-hydroxyphenyl)-3,4-dihydro-2*H*-chromene-3,4,5,7-tetrol (*leucopelargonidin*), (**2.152**) 2-(4-hydroxy-3-methoxyphenyl)-3,4,4a,8a-tetrahydro-2*H*-chromene-3,4,5,7-tetrol (*leucopeonidin*), and (**2.153**) (2*R*,3*S*,4*R*)-2-(3,4-dihydroxyphenyl)-3,4-dihydro-2*H*-chromene-3,4,7-triol (*leuco-fisetinidin*)

2.2.15.2 Deoxyanthocyanidins

Deoxyanthocyanidins are those anthocyanidins that are characterized for the absence of hydroxyl group at carbon-3 of the C-ring. The differences in the structures between the deoxyanthocyanidin, deoxypeonidin, deoxymalvidin, apigeninidin (**2.154**), and luteolinidin (**2.155**) are due to the hydroxylation and methoxylation pattern of ring B (Sousa et al. 2016) as shown in Fig. 2.26. These compounds are more stable to pH-induced color change than the common anthocyanidins and their glycosides. Sorghums are rich in deoxyanthocyanidin (Awika et al. 2004). Deoxyanthocyanidins also occurred as glycosides.

2.2.16 Biflavonyls

Biflavonyls are large phenolic compounds with 30 carbon skeleton. They are usually dimers of flavonols or flavonoids. The most well-known is ginkgetin (Fig. 2.27) present in *Ginko Biloba*, also known as Japanese silver apricot (Vermerris and Nicholson 2008). Several Biflavonyls have been found in genus *Blepharocarya* (Wannan et al. 1985). Ginkgetin (**2.158**) is also found in plant oils. Ginkgetin and isoginkgetin (**2.159**) have been found in plant leaves (Moawad and El Amir 2016).

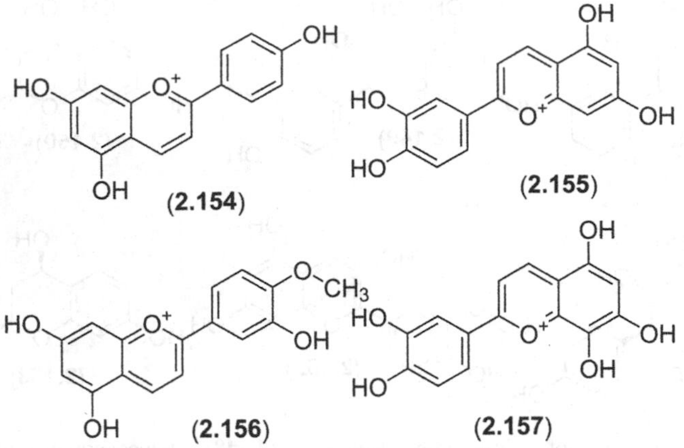

Fig. 2.26 Structures of deoxyanthocyanidins. (**2.154**) 2-(4-hydroxyphenyl)chromenylium-5,7-diol (*apigeninidin*), (**2.155**) 2-(3,4-dihydroxyphenyl)chromenylium-5,7-diol (*luteolinidin*), (**2.156**) 2-(3-hydroxy-4-methoxyphenyl)chromenylium-5,7-diol   (*diosmetinidin*),   and   (**2.157**) 2-(3,4-dihydroxyphenyl)chromenylium-5,7,8-triol (*columnidin*)

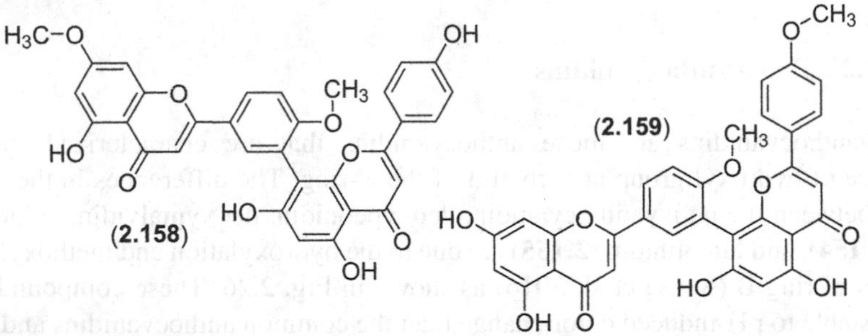

Fig. 2.27 Structures of biflavonyls. (**2.158**) 5,7-Dihydroxy-8-[5-(5-hydroxy-7-methoxy-4--oxochromen-2-yl)-2-methoxyphenyl]-2-(4-hydroxyphenyl)chromen-4-one   (*ginkgetin*),   and (**2.159**)   8-[5-(5,7-dihydroxy-4-oxochromen-2-yl)-2-methoxyphenyl]-5,7-dihydroxy-2-(4-methoxyphenyl)chromen-4-one (*isoginkgetin*)

2.2.17 Betacyanins

Betacyanins are also colored compounds, usually of red color present in *Beta vulgaris* (beets). Chemically they contain heterocyclic nitrogen rings. Important examples are betanidin (**2.160**) and indicaxanthin (**2.161**) as shown in Fig. 2.28. Their absorption spectra are similar to anthocyanins. These natural red pigments from plants are of growing interest as substitutes for synthetic red dyes in the food and pharmaceutical industry (Castellar et al. 2003) and present in *Opuntia* fruits.

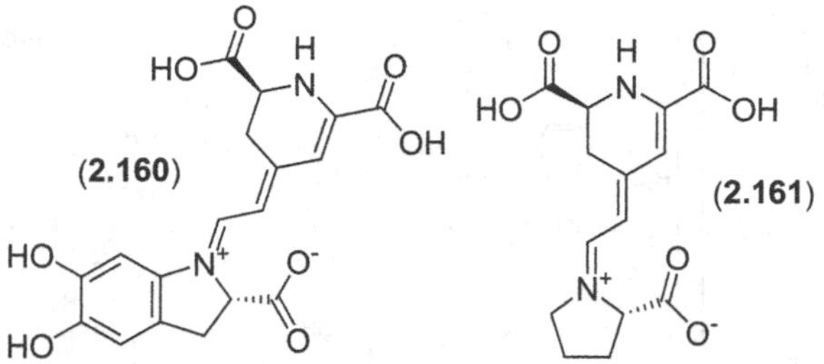

Fig. 2.28 Structures of betacyanins. (**2.160**) (2S)-4-[(E)-2-[(2S)-2-carboxy-5,6-dihydroxy-2,3-dihydroindol-1-yl]ethenyl]-2,3-dihydropyridine-2,6-dicarboxylic acid (*betanidin*), and (**2.161**) (2S)-1-[(2E)-2-[(2S)-2,6-dicarboxy-2,3-dihydro-1H-pyridin-4-ylidene]ethylidene]pyrrolidin-1-ium-2-carboxylate (*indicaxanthin*)

2.2.18 Lignans

Lignans are relatively high molecular weight phenolic compounds which are dimers of lignols. Coumaroyl alcohol (**2.162**), coniferyl alcohol (**2.163**), sinapoyl alcohol (**2.164**) and feruloyl alcohol (**2.165**) are common lignols. Lignans are stereospecific dimers of these cinnamic alcohols (monolignols) bonded at carbon 8 (C8-C8) (Peterson et al. 2010). In the plant, lignans (monolignol dimers) usually occur free or bound to sugars (monosaccharides or disaccharides). The plant lignans most commonly distributed in foods are secoisolariciresinol (**2.166**), matairesinol (**2.167**), syringaresinol (**2.168**), lariciresinol (**2.169**), pinoresinol (**2.170**), enterolactone (**2.171**), and enterodiol (**2.172**) as shown in Fig. 2.29. The most important and widely known lignans present in foods are sesamin (**2.173**), sesaminol, and pinoresinol (Peterson et al. 2010; Zeb et al. 2017).

2.2.19 Lignins

Lignins are high molecular weight polymeric compounds classified as phenolic compounds. They are the second most abundant organic class of compounds present in nature after cellulose. Chemically lignins are biosynthesized from monolignols such as coumaroyl alcohol, coniferyl alcohol, and sinapoyl alcohol and their derivatives. After polymerization, the different lignin subunits are referred to as p-hydroxyphenyl (H), guaiacyl (G), and syringyl (S) residues, depending on whether they originated from p-coumaryl alcohol, coniferyl alcohol, or sinapoyl alcohol, respectively (Vermerris and Nicholson 2008).

The composition of lignin, however, varies from one specie to another. Lignins are biopolymers with structural heterogeneity and there is no define primary structures. Due to the heterogenic structure, it is usually very difficult to measure the degree of polymerization (Laurichesse and Avérous 2014). The molecular masses

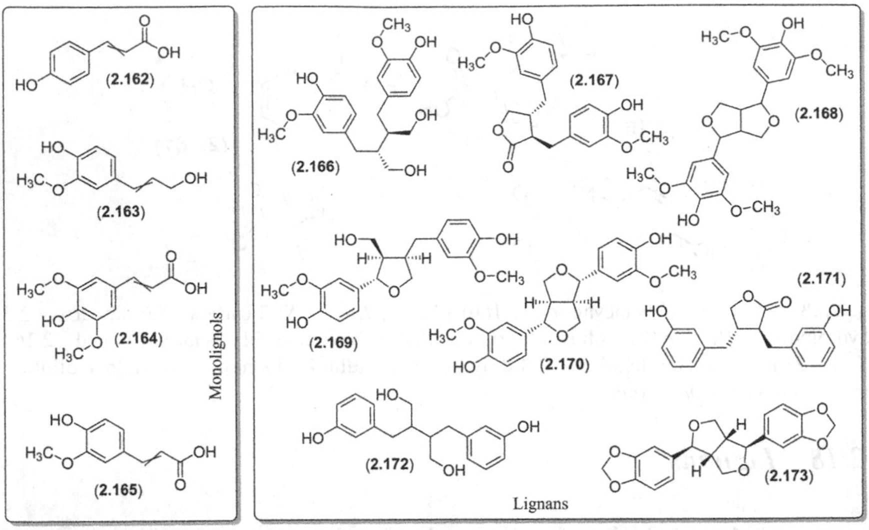

Fig. 2.29 Structures of lignans. (**2.162**) Coumaroyl alcohol, (**2.163**) coniferyl alcohol, (**2.164**) sinapoyl alcohol, (**2.165**) feruloyl alcohol, (**2.166**) (2R,3R)-2,3-bis[(4-hydroxy-3-methoxyphenyl) methyl]butane-1,4-diol (*secoisolariciresinol*), (**2.167**) (3R,4R)-3,4-bis[(4-hydroxy-3-methoxyphenyl) methyl]oxolan-2-one (*Matairesinol*), (**2.168**) 4-[6-(4-hydroxy-3,5-dimethoxyphenyl)-1,3,3a,4,6,6a-hexahydrofuro[3,4-c]furan-3-yl]-2,6-dimethoxyphenol (*Syringaresinol*), (**2.169**) 4-[[(3R,4R,5S)-5-(4-hydroxy-3-methoxyphenyl)-4-(hydroxymethyl)oxolan-3-yl]methyl]-2-methoxyphenol (*Lariciresinol*), (**2.170**) 4-[(3S,3aR,6S,6aR)-6-(4-hydroxy-3-methoxyphenyl)-1,3,3a,4,6,6a-hexahydrofuro[3,4-c]furan-3-yl]-2-methoxyphenol (*Pinoresinol*), (**2.171**) (3R,4R)-3,4-bis[(3-hydroxyphenyl)methyl]oxolan-2-one (*Enterolactone*), (**2.172**) (2R,3R)-2,3-bis[(3-hydroxyphenyl) methyl]butane-1,4-diol (*Enterodiol*), (**2.173**) 5-[(3S,3aR,6S,6aR)-3-(1,3-benzodioxol-5-yl)-1,3,3a,4,6,6a-hexahydrofuro[3,4-c]furan-6-yl]-1,3-benzodioxole (*Sesamin*)

may be more than 10,000. Lignans are not classified as dietary fibers, they share some of the chemical characteristics of lignin, which is insoluble fiber (Davin et al. 2008).

2.2.20 Tannins

Tannins are also high molecular weight phenolic compounds present in plant foods. Tannins were known to us for 200 years. They are widely distributed in many species of the plant kingdom. Tannins have an astringent taste and have extensively studied due to their binding ability to proteins to form precipitates. A Tannin molecule usually requires at least 12 hydroxyl groups and at least five aromatic groups to function as protein binders or tanner. Chemically tannins are classified based on their structure into three main classes as given below: condensed tannins, hydrolyzable tannins, and complex tannins (Khanbabaee and van Ree 2001).

2.2.20.1 Condensed Tannins

Condensed tannins are formed by the condensation of flavans. They are also known as proanthocyanidins, catechol type tannins, non-hydrolyzable tannins, or flavors. They are formed by either oligomerization or polymerization of flavan-3-ol (cate-chin) units (Vermerris and Nicholson 2008). Procyanidin B2 (epicatechin-(4β → 8')-epicatechin, **2.175**) is an example of condensed tannins made of two epicatechin units as shown in Fig. 2.30. The linkage between interflavanyl groups is between C4 of the one unit with C8 of the second unit. The linkage can also be between C4 of one unit and C6 of the second unit. On hydrolysis, they yield differ-ent anthocyanidins. Several types of condensed tannins are present in nature, based on the structure of respective anthocyanidins such as the procyanidins, propelargo-nidins, prodelphinidins, etc. Some condensed tannins are water-soluble due to their hydrophilic nature, whereas high oligomeric or polymeric is not soluble (Hemingway and Karchesy 2012). Condensed tannins are present in a variety of foods such as beans, legumes and seeds (Díaz et al. 2010).

2.2.20.2 Hydrolysable Tannins

These tannins upon hydrolysis yield different phenolic compounds and monosac-charides. They may contain a central core of a monosaccharide or polyhydroxy (polyol) compound which are esterified with gallic acid or ellagic acid. Hydrolyzable tannins may also be classified according to the linkage between monomers, into five types: (1) GOG (and GOGOG), (2) DOG, (3) GOD, (4) D(OG)2 and (5) C-glucosidic type, where G = galloyl, O = oxygen and D = hexahydroxydiphenoyl (Okuda et al. 1993).

Thus, they are classified into gallotannins and ellagitannins.

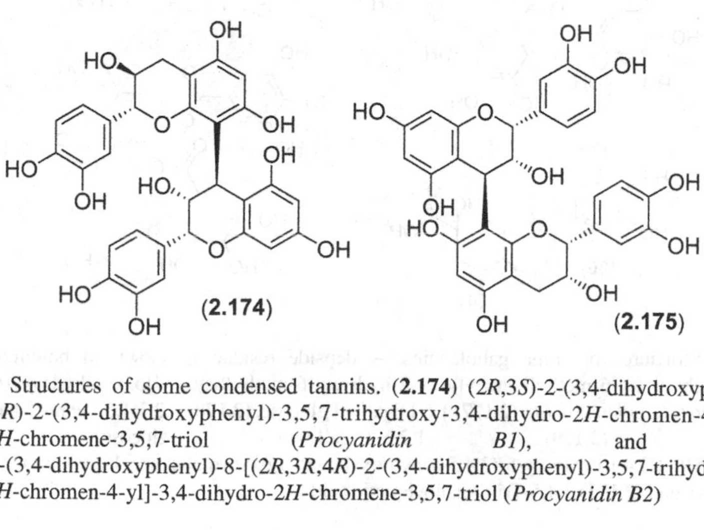

Fig. 2.30 Structures of some condensed tannins. (**2.174**) (2R,3S)-2-(3,4-dihydroxyphenyl)-8-[(2R,3R,4R)-2-(3,4-dihydroxyphenyl)-3,5,7-trihydroxy-3,4-dihydro-2H-chromen-4-yl]-3,4-dihydro-2H-chromene-3,5,7-triol (*Procyanidin B1*), and (**2.175**) (2R,3R)-2-(3,4-dihydroxyphenyl)-8-[(2R,3R,4R)-2-(3,4-dihydroxyphenyl)-3,5,7-trihydroxy-3,4-dihydro-2H-chromen-4-yl]-3,4-dihydro-2H-chromene-3,5,7-triol (*Procyanidin B2*)

Gallotannins Gallotannins are the simplest class of hydrolyzable tannins (HTs). They contain 2–12 gallic acid substituents esterified with a polyol moiety (mostly D-glucose). The glucose is predominately in the β-anomeric configuration. Some rare examples of α-configuration may also be found as natural derivatives. The glucose hydroxyl groups can be partly or totally substituted by phenol units to produce initially β-glucogallin (1-*O*-galloyl-β-D-glucopyranose). They may be found in dimeric, trimeric and tetrameric forms (Smeriglio et al. 2017). Examples are 1,6-digalloyl glucose (**2.177**) and 1,2,6-trigalloyl glucose (**2.179**). An important compound present in the grape is 1,2,3,4,6-penta-*O*-galloyl-β-D-glucose (**2.178**), which is normally considered as the typical high molecular weight example of gallotannins (Fig. 2.31).

The presence of *meta*- or *para*-depside bonds is characteristic of gallotannins. A depside molecule involves esterification of the aromatic hydroxyl group of one aromatic ring with the carboxylic acid moiety of another aromatic ring or another

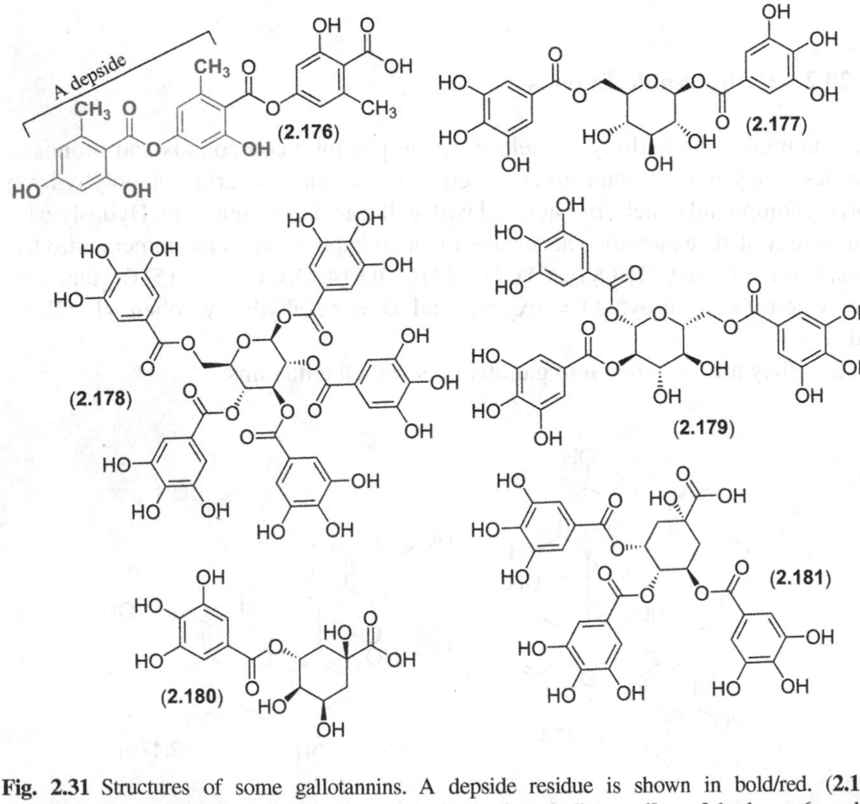

Fig. 2.31 Structures of some gallotannins. A depside residue is shown in bold/red. (**2.176**) 4-[4-(2,4-dihydroxy-6-methylbenzoyl)oxy-2-hydroxy-6-methylbenzoyl]oxy-2-hydroxy-6-methylbenzoic acid (*Gyrophoric acid*), (**2.177**) 1,6-Digalloyl glucose, (**2.178**) 1,2,3,4,6-Penta-*O*-galloyl-β-D-glucose, (**2.179**) 1,2,6-Trigalloyl glucose, (**2.180**) (1*S*,3*R*,4*R*,5*R*)-1,3,4-trihydroxy-5-(3,4,5-trihydroxybenzoyl)oxycyclohexane-1-carboxylic acid (*Theogallin*), and (**2.181**) 3,4,5-Tri-*O*-galloylquinic acid

compound. One important example is Gryphoric acid formed from the hydroxyben-zoic acids. The *Tannase* enzyme is responsible for the formation of a depside bond. However, this bond is more labile than its aliphatic ester bond.

Many other derivatives can also be found, for example, gallotannins with 10 or up to 12 units of gallic acid esterified to a single glucose moiety. They can be found in tannins obtained from sumac (*Rhus semialata*) or oak galls (*Quercus infectoria*). Such tannins had been used since ancient times to tan animal skins to produce leather.

Though D-glucose is the most abundant polyol identified in tannins. Other poly-ols such as glucitol, fructose, shikimic acid, xylose, hamamelose, saccharose, quer-citol, and quinic acid may constitute the core molecules for the subsequent galloylation processes. However, these derivatives are not very common and have only been isolated in maple, chestnut and witch hazel (Smeriglio et al. 2017). Theogallin is chemically trihydroxy benzoic acid glycoside is found in tea is quinic acid-containing gallotannin (Tomás-Barberán and Clifford 2000a, b).

Ellagitannins Ellagitannins (ETs) are classified as a subclass of hydrolyzable tan-nins, however, not all ellagitannins are hydrolyzable in nature. With the help of acids or bases, the ETs are hydrolyzed to yields hexahydroxydiphenic acid (HHDP), which spontaneously lactones to form ellagic acid (EA, **2.182**). The hydrolysis of ETs with hot water and tannase was the main reason to call them hydrolyzable tan-nins. ETs also contain additional C–C bonds between adjacent galloyl moieties in the pentagalloylglucose molecule. This C–C linkage is formed through oxidative coupling between the two adjacent galloyl residues as shown in Fig. 2.32 (Landete 2011). ETs are organized into a complex class of polyphenols categorized by one or more hexahydroxydiphenoyl (HHDP, **2.183**) moieties esterified to a sugar, usually glucose (Khanbabaee and van Ree 2001). The HHDP moieties are made by the oxidative bi-aryl-coupling (C–C coupling) between suitably orientated neighboring galloyl residues.

They can be found in numerous forms: monomeric (e.g. nupharin A, punicalagin (**2.184**), geraniin (**2.186**), eugeniin, davidiin, casuarictin, and corilagin), dimeric (e.g. sanguiin), oligomeric (e.g. agrimoniin, nupharin E, nupharin C and hirtellin A) and *C*-glycosidic (e.g. castalagin (**2.186**), vescalagin (**2.187**), casuarinin, and stachyurin) molecules. Ellagitannins are formed from gallotannins by intermolecu-lar carbon-carbon coupling between at least two galloyl units, yielding hexahy-droxydiphenoyl acid, which in aqueous solution spontaneously lactonizes into ellagic acid. Ellagitannins and ellagic acid have been identified in both fruits as well as in nuts and seeds (Landete 2011). The oxidative coupling commonly involves C-4 and C-6 as in the case of eugeniin. It may be extended to C-2 and C-3 as in the case of casuarictin. Other carbon atoms involved in this reaction are the C-3/C-6, C-2/C-4 or C-1/C-6 pairs. Hexahydroxydiphenoyl acid can undergo esterification not only with glucose but also with hamamelose and terpenoids. Ellagitannins can yield dimers following intermolecular oxidative coupling with other hydrolyzable tannins (as in the case of euphorbins) or high molecular weight oligomers. Based on chemical arrangements, ETs have been classified into two classes. ***Group A***

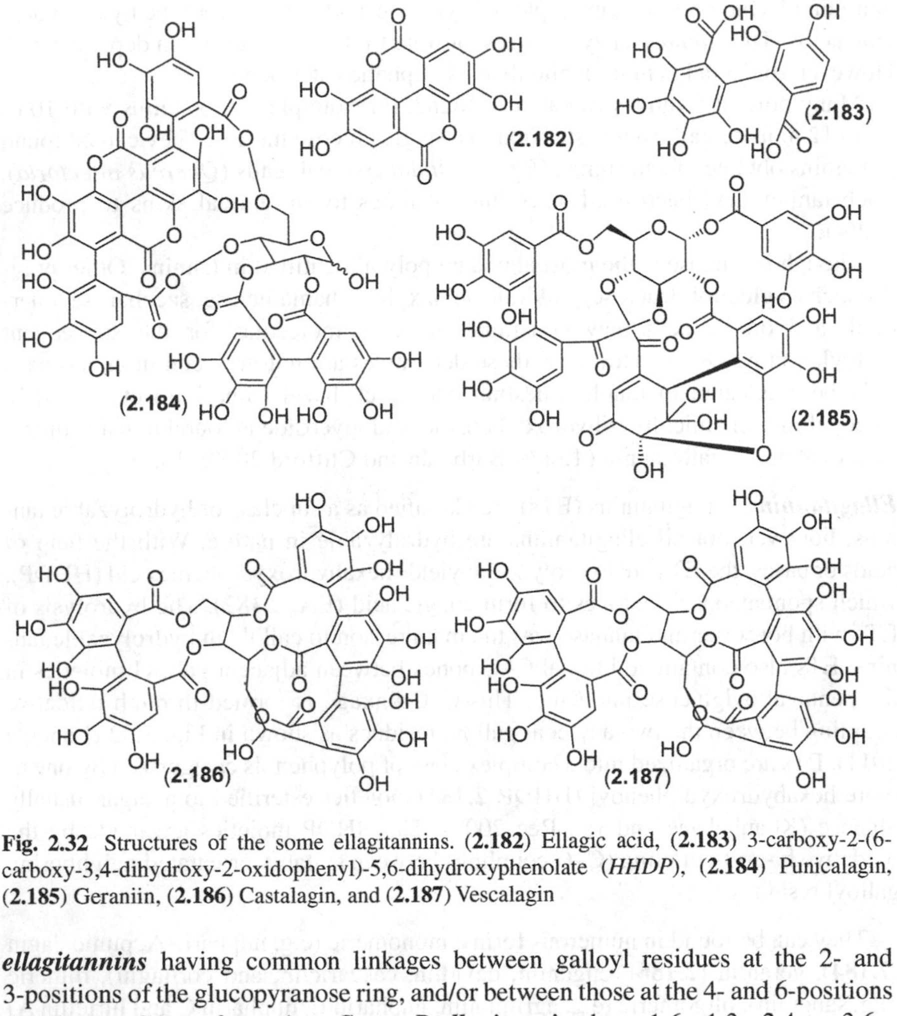

Fig. 2.32 Structures of the some ellagitannins. (**2.182**) Ellagic acid, (**2.183**) 3-carboxy-2-(6-carboxy-3,4-dihydroxy-2-oxidophenyl)-5,6-dihydroxyphenolate (*HHDP*), (**2.184**) Punicalagin, (**2.185**) Geraniin, (**2.186**) Castalagin, and (**2.187**) Vescalagin

ellagitannins having common linkages between galloyl residues at the 2- and 3-positions of the glucopyranose ring, and/or between those at the 4- and 6-positions the glucose C1-conformation. *Group B ellagitannins* have 1,6-, 1,3-, 2,4- or 3,6- C–C linkages can be formed between galloyl residues and glucose (Vermerris and Nicholson 2008). ETs are present in many plant families, and more than 500 molecules have been isolated and identified (Crozier et al. 2009; Clifford and Scalbert 2000; Koponen et al. 2007).

2.2.20.3   Complex Tannins

Complex tannins are formed by the reactions of either gallotannins or ellagitannins with catechins. These compounds are of high molecular weight than its counterpart ellagitannins or gallotannins. They are also called flavono-ellagitannins. Acutissimin A (**2.189**) is an important example of complex tannin formed from ellagitannins and

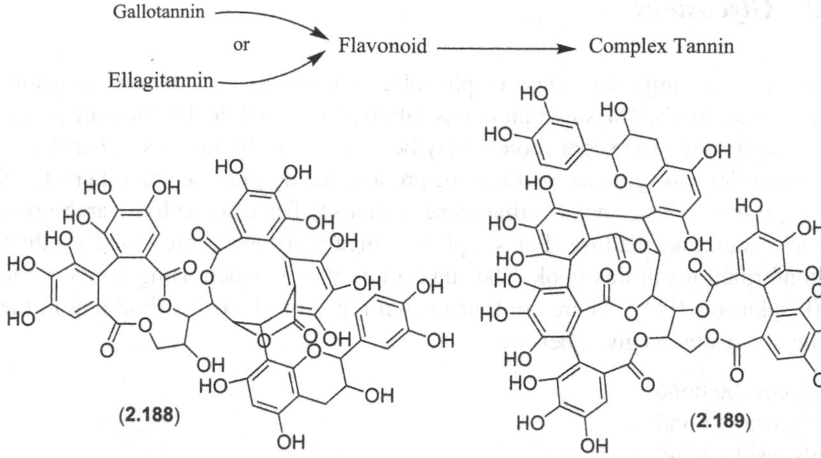

Fig. 2.33 Formation of complex tannins from gallotannins or ellagitannins with flavonoid. Structures of some complex tannins. (**2.188**) Camelliatannin A, and (**2.189**) Acutissimin A

a flavonoid as shown in Fig. 2.33. The flavogallonyl unit is bound to the C1 of glucose with three more hydrolyzable ester bonds. Acutissimin A is formed during the aging process of red wine, whereby the catechin unit originates from the grapes, and the ellagitannin, in this case, vescalagin, originates from the oak barrels (Vermerris and Nicholson 2008). Acutissimin A and other complex tannins present in wine have 250-fold more potent DNA topoisomerase inhibitor than clinically used anticancer drug (Etoposide) (Quideau et al. 2003).

2.2.21 Phlobaphenes

Phlobaphenes are polymeric phenolic compounds present in flowers, leaves, and foods. The chemical nature of phlobaphenes is extremely complex, insoluble in water and soluble in alcohol. It is believed that phlobaphenes are made of polymers of flavan-4-ols such as apiferol, and luteoferol (Vermerris and Nicholson 2008). Physically they are reddish color compounds, imparting color to maize and several fruits. Phlobaphenes have been found to modify pericarp thickness in maize seeds (Landoni et al. 2020).

Phlobaphenes are varying from plant to plant and display a variety of functional groups not seen in the condensed tannins. Douglas-fir phlobaphenes are composed of a mixture of polymeric procyanidins, dihydroquercetin, carbohydrate (glucosyl), and methoxyl moieties. Water insolubility appears to be due to the abundance of methoxy groups (Foo and Karchesy 1989).

2.2.22   Glycosides

Glycosides are an important class of phenolic compounds, heterogenic in nature and found in all foods. The sugar moiety is called glycone while the phenolic part is known as aglycone. The sugar moiety may be a monosaccharide, disaccharides or oligosaccharides. The glycone part may be predominantly glucose, but other sugars are also present. These include rhamnose, rutinose, fructose, xylose, arabinose, glucuronic acid, and maltose. For simplicity, sugars are not represented by their chair conformations in this book. The sugar moiety is attached using a glycosidic bond (Hopkinson 1969). There are four known types of glycosidic bonding in the phenolic glycosides as given below:

- *O*-glycosidic bond
- *C*-glycosidic bond
- *N*-glycosidic bond
- *S*-glycosidic bond
- *C* and *O*-glycosidic bonds

Phenolic glycosides are significant importance in modern medicinal chemistry. Due to the diverse nature of phenolic compounds, the phenolic glycosides are classified according to the classification of their parent phenolic compound.

2.2.22.1 Simple Phenol Glycosides

Simple phenols produce glycosides with a variety of sugar residues like glucose, rhamnose, gentiobiose, ribose, and rutinose. Isoeugenol glycoside, eugenyl rhamnoside, and eugenyl gentiobiose are common examples of simple phenol glycosides present in Eucalyptus (Orihara et al. 1992). Glycosides of cresol, guaiacol and eugenol had been identified in green beans (Dignum et al. 2004). Simple phenol glycosides are usually *O*-glycosides (Fig. 2.34).

2.2.22.2 Hydroxybenzoic Acid and Aldehyde Glycosides

Several glycosides of hydroxybenzoic acids are present in nature in a variety of foods. Vanillin glucoside (**2.196**) formed by the attachment of glucose at C-2 with a hydroxyl group at four positions as shown in Fig. 2.35. It was reported to be present in vanilla beans (Kishor Kumar et al. 2010) and *p*-hydroxybenzaldehyde glucoside (**2.204**) in green beans (Dignum et al. 2004). Galloyl glucoside (**2.197**) had been reported in several vegetables (Romani et al. 2012; Klick and Herrmann 1988).

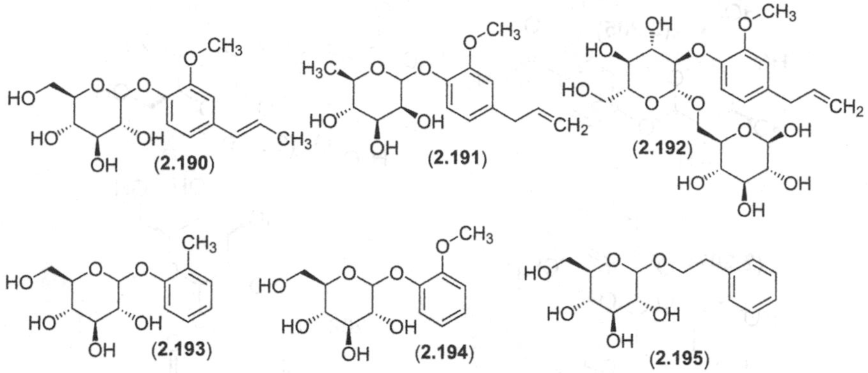

Fig. 2.34 Structures of some of phenol glycosides. (**2.190**) *E*-Isoeugenyl-glucoside, (**2.191**) eugenyl-rhamnoside, (**2.192**) eugenyl gentiobiose, (**2.193**) cresol glucoside, (**2.194**) guaiacol glucoside, and (**2.195**) 2-phenylethanol glucoside

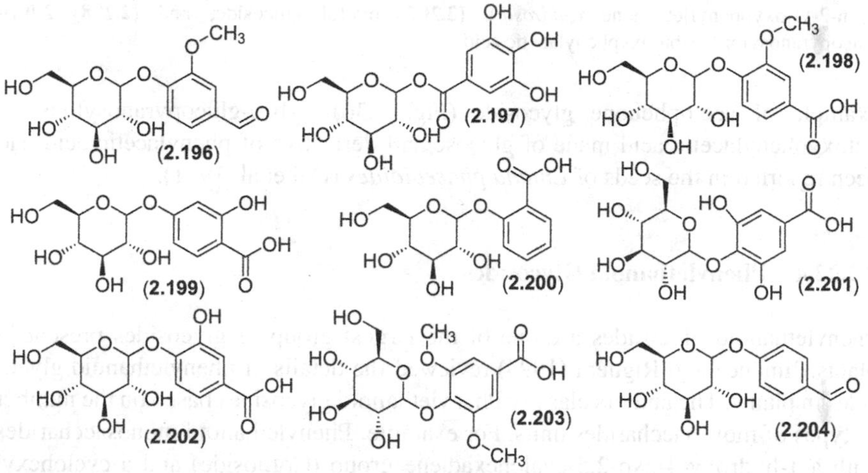

Fig. 2.35 Structures of glycosides of hydroxybenzoic acid and aldehydes. (**2.196**) Vanillin glucoside, (**2.197**) galloyl glucoside, (**2.198**) vanillic acid 4-glucoside, (**2.199**) hydroxybenzoic acid 4-glucoside, (**2.200**) salicylic acid 2-glucoside, (**2.201**) gallic acid 4-glucoside, (**2.202**) protocatechuic acid 4-glucoside, (**2.203**) syringic acid 4-glucoside, and (**2.204**) *p*-hydroxybenzaldehyde glucoside

2.2.22.3 Acetophenone and Phenylacetic Acid Glycosides

Acetophenone glycosides are formed either with bonding with monosaccharide or disaccharide. Androsin (**2.206**), and picein glucosides of acetophenone reported in Thyme along with two other glycosides having apiose sugar moiety and glucose (Wang et al. 1999). The acetophenone glycosides are known for their significant anti-inflammatory activity (Sala et al. 2001). Maltol glycoside (**2.207**) is another

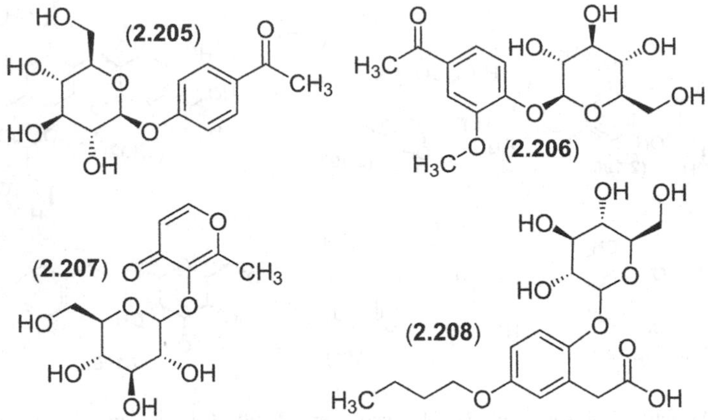

Fig. 2.36 Structures of glycosides of acetophenone and phenylacetic acids. (**2.205**) 1-[4-[(2*S*,3*R*,4*S*,5*S*,6*R*)-3,4,5-trihydroxy-6-(hydroxymethyl)oxan-2-yl]oxyphenyl]ethanone (*picein*), (**2.206**) 1-[3-methoxy-4-[(2*S*,3*R*,4*S*,5*S*,6*R*)-3,4,5-trihydroxy-6-(hydroxymethyl) oxan-2-yl]oxyphenyl]ethanone (*androsin*), (**2.207**) maltol glucoside, and (**2.208**) 2-β-D-glucopyranosyloxy-5-butoxyphenylacetic acid

example of acetophenone glycoside (Fig. 2.36). 2-b-D-glucopyranosyloxy-5--butoxyphenylacetic acid made of glucose and derivative of phenylacetic acid had been reported in the seeds of *Entada phaseoloides* (Dai et al. 1991).

2.2.22.4 Phenylethanoid Glycosides

Phenylethanoid glycosides are one of the largest group of glycosides present in plants. Jiménez and Riguera (1994) reviewed the details of phenylethanoid glycosides in plants. The authors classify Phenylethanoid glycosides based on the number or types of monosaccharides units. For example, Phenylethanoid monosaccharides with a 1-hydroxy-4-oxo-2,5-cyclohexadiene group (Cornoside) and a cyclohexyl ethyl group (Rengynoside A). The other classes of glycosides may contain hydroxy-phenylethyl group bonded to the glucose through the phenolic 4-hydroxy group (Ibotanolide), disaccharides with rhamnose bonded to C-3′ of glucose (Acteoside and its derivatives), disaccharides with the phenylethyl moiety forming a 1,4-dioxane ring (Crenatoside), trisaccharides with rhamnose attached to C-3′ of glucose (Alyssonoside), trisaccharides with rhamnose attached to C-6′ of glucose (Mussatioside), sugar other than glucose (Magnoloside A), and having secoiridoid moiety (Oleoacteoside). Figure 2.37 shows some of the important Phenylethanoid glycosides present in plant foods (Li et al. 2008, 2017).

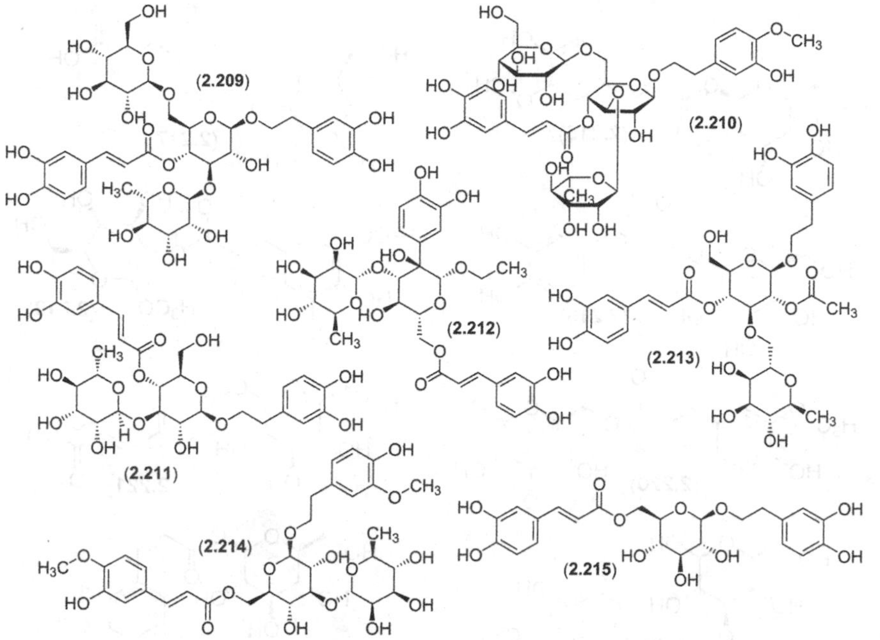

Fig. 2.37 Structures of phenylethanoid glycosides. (**2.209**) Eechinacoside, (**2.210**) cistanoside A, (**2.211**) acteoside, (**2.212**) isoacteoside, (**2.213**) 2′-acetylacteoside, (**2.214**) isomartynoside, and (**2.215**) calceolarioside B

2.2.22.5 Cinnamic Acid Glycosides

The glycosides of cinnamic acids are widely distributed in plant foods. Hydroxycinnamic acid glycosides had been reported in forage legumes (Lu et al. 2000). The cinnamic acids are attached predominately to third position of glycone, but other positions are also occupied. Most of the identified cinnamic acid glycosides are O-glycosides as shown in Fig. 2.38. Feruloyl-1-glucoside, Caffeoyl-1-glucoside had been reported in forage legumes (Lu et al. 2000), whereas sinapoyl-3-glucoside (**2.219**), coumaroyl-4-glucoside, coumaroyl-2-glucoside, and feruloyl glucosides were reported in fruits, vegetables, and cereals (Kylli et al. 2008; Mäkilä et al. 2016). These aglycone parts in cinnamoyl glycosides may be aldehydes, alcohols or esters.

2.2.22.6 Coumarin and Chromone Glycosides

In coumarin glycosides, the sugar moiety is most frequently attached to a hydroxyl group at position C-7 of the coumarin nucleus but it may also be located at position C-4, C-5, C-6, or C-8. Most of the glycosides are O-glycosides, only rare are

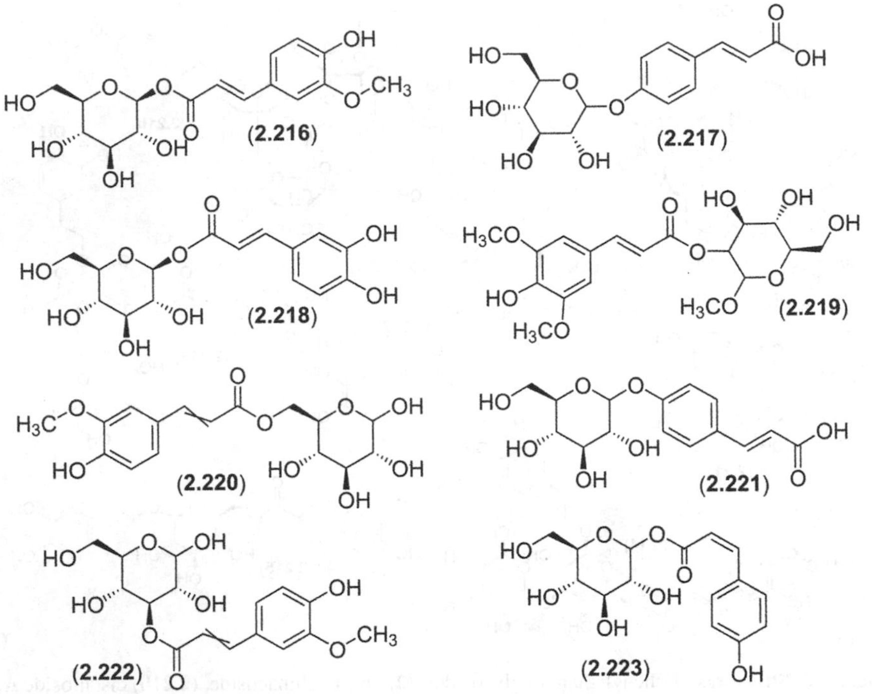

Fig. 2.38 Structures of cinnamic acid glycosides. The crossed double bond represents both E and Z isomers. (**2.216**) 1-O-(E)-Feruloyl-β-D-glucose, (**2.217**) 4-glucopyranosyl-p-coumaric acid, (**2.218**) 1-O-(E)-caffeoyl-β-D-glucoside, (**2.219**) sinapoyl-3-glucoside, (**2.220**) 6-O-feruloyl-glucoside, (**2.221**) (E)-p-coumaric acid 4-O-glucoside, (**2.222**) 3-O-feruloyl-glucoside, and (**2.223**) (Z)-p-coumaroylglucoside

C-glycosides. Most often attached sugars are D-glucose, D-galactose, L-rhamnose, and D-apiose, and more rarely D-xylose and L-arabinose (Bartnik and Facey 2017). Deoxy-sugars are also present. Disaccharides have also been reported as part of the coumarin glycosides, e.g., rutinose (6-O-α-L-rhamnopyranosyl-D-glucopyranose), gentiobiose (6-O-β-D-glucopyranosyl-D-glucopyranose), primeverose (6-O-β-D-xylopyranosyl-D-glucopyranose), and 6-O-8-D-apiosyl-D-glucopyranose. Examples are scopolin (**2.224**), hydroxycnidimoside A (**2.225**), skimmin (**2.226**), 5-hydroxy-2,6,8-trimethylchromone-7-glucoside (**2.227**), and daphnin (**2.228**)as shown in Fig. 2.39 (Petruľová-Poracká et al. 2013). The coumarin and chromone glycosides have been reported in Japanese teas (Yang et al. 2009).

2.2.22.7 Benzophenone and Xanthone Glycosides

Benzophenone and xanthone glycosides are commonly O-glycosides with predominate glycone as glucose. There are C-glycosides present in the plant; one such example is 2,3,4′,5,6-pentahydroxybenzophenone-4-C-glucoside as shown in

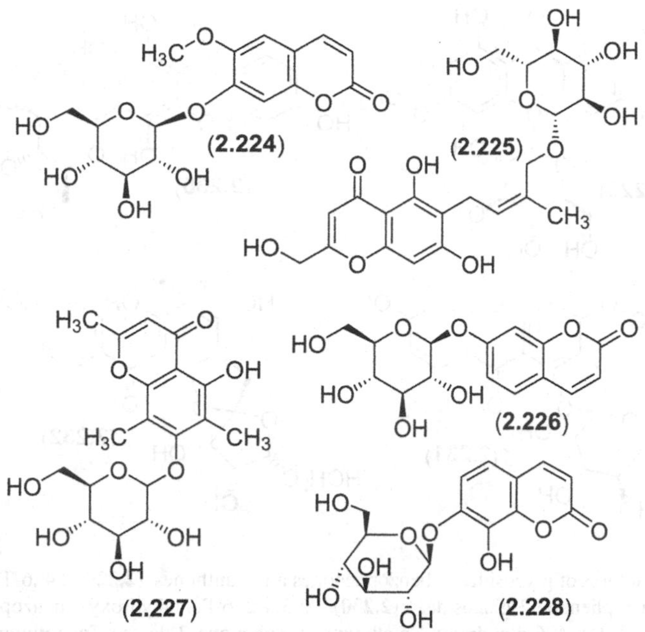

Fig. 2.39 Structures of glycosides of coumarin and chromones. (**2.224**) 6-methoxy-7-[(2S,3R,4S,5S,6R)-3,4,5-trihydroxy-6-(hydroxymethyl)oxan-2-yl]oxychromen-2-one (*Scopolin*), (**2.225**) 5,7-dihydroxy-2-(hydroxymethyl)-6-[(Z)-3-methyl-4-[(2R,3R,4S,5S,6R)-3,4,5-trihydroxy-6-(hydroxymethyl)oxan-2-yl]oxybut-2-enyl]chromen-4-one (*Hydroxycnidimoside A*), (**2.226**) 7-[(2S,3R,4S,5S,6R)-3,4,5-trihydroxy-6-(hydroxymethyl)oxan-2-yl]oxychromen-2-one (*Skimmin*), (**2.227**) 5-hydroxy-2,6,8-trimethyl chromone 7-glucoside, and (**2.228**) 8-hydroxy-7-[(2S,3R,4S,5S,6R)-3,4,5-trihydroxy-6-(hydroxymethyl)oxan-2-yl]oxychromen-2-one (*Daphnin*)

Fig. 2.40, whereas *O*-glycoside is 2,4′,6-Trihydroxy-4-methoxybenzophenone-2--glucoside (**2.229**) present in herbal tea (Schulze et al. 2015). In mango leaves, the glycone moiety is attached using *C* as well as *O*-glycosidic linkage (Pan et al. 2018). The examples of such glycosides are 4′,6-dihydroxy-4-methoxybenzophenone-2-desoxy-fructopyranoside and 4,4′,6-trihydroxybenzophenone-2-desoxy-fructofuranoside (**2.232**). The *C* and *O*-glycosidic bonds form an oxygen heterocycle ring between phenone ring and sugar moiety. The sugar, in this case, is fructose both in furanose as well as pyranose form.

2.2.22.8 Stilbene Glycosides

Stilbene glycosides are one of the diverse classes of stilbenes derivatives (Shen et al. 2009). They may contain one or more of the monosaccharides. For example, astringin is a stilbene glycoside formed by the *O*-glycosidic bond of glucose and astringinin (3′,4′,3,5-Tetrahydroxy-trans-stilbene, **2.233**) present in grape products

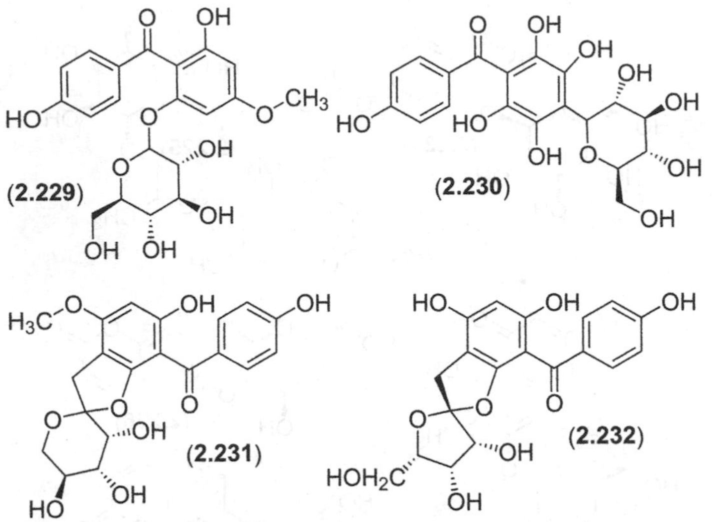

Fig. 2.40 Structures of glycosides of benzophenones and xanthones. (**2.229**) 2,4′,6-Trihydroxy-4-methoxy benzophenone-2-glucoside, (**2.230**) 2,3,4′,5,6-Pentahydroxy benzophenone-4-*C*-glucoside, (**2.231**) 4′,6-dihydroxy-4-methoxybenzophenone-2-desoxy-fructopyranoside, and (**2.232**) 4,4′,6-trihydroxybenzophenone-2-desoxy-fructofuranoside

such as   vine (Vitrac et al. 2005). Its IUPAC name is (2*S*,3*R*,4*S*,5*S*,6*R*)-2-[3-[(*E*)-2-(3,4-dihydroxyphenyl)ethenyl]-5-hydroxyphenoxy]-6-(hydroxymethyl)oxane-3,4,5-triol as shown in Fig. 2.41. Similar is the case of piceid (**2.234**). *Morus alba* had been reported to contain glycosides of resveratrol and hydroxyresveratrol (Piao et al. 2009). These include 4′-glucopyranosyl-2′-hydroxyresveratrol-3-(6-glucopyranosyl-glucopyranoside) with three units of sugars. All sugars are glucose, having two forming a disaccharide unit. Another glycoside with the same backbone was 4′-glucopyranosyl-2′-hydroxyresveratrol 3-(6-apiofuranosyl-glucopyranoside) contain apiofuranose replacing glucose in the former compound. Apiofuranose is attached through C1 to C6 of the glucopyranose.

2.2.22.9 Anthraquinone Glycosides

In anthraquinone glycosides, the sugar is either a single unit, a disaccharide or oligosaccharide. Disaccharides have been more predominant such as gentiobiose and primeverose. The examples of 1-gentiobiose-2-methylol-anthraquinone, 1-gentiobiose-3-hydroxy-2-methyl-anthraquinone,   3-*O*-primeverose-1,8-dihydroxy-2-methyl-anthraquinone, 1-methyl-3-hydroxy-anthraquinone (Wang et al. 2019a), Digiferruginol-1-methylether-11-gentiobioside   (**2.237**),   digiferruginol-11-primeveroside   (**2.238**),   1-hydroxy-5,6-dimethoxy-2-methyl-7--primeverosyloxyanthraquinone (**2.239**),   and   damnacanthol-11-primeveroside

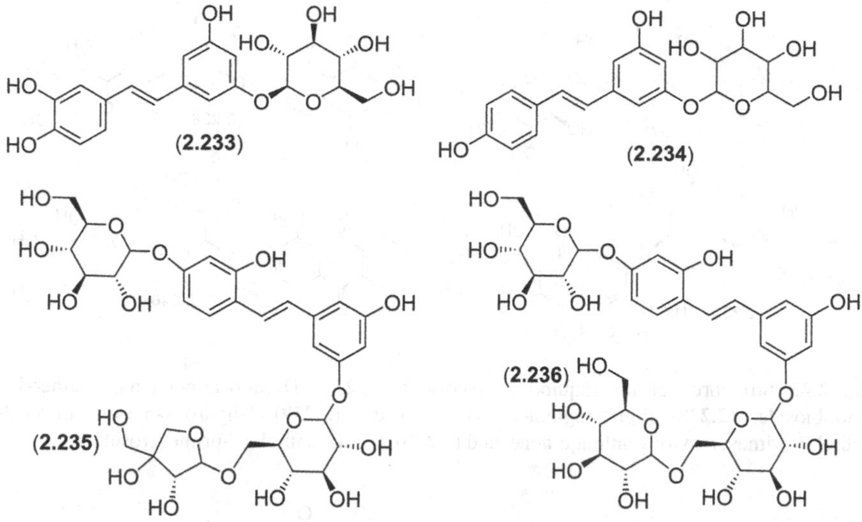

Fig. 2.41 Structures of stilbene glycosides. (**2.233**) (2S,3R,4S,5S,6R)-2-[3-[(E)-2-(3,4-dihydroxyphenyl)ethenyl]-5-hydroxyphenoxy]-6-(hydroxymethyl)oxane-3,4,5-triol (*astringin*), (**2.234**) (2S,3R,4S,5S,6R)-2-[3-hydroxy-5-[(E)-2-(4-hydroxyphenyl)ethenyl]phenoxy]-6-(hydroxymethyl)oxane-3,4,5-triol (*Piceid*), (**2.235**) 4′-glucopyranosyl-2′-hydroxyresveratrol 3-(6-apiofuranosyl-glucopyranoside), and (**2.236**) 4′-glucopyranosyl -2′-hydroxyresveratrol-3-(6-glucopyranosyl-glucopyranoside)

(**2.240**) as shown in Fig. 2.42 (Kamiya et al. 2009). The sugar moiety may be attached to the different sites on the anthraquinone rings.

2.2.22.10 Chalcone Glycosides

Chalcone glycosides contain monosaccharides or disaccharides. Glucose in pyranose form is a common glycone in chalcone glycosides. They are O-glycosidic compound, in which glucose is attached to the position 3, 4, 6, 3′, 4′ and 6′ of the aglycone rings. For example, in neosakuranin (**2.241**), the glucose is bonded to position 6 of aglycone as shown in Fig. 2.43. Neosakuranin had been reported in walnuts, cherry, different fruits (Bhatt et al. 2009), *Brassica Rapa* leaves (Elsayed and Khalil 2018). 4′-glucopyranosyl-4-hydroxy-3′-methoxychalcone, and 4′-glucopyranosyl-3′,4-dimethoxychalcone contains one glucose unit each, whereas 4,4′-di-glucopyranosyl-3′-methoxychalcone (**2.242**) contains two glucose units (Ninomiya et al. 2010). These glucose units are attached directly to the aglycone through C-glycosidic bonds. The glycosides of chalcones exhibited several important medicinal applications such as antioxidant, anti-inflammatory, and antitumor (Debarshi Kar et al. 2017).

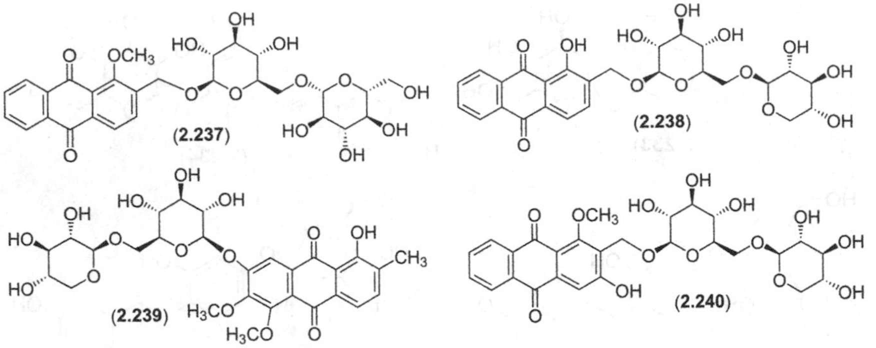

Fig. 2.42 Structures of anthraquinone glycosides. (**2.237**) Digiferruginol-1-methylether-11-gentiobioside, (**2.238**) digiferruginol-11-primeveroside, (**2.239**) 1-hydroxy-5,6-dimethoxy-2-methyl-7-primeverosyloxyanthraquinone, and (**2.240**) damnacanthol-11-primeveroside

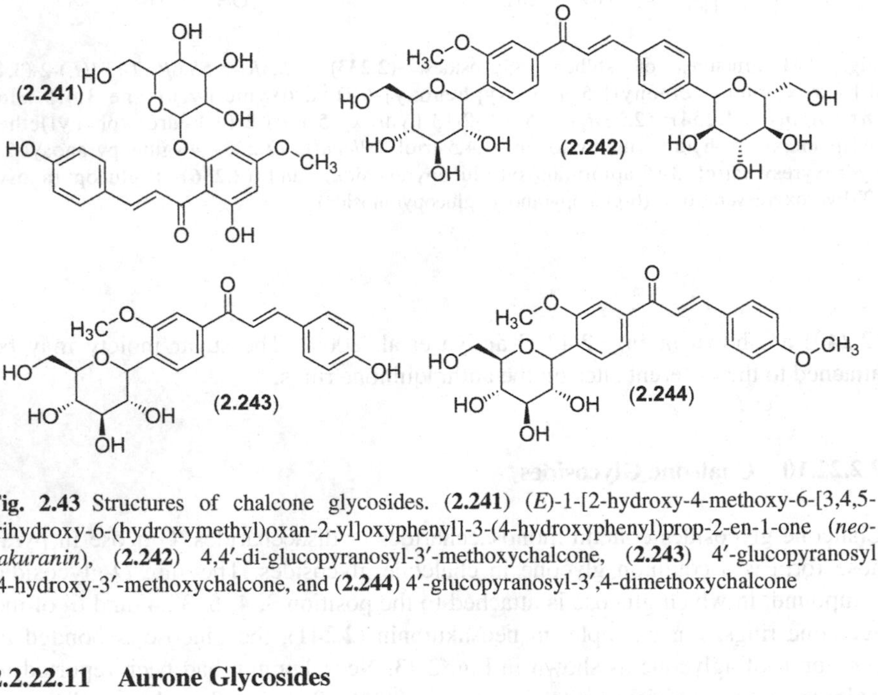

Fig. 2.43 Structures of chalcone glycosides. (**2.241**) (*E*)-1-[2-hydroxy-4-methoxy-6-[3,4,5-trihydroxy-6-(hydroxymethyl)oxan-2-yl]oxyphenyl]-3-(4-hydroxyphenyl)prop-2-en-1-one (*neosakuranin*), (**2.242**) 4,4′-di-glucopyranosyl-3′-methoxychalcone, (**2.243**) 4′-glucopyranosyl-4-hydroxy-3′-methoxychalcone, and (**2.244**) 4′-glucopyranosyl-3′,4-dimethoxychalcone

2.2.22.11 Aurone Glycosides

There are a large number of aurone glycosides, which contain one or more monosaccharide units. Most of these are *O*-glycosides as reported by Boucherle et al. (2017). The authors reported more than 100 aurones and their glycosides present in a variety of plants including foods (Tomás-Barberán and Clifford 2000a, b). Examples are sulfurein (**2.245**), maritimein (**2.246**), leptosin (**2.247**), bidenoside A (**2.248**), aureusin (**2.249**), and cernuoside (**2.250**) as shown in Fig. 2.44.

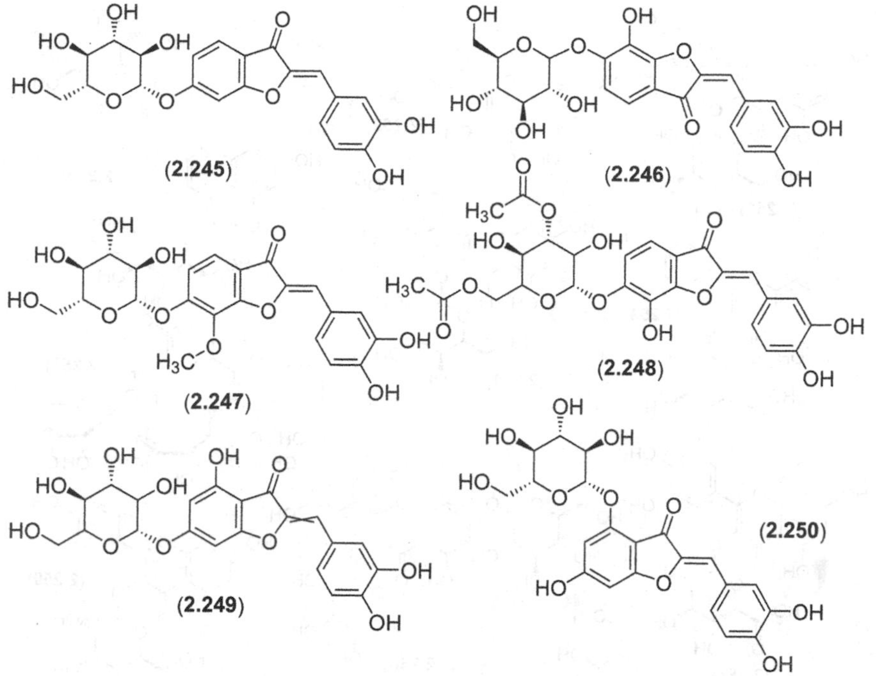

Fig. 2.44 Structures of aurone glycosides. (**2.245**) (2Z)-2-[(3,4-dihydroxyphenyl)methylidene]-6-
[(2S,3R,4S,5S,6R)-3,4,5-trihydroxy-6-(hydroxymethyl)oxan-2-yl]oxy-1-benzofuran-3-one
(*Sulfurein*), (**2.246**) (2Z)-2-[(3,4-dihydroxyphenyl)methylidene]-7-hydroxy-6-[(2S,3R,4S,5S,6R)
-3,4,5-trihydroxy-6-(hydroxymethyl)oxan-2-yl]oxy-1-benzofuran-3-one (*Maritimein*), (**2.247**)
(2Z)-2-[(3,4-dihydroxyphenyl)methylidene]-7-methoxy-6-[(2S,3R,4S,5S,6R)-3,4,5-trihydroxy-6-
(hydroxymethyl)oxan-2-yl]oxy-1-benzofuran-3-one (*Leptosin*), (**2.248**) [(3R,4S,6S)-4-acetyloxy
-6-[[(2Z)-2-[(3,4-dihydroxyphenyl)methylidene]-7-hydroxy-3-oxo-1-benzofuran-6-yl]oxy]-
3,5-dihydroxyoxan-2-yl]methyl acetate (*Bidenoside A*), (**2.249**) 2-[(3,4-dihydroxyphenyl)
methylidene]-4-hydroxy-6-[(2S,4S,5S)-3,4,5-trihydroxy-6-(hydroxymethyl)oxan-2-yl]oxy-1-ben-
zofuran-3-one (*Aureusin*), and (**2.250**) (2Z)-2-[(3,4-dihydroxyphenyl)methylidene]-6-hydroxy-4-
[(2S,3R,4S,5S,6R)-3,4,5-trihydroxy-6-(hydroxymethyl)oxan-2-yl]oxy-1-benzofuran-3-one
(*Cernuoside*)

2.2.22.12 Flavonoid Glycosides

Flavonoid glycosides are one of the most widely distributed class of glycosides in
foods and nearly in all plants. Therefore, flavonoid glycosides will be collectively
discussed. These glycosides may O- or C-glycosides, however, the majority of them
present in foods are O-glycosides. The glycone may be glucose, fructose, galactose,
glucuronide, rutinose, gentiobiose, and pentose (Heim et al. 2002). The glycone
may be a monosaccharide i.e., quercetin-3-glucoside (**2.251**), kaempferol-3-
glucoside (**2.259**), quercetin-4-galactoside or a disaccharide such as in quercetin-3,4′-
diglucoside (**2.255**), quercetin-3-rutinoside (**2.256**), and apigenin-6,8-diglucoside

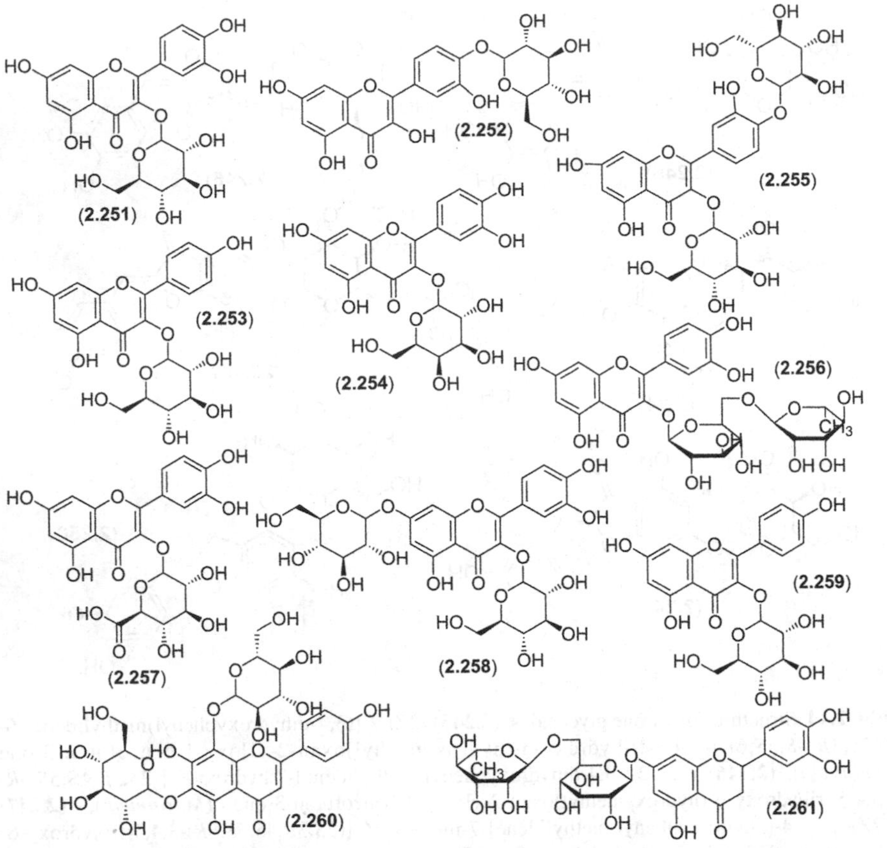

Fig. 2.45 Structures of flavonoid glycosides. (**2.251**) quercetin-3-glucoside, (**2.252**) quercetin-4′-glucoside, (**2.253**) apigenin-3-glucoside, (**2.254**) quercetin-3-galactoside, (**2.255**) quercetin-3,4′-diglucoside, (**2.256**) quercetin-3-rutinoside, (**2.257**) quercetin-3-glucouronide, (**2.258**) quercetin-3,7-diglucoside, (**2.259**) kaempferol-3-glucoside, (**2.260**) apigenin-6,8-diglucoside, and (**2.261**) eriodictyol-7-rutinoside

(**2.260**) (Fig. 2.45). All these glycosides are *O*-glycosides. There are *C*-glycosides such as apigenin-6,8-di-*C*-glucoside in foods.

Glycosides of flavonoids, especially flavonols such as kaempferol, quercetin are widely occurring flavonoids. They demonstrate numerous biological functions, e.g., antioxidant, anti-inflammatory, hepato-protective, cardio-protective, and vasodila- tory effects (Bartnik and Facey 2017). They are present in several vegetables such as carrot, spinach, cauliflower, onion, garlic, ginger, or cabbage. Fruits such as plum, apple, strawberry, and apricot are rich in flavonoid glycosides. They exten- sively studied phenolic compounds in foods. Thus, they are present in nearly all foods especially leafy vegetables are rich sources (Zeb 2015; Ahmad et al. 2020).

2.2.22.13 Anthocyanidins Glycosides

Anthocyanidin glycosides are also known as anthocyanins. They are mostly 3-glycosides, with glycone attached by an O-glycosidic bond. The glycone is usually glucose, galactose and glucuronide, and xylose (Veitch and Grayer 2008). The important examples are cyanidin-3-glucoside (**2.263**), malvidin-3-glucoside (**2.264**), peonidin-3-glucoside (**2.265**), delphinidin-3-glucoside (**2.266**), petunidin-3-glucoside (**2.267**), and cyanidin-3-glucoside and their derivatives (Fig. 2.46). Anthocyanins are one of the most important and extensively studied phenolic compounds in plant foods. They are present in nearly all edible fruits (Einbond et al. 2004) and possess a significant beneficial role in human health (Khoo et al. 2017).

2.2.22.14 Lignan Glycosides

In lignan glycosides, the glycone moiety is either a monosaccharide, disaccharide or oligosaccharide. The first tri and tetra-glycosides of lignans were reported in Japan (Kaneshiro et al. 1991). Significant research had been carried out on the lignan glycosides. In foods such as sesame leaves and oils, lignan glycosides such as sesaminol mono glucoside (**2.268**), sesaminol diglucoside (**2.269**) and triglucoside (**2.270**) as shown in Fig. 2.47. In unroasted *Sesamum indicum* seeds revealed both di- and triglucosides of sesaminol (Shyu and Hwang 2002).

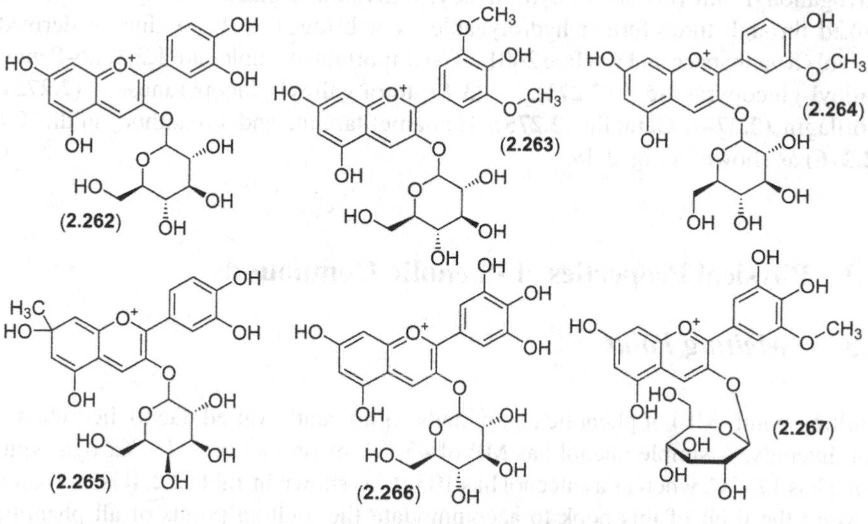

Fig. 2.46 Structures of some anthocyanidin glycosides. (**2.262**) Cyanidin-3-glucoside, (**2.263**) malvidin-3-glucoside, (**2.264**) peonidin-3-glucoside, (**2.265**) 7-methyl-cyanidin-3-galactoside, (**2.266**) delphinidin-3-glucoside, and (**2.267**) petunidin-3-glucoside

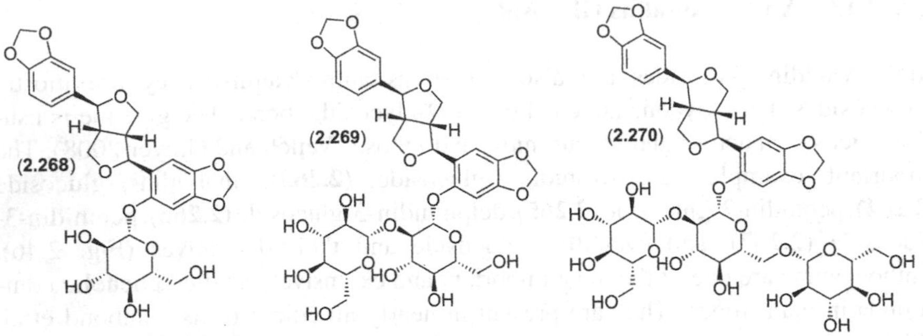

Fig. 2.47 Structures of some lignan glycosides. (**2.268**) Sesaminol glucoside, (**2.269**) sesaminol diglucoside, and (**2.270**) sesaminol triglucoside

2.2.22.15 Tannin Glycosides

In tannins, gallotannins represent the simplest class of hydrolyzable tannins (HTs), containing gallic acid substituents esterified with a D-glucose. The β-anomeric form of glucose is the main configuration, while rare examples of the α-configuration may also be present. The glucose hydroxyl groups can be partly or totally substituted by phenol units to produce initially β-glucogallin (1-galloyl-β-D-glucopyranose) (Smeriglio et al. 2017). Similarly, ellagitannins and complex tannins are glycosides in nature. The glycone moiety may be glucose, fructose, xylose, hamamelose, sucrose, and glucuronic acid. In complex tannins, such as acutissimin A, having a flavogallonyl unit (nonahydroxytriphenoyl unit) bound glucosidically to C-1, and linked through three further hydrolyzable ester bridges to the D-glucose derived polyol (Khanbabaee and van Ree 2001). The important examples are 1,2,3,4,6-Penta-galloyl-glucopyranose (**2.271**), 2,3,4,6-tetra-galloyl-glucopyranose (**2.272**), Corilagin (**2.274**), Geraniin (**2.275**), Hamamelitannin, and Proanthocyanidin C1 (**2.276**) as shown in Fig. 2.48.

2.3   Physical Properties of Phenolic Compounds

2.3.1   Melting Point

Melting point (MP) of phenolic compounds significantly varied due to their chemical diversity. A simple phenol has MP of 43 °C, resorcinol has 111 °C, hydroquinone has 173 °C, whereas a catechol has 105 °C as shown in Table 2.2. It is, however, beyond the limit of this book to accommodate the melting points of all phenolic compounds reported so far. Only those physical properties of phenolic compounds are reported where they are isolated in a pure stable form. The melting point of the

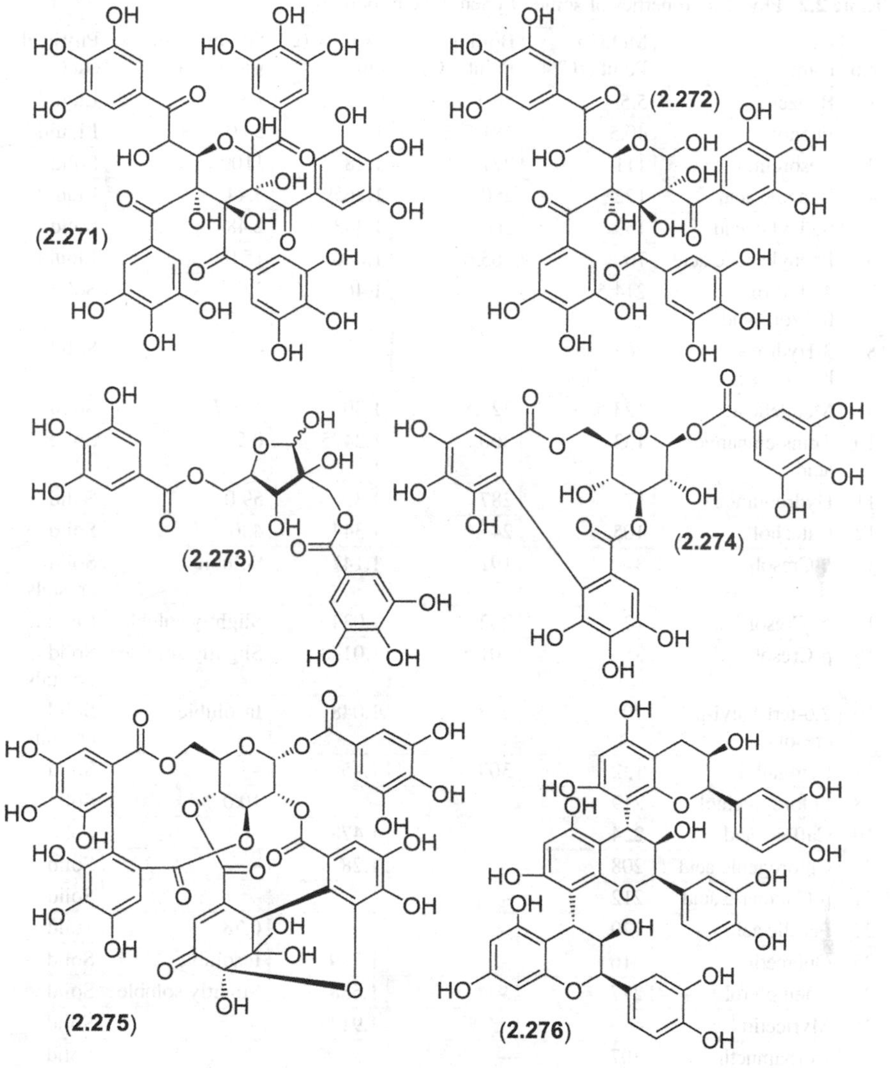

Fig. 2.48 Structures of tannin glycosides. (**2.271**) 1,2,3,4,6-penta-galloyl-glucopyranose, (**2.272**) 2,3,4,6-tetra-galloyl-glucopyranose, (**2.273**) [3,4,5-trihydroxy-4-[(3,4,5-trihydroxybenzoyl)oxy-methyl]oxolan-2-yl]methyl 3,4,5-trihydroxybenzoate (*Hamamelitannin*), (**2.274**) Corilagin, (**2.275**) Geraniin, and (**2.276**) Proanthocyanidin C1

phenolic compounds depends on several structural features including numbers and types of substituents and functional group moiety (Jain et al. 2004). The melting points of high molecular weight phenolic compounds are higher as compared to the simple phenolic compounds.

Table 2.2 Physical properties of some of phenolic compounds

No	Name	Melting Point (°C)	Boiling Point (°C)	Density (g/cm³)	Solubility in water (g/L)	Physical state
1	Benzene	5.5	80.1	0.8765	1.8	Liquid
2	Phenol	40.5	181.7	1.07	83.0	Liquid
3	Resorcinol	111	277	1.28	1100	Solid
4	Benzoic acid	122	250	1.2659	3.44	Liquid
5	Salicylic acid	159	200	1.443	2.48	Solid
6	Phenyl acetic acid	76	265.6	1.0809	15.0	Liquid
7	4-Hydroxy-benzoic acid	214.5	–	1.46	5.0	Solid
8	3-Hydroxy-benzoic acid	202	–	–	–	Solid
9	Mandelic acid	123.5	321.8	1.30	158.7	Solid
10	Trans-cinnamic acid	133	300	1.2475	0.5	Solid
11	Hydroquinone	172	287	1.30	59.0	Solid
12	Catechol	105	245.5	1.344	430	Solid
13	o-Cresol	31	191	1.147	Soluble	Solid crystals
14	m-Cresol	12	202	1.034	Slightly soluble	Liquid
15	p-Cresol	35	201.5	1.0178	Slightly soluble	Solid crystals
16	2,6-tert-butyl-p-cresol	70	265	1.048	Insoluble	Solid crystals
17	Pyrogallol	132	309	1.45	–	Solid
18	Phloroglucinol	219	–	–	10.0	Solid
19	Caffeic acid	224	–	1.478	–	Solid
20	Chlorogenic acid	208	–	1.28	–	Solid
21	p-Coumaric acid	212	–	–	–	Solid
22	Ferulic acid	170	–	–	0.78	Solid
23	Quercetin	316	–	1.799	Insoluble	Solid
24	Kaempferol	277	–	1.688	Slightly soluble	Solid
25	Myricetin	–	–	1.912	–	Solid
26	Isorhamnetin	307	–	–	–	Solid
27	Rutin	242	–	–	125	Solid
28	Apigenin	347	–	–	–	Solid
29	Catechin	176	–	–	–	Solid
30	Diadzein	321	–	–	Insoluble	Solid

2.3.2 Boiling Point

Phenolic compounds generally have higher boiling points as a comparison to other hydrocarbons having equal molecular masses. This is due to the existence of inter-molecular hydrogen bonding between hydroxyl groups of phenol molecules. In

Table 2.3 Colors and flavors of some of phenolic compounds

No	Name	Flavor	Color
1	Hydroxybenzoic acid	Bitter	White
2	Protocatechuic acid	Astringent	Light brown
3	Vanillic acid	Astringent	Light yellow
4	Vanillin	Sweet, Pleasant	White
5	Syringic acid	Bitter	Off-white
6	Sinapic acid	Bitter-sweet	White
7	Ferulic acid	Astringent	Light yellow
8	Cinnamic acid	Cinnamon	White
9	Resveratrol	Bitter	White
10	Catechin	Bitter	White
11	Epicatechin	Bitter-Astringent	White
12	Diadzein	Bitter	Pale yellow
13	Neohesperidin	Bitter-sweet	Yellowish-white
14	Tangretin	Bitter	White
15	Quercetin	Bitter	Yellow
16	Naringin	Bitter	Red purple
17	Pelargonidin	Moderate astringent	Orange-red
18	Cyanidin	Astringent	Red
19	Peonidin	Astringent	Rose-red
20	Delphinidin	Astringent	Blue-violet
21	Petunidin	Astringent	Blue-purple
22	Malvidin	Astringent	Purple

general, the boiling point of phenols increases with an increase in the number of carbon atoms (Norwitz et al. 1986). The substituents have also effects on the boiling points, for example, the boiling point of o-cresol is 191 °C, which is lower than its m-cresol (202 °C) and p-cresol (201.5 °C) (Table 2.2). This means a methyl group at the *ortho-* will boil at a lower temperature than its corresponding *meta-* and *para-* positions. If a butyl group is attached to the para-position the boiling point increased to 265 °C.

2.3.3   Density

Density is a physical parameter, which measured in g/cm^3. Benzene has a density lower than 1 (0.8765 g/cm^3). Almost all phenolic compounds isolated have a density higher than water. The density of phenolic compounds depends on the molecular weight and structural arrangements. In simple phenols like cresol, the position of a single substituent (methyl group) has shown to decrease from o-cresol (1.147 g/cm^3) to m-cresol (1.034 g/cm^3) and p-cresol (1.0178 g/cm^3). High molecular weight phenolic compounds such as quercetin have a higher density (1.799 g/cm^3) than simple phenols (Table 2.2). The density of the phenolic compounds can be used for green extraction technologies (Huang et al. 2019) from plants.

2.3.4   Solubility

Phenolic compounds are usually soluble in water especially simple phenols, while higher phenolic compounds are insoluble in water. For example, phenol is soluble in water, while ellagic acid is insoluble in water as shown in Table 2.2. The following concepts should be remembered while looking for solubility of a phenolic compound:

- The solubility of a phenolic compound in water is administered by the hydroxyl group present. For example, the solubility of benzene is 1.8 g/L, and phenol is 83 g/L.
- The hydroxyl group in a phenolic compound is responsible for the formation of intermolecular hydrogen bonding. Thus, more hydroxyl group means more hydrogen bonding and thus more solubility. For example, the solubility of catechol is significantly higher (430 g/L) than a simple phenol. Thus, hydrogen bonds are formed between water and phenol molecules which make phenol soluble in water.
- The aryl group attached to the hydroxyl group is hydrophobic and decrease the solubility of phenolic compound in water. Thus, the solubility of phenol decreases with the increase in the size of the aryl group.

2.3.5   Physical State

Phenolic compounds exist in different physical states. For example, a phenol, benzoic acid, phenylacetic acid are liquid, while most of the phenolic compounds are solid. The solid form may be crystalline-like o-cresol and p-cresol are crystalline solid as shown in Table 2.2.

2.3.6   Color

Phenolic compounds are either colorless or possess different colors. For example, most of the phenolic compounds are white crystalline solid, however, anthocyanidins and anthocyanins possess different colors (Table 2.3). Coloured phenolic compounds are pelargonidin (orange-red), cyanidin (red), peonidin (rose-red), delphinidin (blue-violet), petunidin (blue-purple), and malvidin (purple).

2.3.7 Flavor and Aroma

The flavor and aroma of the phenolic compounds have been extensively studied in foods. The presence of specific phenolic compounds in foods impart characteristics aroma and flavor to the foods. However, phenolic compounds are usually described as bitter and astringent in flavor as shown in Table 2.3. The flavors can add essential positive attributes as well as unpleasant attributes. These benefits/negatives are also highly concentration-dependent (Soto-Vaca et al. 2012). The intensity and duration of the bitter or taste response of phenolic compounds are altered by the degree of polymerization (Peleg et al. 1999). In humans, distinct bitter taste receptors were activated by one or more of the phenolic compounds. Thus phenolic compounds are characterized by specific flavor and aroma, which can affect the acceptance of specific foods by the consumers (Oliveira et al. 2014).

2.4 Chemical Properties of Phenolic Compounds

Phenolic compounds can undergo several reactions, the details of which is beyond the scope of this book. However, there are several important chemical properties of the phenolic compounds present in foods. These include antioxidant properties, acidity, metal complexation, hydrogen bonding, ester formation, glycosylation, and polymerization. These reactions may be either enzymatic or non-enzymatic or laboratory-based as given below.

2.4.1 Antioxidant Properties

The hydroxyl groups of the phenolic compounds are responsible for the antioxidant activity, whereas the stability of the free radical formed from the antioxidant is stabilized by the aromatic ring. The details of the antioxidant properties have been discussed in Chap. 1. However, it is important to mention that there are several synthetic-free radicals which are used to determine the antioxidant activity. The details of those reactions have been discussed in Chap. 15.

2.4.2 Acidity of Phenolic Compounds

The hydroxyl group(s) in the phenolic compound react with active metals such as sodium, potassium, forming phenoxides or salt. These reactions indicate its acidic nature. In phenol, the sp2 hybridized carbon of the benzene ring attached directly to the hydroxyl group acts as an electron-withdrawing group. Thus, it decreases the electron density on oxygen making hydrogen to be easily released. Due to the delocalization of negative charge on the aromatic ring, phenoxide ions are more stable than alkoxide ions. As a result, phenolic compounds are more acidic than alcohols. In the case of substituted phenols, the acidity decreases if an electron-donating group is attached to the ring while the acidity increases in the case of an electron-withdrawing group. A phenolic compound is considered an acid if it can release a proton (H^+) while in solution. Thus, a phenolic compound (PH) will dissociate to release a H^+ and P^-. The power of dissociation constant Ka is therefore expressed as below:

$$PH \rightleftharpoons P^- + H^+$$

$$K_a = \frac{[P^-][H^+]}{[PH]}$$

The K_a has defined the extent of dissociation of an acid. The pK_a is defined as $- \log 10\ K_a$. The pKa value is used to represent the 50% dissociation of the acid. The pK_a values of most of the phenolic compounds are in the range of 2–14. The Henderson-Hasselbach equation is used to calculate the pH of the phenolic compound.

$$pH = pK_a + \log \frac{[P^-]}{[PH]}$$

This equation is used for the calculation of the pH of a weak acid. Phenolic compounds are usually classed as a weak acid, thus the above equation can be used to calculate its pH. The pH below 7 represents acidity. However, the acidity of aliphatic alcohols is less than aromatic alcohols like phenolic compounds. This is due to the fact that aromatic anion is stabilized by the delocalization of electron and subsequent mesomeric structures as shown in Fig. 2.49. The aromatic anion is termed as phenolate anion.

There are numerous methods used for the determination of dissociation constants of phenolic compounds. Conventionally, potentiometry and UV–vis absorption spectrometry has been the most useful and extensively studied techniques for the determination of equilibrium constants. These methods were used because of

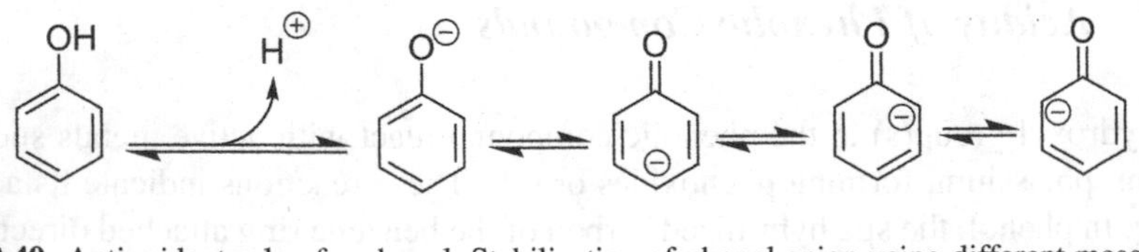

Fig. 2.49 Antioxidant role of a phenol. Stabilization of phenol anion using different mesomeric structures

their accuracy and high reproducibility. Moreover, the use of computer programs for the refinement of equilibrium constants allows the different p*K*a values in polyprotic substances to be determined, even when they are very close (Erdemgil et al. 2007). Alternative techniques for measuring pKa values of phenolic compounds are liquid chromatography (Erdemgil et al. 2007; Sanli et al. 2002)and capillary zone electrophoresis (Poole et al. 2004).

Substituents produce significant effects on the acidity of phenolic compounds. Table 2.4 showed the pKa values of phenolic compounds. In the case of cresol, there was very little difference in the pKa values of ortho and para-cresol, while meta-cresol has the lowest values. Similarly, adding a hydrophobic atom like an alkyl group had reduced the pKa values. For example, Guaiacol (9.93) whereas propyl guaiacol has lesser values (9.85). In case, there is more hydroxyl group, the phenolic compound will have different pKa values, i.e., pKa1, pKa2, and pKa3, etc. (Ragnar et al. 2000; Poole et al. 2004). The difference in pKa values of phenolic compounds is of significant importance during separation by liquid chromatography. The color of the anthocyanins changed with the change in the acidity. Thus, acidity is an important chemical property of phenolic compounds.

2.4.3 Metal Complexation

Metals are electron acceptors and form bonds with electron donor compounds such as phenolic compounds. In nature, phenolic compounds are one of the colored compounds responsible for different colors of foods, after carotenoids. Most of the colors such as blue, red, violet, yellow, and their mixed colors of the foods are due phenolic compounds such as anthocyanins, aurones or as a result of metal chelation of phenolic compounds. In cornflowers, the blue color of the flower leaves is due to the protocyanin complex. Protocyanin complex is formed from one iron, one magnesium and two calcium atoms chelation with anthocyanin. Similarly, Cyanocentaurin is another complex compound formed from an iron complex of 4 mol of cyanidin-3,5-diglucoside, and 3 mol of biflavone glucoside. The cyanidin residues are cyanidin 3-*O*-(6-*O*-succinylglucoside)-5-*O*-glucoside, whereas that of flavone is apigenin-7-*O*-glucuronide-4'-*O*-(6-*O*-malonylglucoside) (Takeda 2006).

In the laboratory, several metals ions such as Ca, Cu, Fe, Sn, and Mg are used to prepare metal complexes. The purpose of metal chelation is diverse. The most important one is the antioxidant activity. For example, Copper chelating ability was found to be highly correlated with FRAP, DPPH and total phenolic contents in Brazilian coffee (Santos et al. 2017). Selective self-assembly of metal chelation in a complex mixture has been used to isolate desirable components for the rapid production of functional substances in foods (Lin et al. 2019).

Table 2.4 Summary of the pKa values of phenolic compounds

No	Name	pKa[a]
1	o-Cresol	10.28
2	m-Cresol	10.08
3	p-Cresol	10.25
4	Guaiacol	9.93
5	Propyl guaiacol	9.85
6	Vanillyl alcohol	9.78
7	Vanillin	7.40
8	Vanillic acid	4.08, 8.54
9	Methyl vanillate	8.30
10	Isovanillic acid	433, 9.9
11	α-Carboxyvanillin	1.60, 7.54
12	Eugenol	10.15
13	Iso-eugenol	9.89
14	α-Hydro-β-oxoferulic acid	2.57, 10.97
15	Ferulic acid	4.56, 9.39
16	Bivanillin	6.16, 10.07
17	Pinoresinol	9.01, 9.76
18	Diguaiacylstilbene	7.27, 10.39
19	Syringol	9.98
20	Syringyl creosol	10.01
21	Syringyl alcohol	9.87
22	Syringaldehyde	7.34
23	Syringic acid	3.86, 7.76
24	Methyl syringate	8.7
25	Trans-sinapaldehyde	8.2
26	Trans-sinapic acid	4.9, 9.2
27	Gallic acid	4.09, 7.3, 12.17
28	Protocatechuic acid	4.19, 7.86, 13.22
29	Gentisic acid	2.98, 10.29
30	Benzoic acid	4.21
31	p-Hydroxybenzoic acid	4.3, 8.68
32	Phenylacetic acid	4.32
33	Resorcinol	9.3
34	Salicylic acid	2.98
35	Caffeic acid	4.3, 8.14, 13.16
36	p-Coumaric acid	4.32, 8.97
37	Ferulic acid	4.27, 8.83

Data are from Ragnar et al. (2000), Sanli et al. (2002), Poole et al. (2004) and Erdemgil et al. (2007)
[a]First value is pKa1, second represents pKa2 and third represents pKa3. The values are based on potentiometric, and spectrometric methods

2.4.4 Hydrogen Bonding

Hydrogen bonding is another chemical property of the hydroxyl group. Hydrogen atom(s) in a phenolic compound is electrostatically attracted by an electronegative atom. The electronegative atoms in phenolic compounds may be either oxygen, nitrogen, fluorine, chlorine or atoms containing free electrons or loan pairs of electrons. This attraction is also referred to as electrostatic attraction or electrostatic bonding. The hydrogen bond is weaker than a covalent and ionic bond, reflecting a higher bond distance between the hydrogen atom and electronegative atom. The bond distance or length depends on several factors, including the structure of the molecule.

Hydrogen bonding in phenolic compounds is usually of two types, i.e., intermolecular and intra-molecular hydrogen bonding. In the intra-molecular hydrogen bonding, a phenolic compound may contain any substituent near a hydroxyl group. For example, an ortho-substituted phenolic compound such as o-cresol is the best example as shown in Fig. 2.50. The intra-molecular hydrogen bonding may be present in the adjacent ring of the same molecule e.g. quercetin. Inter-molecular hydrogen bonding is present between molecules of the same substance or different. For example, the hydrogen bond is formed by hydrogen of one phenol with the oxygen of another phenol molecule.

The bond dissociation energy (BDE) is an important factor for antioxidant property and stability of phenolic compounds. It is measured using a photoacoustic calorimetric method and by density functional theory (DFT) calculations. The relative BDE (O–H) in kcal/mol with respect to phenol had been reported experimentally as follows: 2-methoxy phenol (−4.0), 4-methoxy phenol (−4.9), 2,6-dimethoxyphenol (−10.6), 2,4-dimethoxyphenol (−9.0), 2,4,6-trimethoxyphenol (−13.6), and ubiquinol-0 (−12.0). The intramolecular hydrogen-bond energy in o-methoxy-substituted phenol was −4.3 kcal/mol (de Heer et al. 1999, 2000). The intra-molecular

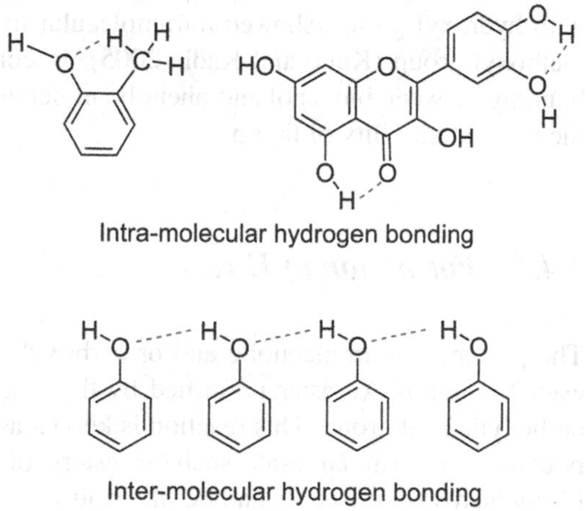

Fig. 2.50 Representations of intra-molecular and inter-molecular hydrogen bonding in phenolic compounds

Intra-molecular hydrogen bonding

Inter-molecular hydrogen bonding

hydrogen-bonded molecule may also form an additional hydrogen bond with solvents having hydrogen bond accepting power. The inter-molecular hydrogen bonding thus increases the solubility and melting point of the phenolic compound (Vermerris and Nicholson 2008). As shown in Table 2.2, the melting point of phenol is 40.1 °C, which increased to 111 °C in resorcinol. A carboxylic acid moiety has a significant effect on melting point, e.g. 3-hydroxybenzoic acid has a high melting point (202 °C) than phenol. Similar is the case of solubility as shown in the above mention table.

The quantitative structure-activity relationship (QSAR) models for predicting the antioxidative capacity of phenolic antioxidants have been studied (Filipović et al. 2015). For this purpose, vitamin C equivalent antioxidative capacity (VCEAC) values various phenolic compounds were reported (Kim and Lee 2004). Thermodynamic and aromaticity properties, natural bond analysis (NBO) and the strength of intramolecular hydrogen bonds were used as independent variables to determine the importance of hydrogen bonding in phenolic compounds (Jeremić et al. 2017). The mathematic modeling revealed a significant influence of internal hydrogen bonds on the antioxidative capacity of the phenolic compounds (Salicylic acid, 4-Hydroxybenzoic acid, 3-Hydroxybenzoic acid, Carvacrol, Butylated hydroxytoluene, Syringic acid, 4-Hydroxyphenylacetic acid, 2-tert-butylhydroquinone, Thymol, Homogentisic acid, Gentisic acid, 3-Hydroxyphenylacetic acid, 2-Hydroxyphenylacetic acid, Butylated hydroxyanisole, Vanillic acid, Protocatechuic acid, 2,3-Dihydroxybenzoic acid, Catechol (**2.282**), Homoprotocatechuic acid, Gallic acid, and Pyrogallol) as shown in Table 2.5. The best correlation with the VCEAC values was achieved within a three-descriptor QSAR model. This model was obtained by including a magnetic aromaticity index. In conclusion, the aromaticity has only secondary effects on the antioxidative capacity.

Hydrogen bonding is a complex phenolic compound such as lignin had been studied using Fourier transformed infrared (FTIR) (Podolyák et al. 2018). In dilute solution non-hydrogen bonded free hydroxyl groups were only observed in the model compounds possessing aliphatic hydroxyl groups. The phenolic compounds with hydroxyl groups showed intramolecular hydrogen bonding with the adjacent methoxyl group (Kubo and Kadla 2005). In conclusion, intermolecular hydrogen bonding between biphenol and phenolic moieties, have a significant contribution to the thermal mobility of lignin.

2.4.5 Formation of Esters

The presence of an alcoholic and or carboxylic acid group is responsible for the ester formation. An ester is formed by the reaction of an alcoholic group with a carboxylic acid group. This reaction is known as esterification. Two phenolic compounds can form an ester such as esters of coumaric acid with quinic acid (5-coumaroylquinic acid), and Ferulic acid with quinic acid (3-feruloylquinic acid).

Table 2.5 Experimental and some theoretical indicators of antioxidative activity of investigated molecules. Reproduced with kind permission of Elsevier (Jeremić et al. 2017)

No	Compound	VCEAC[a] (mg/L)	BDE[b] (kJ/mol)	(IP + PDE) (kJ/mol)	(PA + ETE) (kJ/mol)
1	Salicylic acid	1.4	400.7	581.8	581.8
2	4-Hydroxybenzoic acid	4.8	392.4	573.4	573.5
3	3-Hydroxybenzoic acid	53.7	383.2	564.3	564.3
4	Carvacrol	58	363.1	544.2	544.2
5	Butylated hydroxytoluene	77.4	333.7	514.8	514.8
6	Syringic acid	80.4	351.7	532.8	532.8
7	4-Hydroxyphenylacetic acid	82.8	371	552.1	552.1
8	2-*tert*-Butylhydroquinone	83.9	338.2	519.3	523.9
9	Thymol	85.3	359.6	540.7	540.7
10	Homogentisic acid	87.8	340	521.1	521.1
11	Gentisic acid	90.8	358.5	539.6	539.5
12	3-Hydroxyphenylacetic acid	91.6	374.1	555.2	555.2
13	2-Hydroxyphenylacetic acid	95.1	364.9	546.1	546.1
14	Butylated hydroxyanisole	97.6	337	518.1	518.1
15	Vanillic acid	117.2	371	552.1	552.1
16	Protocatechuic acid	163.2	361	542.1	544.9
17	2,3-Dihydroxybenzoic acid	169.6	362.1	543.2	550.3
18	Catechol	253.1	346.7	527.8	527.8
19	Homoprotocatechuic acid	316.7	316.5	497.6	497.6
20	Gallic acid	324.3	345.7	526.8	526.8
21	Pyrogallol	331.2	332.1	513.2	513.2

BDE bond dissociation energy, *IP* ionization potential, *PDE* proton dissociation enthalpy, *ETC* electron transfer energy, *PA* proton affinity, *VCEAC* vitamin C equivalent antioxidative capacity
[a]Kim and Lee (2004)
[b]Filipović et al. (2015)

Several caffeoylquinic acids are present in foods such as 3-caffeoylquinic acid, 4-caffeoylquinic acid, and 5-caffeoylquinic acid (Nagels et al. 1980). Di-ester of caffeoylquinic acids is also present in foods in the form of 1,5-dicaffeoylquinic acid, 3,4-dicaffeoylquinic acid, and 3,5-dicaffeoylquinic acid as shown in Fig. 2.51. Similarly, gallic acid upon esterification produces ellagic acid. Ester of gallic acids have been identified in several foods and beverages (Newsome et al. 2016). The esters of gallic acid with other phenolic compounds or alkyl esters are of significant antioxidant potential (Rúa et al. 2017).

2.4.6 Glycosylation

Glycosides are formed in foods by condensation and dehydration reactions. Water is released, and glucose is attached to the carboxylic acid group of the phenolic moiety. The glycoside bond thus formed may be either *O*-glycosidic or *C*-glycosidic.

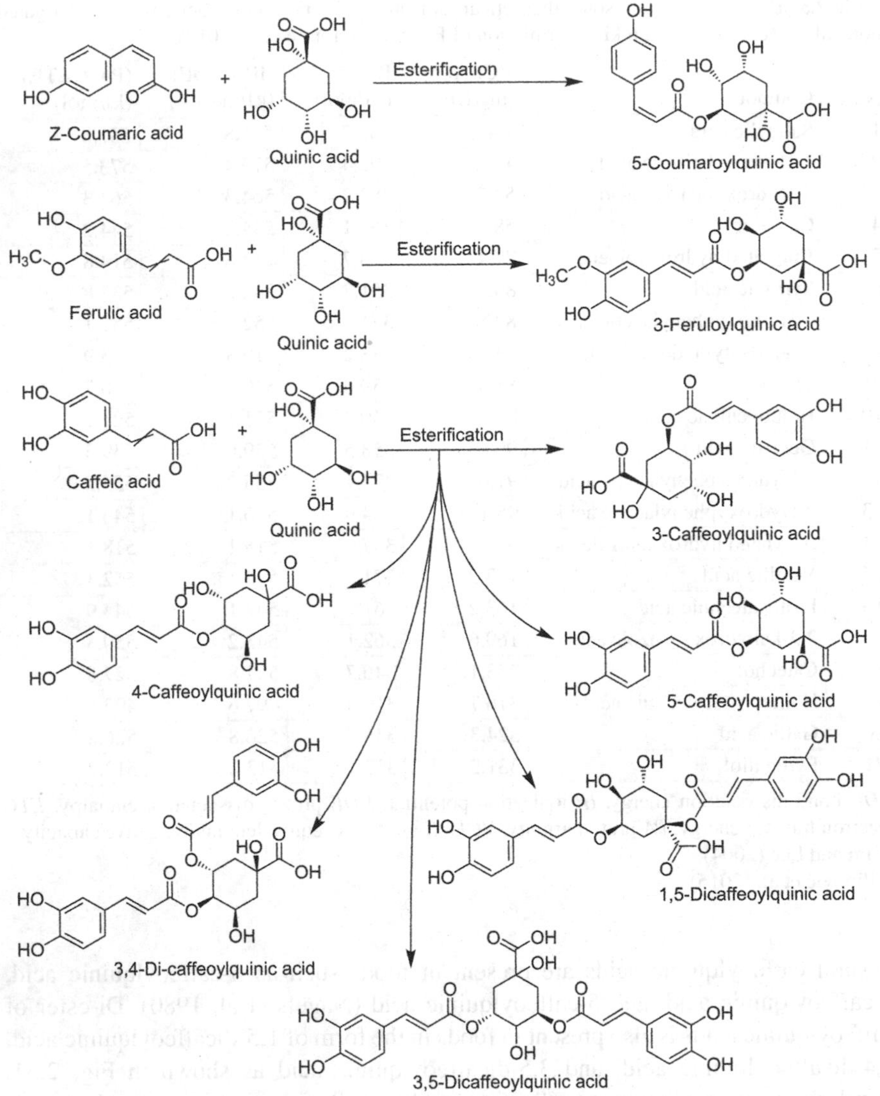

Fig. 2.51 Esterification of Coumaric acid with quinic acid and Ferulic acid with quinic acids. Several esters of caffeic acid with quinic acids are formed and reported in foods

For example, Coumaric acid (**2.34**) upon glycosylation with β-D-glucopyranose (**2.277**) form 1-(4-Coumaroyl)-β-D-glucose (**2.278**) by the loss of water molecule (Fig. 2.52). Such reactions may occur in nature by enzymes or under control conditions in the laboratory.

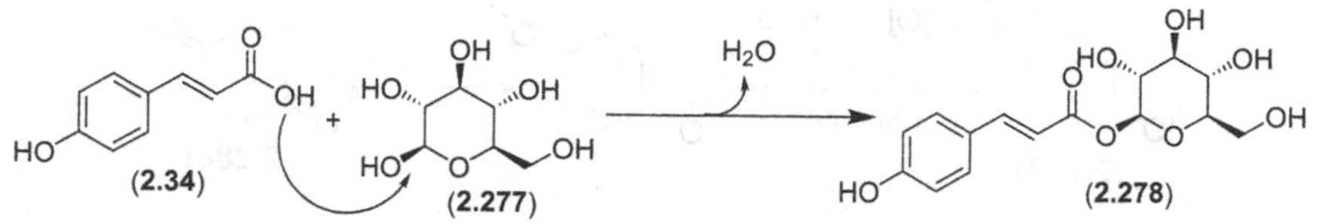

Fig. 2.52 Formation of coumaroyl glucoside from Coumaric acid and glucose. (2.34) Coumaric acid, (2.277) β-D-glucose, and (2.278) 1-(4-Coumaroyl)-β-D-glucose

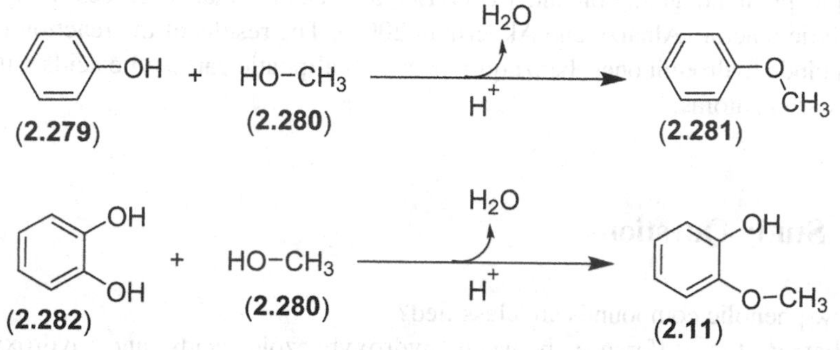

Fig. 2.53 Formation of ethers by phenolic compounds. (2.279) Phenol, (2.280) methanol, (2.281) methoxyphenol, (2.282) catechol, and (2.11) guaiacol

2.4.7 Ether Formation

The phenolic compound reacts with alcohol to form an ether. Ethers of phenolic compounds are commonly present in foods. The simplest reaction is as shown in Fig. 2.53, of the methanol and phenol under acidic medium. A methoxybenzene is formed. Similarly, when catechol reacts with methanol it forms guaiacol. Vanillyl methyl ether is one of the important aroma compounds present in vanilla beans (Pérez-Silva et al. 2006).

2.4.8 Oxidation of Phenolic Compounds

There is a wide range of oxidation reactions of phenolic compounds. Phenolic compounds may be oxidized enzymatically or non-enzymatically. In the enzymatic reaction, the phenolic compound is converted into specialized products. For example, oxidation of 4-hydroxyphenol is catalyzed by the enzyme "Laccase" to form quinone as shown in Fig. 2.54. Further such enzymatic reactions lead to the formation of naphthoquinone in foods. The details will be discussed in Chap. 11.

In the laboratory, phenol is used as a solvent and an important reactant for the synthesis of a diverse class of aromatic compounds. However, some of the products

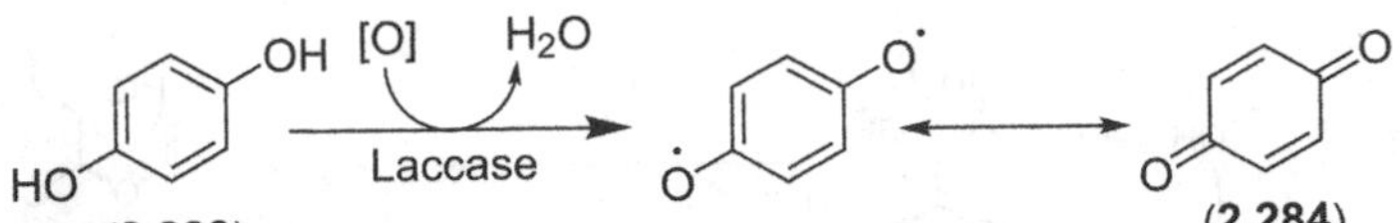

Fig. 2.54 Enzymatic oxidation of phenolic compound using laccase in the presence of oxygen. (**2.283**) p-Hydroxyphenol and (**2.284**) p-quinone

or classes of the compounds are similar to those in biosynthesized in nature. For example, phenol degradation under UV/H_2O_2 advanced oxidation process in a batch photolytic reactor (Alnaizy and Akgerman 2000). The results of the reaction products include hydroquinones, benzoquinones, and aliphatic carboxylic acids with up to six carbon atoms.

2.5 Study Questions

- How phenolic compounds are classified?
- What is the difference between hydroxybenzoic acids and hydroxycinnamic acids?
- Describe the classification of flavonoids.
- What is the structural difference between isoflavones and neoflavones?
- What are glycosides?
- Describe the different classes of glycosides present in plant foods?
- Discuss the structure of chalcone and their glycosides.
- What do you know about the cyanogenic glycosides?
- Describe the physical properties of phenolic compounds.
- Write a detail note on the chemical properties of phenolic compounds.
- Discuss the importance of acidity of phenolic compounds.
- How hydrogen bonding occurred in phenolic compounds?

References

Acosta-Estrada BA, Gutiérrez-Uribe JA, Serna-Saldívar SO (2014) Bound phenolics in foods, a review. Food Chem 152:46–55
Aherne SA, O'Brien NM (2002) Dietary flavonols: chemistry, food content, and metabolism. Nutrition 18(1):75–81
Ahmad S, Zeb A, Ayaz M, Murkovic M (2020) Characterization of phenolic compounds using UPLC–HRMS and HPLC–DAD and anti-cholinesterase and anti-oxidant activities of *Trifolium repens* L. leaves. Eur Food Res Technol 246(3):485–496
Alnaizy R, Akgerman A (2000) Advanced oxidation of phenolic compounds. Adv Environ Res 4(3):233–244

Andrés-Lacueva C, Medina-Remon A, Llorach R, Urpi-Sarda M, Khan N, Chiva-Blanch G, Zamora-Ros R, Rotches-Ribalta M, Lamuela-Raventos RM (2010) Phenolic compounds: chemistry and occurrence in fruits and vegetables. In: de la Rosa LA, Alvarez-Parrilla E, González-Aguilar GA (eds) Fruit and vegetable phytochemicals: chemistry, nutritional value, and stability. Blackwell Publishing, Ames, IA, pp 53–80

Awika JM, Rooney LW, Waniska RD (2004) Properties of 3-deoxyanthocyanins from Sorghum. J Agric Food Chem 52(14):4388–4394

Baggett S, Protiva P, Mazzola EP, Yang H, Ressler ET, Basile MJ, Weinstein IB, Kennelly EJ (2005) Bioactive benzophenones from garcinia xanthochymus fruits. J Nat Prod 68(3):354–360

Barreca D, Gattuso G, Bellocco E, Calderaro A, Trombetta D, Smeriglio A, Laganà G, Daglia M, Meneghini S, Nabavi SM (2017) Flavanones: citrus phytochemical with health-promoting properties. Biofactors 43(4):495–506

Bartnik M, Facey PC (2017) Glycosides. In: Badal S, Delgoda R (eds) Pharmacognosy: fundamentals, applications and strategy. Academic Press, Boston, MA, pp 101–161

Baruah JB (2011) Chemistry of phenolic compounds: state of the art. Nova Science Publishers, New York, NY

Bhatt LR, Bae MS, Kim BM, Oh G-S, Chai KY (2009) A chalcone glycoside from the fruits of Sorbus commixta Hedl. Molecules 14(12):5323–5327

Borges F, Roleira F, Milhazes N, Santana L, Uriarte E (2005) Simple coumarins and analogues in medicinal chemistry: occurrence, synthesis and biological activity. Curr Med Chem 12(8):887–916

Bors W, Michel C (1999) Antioxidant capacity of flavanols and gallate esters: pulse radiolysis studies. Free Radic Biol Med 27(11):1413–1426

Boucherle B, Peuchmaur M, Boumendjel A, Haudecoeur R (2017) Occurrences, biosynthesis and properties of aurones as high-end evolutionary products. Phytochemistry 142:92–111

Castellano G, González-Santander JL, Lara A, Torrens F (2013) Classification of flavonoid compounds by using entropy of information theory. Phytochemistry 93:182–191

Castellar R, Obón JM, Alacid M, Fernández-López JA (2003) Color properties and stability of betacyanins from opuntia fruits. J Agric Food Chem 51(9):2772–2776

Chen J-J, Ting C-W, Hwang T-L, Chen I-S (2009) Benzophenone derivatives from the fruits of Garcinia multiflora and their anti-inflammatory activity. J Nat Prod 72(2):253–258

Cheynier V (2012) Phenolic compounds: from plants to foods. Phytochem Rev 11(2):153–177

Cheynier V, Comte G, Davies KM, Lattanzio V, Martens S (2013) Plant phenolics: recent advances on their biosynthesis, genetics, and ecophysiology. Plant Physiol Biochem 72:1–20

Clifford MN (2000) Miscellaneous phenols in foods and beverages – nature, occurrence and dietary burden. J Sci Food Agric 80(7):1126–1137

Clifford MN, Scalbert A (2000) Ellagitannins–nature, occurrence and dietary burden. J Sci Food Agric 80(7):1118–1125

Covas MI, Miró-Casas E, Fitó M, Farré-Albadalejo M, Gimeno E, Marrugat J, De La Torre R (2003) Bioavailability of tyrosol, an antioxidant phenolic compound present in wine and olive oil, in humans. Drugs Exp Clin Res 29(5–6):203–206

Coward L, Smith M, Kirk M, Barnes S (1998) Chemical modification of isoflavones in soyfoods during cooking and processing. Am J Clin Nutr 68(6):1486S–1491S

Crozier A, Jaganath IB, Clifford MN (2009) Dietary phenolics: chemistry, bioavailability and effects on health. Nat Prod Rep 26(8):1001–1043

Dai J, Kardono LBS, Tsauri S, Padmawinata K, Pezzuto JM, Kinghorn AD (1991) Phenylacetic acid derivatives and a thioamide glycoside from Entada phaseoloides. Phytochemistry 30(11):3749–3752

Davin LB, Jourdes M, Patten AM, Kim K-W, Vassão DG, Lewis NG (2008) Dissection of lignin macromolecular configuration and assembly: comparison to related biochemical processes in allyl/propenyl phenol and lignan biosynthesis. Nat Prod Rep 25(6):1015–1090

De Heer MI, Mulder P, Korth HG, Ingold KU, Lusztyk J (2000) Hydrogen atom abstraction kinetics from intramolecularly hydrogen bonded ubiquinol-0 and other (poly)methoxy phenols. J Am Chem Soc 122(10):2355–2360

Debarshi Kar M, Sanjay Kumar B, Vivek A (2017) Chalcone derivatives: anti-inflammatory potential and molecular targets perspectives. Curr Top Med Chem 17(28):3146–3169

Dias MM, Machado NFL, Marques MPM (2011) Dietary chromones as antioxidant agents—the structural variable. Food Funct 2(10):595–602

Díaz AM, Caldas GV, Blair MW (2010) Concentrations of condensed tannins and anthocyanins in common bean seed coats. Food Res Int 43(2):595–601

Diaz-Muñoz G, Miranda IL, Sartori SK, de Rezende DC, Diaz MAN (2018) Anthraquinones: an overview. In: Rahman A (ed) Studies in natural products chemistry, vol 58. Elsevier, Amsterdam, pp 313–338

Dignum MJW, van der Heijden R, Kerler J, Winkel C, Verpoorte R (2004) Identification of glucosides in green beans of *Vanilla planifolia* Andrews and kinetics of vanilla β-glucosidase. Food Chem 85(2):199–205

Dufossé L (2014) Anthraquinones, the Dr Jekyll and Mr Hyde of the food pigment family. Food Res Int 65:132–136

Dziedzic SZ, Hudson BJF (1983) Polyhydroxy chalcones and flavanones as antioxidants for edible oils. Food Chem 12(3):205–212

Einbond LS, Reynertson KA, Luo X-D, Basile MJ, Kennelly EJ (2004) Anthocyanin antioxidants from edible fruits. Food Chem 84(1):23–28

Elsayed GA, Khalil AK (2018) Facile synthesis of chalcone glycosides isolated from aerial parts of *Brassica rapa* L. Curr Org Synth 15(3):423–429

Erdemgil FZ, Şanli S, Şanli N, Özkan G, Barbosa J, Guiteras J, Beltrán JL (2007) Determination of pKa values of some hydroxylated benzoic acids in methanol–water binary mixtures by LC methodology and potentiometry. Talanta 72(2):489–496

Filipović M, Marković Z, Đorović J, Marković JD, Lučić B, Amić D (2015) QSAR of the free radical scavenging potency of selected hydroxybenzoic acids and simple phenolics. C R Chim 18(5):492–498

Foo LY, Karchesy JJ (1989) Chemical nature of phlobaphenes. In: Hemingway RW, Karchesy JJ, Branham SJ (eds) Chemistry and significance of condensed tannins. Springer, Boston, MA, pp 109–118

Fotsis T, Pepper M, Adlercreutz H, Fleischmann G, Hase T, Montesano R, Schweigerer L (1993) Genistein, a dietary-derived inhibitor of in vitro angiogenesis. Proc Natl Acad Sci 90(7):2690–2694

Fouillaud M, Venkatachalam M, Girard-Valenciennes E, Caro Y, Dufossé L (2016) Anthraquinones and derivatives from marine-derived fungi: structural diversity and selected biological activities. Mar Drugs 14(4):64

Hackman RM, Polagruto JA, Zhu QY, Sun B, Fujii H, Keen CL (2007) Flavanols: digestion, absorption and bioactivity. Phytochem Rev 7(1):195

Harborne J, Simmonds N (1964) The natural distribution of the phenolic aglycones. In: Biochemistry of phenolic compounds. Academic Press, London, pp 77–127

de Heer MI, Korth H-G, Mulder P (1999) Poly methoxy phenols in solution: O–H bond dissociation enthalpies, structures, and hydrogen bonding. J Org Chem 64(19):6969–6975

Heim KE, Tagliaferro AR, Bobilya DJ (2002) Flavonoid antioxidants: chemistry, metabolism and structure-activity relationships. J Nutr Biochem 13(10):572–584

Heleno SA, Martins A, Queiroz MJRP, Ferreira ICFR (2015) Bioactivity of phenolic acids: metabolites versus parent compounds: a review. Food Chem 173:501–513

Hemingway RW, Karchesy JJ (2012) Chemistry and significance of condensed tannins. Springer Science & Business Media, London

Hirose M, Takesada Y, Tanaka H, Tamano S, Kato T, Shirai T (1998) Carcinogenicity of antioxidants BHA, caffeic acid, sesamol, 4-methoxyphenol and catechol at low doses, either alone or in combination, and modulation of their effects in a rat medium-term multi-organ carcinogenesis model. Carcinogenesis 19(1):207–212

Hopkinson SM (1969) The chemistry and biochemistry of phenolic glycosides. Q Rev Chem Soc 23(1):98–124

Hostetler GL, Ralston RA, Schwartz SJ (2017) Flavones: food sources, bioavailability, metabolism, and bioactivity. Adv Nutr 8(3):423–435

Huang H, Wang Z, Aalim H, Limwachiranon J, Li L, Duan Z, Ren G, Luo Z (2019) Green recovery of phenolic compounds from rice byproduct (rice bran) using glycerol based on viscosity, conductivity and density. Int J Food Sci Technol 54(4):1363–1371

Jain A, Yang G, Yalkowsky SH (2004) Estimation of melting points of organic compounds. Ind Eng Chem Res 43(23):7618–7621

Janeczko T, Gładkowski W, Kostrzewa-Susłow E (2013) Microbial transformations of chalcones to produce food sweetener derivatives. J Mol Catal B Enzym 98:55–61

Jeremić S, Radenković S, Filipović M, Antić M, Amić A, Marković Z (2017) Importance of hydrogen bonding and aromaticity indices in QSAR modeling of the antioxidative capacity of selected (poly)phenolic antioxidants. J Mol Graph Model 72:240–245

Jiang N, Doseff AI, Grotewold E (2016) Flavones: from biosynthesis to health benefits. Plants (Basel) 5(2):27

Jiménez C, Riguera R (1994) Phenylethanoid glycosides in plants: structure and biological activity. Nat Prod Rep 11(6):591–606

Kabera JN, Semana E, Mussa AR, He X (2014) Plant secondary metabolites: biosynthesis, classification, function and pharmacological properties. J Pharm Pharmacol 2:377–392

Kamiloglu S, Tomas M, Capanoglu E (2019) Dietary flavonols and O-glycosides. In: Handbook of dietary phytochemicals. Springer Nature Singapore Pte Ltd, Singapore

Kamiya K, Hamabe W, Tokuyama S, Satake T (2009) New anthraquinone glycosides from the roots of *Morinda citrifolia*. Fitoterapia 80(3):196–199

Kaneshiro J, Fukui K, Higuchi H, Nohara T (1991) The first isolation of lignan tri- and tetraglycosides. Chem Pharm Bull 39(6):1623–1625

Kasiotis KM, Pratsinis H, Kletsas D, Haroutounian SA (2013) Resveratrol and related stilbenes: their anti-aging and anti-angiogenic properties. Food Chem Toxicol 61:112–120

Kaur R, Chattopadhyay SK, Tandon S, Sharma S (2012) Large scale extraction of the fruits of *Garcinia indica* for the isolation of new and known polyisoprenylated benzophenone derivatives. Ind Crop Prod 37(1):420–426

Khadem S, Marles RJ (2010) Monocyclic phenolic acids; hydroxy- and polyhydroxybenzoic acids: occurrence and recent bioactivity studies. Molecules 15(11):7985–8005

Khanbabaee K, van Ree T (2001) Tannins: classification and definition. Nat Prod Rep 18(6):641–649

Khoo HE, Azlan A, Tang ST, Lim SM (2017) Anthocyanidins and anthocyanins: colored pigments as food, pharmaceutical ingredients, and the potential health benefits. Food Nutr Res 61(1):1361779–1361779

Kim D-O, Lee CY (2004) Comprehensive study on vitamin C equivalent antioxidant capacity (VCEAC) of various polyphenolics in scavenging a free radical and its structural relationship. Crit Rev Food Sci Nutr 44(4):253–273

Kishor Kumar K, Arul AnanthaKumar A, Ahmad R, Adhikari S, Variyar PS, Sharma A (2010) Effect of gamma-radiation on major aroma compounds and vanillin glucoside of cured vanilla beans (*Vanilla planifolia*). Food Chem 122(3):841–845

Klick S, Herrmann K (1988) Glucosides and glucose esters of hydroxybenzoic acids in plants. Phytochemistry 27(7):2177–2180

Koponen JM, Happonen AM, Mattila PH, Törrönen AR (2007) Contents of anthocyanins and ellagitannins in selected foods consumed in Finland. J Agric Food Chem 55(4):1612–1619

Kostova I, Bhatia S, Grigorov P, Balkansky S, Parmar VS, Prasad AK, Saso L (2011) Coumarins as antioxidants. Curr Med Chem 18(25):3929–3951

Kubo S, Kadla JF (2005) Hydrogen bonding in lignin: a Fourier transform infrared model compound study. Biomacromolecules 6(5):2815–2821

Kylli P, Nousiainen P, Biely P, Sipilä J, Tenkanen M, Heinonen M (2008) Antioxidant potential of hydroxycinnamic acid glycoside esters. J Agric Food Chem 56(12):4797–4805

Landete JM (2011) Ellagitannins, ellagic acid and their derived metabolites: a review about source, metabolism, functions and health. Food Res Int 44(5):1150–1160

Landoni M, Puglisi D, Cassani E, Borlini G, Brunoldi G, Comaschi C, Pilu R (2020) Phlobaphenes modify pericarp thickness in maize and accumulation of the fumonisin mycotoxins. Sci Rep 10(1):1417

Laurichesse S, Avérous L (2014) Chemical modification of lignins: towards biobased polymers. Prog Polym Sci 39(7):1266–1290

Li L, Tsao R, Yang R, Liu C, Young JC, Zhu H (2008) Isolation and purification of phenylethanoid glycosides from *Cistanche deserticola* by high-speed counter-current chromatography. Food Chem 108(2):702–710

Li X-N, Sun J, Shi H, Yu L, Ridge CD, Mazzola EP, Okunji C, Iwu MM, Michel TK, Chen P (2017) Profiling hydroxycinnamic acid glycosides, iridoid glycosides, and phenylethanoid glycosides in baobab fruit pulp (*Adansonia digitata*). Food Res Int 99:755–761

Liggins J, Bluck LJC, Runswick S, Atkinson C, Coward WA, Bingham SA (2000) Daidzein and genistein contents of vegetables. Br J Nutr 84(5):717–725

Likhtenshtein G (2009) Stilbenes: applications in chemistry, life sciences and materials science. Wiley-VCH Verlag GmbH & Co. KGaA, Weinheim

Lin G, Rahim MA, Leeming MG, Cortez-Jugo C, Besford QA, Ju Y, Zhong Q-Z, Johnston ST, Zhou J, Caruso F (2019) Selective metal–phenolic assembly from complex multicomponent mixtures. ACS Appl Mater Interfaces 11(19):17714–17721

Lu Y, Sun Y, Foo LY, McNabb WC, Molan AL (2000) Phenolic glycosides of forage legume *Onobrychis viciifolia*. Phytochemistry 55(1):67–75

Luqman S, Meena A, Singh P, Kondratyuk TP, Marler LE, Pezzuto JM, Negi AS (2012) Neoflavonoids and tetrahydroquinolones as possible cancer chemopreventive agents. Chem Biol Drug Des 80(4):616–624

Lyles JT, Negrin A, Khan SI, He K, Kennelly EJ (2014) In vitro antiplasmodial activity of benzophenones and xanthones from edible fruits of garcinia species. Planta Med 80(08/09):676–681

Maga JA (1978) Simple phenol and phenolic compounds in food flavor. Crit Rev Food Sci Nutr 10(4):323–372

Mah SH (2019) Chalcones in diets. In: Handbook of dietary phytochemicals. Springer Nature Singapore Pte Ltd, Singapore, pp 1–52

Mäkilä L, Laaksonen O, Alanne A-L, Kortesniemi M, Kallio H, Yang B (2016) Stability of hydroxycinnamic acid derivatives, flavonol glycosides, and anthocyanins in black currant juice. J Agric Food Chem 64(22):4584–4598

Moawad A, El Amir D (2016) Ginkgetin or isoginkgetin: the dimethylamentoflavone of *Dioon edule* Lindl. leaves. Eur J Med Plants 16:1–7

Möller B, Herrmann K (1983) Quinic acid esters of hydroxycinnamic acids in stone and pome fruit. Phytochemistry 22(2):477–481

Nagels L, van Dongen W, de Brucker J, de Pooter H (1980) High-performance liquid chromatographic separation of naturally occurring esters of phenolic acids. J Chromatogr A 187(1):181–187

Natella F, Nardini M, Di Felice M, Scaccini C (1999) Benzoic and cinnamic acid derivatives as antioxidants: structure–activity relation. J Agric Food Chem 47(4):1453–1459

Newsome AG, Li Y, van Breemen RB (2016) Improved quantification of free and ester-bound gallic acid in foods and beverages by UHPLC-MS/MS. J Agric Food Chem 64(6):1326–1334

Ninomiya M, Efdi M, Inuzuka T, Koketsu M (2010) Chalcone glycosides from aerial parts of *Brassica rapa* L. 'hidabeni', turnip. Phytochem Lett 3(2):96–99

Nollet LM, Gutierrez-Uribe JA (2018) Phenolic compounds in food: characterization and analysis. CRC Press, Boca Raton, FL

Norwitz G, Nataro N, Keliher PN (1986) Study of the steam distillation of phenolic compounds using ultraviolet spectrometry. Anal Chem 58(3):639–641

Ocakoglu D, Tokatli F, Ozen B, Korel F (2009) Distribution of simple phenols, phenolic acids and flavonoids in Turkish monovarietal extra virgin olive oils for two harvest years. Food Chem 113(2):401–410

Okuda T, Yoshida T, Hatano T (1993) Classification of oligomeric hydrolysable tannins and specificity of their occurrence in plants. Phytochemistry 32(3):507–521

Oliveira LL, Carvalho MV, Melo L (2014) Health promoting and sensory properties of phenolic
compounds in food. Revista Ceres 61:764–779

Orihara Y, Furuya T, Hashimoto N, Deguchi Y, Tokoro K, Kanisawa T (1992) Biotransformation
of isoeugenol and eugenol by cultured cells of *Eucalyptus perriniana*. Phytochemistry
31(3):827–831

Pan J, Yi XM, Zhang SJ, Cheng J, Wang YH, Liu CY, He XJ (2018) Bioactive phenolics from
mango leaves (*Mangifera indica* L.). Ind Crop Prod 111:400–406

Panche AN, Diwan AD, Chandra SR (2016) Flavonoids: an overview. J Nutr Sci 5:e47

Peleg H, Gacon K, Schlich P, Noble AC (1999) Bitterness and astringency of flavan-3-ol monomers, dimers and trimers. J Sci Food Agric 79(8):1123–1128

Peres V, Nagem TJ, de Oliveira FF (2000) Tetraoxygenated naturally occurring xanthones.
Phytochemistry 55(7):683–710

Pérez-Silva A, Odoux E, Brat P, Ribeyre F, Rodriguez-Jimenes G, Robles-Olvera V, García-
Alvarado MA, Günata Z (2006) GC–MS and GC–olfactometry analysis of aroma compounds
in a representative organic aroma extract from cured vanilla (*Vanilla planifolia* G. Jackson)
beans. Food Chem 99(4):728–735

Peterson J, Dwyer J, Adlercreutz H, Scalbert A, Jacques P, McCullough ML (2010) Dietary lignans:
physiology and potential for cardiovascular disease risk reduction. Nutr Rev 68(10):571–603

Petruľová-Poracká V, Repčák M, Vilková M, Imrich J (2013) Coumarins of *Matricaria chamomilla* L.: aglycones and glycosides. Food Chem 141(1):54–59

Piao S-J, Qiu F, Chen L-X, Pan Y, Dou D-Q (2009) New stilbene, benzofuran, and coumarin glycosides from *Morus alba*. Helv Chim Acta 92(3):579–587

Podolyák B, Kun D, Renner K, Pukánszky B (2018) Hydrogen bonding interactions in
poly(ethylene-co-vinyl alcohol)/lignin blends. Int J Biol Macromol 107:1203–1211

Poole SK, Patel S, Dehring K, Workman H, Poole CF (2004) Determination of acid dissociation
constants by capillary electrophoresis. J Chromatogr A 1037(1):445–454

Popova AV, Bondarenko SP, Frasinyuk MS (2019) Aurones: synthesis and properties. Chem
Heterocycl Compd 55(4):285–299

Quideau S, Jourdes M, Saucier C, Glories Y, Pardon P, Baudry C (2003) DNA topoisomerase
inhibitor acutissimin A and other flavano-ellagitannins in red wine. Angew Chem Int Ed
42(48):6012–6014

Ragnar M, Lindgren CT, Nilvebrant N-O (2000) pKa-Values of guaiacyl and syringyl phenols
related to lignin. J Wood Chem Technol 20(3):277–305

Rappoport Z (2004) The chemistry of phenols. John Wiley & Sons, New York, NY

Ribéreau-Gayon P (1972) Plant phenolics. Oliver & Boyd, Edinburgh

Romani A, Campo M, Pinelli P (2012) HPLC/DAD/ESI-MS analyses and anti-radical activity of
hydrolyzable tannins from different vegetal species. Food Chem 130(1):214–221

Rúa J, de Arriaga D, García-Armesto MR, Busto F, del Valle P (2017) Binary combinations of
natural phenolic compounds with gallic acid or with its alkyl esters: an approach to understand
the antioxidant interactions. Eur Food Res Technol 243(7):1211–1217

Saengchantara ST, Wallace TW (1986) Chromanols, chromanones, and chromones. Nat Prod Rep
3:465–475

Sala A, Recio MC, Giner RM, Máñez S, Ríos J-L (2001) New acetophenone glucosides isolated from extracts of *Helichrysum italicum* with antiinflammatory activity. J Nat Prod
64(10):1360–1362

Sanli N, Fonrodona G, Barrón D, Özkan G, Barbosa J (2002) Prediction of chromatographic retention, pKa values and optimization of the separation of polyphenolic acids in strawberries. J
Chromatogr A 975(2):299–309

Santana-Gálvez J, Jacobo-Velázquez DA (2018) Classification of phenolic compounds. In: Nollet
LM, Gutierrez-Uribe JA (eds) Phenolic compounds in food: characterization and analysis.
Taylor & Francis Group, LLC, Boca Raton, FL, pp 3–20

Santana-Gálvez J, Cisneros-Zevallos L, Jacobo-Velázquez DA (2017) Chlorogenic acid: recent advances on its dual role as a food additive and a nutraceutical against metabolic syndrome. Molecules 22(3):358

Santos JS, Alvarenga Brizola VR, Granato D (2017) High-throughput assay comparison and standardization for metal chelating capacity screening: a proposal and application. Food Chem 214:515–522

Schulze AE, Beelders T, Koch IS, Erasmus LM, De Beer D, Joubert E (2015) Honeybush herbal teas (Cyclopia spp.) contribute to high levels of dietary exposure to xanthones, benzophenones, dihydrochalcones and other bioactive phenolics. J Food Compos Anal 44:139–148

Sengupta PK (2017) Pharmacologically active plant flavonols as proton transfer based multiparametric fluorescence probes targeting biomolecules: perspectives and prospects. In: Geddes CD (ed) Reviews in fluorescence 2016. Springer International Publishing, Cham, pp 45–70

Shahidi F, Yeo J (2016) Insoluble-bound phenolics in food. Molecules 21(9):1216

Shen T, Wang X-N, Lou H-X (2009) Natural stilbenes: an overview. Nat Prod Rep 26(7):916–935

Shyu Y-S, Hwang LS (2002) Antioxidative activity of the crude extract of lignan glycosides from unroasted Burma black sesame meal. Food Res Int 35(4):357–365

Skadhauge B, Gruber MY, Thomsen KK, von Wettstein D (1997) Leucocyanidin reductase activity and accumulation of proanthocyanidins in developing legume tissues. Am J Bot 84(4):494–503

Smeriglio A, Barreca D, Bellocco E, Trombetta D (2017) Proanthocyanidins and hydrolysable tannins: occurrence, dietary intake and pharmacological effects. Br J Pharmacol 174(11):1244–1262

Soto-Vaca A, Gutierrez A, Losso JN, Xu Z, Finley JW (2012) Evolution of phenolic compounds from color and flavor problems to health benefits. J Agric Food Chem 60(27):6658–6677

Sousa A, Araújo P, Azevedo J, Cruz L, Fernandes I, Mateus N, de Freitas V (2016) Antioxidant and antiproliferative properties of 3-deoxyanthocyanidins. Food Chem 192:142–148

Stanley WL, Jurd L (1971) Citrus coumarins. J Agric Food Chem 19(6):1106–1110

Sun YN, Li W, Yang SY, Kang JS, Ma JY, Kim YH (2016) Isolation and identification of chromone and pyrone constituents from Aloe and their anti-inflammatory activities. J Funct Foods 21:232–239

Swain T, Bate-Smith E (1962) Flavonoid compounds. In: Florkin M, Mason HS (eds) Comparative biochemistry, vol 3. Elsevier, New York, NY, pp 755–809

Takeda K (2006) Blue metal complex pigments involved in blue flower color. Proc Jpn Acad Ser B Phys Biol Sci 82(4):142–154

Tomás-Barberán FA, Clifford MN (2000a) Flavanones, chalcones and dihydrochalcones – nature, occurrence and dietary burden. J Sci Food Agric 80(7):1073–1080

Tomás-Barberán FA, Clifford MN (2000b) Dietary hydroxybenzoic acid derivatives–nature, occurrence and dietary burden. J Sci Food Agric 80(7):1024–1032

Trouillas P, Marsal P, Siri D, Lazzaroni R, Duroux J-L (2006) A DFT study of the reactivity of OH groups in quercetin and taxifolin antioxidants: the specificity of the 3-OH site. Food Chem 97(4):679–688

Veitch NC, Grayer RJ (2008) Flavonoids and their glycosides, including anthocyanins. Nat Prod Rep 25(3):555–611

Vermerris W, Nicholson RL (2008) Phenolic compound biochemistry. Springer, Dordrecht; London

Vitrac X, Bornet A, Vanderlinde R, Valls J, Richard T, Delaunay J-C, Mérillon J-M, Teissédre P-L (2005) Determination of Stilbenes (δ-viniferin, trans-astringin, trans-piceid, cis- and trans-resveratrol, ε-viniferin) in Brazilian Wines. J Agric Food Chem 53(14):5664–5669

Wang M, Kikuzaki H, Lin C-C, Kahyaoglu A, Huang M-T, Nakatani N, Ho C-T (1999) Acetophenone glycosides from Thyme (Thymus vulgaris L.). J Agric Food Chem 47(5):1911–1914

Wang M, Wang Q, Yang Q, Yan X, Feng S, Wang Z (2019a) Comparison of anthraquinones, iridoid glycosides and triterpenoids in morinda officinalis and morinda citrifolia using UPLC/Q--TOF-MS and multivariate statistical analysis. Molecules 25(1):160

Wang Y, Fong SK, Singh AP, Vorsa N, Johnson-Cicalese J (2019b) Variation of anthocyanins, pro-anthocyanidins, flavonols, and organic acids in cultivated and wild diploid blueberry species. HortScience 54(3):576–585

Wannan BS, Waterhouse JT, Gadek PA, Quinn CJ (1985) Biflavonyls and the affinities of *Blepharocarya*. Biochem Syst Ecol 13(2):105–108

Xie L, Bolling BW (2014) Characterisation of stilbenes in California almonds (*Prunus dulcis*) by UHPLC–MS. Food Chem 148:300–306

Yamane K, Kato Y (2012) Handbook on flavonoids: dietary sources, properties, and health benefits. Nova Science Publishers, Hauppauge, NY

Yang Z, Kinoshita T, Tanida A, Sayama H, Morita A, Watanabe N (2009) Analysis of coumarin and its glycosidically bound precursor in Japanese green tea having sweet-herbaceous odour. Food Chem 114(1):289–294

Zeb A (2015) Phenolic profile and antioxidant potential of wild watercress (*Nasturtium officinale* L.). Springerplus 4:714

Zeb A, Muhammad B, Ullah F (2017) Characterization of sesame (*Sesamum indicum* L.) seed oil from Pakistan for phenolic composition, quality characteristics and potential beneficial properties. J Food Measur Charact 11(3):1362–1369

Zhang J-R, Tolchard J, Bathany K, Langlois d'Estaintot B, Chaudiere J (2018) Production of 3,4-cis- and 3,4-trans-leucocyanidin and their distinct MS/MS fragmentation patterns. J Agric Food Chem 66(1):351–358

Chapter 3
Phenolic Antioxidants in Fruits

Learning Outcomes

After reading this chapter, the reader will be able to:

- Describe the trend in the antioxidant potential of edible fruits.
- Write a detail note on the different hydroxybenzoic acid derivatives in fruits.
- Discuss the composition of phenolic aldehydes, acetophenone, cinnamoyl derivatives and phenylacetic acids in different fruits.
- Understand the types of coumarins, chromones and their derivatives present in fruits.
- Discuss flavonoids, anthocyanins and tannins present in fruits.

3.1 Introduction

Phenolic compounds are widely distributed among the wide range of plant foods. The tremendous growing interest of the scientific community in recent years on phenolic compounds was due to the possible toxicity of synthetic antioxidants. The potential of plant foods to serve as a source of antioxidants to protect against various diseases induced by free radicals has been the prime objective (Shahidi and Ambigaipalan 2015). The scavenging of free radicals especially lipid free radicals by phenolic compounds serves the topmost significant function in the planet earth (Taverne et al. 2018). Besides, several other actions of phenolic antioxidants include inactivating metal catalysts by chelation, hydroperoxides reduction into stable hydroxyl derivatives, and synergistic interacting role with other reducing compounds (Frankel and German 2006; Frankel and Finley 2008). Thus, food manufacturers have been prompted to natural food additives to replace synthetic antioxidants

© Springer Nature Switzerland AG 2021
A. Zeb, *Phenolic Antioxidants in Foods: Chemistry, Biochemistry and Analysis*,
https://doi.org/10.1007/978-3-030-74768-8_3

with ingredients containing natural antioxidative compounds especially phenolic compounds. Therefore, the focus of research on natural phenolic additives has gained significant momentum recently, as they are believed to posing no health risk to consumers (Farag et al. 2003).

Phenolic compounds present in foods include phenolic-acids, hydroxycinnamic-acids, flavonoids, anthocyanins, tannins and lignans as discussed in Chap. 2 are continuously determined. They are the prime sources of natural antioxidants are that may occur in all parts of plant foods. They are present in fruits, vegetables, nuts, seeds, leaves, flours, roots, beverages and their derivative products (Shahidi 2015). Several studies on the biological activities of phenolic compounds, which are potent antioxidants and free radical scavengers have been reported. The literature on this topic is still under consideration and needs to be updated. The types and composition of phenolic compounds present in plant foods regarding the latest scientific knowledge have been discussed in this chapter.

3.2   Phenolic Antioxidants in Fruits

Fruits are rich in natural antioxidants that support lowering the incidence of several degenerative diseases such as cancer, arthritis, arteriosclerosis, heart disease, inflammation, brain dysfunction and acceleration of the aging process (Seifried et al. 2017; Ramana et al. 2018). In 2012, Haminiuk et al. (2012) published a review article on the phenolic compounds in fruits. However, since then tremendous research has been reported. The most abundant antioxidants in fruits are polyphenols; vitamins A, B, C and E, and carotenoids are present to a lesser extent in some fruits (Shahidi and Ambigaipalan 2015).

Polyphenols are present mainly in the ester and glycoside forms existed as flavonoids. The antioxidant potential measured by scavenging of DPPH radicals and iron (III) reducing assays of tropical fruits like guava, papaya, and star fruit is higher as compared to orange (Lim et al. 2007). Some fruits such as banana, star fruit, water apple, langsat, and papaya had been reported to possess weaker primary antioxidant activities than orange, however, they showed powerful secondary antioxidant potential measured by the iron (II) chelating experiment (Lim et al. 2007). The highest content of total flavonols in plum, apples, custard apple, peach, strawberry, and cherry ranged from 70 and 370 mg/100 g of dry weight. The lowest amount was present in avocado, banana, and pear (lower than 4 mg/100 g of dry weight) (García-Alonso et al. 2004).

In another study, the total antioxidant activities of 12 fruits were evaluated with an automated ORAC assay and a peroxyl radical generator (Wang et al. 1996). Based on the edible portion of fruits, strawberry had the highest ORAC activity followed by plum, orange, red grape, kiwi fruit, pink grapefruit, white grape, banana, apple, tomato, pear, and honeydew melon. Whereas strawberry had the highest ORAC followed by plum, orange, pink grapefruit, tomato, kiwi fruit, red grape, white grape, apple, honeydew melon, pear, and banana on a dry weight basis.

According to Vinson et al. (2001), the cranberry had the highest total phenols, followed by red grape, on a fresh weight basis. The authors revealed that citrus fruits had a very low concentration of phenolics and the influence of ascorbic acid to the antioxidants in fruits was minor, except for melon, nectarine, orange, white grape, and strawberry. In the previous work of the authors (Vinson and Hontz 1995), the antioxidant quality of the fruits extracts measured by IC_{50} values showed better levels than the vitamin antioxidants. This was suggested to be due to a synergistic action of the antioxidants present in the extract mixture. According to the total phenol antioxidant index, cherry was most effective followed by red grape, blueberry, strawberry, white grape, cranberry, banana, and apple. Thus, small fruits such as berries are amongst the best sources of polyphenol antioxidants (Sun et al. 2002). These studies concluded that fruits had significantly better quantity and quality of phenolic antioxidants than vegetables.

It is obvious that among fruits berry is consistently ranked amongst the top sources of phenolic compounds. The antioxidant potential of berries is up to four times higher than other fruits, ten times higher than vegetables, and forty times higher than cereals (Halvorsen et al. 2002). Alasalvar and Shahidi (2013) showed that cardiovascular diseases, obesity, various types of cancer, type-2 diabetes, and other chronic diseases were lower in those individuals who regularly consume large amounts of dried fruits (dates, raisins, prunes, figs, acai berries, apples, bananas, black currants, blackberries, cherries, citrus fruits, cranberries, gingers, goji berries, guavas, kiwis, mangoes, mulberries, nectarines, papayas, passion fruits, peaches, pears, pineapples, raspberries, star apples, and strawberries). Thus it is imperative to discuss each class of phenolic compounds (simple phenols, hydroxybenzoic acids, phenolic aldehydes, phenylethanoids, cinnamic acids, and their derivatives, Coumarins, benzophenones, stilbenes, anthraquinones, chalcones, aurones, flavonoids, anthocyanins, lignans, lignins, tannins and glycosides) present in different fruits.

3.2.1   Simple Phenols

Simple phenols are present in fruits. Maga (1978) reviewed the composition of simple phenols in a variety of foods. Simple phenols had been reported to be present in nonalcoholic beverages such as cocoa, coffee and tea, alcoholic beverages (beer, rum, smoked beer, sherry, fusel oil), meat and poultry products (boiled beef, pork liver, chicken broth, fried chicken, eggs), milk and its products (butter, camembert cheese, Lactose-casein, nonfat dry milk, whey powder, cow's milk, Vacherin cheese), nuts and its products (almonds, filberts, Macadamia nuts, peanuts), and vegetables (cauliflower, celery, tomato, and asparagus). Tea contains o-cresol, m-cresol, p-cresol, guaiacol, o-xylenol and thymol (Maga 1978). Catechol is an important member of the simple phenols present in black tea (Sanderson and Grahamm 1973), green tea (Miller et al. 2012) and olive oils (Owen et al. 2000).

Clifford (2000) reviewed miscellaneous phenols in foods and beverages. The author showed that smoked foods are rich in simple phenols. Isoeugenol, cresols, 4-methyl guaiacol and phenol were the highest amount present in smoked foods as shown in Table 3.1. Other simple phenols had trace amounts. The amount can be higher if foods are highly smoked.

Table 3.1 Simple phenolic compounds in smoked foods. Reproduced with kind permission of John Wiley & Sons UK (Clifford 2000)

Compound	Contents (mg/kg)
Phenol	0.1–2.8
Cresols	0.34–3.7
Xylenols	Trace-1.3
Guaiacol	0.5–1.7
4-Methylguaiacol	1.0–3.5
Isoeugenol	1.2–3.9
4-Ethylresorcinol	0.5–1.6
Catechol	Trace-3.6
3-Methoxycatechol	0.2–2.8
Eugenol	2.3
Syringol	Trace
4-Methylsyringol	Trace
4-Ethylsyringol	Trace

3.2.2 Hydroxybenzoic Acids

Hydroxybenzoic acids such as 2-hydroxybenzoic acid (2-HBA), 3-hydroxybenzoic acid (3-HBA), 4-hydroxybenzoic acid (4-HBA), salicylic acid (SA), 3,4,5-trihydroxy benzoic acid (3,4,5-THBA), protocatechuic acid (PTA), vanillic acid (VA), syringic acid (SYA), gallic acid (GA) and gentisic acid (GNA) have been extensively studied in fruits (Kumar and Goel 2019). Soft fruits are rich in hydroxybenzoic acids (Schuster and Herrmann 1985). Among fruits, the highest contents (28 mg/100 g FW) are determined in dark plum, cherry, and one apple variety (Valkea kuulas) (Mattila et al. 2006). Sinapic acid had higher (4.25 mg/100 g FW) in Chinese cabbage, and protocatechuic acid had the highest concentration of all the phenolic acids in white wine (Andrés-Lacueva et al. 2010).

In the year 2000, Tomás-Barberán and Clifford (2000) reviewed the hydroxybenzoic acid derivatives and their burden in the diet. The authors showed that 4-hydroxybenzoic acid, Protocatechuic acid, and gallic acid were the major hydroxybenzoic acid derivatives Blackberry, blackcurrant, whitecurrant, redcurrant, raspberry, and strawberry contain high amounts of these phenolic compounds (Häkkinen et al. 1999a). However, mangosteen fruits were the richest source of 4-HBA (up to 1593 mg/kg), PTA (up to 3812 mg/kg) (Zadernowski et al. 2005, 2009).

The results of Russell et al. (2009) showed that locally produced Scottish fruits had significantly higher levels of phenolic acids (1.61–4.89 g/kg dry weight) than the commonly consumed fruits (0.07–0.22 g/kg dry weight) as shown in Table 3.2. These results revealed that Scottish fruits are highly varied in their composition with strawberries, raspberries, and blackcurrants. They have almost similar profiles and all having gallic acid as the principal phenolic compound with 29%, 48% and 31%, respectively. In these fruits, gallic acid was mostly found conjugated to other plant

Table 3.2 Some major sources of common hydroxybenzoic acids in fruits (mg/kg). Redrawn with kind permission of Elsevier (Russell et al. 2009)

Sample	Raspberry	Gooseberry	Blackcurrant	Strawberry	Banana	Apple	Pear	Grapes	Orange
Free Hydroxybenzoic acids									
Gallic acid	5.73	1.27	5.63	1.67	0.02	0.11	n.d	0.17	n.d
Protocatechuic	3.83	27.22	10.4	39.0	0.16	3.37	0.45	1.67	2.53
p-Hydroxybenzoic	33.31	n.d.	4.33	193.9	n.d	34.09	0.48	1.20	17.4
Gentisic acid	n.d.	2.01	n.d.	30.8	n.d	n.d	n.d	n.d	0.20
Vanillic acid	24.58	n.d	15.0	98.4	0.30	n.d	n.d	n.d	4.86
Syringic acid	107.51	n.d.	n.d.	n.d.	0.21	1.10	n.d	n.d	2.53
p-Coumaric acid	n.d.	n.d.	19.8	n.d.	n.d.	11.65	n.d	n.d	6.65
Sinapic acid	36.89	n.d.	n.d.	450.3	n.d.	13.42	0.96	n.d	17.3
Salicylic acid	7.64	n.d.	62.1	n.d.	n.d.	n.d	0.37	n.d	2.37
Conjugated Hydroxybenzoic acids									
Gallic acid	1669.2	43.59	999.1	1416.0	3.22	0.44	0.45	1.29	n.d
Protocatechuic	273.26	389.8	355.9	76.1	n.d.	28.5	2.65	21.17	1.8
p-Hydroxybenzoic	676.31	120.7	259.1	429.5	1.62	n.d	n.d	6.94	15.1
Gentisic acid	n.d.	41.2	77.3	89.8	0.64	2.29	5.81	2.99	0.92
Vanillic acid	94.34	48.2	83.4	47.3	1.04	5.21	7.98	0.22	13.74
Syringic acid	5.91	n.d.	10.27	n.d.	3.54	n.d	33.03	n.d	0.59
p-Coumaric acid	224.6	504.8	591.3	1107.6	3.47	2.00	5.84	16.3	20.9
Sinapic acid	42.9	n.d.	37.3	n.d.	0.27	6.34	34.87	2.93	28.9
Salicylic acid	25.1	n.d.	55.8	93.2	3.93	6.73	1.93	9.50	16.9

Values are means and n.d. represents not detected

components. In comparison to other fruits, phenolic acids that occurred in mango and blueberry were significantly higher (Gu et al. 2019). Gallic acid and protocate-chuic acid were the major phenolic acids in mangos.

Syringic acids and derivative of *p*-hydroxybenzoic acid, *o*- and *m*-coumaric, gal-lic, *m*- and *p*-hydroxybenzoic, protocatechuic and vanillic acids had been reported in bilberry juice (Ancillotti et al. 2016). Ayaz et al. (2005) also reported phenolic acids in blueberry. Similarly, 15 phenolic acids, namely benzoic, *o*-hydroxybenzoic, cinnamic, *m*-hydroxybenzoic, *p*-hydroxybenzoic, *p*-hydroxyphenyl acetic, phthalic, 2,3-dihydroxybenzoic, vanillic, *o*-hydroxycinnamic, 2,4-dihydroxybenzoic, *p*-cou-maric, ferulic, caffeic, and sinapic acids were identified in cranberry fruit in the free and bound forms (Zuo et al. 2002). Gallic acid (58.1–108.8 mg/kg), protocatechuic acid (34.9–316.6 mg/kg), protocatechuic aldehyde (81.0–87.9 mg/kg), *p*-hydroxy-benzoic acid (30.9–152.7 mg/kg), and vanillic acid (23.1–82.0 mg/kg) had been reported in the fruits of three mulberry cultivars from Pakistan (Memon et al. 2010). Melon fruit from Tunisia was reported to contain gallic acid (120.7 mg/kg), proto-catechuic acid (34.6 mg/kg), isovanillic acid (237 mg/kg), and 3-HBA (334.5 mg/kg) (Mallek-Ayadi et al. 2017). Gallic acid, hydroxybenzoic acid and its derivatives have been reported in Melon seeds (Zeb 2016). The total phenolic content measured in melon both with Folin Ciocaltaeu reagent and HPLC showed promising results as shown in Fig. 3.1. High amounts of phenolic compounds were reported in water followed by methanol-water mixture.

A recent study showed that both free and conjugated 2-HBA, 3-HBA, 4-HBA, 2,3-DHBA, vanillic acid, and sinapic acids were present in the wild fruits of olive, Jujube, and common fig (Ahmad et al. 2016). This study concluded that wild fruits are also good sources of phenolic antioxidants. Similarly, tomato fruits were rich in Protocatechuic acid and gallic acid along with other phenolic compounds, which helps in protection against UV irradiation (Liu et al. 2018). Hydroxybenzoic acid, vanillic acid, and protocatechuic acid were also reported in Chinese citrus fruits (Sun et al. 2019). In Brazilian red and yellow araçá (*Psidium cattleianum* Sabine) fruits, gallic acid was present in the range of 7.7–34.1 mg/kg, 2,4-dihydroxybenzoic (none-7.5 mg/kg), and a small amount of sinapic acid (Mallmann et al. 2020) amongst other phenolic compounds.

3.2.3 Phenolic Aldehydes

Phenolic aldehydes such as vanillin and its derivatives are present in vanilla fruits (Jadhav et al. 2009; Sharma et al. 2006). Vanillin had also been determined in vari-ous citrus juices in the low parts-per-million (ppm) ranges. The measured amounts were 0.20 ppm, 0.35 ppm, 0.41 ppm, 0.35 ppm, and 0.60 ppm, in orange, tangerine, lemon, lime, and grapefruit juices, respectively (Goodner et al. 2000). It had been observed that pasteurization of grapefruit juices produced an average of 15% increase in the amounts of vanillin. Vanillin is used as a flavoring agent and has a

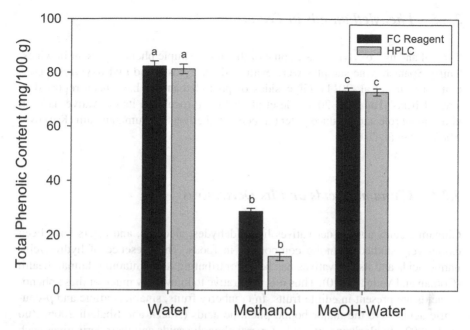

Fig. 3.1 Total phenolic acids (TPC) of different extracts and analytical methods of melon seeds. Different letters (a–c) in the same procedure represent significant difference at $p < 0.05$ (Holm-Sidak method). (Adopted from Zeb 2016)

strong antioxidant and anti-microbial property (Sinha et al. 2008). It has the potential to enhance the quality attributes of strawberry and other juices (Cassani et al. 2016).

3.2.4   Acetophenone and Phenylacetic Acids

Acetophenones had been identified in fruits. Fruits of *Evodia merrillii* had been found to contain 4-(1′-geranloxy)-2,6-β-trihydroxy-3-dimethylallylacetophenone, and 2-(1′-geranloxy)-4,6-β-trihydroxy-acetophenone (Lin et al. 1993). Similarly, in mango fruit, acetophenone and phenylacetaldehyde having sweet and floral flavors were found with an amount of 3.36 μg/kg (MacLeod and de Troconis 1982). Acetophenones and phenylacetic acid derivatives have been identified as aroma components of fruits (Zhu et al. 2018). 3,4-Dihydroxyphenylacetic acid and 4-hydroxyphenyl acetic acid had been identified in olive fruits (Cabrini et al. 2001). In Osage orange (*Maclura pomifera*) fruit, 3,4-dihydroxybenzoic acid, 2,3,4-trihydroxy benzoic acid have been identified (Su et al. 2017). Melon fruit was recently found to contain 3.27 mg/kg of phenylacetic acid along with other phenolic compounds (Mallek-Ayadi et al. 2017).

3.2.5 Phenylethanoids in Fruits

Tyrosol and hydroxytyrosol are some of the major simple phenols present in wines. Thirty Spanish wine samples were evaluated for Tyrosol and hydroxytyrosol contents (Piñeiro et al. 2011). Glycosides of phenylethanoids have been reported in several fruits (Hu et al. 2018; Ge et al. 2017). Tyrosol and its derivative have an antioxidant role and produce potential beneficial effects on human health (Karković Marković et al. 2019).

3.2.6 Cinnamic Acids and Its Derivatives

Cinnamic acids and its derivatives like aldehydes, alcohols, and esters have been extensively studied phenolic compounds in foods. The presence of hydroxycinnamic acids and its derivatives has been contributing to maintaining human health (Coman and Vodnar 2020). Thus it is imperative to know how much of these phenolic acids are present in edible fruits. In cranberry fruits, sinapic, caffeic and p-coumaric acids were the major bound phenolic acids (Naczk and Shahidi 2006; Zuo et al. 2002). In blackcurrant seeds, several phenolic acids and their derivatives such as caffeic, ferulic, p-coumaric, gallic, protocatechuic and p-hydroxybenzoic acids as well as 1-cinnamoyl-β-D-glucoside and 1-p-coumaroyl-β-D-glucoside had been reported (Lu et al. 2002). Similarly, several phenolic acids have been identified in blackberries. These include caffeic, ferulic, p-coumaric and ellagic acids (Sellappan et al. 2002). Several research reports have indicated that sweet cherries are rich in phenolic acids. Mulabagal et al. (2009) reported hydroxycinnamic acids, such as caffeic and coumaric acids, with antioxidant and anti-inflammatory properties.

The skins of blueberries consist mainly of cinnamic acids and flavonol glycosides, with a minor amount of gallic acid and syringic acid. The flesh is a rich source of cinnamic acids, where chlorogenic acid (3-caffeoylquinic acid) was the key phenolic compound. The seed fraction also has cinnamic acids and flavonol glycosides (Lee and Wrolstad 2004). Grape berries and their skins contain phenolic acids such as caftaric acid (trans-caffeoyl tartaric acid), coutaric acid (p-coumaryl tartaric acid) and trans-fertaric acid (Cantos et al. 2002; Souquet et al. 2000). In Juneberry, hydroxycinnamic acids such as cis-5-caffeoylquinic acid (88.5 mg/kg), 5-coumaroylquinic acids (23–28 mg/kg), chlorogenic acid (35–94 mg/kg) along with cinnamic acids glycosides were reported recently (Mikulic-Petkovsek et al. 2020). The fruits of Juneberry was thus recommended to be used as a source of important bioactive components in various food products, such as tea, juice, food supplements, and alcoholic beverages.

In Chinese mulberry (Morus alba) fruits, 50 polyphenolic compounds were identified including phenolic acids such as p-hydroxycinnamic acid, ferulic acid, cis-cinnamic acid, cis-p-hydroxycinnamic acid. These compounds have been reported to possess strong antioxidant activity as compared to ascorbic acid (Xu et al. 2020). Similarly, Chilean Berry Fruits: Murta (*Ugni molinae* Turcz) and Calafate (*Berberis buxifolia* Lam.) also contain phenolic acids and have strong antioxidant activity (Fredes et al. 2020).

In apples, hydroxycinnamic acids may play a significant role in the flesh. The *p*-coumaroylquinic acid, 5-caffeoylquinic acid, and 4-caffeoylquinic acid have been reported along with other polyphenols. It was reported that 5-caffeoylquinic and *p*-coumaroylquinic acids showed the highest sensitivity to the treatment of pulsed electric field (PEF). Their amounts declined by 88–95% (*p*-coumaroylquinic acid) and 86–91% (5-caffeoylquinic acid) with PEF of 1.8 and 7.3 kJ/kg as compared to untreated apple. In a study of different cultivars (Fuji, Pink Lady and Golden Delicious) of apples from the USA showed 11 hydroxycinnamic acids (Lee et al. 2017). These include chlorogenic acid, caffeic acid, rosmarinic acid, coumaroylquinic acid, Ferulic acid and sinapic acids and its derivatives. The levels of average phenolic acids in all apple varieties were significantly different on a dry weight basis in the pulp as compared to peel. Similarly, the levels of chlorogenic, and caffeic acid were predominant in all varieties. Pulp have significantly higher amounts of these phenolic acids as compared to the peel. For example, levels of total chlorogenic acid ranged from 157.0 to 745.8 mg/kg DW (pulp) and 63.7–449.0 mg/kg DW was in the peel. Levels of caffeic acid ranged from 44.3 to 287.5 mg/kg DW in pulp and 1.4–125 mg/kg DW in the peel. These two phenolic acids constitute between 59% and 68% in peel and 75–83% in the pulp of all apples as total phenolic acids (Lee et al. 2017).

Spanish and Italian "*Golden Delicious*" apples were evaluated for phenolic composition. The derivatives of hydroxycinnamic acids (HA) were present in significant concentrations in both apples, representing approximately 23% of total apple polyphenols. Chlorogenic acid was the major HA found in the Spanish and Italian untreated apples (626.0 µg/g DW and 592.6 µg/g DW, respectively). The apple samples also contain neochlorogenic acid and cryptochlorogenic acid. Cryptochlorogenic acid showed a higher value in the Spanish apples than in Italian apple (63.5 µg/g DW vs. 45.1 µg/g DW). However, neochlorogenic acid content was higher in the untreated Italian apples than in Spanish (41.2 µg/g DW vs. 33.7 µg/g DW). Other hydroxycinnamic acid derivative found in both apples identified as coumaroyl quinic acid showed similar concentration in both apples (34.7 µg/g DW and 35.6 µg/g DW, respectively) (Fernández-Jalao et al. 2019). Furthermore, it had been reported that high-pressure processing of both apples showed that Italian apples exhibited higher antioxidant activity than Spanish apples. Guyot et al. (1998) showed that chlorogenic acid was the major hydrocinnamic acid present in apple fruit accounting for up to 87% of the total phenolic compounds.

Similarly, 4-caffeoylquinic acid had shown sensitivity under the same PEF doses with 38–40% reduction in its composition (Ribas-Agustí et al. 2019). In Italian peaches, the amount of neochlorogenic acid and chlorogenic acids varied between cultivars and flesh and skins (Ceccarelli et al. 2016). The amounts of chlorogenic acid were higher than neochlorogenic acid in all cultivars. The exocarp (peel/skin) was rich in chlorogenic acid as shown in Table 3.3. Different Prunus smoothies (*Prunus cerasus, Prunus persica, Prunus armeniaca*) were evaluated for their

Table 3.3 Hydroxycinnamic acids: neochlorogenic and chlorogenic acid (mg/kg FW) in exocarp (peel) and mesocarp (flesh) of Italian peaches. Reproduced with kind permission of Taylor & Francis (Ceccarelli et al. 2016)

	Neochlorogenic acid		Chlorogenic acid	
Cultivar	Exocarp	Mesocarp	Exocarp	Mesocarp
White flesh				
Bea	6	20	75	53
Caldesi 2000	20	10	117	10
Cesarini	18	55	121	38
Iris Rosso	27	15	158	34
Isabella d'Este	38	23	206	52
Yellow flesh				
Gilda Rossa	5	11	80	21
Jonia	22	33	125	110
Lolita	12	9	140	30
Roberta	17	12	90	9
Romea	28	12	121	26

Data expressed as means of n = 3. Data has been converted to mg/kg from mg/100 g

polyphenols profile and determine the correlation between the polyphenols, and antioxidant capacity (Nowicka et al. 2016). The prepared products were found to contain 3-*p*-feruloylquinic acid, neochlorogenic acid, chlorogenic acid, 3-*p*-coumarylquinc acid, 4-caffeoylquinic acid, 3-caffeoylshikimic acid and 3-*p*-feruloylquinic acid. In Chinese peaches, chlorogenic acid ranges from 5.2 to 107.9 mg/kg and Neochlorogenic acid from 0.8 to 74.5 mg/kg (Ding et al. 2020).

In the peel and pulp extracts of quince (*Cydonia oblonga* Mill.), six hydroxycinnamic acids i.e., 3-caffeoylquinic acid (3-CQA), 4-coumaroylquinic acid (HC1), 4-caffeoylquinic acid (4-CQA), 5-caffeoylquinic acid (5-CQA), a derivative of *p*-coumaroylquinic acid (HC2) and 3,5-dicaffeoylquinic acid (3,5-diCQA) were found. The most abundant hydroxycinnamic acid was 5-CQA with 259.1–481.4 mg/kg FW in peel and 97.33–217.36 mg/kg in quince pulp (Stojanović et al. 2017).

The baobab fruit popularly called monkey's bread widely consumed as a portion of basic food in many African countries. The fruit has an oblong-cylindrical shape having black seeds embedded in a white and chalky pulp. The fruit is a rich source of phenolic antioxidants, where total flavonoids are higher than total phenolic contents. It contains *p*-hydroxycinnamic acid, caffeic acid and chlorogenic acid (Ismail et al. 2019). Cinnamic acid, chlorogenic acid, sinapine, ferulic acid, its 4-glycoside, 3-feruloylquinic acid, 3-methyrosmarinic acid, along with other phenolic compounds were reported in the palm fruits such as jelly palm (*Butia ordorata*) and fishtail palm (*Caryota uren*) (Ma et al. 2019).

In wild berries like sea buckthorn, hawthorn, blackthorn, blackberry, cornelian cherry, dog rose berries, and bird cherries, phenolic acids such vanillic acid, chlorogenic acid, caffeic acid, syringic acid, coumaric acid, ferulic acid, sinapic acid, and salicylic acid were quantified along with other phenolic compounds (Cosmulescu et al. 2017). The high antioxidant activities have been attributed to the presence of phenolic acids and other polyphenols in the wild berries.

The peel of two Australian mangoes (*Mangifera indica*) cultivars (Keitt and Kensington Pride) revealed several polyphenols including cinnamic acid, 1,5-dicaffeoylquinic acid, feruloyl tartaric acid, caffeic acid 3-glucuronide, 3-caffeoylquinic acid, isoferulic acid, ferulic acid 4-glucoside, *p*-coumaric acid 4-glucoside, chicoric acid, sinapic acid, ferulic acid 4-sulfate, ferulic acid 4-glucuronide, sinapine, 3-*p*-coumaroylquinic acid, verbascoside and *p*-coumaroyl tartaric acid (Peng et al. 2019). Keitt peel contained higher concentrations of total phenolic compounds and thus higher antioxidant capacity in DPPH, FRAP, and ABTS assays as compared to K&P peel.

A recent study by Song et al. (2020) showed that hydroxycinnamic acids were the main phenolic acids in Chinese fruits, accounting for 52.2–88.7% of total phenolic acids (TPA) contents. The authors showed that predominant hydroxycinnamic acids were isoferulic acid (15.5–31.3%) in fruits. Other hydroxycinnamic acids were ferulic acid, caffeic acid, syringic acid, sinapic acid and chlorogenic acid (Fig. 3.2).

The phenolic acids were also identified in the fruit of seven Saskatoon berry genotypes and particular fruit parts using the LC/MS technique (Lachowicz et al.

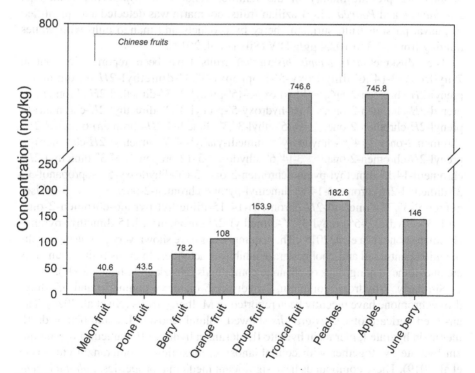

Fig. 3.2 Total phenolic acids in different fruits. Data has been converted to mg/kg from original and updated. (The data from the Chinese fruits has been reproduced with kind permission of Elsevier, Song et al. 2020)

2017, 2019). The contents of hydroxycinnamic acids significantly varied depending on the analyzed part of berries. Their total content ranged from 2020 to 11,690 mg/ kg DW (Lachowicz et al. 2020). In Brazilian red and yellow araçá (*Psidium cattleianum* Sabine) fruits, 3-caffeoyl-shikimic acid was present in range up to 87.1 mg/ kg, *p*-coumaroyl-feruloyl glycerol I (none-26.6 mg/kg), 3-feruloylquinic acid (none-25.8 mg/kg), *p*-coumaroyl-feruloyl glycerol II (none-51.7 mg/kg), 4-caffeoyl-shikimic acid (none-19.2 mg/kg), 5-caffeoyl-shikimic acid (none-15.6 mg/ kg), and 5-feruloylquinic acid (none-30.7 mg/kg) (Mallmann et al. 2020) amongst other phenolic compounds.

3.2.7 Coumarins and Chromones

Coumarins and chromones have been identified in fruits, but they had been reported to have significant importance in medicinal chemistry (Peng et al. 2013). Among the examples of Coumarins, scopoletin present in cassava fruits has significant health implications (Obidoa and Obasi 1991). Coumarin is present along with its glycosides having a sweet taste in Japanese green tea (Yang et al. 2009b). These compounds are present mainly in the families *Asteraceae*, *Apiaceae*, *Fabaceae*, *Lamiaceae*, and *Poaceae*. In Brazilian fruits, coumarin was detected and quantified in guava, passion fruit, surinam cherry by-products and mango pulp with values ranging from 57.3 to 102.4 µg/g DW (Silva et al. 2014).

The Muskmelon (*Cucumis bisexualis*) fruits have been reported to contain 7-hydroxy-3-(4′,6′-dihydroxy-5′-isopropyl-3″,3″-dimethyl-2*H*-chromen)-6-prenyl-2*H*-chromen-2-one, 7-hydroxy-3-(5′-prenyl-3″,3″-dimethyl-2*H*-chromen)-6-prenyl-2*H*-chromen-2-one, 3-(6′-hydroxy-5′-prenyl-3″,3″-dimethyl-2*H*-chromen)-6-prenyl-2*H*-chromen-2-one, 3-(5′-ethyl-3″,3″-dimethyl-2*H*-chromen)-6-prenyl-2*H*-chromen-2-one, 3-(4′,6′-dihydroxy-5′-dimeth-ylallyl-3″,3″-dimethyl-2*H*-chromen)-6-prenyl-2*H*-chromen-2-one, 3-[4′,6′-dihydroxy-5′-(2-propenyl)-3″,3″-dimethyl-2*H*-chromen]-14,15-dimethyl-pyrano-chromen-2-one, 3-(6′-dihydroxy-5′-isopropanol-3″, 3″-dimethyl-2*H*-chromen)-14,15-dimethyl-pyrano-chromen-2-one, 3-(5′-isopentenol-3″,3″-dimethyl-2*H*-chromen)-14,15-dimethyl-pyrano-chromen-2-one, 3-(4′,6′-dihydroxy-5′-prenyl-3″,3″-dimethyl-2*H*-chromen)-14,15-dimethyl-pyrano-chromen-2-one (Ma et al. 2018). These compounds have shown very promising results of antioxidants, and anti-cholinesterase inhibition activities. In citrus fruits, coumarins are one of the major groups of phenolic compounds (Naczk and Shahidi 2006).

Similarly, 7-hydroxy coumarin, 7-hydroxy-6-methoxy coumarin and 5,7-dihydroxychromone have recently been reported in Mulberry fruits (Xu et al. 2020). The air-dried pericarp of *Citrus grandis* showed columbianoside I, columbianoside II, meranzin hydrate I, meranzin hydrate II, meranzin hydrate III, paniculin III, meranzin hydrate IV, together with columbianoside, mexoticin, isomeranzin etc. (Tian et al. 2019). These compounds have significant medicinal properties. *Solanum incanum* with the common English name of bitter apple is a perennial, wild shrub distributed mainly in Africa, Middle East, India, Australia, and Taiwan. The fruit of bitter apple contains fraxinol-6-D-galactopyranoside (Kaunda and Zhang 2020). *Clausena indica* fruit is commonly used for food ingredients. The fruit of *C. indica* contains coumarins like Seselin, Braylin, and Dentatin (Hoang Anh et al. 2020).

3.2.8 Benzophenones and Xanthones

In fruits, benzophenones and xanthones are widely distributed in specific classes of plants such as Garcinia. Garcinia is a plant genus of the Guttiferae family containing several species. The edible fruit is rich in phenolic compounds including benzophenones. *Garcinia mangostana* (mangosteen) is a tropical fruit with reddish-purple skin that contains about 80% of xanthones such as α-, β-, and γ-mangostin (El-Seedi et al. 2020). In *Garcinia xanthochymus* fruits, benzophenones such as guttiferone H, aristophenone A, guttiferone E, and gambogenone were identified (Baggett et al. 2005). In another study, garciyunnanin A, garciyunnanin A, garciyunnanin B, oblongifolin C were identified in *Garcinia yunnanensis*, another specie of Garcinia (Xu et al. 2008). The fruits of *Garcinia cambogia* also showed benzophenones such as guttiferone M, guttiferone N, and Oxy-guttiferone K (Masullo et al. 2008). Similarly, guttiferone A, guttiferone K, xanthochymol, guttiferone E, cycloxanthochymol, isoxanthochymol, and gambogenone were identified in the fruits of *Garcinia livingstonei* (Yang et al. 2010). In other species of Garcinia such as *Garcinia griffithii* and *Garcinia mangostana*, xanthones and benzophenones were identified (Nguyen et al. 2005). α-Mangostin, cowanin, cowanol, garcinone D, β-mangostin, fuscaxanthone C, fuscaxanthone I, kaennacowanol A, jacareubin, fuscaxanthone A, and 1-isomagostin had been reported in *Garcinia cowa* (Ruan et al. 2017; Raksat et al. 2020). The benzophenones isolated from Garcinia species have shown significant DPPH radical scavenging activity and antiplasmodial activity (Lyles et al. 2014).

3.2.9 Stilbenes

Stilbenes were identified in a limited number of foods, however, due to the significant medicinal and health applications, they are the focus of modern research. Recently several fruits of the Umbelliferae family have been reported to contain stilbenes (Serag et al. 2020). Berries are a rich source of stilbene as reviewed by Błaszczyk et al. (2019). Several stilbenes, namely *E*- and *Z*-resveratrol (3,5,4-trihydroxystilbene), *E*- and *Z*-piceids (3-glucosides of resveratrol), *E*- and *Z*-astringins (3-glucosides of 3-hydroxyresveratrol), *E*- and *Z*-resveratrolosides (4-glucosides of resveratrol) and pterostilbene (a dimethylated derivative of stilbene) had been reported in both grape leaves and berries (Jeandet et al. 2002; Naczk and Shahidi 2006).

In bilberries and blueberries, the presence of resveratrol had been reported (Lyons et al. 2003). It has been observed that Bilberry (*Vaccinium myrtillus* L.), both cultivated and wild were evaluated to contain resveratrol from 0.77 to 6.8 mg/kg (Može et al. 2011; Rimando and Cody 2005; Ehala et al. 2005). The Vaccinium fruits have been reported to possess antioxidant properties. The resveratrol amount was 30 mg/kg in Cowberry, 19.2 mg/kg in Cranberry, 15.7 mg/kg in redcurrant (Ehala et al. 2005). The wild and cultivated Deerberry (*Vaccinium stamineum* L.) contains resveratrol (Rimando and Cody 2005). Very high amounts of resveratrol and piceid were reported Passion fruit (*Passiflora edulis*) (Kawakami et al. 2014). In tomato fruit, *E*-resveratrol were higher (15.3 mg/kg) than Z-resveratrol (2.71 mg/kg) (Ragab et al. 2006). Structures of some of the stilbenes identified in fruits are shown in Fig. 3.3.

In fruits, grape contains several stilbenes, among which resveratrol and piceid are the fingerprinting compounds as shown in Table 3.4. The grapefruit from different cultivated countries and cultivars showed resveratrol, piceid, and their isomers with the varied composition (Moriartry et al. 2001; Bavaresco et al. 2002; Rimando and Cody 2005; Ragab et al. 2006; Sun et al. 2006; Guerrero et al. 2010a; Viñas et al. 2011). These compounds were also reported in wild grape (Kiselev et al. 2017). However, there were other stilbenes reported in the Southeast China grape (*Vitis chunganesis*). These were Chunganenol, amurensin B and G, gnetin H, viniferin, vitisin A, hopeaphenol together with resveratrol (He et al. 2009a, b). In the last decade, the rise of high tech separation and identification equipment such as UPLC-MS/MS has increased the identification of new stilbene identification. For

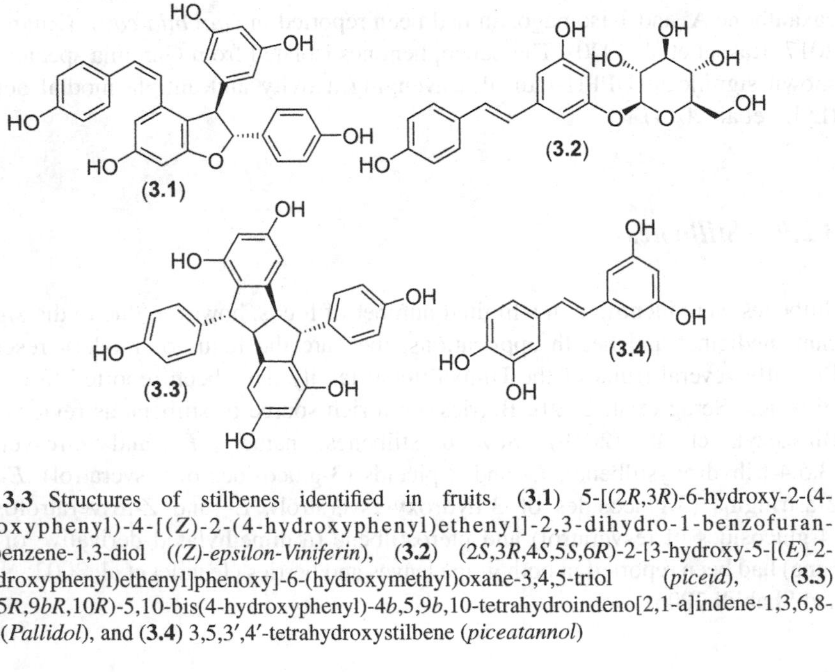

Fig. 3.3 Structures of stilbenes identified in fruits. (**3.1**) 5-[(2*R*,3*R*)-6-hydroxy-2-(4-hydroxyphenyl)-4-[(*Z*)-2-(4-hydroxyphenyl)ethenyl]-2,3-dihydro-1-benzofuran-3-yl]benzene-1,3-diol (*(Z)-epsilon-Viniferin*), (**3.2**) (2*S*,3*R*,4*S*,5*S*,6*R*)-2-[3-hydroxy-5-[(*E*)-2-(4-hydroxyphenyl)ethenyl]phenoxy]-6-(hydroxymethyl)oxane-3,4,5-triol   (*piceid*),   (**3.3**) (4*b*R,5*R*,9*b*R,10*R*)-5,10-bis(4-hydroxyphenyl)-4*b*,5,9*b*,10-tetrahydroindeno[2,1-a]indene-1,3,6,8-tetrol (*Pallidol*), and (**3.4**) 3,5,3′,4′-tetrahydroxystilbene (*piceatannol*)

Table 3.4 Stilbene contents present in different fruits

Sample	Cultivars/conditions	Stilbenes	Amount (µg/g)	Amount based on	References
Bilberry (*Vaccinium myrtillus* L.)	–	*E*-Resveratrol	2.0	FW	Može et al. (2011)
Bilberry (*Vaccinium myrtillus* L.)	Wild	Resveratrol	0.77	DW	Rimando and Cody (2005)
Bilberry (*Vaccinium myrtillus* L.)		*E*-Resveratrol	6.8	FW	Ehala et al. (2005)
Blueberries (*Vaccinium corymbosum* L.)		*E*-Resveratrol	4.0	FW	Može et al. (2011)
Highbush blueberry (*Vaccinium corymbosum* L.)	Bluecrop and Wild/*from conventional and sustainable farming*	*Conventional farming*		DW	Rimando and Cody (2005)
		Resveratrol	0.85		
		Piceatannol	0.42		
		Sustainable farming			
		Resveratrol	0.32		
		Piceatannol	0.19		
		Wild			
		Resveratrol	1.07		
Rabbiteye blueberry (*Vaccinium ashei* Reade)	Tifblue	*Tifblue*		DW	Rimando and Cody (2005)
		Resveratrol	0.11		
		Pterostilbene	0.15		
	Climax	*Climax*			
		Resveratrol	0.39		
		Pterostilbene	Nd		
	Premier	*Premier*			
		Resveratrol	0.1		
Cowberry (*Vaccinium Vitis-idaea* L.)		*E*-Resveratrol	30.0	FW	Ehala et al. (2005)
Cranberry (*Vaccinium oxycoccos*)		*E*-Resveratrol	19.2	FW	Ehala et al. (2005)
Red currant (*Ribes rubrum* L.)		*Red currant*		FW	Ehala et al. (2005)
		E-Resveratrol	15.7		
Black currant (*Ribes nigrum* L.)		*Black currant*			
		E-Resveratrol	Nd		

(continued)

Table 3.4 (continued)

Sample	Cultivars/conditions	Stilbenes	Amount (μg/g)	Amount based on	References
Deerberry (*Vaccinium stamineum* L.)	Wild	*Wild*		DW	Rimando and Cody (2005)
		Resveratrol	0.20		
	SHF3A-2-108	*SHF3A-2-108*			
		Resveratrol	0.12		
		Pterostilbene	0.52		
	B-76	*B-76*			
		Resveratrol	0.50		
		Piceatannol	0.19		
Table grapes (*Vitis vinifera* L.)		Resveratrol		FW	Moriartry et al. (2001)
Californian	Black Corinth	*Black Corinth cultivar*			
		Non-irradiated fruits:	0–25.1		
		Irradiated fruits:	0.9–33.2		
	Flame Seedless, skin, irradiation	*Flame Seedless cultivar*			
		Non-irradiated fruits	1–13.2		
		Irradiated fruits	1.8–57.3		
Grapes (*Vitis vinifera* L.)	Cabernet sauvignon	*E*-Resveratrol	0.29	FW	Bavaresco et al. (2002)
		E-Piceid	0.097		
		Piceatannol	0.052		
		Z-Resveratrol	Nd		
		Z-Piceid	Nd		
Grapes (*Vitis vinifera* L.)	Cabernet	*Cabernet*		DW	Rimando and Cody (2005)
		Resveratrol	2.48		
	Pinot Noir	*Pinot Noir*			
		Resveratrol	5.75		
	Merlot	*Merlot*			
		Resveratrol	6.36		
	Table grapes	*Table grapes*			
		Resveratrol	6.47		
Table grapes (*Vitis vinifera* L.)	Seedless red	*Z*-Resveratrol	20	DW	Ragab et al. (2006)
		E-Resveratrol	26–80		
		Z-Piceid	30		
		E-Piceid	50		
Grapes (*Vitis vinifera* L.)	Castelao, Syrah	*E*-Resveratrol	~10–100	FW	Sun et al. (2006)
		E-Piceid	~10–140		
		Z-Piceid	~20–200		

(continued)

Table 3.4 (continued)

Sample	Cultivars/conditions	Stilbenes	Amount (µg/g)	Amount based on	References
Southeast China grape (*Vitis chunganensis*)		Chunganenol	8.57	FW	He et al. (2009a)
		Amurensin B	242.8		
		Gnetin H	131.4		
		ε-Viniferin	154.2		
		Amurensin G	262.9		
		Vitisin A	191.4		
		Hopeaphenol	88.6		
		Resveratrol	48.6		
Southeast China grape (*Vitis chunganensis*)		Hopeaphenol	84.4	DW	He et al. (2009b)
		Amurensin G	148.8		
		Vitisin A	382.4		
Grapes (*Vitis vinifera* L.)	Red and white	*Red grapes*		FW	Viñas et al. (2011)
		E-Resveratrol	0.029		
		Z-Resveratrol	0.0028		
		Piceatannol	0.024		
		White grapes			
		E-Resveratrol	0.006		
		Z-Resveratrol	Nd		
		Piceatannol	0.002		
Grapes	(*3 Vitis vinifera sylvestris*, 7 *Vitis vinifera sativa*, 2 Hybrid Direct Producers)	*Merlot: non-UV, UV-C treatment*		FW	Guerrero et al. (2010b)
		E-Resveratrol	2.8, 9.7		
		Piceatannol	0.48, 2.55		
		Viniferins	0.29, 2.39		
		Syrah: non-UV, UV-C treatment			
		E-Resveratrol	3.56, 19.56		
		Piceatannol	0.46, 0.31		
		Viniferins	Nd, 1.49		
		Tempranillo: non-UV, UV-C treatment			
		E-Resveratrol	0.31, 3.78		
		Piceatannol	Nd, 1.10		
		Viniferins	0.14, 1.19		

(continued)

Table 3.4 (continued)

Sample	Cultivars/conditions	Stilbenes	Amount (µg/g)	Amount based on	References
Grapes (*Vitis vinifera* L.)	Red and White	*Red grapes*		FW	Viñas et al. (2011)
		E-Resveratrol	1.64		
		Z-Resveratrol	0.41		
		Piceatannol	0.37		
		White grapes			
		E-Resveratrol	0.23		
		Z-Resveratrol	0.082		
		Piceatannol	0.043		
Grapes (*Vitis vinifera* L.)		Piceatannol	0.08	FW	Vrhovsek et al. (2012)
		E-Piceid	0.27		
		Z-Piceid	0.02		
		Astringin	0.69		
		Isorhapontin	0.08		
		E-ε-Viniferin	0.32		
		Pallidol	0.04		
		Isohopeaphenol	0.19		
Wild grape (*Vitis wilsonae*)		Wilsonol A	1.87	DW	Jiang et al. (2012)
		Wilsonol B	0.72		
		Wilsonol C	1.13		
		Diviniferin B	3.03		
		Pallidol	0.75		
		ε-Viniferin	0.23		
		Ampelopsin B	1.13		
		Ampelopsin D	0.32		
		Miyabenol C	3.75		
		Dividol A	1.9		
		Hopeaphenol	6.13		
		Gnetin H	11.7		
		Heyneanol A	4.3		
		Ampelopsin G	1.87		
		Amurensin G	2.78		
		Visitin E	4.02		

(continued)

Table 3.4   (continued)

Sample	Cultivars/conditions	Stilbenes	Amount (µg/g)	Amount based on	References
Red grapes (*Vitis vinifera*)	Raboso Piave, Primitivo	E-Astringin	0.88	FW	Flamini et al. (2013)
		E-Piceid	2.34		
		Z-Astringin	0.12		
		Piceatannol	0.18		
		Z-Piceid	1.78		
		Pallidol	0.36		
		Pallidol-3-glucoside	0.19		
		Parthenocissin A	0.13		
		E-Resveratrol	1.14		
		Caraphenol B	0.11		
		Hopheapenol	0.10		
		Ampelopsin	0.54		
		Z-ε-Viniferin	0.38		
		E-ε-Viniferin	0.70		
		Z-Miyabenol C	0.24		
		E-Miyabenol C	1.36		
			0.068		
Grapes (*Vitis vinifera* L.)	21 red cultivars, skin	E-Resveratrol	19–508	FW	Vincenzi et al. (2013)
		E-Piceid	Nd to 551		
		Z-Piceid	Nd to 1		
		E-Piceatannol	Nd to 72.1		
Commercial Grapes (*Vitis vinifera*)	Mencía	E-Resveratrol	2.10–3.28	DW	Vilanova et al. (2015)

(continued)

Table 3.4 (continued)

Sample	Cultivars/conditions	Stilbenes	Amount (µg/g)	Amount based on	References
Grapes (*Vitis vinifera*)	Negro Amaro	*Non-infected, infected grapes*		FW	Flamini et al. (2013)
		E-Resveratrol	1.83, 3.89		
		Piceatannol	0.65, 0.78		
		Z-Piceid	0.78, 0.71		
		E-Piceid	1.71, 0.92		
		E-Astringin	1.08, 0.50		
		Z-Astringin	0.05, 0.05		
		Pallidol	0.70, 0.59		
		Z-ε-Viniferin	0.29, 0.25		
		ω-Viniferin:	0.88, 1.99		
		Z-Miyabenol C	0.04, 0.05		
		E-Miyabenol C	0.66, 1.46		
			0.21, 0.57		
Grapes (*Vitis vinifera*)	Raboso Piave	*Fresh, dried*		FW, DW	Rosso et al. (2016)
		α-Viniferin	0.79, 1.0		
		Z-Piceid	4.0, 5.3		
		E-Astringin	1.44, 1.9		
		Pallidol-3-glucoside	0.91, 1.2		
		Piceatannol	1.95, 2.5		
		Pallidol	1.02, 1.36		
		Parthenocissin A	0.56, 0.6		
		Z-ε-Viniferin	1.6, 2.1		
		E-ε-Viniferin	5.8, 7.73		
		δ-Viniferins	0.81, 1.08		
		Z-Miyabenol C	0.61, 0.82		
		E-Miyabenol C	5.5, 7.35		
		E-Piceid	3.84, 5.13		
		E-Resveratrol	4.93, 6.58		
			0.13, 0.17		

(continued)

Table 3.4 (continued)

Sample	Cultivars/conditions	Stilbenes	Amount (µg/g)	Amount based on	References
Grapes (*Vitis vinifera* L.)	Crimson Seedless	*E*-Resveratrol	121.62	FW	Guerrero et al. (2016)
		Z-Piceid	61.22		
		E-Piceid	15.81		
		E-Piceatannol	14.7		
		ε-Viniferin	11.25		
		ω-Viniferin	5.8		
		Isohopeaphenol	4.27		
		Stilbenoid	234.67		
Grapes (*Vitis vinifera* L.)	28 grape interspecific hybrids	*E*-Piceid	71	DW	Samoticha et al. (2017)
		Rielsing	25		
		Hibernal	19		
Vitis amurensis Rupr.	Wild	*E*-Piceid	55	DW	Kiselev et al. (2017)
		Z-Piceid	127		
		E-Resveratrol	14		
		Z-ε-Viniferin	7		
		E-ε-Viniferin	46		
		E-δ-Viniferin	Nd		
Lingonberry (*Vaccinium vitis-idaea* L.)	Wild	Resveratrol	5.88	DW	Rimando and Cody (2005)
Passion fruit (*Passiflora edulis*)	seeds	Piceatannol	85–400	DW	Kawakami et al. (2014)
		Scirpusin B	54–700		
Tomato (*Lycopersicon esculentum* Mill.)	MicroTom, Beafsteak, UglyRipe, Heirloo, PlumTom	*MicroTom*			Ragab et al. (2006)
		Z-Resveratrol	2.71		
		E-Resveratrol	15.30		
		Z-Piceid	0.26		
		E-Piceid	0.10		
		Beafsteak			
		UglyRipe			
		E-Resveratrol	Nd		
		Heirloom	0.38		
		Z-Resveratrol	0.11		
		E-Resveratrol	1.75		
		PlumTom			
		E-Resveratrol	0.34		

(continued)

Table 3.4 (continued)

Sample	Cultivars/conditions	Stilbenes	Amount (μg/g)	Amount based on	References
Table grapes (*Vitis vinifera* L.)	Dominga (D), Superior Seedless (SS), Autumn Royal (AR), Red Globe (RG)	D:		FW	Sanchez-Ballesta et al. (2020)
		E-Piceatannol	0.93		
		Z-Miyabenol	0.25		
		Pallidol	0.01		
		SS:			
		E-Piceatannol	0.83		
		Z-Miyabenol	0.10		
		Pallidol	0.02		
		AR:			
		E-Piceatannol	1.78		
		Z-Miyabenol	4.2		
		Pallidol	11.1		
		RG:			
		E-Piceatannol	5.13		
		Z-Miyabenol	1.63		
		Pallidol			

Nd not detected, *FW* fresh weight, *DW* dry weight

example, astringin, isorhapontin, viniferin (**3.1**), pallidal (**3.3**), piceatannol (**3.4**), and isohopephenol were the major compounds in the later studies (Vrhovsek et al. 2012; Jiang et al. 2012; Guerrero et al. 2016; Samoticha et al. 2017). In addition to these compounds, isomers of Miyabenol C, were also identified to have a concentration of less than 2 mg/kg (Flamini et al. 2013). Similarly, Parthenocissin A and Pallidol-3-glucoside were new compounds identified in grape cultivar (Raboso Piave) (Rosso et al. 2016). A recent study showed that *E*-piceatannol has higher composition than Z-miyabenol and pallidol in different varieties of table grape (Sanchez-Ballesta et al. 2020).

3.2.10 Anthraquinones

Anthraquinones have been rarely been found in widely consumed fruits. *Morinda citrifolia* widely known as noni is a Rubiaceous plant distributed across the tropical areas of the world. Its fruits have been traditionally used as food and for the prevention or improvement of diversified health problems. It contains several anthraquinone derivatives (Deng et al. 2009). Due to the toxicity of anthraquinones, the plant is not usually preferred as foods (Shalan et al. 2017). However, a recent study of Wang et al. (2019) showed that the fruits of *Morinda citrifolia* and *Morinda Officinalis* containing anthraquinones and other polyphenols have no in-vitro toxicity to normal cells.

3.2.11   Flavonoids

Flavonoids are one of the highly important and widely distributed classes of phenolic compounds present in foods. Among the fruits, berries are rich sources of flavonoids. Flavonoids such as quercetin, kaempferol and Myricetin had been reported in apples, plum, apricot, strawberry and mulberry was reported by Sultana et al. (2008). The highest among the flavonoids was Myricetin. For example, fresh bilberries were a good source of flavonols. The flavonol content of fresh bilberries was about 41 mg/kg (Shahidi and Ambigaipalan 2015). The fruit of blackcurrant serves as a rich source of phenolic compounds including flavonoids (Costantino et al. 1992; Naczk et al. 2005). Unlike other berries, in blackcurrant the flavonoids are dominated by myricetin, followed by quercetin and kaempferol (Häkkinen et al. 1999a; Mikkonen et al. 2001; Vuorinen et al. 2000). Total flavonoids were 20.9 mg/g (DW) in blackberries (Chen et al. 2016). On a fresh weight basis, the order of total polyphenols follows the order of dried cranberries > frozen cranberries > cranberry sauce > jellied sauce. The frozen cranberries and dried cranberries provide the most antioxidants distantly followed by mixed and 27% cranberry juice and other cranberry products (Vinson et al. 2008). Raspberries had the highest antioxidant capacity followed by strawberries, kiwi, broccoli, leek, apple, and, finally, tomato (Beekwilder et al. 2005). The antioxidant activity of the raspberry fruit was correlated to its total phenolic and flavonoid contents (Liu et al. 2002). In melon seeds, quercetin-3-rutinoside, and catechin have been reported (Zeb 2016).

The apple skin had been found to contain approximately 46% of the total phenolic compounds in apples (McGhie et al. 2005). The major flavonoids present in apples are flavan-3-ols/procyanidins (mainly catechin, epicatechin, and procyanidins B1 and B2), flavonols (quercetin galactoside, glucoside, rhamnoside, arabinoside, and xyloside), phloridzin and phloretin-2'-xyloglucoside (Tsao et al. 2005).

Grape berries and their skins contain flavonols such as quercetin-3-glucuronide, quercetin-3-glucoside, and myricetin-3-glucuronide (Souquet et al. 2000) as well as flavanonols, such as astilbin (dihydroquercetin-3-rhamnoside) and engeletin (dihydrokaempferol-3-rhamnoside) as shown in Fig. 3.4 (Lu and Foo 1999). Grape seeds are a rich source of monomeric phenolic compounds such as (+)-catechins, (−)-epicatechin, (+)-gallocatechin, (−)-epigallocatechin, and their dimeric, trimeric, and tetrameric proanthocyanidins (de Freitas et al. 1998).

The results of Bulgarian fruits had shown to contain several phenolic antioxidants. The quercetin in unpeeled red apples were 15.9 mg/kg, unpeeled green apples (13.9 mg/kg), blackberry (27.0 mg/kg), blueberry (99.2 mg/kg), cherry (25.2 mg/kg), fig (8.7 mg/kg), black grape (23.2 mg/kg), white grape (15.6 mg/kg), peach (34.1 mg/kg), unpeeled pear (5.9 mg/kg), plum (23.4 mg/kg), raspberry (16.0 mg/kg), sour berry (10.8 mg/kg), and strawberry (10.8 mg/kg) (Tsanova-Savova et al. 2018).

In 90 cultivars of strawberries, the major part of polyphenols is anthocyanin glycosides with pelargonidin-3-glucoside as a predominate compound (Nowicka et al. 2019). They also contain minor amounts of cyanidin-3-glucoside,

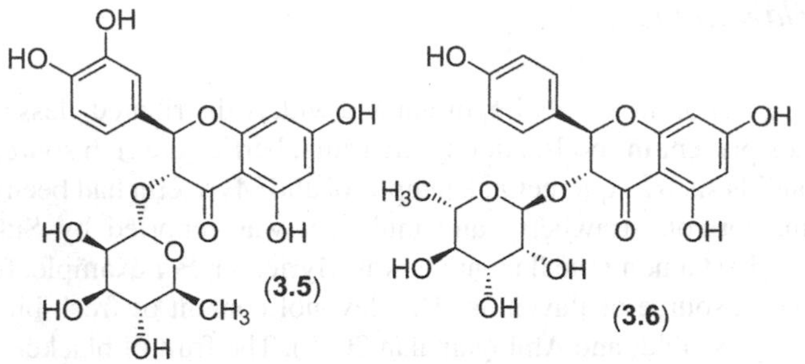

Fig. 3.4 Structures of flavonoids identified in grapes. (**3.5**) Astilbin, and (**3.6**) Engeletin

pelargonidin-3-rutinoside,   cyaniding-3-O-(6″malonyl)glucoside,   and
pelargonidin-3-(6″malonyl)glucoside (Aaby et al. 2012).

Polymethoxyflavones and their derivatives like hydroxy polymethoxyflavones
are distinctive compounds present exclusively in fruits of the citrus genus, especially in the citrus peels (Li et al. 2009). The most abundant polymethoxyflavones
are tangeretin and nobiletin present in citrus peels (Li et al. 2009). However, Marín
and Del Río (2001) showed that diosmin (4′methoxy-5,7,3′-trihydroxyflavone-7-
rutinoside,   **3.7**)   and   neodiosmin   (4′methoxy-5,7,3′-trihydroxyflavone-7-
neohepseridoside) are the predominant glycosylated flavones identified in citrus
fruits.   Nobiletin   (5,6,7,8,3′,4′-hexamethoxyflavone,   **3.8**)   and   sinensetin
(5,6,7,3′,4′-pentamethoxyflavone, **3.9**) have been identified in orange, whereas tangeretin, 3,5,6,7,8,3′,4′-heptamethoxyflavone, 5,7,8,4′-tetramethoxyflavone, and
5,7,8,3′,4′-pentamethoxyflavone exist in grapefruits (Naczk and Shahidi 2006; Li
et al. 2009). The structures of some of these compounds are shown in Fig. 3.5.

In Romanian wild fruits, Myricetin and quercetin were the major flavonoids.
Myricetin was 166.3 mg/kg in Sea buckthorn berries, 14.3 mg/kg in Hawthorn berries, 26.6 mg/kg in wild European blackberries, 305.4 mg/kg in Cornelian cherries,
74.4 mg/kg in blackthorn berries, 19.1 mg/kg in dog rose berries, and 53.3 mg/kg in
bird cherries (Cosmulescu et al. 2017). Similarly, quercetin was 35.4 mg/kg in Sea
buckthorn berries, 57.2 mg/kg in Hawthorn berries, 68.3 mg/kg in wild European
blackberries, 33.6 mg/kg in Cornelian cherries, 65.5 mg/kg in blackthorn berries,
49.9 mg/kg in dog rose berries and 68.8 mg/kg in bird cherries.

Three classes of flavonoids occur in citrus fruits i.e., flavanones, flavones and
flavonols (Calabro et al. 2004). Recent studies had shown that the flavonoids extracts
of citrus fruits represent a significant source of phenolic antioxidants with potential
prophylactic properties for the development of functional foods (Ramful et al.
2010). Limonoids are a group of highly oxygenated triterpenoids found in citrus
that was also reported as an anticancer agent (Hasegawa et al. 1999). In red and yellow araçá (*Psidium cattleianum* Sabine) fruits from Brazil, luteolin 7-glucuronide
was present in the range up to 18.4 mg/kg, quercetin hexoside (none-23.6 mg/kg),
and quercetin glucuronide (none-38.5 mg/kg) (Mallmann et al. 2020) amongst other
phenolic compounds.

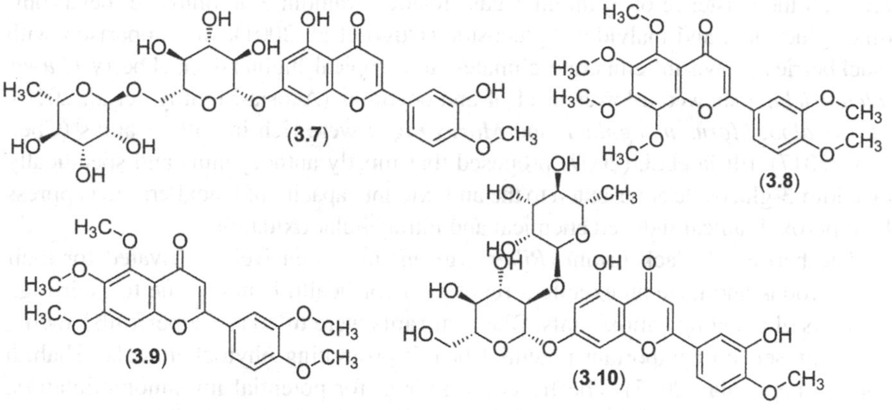

Fig. 3.5 Structures of flavonoids identified in citrus fruits. (**3.7**) Diosmin, and (**3.8**) nobiletin, (**3.9**) sinensetin, and (**3.10**) neodiosmin

3.2.12 Anthocyanidins and Anthocyanins

Fruits are rich sources of anthocyanidins, among which berries contain a high quantity of anthocyanidins. Among the different berry varieties, the highest amount of anthocyanins in bilberry contains for cyanidin-3-glycoside, delphinidin-3-glycoside, and malvidin-3-glycoside (Du et al. 2004). The anthocyanosides in Bilberry demonstrating all possible combinations of five anthocyanidins (cyanidin, delphinidin, peonidin, petunidin, malvidin) with three sugar moieties like 3-arabinoside, 3-glucoside, 3-galactoside were separated using HPLC (Madhavi et al. 1998). Du et al. (2004) reported delphinidin-3-sambubioside, and cyanidin-3-sambubioside in bilberry. Similarly, the commercial extract of bilberry had been reported to contain mostly glycosides of delphinidin and cyanidin (Baj et al. 1983). These anthocyanidins had been used to treat several microcirculation diseases resulting from capillary fragility and have been used to maintain normal vascular permeability (Wang et al. 1997). An *in-vitro* study revealed that the use of bilberry and bilberry extracts had been useful as functional foods and food supplements designed for the prevention of chronic diseases associated with oxidative stress (Valentová et al. 2007).

Blackberries are also a rich source of anthocyanins and other polyphenolic antioxidants (Siriwoharn et al. 2004). The anthocyanins of blackberries have been extensively studied and it was revealed that they are solely cyanidin-based phenolic compounds. In blackberries, five anthocyanins were reported. They were cyanidin-3-glucoside as a major anthocyanin, cyanidin-3-rutinoside, malonic acid acylated cyanidin-3-glucoside and a derivative of xylose-cyanidin (Siriwoharn et al. 2004). In evergreen blackberry, cyanidin-3-dioxalyl-glucoside, a novel zwitterionic anthocyanin was isolated (Stintzing et al. 2002). Similarly, cyanidin-3-(6'-*p*-coumaryl)glucoside was identified in Marionberries and cyanidin-3-arabinoside in evergreen blackberries (Wada and Ou 2002). Another study on a commercial Italian blackberry reported the existence of cyanidin-3-galactoside, cyanidin-3-arabinoside, pelargoni-din-3-glucoside, and malvidin-3-glucoside (Dugo et al. 2001). In comparison with blackberries grown in temperate climates, the tropical highland blackberry (*Rubus adenotrichus*) showed a lower level of anthocyanins (Acosta-Montoya et al. 2010). *Morus alba*, *Morus mongolica*, and *Morus nigra* were rich in anthocyanins (Chen et al. 2017). Elisia et al. (2007) proposed that mostly anthocyanins and specifically cyanidin-3-glucoside contributed to the antioxidant capacity of blackberry to suppress both peroxyl radical-induced chemical and intracellular oxidation.

The berries of blackcurrant (*Ribes nigrum*) are extensively cultivated for their use in foods and have an excellent reputation for health benefits due to their high contents of phenolic antioxidants. Black currants were titled as "superfruits" owing to the presence of important potential health-promoting phytochemicals (Shahidi and Ambigaipalan 2015). The fruits are known for potential immunomodulatory, antimicrobial and anti-inflammatory actions. They inhibit low-density lipoproteins and thus reduce the risk of cardiovascular diseases (Nour et al. 2014). Anthocyanins of 11 anthocyanidins such as delphinidin, cyanidin, malvidin, petunidin, and peoni-din had been reported in blackcurrant, with the main components being delphinidin-3-glucoside, delphinidin-3-rutinoside and cyanidin-3-rutinose (Borges et al. 2010). The HPLC-MS analyses of ethanolic extracts of black currants showed delphinidin-3-glucoside, delphinidin-3-rutinoside, cyanidin-3-glucoside, cyanidin-3-rutinoside, petunidin-3-rutinoside, pelargonidin-3-rutinoside, peonidin-3-rutinoside, petuni-din-3-(6-coumaroyl)-glucoside and cyanidin-3-(6-coumaroyl)-glucoside (Nour et al. 2014). The Southeastern U.S. range blackberry cultivars showed strong anti-oxidant and anti-inflammatory activities due to the presence of anthocyanins (Srivastava et al. 2010). Another study of berries suggested that blackcurrants had the highest antioxidant activity in the FRAP assay followed by blueberries, raspber-ries, and red currants, and the lowest was noted for cranberries (Borges et al. 2010). However, several previous reports had confirmed that blueberries contain high anti-oxidant activity compared with other fruits (Vinson et al. 2001; Prior et al. 1998). The total anthocyanin and phenolic contents were responsible for the high antioxi-dant capacity of whole blueberries. In blueberries, the anthocyanins are concen-trated in the skin, and antioxidant properties are mostly because of the skin (Lee and Wrolstad 2004). Fifteen anthocyanins were reported in blueberries, many of which exhibited substantial antioxidant activity. Oligomeric B-type procyanidins from dimers to octamers have been identified in blueberry (Prior et al. 1998). The pro-cessing wastes of blueberry have high total phenols and antioxidant activity and are a good source of natural colorant and nutraceuticals (Lee et al. 2005).

Cranberries had been reported to contain the highest amount of phenolic antioxi-dants than any commonly consumed fruit (Vinson et al. 2001). It was the main reason that it was approved for the treatment of urinary tract infection (Vinson et al. 2008; Koradia et al. 2019). The flavonoids in cranberry include anthocyanins, pro-anthocyanidins, and flavonols (Gregoire et al. 2007). Galactoside and arabinoside largely form conjugates of cyanidin and peonidin in cranberry (Singh et al. 2009). Proanthocyanidin isolates contain epicatechin (monomer), dimer (epicatechin-(4β → 8, 2β → O → 7)-epicatechin), and trimer (epicatechin-(4β → 8)-epicatechin-(4β → 8, 2β → O → 7)-epicatechin) (Singh et al. 2009). In cranberry flavonols such as myricetin-3-galactoside, myricetin-3-arabino-furanoside, quercetin-3-galactoside, quercetin-3-glucoside, quercetin-3-rhamno-spyranoside, and quercetin-3-(6″-p-benzoyl)-galactoside (Singh et al. 2009). Free flavonoids such as myricetin, quercetin, and kaempferol had been identified in a small percentage of the total flavonol content in cranberry (Häkkinen et al. 1999b). When both flavonol glycosides and cyanidin-3-galactoside from the whole cranberry were compared to vitamin E, the former was more effective in scavenging of free radicals and preventing oxidation of LDL (low density lipoproteins).

Red raspberries (*Rubus idaeus* L.) possess strong antioxidant activity, due to the presence of high levels of anthocyanins and other phenolic compounds (Kafkas et al. 2008). The major anthocyanin was cyanidin-3-sophoroside in raspberry with smaller quantities of other anthocyanins, including cyanidin-3-(2-glucosylrutino-side), cyanidin-3-glucoside, cyanidin-3-rutinoside, pelargonidin-3-sophoroside, pelargonidin-3(2-glucosylrutinoside), and pelargonidin-3-glucoside (Borges et al. 2010). They also contain quercetin and kaempferol based flavonol conjugates and trace levels of (−)-epicatechin and a procyanidin dimer (Mullen et al. 2004). Black raspberries have much higher levels of anthocyanins (6870 mg/kg, fw) than red raspberries, which contribute to their darker coloration, and the majority of these are also cyanidins (Rao and Snyder 2010).

Anthocyanin contents of Bilberry fruits collected from different areas of the Italian Northern Apennines were reported by (Benvenuti et al. 2018). The total content of anthocyanins varied from 582.4 mg/100 g to 795.2 mg/100 g (FW). These include glycosides of delphinidin, cyanidin, petunidin, peonidin and malvidin as shown in Table 3.5.

In ninety cultivars of strawberries, the mean values of total flavonoids were 53.1 mg/kg in 2011 and 54.7 mg/kg in 2012. The quercetin-3-glucuronide was accounted for 48.8% of the total flavonoids, whereas others such as isorhamne-tin-3-glucuronide, and kaempferol-3-glucuronide were also present (Nowicka et al. 2019).

In grapes, the flavonoids include flavan-3-ols (catechin), flavonols (quercetin) and anthocyanins. In the grape berry, the anthocyanins are confined in the skins (Palomino et al. 2000), whereas flavan-3-ols are present both in the skins and in the seeds (Yang et al. 2009a). Grapes are one of the major dietary sources of anthocyanins, responsible for a wide range of colors of black, red and purple grapes; however, these phenolics are absent in white grapes. In particular, anthocyanins mostly accumulate in the skins, whereas procyanidins are located in the seeds. The anthocyanins in grape skins are predominately 3-glucosides of malvidin, cyanidin, delphinidin, peonidin and petunidin. Malvidin, the reddest of all anthocyanins, is the major one in dark red vinifera grapes, with higher proportions of cyanidin in red grapes. The major anthocyanins in Concord grapes are cyanidin-3-monoglucoside, and delphinidin-3-monoglucoside (Yang et al. 2009a). However, in different grape cultivars, anthocyanin and phenolic compounds either alone or in combination, are responsible for the antioxidant activity (Orak 2007). Díaz-Mula et al. (2019) showed

Table 3.5 Anthocyanin contents of Bilberry fruits collected from different areas of the Italian Northern Apennines. Reproduced with kind permission of Springer Nature (Benvenuti et al. 2018)

Compound	Samples of Bilberry fruits (mg/100 g)		
	1	2	3
Delphinidin-3-galactoside	117.0 ± 10.3	99.4 ± 7.5	127.1 ± 2.9
Delphinidin-3-glucoside	123.8 ± 3.9	95.7 ± 1.3	119.0 ± 6.1
Cyanidin-3-galactoside	50.9 ± 8.6	44.6 ± 0.6	56.3 ± 5.1
Delphinidin-3-arabinoside	91.3 ± 4.3	77.0 ± 5.8	97.4 ± 5.1
Cyanidin-3-glucoside	58.1 ± 3.4	45.7 ± 1.6	55.0 ± 4.2
Petunidin-3-galactoside	38.7 ± 4.0	28.4 ± 0.4	33.1 ± 2.2
Cyanidin-3-arabinoside	36.3 ± 4.9	30.7 ± 1.0	37.1 ± 1.6
Petunidin-3-glucoside	82.8 ± 1.2	56.9 ± 2.9	65.9 ± 1.3
Petunidin-3-arabinoside	23.1 ± 1.5	16.7 ± 0.4	20.1 ± 0.4
Peonidin-3-glucoside	36.4 ± 2.9	22.8 ± 2.4	24.8 ± 6.0
Malvidin-3-galactoside	26.8 ± 3.9	14.4 ± 1.4	19.7 ± 2.4
Malvidin-3-glucoside	92.4 ± 4.1	50.0 ± 3.1	56.1 ± 4.4
Malvidin-3-arabinoside	17.6 ± 1.7	Trace	Trace
Total anthocyanins	795.2 ± 57.7	582.4 ± 28.6	712.1 ± 20.8

the anthocyanins composition of pomegranate fruits. The fruits were a rich mixture of flavan-3-ols.

Anthocyanins are also present in sweet cherries and tart cherries. In bing and other sweet cherry cultivars, cyanidin-3-rutinoside, cyanidin-3-glucoside, peonidin-3-rutinoside, peonidin-3-glucoside, and pelargonidin-3-rutinoside had been reported (Chaovanalikit and Wrolstad 2004). Similarly, cyanidin-3-glucoside, cyanidin-3-rutinoside, cyanidin-3-glucosylrutinoside, cyanidin-3-sophoroside, pelargonidin-3-glucoside, peonidin-3-rutinoside and cyanidin-3-arabinosylrutinoside were reported in sour cherries (Chaovanalikit and Wrolstad 2004). The fruits of tart cherry contain cyanidin 3-glucosylrutinoside, cyanidin 3-rutinoside, cyanidin sophoroside, peonidin 3-glucoside; kaempferol, quercetin and isorhamnetin and their derivatives (Kirakosyan et al. 2010). Cherries have been found to contain higher levels of anthocyanins as compared to several other fruits. It was, therefore, implicit that regular consumption of cherries is beneficial in reducing risk factors for heart diseases, diabetes and certain types of cancers (Mulabagal et al. 2009).

Apples are an excellent source of several phenolic compounds and also possess high total antioxidant capacity (Biedrzycka and Amarowicz 2008). The amounts of total phenolic compounds are higher in the peel than in the flesh and the antioxidant and antiproliferative activities of unpeeled apples were greater than those of peeled apples (Burda et al. 1990). Catechins, procyanidins, phloridzin, phloretin glycosides are present in the flesh of the apple. However, in addition to these compounds, the peel also contains flavonoids not found in the flesh, such as quercetin glycosides (Wolfe et al. 2003). Quercetin glycosides and cyanidin-3-galactoside had been reported in the extract of apple skin, which was not found in the flesh (Huber and Rupasinghe 2009). The anthocyanins present in apples are cyanidin-3-galactoside, procyanidins (mainly catechin, epicatechin, and procyanidins B1 and B2) (Tsao et al. 2005). In apples, flavan-3-ols were the maximum contributor of individual antioxidant activity, followed by quercetin glycosides, chlorogenic acid, cyanidin-3-galactoside, and phloridzin (van der Sluis et al. 2002). However, Chinnici et al. (2004) reported the order of antioxidant activity of individual phenolic compounds as quercetin glycosides > procyanidins > chlorogenic > phloridzin. The pomace of apple is a valuable source of anthocyanins relevant to its antioxidant activity. The antioxidant activity of apple pomace can be forecasted by the amounts of phloridzin, procyanidin B2, rutin + isoquercitrin, protocatechuic acid and hyperin (García et al. 2009).

In citrus fruits, like sweet oranges, narirutin and hesperidin, hesperetin 7-rutinoside were present, while in sour oranges naringin, neohesperidin and hesperetin 7-neohesperidoside had been identified (Kanes et al. 1993). In Italian blood oranges, cyanidin-3-glycoside, cyanidin-3-(6″-malonyl)-glycoside and cyanidin-3-rhamnoside were the major anthocyanins (Maccarone et al. 1998). In blood orange, cyanidin-3-glycoside and cyanidin-3-(6″-malonyl)-glycoside were reported in high amounts (Lee 2002).

Pomegranates are a rich source of anthocyanins. Anthocyanins like cyanidin-3-glucoside, delphinidin-3-glucoside, cyanidin-3,5-diglucoside, delphinidin-3,5-diglucoside and pelargonidin-3-glucoside had been reported in pomegranate fruits (Noda et al. 2002). Delphinidin-3-glucoside, delphinidin-3,5-diglucoside, cyanidin-3-glucoside, cyanidin-3,5-diglucoside, pelargonidin-3-glucoside, and pelargonidin-3,5-diglucoside were present in seed coat of pomegranate (Du et al. 1975). In red and yellow araçá (*Psidium cattleianum* Sabine) fruits from Brazil, Vescalagin was present in range up to 77.1 mg/kg, castalagin (none-85.5 mg/kg), proanthocyanidin dimer B I (none-132.3 mg/kg), cyanidin-3-hexoside (none-92.9 mg/kg), brevifolin (none-20.9 mg/kg), delphinidin (none-21.0 mg/kg), myricetin pentoside II (none-5.8 mg/kg), and cyanidin (none-32.5 mg/kg) (Mallmann et al. 2020), amongst other phenolic compounds.

The peel and flesh of apricot fruit contain four main groups of phenolic compounds, procyanidins, hydroxycinnamic acid derivatives, flavonol, and anthocyanins (Ruiz et al. 2005). Similarly, in the skin and flesh of the different cultivars of apricot fruits, procyanidins B1, B2, and B4, and cyanidin 3-rutinoside, and cyanidin-3-glucoside were reported along with other phenolic compounds (Han et al. 2013).

3.2.13   Tannins

Pomegranates are a rich source of hydrolyzable tannins. Several studies have shown anthocyanins, ellagitannins, gallotannins, non-colored flavonoids, and lignans amongst other phenolic compounds in fruits of pomegranate (Bonzanini et al. 2009). Hydrolyzable tannins are the most abundant polyphenols and antioxidant compounds in pomegranates. They include gallotannins, ellagitannins and galloyl esters, for example, punicalagin and punicalin (Madrigal-Carballo et al. 2009). Among the 29 local and domesticated accessions of grapes from Israel, four major hydrolyzable tannins, including punicalagin, punicalins, gallagic acid, and ellagic acid, were identified (Tzulker et al. 2007; Bar-Ya'akov et al. 2019).

A unique phytochemical profile rich in hydrolyzable tannins (ellagitannins) and anthocyanins had been reported in raspberries that distinguishes them from other berries and fruits (Rao and Snyder 2010). A recent study by Abel Sánchez-Velázquez et al. (2019) showed the tannins profile of two wild blackberries from the Northwest of Mexico, i.e., *Rubus liebmannii* and *Rubus palmeri*. These include epicatechin, and isomers of Pedunculagin, Casuarictin, Potentillin, Sanguiin H-2, Lambertianin B, Lambertianin C, Sanguiin H-10, Nobotanin A, Malabathrin B, Sanguiin H-6 and Lambertianin A. Similarly, 16 tannins including ellagitannins, gallotannins and pro-anthocyanidins and its related compounds were identified in 90 cultivars of straw-berries (Nowicka et al. 2019). Ellagitannins were the major class of compounds having agrimoniin as the predominant compound with concentration ranges from 122.8 to 149.6 mg/kg. The agrimoniin in 27 cultivars studied was reported to be an average of 88.0 mg/kg by Aaby et al. (2012), while in six cultivars an average of 118.9 mg/kg as reported by Gasperotti et al. (2013). The seeds of melon fruits also revealed ellagic derivatives including ellagitannins (Zeb 2016).

In chestnut (*Castanea sativa* Mill.) fruits, high amounts of tannins have recently been reported having significant anti-inflammatory properties (Piazza et al. 2019). In fruits of *Myrciaria trunciflora* and *M. jaboticaba* several hydrolyzable tannins such as HHDP-galloylglucose, bis-HHDP-glucose, pedunculagin, trisgalloyl HHDP glucose, HHDP-digalloylglucose, castalin, epicatechin, bis-HHDP-galloylglucose (casuarinin), bis-HHDP-galloylglucose isomer (casuarictin), digalloylglucose, tri-galloylglucose, castalagin, vescalagin, tetragalloylglucose, galloyl-castalagin, pen-tagalloylglucose and their isomers. The total amount of hydrolyzable tannins in *Myrciaria trunciflora* was 3315.4 mg/100 g, whereas 788.5 mg/100 g in *M. jaboti-caba* fruits (Quatrin et al. 2019). Phenolic compounds and tannins have been identi-fied in different apple cultivars showed significant antioxidant activity in the peel as compared to whole fruit and flesh portion (Bahukhandi et al. 2019). Khundamri et al. (2020) studied the depolymerization products of condensed tannins in the pericarp of the mangosteen fruit. The authors showed that the pericarp of the fruit has condensed tannins in the form of mainly dimers and trimers.

3.2.14 Glycosides

Phenolic compounds in fruits are present mostly in glycosides form apart from the anthocyanins. They are present in all fruits and thus it is not possible to accommo-date them here. Flavonoids occurred in the forms of *O*-glycoside or *C*-glycoside forms in fruits. The dietary flavonoid *C*-glycosides have received less attention than their corresponding *O*-glycosides (Xiao et al. 2016). Among the flavonoid *C*-glycosides, flavone *C*-glycosides, particularly vitexin, isoorientin, orientin, isovi-texin and their multiglycosides are more frequently reported. Citrus fruits can be considered as a major source of *C*-glycosylflavones (Bucar et al. 2020).

The glycosides of flavonol such as myricetin-3-rutinoside, myricetin-3-gluco-side, quercetin-3-rutinoside, quercetin-3-glucoside and kaempferol-3-glucoside have been identified in blackcurrant seeds (Lu et al. 2002). Hesperidin, narirutin and didymin (isosakuranetin 7-rutinoside) are the predominant flavanone glycosides in the navel (Gil-Izquierdo et al. 2002) and blood oranges (Pannitteri et al. 2017).

3.3 Study Questions

- Describe the trend in the antioxidant potential of edible fruits.
- Write a detail note on the different hydroxybenzoic acid derivatives in fruits.
- What are the composition of phenolic aldehydes, acetophenone and phenylacetic acids in different fruits?
- Write a detail note on the cinnamoyl derivatives in fruits.
- What are the types of coumarins, chromones and their derivatives present in fruits?
- What do you know about the stilbenes in fruits?
- Discuss a detail composition of flavonoids in fruits.
- Describe the anthocyanins in fruits.
- Are tannins present in fruits?
- Explain a detail note of phenolic glycosides in fruits.

References

Aaby K, Mazur S, Nes A, Skrede G (2012) Phenolic compounds in strawberry (Fragaria x anan-assa Duch.) fruits: composition in 27 cultivars and changes during ripening. Food Chem 132(1):86–97

Abel Sánchez-Velázquez O, Montes-Ávila J, Milán-Carrillo J, Reyes-Moreno C, Mora-Rochin S, Cuevas-Rodríguez E-O (2019) Characterization of tannins from two wild blackberries (Rubus spp) by LC–ESI–MS/MS, NMR and antioxidant capacity. J Food Measur Charact 13(3):2265–2274

Acosta-Montoya Ó, Vaillant F, Cozzano S, Mertz C, Pérez AM, Castro MV (2010) Phenolic content and antioxidant capacity of tropical highland blackberry (Rubus adenotrichus Schltdl.) during three edible maturity stages. Food Chem 119(4):1497–1501

Ahmad N, Zuo Y, Lu X, Anwar F, Hameed S (2016) Characterization of free and conjugated phenolic compounds in fruits of selected wild plants. Food Chem 190:80–89

Alasalvar C, Shahidi F (2013) Composition, phytochemicals, and beneficial health effects of dried fruits: an overview. In: Dried fruits: phytochemicals and health effects. Oxford, Wiley-Blackwell, pp 1–18

Ancillotti C, Ciofi L, Pucci D, Sagona E, Giordani E, Biricolti S, Gori M, Petrucci WA, Giardi F, Bartoletti R (2016) Polyphenolic profiles and antioxidant and antiradical activity of Italian berries from Vaccinium myrtillus L. and Vaccinium uliginosum L. subsp. gaultherioides (Bigelow) SB Young. Food Chem 204:176–184

Andrés-Lacueva C, Medina-Remon A, Llorach R, Urpi-Sarda M, Khan N, Chiva-Blanch G, Zamora-Ros R, Rotches-Ribalta M, Lamuela-Raventos RM (2010) Phenolic compounds:

chemistry and occurrence in fruits and vegetables. In: de la Rosa LA, Alvarez-Parrilla E, González-Aguilar GA (eds) Fruit and vegetable phytochemicals: chemistry, nutritional value, and stability. Blackwell Publishing, Ames, IA, pp 53–80

Ayaz FA, Hayirlioglu-Ayaz S, Gruz J, Novak O, Strnad M (2005) Separation, characterization, and quantitation of phenolic acids in a little-known blueberry (*Vaccinium arctostaphylos* L.) fruit by HPLC-MS. J Agric Food Chem 53(21):8116–8122

Baggett S, Protiva P, Mazzola EP, Yang H, Ressler ET, Basile MJ, Weinstein IB, Kennelly EJ (2005) Bioactive benzophenones from *Garcinia xanthochymus* fruits. J Nat Prod 68(3):354–360

Bahukhandi A, Dhyani P, Jugran AK, Bhatt ID, Rawal RS (2019) Total phenolics, tannins and antioxidant activity in twenty different apple cultivars growing in West Himalaya, India. Proc Natl Acad Sci India Sec B Biol Sci 89(1):71–78

Baj A, Bombardelli E, Gabetta B, Martinelli E (1983) Qualitative and quantitative evaluation of Vaccinium myrtillus anthocyanins by high-resolution gas chromatography and high-performance liquid chromatography. J Chromatogr A 279:365–372

Bar-Ya'akov I, Tian L, Amir R, Holland D (2019) Primary metabolites, anthocyanins, and hydrolyzable tannins in the pomegranate fruit. Front Plant Sci 10:620–620

Bavaresco L, Fregoni M, Trevisan M, Mattivi F, Vrhovsek U, Falchetti R (2002) The occurrence of the stilbene piceatannol in grapes. Vitis 44(3):133–136

Beekwilder J, Hall RD, De Vos C (2005) Identification and dietary relevance of antioxidants from raspberry. Biofactors 23(4):197–205

Benvenuti S, Brighenti V, Pellati F (2018) High-performance liquid chromatography for the analytical characterization of anthocyanins in *Vaccinium myrtillus* L. (bilberry) fruit and food products. Anal Bioanal Chem 410(15):3559–3571

Biedrzycka E, Amarowicz R (2008) Diet and health: apple polyphenols as antioxidants. Food Rev Int 24(2):235–251

Błaszczyk A, Sady S, Sielicka M (2019) The stilbene profile in edible berries. Phytochem Rev 18(1):37–67

Bonzanini F, Bruni R, Palla G, Serlataite N, Caligiani A (2009) Identification and distribution of lignans in *Punica granatum* L. fruit endocarp, pulp, seeds, wood knots and commercial juices by GC–MS. Food Chem 117(4):745–749

Borges G, Degeneve A, Mullen W, Crozier A (2010) Identification of flavonoid and phenolic antioxidants in black currants, blueberries, raspberries, red currants, and cranberries. J Agric Food Chem 58(7):3901–3909

Bucar F, Xiao JB, Ochensberger S (2020) Flavonoid C-glycosides in Diets. In: Xiao J, Sarker SD, Asakawa Y (eds) Handbook of dietary phytochemicals. Springer, Singapore, pp 1–37

Burda S, Oleszek W, Lee CY (1990) Phenolic compounds and their changes in apples during maturation and cold storage. J Agric Food Chem 38(4):945–948

Cabrini L, Barzanti V, Cipollone M, Fiorentini D, Grossi G, Tolomelli B, Zambonin L, Landi L (2001) Antioxidants and total peroxyl radical-trapping ability of olive and seed oils. J Agric Food Chem 49(12):6026–6032

Calabro M, Galtieri V, Cutroneo P, Tommasini S, Ficarra P, Ficarra R (2004) Study of the extraction procedure by experimental design and validation of a LC method for determination of flavonoids in *Citrus bergamia* juice. J Pharmaceut Biomed 35(2):349–363

Cantos E, Espin JC, Tomás-Barberán FA (2002) Varietal differences among the polyphenol profiles of seven table grape cultivars studied by LC–DAD–MS–MS. J Agric Food Chem 50(20):5691–5696

Cassani L, Tomadoni B, Viacava G, Ponce A, Moreira MR (2016) Enhancing quality attributes of fiber-enriched strawberry juice by application of vanillin or geraniol. LWT Food Sci Technol 72:90–98

Ceccarelli D, Simeone AM, Nota P, Piazza MG, Fideghelli C, Caboni E (2016) Phenolic compounds (hydroxycinnamic acids, flavan-3-ols, flavonols) profile in fruit of Italian peach varieties. Plant Biosyst 150(6):1370–1375

Chaovanalikit A, Wrolstad R (2004) Anthocyanin and polyphenolic composition of fresh and processed cherries. J Food Sci 69(1):FCT73–FCT83

Chen H, Pu J, Liu D, Yu W, Shao Y, Yang G, Xiang Z, He N (2016) Anti-inflammatory and anti-nociceptive properties of flavonoids from the fruits of black mulberry (*Morus nigra* L.). PLoS One 11(4):e0153080–e0153080

Chen H, Yu W, Chen G, Meng S, Xiang Z, He N (2017) Antinociceptive and antibacterial properties of anthocyanins and flavonols from fruits of black and non-black mulberries. Molecules 23(1)

Chinnici F, Bendini A, Gaiani A, Riponi C (2004) Radical scavenging activities of peels and pulps from cv. Golden Delicious apples as related to their phenolic composition. J Agric Food Chem 52(15):4684–4689

Clifford MN (2000) Miscellaneous phenols in foods and beverages – nature, occurrence and dietary burden. J Sci Food Agric 80(7):1126–1137

Coman V, Vodnar DC (2020) Hydroxycinnamic acids and human health: recent advances. J Sci Food Agric 100(2):483–499

Cosmulescu S, Trandafir I, Nour V (2017) Phenolic acids and flavonoids profiles of extracts from edible wild fruits and their antioxidant properties. Int J Food Prop 20(12):3124–3134

Costantino L, Albasini A, Rastelli G, Benvenuti S (1992) Activity of polyphenolic crude extracts as scavangers of superoxide radicals and inhibitors of xanthine oxidase. Planta Med 58(04):342–344

Deng S, West BJ, Jensen CJ, Basar S, Westendorf J (2009) Development and validation of an RP-HPLC method for the analysis of anthraquinones in noni fruits and leaves. Food Chem 116(2):505–508

Díaz-Mula HM, Tomás-Barberán FA, García-Villalba R (2019) Pomegranate fruit and juice (cv. Mollar), rich in ellagitannins and anthocyanins, also provide a significant content of a wide range of proanthocyanidins. J Agric Food Chem 67(33):9160–9167

Ding T, Cao K, Fang W, Zhu G, Chen C, Wang X, Wang L (2020) Evaluation of phenolic components (anthocyanins, flavanols, phenolic acids, and flavonols) and their antioxidant properties of peach fruits. Sci Hort Amst 268:109365

Du C, Wang P, Francis F (1975) Anthocyanins of pomegranate, *Punica granatum*. J Food Sci 40(2):417–418

Du Q, Jerz G, Winterhalter P (2004) Isolation of two anthocyanin sambubiosides from bilberry (*Vaccinium myrtillus*) by high-speed counter-current chromatography. J Chromatogr A 1045(1–2):59–63

Dugo P, Mondello L, Errante G, Zappia G, Dugo G (2001) Identification of anthocyanins in berries by narrow-bore high-performance liquid chromatography with electrospray ionization detection. J Agric Food Chem 49(8):3987–3992

Ehala S, Vaher M, Kaljurand M (2005) Characterization of phenolic profiles of Northern European berries by capillary electrophoresis and determination of their antioxidant activity. J Agric Food Chem 53(16):6484–6490

Elisia I, Hu C, Popovich DG, Kitts DD (2007) Antioxidant assessment of an anthocyanin-enriched blackberry extract. Food Chem 101(3):1052–1058

El-Seedi HR, Salem MA, Khattab OM, El-Wahed AA, El-Kersh DM, Khalifa SAM, Saeed A, Abdel-Daim MM, Hajrah NH, Alajlani MM, Halabi MF, Jassbi AR, Musharraf SG, Farag MA (2020) Dietary xanthones. In: Xiao J, Sarker SD, Asakawa Y (eds) Handbook of dietary phytochemicals. Springer, Singapore, pp 1–22

Farag R, El-Baroty G, Basuny AM (2003) Safety evaluation of olive phenolic compounds as natural antioxidants. Int J Food Sci Nutr 54(3):159–174

Fernández-Jalao I, Sánchez-Moreno C, De Ancos B (2019) Effect of high-pressure processing on flavonoids, hydroxycinnamic acids, dihydrochalcones and antioxidant activity of apple 'Golden Delicious' from different geographical origin. Innovative Food Sci Emerg Technol 51:20–31

Flamini R, De Rosso M, De Marchi F, Dalla Vedova A, Panighel A, Gardiman M, Maoz I, Bavaresco L (2013) An innovative approach to grape metabolomics: stilbene profiling by suspect screening analysis. Metabolomics 9(6):1243–1253

Frankel EN, Finley JW (2008) How to standardize the multiplicity of methods to evaluate natural antioxidants. J Agric Food Chem 56(13):4901–4908

Frankel EN, German JB (2006) Antioxidants in foods and health: problems and fallacies in the field. J Sci Food Agric 86(13):1999–2001

Fredes C, Parada A, Salinas J, Robert P (2020) Phytochemicals and traditional use of two southernmost Chilean berry fruits: Murta (*Ugni molinae* Turcz) and Calafate (*Berberis buxifolia* Lam.). Foods 9(1):54

de Freitas VA, Glories Y, Bourgeois G, Vitry C (1998) Characterisation of oligomeric and polymeric procyanidins from grape seeds by liquid secondary ion mass spectrometry. Phytochemistry 49(5):1435–1441

García YD, Valles BS, Lobo AP (2009) Phenolic and antioxidant composition of by-products from the cider industry: apple pomace. Food Chem 117(4):731–738

García-Alonso MA, de Pascual-Teresa S, Santos-Buelga C, Rivas-Gonzalo JC (2004) Evaluation of the antioxidant properties of fruits. Food Chem 84(1):13–18

Gasperotti M, Masuero D, Guella G, Palmieri L, Martinatti P, Pojer E, Mattivi F, Vrhovsek U (2013) Evolution of ellagitannin content and profile during fruit ripening in Fragaria spp. J Agric Food Chem 61(36):8597–8607

Ge L, Zhang W, Zhou G, Ma B, Mo Q, Chen Y, Wang Y (2017) Nine phenylethanoid glycosides from Magnolia officinalis var. biloba fruits and their protective effects against free radical-induced oxidative damage. Sci Rep 7(1):45342

Gil-Izquierdo A, Gil MI, Ferreres F (2002) Effect of processing techniques at industrial scale on orange juice antioxidant and beneficial health compounds. J Agric Food Chem 50(18):5107–5114

Goodner KL, Jella P, Rouseff RL (2000) Determination of vanillin in orange, grapefruit, tangerine, lemon, and lime juices using GC–olfactometry and GC–MS/MS. J Agric Food Chem 48(7):2882–2886

Gregoire S, Singh A, Vorsa N, Koo H (2007) Influence of cranberry phenolics on glucan synthesis by glucosyltransferases and Streptococcus mutans acidogenicity. J Appl Microbiol 103(5):1960–1968

Gu C, Howell K, Dunshea FR, Suleria HAR (2019) LC-ESI-QTOF/MS Characterisation of phenolic acids and flavonoids in polyphenol-rich fruits and vegetables and their potential antioxidant activities. Antioxidants 8(9):405

Guerrero RF, Puertas B, Fernández MI, Piñeiro Z, Cantos-Villar E (2010a) UVC-treated skin-contact effect on both white wine quality and resveratrol content. Food Res Int 43(8):2179–2185

Guerrero RF, Puertas B, Fernández MI, Palma M, Cantos-Villar E (2010b) Induction of stilbenes in grapes by UV-C: Comparison of different subspecies of Vitis. Innov Food Sci Emerg Technol 11(1):231–238

Guerrero RF, Cantos-Villar E, Puertas B, Richard T (2016) Daily preharvest UV-C light maintains the high stilbenoid concentration in grapes. J Agric Food Chem 64(25):5139–5147

Guyot S, Marnet N, Laraba D, Sanoner P, Drilleau JF (1998) Reversed-phase HPLC following thiolysis for quantitative estimation and characterization of the four main classes of phenolic compounds in different tissue zones of a French cider apple variety (*Malus domestica* var. Kermerrien). J Agric Food Chem 46(5):1698–1705

Häkkinen S, Heinonen M, Kärenlampi S, Mykkänen H, Ruuskanen J, Törrönen R (1999a) Screening of selected flavonoids and phenolic acids in 19 berries. Food Res Int 32(5):345–353

Häkkinen SH, Kärenlampi SO, Heinonen IM, Mykkänen HM, Törrönen AR (1999b) Content of the flavonols quercetin, myricetin, and kaempferol in 25 edible berries. J Agric Food Chem 47(6):2274–2279

Halvorsen BL, Holte K, Myhrstad MC, Barikmo I, Hvattum E, Remberg SF, Wold A-B, Haffner K, Baugerød H, Andersen LF (2002) A systematic screening of total antioxidants in dietary plants. J Nutr 132(3):461–471

Haminiuk CW, Maciel GM, Plata-Oviedo MS, Peralta RM (2012) Phenolic compounds in fruits– an overview. Int J Food Sci Technol 47(10):2023–2044

Han ZP, Liu RL, Cui HY, Zhang ZQ (2013) Microwave-assisted extraction and Lc/Ms analysis of phenolic antioxidants in sweet apricot (*Prunus Armeniaca* L.) kernel skins. J Liq Chromatogr Relat Technol 36(15):2182–2195

Hasegawa S, Lam LK, Miller EG (1999) Citrus limonoids: biochemistry and possible importance to human nutrition. In: Phytochemicals and phytopharmaceuticals. AOCS Press, New York, pp 79–94

He S, Jiang L, Wu B, Li C, Pan Y (2009a) Chunganenol: an unusual antioxidative resveratrol hexamer from *Vitis chunganensis*. J Org Chem 74(20):7966–7969

He S, Lu Y, Jiang L, Wu B, Zhang F, Pan Y (2009b) Preparative isolation and purification of antioxidative stilbene oligomers from *Vitis chunganensis* using high-speed counter-current chromatography in stepwise elution mode. J Sep Sci 32(14):2339–2345

Hoang Anh L, Xuan TD, Dieu Thuy NT, Quan NV, Trang LT (2020) Antioxidant and α-amylase inhibitory activities and phytocompounds of *Clausena indica* Fruits. Medicines 7(3):10

Hu F, Liao X, Chen Z (2018) Determination of three phenylethanoid glycosides in Osmanthus fragrans fruits by high-performance liquid chromatography with fluorescence detection. J Sep Sci 41(21):3995–4000

Huber G, Rupasinghe H (2009) Phenolic profiles and antioxidant properties of apple skin extracts. J Food Sci 74(9):C693–C700

Ismail BB, Pu Y, Guo M, Ma X, Liu D (2019) LC-MS/QTOF identification of phytochemicals and the effects of solvents on phenolic constituents and antioxidant activity of baobab (*Adansonia digitata*) fruit pulp. Food Chem 277:279–288

Jadhav D, Rekha B, Gogate PR, Rathod VK (2009) Extraction of vanillin from vanilla pods: a comparison study of conventional soxhlet and ultrasound assisted extraction. J Food Eng 93(4):421–426

Jeandet P, Douillet-Breuil A-C, Bessis R, Debord S, Sbaghi M, Adrian M (2002) Phytoalexins from the Vitaceae: biosynthesis, phytoalexin gene expression in transgenic plants, antifungal activity, and metabolism. J Agric Food Chem 50(10):2731–2741

Jiang L, He S, Sun C, Pan Y (2012) Selective $1O_2$ quenchers, oligostilbenes, from *Vitis wilsonae*: structural identification and biogenic relationship. Phytochemistry 77:294–303

Kafkas E, Özgen M, Özoğul Y, Türemiş N (2008) Phytochemical and fatty acid profile of selected red raspberry cultivars: a comparative study. J Food Qual 31(1):67–78

Kanes K, Tisserat B, Berhow M, Vandercook C (1993) Phenolic composition of various tissues of Rutaceae species. Phytochemistry 32(4):967–974

Karković Marković A, Torić J, Barbarić M, Jakobušić Brala C (2019) Hydroxytyrosol, tyrosol and derivatives and their potential effects on human health. Molecules 24(10):2001

Kaunda JS, Zhang Y-J (2020) Chemical constituents from the fruits of *Solanum incanum* L. Biochem Syst Ecol 90:104031

Kawakami S, Kinoshita Y, Maruki-Uchida H, Yanae K, Sai M, Ito T (2014) Piceatannol and its metabolite, isorhapontigenin, induce SIRT1 expression in THP-1 human monocytic cell line. Nutrients 6(11):4794–4804

Khundamri N, Aouf C, Fulcrand H, Dubreucq E, Tanrattanakul V (2020) Condensed tannins in mangosteen pericarps determined from ultra-performance liquid chromatography-mass spectrometry. Songklanakarin J Sci Technol 42(1)

Kirakosyan A, Seymour EM, Noon KR, Llanes DEU, Kaufman PB, Warber SL, Bolling SF (2010) Interactions of antioxidants isolated from tart cherry (*Prunus cerasus*) fruits. Food Chem 122(1):78–83

Kiselev KV, Aleynova OA, Grigorchuk VP, Dubrovina AS (2017) Stilbene accumulation and expression of stilbene biosynthesis pathway genes in wild grapevine *Vitis amurensis* Rupr. Planta 245(1):151–159

Koradia P, Kapadia S, Trivedi Y, Chanchu G, Harper A (2019) Probiotic and cranberry supplementation for preventing recurrent uncomplicated urinary tract infections in premenopausal women: a controlled pilot study. Expert Rev Anti-Infect Ther 17(9):733–740

Kumar N, Goel N (2019) Phenolic acids: natural versatile molecules with promising therapeutic applications. Biotechnol Rep (Amst) 24:e00370

Lachowicz S, Oszmiański J, Seliga Ł, Pluta S (2017) Phytochemical composition and antioxidant capacity of seven saskatoon berry (*Amelanchier alnifolia* Nutt.) Genotypes Grown in Poland. Molecules 22(5):853

Lachowicz S, Oszmiański J, Wiśniewski R, Seliga Ł, Pluta S (2019) Chemical parameters profile analysis by liquid chromatography and antioxidative activity of the Saskatoon berry fruits and their components. Eur Food Res Technol 245(9):2007–2015

Lachowicz S, Seliga Ł, Pluta S (2020) Distribution of phytochemicals and antioxidative potency in fruit peel, flesh, and seeds of Saskatoon berry. Food Chem 305:125430

Lee HS (2002) Characterization of major anthocyanins and the color of red-fleshed budd blood orange (*Citrus sinensis*). J Agric Food Chem 50(5):1243–1246

Lee J, Wrolstad R (2004) Extraction of anthocyanins and polyphenolics from blueberry processing waste. J Food Sci 69(7):564–573

Lee J, Durst RW, Wrolstad RE (2005) Determination of total monomeric anthocyanin pigment content of fruit juices, beverages, natural colorants, and wines by the pH differential method: collaborative study. J AOAC Int 88(5):1269–1278

Lee J, Chan BL, Mitchell AE (2017) Identification/quantification of free and bound phenolic acids in peel and pulp of apples (Malus domestica) using high resolution mass spectrometry (HRMS). Food Chem 215:301–310

Li S, Pan M-H, Lo C-Y, Tan D, Wang Y, Shahidi F, Ho C-T (2009) Chemistry and health effects of polymethoxyflavones and hydroxylated polymethoxyflavones. J Funct Foods 1(1):2–12

Lim YY, Lim TT, Tee JJ (2007) Antioxidant properties of several tropical fruits: a comparative study. Food Chem 103(3):1003–1008

Lin L-C, Chou C-J, Chen K-T, Chen C-F (1993) Two new acetophenones from fruits of Evodia merrillii. J Nat Prod 56(6):926–928

Liu M, Li XQ, Weber C, Lee CY, Brown J, Liu RH (2002) Antioxidant and antiproliferative activities of raspberries. J Agric Food Chem 50(10):2926–2930

Liu C, Zheng H, Sheng K, Liu W, Zheng L (2018) Effects of postharvest UV-C irradiation on phenolic acids, flavonoids, and key phenylpropanoid pathway genes in tomato fruit. Sci Hort Amst 241:107–114

Lu Y, Foo LY (1999) The polyphenol constituents of grape pomace. Food Chem 65(1):1–8

Lu Y, Foo LY, Wong H (2002) Nigrumin-5-p-coumarate and nigrumin-5-ferulate, two unusual nitrile-containing metabolites from black currant (*Ribes nigrum*) seed. Phytochemistry 59(4):465–468

Lyles JT, Negrin A, Khan SI, He K, Kennelly EJ (2014) In vitro antiplasmodial activity of benzophenones and xanthones from edible fruits of garcinia species. Planta Med 80(08/09):676–681

Lyons MM, Yu C, Toma R, Cho SY, Reiboldt W, Lee J, van Breemen RB (2003) Resveratrol in raw and baked blueberries and bilberries. J Agric Food Chem 51(20):5867–5870

Ma Q-G, Wei R-R, Yang M, Huang X-Y, Wang F, Sang Z-P, Liu W-M, Yu Q (2018) Molecular characterization and bioactivity of coumarin derivatives from the fruits of *Cucumis bisexualis*. J Agric Food Chem 66(22):5540–5548

Ma C, Dunshea FR, Suleria HAR (2019) LC-ESI-QTOF/MS characterization of phenolic compounds in palm fruits (jelly and fishtail palm) and their potential antioxidant activities. Antioxidants 8(10):483

Maccarone E, Arena E, Rapisarda P, Fanella F, Mondello L (1998) Cyanidin-3-(6′-malonyl)-beta-glucoside. One of the major anthocyanins in blood orange juice. Ital J Food Sci 10:367–372

MacLeod AJ, de Troconis NG (1982) Volatile flavour components of mango fruit. Phytochemistry 21(10):2523–2526

Madhavi D, Bomser J, Smith M, Singletary K (1998) Isolation of bioactive constituents from *Vaccinium myrtillus* (bilberry) fruits and cell cultures. Plant Sci 131(1):95–103

Madrigal-Carballo S, Rodriguez G, Krueger C, Dreher M, Reed J (2009) Pomegranate (*Punica granatum*) supplements: authenticity, antioxidant and polyphenol composition. J Funct Foods 1(3):324–329

Maga JA (1978) Simple phenol and phenolic compounds in food flavor. Crit Rev Food Sci Nutr 10(4):323–372

Mallek-Ayadi S, Bahloul N, Kechaou N (2017) Characterization, phenolic compounds and functional properties of Cucumis melo L. peels. Food Chem 221:1691–1697

Mallmann LP, Tischer B, Vizzotto M, Rodrigues E, Manfroi V (2020) Comprehensive identification and quantification of unexploited phenolic compounds from red and yellow araçá (*Psidium cattleianum* Sabine) by LC-DAD-ESI-MS/MS. Food Res Int 131:108978

Marín FR, Del Río JA (2001) Selection of hybrids and edible Citrus species with a high content in the diosmin functional compound. Modulating effect of plant growth regulators on contents. J Agric Food Chem 49(7):3356–3362

Masullo M, Bassarello C, Suzuki H, Pizza C, Piacente S (2008) Polyisoprenylated benzophenones and an unusual polyisoprenylated tetracyclic xanthone from the fruits of *Garcinia cambogia*. J Agric Food Chem 56(13):5205–5210

Mattila P, Hellstrom J, Torronen R (2006) Phenolic acids in berries, fruits, and beverages. J Agric Food Chem 54(19):7193–7199

McGhie TK, Hunt M, Barnett LE (2005) Cultivar and growing region determine the antioxidant polyphenolic concentration and composition of apples grown in New Zealand. J Agric Food Chem 53(8):3065–3070

Memon AA, Memon N, Luthria DL, Bhanger MI, Pitafi AA (2010) Phenolic acids profiling and antioxidant potential of mulberry (*Morus laevigata* W., *Morus nigra* L., *Morus alba* L.) leaves and fruits grown in Pakistan. Pol J Food Nutr Sci 60(1):25–32

Mikkonen TP, Määttä KR, Hukkanen AT, Kokko HI, Törrönen AR, Kärenlampi SO, Karjalainen RO (2001) Flavonol content varies among black currant cultivars. J Agric Food Chem 49(7):3274–3277

Mikulic-Petkovsek M, Koron D, Rusjan D (2020) The impact of food processing on the phenolic content in products made from juneberry (*Amelanchier lamarckii*) fruits. J Food Sci 85(2):386–393

Miller RJ, Jackson KG, Dadd T, Nicol B, Dick JL, Mayes AE, Brown AL, Minihane AM (2012) A preliminary investigation of the impact of catechol-*O*-methyltransferase genotype on the absorption and metabolism of green tea catechins. Eur J Nutr 51(1):47–55

Moriartry JM, Harmon R, Weston LA, Bessis R, Breuil A-C, Adrian M, Jeandet P (2001) Resveratrol content of two Californian table grape cultivars. Vitis 40(1):43–44

Može S, Polak T, Gasperlin L, Koron D, Vanzo A, Poklar Ulrih N, Abram V (2011) Phenolics in Slovenian bilberries (Vaccinium myrtillus L.) and blueberries (*Vaccinium corymbosum* L.). J Agric Food Chem 59(13):6998–7004

Mulabagal V, Lang GA, DeWitt DL, Dalavoy SS, Nair MG (2009) Anthocyanin content, lipid peroxidation and cyclooxygenase enzyme inhibitory activities of sweet and sour cherries. J Agric Food Chem 57(4):1239–1246

Mullen W, Boitier A, Stewart AJ, Crozier A (2004) Flavonoid metabolites in human plasma and urine after the consumption of red onions: analysis by liquid chromatography with photodiode array and full scan tandem mass spectrometric detection. J Chromatogr A 1058(1–2):163–168

Naczk M, Shahidi F (2006) Phenolics in cereals, fruits and vegetables: occurrence, extraction and analysis. J Pharm Biomed Anal 41(5):1523–1542

Naczk M, Amarowicz R, Zadernowski R, Pegg R, Shahidi F (2005) Antioxidant capacity of phenolic extracts from wild blueberry leaves and fruits. Food Flavor Chem Explor 21 Cent (300):293–303

Nguyen L-HD, Venkatraman G, Sim K-Y, Harrison LJ (2005) Xanthones and benzophenones from *Garcinia griffithii* and *Garcinia mangostana*. Phytochemistry 66(14):1718–1723

Noda Y, Kaneyuki T, Mori A, Packer L (2002) Antioxidant activities of pomegranate fruit extract and its anthocyanidins: delphinidin, cyanidin, and pelargonidin. J Agric Food Chem 50(1):166–171

Nour V, Trandafir I, Cosmulescu S (2014) Antioxidant capacity, phenolic compounds and minerals content of blackcurrant (*Ribes nigrum* L.) leaves as influenced by harvesting date and extraction method. Ind Crop Prod 53:133–139

Nowicka P, Wojdyło A, Samoticha J (2016) Evaluation of phytochemicals, antioxidant capacity, and antidiabetic activity of novel smoothies from selected Prunus fruits. J Funct Foods 25:397–407

Nowicka A, Kucharska AZ, Sokół-Łętowska A, Fecka I (2019) Comparison of polyphenol content and antioxidant capacity of strawberry fruit from 90 cultivars of Fragaria×ananassa Duch. Food Chem 270:32–46

Obidoa O, Obasi SC (1991) Coumarin compounds in cassava diets: 2 health implications of scopoletin in gari. Plant Foods Hum Nutr 41(3):283–289

Orak HH (2007) Total antioxidant activities, phenolics, anthocyanins, polyphenoloxidase activities of selected red grape cultivars and their correlations. Sci Hort Amst 111(3):235–241

Owen RW, Mier W, Giacosa A, Hull WE, Spiegelhalder B, Bartsch H (2000) Phenolic compounds and squalene in olive oils: the concentration and antioxidant potential of total phenols, simple phenols, secoiridoids, lignansand squalene. Food Chem Toxicol 38(8):647–659

Palomino O, Gomez-Serranillos M, Slowing K, Carretero E, Villar A (2000) Study of polyphenols in grape berries by reversed-phase high-performance liquid chromatography. J Chromatogr A 870(1–2):449–451

Pannitteri C, Continella A, Cicero LL, Gentile A, La Malfa S, Sperlinga E, Napoli E, Strano T, Ruberto G, Siracusa L (2017) Influence of postharvest treatments on qualitative and chemical parameters of Tarocco blood orange fruits to be used for fresh chilled juice. Food Chem 230:441–447

Peng X-M, Damu GLV, Zhou H (2013) Current developments of coumarin compounds in medicinal chemistry. Curr Pharm Design 19(21):3884–3930

Peng D, Zahid HF, Ajlouni S, Dunshea FR, Suleria HAR (2019) LC-ESI-QTOF/MS profiling of Australian mango peel by-product polyphenols and their potential antioxidant activities. Processes 7(10):764

Piazza S, Sangiovanni E, Vrhovsek U, Fumagalli M, Khalilpour S, Masuero D, Colombo L, Mattivi F, Fabiani ED, Dell'Agli M (2019) Tannins from Chestnut (Castanea sativa Mill.) leaves and fruits show promising in vitro antiinflammatory properties in gastric epithelial cells. Planta Med 85(18):388

Piñeiro Z, Cantos-Villar E, Palma M, Puertas B (2011) Direct liquid chromatography method for the simultaneous quantification of hydroxytyrosol and tyrosol in red wines. J Agric Food Chem 59(21):11683–11689

Prior RL, Cao G, Martin A, Sofic E, McEwen J, O'Brien C, Lischner N, Ehlenfeldt M, Kalt W, Krewer G (1998) Antioxidant capacity as influenced by total phenolic and anthocyanin content, maturity, and variety of Vaccinium species. J Agric Food Chem 46(7):2686–2693

Quatrin A, Pauletto R, Maurer LH, Minuzzi N, Nichelle SM, Carvalho JFC, Maróstica MR, Rodrigues E, Bochi VC, Emanuelli T (2019) Characterization and quantification of tannins, flavonols, anthocyanins and matrix-bound polyphenols from jaboticaba fruit peel: a comparison between Myrciaria trunciflora and M. jaboticaba. J Food Compos Anal 78:59–74

Ragab AS, Van Fleet J, Jankowski B, Park J-H, Bobzin SC (2006) Detection and quantitation of resveratrol in tomato fruit (Lycopersicon esculentum Mill.). J Agric Food Chem 54(19):7175–7179

Raksat A, Phukhatmuen P, Yang J, Maneerat W, Charoensup R, Andersen RJ, Wang YA, Pyne SG, Laphookhieo S (2020) Phloroglucinol benzophenones and xanthones from the leaves of Garcinia cowa and their nitric oxide production and α-glucosidase inhibitory activities. J Nat Prod 83:164

Ramana KV, Reddy A, Majeti N, Singhal SS (2018) Therapeutic potential of natural antioxidants. Oxidative Med Cell Longev 2018:9471051

Ramful D, Bahorun T, Bourdon E, Tarnus E, Aruoma OI (2010) Bioactive phenolics and antioxidant propensity of flavedo extracts of Mauritian citrus fruits: potential prophylactic ingredients for functional foods application. Toxicology 278(1):75–87

Rao AV, Snyder DM (2010) Raspberries and human health: a review. J Agric Food Chem 58(7):3871–3883

Ribas-Agustí A, Martín-Belloso O, Soliva-Fortuny R, Elez-Martínez P (2019) Enhancing hydroxycinnamic acids and flavan-3-ol contents by pulsed electric fields without affecting quality attributes of apple. Food Res Int 121:433–440

Rimando AM, Cody R (2005) Determination of stilbenes in blueberries. LCGC N Am 23(11):1192–1200

Rosso MD, Soligo S, Panighel A, Carraro R, Vedova AD, Maoz I, Tomasi D, Flamini R (2016) Changes in grape polyphenols (V. vinifera L.) as a consequence of post-harvest withering by high-resolution mass spectrometry: Raboso Piave versus Corvina. J Mass Spectrom 51(9):750–760

Ruan J, Zheng C, Liu Y, Qu L, Yu H, Han L, Zhang Y, Wang T (2017) Chemical and biological research on herbal medicines rich in xanthones. Molecules 22(10):1698

Ruiz D, Egea J, Gil MI, Tomás-Barberán FA (2005) Characterization and quantitation of phenolic compounds in new apricot (Prunus armeniaca L.) varieties. J Agric Food Chem 53(24):9544–9552

Russell WR, Labat A, Scobbie L, Duncan GJ, Duthie GG (2009) Phenolic acid content of fruits commonly consumed and locally produced in Scotland. Food Chem 115(1):100–104

Samoticha J, Wojdyło A, Golis T (2017) Phenolic composition, physicochemical properties and antioxidant activity of interspecific hybrids of grapes growing in Poland. Food Chem 215:263–273

Sanchez-Ballesta MT, Alvarez I, Escribano MI, Merodio C, Romero I (2020) Effect of high CO_2 levels and low temperature on stilbene biosynthesis pathway gene expression and stilbenes production in white, red and black table grape cultivars during postharvest storage. Plant Physiol Biochem 151:334–341

Sanderson GW, Grahamm HN (1973) Formation of black tea aroma. J Agric Food Chem 21(4):576–585

Schuster B, Herrmann K (1985) Hydroxybenzoic and hydroxycinnamic acid derivatives in soft fruits. Phytochemistry 24(11):2761–2764

Seifried RM, Harrison E, Seifried HE (2017) Antioxidants in health and disease. In: Nutrition in the prevention and treatment of disease. Elsevier, New York, NY, pp 321–346

Sellappan S, Akoh CC, Krewer G (2002) Phenolic compounds and antioxidant capacity of Georgia-grown blueberries and blackberries. J Agric Food Chem 50(8):2432–2438

Serag A, Baky MH, Döll S, Farag MA (2020) UHPLC-MS metabolome based classification of umbelliferous fruit taxa: a prospect for phyto-equivalency of its different accessions and in response to roasting. RSC Adv 10(1):76–85

Shahidi F (2015) Handbook of antioxidants for food preservation. Woodhead Publishing, Cambridge

Shahidi F, Ambigaipalan P (2015) Phenolics and polyphenolics in foods, beverages and spices: antioxidant activity and health effects–a review. J Funct Foods 18:820–897

Shalan NAAM, Mustapha NM, Mohamed S (2017) Chronic toxicity evaluation of *Morinda citrifolia* fruit and leaf in mice. Regul Toxicol Pharmacol 83:46–53

Sharma A, Verma SC, Saxena N, Chadda N, Singh NP, Sinha AK (2006) Microwave- and ultra-sound-assisted extraction of vanillin and its quantification by high-performance liquid chromatography in *Vanilla planifolia*. J Sep Sci 29(5):613–619

Silva LMR, Figueiredo EAT, Ricardo NMPS, Vieira IGP, Figueiredo RW, Brasil IM, Gomes CL (2014) Quantification of bioactive compounds in pulps and by-products of tropical fruits from Brazil. Food Chem 143:398–404

Singh AP, Wilson T, Kalk AJ, Cheong J, Vorsa N (2009) Isolation of specific cranberry flavonoids for biological activity assessment. Food Chem 116(4):963–968

Sinha AK, Sharma UK, Sharma N (2008) A comprehensive review on vanilla flavor: extraction, isolation and quantification of vanillin and others constituents. Int J Food Sci Nutr 59(4):299–326

Siriwoharn T, Wrolstad RE, Finn CE, Pereira CB (2004) Influence of cultivar, maturity, and sampling on blackberry (Rubus L. Hybrids) anthocyanins, polyphenolics, and antioxidant properties. J Agric Food Chem 52(26):8021–8030

van der Sluis AA, Dekker M, Skrede G, Jongen WM (2002) Activity and concentration of polyphenolic antioxidants in apple juice. 1. Effect of existing production methods. J Agric Food Chem 50(25):7211–7219

Song H, Zhang L, Wu L, Huang W, Wang M, Zhang L, Shao Y, Wang M, Zhang F, Zhao Z, Mei X, Li T, Wang D, Liang Y, Li J, Xu T, Zhao Y, Zhong Y, Chen Q, Lu B (2020) Phenolic acid profiles of common food and estimated natural intake with different structures and forms in five regions of China. Food Chem 321:126675

Souquet J-M, Labarbe B, Le Guernevé C, Cheynier V, Moutounet M (2000) Phenolic composition of grape stems. J Agric Food Chem 48(4):1076–1080

Srivastava A, Greenspan P, Hartle DK, Hargrove JL, Amarowicz R, Pegg RB (2010) Antioxidant and anti-inflammatory activities of polyphenolics from southeastern US range blackberry cultivars. J Agric Food Chem 58(10):6102–6109

Stintzing FC, Stintzing AS, Carle R, Wrolstad RE (2002) A novel zwitterionic anthocyanin from evergreen blackberry (Rubus laciniatus Willd.). J Agric Food Chem 50(2):396–399

Stojanović BT, Mitić SS, Stojanović GS, Mitić MN, Kostić DA, Paunović DĐ, Arsić BB, Pavlović AN (2017) Phenolic profiles and metal ions analyses of pulp and peel of fruits and seeds of quince (*Cydonia oblonga Mill.*). Food Chem 232:466–475

Su Z, Wang P, Yuan W, Grant G, Li S (2017) Phenolics from the Fruits of *Maclura pomifera*. Nat Prod Commun 12(11):1934578X1701201122

Sultana B, Anwar F, Iqbal S (2008) Effect of different cooking methods on the antioxidant activity of some vegetables from Pakistan. Int J Food Sci Technol 43(3):560–567

Sun J, Chu Y-F, Wu X, Liu RH (2002) Antioxidant and antiproliferative activities of common fruits. J Agric Food Chem 50(25):7449–7454

Sun B, Ribes AM, Leandro MC, Belchior AP, Spranger MI (2006) Stilbenes: quantitative extraction from grape skins, contribution of grape solids to wine and variation during wine maturation. Anal Chim Acta 563(1–2):382–390

Sun Y, Tao W, Huang H, Ye X, Sun P (2019) Flavonoids, phenolic acids, carotenoids and antioxidant activity of fresh eating citrus fruits, using the coupled in vitro digestion and human intestinal HepG2 cells model. Food Chem 279:321–327

Taverne YJ, Merkus D, Bogers AJ, Halliwell B, Duncker DJ, Lyons TW (2018) Reactive oxygen species: radical factors in the evolution of animal life: a molecular timescale from Earth's earliest history to the rise of complex life. BioEssays 40(3)

Tian D, Wang F, Duan M, Cao L, Zhang Y, Yao X, Tang J (2019) Coumarin analogues from the *Citrus grandis* (L.) osbeck and their hepatoprotective activity. J Agric Food Chem 67(7):1937–1947

Tomás-Barberán FA, Clifford MN (2000) Dietary hydroxybenzoic acid derivatives–nature, occurrence and dietary burden. J Sci Food Agric 80(7):1024–1032

Tsanova-Savova S, Ribarova F, Petkov V (2018) Quercetin content and ratios to total flavonols and total flavonoids in Bulgarian fruits and vegetables. Bulg Chem Commun 50(1):69–73

Tsao R, Yang R, Xie S, Sockovie E, Khanizadeh S (2005) Which polyphenolic compounds contribute to the total antioxidant activities of apple? J Agric Food Chem 53(12):4989–4995

Tzulker R, Glazer I, Bar-Ilan I, Holland D, Aviram M, Amir R (2007) Antioxidant activity, polyphenol content, and related compounds in different fruit juices and homogenates prepared from 29 different pomegranate accessions. J Agric Food Chem 55(23):9559–9570

Valentová K, Ulrichová J, Cvak L, Šimánek V (2007) Cytoprotective effect of a bilberry extract against oxidative damage of rat hepatocytes. Food Chem 101(3):912–917

Vilanova M, Rodríguez I, Canosa P, Otero I, Gamero E, Moreno D, Talaverano I, Valdés E (2015) Variability in chemical composition of Vitis vinifera cv Mencía from different geographic areas and vintages in Ribeira Sacra (NW Spain). Food Chem 169:187–196

Viñas P, Martínez-Castillo N, Campillo N, Hernández-Córdoba M (2011) Directly suspended droplet microextraction with in injection-port derivatization coupled to gas chromatography–mass spectrometry for the analysis of polyphenols in herbal infusions, fruits and functional foods. J Chromatogr A 1218(5):639–646

Vincenzi S, Tomasi D, Gaiotti F, Lovat L, Giacosa S, Torchio F, Segade SR, Rolle L (2013) Comparative study of the resveratrol content of twenty-one Italian red grape varieties. South Afr J Enol Vitic 34(1):30–35

Vinson JA, Hontz BA (1995) Phenol antioxidant index: comparative antioxidant effectiveness of red and white wines. J Agric Food Chem 43(2):401–403

Vinson JA, Su X, Zubik L, Bose P (2001) Phenol antioxidant quantity and quality in foods: fruits. J Agric Food Chem 49(11):5315–5321

Vinson JA, Bose P, Proch J, Al Kharrat H, Samman N (2008) Cranberries and cranberry products: powerful in vitro, ex vivo, and in vivo sources of antioxidants. J Agric Food Chem 56(14):5884–5891

Vrhovsek U, Masuero D, Gasperotti M, Franceschi P, Caputi L, Viola R, Mattivi F (2012) A versatile targeted metabolomics method for the rapid quantification of multiple classes of phenolics in fruits and beverages. J Agric Food Chem 60(36):8831–8840

Vuorinen H, Määttä K, Törrönen R (2000) Content of the flavonols myricetin, quercetin, and kaempferol in Finnish berry wines. J Agric Food Chem 48(7):2675–2680

Wada L, Ou B (2002) Antioxidant activity and phenolic content of Oregon caneberries. J Agric Food Chem 50(12):3495–3500

Wang H, Cao G, Prior RL (1996) Total antioxidant capacity of fruits. J Agric Food Chem 44(3):701–705

Wang H, Cao G, Prior RL (1997) Oxygen radical absorbing capacity of anthocyanins. J Agric Food Chem 45(2):304–309

Wang M, Wang Q, Yang Q, Yan X, Feng S, Wang Z (2019) Comparison of anthraquinones, iridoid glycosides and triterpenoids in *Morinda officinalis* and *Morinda citrifolia* using UPLC/Q-TOF-MS and multivariate statistical analysis. Molecules 25(1):160

Wolfe K, Wu X, Liu RH (2003) Antioxidant activity of apple peels. J Agric Food Chem 51(3):609–614

Xiao J, Capanoglu E, Jassbi AR, Miron A (2016) Advance on the Flavonoid C-glycosides and Health Benefits. Crit Rev Food Sci Nutr 56(suppl 1):S29–S45

Xu G, Feng C, Zhou Y, Han Q-B, Qiao C-F, Huang S-X, Chang DC, Zhao Q-S, Luo KQ, Xu H-X (2008) Bioassay and ultraperformance liquid chromatography/mass spectrometry guided isolation of apoptosis-inducing benzophenones and xanthone from the pericarp of *Garcinia yunnanensis* Hu. J Agric Food Chem 56(23):11144–11150

Xu X, Huang Y, Xu J, He X, Wang Y (2020) Anti-neuroinflammatory and antioxidant phenols from mulberry fruit (*Morus alba L.*). J Funct Foods 68:103914

Yang J, Martinson TE, Liu RH (2009a) Phytochemical profiles and antioxidant activities of wine grapes. Food Chem 116(1):332–339

Yang Z, Kinoshita T, Tanida A, Sayama H, Morita A, Watanabe N (2009b) Analysis of coumarin and its glycosidically bound precursor in Japanese green tea having sweet-herbaceous odour. Food Chem 114(1):289–294

Yang H, Figueroa M, To S, Baggett S, Jiang B, Basile MJ, Weinstein IB, Kennelly EJ (2010) Benzophenones and biflavonoids from *Garcinia livingstonei fruits*. J Agric Food Chem 58(8):4749–4755

Zadernowski R, Naczk M, Nesterowicz J (2005) Phenolic acid profiles in some small berries. J Agric Food Chem 53(6):2118–2124

Zadernowski R, Czaplicki S, Naczk M (2009) Phenolic acid profiles of mangosteen fruits (*Garcinia mangostana*). Food Chem 112(3):685–689

Zeb A (2016) Phenolic profile and antioxidant activity of melon (*Cucumis Melo L.*) seeds from Pakistan. Foods 5(4):67

Zhu J, Wang L, Xiao Z, Niu Y (2018) Characterization of the key aroma compounds in mulberry fruits by application of gas chromatography–olfactometry (GC-O), odor activity value (OAV), gas chromatography-mass spectrometry (GC–MS) and flame photometric detection (FPD). Food Chem 245:775–785

Zuo Y, Wang C, Zhan J (2002) Separation, characterization, and quantitation of benzoic and phenolic antioxidants in American cranberry fruit by GC–MS. J Agric Food Chem 50(13):3789–3794

Chapter 4
Phenolic Antioxidants in Vegetables

Learning Outcomes

After reading this chapter, the reader will be able to:

- Have detailed knowledge of different types of hydroxybenzoic acids present in vegetables.
- Understand the different types of cinnamoyl derivatives present in vegetables.
- Discuss the composition of coumarins and chromones present in vegetables.
- Describe flavonoids, anthocyanins, and their derivatives present in vegetables.

4.1 Introduction

The consumption of vegetables is associated with a reduced risk of certain chronic diseases as evident from the epidemiological studies. The increase in the consumption of vegetables having high levels of phenolic antioxidants has been recommended to prevent chronic diseases in the human body induced by oxidative stress (Chu et al. 2002). In vegetables, phenolic compounds are present in both free as well as bound forms. Bound phenolic compounds are in the form of β-glycosides. During digestion, these glycosides may survive the acidic environment of the human stomach and then in small intestine digestion. In the colon, they are released and exert their beneficial effects (Shahidi and Yeo 2016). Chu et al. (2002, 2000) showed that vegetables contain high amounts (more than 60%) of free phenolic compounds as compared to bound phenolic compounds. The authors showed that Broccoli contained the highest total phenolic content, followed by spinach, yellow onion, red pepper, carrot, cabbage, potato, lettuce, celery, and cucumber. The total antioxidant

© Springer Nature Switzerland AG 2021   131
A. Zeb, *Phenolic Antioxidants in Foods: Chemistry, Biochemistry and Analysis*,
https://doi.org/10.1007/978-3-030-74768-8_4

activity of Red pepper was highest, followed by broccoli, carrot, spinach, cabbage, yellow onion, celery, potato, lettuce, and cucumber.

The antioxidant activities and total phenolic contents of 56 commonly consumed vegetables revealed that the highest antioxidant activities and phenolic contents were found in Chinese toon bud, loosestrife, perilla leaf, cowpea, caraway, lotus root, sweet potato leaf, soybean (green), pepper leaf, ginseng leaf, chives, and broccoli, whilst the values were very low in marrow squash and eggplant (purple) (Deng et al. 2013). Morales et al. (2012) studied the antioxidant properties of wild and cultivated Spanish vegetables. Among the leafy vegetables, *Silene vulgaris* and *Apium nodiflorum* showed the highest antioxidant activity and antioxidants composition. In the wild asparagus, the highest antioxidant activity was obtained in *Humulus lupulus*. Similarly, 11 vegetables from India were screened for antioxidant potential, with *Ipomoea reptans* has good activity amongst all (Dasgupta and De 2007). Vegetables are thus contributing toward antioxidant capacity that is extensively studied. The phenolic compounds reported so far in the vegetables are given below.

4.2 Phenolic Antioxidants in Vegetables

4.2.1 Simple Phenols

Simple phenols have been a rare focus of research in vegetables. In vegetables, the total phenolic contents are measured as gallic acid equivalents. The total phenolic contents in vegetables have shown a strong correlation with the antioxidant activity such as ABTS and DPPH activities (Khanam et al. 2012; Branca et al. 2018). Beetroot peel had the second-highest amounts of total phenols on a dry weight basis (Kähkönen et al. 1999). Phloroglucinol had been found in seaweeds (Pádua et al. 2015). It has shown strong effects along with carotenoids on certain cancers. Recently, 1,2-dihydroxybenzene has been reported in the range of 95.5–234.0 mg/kg in onions (Ozcan et al. 2018).

4.2.2 Hydroxybenzoic Acids and Aldehydes

Hydroxybenzoic acids and their derivatives have been found in nearly all vegetables. Tomás-Barberán and Clifford (2000) reviewed the dietary hydroxybenzoic acids, their occurrence in foods. The authors pointed out the need for comprehensive compositional data. Since then, several studies had reported the composition of hydroxybenzoic acids in a variety of foods including vegetables. Watercress had been found rich in gallic acid derivatives, hydroxybenzoic acid, and sinapic acid (Zeb 2015). A study of seven Chinese vegetables showed the presence of free phenolic acids like hydroxybenzoic acids, vanillic acid, p-hydroxybenzoic acid, and protocatechuic acid (Gao et al. 2017). Similarly, another study on 44 Chinese vegetables namely zucchini, pumpkin, wax gourd, towel gourd, cucumber, chili pepper, pepper, green pepper, tomato, eggplant, broccoli, cauliflower, day lily, soybean sprout, pea, green pea, hyacinth bean, snow pea, cowpea, phaseolus vulgaris, crown daisy chrysanthemum, lettuce, edible amaranth, purple cabbage, garlic, onion, Chinese chives, scallion, caraway, spring rape, leaf lettuce, celery, spinach, baby cabbage, Chinese cabbage, asparagus lettuce, bamboo shoots, cane shoots, fresh ginger, lotus root, carrot, white radich, radish, and green radish. The results showed that gallic acid, 3,5-dihydroxybenzoic acid, p-hydroxybenzoic acid, 2,3,4-trihydroxybenzoic acid, gentisic acid, vanillic acid, protocatechuic acid were presented by Zhang et al. (2019). Hydroxybenzoic acid was present in forty vegetables, and 3,5-dihydroxybenzoic acid in 24 vegetables. In nearly all of the selected vegetables, free phenolic acids were higher as compared to bound phenolic acids as shown in Fig. 4.1.

A recent study by Song et al. (2020) revealed 116 food samples from five regions of China. The results showed that fruit vegetables has total phenolic acids as 1.6 mg/kg, stem vegetables (4.9 mg/kg), root vegetables (5.0 mg/kg), leafy vegetables (61.6 mg/kg), flower vegetables (922.7 mg/kg), bean vegetables (986.4 mg/kg), yam (98.8 mg/kg), potato (145.2 mg/kg), sweet potato (300.5 mg/kg) (Song et al. 2020) as shown in Fig. 4.2. Hydroxybenzoic acid, p-hydroxybenzaldehyde, vanillin, and protocatechuic acid were reported to be present in the Chinese wild rice (*Zizania latifolia*) during different stages of germination (Chu et al. 2020). The amounts of these phenolic compounds have been found to increase significantly during germination.

In fresh frozen leafy vegetables such as Komatsuna, Mizuna, Pok Choi, Mitsuba, spinach, and lettuce from Japan, several hydroxybenzoic acids were reported. These include gallic acid (none-3.16 mg/kg), vanillic acid (6.6–8.43 mg/kg), syringic acid (none-7.7 mg/kg), p-hydroxybenzoic acid (none-3.6 mg/kg), and salicylic acid (none-117.3 mg/kg) (Khanam et al. 2012). Another study from Australia showed the presence of protocatechuic acid in garlic and ginger (Gu et al. 2019). Onions from Turkey showed gallic acid, 3,4-dihydroxybenzoic acid, and syringic acid (Ozcan et al. 2018).

4.2.3   Cinnamic Acids, Aldehydes, Esters, and Alcohols

Cinnamoyl derivatives such as hydroxycinnamic acids, cinnamaldehydes, esters, and cinnamoyl alcohols are one of the major classes of phenolic antioxidants present in all vegetables. Mattila and Hellström (2007) reported phenolic acids contents in 14 potatoes and 45 other vegetables. The derivatives of chlorogenic acid were major soluble phenolic acids, whereas the most dominant aglycone was caffeic acid in all vegetables. Among the vegetables, brassica vegetables are well-known for their special taste and composition. These include a different genus of cabbage

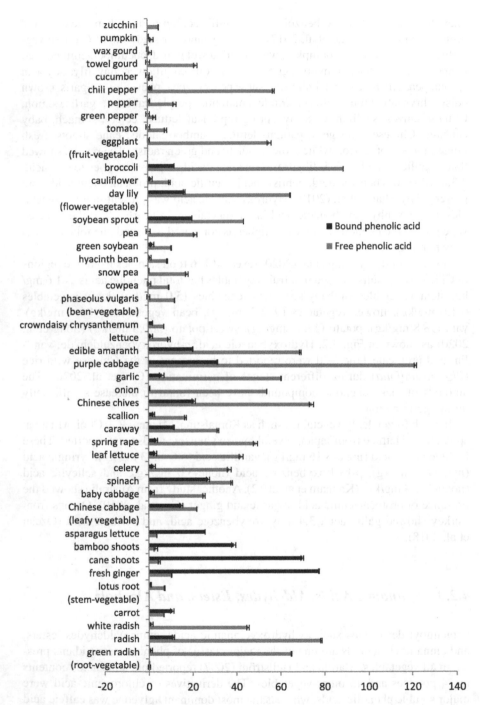

Fig. 4.1 Concentrations and comparisons of free and bound phenolic acids in six categories of vegetables (µg/g DW). (Reproduced with kind permission of Elsevier, Zhang et al. 2019)

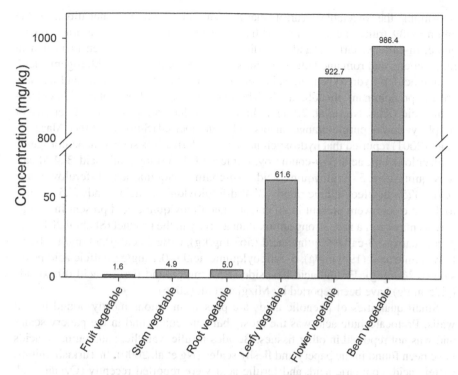

Fig. 4.2 Phenolic acids in different in six categories of vegetables. (Data has been converted to mg/kg from the source, which has been reproduced with the kind permission of Elsevier, Song et al. 2020)

(white, red, savoy, swamp, Chinese), broccoli, cauliflower, Brussels sprouts, and kale, which are consumed all over the world (Podsedek et al. 2006). Broccoli is a source of flavonol and hydroxycinnamoyl derivatives. In broccoli, 1-sinapoyl-2-feruloyl-gentiobiose, 1,2-di-feruloyl gentiobiose, 1,2,2′-trisinapoylgentiobiose, and neochlorogenic acid were identified as the predominant hydroxycinnamoyl acids (Soengas et al. 2009; Hosny et al. 2018). In addition, 1,2′-disinapoyl-2-feruloylgentiobiose and 1,2-disinapoylgentiobiose, 1-sinapoyl-2,2-diferuloyl gentiobiose, isomers of 1,2,2′-trisinapoylgentiobiose, and chlorogenic acid were found in broccoli (Fernández-León et al. 2012). Due to the presence of high amounts of phenolic compounds and vitamin C in brassica vegetables (kale, broccoli, cabbage, Brussels sprouts, and cauliflower), they showed a considerable antioxidant and medicinal activity (Idrees et al. 2019). The highest antioxidant activity was reported for kale, which was attributed to its carotenoids composition (Sikora et al. 2012). After the kale with the highest antioxidant activity against hydroxyl radicals, Brussels sprouts are the second vegetable followed by alfalfa sprouts, beets, spinach, and broccoli flowers (Cao et al. 1996). A recent study of the leaves of 14 amaranth genotypes showed that caffeoylaldaric and isocitric acid esters were the most abundant hydroxycinnamic acid derivatives (Schröter et al. 2018).

Among the vegetables, carrots have been placed tenth in nutritional value amongst 39 fruits and vegetables (Sharma et al. 2012). Caffeoyltartaric, chlorogenic, di-caffeoyl tartaric, and 3',5'-dicaffeoylquinic acids had been identified in red, iceberg, and romaine lettuce (Cantos et al. 2001). In carrots, chlorogenic acid, caffeic acid, p-hydroxybenzoic acid, ferulic acid, and other cinnamic acid isomers are the predominant phenolic acids. The main amongst all phenolic acids chlorogenic acid (Alasalvar et al. 2001). A high antioxidant composition and activity of purple-yellow or purple-orange carrots had been reported (Sun et al. 2009). Alasalvar et al. (2001) reported that hydroxycinnamic acid derivatives such as neochlorogenic acid, chlorogenic acid, 3'-p-coumaroylquinic acid, 3'-feruloyquinic acid, 3',4'-dicaffeoylquinic acid, 5'-feruloyquinic acid, 5'-p-coumaroylquinic acid, 4'-feruloylquinic acid, 3',5'-dicaffeoylquinic acid, 3',4'-diferuloylquinic acid, and 3',5'-diferuloylquinic acid were present in all carrots. Dicaffeoylquinic acid present in orange carrots may exert a very strong antioxidant activity in the product (Shahidi 2015). In purple carrots, 3-caffeoylquinic acid (567 mg/kg), caffeic acid (0.53 mg/g), 4-caffeoylquinic acid (19.9 mg/g), 5-caffeoylquinic acid (80.1 mg/g), caffeic acid hexoside (0.76 mg/g), Ferulic acid hexoside (5.51 mg/g), and Ferulic acid dihexoside (5.25 mg/g) have been reported by Mizgier et al. (2016).

Small quantities of phenolic acids are present in onions mostly bound to cell walls. Protocatechuic acid was the most abundant compound in the papery scales and was not reported in other tissues. Besides ferulic, vanillic, and coumaric acids have been found in the papery and fleshy scales (Ng et al. 2000). In Turkish onions, caffeic acid, coumaric acid, and ferulic acid were reported recently (Ozcan et al. 2018). Another member of the onion family (*Allium tripedale*) also contains several cinnamic acids and their derivatives (Chehri et al. 2018).

In potato tubers, chlorogenic acid constituting up to 90% of its total phenolic content. In the potato peel, 50% of the phenolic compounds are present and the adjoining tissues (Friedman 1997). Chlorogenic acid, gallic acid, protocatechuic acid, caffeic acid, and quercetin had been identified in potato peel (Mohdaly et al. 2013). Other phenolic compounds in potatoes include ferulic acid, p-coumaric acid as well as small amounts of flavonoids (Nara et al. 2006).

In sweet potatoes, chlorogenic, isochlorogenic, and cinnamic acids, had been identified amongst other phenolic compounds by Oki et al. (2002). Similarly, Philpott et al. (2003) reported that hydroxycinnamic acids were the major antioxidant components of sweet potatoes. The high antioxidative capacity of the sweet potato extracts was attributed to the presence of phenolic acids, including chlorogenic, isochlorogenic, 4-caffeoylquinic, and neochlorogenic acids (Rumbaoa et al. 2009). In the peel and flesh of the potato, three caffeoylquinic acid isomers such as caffeic acid, 3,5-dicaffeoylquinic acid, and N-[2-hydroxy-2-(4-hydroxyphenyl) ethyl] ferulamide had been identified as the main phenolic compounds (Lopez-Cobo et al. 2014).

The sweet potato cultivars with purple flesh had higher antioxidant activity than those with white, yellow, or orange flesh (Furuta et al. 1998). Another study showed a better radical scavenging activity, reducing power, and oxidation inhibition than α-tocopherol and iron-chelating capacity than EDTA (ethylenediaminetetraacetic acid), which makes sweet potato a viable alternative source for antioxidants (Rumbaoa et al. 2009). In the extract, sweet potato amino acids may have synergistic antioxidant effects with phenolic compounds (Shih et al. 2009). In leaves of sweet potato, several cinnamic acid derivatives had been reported with composition order to 3,5-di-caffeoylquinic acid > 4,5-di-caffeoylquinic acid > chlorogenic acid > 3,4-di-O-caffeoylquinic acid > 3,4,5-tri-caffeoylquinic acid > caffeic acid. Among these, 3,4,5-tri-caffeoylquinic acid and 4,5-di-caffeoylquinic acid were the major compounds in sweet potato (Islam et al. 2002). The leaves of different varieties of sweet potato had also shown strong antioxidant activity (Hue et al. 2012). Coumaric acid, caftaric acid, and caffeoymalic acid had been reported in watercress leave (Zeb 2015). In another study, caffeic acid (0.64 mg/kg), caffeoyl hexose (0.65 mg/kg), 3-caffeoylquinic acid (3.22 mg/kg), caftaric acid (17.3 mg/kg), coumaroyl derivatives (1.46 mg/kg), 5-caffeoylquinic acid (6.51 mg/kg), 4-caffeoylquinic acid (12.8 mg/kg), 4-feruloylquinic acid (13.4 mg/kg), 3,4-dicaffeoylquinic acid (3.12 mg/kg), and caffeoyl feruloylquinic acid (11.7 mg/kg) had been reported in watercress leave (Zeb and Habib 2018).

Beetroot is a commonly used vegetable. In the peels of beetroot, the key compound identified was ferulic acid ester fructofuranosyl-(6-E-feruloylglucopyranoside) (Kujala et al. 2000). Besides, p-hydroxybenzoic, cis- and trans-ferulic, trans-coumaric, and vanillic acids, as well as p-hydroxybenzaldehyde and vanillin have been reported in beetroot (Chhikara et al. 2019). In Detroit Dark Red beetroots, 4-hydroxybenzoic acid (12 mg/kg), chlorogenic acid (18 mg/kg), caffeic acid (37 mg/kg), and catechin hydrate (47 mg/kg), whereas extract of hairy root cultures contains 4-hydroxybenzoic acid (39.6 mg/kg), caffeic acid (20.3 mg/kg), and catechin hydrate (37.2 mg/kg) (Georgiev et al. 2010). The Detroit beetroot pomace extract had been shown to contain ferulic acid (1325.2 mg/kg), vanillic acid (51.2 mg/kg), p-hydroxybenzoic acid (11.3 mg/kg), caffeic acid (71.1 mg/kg), Protocatechuic acid (54.2 mg), and catechin (379.6 mg/kg) (Chhikara et al. 2019). When conventional and organic farming is considered, the hydroxycinnamic acids such as chlorogenic acid, caffeic acid, and Coumaric acid, are higher in conventionally farmed beetroots as compared to organic farmed (Carrillo et al. 2019).

4.2.4 Coumarins and Chromones

In vegetables, coumarins are widely distributed among the *Umbelliferae* family, which includes carrots, coriander, celery, fennel, and parsley (Cherng et al. 2008; Zobel 1997). Among the Coumarins, apigravin, apimentin, apimoside, bergapten, celerin, isopimpinellin, and xanthoxin have been reported in these vegetables (Lobo et al. 2018). Coumarins such as daphnetin, scopoletin, and umbelliferone have been also found in peas, whereas aesculetin, scopoletin, and umbelliferone in maize, and aesculin, scopoletin, and umbelliferone in potatoes and sweet potatoes (Mäder et al. 2009). Simple coumarins i.e., esculetin and scopoletin is present in carrots, but in some cases, the roots seem to be free of the furanocoumarins (Mercier et al. 1993). Coumarins have been reported in vegetables like soybean and broccoli (Sarker and Nahar 2019). In carrot tissues, 6-methoxymellein and 6-hydroxymellein were the major coumarins (Naczk and Shahidi 2006). These Coumarins had been found to mainly gather in the periderm tissue of carrots (Talcott and Howard 1999).

4.2.5 Flavonoids

Vegetables are rich sources of flavonoids. They are present in free and bound form in vegetables and are affected by different processing conditions (Ahmed and Eun 2018). In vegetables like broccoli florets, glycosides of quercetin and kaempferol had been reported (Price et al. 1998). Spinach leaves also contain quercetin and myricetin as major flavonoids (Chu et al. 2000). Another study by Dehkharghanian et al. (2010) showed quercetin, kaempferol, myricetin, apigenin, and luteolin in fresh leaves of spinach. The flavonoids composition such as myricetin, quercetin, and kaempferol in nine vegetables were studied. Myricetin was higher in cauli-flower, spinach, and turnip, whereas kaempferol was highest in spinach amongst the 9 vegetables analyzed (Sultana and Anwar 2008). Similarly, glycosides of isoquer-citrin, kaempferol-3-glucoside, and kaempferol diglucoside were also reported in broccoli (Bors et al. 2014). In another study, Nielsen et al. (1993) reported the presence of derivatives of kaempferol and quercetin in the form of 3-sophoroside-7-glucosides amongst other phenolic compounds in cabbage leaves. The authors also reported the acylated esters of the aglycone with either sinapic, ferulic, or caffeic acid and unmodified kaempferol tetraglucosides. Several derivatives of kaempferol and myricetin have been reported in Brassica vegetables. However, the review by Podsędek (2007) showed that myricetin was not reported in broccoli, white cabbage, purple cabbage, and cauliflower. The hydrolyzed extracts of different Brassica vegetables, except for broccoli showed the apigenin and luteolin as flavonoids (Bahorun et al. 2004). Apigenin and quercetin were the major aglycones in several vegetables, the highest amount the flavonoids were of myricetin and quercetin (Shahidi and Ambigaipalan 2015; Chu et al. 2000). Quercetin-3-(cafferoyldiglucoside)-7-glucoside, and kaempferol-3-(caffeoyldiglucoside)-7-rhamnoside had been reported in watercress leaves (Zeb 2015).

In *Spinacia oleracea* leaves, three important flavonoids glycosides had been reported namely jaceidin 4'-glucuronide; 5,3',4'-trihydroxy-3-methoxy-6:7-methylenedioxyflavone 4'-glucuronide and 5,4'-dihydroxy-3,3'-dimethoxy-6:7-methylenedioxyflavone 4'-glucuronide (Aritomi and Kawasaki 1984). Similarly, from spinach leaves extracts, Ferreres et al. (1997) isolated and identified flavonoids namely spinacetin-3-glucopyranosyl(1 → 6)-[apiofuranosyl(1 → 2)]-glucopyranoside, patuletin-3-(2'feruloylglucopyranosyl)(1 → 6)-[apiofuranosyl(1 → 2)]-glucopyranoside, spinacetin-3-(2'-p-coumaroylglucopyranosyl)(1 → 6)-[apiofurano-syl(1 → 2)]-glucopyranoside, spinacetin-3-(2'feruloylglucopyranosyl)(1 → 6)-[apio-furanosyl(1 → 2)]-glucopyranoside and spinacetin-3-(2'feruloylglucopyranosyl)

$(1 \rightarrow 6)$-glucopyranoside. The antioxidants isolated in purified forms from spinach leaves were water-soluble having glucorinated flavonoid structures that showed strong antioxidant capacities (Bergman et al. 2003). Apigenin, kaempferol, and their glyco-sides were recently reported to be present in spinach leaves extract (Zeb and Ullah 2019). The spinach extract rich in phenolic compounds was also found to extend the thermal stability of sunflower oil during frying. Similarly, Lee et al. (2002) showed that spinach powder minimizes the lipid peroxidation in the fried products. When feru-lic acid, caffeic acid, and epigallocatechin-3-gallate were combined with the antioxi-dants from spinach leaves, the synergistic actions reduced the generation of reactive oxygen species (Hait-Darshan et al. 2009; Zeb 2019a).

Lettuce is another good source of important flavonoids. In green leafy lettuce, quercetin-3-galactoside, quercetin-3-glucoside, quercetin-3-glucuronide, querce-tin-3-(6-malonyl)glucoside, and quercetin-3-rhamnoside, and luteolin-7-glucuro-nide were reported along with other phenolic compounds detected in green-leaf lettuce (Pérez-López et al. 2014). Similarly, a higher levels of flavonoids had been reported in red lettuce than green lettuce varieties (Crozier et al. 2009; Naczk and Shahidi 2006).

In different salads vegetables, like *Lactuca sativa, Cichorium intybus, Plantago coronopus, Eruca sativa*, and *Diplotaxis tenuifolia*, flavonoids such as quercetin, kaempferol, luteolin, apigenin, and chrysoeriol were reported (Heimler et al. 2007). Among these vegetables, *C. intybus* has a higher antioxidant activity as compared to other vegetables studied. Recently, in chicory samples from different origins, flavo-noids glycosides such as quercetin-3-feruloylsophoroside, quercetin-3-glucoside, quercetin-di-glucoside,   kaempferol-3-sophoroside,   luteolin-7-glucoside, quercetin-3-rutinoside, and kaempferol-3-glucoside were identified (Zeb 2019b). The composition of these phenolic compounds has been reported to be affected by the source of irrigation waters (Zeb 2019b). The edible leaves of *Oxalis corniculata* have been found to contain kaempferol-3-(p-coumaroyldiglucoside)-7-glucoside, kaempferol-3-glucuronide,   quercetin-3-(caffeoyldiglucoside)-7-glucoside,   and kaempferol-3-sophorotioside (Zeb and Imran 2019). These flavonoids are affected by climatic conditions.

Tomato is an important source of different antioxidants such as ascorbic acid, carotenoids, flavonoids, and phenolic compounds. Tomato and its products are rich in flavonoids and possess antioxidant and health benefits (Zeb and Haq 2016). In tomatoes, conjugates of quercetin especially quercetin-3-rhamnosylglucoside are the major flavonols, whereas kaempferol conjugates are present in lower amounts (Crozier et al. 1997). Kaempferol, quercetin, Myricetin, and quercetin-3-rhamnosylglucoside were identified in tomato using micellar electrokinetic capil-lary electrophoresis (Marti et al. 2017). Silva Souza et al. (2020) showed that purple tomato has been shown to contain kaempferol and quercetin-3-rhamnosylglucoside as compared to orange tomato. Recently Bakir et al. (2020) studied 50 landraces of tomato from Turkey. Among the flavonoid glycosides, quercetin-3-rutinoside, quercetin-hexose-deoxyhexose, quercetin-3-rutinoside-7-glucoside, kaempferol-3-rutinoside have been reported. Valdez-Morales et al. (2014) reported antioxidant and antimutagenic activities of the peel and seeds of different tomato types such as grape, cherry, bola, and saladette.

Among the vegetables, onion is one of the richest sources of flavonoids (Slimestad et al. 2007). Onions contain quercetin, kaempferol, myricetin, and catechin, which are responsible for high antioxidant activity (Perez-Gregorio et al. 2014). Rhodes and Price (1996) reported that 80% of the flavonoids in onions were only monoglucoside and diglucoside derivatives of quercetin. However, as compared to other vegetables, the amounts of quercetin glucosides in onions are much higher (Shahidi and Naczk 2003). Higher amounts of quercetin had been reported in yellow onions than red onions, whereas the lowest amounts were in pink and white onions (Patil et al. 1995). However, Gokce et al. (2010) suggested that antioxidant activities of red onions were higher than yellow and white onions though yellow onions had higher phenolic contents. In onion, the major flavonoids are quercetin, quercetin-3-glucoside, quercetin-4′-glucoside, and quercetin-3,4′-di-glucoside. Among these, quercetin-4′-glucoside contributed to the highest antioxidant capacity (Zielinska et al. 2008).

When antioxidant activities of extract of onions were compared with garlic extract, higher activity was reported for onions than garlic, whereas red onions had higher activity than yellow onions (Nuutila et al. 2003). Similarly, Albishi et al. (2013) also reported that skins of pearl, red, yellow, and white onions contain six times higher phenolic contents and antioxidant activities as compared to that of their flesh counterparts. Recently, Moreno-Rojas et al. (2018) showed that quercetin, quercetin-3-glucoside, kaempferol-3-rutinoside, quercetin-3-rhamnosylglucoside, apigenin, isorhamnetin, and luteolin were identified in fresh shallot onions and black onions. The authors showed that in black onion, free quercetin was 144 mg/kg FW, representing 89.6% of the total flavonoids. In fresh onions, free quercetin was 87 mg/kg FW, followed by quercetin-4-glucoside (39.4 mg/kg FW) and two isomers of quercetin-diglucoside (63 mg/kg FW) representing 94.6% of the total flavonoids. Similarly, quercetin 3-glucoside, quercetin, luteolin, and kaempferol were reported significantly higher amounts in different varieties of onion skins (Sagar et al. 2020).

It is well-known that in asparagus, flavonoids are the most abundant phenolic compounds. Among flavonoids, quercetin-3-rutinoside was the predominant phenolic antioxidant (Sun et al. 2005). Fuentes-Alventosa et al. (2008) reported that *triguero* asparagus flavonoids include quercetin being the major flavonoid, followed by isorhamnetin and kaempferol. The authors also identified quercetin-3-rutinoside, kaempferol-3-rutinoside, isorhamnetin-3-rutinoside, and isorhamnetin-3-glucoside in asparagus.

Several studies on celery showed flavonoid glycosides such as 7-apiosylglucosides and 7-glucosides of luteolin, apigenin, and chrysoeriol amongst the other phenolic compounds (Garg and Garg 1980). Lin et al. (2007) reported the presence of ten glycosylated flavone malonates in celery, Chinese celery, and celery seeds. Kaempferol-3-glucoside, kaempferol-3-glucuronide, and kaempferol-3-(6-malonyl) glucoside were reported in endive varieties. The kaempferol 3-(6-malonyl) glucoside was identified for the first time in the endive by DuPont et al. (2000). These authors also showed that kaempferol-3-glucuronide covered over 70% of the total flavonoids in the endive and lettuce.

In Bulgarian vegetables, quercetin was present okra (200.3 mg/kg), tomatoes (14.2 mg/kg), red pepper (14.9 mg/kg), green pepper (102.7 mg/kg), lettuce (153.9 mg/kg), Brussels sprouts (26.3 mg/kg), broccoli (29.4 mg/kg), red onions (452.4 mg/kg), white onion (204.1 mg/kg), spring onion (103.2 mg/kg), green beans (21.3 mg/kg), and yellow beans (22.9 mg/kg) (Tsanova-Savova et al. 2018).

4.2.6 Anthocyanins in Vegetables

Anthocyanins are one of the most widely studied phenolic compounds in vegetables. The specific coloration of vegetables has been attributed to the anthocyanin contents. Studies in vegetables are usually based on the total anthocyanin contents rather than detailed identification (Georgiadou et al. 2018). The red pigmentation of red cabbage is known to be due to anthocyanins. Fifteen different anthocyanins in the form of acyl glycosides of cyanidin have been identified in red cabbage (Podsędek 2007). Similarly, 20 different anthocyanins having the main structure of cyanidin-3-diglucoside-5-glucosides had been reported in red cabbages (Wiczkowski et al. 2014). Another study reported 21 anthocyanins namely peonidin hexoside, rosinidin hexoside, delphinidin malonylhexoside, cyanidin-3-glucoside, delphinidin malonyl-malonylhexoside, peonidin, cyanidin acetylhexoside, pelargonidin acetylhexoside, procyanidin A1, procyanidin A2, procyanidin B2, procyanidin B3, delphinidin, pelargonidin, delphinidin-3-glucoside, cyanidin-3-rutinoside, malvidin-3,5-diglucoside, pelargonin, petunidin-3-glucoside, pelargonidin-3-glucoside, and cyanidin (Zhuang et al. 2019).

In newly developed *Brassica napus* lines, the anthocyanins were cyanidin-3-diglucoside-5-glucoside (none-590 mg/kg), cyanidin-3-(sinapoyl)-diglucoside-5-glucoside (none-1630 mg/kg), cyanidin-3-sophoroside-5-malonylglucoside (none-70 mg/kg), malvidin-3-(p-coumaroyl)glucoside pyruvic derivative (none-67 mg/kg), pelargonidin-3-(6-acetyl)-glucoside (none-262 mg/kg), and cyanidin-3-(p-coumaroyl) diglucoside-5-glucoside (none-200 mg/kg) (Goswami et al. 2018). Similarly, Zhu et al. (2018) reported several anthocyanins namely cyanidin-3-diglucoside-5-glucoside, cyanidin-3-(sinapoyl)-diglucoside-5-glucoside, cyanidin-3-(feruloyl)-glucoside-5-glucoside, cyanidin-3-(feruloyl)-diglucoside-5-glucoside, cyanidin-3-(sinapoyl)-diglucoside-5-glucoside, cyanidin-3-(sinapoyl)(feruloyl)-diglucoside-5-glucoside, cyanidin-3-(sinapoyl)(sinapoyl)-diglucoside-5-glucoside, and cyanidin-3-(sinapoyl)(p-coumaroyl)-diglucoside-5-glucoside in kale.

Five reddish-purple vegetables obtained from the local market in Romania (onion, eggplant, sweet potato, eggplant, and chicory) were studied for phenolic composition including anthocyanins. The anthocyanins in red chicory were only cyanidin-3-glucoside. In red onion there were four anthocyanins i.e., cyanidin-3-glucoside and cyanidin-3-laminaribioside, cyanidin-3-(6″-malonyl-glucoside) and cyanidin-3-(6″-

malonyl-laminaribioside). In the eggplant sample, delphinidin-3-rutinoside was the only anthocyanin. Similarly, in sweet potato, delphinidin-3-rutinoside, peonidin-3-glucoside, cyanidin-3-*p*-hydroxybenzoylsophoroside-5-glucoside, peonidin-3-*p*-hydroxybenzoylsophoroside-5-glucoside, cyanidin-3-caffeoylsophoroside-5-glucoside, peonidin-3-caffeoylsophoroside-5-glucoside, cyanidin-3-caffeoyl-*p*-hydroxybenzoylsophoroside-5-glucoside, peonidin-3-dicaffeoylsophoroside-5-glucoside, peonidin-3-caffeoyl-*p*-hydroxybenzoylsophoroside-5-glucoside, and peonidin-3-caffeoyl-feruloylsophoroside-5-glucoside (Frond et al. 2019). The antioxidant activities of these vegetables were in the order of red chicory > purple sweet potato > black carrot > eggplant > red onion.

In red-leaf lettuce varieties, cyanidin-3-glucoside and cyanidin-3-(6-malonyl) glucoside were reported (Pérez-López et al. 2014). In onions, anthocyanins are predominate in the edible bulb of red onions than the bulbs of white or sweet yellow onions (Rhodes and Price 1996). Anthocyanins such as peonidin-3-glucoside, cyanidin-3-glucoside, and cyanidin-3-arabinoside and their malonylated derivatives, cyanidin-3-laminariobioside, and delphinidin and petunidin derivatives have been reported in the skins and outer fleshy layer of red onions (Donner et al. 1997). Similarly, in the whole red onion, Fossen et al. (1996) also identified cyanidin derivatives such as 3-(6″-malonyl-3″-glucosylglucoside), 3-(3″,6″dimalonylglucoside), 3-(6″-malonylglucoside), 3-(3″-malonylglucoside), 3-(3″-glucosylglucoside) and 3-glucoside. Previous study (Gennaro et al. 2002) showed that in whole red onion, the derivatives of cyanidin and delphinidin comprise over 50% and 30% of total anthocyanins, respectively.

In potato with purple-fleshed, petunidin and malvidin-3-rutinoside-5-glycosides were acylated with *p*-coumaric and ferulic acid. In potato with red-fleshed, pelargonidin and peonidin-3-rutinoside-5-glycosides were acylated with *p*-coumaric and ferulic acid (Reyes et al. 2005). In potatoes with purple flesh had a higher antioxidant activity due to the presence of a large quantity of anthocyanidins and their hydroxylated derivatives such as malvidin (Lachman et al. 2008). Similarly, 13 acylated anthocyanins were reported in purple-fleshed sweet potato from China, having cyanidin-3-caffeoyl-vanilloyl sophoroside-5-glucoside and peonidin-3-caffeoyl-vanilloyl sophoroside-5-glucoside as major compounds (He et al. 2016). In flours made from sweet potatoes had a higher antioxidant activity due to their high amounts of anthocyanins amongst other phenolic compounds (Huang et al. 2006). In purple-fleshed sweet potatoes, the major anthocyanins were cyanidin-3-dicaffeoyl sophoroside-5-glucoside, cyanidin-3-caffeoyl-*p*-coumarylsophoroside-5-glucoside, and peonidin-3-caffeoyl-*p*-hydroxybenzoyl sophoroside-5-glucoside, comprising 57.27% of the total anthocyanin contents (Liao et al. 2019).

In most of the varieties of red beetroot, 5-glucopyranosyl betanidine was the major coloring compound, which comprises 75–90% of the total coloring substances (Kujala et al. 2000). Other studies reported betanidin, isobetanidin, isbetanin (5-glucopyranosylisobetanidin), prebetanin (betanin 6′-sulphate), neobetanin (5-glucopyranosylneobetanidin), amaranthin, lampranthin I, and lampranthin II in different parts and cell cultures of red beetroot (Bokern et al. 1991). The beetroot has shown strong antioxidant activity, which was attributed to the anthocyanin contents (Carrillo et al. 2019).

4.3 Study Questions

- Describe in detail the types and hydroxybenzoic acids present in vegetables.
- What are the different types of cinnamoyl derivatives present in vegetables?
- Write a short note on coumarins and chromones in vegetables.
- What do you know about the flavonoids present in vegetables?
- Are anthocyanins and their derivatives are present in vegetables?

References

Ahmed M, Eun JB (2018) Flavonoids in fruits and vegetables after thermal and nonthermal processing: a review. Crit Rev Food Sci Nutr 58(18):3159–3188

Alasalvar C, Grigor JM, Zhang D, Quantick PC, Shahidi F (2001) Comparison of volatiles, phenolics, sugars, antioxidant vitamins, and sensory quality of different colored carrot varieties. J Agric Food Chem 49(3):1410–1416

Albishi T, John JA, Al-Khalifa AS, Shahidi F (2013) Antioxidative phenolic constituents of skins of onion varieties and their activities. J Funct Foods 5(3):1191–1203

Aritomi M, Kawasaki T (1984) Three highly oxygenated flavone glucuronides in leaves of *Spinacia oleracea*. Phytochemistry 23(9):2043–2047

Bahorun T, Luximon-Ramma A, Crozier A, Aruoma OI (2004) Total phenol, flavonoid, proanthocyanidin and vitamin C levels and antioxidant activities of Mauritian vegetables. J Sci Food Agric 84(12):1553–1561

Bakir S, Capanoglu E, Hall RD, de Vos RCH (2020) Variation in secondary metabolites in a unique set of tomato accessions collected in Turkey. Food Chem 317:126406

Bergman M, Perelman A, Dubinsky Z, Grossman S (2003) Scavenging of reactive oxygen species by a novel glucurinated flavonoid antioxidant isolated and purified from spinach. Phytochemistry 62(5):753–762

Bokern M, Heuer S, Wray V, Witte L, Macek T, Vanek T, Strack D (1991) Ferulic acid conjugates and betacyanins from cell cultures of *Beta vulgaris*. Phytochemistry 30(10):3261–3265

Bors MD, Socaci S, Tofana M, Muresan V, Pop AV, Nagy M, Vlaic R (2014) Determination of total phenolics, flavonoids and antioxidant capacity of methanolic extracts of some brassica seeds. Bull Univ Agric Sci Veterin Med Cluj Napoca Food Sci Technol 71:205–206

Branca F, Chiarenza GL, Cavallaro C, Gu HH, Zhao ZQ, Tribulato A (2018) Diversity of Sicilian broccoli (*Brassica oleracea var.* italica) and cauliflower (*Brassica oleracea* var. botrytis) landraces and their distinctive bio-morphological, antioxidant, and genetic traits. Genet Resour Crop Ev 65(2):485–502

Cantos E, Espín JC, Tomás-Barberán FA (2001) Effect of wounding on phenolic enzymes in six minimally processed lettuce cultivars upon storage. J Agric Food Chem 49(1):322–330

Cao G, Sofic E, Prior RL (1996) Antioxidant capacity of tea and common vegetables. J Agric Food Chem 44(11):3426–3431

Carrillo C, Wilches-Pérez D, Hallmann E, Kazimierczak R, Rembiałkowska E (2019) Organic versus conventional beetroot. Bioactive compounds and antioxidant properties. LWT Food Sci Technol 116:108552

Chehri Z, Zolfaghari B, Sadeghi Dinani M (2018) Isolation of cinnamic acid derivatives from the bulbs of *Allium tripedale*. Adv Biomed Res 7:60–60

Cherng J-M, Chiang W, Chiang L-C (2008) Immunomodulatory activities of common vegetables and spices of Umbelliferae and its related coumarins and flavonoids. Food Chem 106(3):944–950

Chhikara N, Kushwaha K, Sharma P, Gat Y, Panghal A (2019) Bioactive compounds of beetroot and utilization in food processing industry: a critical review. Food Chem 272:192–200

Chu YH, Chang CL, Hsu HF (2000) Flavonoid content of several vegetables and their antioxidant activity. J Sci Food Agric 80(5):561–566

Chu Y-F, Sun J, Wu X, Liu RH (2002) Antioxidant and antiproliferative activities of common vegetables. J Agric Food Chem 50(23):6910–6916

Chu C, Du Y, Yu X, Shi J, Yuan X, Liu X, Liu Y, Zhang H, Zhang Z, Yan N (2020) Dynamics of antioxidant activities, metabolites, phenolic acids, flavonoids, and phenolic biosynthetic genes in germinating Chinese wild rice (*Zizania latifolia*). Food Chem 318:126483

Crozier A, Lean MEJ, McDonald MS, Black C (1997) Quantitative analysis of the flavonoid content of commercial tomatoes, onions, lettuce, and celery. J Agric Food Chem 45(3):590–595

Crozier A, Jaganath IB, Clifford MN (2009) Dietary phenolics: chemistry, bioavailability and effects on health. Nat Prod Rep 26(8):1001–1043

Dasgupta N, De B (2007) Antioxidant activity of some leafy vegetables of India: a comparative study. Food Chem 101(2):471–474

Dehkharghanian M, Adenier H, Vijayalakshmi MA (2010) Study of flavonoids in aqueous spinach extract using positive electrospray ionisation tandem quadrupole mass spectrometry. Food Chem 121(3):863–870

Deng G-F, Lin X, Xu X-R, Gao L-L, Xie J-F, Li H-B (2013) Antioxidant capacities and total phenolic contents of 56 vegetables. J Funct Foods 5(1):260–266

Donner H, Gao L, Mazza G (1997) Separation and characterization of simple and malonylated anthocyanins in red onions, *Allium cepa* L. Food Res Int 30(8):637–643

DuPont MS, Mondin Z, Williamson G, Price KR (2000) Effect of variety, processing, and storage on the flavonoid glycoside content and composition of lettuce and endive. J Agric Food Chem 48(9):3957–3964

Fernández-León M, Fernández-León A, Lozano M, Ayuso M, González-Gómez D (2012) Identification, quantification and comparison of the principal bioactive compounds and external quality parameters of two broccoli cultivars. J Funct Foods 4(2):465–473

Ferreres F, Castañer M, Tomás-Barberán FA (1997) Acylated flavonol glycosides from spinach leaves (*Spinacia oleracea*). Phytochemistry 45(8):1701–1705

Fossen T, Andersen OM, Ovstedal DO, Pedersen AT, Raknes A (1996) Characteristic anthocyanin pattern from onions and other Allium spp. J Food Sci 61(4):703–706

Friedman M (1997) Chemistry, biochemistry, and dietary role of potato polyphenols. A review. J Agric Food Chem 45(5):1523–1540

Frond AD, Iuhas CI, Stirbu I, Leopold L, Socaci S, Andreea S, Ayvaz H, Andreea S, Mihai S, Diaconeasa Z, Carmen S (2019) Phytochemical characterization of five edible purple-reddish vegetables: anthocyanins, flavonoids, and phenolic acid derivatives. Molecules 24(8):1536

Fuentes-Alventosa J, Jaramillo S, Rodriguez-Gutierrez G, Cermeño P, Espejo J, Jiménez-Araujo A, Guillén-Bejarano R, Fernández-Bolaños J, Rodríguez-Arcos R (2008) Flavonoid profile of green asparagus genotypes. J Agric Food Chem 56(16):6977–6984

Furuta S, Suda I, Nishiba Y, Yamakawa O (1998) High tert-butylperoxyl radical scavenging activities of sweet potato cultivars with purple flesh. Food Sci Technol Int 4(1):33–35

Gao Y, Ma S, Wang M, Feng XY (2017) Characterization of free, conjugated, and bound phenolic acids in seven commonly consumed vegetables. Molecules 22(11):1878

Garg BK, Garg OP (1980) Sodium-carbonate and bicarbonate induced changes in growth, chlorophyll, nucleic-acids and protein contents in leaves of *pisum-sativum*. Photosynthetica 14(4):594–598

Gennaro L, Leonardi C, Esposito F, Salucci M, Maiani G, Quaglia G, Fogliano V (2002) Flavonoid and carbohydrate contents in Tropea red onions: effects of homelike peeling and storage. J Agric Food Chem 50(7):1904–1910

Georgiadou EC, Goulas V, Majak I, Ioannou A, Leszczyńska J, Fotopoulos V (2018) Antioxidant potential and phytochemical content of selected fruits and vegetables consumed in Cyprus. Biotechnol Food Sci 82(1):3–14

Georgiev VG, Weber J, Kneschke E-M, Denev PN, Bley T, Pavlov AI (2010) Antioxidant activity and phenolic content of betalain extracts from intact plants and hairy root cultures of the red beetroot *Beta vulgaris* cv. Detroit dark red. Plant Foods Hum Nutr 65(2):105–111

Gokce AF, Kaya C, Serce S, Ozgen M (2010) Effect of scale color on the antioxidant capacity of onions. Sci Hort Amst 123(4):431–435

Goswami G, Nath UK, Park J-I, Hossain MR, Biswas MK, Kim H-T, Kim HR, Nou I-S (2018) Transcriptional regulation of anthocyanin biosynthesis in a high-anthocyanin resynthesized *Brassica napus* cultivar. J Biol Res Thessaloniki 25(1):19

Gu C, Howell K, Dunshea FR, Suleria HAR (2019) LC-ESI-QTOF/MS Characterisation of phenolic acids and flavonoids in polyphenol-rich fruits and vegetables and their potential antioxidant activities. Antioxidants 8(9):405

Hait-Darshan R, Grossman S, Bergman M, Deutsch M, Zurgil N (2009) Synergistic activity between a spinach-derived natural antioxidant (NAO) and commercial antioxidants in a variety of oxidation systems. Food Res Int 42(2):246–253

He W, Zeng M, Chen J, Jiao Y, Niu F, Tao G, Zhang S, Qin F, He Z (2016) Identification and quantitation of anthocyanins in purple-fleshed sweet potatoes cultivated in China by UPLC-PDA and UPLC-QTOF-MS/MS. J Agric Food Chem 64(1):171–177

Heimler D, Isolani L, Vignolini P, Tombelli S, Romani A (2007) Polyphenol content and antioxidative activity in some species of freshly consumed salads. J Agric Food Chem 55(5):1724–1729

Hosny H, El Gohary N, Saad E, Handoussa H, El Nashar RM (2018) Isolation of sinapic acid from broccoli using molecularly imprinted polymers. J Sep Sci 41(5):1164–1172

Huang Y-C, Chang Y-H, Shao Y-Y (2006) Effects of genotype and treatment on the antioxidant activity of sweet potato in Taiwan. Food Chem 98(3):529–538

Hue S-M, Boyce AN, Somasundram C (2012) Antioxidant activity, phenolic and flavonoid contents in the leaves of different varieties of sweet potato (*'Ipomoea batatas'*). Aust J Crop Sci 6(3):375

Idrees N, Tabassum B, Sarah R, Hussain MK (2019) Natural compound from genus brassica and their therapeutic activities. In: Akhtar MS, Swamy MK, Sinniah UR (eds) Natural bio-active compounds: volume 1: production and applications. Springer, Singapore, pp 477–491

Islam MS, Yoshimoto M, Yahara S, Okuno S, Ishiguro K, Yamakawa O (2002) Identification and characterization of foliar polyphenolic composition in sweet potato (*Ipomoea batatas* L.) genotypes. J Agric Food Chem 50(13):3718–3722

Kähkönen MP, Hopia AI, Vuorela HJ, Rauha J-P, Pihlaja K, Kujala TS, Heinonen M (1999) Antioxidant activity of plant extracts containing phenolic compounds. J Agric Food Chem 47(10):3954–3962

Khanam UKS, Oba S, Yanase E, Murakami Y (2012) Phenolic acids, flavonoids and total antioxidant capacity of selected leafy vegetables. J Funct Foods 4(4):979–987

Kujala TS, Loponen JM, Klika KD, Pihlaja K (2000) Phenolics and betacyanins in red beetroot (*Beta vulgaris*) root: distribution and effect of cold storage on the content of total phenolics and three individual compounds. J Agric Food Chem 48(11):5338–5342

Lachman J, Hamouz K, Orsák M, Pivec V, Dvořák P (2008) The influence of flesh colour and growing locality on polyphenolic content and antioxidant activity in potatoes. Sci Hort Amst 117(2):109–114

Lee J, Lee S, Lee H, Park K, Choe E (2002) Spinach (*Spinacia oleracea*) powder as a natural food-grade antioxidant in deep-fat-fried products. J Agric Food Chem 50(20):5664–5669

Liao M, Zou B, Chen J, Yao Z, Huang L, Luo Z, Wang Z (2019) Effect of domestic cooking methods on the anthocyanins and antioxidant activity of deeply purple-fleshed sweet potato GZ9. Heliyon 5(4):e01515

Lin L-Z, Lu S, Harnly JM (2007) Detection and quantification of glycosylated flavonoid malonates in celery, Chinese celery, and celery seed by LC-DAD-ESI/MS. J Agric Food Chem 55(4):1321–1326

Lobo M, Hounsome N, Hounsome B (2018) Biochemistry of vegetables: secondary metabolites in vegetables—terpenoids, phenolics, alkaloids, and sulfur-containing compounds. In: Siddiq

M, Uebersax MA (eds) Handbook of vegetables and vegetable processing, vol 1. John Wiley & Sons Ltd, Oxford, p 47

Lopez-Cobo A, Gomez-Caravaca AM, Cerretani L, Segura-Carretero A, Fernandez-Gutierrez A (2014) Distribution of phenolic compounds and other polar compounds in the tuber of *Solanum tuberosum* L. by HPLC-DAD-q-TOF and study of their antioxidant activity. J Food Compos Anal 36(1–2):1–11

Mäder J, Rawel H, Kroh LW (2009) Composition of phenolic compounds and glycoalkaloids α-solanine and α-chaconine during commercial potato processing. J Agric Food Chem 57(14):6292–6297

Marti R, Valcarcel M, Herrero-Martinez JM, Cebolla-Cornejo J, Rosello S (2017) Simultaneous determination of main phenolic acids and flavonoids in tomato by micellar electrokinetic capillary electrophoresis. Food Chem 221:439–446

Mattila P, Hellström J (2007) Phenolic acids in potatoes, vegetables, and some of their products. J Food Compos Anal 20(3):152–160

Mercier J, Ponnampalam R, Bérard L, Arul J (1993) Polyacetylene content and UV-induced 6-methoxymellein accumulation in carrot cultivars. J Sci Food Agric 63(3):313–317

Mizgier P, Kucharska AZ, Sokół-Łętowska A, Kolniak-Ostek J, Kidoń M, Fecka I (2016) Characterization of phenolic compounds and antioxidant and anti-inflammatory properties of red cabbage and purple carrot extracts. J Funct Foods 21:133–146

Mohdaly AAA, Hassanien MFR, Mahmoud A, Sarhan MA, Smetanska I (2013) Phenolics extracted from potato, sugar beet, and sesame processing by-products. Int J Food Prop 16(5):1148–1168

Morales P, Carvalho AM, Sánchez-Mata MC, Cámara M, Molina M, Ferreira ICFR (2012) Tocopherol composition and antioxidant activity of Spanish wild vegetables. Genet Resour Crop Evol 59(5):851–863

Moreno-Rojas JM, Moreno-Ortega A, Ordóñez JL, Moreno-Rojas R, Pérez-Aparicio J, Pereira-Caro G (2018) Development and validation of UHPLC-HRMS methodology for the determination of flavonoids, amino acids and organosulfur compounds in black onion, a novel derived product from fresh shallot onions (*Allium cepa* var. aggregatum). LWT Food Sci Technol 97:376–383

Naczk M, Shahidi F (2006) Phenolics in cereals, fruits and vegetables: occurrence, extraction and analysis. J Pharm Biomed Anal 41(5):1523–1542

Nara K, Miyoshi T, Honma T, Koga H (2006) Antioxidative activity of bound-form phenolics in potato peel. Biosci Biotechnol Biochem 70(6):1489–1491

Ng A, Parker ML, Parr AJ, Saunders PK, Smith AC, Waldron KW (2000) Physicochemical characteristics of onion (*Allium cepa* L.) tissues. J Agric Food Chem 48(11):5612–5617

Nielsen JK, Olsen CE, Petersen MK (1993) Acylated flavonol glycosides from cabbage leaves. Phytochemistry 34(2):539–544

Nuutila AM, Puupponen-Pimia R, Aarni M, Oksman-Caldentey KM (2003) Comparison of antioxidant activities of onion and garlic extracts by inhibition of lipid peroxidation and radical scavenging activity. Food Chem 81(4):485–493

Oki T, Masuda M, Furuta S, Nishiba Y, Terahara N, Suda I (2002) Involvement of anthocyanins and other phenolic compounds in radical-scavenging activity of purple-fleshed sweet potato cultivars. J Food Sci 67(5):1752–1756

Ozcan MM, Dogu S, Uslu N (2018) Effect of species on total phenol, antioxidant activity and phenolic compounds of different wild onion bulbs. J Food Measur Charact 12(2):902–905

Pádua D, Rocha E, Gargiulo D, Ramos AA (2015) Bioactive compounds from brown seaweeds: phloroglucinol, fucoxanthin and fucoidan as promising therapeutic agents against breast cancer. Phytochem Lett 14:91–98

Patil BS, Pike LM, Yoo KS (1995) Variation in the quercetin content in different colored onions (*Allium-Cepa* L). J Am Soc Hortic Sci 120(6):909–913

Perez-Gregorio MR, Regueiro J, Simal-Gandara J, Rodrigues AS, Almeida DPF (2014) Increasing the added-value of onions as a source of antioxidant flavonoids: a critical review. Crit Rev Food Sci Nutr 54(8):1050–1062

Pérez-López U, Pinzino C, Quartacci MF, Ranieri A, Sgherri C (2014) Phenolic composition and related antioxidant properties in differently colored lettuces: a study by Electron Paramagnetic Resonance (EPR) kinetics. J Agric Food Chem 62(49):12001–12007

Philpott M, Gould KS, Markham KR, Lewthwaite SL, Ferguson LR (2003) Enhanced coloration reveals high antioxidant potential in new sweet potato cultivars. J Sci Food Agric 83(10):1076–1082

Podsędek A (2007) Natural antioxidants and antioxidant capacity of Brassica vegetables: a review. LWT Food Sci Technol 40(1):1–11

Podsedek A, Sosnowska D, Redzynia M, Anders B (2006) Antioxidant capacity and content of *Brassica oleracea* dietary antioxidants. Int J Food Sci Technol 41:49–58

Price KR, Casuscelli F, Colquhoun IJ, Rhodes MJ (1998) Composition and content of flavonol glycosides in broccoli florets (*Brassica oleracea*) and their fate during cooking. J Sci Food Agric 77(4):468–472

Reyes L, Miller J, Cisneros-Zevallos L (2005) Antioxidant capacity, anthocyanins and total phenolics in purple-and red-fleshed potato (*Solanum tuberosum* L.) genotypes. Am J Potato Res 82(4):271

Rhodes MJC, Price KR (1996) Analytical problems in the study of flavonoid compounds in onions. Food Chem 57(1):113–117

Rumbaoa RGO, Cornago DF, Geronimo IM (2009) Phenolic content and antioxidant capacity of Philippine sweet potato (*Ipomoea batatas*) varieties. Food Chem 113(4):1133–1138

Sagar NA, Pareek S, Gonzalez-Aguilar GA (2020) Quantification of flavonoids, total phenols and antioxidant properties of onion skin: a comparative study of fifteen Indian cultivars. J Food Sci Technol 57:2423

Sarker SD, Nahar L (2019) Dietary coumarins. In: Xiao J, Sarker SD, Asakawa Y (eds) Handbook of dietary phytochemicals. Springer, Singapore, pp 1–56

Schröter D, Baldermann S, Schreiner M, Witzel K, Maul R, Rohn S, Neugart S (2018) Natural diversity of hydroxycinnamic acid derivatives, flavonoid glycosides, carotenoids and chlorophylls in leaves of six different amaranth species. Food Chem 267:376–386

Shahidi F (2015) Handbook of antioxidants for food preservation. Woodhead Publishing, Cambridge

Shahidi F, Ambigaipalan P (2015) Phenolics and polyphenolics in foods, beverages and spices: antioxidant activity and health effects–a review. J Funct Foods 18:820–897

Shahidi F, Naczk M (2003) Phenolics in food and nutraceuticals. CRC press, Boca Raton, FL

Shahidi F, Yeo J (2016) Insoluble-bound phenolics in food. Molecules 21(9):1216

Sharma KD, Karki S, Thakur NS, Attri S (2012) Chemical composition, functional properties and processing of carrot—a review. J Food Sci Technol 49(1):22–32

Shih M-C, Kuo C-C, Chiang W (2009) Effects of drying and extrusion on colour, chemical composition, antioxidant activities and mitogenic response of spleen lymphocytes of sweet potatoes. Food Chem 117(1):114–121

Sikora E, Cieslik E, Filipiak-Florkiewicz A, Leszczynska T (2012) Effect of hydrothermal processing on phenolic acids and flavonols contents in selected brassica vegetables. Acta Sci Pol Technol Aliment 11(1):45–51

Silva Souza MA, Peres LEP, Freschi JR, Purgatto E, Lajolo FM, Hassimotto NMA (2020) Changes in flavonoid and carotenoid profiles alter volatile organic compounds in purple and orange cherry tomatoes obtained by allele introgression. J Sci Food Agric 100(4):1662–1670

Slimestad R, Fossen T, Vagen IM (2007) Onions: a source of unique dietary flavonoids. J Agric Food Chem 55(25):10067–10080

Soengas P, Cartea E, Lema M, Velasco P (2009) Effect of regeneration procedures on the genetic integrity of *Brassica oleracea* accessions. Mol Breed 23(3):389–395

Song H, Zhang L, Wu L, Huang W, Wang M, Zhang L, Shao Y, Wang M, Zhang F, Zhao Z, Mei X, Li T, Wang D, Liang Y, Li J, Xu T, Zhao Y, Zhong Y, Chen Q, Lu B (2020) Phenolic acid profiles of common food and estimated natural intake with different structures and forms in five regions of China. Food Chem 321:126675

Sultana B, Anwar F (2008) Flavonols (kaempferol, quercetin, myricetin) contents of selected fruits, vegetables and medicinal plants. Food Chem 108(3):879–884

Sun T, Tang J, Powers JR (2005) Effect of pectolytic enzyme preparations on the phenolic composition and antioxidant activity of asparagus juice. J Agric Food Chem 53(1):42–48
Sun T, Simon PW, Tanumihardjo SA (2009) Antioxidant phytochemicals and antioxidant
   capacity of biofortified carrots (*Daucus carota* L.) of various colors. J Agric Food Chem
   57(10):4142–4147
Talcott S, Howard L (1999) Determination and distribution of 6-methoxymellein in fresh and processed carrot puree by a rapid spectrophotometric assay. J Agric Food Chem 47(8):3237–3242
Tomás-Barberán FA, Clifford MN (2000) Flavanones, chalcones and dihydrochalcones – nature,
   occurrence and dietary burden. J Sci Food Agric 80(7):1073–1080
Tsanova-Savova S, Ribarova F, Petkov V (2018) Quercetin content and ratios to total flavonols and
   total flavonoids in Bulgarian fruits and vegetables. Bulg Chem Commun 50(1):69–73
Valdez-Morales M, Espinosa-Alonso LG, Espinoza-Torres LC, Delgado-Vargas F, Medina-Godoy
   S (2014) Phenolic content and antioxidant and antimutagenic activities in tomato peel, seeds,
   and byproducts. J Agric Food Chem 62(23):5281–5289
Wiczkowski W, Topolska J, Honke J (2014) Anthocyanins profile and antioxidant capacity of red
   cabbages are influenced by genotype and vegetation period. J Funct Foods 7:201–211
Zeb A (2015) Phenolic profile and antioxidant potential of wild watercress (*Nasturtium officinale*
   L.). Springerplus 4:714
Zeb A (2019a) Chemistry of interactions in frying. In: Food frying: chemistry, biochemistry and
   safety, vol 1. John Wiley & Sons, Oxford, pp 175–205
Zeb A (2019b) Chemo-metric analysis of the polyphenolic profile of *Cichorium intybus* L. leaves
   grown on different water resources of Pakistan. J Food Meas Charact 13(1):728–734
Zeb A, Habib A (2018) Lipid oxidation and changes in the phenolic profile of watercress
   (*Nasturtium officinale* L.) leaves during frying. J Food Meas Charact 12(4):2677–2684
Zeb A, Haq I (2016) The protective role of tomato powder in the toxicity, fatty infiltration and
   necrosis induced by oxidized tallow in rabbits. J Food Biochem 40:428–435
Zeb A, Imran M (2019) Carotenoids, pigments, phenolic composition and antioxidant activity of
   Oxalis corniculata leaves. Food Biosci 32:100472
Zeb A, Ullah F (2019) Effects of spinach leaf extracts on quality characteristics and phenolic pro-
   file of sunflower oil. Eur J Lipid Sci Technol 121(1):1800325
Zhang L, Li Y, Liang Y, Liang K, Zhang F, Xu T, Wang M, Song H, Liu X, Lu B (2019)
   Determination of phenolic acid profiles by HPLC-MS in vegetables commonly consumed in
   China. Food Chem 276:538–546
Zhu P, Tian Z, Pan Z, Feng X (2018) Identification and quantification of anthocyanins in different
   coloured cultivars of ornamental kale (*Brassica oleracea* L. var. acephala DC). J Hortic Sci
   Biotechnol 93(5):466–473
Zhuang H, Lou Q, Liu H, Han H, Wang Q, Tang Z, Ma Y, Wang H (2019) Differential regulation
   of anthocyanins in green and purple turnips revealed by combined de novo transcriptome and
   metabolome analysis. Int J Mol Sci 20(18):4387
Zielinska D, Wiczkowski W, Piskula MK (2008) Determination of the relative contribution of
   quercetin and its glucosides to the antioxidant capacity of onion by cyclic voltammetry and
   spectrophotometric methods. J Agric Food Chem 56(10):3524–3531
Zobel AM (1997) Coumarins in fruit and vegetables. In: Tomás-Barberán FA, Robins RJ (eds)
   Phytochemistry of fruit and vegetables. Oxford Science, Oxford, pp 73–204

Chapter 5
Phenolic Antioxidants in Cereals

5.1   Introduction

Cereals are normal staples for human nutrition. Major cereals having important global production are maize, rice, wheat, barley, and sorghum. Minor cereals having local significance comprise millets, oats, rye, and others. These cereals are naturally pigmented and may be black, purple, blue, pink, red, and brown (Zhu 2018). Phenolic antioxidants are present in cereals and legumes. In healthy diets, whole grains are usually recommended due to the presence of important fibers and several antioxidants. In cereals and grains, the outer layers (husk, pericarp, testa, and aleurone cells) contain high amounts of phenolic compounds than the inside part or endosperm layers (Kähkönen et al. 1999). For example, free and esterified phenolic acids were found in wheat, corn, rice, and oat (Sosulski et al. 1982). The antioxidant properties of wheat extracts were due to the presence of phenolic compounds, which seem to play a powerful antioxidants role, using their properties of radical scavenging and/or chelation of metals (Liyana-Pathirana et al. 2006). However, the barley also contained considerable amounts of phenolic antioxidants. These antioxidants had effective properties of scavenging DPPH, peroxyl, and hydroxyl radicals. They also effectively control LDL-cholesterol oxidation, thereby possess an excellent

© Springer Nature Switzerland AG 2021   149
A. Zeb, *Phenolic Antioxidants in Foods: Chemistry, Biochemistry and Analysis*,
https://doi.org/10.1007/978-3-030-74768-8_5

perspective in the development of antioxidants rich nutraceuticals (Madhujith and Shahidi 2006).

5.2   Phenolic Antioxidants in Cereals

5.2.1   Hydroxybenzoic Acids and Aldehydes in Cereals

Hydroxybenzoic acids and their aldehyde derivatives had been identified in nearly all cereals and legumes. They are present either in the outer layer/skin or inside layers. For example, sinapic acid had been reported to be present in the pericarp and germ of wheat, whereas isoferulic acid was accumulated in the germ of purple barley (Ndolo and Beta 2014). Additionally, in the pericarp, a large amount of syringic and vanillic acids had been reported, whereas sinapic acid was concentrated in the aleurone layer of yellow corn. Corn had the highest total phenolic contents i.e., 15.5μmol of gallic acid equivalent per gram of grain among tested grains, followed by wheat (7.99μmol of gallic acid eq/g of grain), oats (6.53μmol of gallic acid eq/g of grain), and rice (5.56μmol of gallic acid eq/g of grain) as reported by Adom and Liu (2002).

In kernels of wheat, several phenolic compounds, specifically ferulic, vanillic, gentisic, caffeic, salicylic, syringic, and sinapic acids as well as vanillin and syringaldehyde, has been identified (Liyana-Pathirana et al. 2006). In rye grain and its products benzoic acid and its derivatives, vanillic acid, syringic acid, and *para*-hydroxybenzoic acid had been reported (Andreasen et al. 2000). Watanabe et al. (1997) reported protocatechuic acid and 3,4-dihydroxybenzaldehyde from buckwheat hulls.

In wild rice, *p*-coumaric, vanillic, syringic, and *p*-hydroxybenzoic acids, along with two phenolic aldehydes, *p*-hydroxybenzaldehyde and vanillin were reported in both soluble and insoluble forms (Qiu et al. 2010). Wild rice had been found to have 30 times higher antioxidant activity than white rice. The individual phenolic acids such as gallic acid, protocatechuic acid, 2,5-dihydroxybenzoic acid, ferulic acid, sinapic acid present in 18 rice cultivars from China had significantly positive correlations with antioxidant capacity as reported by Pang et al. (2018). The authors also found that protocatechuic acid and 2,5-dihydroxybenzoic acid were not present in the samples of white rice. Similarly, Thai rice contains protocatechuic acid, vanillic acid, and sinapic acid (Peanparkdee et al. 2019).

The antioxidant activity of wild rice was 30-times higher than that of white rice (Qiu et al. 2009). The higher phenolic contents (2.51–3.59 mg/g) were reported in rice bran as compared to wheat bran (Iqbal et al. 2005). In non-pigmented, red, and black rice cultivars, the hydroxybenzoic acid derivatives were reported such as protocatechuic acid, 2,5-dihydroxybenzoic acid, *p*-hydroxybenzoic acid, vanillic acid, and syringic acid amongst other phenolic compounds (Shao et al. 2018). A higher total bound phenolic acids had been reported in black rice than white and red rice, whereas white rice at all stages of development after flowering. In red rice and black rice had relatively higher amounts of protocatechuic acid during the first week of development (14.1 mg/kg) and maturity (44.8 mg/kg), respectively. Similarly, vanillic acid (24–54 mg/kg) was only present in black rice at maturity (Shao et al. 2014b). The changes in the gallic acid, 4-hydroxybenzoic acid, vanillic acid, syringic acid, and other hydroxycinnamic acids were associated with enzyme activity in the shoot and rice during germination (Cho and Lim 2018). In black rice, drying temperature reduces the levels of free phenolic acids like gallic acid (Lang et al. 2019).

In grains, sorghum is the world's fifth largest produced cereal and also a good source of important phenolic antioxidants (Vanamala et al. 2018). The main hydroxybenzoic acids reported in sorghum grain are gallic, vanillic, protocatechuic, p-hydroxybenzoic, syringic, and sinapic acids (Vanamala et al. 2018; Xiong et al. 2019). They are present in the pericarp, endosperm, and testa of the grain. Protocatechuic, p-hydroxybenzoic, p-coumaric, and ferulic acids are found in the free and bound forms (Dykes 2019a). However, in some sorghum varieties, vanillic acid is found in the free and/or bound forms. Gallic acid had been reported to be present only in the bound form (Shahidi and Yeo 2016). Sinapic acid was reported along with other phenolic acids in all pearling fine and pearled kernel fractions of two sorghum genotypes namely white (PR6E6) and red (PR6E14) (Luthria and Liu 2013). Recently, 4-hydroxybenzoic acid, gallic acid, protocatechuic acid, sinapic acid, and vanillic acid has been reported in two Polish Sorghum cultivars (red and white) (Przybylska-Balcerek et al. 2019). These compounds have shown strong antioxidant activities. In the Chinese red sorghum variety (Ji Liang No. 1), p-hydroxybenzoic acid was the only hydroxybenzoic acid reported in free form (3.41 mg/kg), soluble conjugated form (20.3 mg/kg) and as insoluble bound form (21.5 mg/kg) (Zhang et al. 2019b).

5.2.2   Cinnamic Acids, Aldehydes, Esters, and Alcohols in Cereals

Cinnamic acids and their derivatives like cinnamaldehydes, esters, and cinnamoyl alcohols have been thoroughly studied in cereals and legumes. Cinnamic acids and their derivatives may be present in the outer layers, where they are mainly esterified to the arabinose side groups of arabinoxylans. Some studies have shown that cinnamic acids are present mostly in the aleurone layer and endosperm (Yu et al. 2001). Naczk and Shahidi (2006) reported that the highest amounts of phenolic acids are in the aleurone layer of cereal grains, but are also found in embryos and seed coat. In barley grains, salicylic acid, p-hydroxybenzoic acid, vanillic acid, protocatechuic acid, o-, m- and p-coumaric acids, syringic acid, ferulic acid, and sinapic acid had been identified in free forms (Madhujith and Shahidi 2006; Naczk and Shahidi 2006).

The purified monomeric and dimeric hydroxycinnamates and of phenolic extracts from rye (whole grain, bran, and flour) were investigated for antioxidant activities using an *in-vitro* copper-catalyzed human LDL oxidation assay. The 4,8-diferulic acid dehydrodimer was the most abundant compound in the rye had a slightly better antioxidant activity than ferulic acid and *p*-coumaric acid. The antioxidant activity decreased in the order of caffeic acid > sinapic acid > ferulic acid > *p*-coumaric acid (Andreasen et al. 2001). Good antioxidant activity in terms of their ability to inhibit hydroperoxide formation in bulk methyl linoleate and methyl linoleate emulsion had been reported for steryl ferulate extracts obtained from wheat or rye bran (Nyström et al. 2005). In rice, the sterols are 4,4-dimethylsterols with two methyl groups on carbon-4, whereas the sterols in other cereals are principally desmethyl-sterols that have no methyl groups in their C-4 position. In wild rice, the most insoluble form of ferulic acid had been reported as the most abundant phenolic acid (up to 355 mg/kg) followed by sinapic acid (Irakli et al. 2012). In the cell wall of the rice grain, phenolic acid dehydrodimers are present as insoluble fractions contained diferulic acids and disinapic acids (Qiu et al. 2010). In rice, phenolic acids are also present in bound form with lipids such as phytosterols. For example, cycloartenyl ferulate (**5.1**), 24-methylenecycloartanyl ferulate (**5.2**), and compestanyl ferulate (**5.3**) are major steryl ferulates as shown in Fig. 5.1.

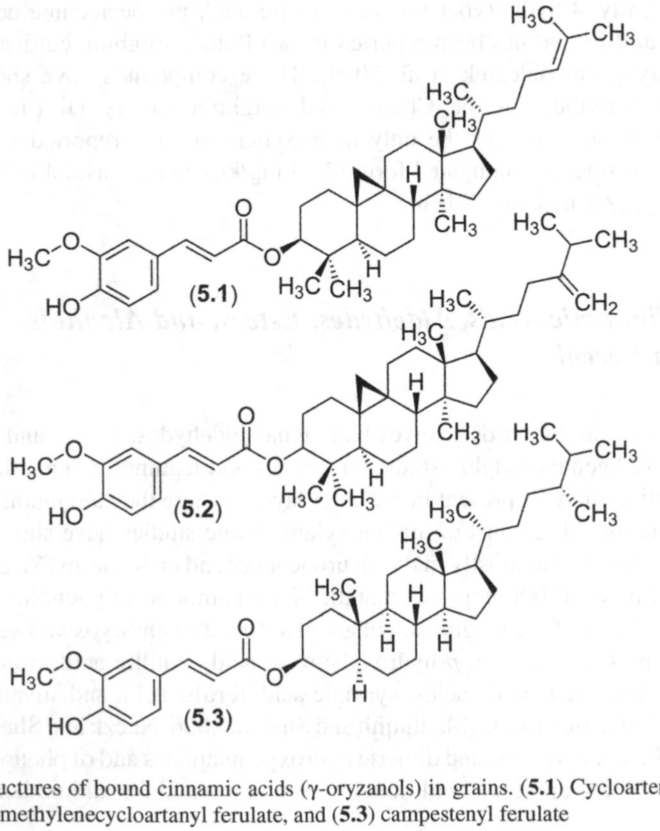

Fig. 5.1 Structures of bound cinnamic acids (γ-oryzanols) in grains. (**5.1**) Cycloartenyl ferulate, and (**5.2**) 24-methylenecycloartanyl ferulate, and (**5.3**) campestenyl ferulate

The steryl ferulates (γ-oryzanol) of rice possess good antioxidant activity especially DPPH radicals (Zhang et al. 2019a). A recent study of 223 landraces of rice from South Korea was studied for oryzanol contents. The total γ-oryzanol contents ranging from 6.9 to 514.0 mg/kg. The minor γ-oryzanols were Δ^7-stigmastenyl ferulate, stigmasteryl ferulate, Δ^7-campestenyl ferulate, Δ^7-sitostenyl ferulate, sitosteryl ferulate, compestanyl ferulate, and sitostanyl ferulate (Cho et al. 2019). The oryzanols have been shown to possess good antioxidant activity.

In wheat grains, campestanyl ferulate and sitostanyl ferulate were the key components of steryl ferulates. The bran fractions of the rye and wheat have shown to contain comparably higher amounts of steryl ferulate than corn (Hakala et al. 2002). Ferulic acid dehydrodimers are also present in wheat bran, which strengthen the aleurone walls during the maturation of wheat grain by the formation of bridges between two arabinoxylan chains. These dehydrodimers are formed by the peroxidase catalyzed oxidative coupling of ferulic acid (Naczk and Shahidi 2006). Five phenolic acids such as ferulic, sinapic, p-coumaric, vanillic, and 4-hydroxybenzoic, and cis-isomers of ferulic and sinapic acid were determined in 12 wheat genotypes of 4 different grain colors i.e., standard red, yellow endosperm, purple pericarp, and blue aleurone (Paznocht et al. 2020).

In Brazilian rice cultivars such as brown, black, and red rice from IRGA 417, IAC-600, and MPB-10 genotypes, respectively contained coumaric acid, ferulic acid, and caffeic acid in free and bound forms (Ziegler et al. 2018). Similarly, five grain varieties i.e., red, brown, black, brown (sushi), and white (sushi) from an Asian market in Seville, Spain were studied for phenolic acid contents. Caffeic acid, chlorogenic acids, coumaric acid, ferulic acid, and sinapic acids were higher in red and brown rice grains as compared to others (Table 5.1). In black rice, only coumaric acid and ferulic acid were detected (Setyaningsih et al. 2019). In brown rice, *Trans*-ferulic acid was the major phenolic acid ranging from 1.61 to 3.748 mg/kg. The percentage of *trans*-ferulic acid in bound fraction ranged from 96.4 to 99.2% (Gong et al. 2017).

In barley, chlorogenic and protocatechuic acids had been reported (Yu et al. 2005), whereas another study showed that ferulic acid was the major free phenolic

Table 5.1 The level of cinnamic acid derivatives in rice grain samples obtained from the local Asian market at Seville, Spain. Reproduced with kind permission of Elsevier (Setyaningsih et al. 2019)

Phenolics (mg/kg)	Rice varieties				
	Red	Black	Brown	Brown (Sushi)	White (Sushi)
Caffeic acid	7.20	ND	7.20	3.45	0.94
Chlorogenic acid	3.98	ND	2.33	2.13	1.05
p-Coumaric acid	5.45	3.48	4.11	1.51	0.49
Ferulic acid	12.16	2.25	22.68	10.66	4.02
Sinapic acid	0.64	ND	1.24	TR	TR
Isoferulic acid	1.91	ND	1.80	2.03	TR

Note. *ND* not detected due to less than LOD. *TR* trace due to the concentration less than LOQ but higher than LOD acid in seeds (Nordkvist et al. 1984) and bran of barley (Renger and Steinhart 2000). The total phenolic contents in defatted barley flour ranged from 0.81 to 1.38 mg of ferulic acid eq/g. Madhujith and Shahidi (2009) reported higher antioxidant and antiradical activities of insoluble-bound phenolic fraction than those of soluble conjugates and free phenolic fractions. Similarly, bound phenolics of syringic, and *p*-coumaric acids had been reported in bran–aleurone fraction of buckwheat (Shahidi 2015).

In corn, insoluble bound phenolic acids constitute the major fraction of phenolic acids present (Shahidi and Ambigaipalan 2015). Hydroxycinnamic acids were linked covalently to amine functionalities, such as feruoylputrescine, *p*-coumarylputrescine, diferuloylputrescine, di-*p*-coumarylputrescine, *p*-coumarylspermidine, diferuloylspermidine, and diferuloylspermine and present in the embryo and aleurone layer of corn (Sen et al. 1994).

Caffeic acids were the major free phenolic acid, whereas bound forms of ferulic acid, caffeic acid, and coumaric acid were the major compounds in finger millets (Subba Rao and Muralikrishna 2002). Chandrasekara and Shahidi (2010) had shown that ferulic acid and *p*-coumaric acids were the major bound phenolic compounds present in several varieties of millets such as kodo, finger, foxtail, proso, pearl, and little millets. Amongst other phenolic compounds, coumaric, syringic and vanillic acids were predominant phenolic compounds in finger millets (Viswanath et al. 2009).

5.2.3 Coumarins and Chromones in Cereals

In breakfast cereals obtained from the local Danish market showed that coumarins were below the range approved by the European Union regulation (EC) No 1334/2008, ranging between 0.9 and 10.0 mg/kg with a mean coumarin content of 3.3 mg/kg (Ballin and Sørensen 2014). Plant foods of the families Apiaceae, Asteraceae, and Rutaceae are major sources of naturally occurring coumarins (Sarker and Nahar 2019).

5.2.4 Flavonoids in Cereals

Cereals are rich in different flavonoids (Schendel 2019). The flavonoids of cereals had shown significant antioxidant activity measured using Trolox equivalent antioxidant capacity (TEAC), 2,2-diphenyl-1-picrylhydrazyl radical scavenging, reducing power, oxygen radical absorbance capacity (ORAC), inhibition of oxidation of human low-density lipoprotein (LDL) cholesterol and DNA, Rancimat, inhibition of photochemiluminescence (PCL), and iron(II) chelation activity (Van Hung 2016). In wheat 5,7,4-trihydroxy 3,5-dimethoxyflavone was found to be the major flavone. In wheat-bran, two *C*-glycosylflavones, namely 6-*C*-pentosyl-8-*C*-hexosylapigenin and 6-C-hexosyl-8-C-pentosylapigenin were isolated (Naczk and Shahidi 2006). The consumption of wheat with bran in the form of whole-grain may provide beneficial health effects due to the fact that phenolic compounds have been concentrated in the bran having higher antioxidant activity than other milling fractions (Liyana-Pathirana and Shahidi 2007a, b).

In grains of barley, several flavanols such as catechin, and epicatechin had been reported (Goupy et al. 1999). A novel flavone C-glycoside i.e., 2″-O-glycosylisovitexin was isolated from green barley leaves and reported that its antioxidative activity was similar to that exhibited by α-tocopherol (Osawa et al. 1992). A study showed that bread made by replacing 40% of wheat flour with barley flour had increased the antioxidant properties of the bread (Holtekjølen et al. 2008). Buckwheat is a pseudo-cereal and its seed serves as a rich source of flavonoids. In immature buckwheat seeds, several flavonoids such as quercetin-3-rutinoside, hyperin, quercitrin, quercetin, vitexin, isovitexin, orientin, isoorientin had been reported (Guo et al. 2012). In the ethanolic extracts of buckwheat groats, flavonoids such as epicatechin, catechin 7-O-β-D-glucopyranoside, epicatechin 3-O-p-hydroxybenzoate, and epicatechin 3-O-(3,4-di-O-methyl)gallate were identified (Dietrych-Szostak and Oleszek 1999). In hydrolysates of aleurone tissues of corn, monomeric flavan-3,4-diols, such as kaempferol and quercetin, had been reported (Shahidi and Ambigaipalan 2015).

Flavonoids such as orientin, isoorientin, vitexin, isovitexin, saponarin, violanthin, lucenin-1, and tricin had been reported in the leaves of finger millet (Dykes and Rooney 2006). The tiny millet grain showed to have different colors, usually a dark brown seed coat, which is rich in polyphenols compared to other continental cereals such as barley, rice, maize, and wheat. Finger millet seed coat is a good source of polyphenols with significantly higher antioxidant activity compared to whole flour (Viswanath et al. 2009). Two flavonoids were isolated from an ethanolic extract of the grain of Japanese barnyard millet (cv. Kurohie) using preparative high-performance liquid chromatography were luteolin, and tricin (Watanabe 1999).

In red sorghums, the pericarp contains luteoforol and apiforol, which are produced from flavanones i.e., naringenin and eriodictyol, and maybe precursors of sorghum anthocyanidins (Dykes and Rooney 2006). Other flavonoids isolated and identified in sorghum grains include apigenin and luteolin, which are the main flavons in tan-pigmented sorghums. Flavanones, eriodictyol, eriodictyol-5-glucoside, kaempferol 3-rutinoside-7-glucuronide, and the dihydroflavonols taxifolin, and taxifolin 7-glucoside had also been reported (Dykes and Rooney 2006). Recently, luteolinidin, apigeninidin, luteolin, apigenin, taxifolin, and naringenin have been reported in both free and bound forms in eight different sorghum grains from China (Shen et al. 2018). The free flavonoid fractions showed higher DPPH and FRAP values than their corresponding bound fractions. Kil et al. (2009) suggested that due to the antioxidant and antimicrobial properties, sorghum could be used as a natural ingredient with biological function in the food industry.

Rice is rich in several flavonoids (Das et al. 2017; Wang et al. 2018). In pigmented and non-pigmented rice cultivars, naringenin, apigenin, and quercetin have been reported in the ranges of 40–400, 80–1830, and 40–1200 mg/kg, respectively (Chutipaijit and Sutjaritvorakul 2018). Phenolic compounds in the free fraction of pigmented rice had higher antioxidant capacity as compared to those in the bound form, while the non-pigmented rice cultivars showed the opposite trend (Irakli et al. 2016). Nineteen different C-glycosidic flavonoids were identified in yellow grains mutants included isoorientin, isoorientin-2″-glucoside, luteolin-6,8-di-C-hexoside, luteolin-6-C-pentosyl-8-C-hexoside, apigenin-6,8-di-C-hexoside, chrysoeriol-6,8-di-C-hexoside, isoorientin-2″-glucoside, and vitexin-2″-glucoside were identified based on UHPLC-DAD-ESI-Q-TOF-MS analysis (Kim et al. 2018). Similarly, Chinese wild rice varieties showed quercetin, rutin, and procyanidin B1 and B2 (Chu et al. 2020).

The flour of oat has been shown to consist of apigenin, luteolin, and tricin (Shahidi and Ambigaipalan 2015). The phenolic compounds in oat have been shown to synergistically interact with vitamin-C to protect LDL during oxidation (Chen et al. 2004). These phenolic compounds have also been shown to possess anti-inflammatory and anti-hypertensive activity (Chu et al. 2013).

5.2.5 Anthocyanins in Cereals

Anthocyanins are glycosides of cyanidins and are present in a variety of cereals. Several review articles showed the anthocyanin contents in important cereals (Escribano-Bailón et al. 2004; Shahidi and Ambigaipalan 2015; Das et al. 2017). A summary of those studies along with the recent advances in the anthocyanin contents in cereals has been presented. In whole grain of finger millet, Apigeninidin, and luteolinidin-type anthocyanins were reported.

The grains of the colored barley contained at least eight pigmented compounds that were identified as derivatives of cyanidin, delphinidin, and pelargonidin, as well as cyanidin-3-arabinoside, and cyanidin-3-glucoside (Madhujith and Shahidi 2009). Similarly, another study (Goupy et al. 1999) reported dimeric prodelphinidin B3, procyanidin B3, as well as trimeric procyanidin C2, and three trimeric prodelphinidins in barley.

Anthocyanins such as cyanidin-3-glucoside, pelargonidin-3-glucoside, and peonidin-3-glucoside had been found in Chinese purple corn (Yang et al. 2009). Mohsen and Ammar (2009) reported that the ethanolic extract of corn tassels was successfully utilized to retard the oxidation of sunflower oil. The corncob was found to be a good source of phenolic antioxidants that can be used in nutraceutical and functional food applications and to protect vegetable oil (Sultana et al. 2007). Maize has been reported to have several colored cultivars as shown in Fig. 5.2. These include blue aleurone (A), pink aleurone (B), blue aleurone without acylated anthocyanins (C), pericarp without condensed forms (D), pericarp with condensed forms (E), phlobabaphenes (F), and bronze pigment (G) (Paulsmeyer et al. 2017).

Anthocyanin pigments are mainly present in the aleurone layers of blue maize kernel which has thin and colorless pericarp as shown in Fig. 5.3a (Paulsmeyer et al. 2017). In contrast, pigments are mostly concentrated in the pericarp in purple maize kernel, which tends to be thick (Fig. 5.3b). No maize genotype with endosperms

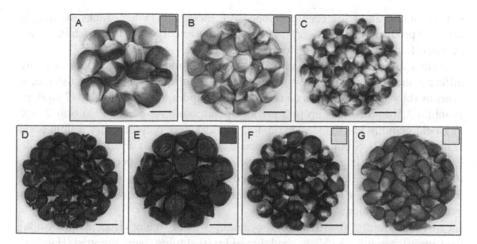

Fig. 5.2 Kernel samples from each major pigment category; blue aleurone (**A**), pink aleurone (**B**), blue aleurone without acylated anthocyanins (**C**), pericarp without condensed forms (**D**), pericarp with condensed forms (**E**), phlobabaphenes (**F**), and bronze pigment (**G**). (Reproduced with kind permission of the American Chemical Society, Paulsmeyer et al. 2017)

Fig. 5.3 Micrographs showing pigmentation in the major anthocyanin-producing tissues of the maize kernel: pericarp (**A**) and aleurone (**B**). (Reproduced with kind permission of the American Chemical Society, Paulsmeyer et al. 2017)

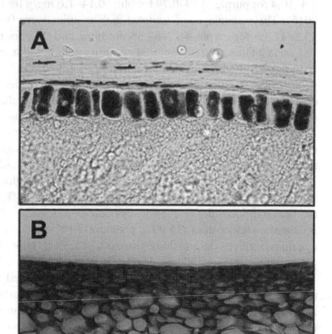

rich in anthocyanins has been developed yet, though this cereal ranks the first in terms of production quantity (Zhu 2018). The maize cob can also be rich in anthocyanins (Lao and Giusti 2016; Hong et al. 2020).

Anthocyanin composition and concentrations in the kernels and cobs of many different maize genotypes have been identified and quantified during the recent years as shown in Table 5.2. In colored maize, cyanidin-3-glucoside (0–547 mg/kg), cyanidin-3,5-diglucoside (up to 5 mg/kg), pelargonidin-3-glucoside (up to 2 mg/kg), pelargonidin-3,5-diglucoside (up to 2 mg/kg) were present in dry weight basis (Zilic et al. 2012). Similarly, in waxy maize cultivars (n = 3), cyanidin-3-glucoside (0–380µg/g), pelargonidin-3-glucoside (0–99µg/g), peonidin-3-glucoside (0–133µg/g), cyanidin-3-(6″-malonylglucoside) (0–279µg/g), pelargonidin-3-(6″-malonylglucoside) (0–20µg/g), peonidin-3-(6″-malonylglucoside) (0–26µg/g), cyanidin-3-(3″,6″-dimalonylglucoside) (1–183µg/g), cyanidin-3-(6″-succinylglucoside) (0–101µg/g), cyanidin-3-(3″,6″-malonylsuccinylglucoside) (0–15µg/g), peonidin-3-(6″-succinylglucoside) (0–44µg/g) were reported (Harakotr et al. 2014, 2015). In blue, red, purple, and red/blue maize, cyanidin-3-glucoside (0–0.16 mg/g) for red samples, 4.8–16.4 mg/g for purple, 0.26–0.64 mg/g for blue, 0.13–0.84 mg/g for red/blue; cyanidin-3,6″-malonyl-glucoside (0–0.17 for red, 2.3–10.4 for purple, 0.39–0.79 for blue, 0.14–1.6 mg/g for red/blue) (Collison et al. 2015). The percentages of acylated anthocyanins were 0–43.4% for red samples, 35.2–47.3% for purple, 56.2–59.3% for blue, and 61% for red/blue maize.

Lao and Giusti (2016) reported cyanidin, pelargonidin, peonidin, catechin-(4,8)-cyanidin-3,5-diglucoside, cyanidin-3,5-diglucoside, cyanidin-3-glucoside, pelargonidin-3-glucoside, cyanidin-3-malonylglucoside, peonidin-3-glucoside, cyanidin-3-(6″-malonylglucoside), cyanidin-3-succinylglucoside, pelargonidin-3-(6″-malonylglucoside), cyanidin-3-dimalonylglucoside, peonidin-3-(6″-malonylglucoside) in 14 purple maize cultivars. Another study showed the percent of individual anthocyanins in blue maize, whole flour is grown in 3 different locations (Nankar et al. 2016). These include cyanidin-3-glucoside (62%), pelargonidin-3-glucoside (14%), cyanidin-3-disuccinylglucoside (5%), cyanidin-3-succinylglucoside (4%), and peonidin-3-glucoside (3.4%). Similarly, cyanidin-3-(6″-malonylglucoside) (46.5%), cyanidin-3-glucoside (16.7%), cyanidin-3-(3″, 6″-dimalonylglucoside) (15.9%), peonidin-3-(6″-ethylmalonylglucoside) (10.6%), cyanidin-3-glucoside-2-malonylglucoside (5.3%), peonidin-3-glucoside (3.4%), peonidin-3-(dimalonyl glucoside) (1.6%) were also reported in blue maize (Camelo-Méndez et al. 2016). Recently, Hong et al. (2020) reported that glucosides of cyanidin and pelargonidin were the major compounds in purple and red maize. Total anthocyanin concentration reached 51 mg/g at the optimum sweetcorn eating stage. Cyanidin-3-malonylglucoside, four isomers of pelargonidin-3-malonylglucoside, and two to three isomers each of cyanidin-3-dimalonylglucoside, peonidin-3-malonylglucoside, and pelargonidin-3-dimalonylglucoside were also reported. Flavones have been reported to stabilize pelargonidin from corn (Chatham et al. 2020). The presence of these compounds in corn is responsible for higher antioxidant activity (Syedd-León et al. 2020).

Table 5.2 Anthocyanins composition in diverse colored cereals. Updated with latest literature and part of the table has been reproduced with kind permission of Elsevier (Zhu 2018)

Cereal type	Sample type and number of genotypes (n)	Extraction solvent	Quantification and identification methods	Total anthocyanin content	Anthocyanin (composition)	References
Maize	Colored, n = 12	Methanol with 1N HCl (85:15)	HPLC-MS, total anthocyanin method, UVV	3–696 mg CGE/kg, db	Cyanidin-3-glucoside (0–547 mg/kg, db), cyanidin-3,5-diglucoside (up to 5 mg/kg, db), pelargonidin-3-glucoside (up to 2 mg/kg, db), pelargonidin-3,5-diglucoside (up to 2 mg/kg, db)	Zilic et al. (2012)
Maize	Waxy maize, n = 12 (samples from 2 maturation stages: milk and mature)	Aqueous acetone (70%) with HCl (0.01%)	pH differential, HPLC-MS, UVV	0–1439µg CGE/g, db	Cyanidin-3-glucoside (0–380µg/g, db), pelargonidin-3-glucoside (0–99µg/g, db), peonidin-3-glucoside (0–133µg/g, db), cyanidin-3-(6″-malonylglucoside) (0–279µg/g, db), pelargonidin-3-(6″-malonylglucoside) (0–20µg/g, db), peonidin 3-(6″-malonylglucoside) (0–26µg/g, db), cyanidin-3-(3″,6″-dimalonylglucoside) (1–183µg/g, db), cyanidin-3-(6″-succinylglucoside) (0–101µg/g, db), cyanidin-3-(3″,6″-malonylsuccinylglucoside) (0–15µg/g, db), peonidin-3-(6″-succinylglucoside) (0–44µg/g, db)	Harakotr et al. (2014)
Maize	Waxy maize, n = 49	Aqueous acetone (70%) with HCl (0.01%)	pH differential	0–1063µg CGE/g, db	Cyanidin-3-glucoside	Harakotr et al. (2015)

(continued)

Table 5.2 (continued)

Cereal type	Sample type and number of genotypes (n)	Extraction solvent	Quantification and identification methods	Total anthocyanin content	Anthocyanin (composition)	References
Maize	Blue, red, purple, and red/blue maize, whole flour, n = 24, samples grown in 2 different environments in Texas	Methanol with HCl (1%)	HPLC-MS, UVV, pH differential	Unit: μg CGE/g, db; 9–127 for red samples, 891–3312 for purple samples, 0–540 for blue samples, 46–368 for red/blue samples	Unit: mg/g, cyanidin-3-glucoside (0–0.16 for red samples, 4.8–16.4 for purple, 0.26–0.64 for blue, 0.13–0.84 for red/blue), cyanidin-3,6″-malonyl-glucoside (0–0.17 for red, 2.3–10.4 for purple, 0.39–0.79 for blue, 0.14–1.6 for red/blue); percentages of acylated anthocyanins (0–43.4% for red samples, 35.2–47.3% for purple, 56.2–59.3% for blue, and 61% for red/blue)	Collison et al. (2015)
Maize	Purple maize cob, n = 14	Acetone (70%) with HCl (0.01%)	HPLC-MS, HPLC of intact and HCl-hydrolysed anthocyanins, pH differential, total anthocyanin	Unit: mg CGE/g; 4.3–117 (total anthocyanins method); 3.1–100 (pH differential); 3.1–98.1 (HPLC of intact anthocyanins); 1.1–56.1 (HPLC of HCl-hydrolysed anthocyanins)	Cyanidin, pelargonidin, peonidin, catechin-(4,8)-cyanidin-3,5-diglucoside, cyanidin 3,5-diglucoside, cyanidin-3-glucoside, pelargonidin-3-glucoside, cyanidin-3-malonylglucoside, peonidin-3-glucoside, cyanidin-3-(6″-malonylglucoside), cyanidin-3-succinylglucoside, pelargonidin-3-(6″-malonylglucoside), cyanidin-3-dimalonylglucoside, peonidin-3-(6″-malonylglucoside)	Lao and Giusti (2016)
Maize	Blue maize, whole flour, n = 8, grown in 3 different locations	Acidified methanol	HPLC-MS, UVV, pH differential method	0.04–0.88 g CGE/kg kernel tissue (pH differential)	Cyanidin-3-glucoside (62%), pelargonidin-3-glucoside (14%), cyanidin-3-disuccinylglucoside (5%), cyanidin-3-succinylglucoside (4%), peonidin-3-glucoside (3.4%)	Nankar et al. (2016)

Cereal type	Sample type and number of genotypes (n)	Extraction solvent	Quantification and identification methods	Total anthocyanin content	Anthocyanin (composition)	References
Maize	Blue maize, whole flour, n = 1	Methanol (75%) acidified with 1N HCl (5%)	HPLC-MS, UVV	-	Cyanidin-3-(6″-malonylglucoside) (46.5%), cyanidin-3-glucoside (16.7%), cyanidin 3-(3″, 6″-dimalonylglucoside) (15.9%), peonidin-3-(6″-ethylmalonylglucoside) (10.6%), cyanidin-3-glucoside-2-malonylglucoside (5.3%), peonidin-3-glucoside (3.4%), peonidin-3-(dimalonyl glucoside) (1.6%)	Camelo-Méndez et al. (2016)
Maize	Pigmented whole grain flour, n = 398	Formic acid aqueous solution (2%)	HPLC, UVV	23–252 mg CGE/kg (average values of 5 groups differing in colour features)	Cyanidin-3-glucoside (3–57%), pelargonidin-3-glucoside (2–15%), peonidin-3-glucoside (1–12%), acylated anthocyanins (8–63%)	Paulsmeyer et al. (2017)
Rice	Black, red, n = 2	Methanol with formic acid (1%)	HPLC-MS/MS, UVV	3.5 mg CGE/g for black rice, 4.3 CGE µg/g for red rice, 0 for brown rice	Unit: CGE µg/g, cyanidin-3,5-diglucoside (20 for black), cyanidin-3-glucoside (2857 for black and 3 for red), cyanidin-3-(6″-p-coumaryl) glucoside (57 for black and 1.3 for red), pelargonidin-3-glucoside (8 for black), peonidin-3-O-glucoside (500 for black), peonidin-3-(6″-p-coumaryl)glucoside (23 for black), cyanidin-3-arabidoside (9 for black)	Pereira-Caro et al. (2013b)
Rice	Black-purple rice, n = 1	Methanol with formic acid (1%)	HPLC-MS/MS, UVV	1.4 CGE mg/g	Unit: cyanidin-3-glucoside equivalent µg/g, cyanidin-3,5-diglucoside (13), cyanidin-3-glucoside (1239), cyanidin-3-(6″-p-coumaryl) glucoside (4), pelargonidin-3-glucoside (6), peonidin-3-glucoside (131), peonidin-3-(6″-p-coumaryl)glucoside (4), cyanidin-3-arabidoside (3)	Pereira-Caro et al. (2013a)

(continued)

Table 5.2 (continued)

Cereal type	Sample type and number of genotypes (n)	Extraction solvent	Quantification and identification methods	Total anthocyanin content	Anthocyanin (composition)	References
Rice	Black, whole grain, embryo, bran, endosperm, n = 1	Methanol acidified with HCl	HPLC-MS/MS, UVV, pH differential	0.87 mg CGE/g for whole grain, 0.34 for embryo, 0 for endosperm, and 6.3 for bran	Unit: CGE mg/g, cyanidin-3-glucoside (0.52 for whole grain, 0.12 for embryo, 3.6 for bran, and 0 for endosperm), peonidin-3-glucoside (0.14 for whole grain, 0.09 for embryo, 0.7 for bran, and 0 for endosperm), cyanidin-3-rutinoside (0.06 for whole grain, 0 for embryo, 0.3 for bran, and 0 for endosperm)	Shao et al. (2014a, b)
Rice	Black, whole grain, n = 1	Ethanol (50%) solution with HCl (0.5%)	HPLC-MS, UVV	417 mg CGE/g extract	Unit, mg CGE/g extract, cyanidin-3,5-diglucoside (7.5), cyanidin-3-gentiobioside (7.6), cyanidin-3-glucoside (268), cyanidin-3-sambubioside (7.6), cyanidin-3-rutinoside (7.9), peonidin-3-glucoside (23.8), cyanidin derivative (7), cyanidin (80), peonidin (7)	Hao et al. (2015)
Rice	Whole grain and polished, n = 5	HCL acidified aqueous methanol	HPLC, UVV	Up to 1 mg/g of grain, db, for polished and engineered genotype	Cyanidin-3-glucoside, peonidin-3-glucoside	Zhu et al. (2017)
Rice	Whole grains, n = 7	Methanol (85:15 v/v)	HPLC, PDA	-	Cyanidin-3-glucoside, peonidin-3-glucoside	Melini et al. (2019)
Wheat	Blue wheat	HCL acidified aqueous methanol	HPLC-MS, UVV	-	Delphinidin-3-glucoside (135 mg/kg), cyanidin-3-glucoside (86 mg/kg), delphinidin-3-rutinoside (65 mg/kg), cyanidin-3-rutinoside (5.3 mg/kg)	Abdel-Aal et al. (2008)

Cereal type	Sample type and number of genotypes (n)	Extraction solvent	Quantification and identification methods	Total anthocyanin content	Anthocyanin (composition)	References
Wheat	Blue wheat, whole grain flour	MeOH (80%) containing 0.3% TFA	HPLC-MS, UVV, NMR	–	Delphinidin-3-glucoside (9.3 mg/kg), delphinidin-3-rutinoside (33.1 mg/kg), cyanidin-3-glucoside (4.1 mg/kg), cyanidin-3-rutinoside (18.5 mg/kg)	Tyl and Bunzel (2012)
Wheat	Whole grain, purple wheat, n = 1	HCl acidified methanol	HPLC, UVV	–	Cyanidin-3-glucoside (43%), peonidin-3-glucoside (40%), malvidin-3-galactoside (17%)	Chen et al. (2013)
Wheat	Blue-aleurone bread wheat (T. aestivum), n = 4	Methanol (85%) acidified with HCl (1N)	HPLC, UVV, pH differential	83–174µg/g	Unit, mg/kg, cyanidin-3-glucoside (1.2–7.1), cyanidin-3-rutinoside (7.1–11), delphinidin-3-glucoside (9.9–18.8), delphinidin-3-rutinoside (31.8–35.9), peonidin-3-glucoside (0.8–1.0), peonidin-3-galactoside (1.9), peonidin-3-arabinoside (0.9–3.5), malvidin-3-glucoside (10.3–13.8)	Ficco et al. (2014)
Durum wheat	Red (n = 43)- and purple (n = 27)-pericarp durum	Methanol acidified with HCl (1N) (85:15)	HPLC, UVV, pH differential	Purple (8–50µg/g), red (1–25µg/g)	Unit, mg/kg, Cyanidin-3-glucoside (0.59–10.23), peonidin-3-galactoside (0.06–3.87), malvidin-3-glucoside (0.06–2.35)	Ficco et al. (2014)

(continued)

Table 5.2 (continued)

Cereal type	Sample type and number of genotypes (n)	Extraction solvent	Quantification and identification methods	Total anthocyanin content	Anthocyanin (composition)	References
Various wheat species	Whole grain flour, 1 genotype for each species	Acidified ethanol or methanol	Total anthocyanin	11.6 mg/kg (db) for *T. monococcum* ssp. *monococcum*, 43 mg/kg (db) for *T. monococcum* ssp. *thaoudar*, 17.4 mg/kg (db) for *T. monococcum* ssp. *aegilopoides*, 15.3 mg/kg (db) for *T. urartu*, 2.9 mg/kg (db) for *T. turgidum* ssp. *dicoccum*, 2.6 mg/kg (db) for *T. aestivum* ssp. *spelta*	Cynidin-3-glucoside , ..	Brandolini et al. (2015)
Wheat	Purple wheat, whole flour, flakes, bran, and products	Methanol acidified with 1.0N HCl (85:15, v/v)	HPLC, UV, UVV	126.8 mg/kg for whole flour, 120 mg/kg for flakes, and 806 mg/kg for bran	Cyanidin-3-glucoside, cyanidin-3-rutinoside, pelargonidin-3-glucoside	Gamel et al. (2020)
Barley	Whole grain, n = 127 (black, blue, purple)	Methanol (80%) with HCl (0.1%)	HPLC, UVV	60–350 mg/kg (average values of 7 groups) differing in color features	Delphinidin-3-glucoside (15–168 mg/kg), cyanidin-3-glucoside (9–240 mg/kg), pelargonidin-3-glucoside (12–42 mg/kg), peonidin-3-glucoside (3–38 mg/kg), malvidin 3-glucoside (0.6–39 mg/kg)	Kim et al. (2007)
Barley	Whole grain black, n = 1	Ethanol (85%) with 1N HCl	HPLC, pH differential	0.27 mg/g (pH differential)	Cyanidin-3-glucoside (2.52 mg/100 g), delphinidin-3-glucoside (2.13 mg/100 g), petunidin-3-glucoside (29 mg/100 g)	Yao et al. (2010)

Cereal type	Sample type and number of genotypes (n)	Extraction solvent	Quantification and identification methods	Total anthocyanin content	Anthocyanin (composition)	References
Barley	Whole grain (hulless) flour, refined flour, and bran, n = 4	Acetone extraction, chloroform partition	Total anthocyanin, HPLC-MS, UVV	3–679 mg/kg for whole flour, 5–1655 mg/kg for bran, 2–257 mg/kg for refined flour	Cyanidin-3-glucoside, pelargonidin 3-glucoside, peonidin-3-glucoside, cyanidin-3-(6″-succinyl) glucoside, peonidin-3-(6″-succinyl) glucoside, and some un-identified cyanidin- and peonidin-derivatives were recorded in purple barley, whereas black and yellow genotypes contained only 1 peonidin-derivative	Lee et al. (2013)
Barley	Winter blue barley, n = 5, harvested in different years	Methanol (85%) with 1% TFA	HPLC-MS/MS, UVV, total anthocyanin	47–84 mg/kg (total anthocyanin)	Delphinidin-3-glucoside, cyanidin-3-glucoside, delphinidin-3-malonylglucoside, cyanidin-3-malonylglucoside	Diczházi and Kursinszki (2014)
Barley	Tibetan hulless barley-purple barley, blue barley, and white barley	Methanol (80%) with HCl (0.1%)	LC-MS, UVV total anthocyanin	466.2 mg/kg in purple, 82 mg/kg in blue, none in white	Cyanidin-3-glucoside	Zhang et al. (2019a, b)
Barley	Whole grain, 17 colored cultivars	Methanol (85%) acidified with HCl (1N)	HPLC, PDA, UVV total anthocyanin	Average of 1337 mg/kg in 17 cultivars	Malvidin 3-glucoside (51.8%), delphinidin-3-glucoside (12.7%), cyanidin (5.0%), both petunidin and peonidin-3-glucoside (~2.5%), both delphinidin-3-rutinoside anthocyanidin-3-glucoside (~1.4%)	Suriano et al. (2019)
Barley	Purple highland barley bran	Ultrasonic extractions	LC-MS	124,374 mg/kg	Cyanidin malonyl glucoside, cyanidin-3-galactoside, cyanidin acetyl galactoside, cyanidin di-glucoside, pelargonidin-3-glucoside, peonidin glucoside	Zhang et al. (2020)

(continued)

Table 5.2 (continued)

Cereal type	Sample type and number of genotypes (n)	Extraction solvent	Quantification and identification methods	Total anthocyanin content	Anthocyanin (composition)	References
Sorghum	Whole grain flour, n = 13	Methanol with 1% HCl	HPLC, UVV	0–680µg/g (3-deoxyanthocyanin content)	Luteolinidin (0–282µg/g), apigeninidin (0–166µg/g), 5-methoxyluteolinidin (0–154µg/g), 7-methoxyapigeninidin (0–137µg/g)	Dykes et al. (2009)
Sorghum	Leaf sheath, n = 1	A mixture of 1,3-ethanol and butanediol	UVV, MS, NMR	–	Pyrano-apigeninidin 4-vinylphenol	Khalil et al. (2010)
Sorghum	Leaf sheath, n = 6	Methanol and HCl mixture (85:15)	HPLC, UVV, total anthocyanin	14–35 CGE mg/g, db	Apigeninidin (17–46 mg/g), luteolinidin (0.4–2.4 mg/g), malvidin (0.6–1.0 mg/g)	Kayodé et al. (2011)
Sorghum	Leaf sheath, n = 6	Aqueous methanol (50%)	HPLC, UV–vis, MS	-	Apigeninidin-flavene dimer, apigenin-7-O-methylflavene dimer	Geera et al. (2012)
Sorghum	Leaf, n = 1 (mutant)	Acidified methanol (1% HCl)	pH differential method, HPLC, UV–vis	Unit: luteolinidin equivalents, 10.05 mg/g (db) (pH differential); 2612µg/g, db (HPLC)	Luteolidin (1768µg/g, db), apigeninidin (421µg/g, db), 7-O-methyl luteolidin (336µg/g, db), 5-O-methyl luteolidin (31µg/g, db), 7-O-methyl apigeninidin (39µg/g, db), 5,7-dimethyl apigeninidin (17µg/g, db)	Petti et al. (2014)
Finger millet	Whole grain flour, n = 1	Ethyl acetate	UV–vis based method for qualitative analysis	-	Apigeninidin-type anthocyanin, luteolinidin-type anthocyanin	Siwela et al. (2010)

CGE cyanidin 3-glucoside equivalent, *HPLC-MS* high-performance liquid chromatography-mass spectrometry, *UVV* ultraviolet–visible spectroscopy. *PDA* photodiode array detector, *db* dry weight basis

Rice is rich in important anthocyanins amongst other phenolic compounds (Goufo and Trindade 2014). In rice like black rice is a rich source of cyanidin 3-glucoside and peonidin-3-glucoside, which possess antioxidative and anti-inflammatory activities (Kong and Lee 2010). For example, in black and red rice, cyanidin-3,5-diglucoside, cyanidin-3-glucoside, cyanidin-3-(6″-*p*-coumaryl)glucoside, pelargonidin-3-glucoside, peonidin-3-glucoside, peonidin-3-(6″-*p*-coumaryl) glucoside, and cyanidin-3-arabidoside had been reported (Pereira-Caro et al. 2013b). Similarly, in black-purple rice, cyanidin-3,5-diglucoside, cyanidin-3-glucoside, cyanidin-3-(6″-p-coumaryl)glucoside, pelargonidin-3-glucoside, peonidin-3-glucoside, peonidin-3-(6″-*p*-coumaryl)glucoside, and cyanidin-3-arabidoside had been reported by Pereira-Caro et al. (2013a) as shown in Table 5.2. In black rice, whole grain, embryo, bran, and endosperm were studied for anthocyanins (Shao et al. 2014a). The amount of cyanidin 3-glucoside (0.52 for whole grain, 0.12 for embryo, 3.6 for bran, and 0 mg/g for endosperm), peonidin 3-glucoside (0.14 for whole grain, 0.09 for embryo, 0.7 for bran, and 0 mg/g for endosperm), and cyanidin-3-rutinoside (0.06 for whole grain, 0 for embryo, 0.3 for bran, and 0 mg/g for endosperm). Similarly, in black whole grain, cyanidin 3,5-diglucoside, cyanidin 3-gentiobioside, cyanidin-3-glucoside, cyanidin-3-sambubioside, cyanidin-3-rutinoside, peonidin-3-glucoside, cyanidin, and peonidin were reported (Hao et al. 2015). The whole grain and polished grains showed cyaniding-3-glucoside and peonidin-3-glucoside as major anthocyanins (Zhu et al. 2017). The pigmented rice samples from Italy and France showed cyanidin-3-glucoside accounted for more than 87% of black rice (Melini et al. 2019). In different types of rice, these anthocyanins impart characteristic higher antioxidant activity (Wang et al. 2020).

Among cereals, wheat is also rich in several important anthocyanins, which act as antioxidants and provide protection to the wheat grain and seedlings (Shoeva et al. 2017). For example, in blue wheat grains, delphinidin-3-glucoside (135 mg/kg), cyanidin 3-glucoside (86 mg/kg), delphinidin 3-rutinoside (65 mg/kg), and cyanidin 3-rutinoside (5.3 mg/kg) had been reported (Abdel-Aal et al. 2008). Similarly, Tyl and Bunzel (2012) reported delphinidin-3-glucoside (9.3 mg/kg), delphinidin-3-rutinoside (33.1 mg/kg), cyanidin-3-glucoside (4.1 mg/kg), cyanidin-3-rutinoside (18.5 mg/kg) had been reported in blue wheat grains and whole-grain flour. The percent anthocyanins in the whole grain of purple wheat contain cyanidin-3-glucoside (43%), peonidin-3-glucoside (40%), and malvidin-3-galactoside (17%) (Chen et al. 2013). In blue-aleurone bread wheat, cyanidin-3-glucoside (1.2–7.1 mg/kg), cyanidin-3-rutinoside (7.1–11 mg/kg), delphinidin-3-glucoside (9.9–18.8 mg/kg), delphinidin-3-rutinoside (31.8–35.9 mg/kg), peonidin-3-glucoside (0.8–1.0 mg/kg), peonidin-3-galactoside (1.9 mg/kg), peonidin-3-arabinoside (0.9–3.5 mg/kg), and malvidin-3-glucoside (10.3–13.8 mg/kg) had been reported (Ficco et al. 2014). In durum wheat, only cyanidin-3-glucoside (0.59–10.2 mg/kg), peonidin-3-galactoside (0.06–3.87 mg/kg), and malvidin-3-glucoside (0.06–2 mg/kg) were present. Several species of wheat whole grains were reported to contain total anthocyanins on a dry weight basis by Brandolini et al. (2015). These include 11.6 mg/kg for *Triticum monococcum* ssp. *monococcum*,

43 mg/kg for *T. monococcum* ssp. *thaoudar*, 17.4 mg/kg for *T. monococcum* ssp. *aegilopoides*, 15.3 mg/kg for *T. urartu*, 2.9 mg/kg for *T. turgidum* ssp. *dicoccum*, 2.6 mg/kg for *T. aestivum* ssp. *Spelta.* Recently, Gamel et al. (2020) reported anthocyanins in purple wheat whole flour, flakes, bran, and food products. The authors found cyanidin 3-glucoside, cyanidin 3-rutinoside, and pelargonidin-3-glucoside as major compounds. The authors showed that bran has highest amounts of total anthocyanins (806 mg/kg) followed by 126.8 mg/kg for whole flour, and 120 mg/kg for flakes, which corresponds to the high ABTS, DPPH and peroxy radical inhibitions in the same order.

Barley is one of the most important cereal crops grown and cultivated since ancient times. The barley grains are rich in several important antioxidants including phenolic compounds (Idehen et al. 2017). The whole grains of black, blue and purple cultivars of barley had been reported to contain delphinidin-3-glucoside (15–168 mg/kg), cyanidin-3-glucoside (9–240 mg/kg), pelargonidin-3-glucoside (12–42 mg/kg), peonidin-3-glucoside (3–38 mg/kg), and malvidin-3-glucoside (0.6–39 mg/kg). Another study by Yao et al. (2010) reported cyanidin-3-glucoside (25.2 mg/kg), delphinidin-3-glucoside (21.3 mg/kg), and petunidin-3-glucoside (290 mg/kg) in whole grain of black cultivar. Similarly, in whole grain flour, refined flour, and bran, cyanidin-3-glucoside, pelargonidin-3-glucoside, peonidin-3-glucoside, cyanidin-3-(6″-succinyl) glucoside, peonidin-3-(6″-succinyl) glucoside, and some un-identified cyanidin- and peonidin-derivatives were recorded in purple barley, whereas black and yellow genotypes contained only one peonidin-derivative (Lee et al. 2013). Diczházi and Kursinszki (2014) reported that in winter blue barley harvested in different years, delphinidin-3-glucoside, cyanidin-3-glucoside, delphinidin-3-malonylglucoside, and cyanidin-3-malonylglucoside were the predominate anthocyanins. A recent study by Suriano et al. (2019) of the seventeen colored genotypes of barley grains consist of 51.8% of malvidin-3-glucoside, 12.7% of delphinidin-3-glucoside, 5.0% of cyanidin, ~2.5% of both petunidin and peonidin-3-glucoside, and ~1.4% of both delphinidin-3-rutinoside, anthocyanidin-3-glucoside, with the remaining 23% was unidentified anthocyanins. Similarly, the bran of purple highland barley from China consists of cyanidin malonyl glucoside, cyanidin-3-galactoside, cyanidin acetyl galactoside, cyanidin di-glucoside, pelargonidin-3-glucoside, peonidin glucoside (Zhang et al. 2020). These compounds showed significant antioxidant activity.

In Sorghum grains, significant amounts of anthocyanins have been reported. In addition to common anthocyanins, sorghum anthocyanins are unique as they do not contain the hydroxyl group in the 3-position of the *C*-ring and thus are called 3-deoxyanthocyanins (Dykes and Rooney 2006). In sorghums, the proanthocyanidins are the B-type with epicatechin as extension units and catechin as terminal units (Gu et al. 2003). The antioxidant properties of the 3-deoxyanthocyanidins are similar to those of the anthocyanins (Awika et al. 2004). In sorghum, the two common 3-deoxyanthocyanidins that are the yellow is apigeninidin, and the orange one is luteolinidin. Other 3-deoxyanthocyanins identified in sorghum grains include apigeninidin-5-glucoside, luteolinidin-5-glucoside, 5-methoxyluteolinidin, 5-methoxyluteolinidin-7-glucoside, 7-methoxyapigeninidin, 7-methoxyapigeninidin-5-glucoside, 5-methoxyapigeninidin and 7-methoxyluteolinidin (Dykes and Rooney 2006). In whole grain flour, luteolinidin (0–282 mg/kg), apigeninidin (0–166 mg/kg), 5-methoxyluteolinidin (0–154 mg/kg), and 7-methoxyapigeninidin (0–137 mg/kg) (Dykes et al. 2009). In a similar study, apigeninidin (17–46 mg/g), luteolinidin (0.4–2.4 mg/g), and malvidin (0.6–1.0 mg/g) were reported in leaf sheath (Kayodé et al. 2011). They also contain apigeninidin-flavene dimer, and apigenin-7-methylflavene dimer (Geera et al. 2012), and methyl derivatives of anthocyanins (Petti et al. 2014).

5.2.6 Tannins in Cereals

In cereals, tannins are made of different types of phenolic residues depending upon soluble or insoluble nature. For example, pelargonidin and cyanidin structures had been confirmed in soluble condensed tannins (Shahidi and Ambigaipalan 2015). A mixture of proanthocyanidins having various degrees of polymerization had been found in condensed tannins of buckwheat (Watanabe et al. 1997). The antioxidant activity of buckwheat seeds proved to be higher as compared to oats, barley, buckwheat straws, and hulls (Holasova et al. 2002). In human hepatoma HepG2 cells, two types of buckwheat sprouts i.e., Tartary buckwheat (TBS) and common buckwheat (CBS) extracts revealed a significant decline in the production of intracellular peroxides and eliminate the intracellular superoxide anions. The TBS was more effective than CBS in reducing the cellular oxidative stress, which may be due to its higher quercetin-3-rutinoside contents (Liu et al. 2008). This was further confirmed in a study by Lin et al. (2009) by replacing wheat flour (15%) in the bread formula with common buckwheat flour, which showed enhanced functional components such as quercetin and quercetin-3-rutinoside and its antioxidant activity. These phenolic compounds in buckwheat also possess antimicrobial activity as reported by Chitarrini et al. (2014). These authors also suggested that buckwheat antioxidants could be considered as markers for tolerance against mycotoxigenic pathogens and can be used for improving food safety. Phenolic extracts from the sprouts of buckwheat also have been shown to exhibit strong antioxidant activity (Shahidi and Ambigaipalan 2015). Different processing treatments such as cooking, fermentation, and germination have been found to reduce the tannin contents and significantly increases the total phenol contents (Siwatch et al. 2019).

In sorghum, tannins are readily associated with sorghum proteins. The varieties with a pigmented testa had been reported to contain condensed tannins. The tannins in sorghum are excellent antioxidants, which slow hydrolysis in foods, produce naturally dark-colored products, and increase the dietary fiber levels of food products (Dykes and Rooney 2006; Dykes 2019b). The condensed tannins present in sorghum grains possess remarkable structural diversity and also exhibited strong antioxidant activity (Jiang et al. 2020).

Millets are also rich sources of tannins, which can act as antioxidants (Kaur et al. 2019). Recently Xiang et al. (2019) reported that the white variety had significantly lower levels of total condensed tannins (TCT) than the colored finger millet varieties. The brown variety contained the highest TCT (83.5 mg CE/100 g, DW), which was 2.6 times that of the white variety. Reddish and red finger millets also contained more than two times the TCT of white finger millet. Previously Chandrasekara and Shahidi (2010) showed that finger millet contained significantly higher TCT than all other millet types, including foxtail, proso, pearl, kodo, and little millets.

Oat is a good source of several phenolic compounds that exhibit antioxidant activity. It has been found that the scavenging free-radicals of oat bran is less effective as compared to other cereal brans such as wheat, barley, and rye (Shahidi and Ambigaipalan 2015). However, the soluble fiber fraction of oat had shown a greater antioxidant capacity than wheat, barley, and rye (Lehtinen and Laakso 1997).

5.3   Study Questions

- Discuss the hydroxybenzoic acid and its derivatives in cereals.
- What do you know about the composition of cinnamic acid and its derivatives in cereals?
- What is γ-oryzanol? Describe their structure and composition in cereals.
- Write a detailed note on flavonoids in cereals.
- Explain the distribution and role of anthocyanins in cereals.
- What are the names of cereals in which tannins have been reported?

References

Abdel-Aal E-SM, Abou-Arab AA, Gamel TH, Hucl P, Young JC, Rabalski I (2008) Fractionation of blue wheat anthocyanin compounds and their contribution to antioxidant properties. J Agric Food Chem 56(23):11171–11177

Adom KK, Liu RH (2002) Antioxidant activity of grains. J Agric Food Chem 50(21):6182–6187

Andreasen MF, Christensen LP, Meyer AS, Hansen A (2000) Content of phenolic acids and ferulic acid dehydrodimers in 17 rye (Secale cereale L.) varieties. J Agric Food Chem 48(7):2837–2842

Andreasen MF, Kroon PA, Williamson G, Garcia-Conesa MT (2001) Intestinal release and uptake of phenolic antioxidant diferulic acids. Free Radic Biol Med 31(3):304–314

Awika JM, Rooney LW, Waniska RD (2004) Properties of 3-deoxyanthocyanins from Sorghum. J Agric Food Chem 52(14):4388–4394

Ballin NZ, Sørensen AT (2014) Coumarin content in cinnamon containing food products on the Danish market. Food Control 38:198–203

Brandolini A, Hidalgo A, Gabriele S, Heun M (2015) Chemical composition of wild and feral diploid wheats and their bearing on domesticated wheats. J Cereal Sci 63:122–127

Camelo-Méndez GA, Agama-Acevedo E, Sanchez-Rivera MM, Bello-Pérez LA (2016) Effect on in vitro starch digestibility of Mexican blue maize anthocyanins. Food Chem 211:281–284

Chandrasekara A, Shahidi F (2010) Content of insoluble bound phenolics in millets and their contribution to antioxidant capacity. J Agric Food Chem 58(11):6706–6714

Chatham LA, Howard JE, Juvik JA (2020) A natural colorant system from corn: flavone-anthocyanin copigmentation for altered hues and improved shelf life. Food Chem 310:125734

Chen CY, Milbury PE, Kwak HK, Collins FW, Samuel P, Blumberg JB (2004) Avenanthramides and phenolic acids from oats are bioavailable and act synergistically with vitamin C to enhance hamster and human LDL resistance to oxidation. J Nutr 134(6):1459–1466

Chen W, Müller D, Richling E, Wink M (2013) Anthocyanin-rich purple wheat prolongs the life span of *Caenorhabditis elegans* probably by activating the DAF-16/FOXO transcription factor. J Agric Food Chem 61(12):3047–3053

Chitarrini G, Nobili C, Pinzari F, Antonini A, De Rossi P, Del Fiore A, Procacci S, Tolaini V, Scala V, Scarpari M, Reverberi M (2014) Buckwheat achenes antioxidant profile modulates *Aspergillus flavus* growth and aflatoxin production. Int J Food Microbiol 189:1–10

Cho D-H, Lim S-T (2018) Changes in phenolic acid composition and associated enzyme activity in shoot and kernel fractions of brown rice during germination. Food Chem 256:163–170

Cho Y-H, Lim S-Y, Rehman A, Farooq M, Lee D-J (2019) Characterization and quantification of γ-oryzanol in Korean rice landraces. J Cereal Sci 88:150–156

Chu Y-F, Wise ML, Gulvady AA, Chang T, Kendra DF, Van Klinken BJ-W, Shi Y, O'Shea M (2013) In vitro antioxidant capacity and anti-inflammatory activity of seven common oats. Food Chem 139(1–4):426–431

Chu C, Du Y, Yu X, Shi J, Yuan X, Liu X, Liu Y, Zhang H, Zhang Z, Yan N (2020) Dynamics of antioxidant activities, metabolites, phenolic acids, flavonoids, and phenolic biosynthetic genes in germinating Chinese wild rice (*Zizania latifolia*). Food Chem 318:126483

Chutipaijit S, Sutjaritvorakul T (2018) Comparative study of total phenolic compounds, flavonoids and antioxidant capacities in pigmented and non-pigmented rice of indica rice varieties. J Food Measur Charact 12(2):781–788

Collison A, Yang L, Dykes L, Murray S, Awika JM (2015) Influence of genetic background on anthocyanin and copigment composition and behavior during thermoalkaline processing of maize. J Agric Food Chem 63(22):5528–5538

Das G, Patra JK, Choi J, Baek K-H (2017) Rice grain, a rich source of natural bioactive compounds. Pak J Agric Sci 54(3):671–682

Diczházi I, Kursinszki L (2014) Anthocyanin content and composition in winter blue barley cultivars and lines. Cereal Chem 91(2):195–200

Dietrych-Szostak D, Oleszek W (1999) Effect of processing on the flavonoid content in buckwheat (*Fagopyrum esculentum* Möench) grain. J Agric Food Chem 47(10):4384–4387

Dykes L (2019a) Sorghum phytochemicals and their potential impact on human health. In: Zhao Z-Y, Dahlberg J (eds) Sorghum: methods and protocols. Springer, New York, pp 121–140

Dykes L (2019b) Tannin analysis in sorghum grains. In: Zhao Z-Y, Dahlberg J (eds) Sorghum: methods and protocols. Springer, New York, pp 109–120

Dykes L, Rooney LW (2006) Sorghum and millet phenols and antioxidants. J Cereal Sci 44(3):236–251

Dykes L, Seitz LM, Rooney WL, Rooney LW (2009) Flavonoid composition of red sorghum genotypes. Food Chem 116(1):313–317

Escribano-Bailón MT, Santos-Buelga C, Rivas-Gonzalo JC (2004) Anthocyanins in cereals. J Chromatogr A 1054(1–2):129–141

Ficco DB, De Simone V, Colecchia SA, Pecorella I, Platani C, Nigro F, Finocchiaro F, Papa R, De Vita P (2014) Genetic variability in anthocyanin composition and nutritional properties of blue, purple, and red bread (*Triticum aestivum* L.) and durum (*Triticum turgidum* L. ssp. turgidum convar. durum) wheats. J Agric Food Chem 62(34):8686–8695

Gamel TH, Wright AJ, Pickard M, Abdel-Aal E-SM (2020) Characterization of anthocyanin-containing purple wheat prototype products as functional foods with potential health benefits. Cereal Chem 97(1):34–38

Geera B, Ojwang LO, Awika JM (2012) New highly stable dimeric 3-deoxyanthocyanidin pigments from Sorghum bicolor leaf sheath. J Food Sci 77(5):C566–C572

Gong ES, Luo SJ, Li T, Liu CM, Zhang GW, Chen J, Zeng ZC, Liu RH (2017) Phytochemical profiles and antioxidant activity of brown rice varieties. Food Chem 227:432–443

Goufo P, Trindade H (2014) Rice antioxidants: phenolic acids, flavonoids, anthocyanins, proanthocyanidins, tocopherols, tocotrienols, gamma-oryzanol, and phytic acid. Food Sci Nutr
2(2):75–104

Goupy P, Hugues M, Boivin P, Amiot MJ (1999) Antioxidant composition and activity of barley
(*Hordeum vulgare*) and malt extracts and of isolated phenolic compounds. J Sci Food Agric
79(12):1625–1634

Gu L, Kelm MA, Hammerstone JF, Beecher G, Holden J, Haytowitz D, Prior RL (2003) Screening
of foods containing proanthocyanidins and their structural characterization using LC-MS/MS
and thiolytic degradation. J Agric Food Chem 51(25):7513–7521

Guo X-D, Wu C-S, Ma Y-J, Parry J, Xu Y-Y, Liu H, Wang M (2012) Comparison of milling fractions
of tartary buckwheat for their phenolics and antioxidant properties. Food Res Int 49(1):53–59

Hakala P, Lampi A-M, Ollilainen V, Werner U, Murkovic M, Wähälä K, Karkola S, Piironen V
(2002) Steryl phenolic acid esters in cereals and their milling fractions. J Agric Food Chem
50(19):5300–5307

Hao J, Zhu H, Zhang Z, Yang S, Li H (2015) Identification of anthocyanins in black rice (*Oryza
sativa* L.) by UPLC/Q-TOF-MS and their in vitro and in vivo antioxidant activities. J Cereal
Sci 64:92–99

Harakotr B, Suriharn B, Tangwongchai R, Scott MP, Lertrat K (2014) Anthocyanins and antioxidant activity in coloured waxy corn at different maturation stages. J Funct Foods 9:109–118

Harakotr B, Suriharn B, Scott MP, Lertrat K (2015) Genotypic variability in anthocyanins,
total phenolics, and antioxidant activity among diverse waxy corn germplasm. Euphytica
203(2):237–248

Holasova M, Fiedlerova V, Smrcinova H, Orsak M, Lachman J, Vavreinova S (2002) Buckwheat—
the source of antioxidant activity in functional foods. Food Res Int 35(2–3):207–211

Holtekjølen AK, Bævre A, Rødbotten M, Berg H, Knutsen SH (2008) Antioxidant properties and
sensory profiles of breads containing barley flour. Food Chem 110(2):414–421

Hong H, Netzel M, O'Hare T (2020) Optimisation of extraction procedure and development of
LC–DAD–MS methodology for anthocyanin analysis in anthocyanin-pigmented corn kernels.
Food Chem 319:126515

Idehen E, Tang Y, Sang S (2017) Bioactive phytochemicals in barley. J Food Drug Anal
25(1):148–161

Iqbal S, Bhanger M, Anwar F (2005) Antioxidant properties and components of some commercially available varieties of rice bran in Pakistan. Food Chem 93(2):265–272

Irakli MN, Samanidou VF, Biliaderis CG, Papadoyannis IN (2012) Simultaneous determination of
phenolic acids and flavonoids in rice using solid-phase extraction and RP-HPLC with photodiode array detection. J Sep Sci 35(13):1603–1611

Irakli MN, Samanidou VF, Katsantonis DN, Biliaderis CG, Papadoyannis IN (2016) Phytochemical
profiles and antioxidant capacity of pigmented and non-pigmented genotypes of rice (*Oryza
sativa* L.). Cereal Res Commun 44(1):98–110

Jiang Y, Zhang H, Qi X, Wu G (2020) Structural characterization and antioxidant activity of condensed tannins fractionated from sorghum grain. J Cereal Sci 92:102918

Kähkönen MP, Hopia AI, Vuorela HJ, Rauha J-P, Pihlaja K, Kujala TS, Heinonen M (1999)
Antioxidant activity of plant extracts containing phenolic compounds. J Agric Food Chem
47(10):3954–3962

Kaur P, Purewal SS, Sandhu KS, Kaur M, Salar RK (2019) Millets: a cereal grain with potent
antioxidants and health benefits. J Food Measur Charact 13(1):793–806

Kayodé AP, Nout MR, Linnemann AR, Hounhouigan JD, Berghofer E, Siebenhandl-Ehn S (2011)
Uncommonly high levels of 3-deoxyanthocyanidins and antioxidant capacity in the leaf sheaths
of dye sorghum. J Agric Food Chem 59(4):1178–1184

Khalil A, Baltenweck-Guyot R, Ocampo-Torres R, Albrecht P (2010) A novel symmetrical
pyrano-3-deoxyanthocyanidin from a Sorghum species. Phytochem Lett 3(2):93–95

Kil HY, Seong ES, Ghimire BK, Chung I-M, Kwon SS, Goh EJ, Heo K, Kim MJ, Lim JD, Lee
D (2009) Antioxidant and antimicrobial activities of crude sorghum extract. Food Chem
115(4):1234–1239

Kim M-J, Hyun J-N, Kim J-A, Park J-C, Kim M-Y, Kim J-G, Lee, S-J, Chun, S-C, Chung, I-M (2007) Relationship between phenolic compounds, anthocyanins content and antioxidant activity in colored barley germplasm. J Agric Food Chem 55(12):4802–4809

Kim B, Woo S, Kim M-J, Kwon S-W, Lee J, Sung SH, Koh H-J (2018) Identification and quantification of flavonoids in yellow grain mutant of rice (*Oryza sativa* L.). Food Chem 241:154–162

Kong S, Lee J (2010) Antioxidants in milling fractions of black rice cultivars. Food Chem 120(1):278–281

Lang GH, Lindemann IS, Ferreira CD, Hoffmann JF, Vanier NL, de Oliveira M (2019) Effects of drying temperature and long-term storage conditions on black rice phenolic compounds. Food Chem 287:197–204

Lao F, Giusti MM (2016) Quantification of purple corn (*Zea mays* L.) anthocyanins using spectrophotometric and HPLC approaches: method comparison and correlation. Food Anal Methods 9(5):1367–1380

Lee C, Han D, Kim B, Baek N, Baik BK (2013) Antioxidant and anti-hypertensive activity of anthocyanin-rich extracts from hulless pigmented barley cultivars. Int J Food Sci Technol 48(5):984–991

Lehtinen P, Laakso S (1997) Antioxidative-like effect of different cereals and cereal fractions in aqueous suspension. J Agric Food Chem 45(12):4606–4611

Lin L-Y, Liu H-M, Yu Y-W, Lin S-D, Mau J-L (2009) Quality and antioxidant property of buckwheat enhanced wheat bread. Food Chem 112(4):987–991

Liu C-L, Chen Y-S, Yang J-H, Chiang B-H (2008) Antioxidant activity of tartary (*Fagopyrum tataricum* (L.) Gaertn.) and common (*Fagopyrum esculentum* Moench) buckwheat sprouts. J Agric Food Chem 56(1):173–178

Liyana-Pathirana CM, Shahidi F (2007a) Antioxidant and free radical scavenging activities of whole wheat and milling fractions. Food Chem 101(3):1151–1157

Liyana-Pathirana CM, Shahidi F (2007b) The antioxidant potential of milling fractions from breadwheat and durum. J Cereal Sci 45(3):238–247

Liyana-Pathirana C, Dexter J, Shahidi F (2006) Antioxidant properties of wheat as affected by pearling. J Agric Food Chem 54(17):6177–6184

Luthria DL, Liu K (2013) Localization of phenolic acids and antioxidant activity in sorghum kernels. J Funct Foods 5(4):1751–1760

Madhujith T, Shahidi F (2006) Optimization of the extraction of antioxidative constituents of six barley cultivars and their antioxidant properties. J Agric Food Chem 54(21):8048–8057

Madhujith T, Shahidi F (2009) Antioxidant potential of barley as affected by alkaline hydrolysis and release of insoluble-bound phenolics. Food Chem 117(4):615–620

Melini V, Panfili G, Fratianni A, Acquistucci R (2019) Bioactive compounds in rice on Italian market: pigmented varieties as a source of carotenoids, total phenolic compounds and anthocyanins, before and after cooking. Food Chem 277:119–127

Mohsen SM, Ammar AS (2009) Total phenolic contents and antioxidant activity of corn tassel extracts. Food Chem 112(3):595–598

Naczk M, Shahidi F (2006) Phenolics in cereals, fruits and vegetables: occurrence, extraction and analysis. J Pharm Biomed Anal 41(5):1523–1542

Nankar AN, Dungan B, Paz N, Sudasinghe N, Schaub T, Holguin FO, Pratt RC (2016) Quantitative and qualitative evaluation of kernel anthocyanins from southwestern United States blue corn. J Sci Food Agric 96(13):4542–4552

Ndolo VU, Beta T (2014) Comparative studies on composition and distribution of phenolic acids in cereal grain botanical fractions. Cereal Chem 91(5):522–530

Nordkvist E, Salomonsson AC, Åman P (1984) Distribution of insoluble bound phenolic acids in barley grain. J Sci Food Agric 35(6):657–661

Nyström L, Mäkinen M, Lampi A-M, Piironen V (2005) Antioxidant activity of steryl ferulate extracts from rye and wheat bran. J Agric Food Chem 53(7):2503–2510

Osawa T, Katsuzaki H, Hagiwara Y, Hagiwara H, Shibamoto T (1992) A novel antioxidant isolated from young green barley leaves. J Agric Food Chem 40(7):1135–1138

Pang Y, Ahmed S, Xu Y, Beta T, Zhu Z, Shao Y, Bao J (2018) Bound phenolic compounds and antioxidant properties of whole grain and bran of white, red and black rice. Food Chem 240:212–221

Paulsmeyer M, Chatham L, Becker T, West M, West L, Juvik J (2017) Survey of anthocyanin composition and concentration in diverse maize germplasms. J Agric Food Chem 65(21):4341–4350

Paznocht L, Kotíková Z, Burešová B, Lachman J, Martinek P (2020) Phenolic acids in kernels of different coloured-grain wheat genotypes. Plant Soil Environ 66(2):57–64

Peanparkdee M, Patrawart J, Iwamoto S (2019) Effect of extraction conditions on phenolic content, anthocyanin content and antioxidant activity of bran extracts from Thai rice cultivars. J Cereal Sci 86:86–91

Pereira-Caro G, Cros G, Yokota T, Crozier A (2013a) Phytochemical profiles of black, red, brown, and white rice from the Camargue region of France. J Agric Food Chem 61(33):7976–7986

Pereira-Caro G, Watanabe S, Crozier A, Fujimura T, Yokota T, Ashihara H (2013b) Phytochemical profile of a Japanese black–purple rice. Food Chem 141(3):2821–2827

Petti C, Kushwaha R, Tateno M, Harman-Ware AE, Crocker M, Awika J, DeBolt S (2014) Mutagenesis breeding for increased 3-deoxyanthocyanidin accumulation in leaves of *Sorghum bicolor* (L.) Moench: a source of natural food pigment. J Agric Food Chem 62(6):1227–1232

Przybylska-Balcerek A, Frankowski J, Stuper-Szablewska K (2019) Bioactive compounds in sorghum. Eur Food Res Technol 245(5):1075–1080

Qiu Y, Liu Q, Beta T (2009) Antioxidant activity of commercial wild rice and identification of flavonoid compounds in active fractions. J Agric Food Chem 57(16):7543–7551

Qiu Y, Liu Q, Beta T (2010) Antioxidant properties of commercial wild rice and analysis of soluble and insoluble phenolic acids. Food Chem 121(1):140–147

Renger A, Steinhart H (2000) Ferulic acid dehydrodimers as structural elements in cereal dietary fibre. Eur Food Res Technol 211(6):422–428

Sarker SD, Nahar L (2019) Dietary coumarins. In: Xiao J, Sarker SD, Asakawa Y (eds) Handbook of dietary phytochemicals. Springer, Singapore, pp 1–56

Schendel RR (2019) Phenol content in sprouted grains. In: Feng H, Nemzer B, DeVries JW (eds) Sprouted grains. AACC International Press, St. Paul, pp 247–315

Sen A, Bergvinson D, Miller SS, Atkinson J, Fulcher RG, Arnason JT (1994) Distribution and microchemical detection of phenolic acids, flavonoids, and phenolic acid amides in maize kernels. J Agric Food Chem 42(9):1879–1883

Setyaningsih W, Saputro IE, Carrera CA, Palma M (2019) Optimisation of an ultrasound-assisted extraction method for the simultaneous determination of phenolics in rice grains. Food Chem 288:221–227

Shahidi F (2015) Handbook of antioxidants for food preservation. Woodhead Publishing, Cambridge

Shahidi F, Ambigaipalan P (2015) Phenolics and polyphenolics in foods, beverages and spices: antioxidant activity and health effects—a review. J Funct Foods 18:820–897

Shahidi F, Yeo J (2016) Insoluble-bound phenolics in food. Molecules 21(9):1216

Shao Y, Xu F, Sun X, Bao J, Beta T (2014a) Identification and quantification of phenolic acids and anthocyanins as antioxidants in bran, embryo and endosperm of white, red and black rice kernels (*Oryza sativa* L.). J Cereal Sci 59(2):211–218

Shao Y, Xu F, Sun X, Bao J, Beta T (2014b) Phenolic acids, anthocyanins, and antioxidant capacity in rice (*Oryza sativa* L.) grains at four stages of development after flowering. Food Chem 143:90–96

Shao Y, Hu Z, Yu Y, Mou R, Zhu Z, Beta T (2018) Phenolic acids, anthocyanins, proanthocyanidins, antioxidant activity, minerals and their correlations in non-pigmented, red, and black rice. Food Chem 239:733–741

Shen S, Huang R, Li C, Wu W, Chen H, Shi J, Chen S, Ye X (2018) Phenolic compositions and antioxidant activities differ significantly among sorghum grains with different applications. Molecules 23(5):1203

Shoeva O, Gordeeva E, Arbuzova V, Khlestkina E (2017) Anthocyanins participate in protection of wheat seedlings from osmotic stress. Cereal Res Commun 45(1):47–56

Siwatch M, Yadav RB, Yadav BS (2019) Influence of processing treatments on nutritional and physicochemical characteristics of buckwheat (*Fagopyrum esculentum*). Curr Nutr Food Sci 15(4):408–414

Siwela M, Taylor JR, de Milliano WA, Duodu KG (2010) Influence of phenolics in finger millet on grain and malt fungal load, and malt quality. Food Chem 121(2):443–449

Sosulski F, Krygier K, Hogge L (1982) Free, esterified, and insoluble-bound phenolic acids. 3. Composition of phenolic acids in cereal and potato flours. J Agric Food Chem 30(2):337–340

Subba Rao MV, Muralikrishna G (2002) Evaluation of the antioxidant properties of free and bound phenolic acids from native and malted finger millet (ragi, *Eleusine coracana* Indaf-15). J Agric Food Chem 50(4):889–892

Sultana B, Anwar F, Przybylski R (2007) Antioxidant potential of corncob extracts for stabilization of corn oil subjected to microwave heating. Food Chem 104(3):997–1005

Suriano S, Savino M, Codianni P, Iannucci A, Caternolo G, Russo M, Pecchioni N, Troccoli A (2019) Anthocyanin profile and antioxidant capacity in coloured barley. Int J Food Sci Technol 54(7):2478–2486

Syedd-León R, Orozco R, Álvarez V, Carvajal Y, Rodríguez G (2020) Chemical and antioxidant charaterization of native corn germplasm from two regions of Costa Rica: a conservation approach. Int J Food Sci 2020:2439541

Tyl CE, Bunzel M (2012) Antioxidant activity-guided fractionation of blue wheat (UC66049 *Triticum aestivum* L.). J Agric Food Chem 60(3):731–739

Van Hung P (2016) Phenolic compounds of cereals and their antioxidant capacity. Crit Rev Food Sci Nutr 56(1):25–35

Vanamala JKP, Massey AR, Pinnamaneni SR, Reddivari L, Reardon KF (2018) Grain and sweet sorghum (*Sorghum bicolor* L. Moench) serves as a novel source of bioactive compounds for human health. Crit Rev Food Sci Nutr 58(17):2867–2881

Viswanath V, Urooj A, Malleshi N (2009) Evaluation of antioxidant and antimicrobial properties of finger millet polyphenols (*Eleusine coracana*). Food Chem 114(1):340–346

Wang W, Li Y, Dang P, Zhao S, Lai D, Zhou L (2018) Rice secondary metabolites: structures, roles, biosynthesis, and metabolic regulation. Molecules 23(12):3098

Wang Y, Zhao L, Zhang R, Yang X, Sun Y, Shi L, Xue P (2020) Optimization of ultrasound-assisted extraction by response surface methodology, antioxidant capacity, and tyrosinase inhibitory activity of anthocyanins from red rice bran. Food Sci Nutr 8(2):921–932

Watanabe M (1999) Antioxidative phenolic compounds from Japanese barnyard millet (*Echinochloa utilis*) grains. J Agric Food Chem 47(11):4500–4505

Watanabe M, Ohshita Y, Tsushida T (1997) Antioxidant compounds from buckwheat (*Fagopyrum esculentum* Möench) hulls. J Agric Food Chem 45(4):1039–1044

Xiang J, Apea-Bah FB, Ndolo VU, Katundu MC, Beta T (2019) Profile of phenolic compounds and antioxidant activity of finger millet varieties. Food Chem 275:361–368

Xiong Y, Zhang P, Warner RD, Fang Z (2019) Sorghum grain: from genotype, nutrition, and phenolic profile to its health benefits and food applications. Compr Rev Food Sci Food Saf 18(6):2025–2046

Yang J, Liu RH, Halim L (2009) Antioxidant and antiproliferative activities of common edible nut seeds. LWT Food Sci Technol 42(1):1–8

Yao Y, Sang W, Zhou M, Ren G (2010) Antioxidant and α-glucosidase inhibitory activity of colored grains in China. J Agric Food Chem 58(2):770–774

Yu J, Vasanthan T, Temelli F (2001) Analysis of phenolic acids in barley by high-performance liquid chromatography. J Agric Food Chem 49(9):4352–4358

Yu J, Ahmedna M, Goktepe I (2005) Effects of processing methods and extraction solvents on concentration and antioxidant activity of peanut skin phenolics. Food Chem 90(1–2):199–206

Zhang L, Zhang T, Chang M, Lu M, Liu R, Jin Q, Wang X (2019a) Effects of interaction between
α-tocopherol, oryzanol, and phytosterol on the antiradical activity against DPPH radical. LWT
Food Sci Technol 112:108206

Zhang Y, Li M, Gao H, Wang B, Tongcheng X, Gao B, Yu L (2019b) Triacylglycerol, fatty acid,
and phytochemical profiles in a new red sorghum variety (Ji Liang No. 1) and its antioxidant
and anti-inflammatory properties. Food Sci Nutr 7(3):949–958

Zhang Y, Lin Y, Huang L, Tekliye M, Rasheed HA, Dong M (2020) Composition, antioxidant, and
anti-biofilm activity of anthocyanin-rich aqueous extract from purple highland barley bran.
LWT Food Sci Technol 125:109181

Zhu F (2018) Anthocyanins in cereals: composition and health effects. Food Res Int 109:232–249

Zhu Q, Yu S, Zeng D, Liu H, Wang H, Yang Z, Xie X, Shen R, Tan J, Li H (2017) Development
of "purple endosperm rice" by engineering anthocyanin biosynthesis in the endosperm with a
high-efficiency transgene stacking system. Mol Plant 10(7):918–929

Ziegler V, Ferreira CD, Hoffmann JF, Chaves FC, Vanier NL, de Oliveira M, Elias MC (2018)
Cooking quality properties and free and bound phenolics content of brown, black, and red rice
grains stored at different temperatures for six months. Food Chem 242:427–434

Zilic S, Serpen A, Akillioglu G, Gokmen V, Vancetovic J (2012) Phenolic compounds, carotenoids,
anthocyanins, and antioxidant capacity of colored maize (*Zea mays* L.) kernels. J Agric Food
Chem 60(5):1224–1231

Chapter 6
Phenolic Antioxidants in Legumes and Nuts

Learning Objectives

In this chapter, the reader will be able to:

- Describe in detail the hydroxybenzoic acid derivatives in legumes and nuts.
- Understand cinnamic acid and its derivatives present in legumes and nuts.
- Discuss in detail the composition of flavonoids in legumes.
- Understand different anthocyanins present in legumes.
- Learn about tannins present in legumes and stilbenes in nuts.
- Describe in detail the flavonoids and anthocyanins present in nuts.

6.1 Introduction

Phenolic antioxidants are present in all types of legumes and nuts. There is a vast amount of literature available regarding the phenolic compounds in both legumes and nuts, however, in this chapter, only those studies will be included which focused on edible legumes and nuts. Both legumes and nuts are important components of our daily foods due to their significant nutritional values. In this chapter, phenolic antioxidants such as derivatives of hydroxybenzoic acids, cinnamic acids, flavonoids, catechins, proanthocyanins, and anthocyanins have been discussed.

6.2 Phenolic Antioxidants in Legumes

Legumes are one of the most nutritious foods widely consumed in the world. There are thousands of legume species occurred in nature, but beans are more commonly consumed by the human than any other. In some countries such as Mexico and

© Springer Nature Switzerland AG 2021   177
A. Zeb, *Phenolic Antioxidants in Foods: Chemistry, Biochemistry and Analysis*,
https://doi.org/10.1007/978-3-030-74768-8_6

Brazil, beans are the primary source of proteins in human diets. As half the grain legumes consumed worldwide are common beans, they represent the species of choice for the study of grain legume nutrition (Broughton et al. 2003). Important legumes are chickpea, lentil, cowpea, and green peas. Legumes are different than cereals due to the fact, that legumes belong to the family Fabaceae, whereas cereals are mostly from the Poaceae family. Other differences may include chemical nature and their responses toward nitrogen availability (Adams et al. 2018). Amarowicz and Pegg (2008) had reviewed legumes as a source of natural antioxidants. Similarly, Singh et al. (2017a) also reviewed the phenolic composition and antioxidant activity of legumes grains. The major polyphenolic compounds of legumes consist mainly of tannins, phenolic acids, and flavonoids. Phenolic compounds are usually associated with cell walls in legumes. However, the antioxidant activities such as DPPH, ABTS, FRAP, and ORAC were higher in the whole legume as compared to cell wall extracts (Salawu et al. 2014).

6.2.1 Hydroxybenzoic Acids

Hydroxybenzoic acids have been identified in several legumes. For example, broad beans, chickpeas, yellow split peas, green split peas, large lentils, small lentils, giant beans, elephant beans, large beans, medium beans, small beans, black-eyed beans, pinto beans, and white lupines had been reported to contain gallic acid, *p*-hydroxybenzoic acid, syringic acid, vanillic acid, and protocatechuic acids along with other phenolic compounds (Kalogeropoulos et al. 2010). In Marama bean seed coats, gallic acid (1.98 mg/g), homogentisic acid (0.49 mg/g), and *p*-hydroxybenzoic acid (1.76 mg/g) were reported (Shelembe et al. 2012).

The concentrations of vanillic, *p*-hydroxybenzoic, and protocatechuic acids vary in beans, peas, and lentils (López et al. 2013). Raw lentils had a total of 5.69 mg/kg of hydroxybenzoic acids which includes dihydroxybenzoic acid, vanillic acid glycoside, protocatechuic aldehyde, and *p*-hydroxybenzoic acid (Aguilera et al. 2010). Dihydroxybenzoic acid (3.68 mg/kg) and *p*-hydroxybenzoic acid (1.90 mg/kg) were common phenolic acids in lentils. Similarly, Alshikh et al. (2015) reported the presence of gallic acid, methyl vanillate, and protocatechuic acid in lentil cultivars. In eleven lentil cultivars, gallic acid (90.9–136.8 mg/kg), protocatechuic acid (20.2–37.7 mg/kg), 2,3,4-trihydroxybenzoic acid (16.9–29.2 mg/kg), protocatechualdehyde (3.6–12.1 mg/kg), *p*-hydroxybenzoic acid (15.7–44.9 mg/kg) and vanillic acid (0.59–3.22 mg/kg) were reported (Xu and Chang 2010).

The colorless Cannellini and colored Pinto beans had been found to contain a total of 21.6 and 84.9 mg/kg of hydroxybenzoic acids, respectively (Aguilera et al. 2011). In Pinto beans, the identified compounds were salicylic acid (44.9 mg/kg), vanillic acid (17.0 mg/kg), *p*-hydroxybenzoic acid (12.2 mg/kg), *p*-hydroxyphenyl acetic acid (8.42 mg/kg), and protocatechuic acid (2.40 mg/kg). In Cannellini beans, the amounts of vanillic acid were 10.7 mg/kg, *p*-hydroxyphenyl acetic acid (6.92 mg/kg), and *p*-hydroxybenzoic acid (4.30 mg/kg). Similarly, Amarowicz and Pegg (2008) showed that the crude extract of adzuki bean contains protocatechuic acid

(67.6 mg/kg) and protocatechuic aldehyde (7.71 mg/kg), while in the acetone extract of red lentil, gallic aldehyde (13.45 mg/kg) and p-hydroxybenzoic acid (73.46 mg/kg) was reported (Amarowicz et al. 2009). The main free hydroxybenzoic acids reported in black cowpea seed coats were gallic acid was 27 mg/kg, protocatechuic was 18.9 mg/kg and p-hydroxybenzoic was 5.81 mg/kg (Gutiérrez-Uribe et al. 2011). Similarly, in alkaline hydrolyzable phenolic fractions of dark-colored cranberry beans, seed coats, the amount of protocatechuic acid was 217 mg/kg, and p-hydroxybenzoic acid (239 mg/kg), whereas they were not present in a similar fraction of light-colored cranberry beans seed coats (Chen et al. 2015a). Similarly, the authors also reported that alkaline hydrolyzable phenolic fractions of regular darkening cultivars of whole cranberry beans showed p-hydroxybenzoic acid in the range of 29.08–71.26 mg/kg as the main hydroxybenzoic acid. Wang et al. (2016) reported that gallic and protocatechuic are the common phenolic acids present in beans. Similarly, red sword bean and black sword bean coats had been reported to contain 9.68 and 5.0 mg/g of gallic acid, respectively (Gan et al. 2016). In soybean seeds, benzoic acid (57 μg/g) and protocatechuic (44 mg/kg) acids were the main phenolic acids (Chung et al. 2011). The amounts of p-hydroxybenzoic acid were 10.33 and 10 mg/kg in kidney beans and its sprouts, respectively (Dueñas et al. 2015). Similarly, Dueñas et al. (2016) reported that dark beans had 7.3 and 20.4 mg/kg of total hydroxybenzoic acids such as protocatechuic acid, protocatechuic aldehyde, p-hydroxybenzoic acid, p-hydroxybenzoic aldehyde in soluble and insoluble dietary fiber, respectively. The authors also reported raw lentils to have a high level of these hydroxybenzoic acids in insoluble (73.2 mg/kg) than soluble (2 mg/kg) dietary fiber.

Khang et al. (2016) studied six beans and peanuts for phenolic composition on a dry weight basis. Gallic acid (21.3 mg/kg) was present only in soybeans, the highest amounts of protocatechuic acid were present in soybean as compared to other legumes. p-Hydroxybenzoic acid has been reported in all selected legumes except while cowpeas. Other derivatives reported were vanillic acids (none to 189.2 mg/kg), syringic acid (none to 327.5 mg/kg), vanillin (none to 28.9 mg/kg) and benzoic acid (none to 636.0 mg/kg) (Khang et al. 2016).

Recently, Magalhães et al. (2017) reported the phenolic profile of 29 mature raw varieties of grain legume seeds such as chickpeas, field peas, faba beans, common vetch, and lupins produced in Europe. In chickpea varieties, the amount of p-hydroxybenzoic acid ranged from 19.2 to 60.5 mg/kg, syringic acid was only present in one variety i.e., Elmo, whereas gentisic acid ranged from 8.1 to 26 mg/kg. In field peas, protocatechuic acid ranged from 12.1 to 163.5 mg/kg, and p-hydroxybenzoic acid (45.4–101.7 mg/kg). In faba beans, only gallic acid (22.9–138.2 mg/kg) was reported. Similarly, in white lupin, narrow-leafed lupin, and yellow lupin, protocatechuic acid was the only identified hydroxybenzoic acid derivative. Recently six legumes such as black beans, black lentils, pinto beans, Ruviotto beans, black chickpeas, and chickpeas were studied for phenolic compounds and processing effects (Giusti et al. 2019). In Australian sweet lupin seeds, protocatechuic acid was the main hydroxybenzoic acid reported (Zhong et al. 2019). Protocatechuic acid, protocatechuic aldehyde, p-hydroxybenzoic acid, and syringic acids have been reported in lentils, black soybean, and black turtle bean (Zhang et al. 2018). The identified phenolic compounds were gallic acid, 4-hydroxybenzoic acid, and syringic acid along with other phenolic compounds. In Portuguese unexplored legume germplasm, several hydroxybenzoic acids were reported such as vanillic acid, protocatechuic acid, p-hydroxybenzoic acid, protocatechuic acid-4-glucoside, p-hydroxybenzoic acid-4-glucoside, and gentisic acid (Mecha et al. 2019). Syringic acid contents in legumes such as green beans, red beans, black beans, and soybean from five different regions of China showed a composition of 28.6–72.4% (Song et al. 2020).

6.2.2 Hydroxycinnamic Acids

Hydroxycinnamic acids are present in grain legume seeds. These include caffeic acid, p-coumaric acid, *trans*-ferulic acid, sinapic acid, and chlorogenic acids. The levels of hydroxycinnamic acids vary in legumes such as beans, peas, and lentils (Amarowicz and Pegg 2019). *Trans*-ferulic acid had been reported in beans, peas, and lentils and *cis*-ferulic acid only in beans. Similarly, in lentils and peas, *trans-p*-coumaric acid had been reported, and *cis-p*-coumaric acid only in peas (López-Amorós et al. 2006). In 15 dry edible bean varieties, the mean ferulic acid, p-coumaric acid, and sinapic acid were reported as 178, 63, and 70 mg/kg, respectively (Luthria and Pastor-Corrales 2006). Lentils had been reported to contain 3.76 mg/kg of hydroxycinnamic acids, which include *trans-p*-coumaric, *cis-p*-coumaric and hydroxy acids (Aguilera et al. 2010). Similarly, *trans-p*-coumaroyl malic acid (10.02 mg/kg), *trans-p*-coumaroyl glycolic acid (2.88 mg/kg), *trans-p*-coumaric acid (5.74 mg/kg) were reported in raw lentils as main hydroxycinnamic acids (Dueñas et al. 2007). Another study by Xu and Chang (2010) showed that sinapic acid was in the range of 1099–2217 mg/kg and chlorogenic acid (159–213 mg/kg) as the main cinnamic acid derivatives detected in eleven lentil cultivars.

The beans of Cannellini and Pinto had shown to contain 23.5 and 36.3 mg/kg of hydroxycinnamic acids, respectively with the majority of them being the aldaric derivatives of ferulic and p-coumaric free acids (Aguilera et al. 2011). Additionally, the key hydroxycinnamic acids reported in Cannellini and Pinto beans were *trans*-ferulic acid having a concentration of 8.9 and 11.8 mg/kg, respectively. Alshikh et al. (2015) reported caffeic, ferulic, p-coumaric, and sinapic acids in different lentil cultivars. In another study, Amarowicz et al. (2009) reported *trans*-ferulic, *trans-p*-coumaric, and sinapic acids in red lentils. *Trans-p*-coumaric acid, *trans*-ferulic acid, ferulic acid, caffeic acid, and sinapic acid had been reported in hull portions of legume seeds (Amarowicz and Pegg 2008). Similarly, in the crude extract of adzuki bean, the *trans-p*-coumaric acid had 31.3 mg/kg and *trans-p*-coumaroyl malic acid 4.57 mg/kg (Amarowicz and Pegg 2008), while in the extract of red lentil, trans-p-coumaric acid had the amount of 38.8 mg/kg, and sinapic acid as 0.06 mg/kg (Amarowicz et al. 2009). Similarly, trans-p-coumaric acid (37.3 mg/kg), *trans-p*-coumaric acid derivative (6.4 mg/kg), and trans-ferulic acid (10.1 mg/kg) had been reported in green lentil (Amarowicz et al. 2010). In soybean seed, the amount of ferulic was 95 mg/kg as a main hydroxycinnamic acid (Chung et al. 2011). Similarly, in black cowpea seeds, ferulic acid was 26.2 mg/kg, and coumaric acid was 1.25 mg/kg as free phenolic acids (Gutiérrez-Uribe et al. 2011). In beans such as white kidney bean and round purple bean, aldaric derivatives of ferulic acid, p-coumaric acid, and sinapic acid were the main phenolic acids (García-Lafuente et al. 2014). Kidney beans are a rich source of hydroxycinnamic acids in both free forms and their glycosides accounting for up to 50% of the total phenolic contents (Dueñas et al. 2015). The most active hydrolysates of pinto beans contain p-coumaric acid (1 mg/kg), sinapic acid (0.21 mg/kg), ferulic acid (28.2 mg/kg), catechin (41.2 mg/kg) and epi-catechin (15.2 mg/kg) (Garcia-Mora et al. 2015).

Khang et al. (2016) showed that caffeic acid was only present in white cowpeas among the five beans (black beans, mung bean, adzuki beans, soybeans, and white cowpeas), whereas syringic acid, p-coumaric acid, cinnamic acid, and sinapic acid were detected in all selected beans. Dark beans have 1.2 and 1.3 mg/kg of hydroxy-cinnamic compounds in soluble and insoluble dietary fractions, respectively (Dueñas et al. 2016). Ferulic and p-coumaric are the two main hydroxycinnamic acids commonly found in selected beans (Wang et al. 2016). In mung beans, caffeic acid, p-coumaric acid, ferulic acid, and sinapic acid had been reported in the hull, cotyledon, and whole beans (Singh et al. 2017b).

Trans-cinnamic acid, 2-hydroxycinnamic acid, and p-coumaric acid were the major hydroxycinnamic acids present in several legumes seeds from central Balkans (Malenčić et al. 2018). Zhang et al. (2018) studied four cultivars each of lentils, black soybean, and black turtle beans, and found that the amounts of coumaric acid were 0.54–2865 mg/kg in lentils, 2.6–1998 mg/kg in black soybean and not detected in black turtle beans. Ferulic acid was only reported in black turtle beans in the range of 3.19–3897 mg/kg. Sinapic acid was reported after the thermal treatment of these beans, whereas coumaric acid was detected in black turtle beans (Zhang and Chang 2019). In Australian sweet lupin seeds, several hydroxybenzoic acid deriva-tives have been reported by Zhong et al. (2019). These include cinnamic acid gluco-side, coumaric acid glucoside, ferulic acid, and its glucoside. Similarly, ferulic acid, gallic acid, and p-coumaric acid were the major hydroxycinnamic acids present in black beans (Damián-Medina et al. 2020).

Recently Madrera and Valles (2020) studied 17 accessions of dry beans (*Phaseolus vulgaris* L.) for phenolic composition. In addition to other phenolic compounds, the authors were able to separate 15 hydroxycinnamic acid derivatives as shown in Fig. 6.1. The majority of the compounds were aldaric derivatives of coumaric acid, ferulic acid, and sinapic acid. The identified compounds were O-p-coumaryl aldaric I, O-feruloyl aldaric I, O-p-coumaryl aldaric II, O-feruloyl aldaric II, O-p-coumaryl aldaric III, O-sinapyl aldaric I, O-feruloyl aldaric III, O-feruloyl aldaric IV, O-sinapyl aldaric II, O-sinapyl aldaric III, O-feruloyl aldaric V, O-sinapyl aldaric IV, p-coumaric acid, ferulic acid, and sinapic acid. These authors showed significant variations among the different accessions as given in Table 6.1. Similarly, Mecha et al. (2020) studied phenolic composition including hydroxycinnamic acid derivatives in common beans using ultra-high performance liquid chromatography-quadrupole-time of flight-MS (UPLCQ-TOF-MS). The authors showed that caffeic

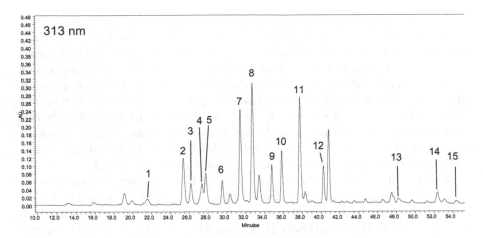

Fig. 6.1 Separation of phenolic acid derivatives in a dry bean extract of the Red Mexican variety. (1) *O-p*-coumaryl aldaric I, (2) *O*-feruloyl aldaric I, (3) *O-p*-coumaryl aldaric II, (4) *O*-feruloyl aldaric II, (5) *O-p*-coumaryl aldaric III, (6) *O*-sinapyl aldaric I, (7) *O*-feruloyl aldaric III, (8) *O*-feruloyl aldaric IV, (9) *O*-sinapyl aldaric II, (10) *O*-sinapyl aldaric III, (11) *O*-feruloyl aldaric V, (12) *O*-sinapyl aldaric IV, (13) *p*-coumaric acid, (14) ferulic acid, and (15) sinapic acid. (Reproduced with kind permission of Elsevier, Madrera and Valles 2020)

acid, *p*-coumaric acid, *trans*-ferulic acid, and sinapic acid were the major cinnamic acids. They also reported the availability and absorption of the studied compounds in human volunteers. It was found that 50% of the total amount of metabolites, such as 4-methylcatechol-sulfate and dihydrocaffeic acid-3-sulfate, was excreted after 8 h post-consumption, indicating colonic bacterial metabolism of the phenolic compounds.

6.2.3   Esters of Cinnamic Acids

In legumes, cinnamoyl esters have been reported including chlorogenic acids as a major component. The highest known amounts of cinnamates such as feruloylquinic acids and caffeoylquinic acids had been reported in coffee beans (Clifford 1999; Clifford and Wight 1976). In green coffee beans, *trans*-3-caffeoylquinic acid, *trans*-5-caffeoylquinic acid, *trans*-4-caffeoylquinic acid, and *cis*-5-caffeoylquinic acid had been reported (Clifford et al. 2008). Apart from coffee beans, esters of cinnamic acids have been reported in other legumes (Upadhyay and Mohan Rao 2013). Esters of ferulic acid, coumaric acid, sinapic acid, and cinnamic acid had been reported in four cultivars of dry beans (*Phaseolus Vulgaris*) (Drumm et al. 1990). In mung beans, 3-caffeoylquinic acid and 5-caffeoylquinic acid have been reported (Ganesan and Xu 2018). The concentration of 3-caffeoylquinic acid in the whole mung beans using different solvent extractions ranged from 77.9 to 136.4 mg/kg (Singh et al. 2017b).

Table 6.1 Content of hydroxycinnamic acid (mg/kg) derivatives in 17 accessions of dry beans (*Phaseolus vulgaris* L.). Redrawn with kind permission of Elsevier (Madrera and Vallés 2020)

	MDRK	Xana	Cornell 49 242	BAT93	Red Mexican 35	Maravilla de Venecia	Garrafal oro	Buenos Aires	Contender	Black Turtle Soup	Don Timoteo	Borloto Rosso	Lamon	Redlands Green Leaf C	Tendergreen	Primel	MG38
O-Feruloyl aldaric I	17.4	17.0	13.1	13.1	20.3	22.2	21.5	32.3	28.3	32.5	22.3	18.5	18.4	22.9	27.3	29.1	45.8
O-*p*-Coumaryl aldaric II	5.2	8.0	1.9	7.3	6.2	4.5	18.4	8.7	5.5	16.1	5.6	2.4	5.2	3.0	10.9	6.2	13.6
O-Sinapyl aldaric I	3.5	2.7	7.2	7.0	9.4	8.2	5.0	3.6	3.8	9.5	8.3	4.7	3.6	3.0	3.7	4.1	1.9
O-Feruloyl aldaric III	30.8	28.9	23.0	45.5	37.2	38.6	37.1	55.7	49.5	56.9	39.3	33.3	32.9	40.5	49.0	47.8	80.6
O-Feruloyl aldaric IV	39.9	39.6	28.4	61.6	46.6	53.7	54.4	80.5	66.8	73.4	50.4	46.4	45.6	55.0	63.2	64.7	116.1
O-Sinapyl aldaric II	5.7	4.4	10.8	10.3	15.0	12.5	7.6	5.9	5.9	14.3	12.6	7.0	5.7	4.9	5.9	5.7	2.6
O-Sinapyl aldaric III	7.4	5.0	14.3	14.3	19.8	17.4	10.8	8.4	7.5	18.5	17.5	12.1	8.8	7.1	8.4	8.1	2.8
O-Feruloyl aldaric V	35.0	41.6	27.6	48.0	38.9	36.9	46.8	60.1	46.3	63.0	33.0	34.6	32.0	37.9	36.6	60.1	87.7
O-Sinapyl aldaric IV	6.9	4.1	12.2	9.0	13.3	12.0	8.7	4.8	3.9	12.6	10.0	6.0	4.5	4.0	5.0	6.1	0.9
p-Coumaric acid	1.6	2.5	0.4	1.3	0.9	0.4	1.1	0.9	1.3	1.6	<loq	1.0	0.6	0.6	0.9	1.7	1.2
Ferulic acid	9.1	9.6	10.1	7.8	6.0	3.7	3.8	6.2	7.9	11.0	2.9	8.9	4.6	3.0	3.6	7.2	7.5
Sinapic acid	1.2	0.9	2.9	1.5	1.7	1.3	1.2	1.0	0.9	0.5	0.7	1.2	0.6	0.6	0.5	0.8	<loq

loq limit of quantification

6.2.4   Stilbenes

Stilbenes such as trans-stilbenes and resveratrol have been reported in a few legumes. For example, in the hull extracts of mung beans, resveratrol was present in the range of 9.0–30.5 mg/kg, cotyledon has mean values of 7.22 mg/kg and the whole grain contain 6.81–11.6 mg/kg (Singh et al. 2017b). Trans-stilbene was also present in the hull, cotyledon, and whole mung beans. In lentils, chickpea, and beans, one stilbene glycoside was identified to be resveratrol glucoside using untargeted LC-MS metabolomics (Llorach et al. 2019).

6.2.5   Flavonoids

In legumes, flavonoids are the key phenolic antioxidants. The raw lentils revealed myricetin-3-rhamnoside (5.95 mg/kg) and quercetin-3-rutinoside (5.24 mg/kg) as the main flavonol glycosides. Whereas apigenin-7-apiofuranosyl glucoside (6.20 mg/kg), apigenin-7-glucoside (1.87 mg/kg), and luteolin-7-glucoside (1.29 mg/kg), were the main flavone glycosides (Dueñas et al. 2007). Seventeen percent of the total identified phenols in raw lentils had been reported to be flavonols and dihydroflavonols (Aguilera et al. 2010). The glycosides of kaempferol were the key flavonols reported in raw lentils with the most common were kaempferol-3-rutinoside (5.95 mg/kg) and kaempferol glucoside (3.66 mg/kg). Similarly, in six cultivars of lentil, kaempferol dirutinoside was the major flavonol (Alshikh et al. 2015). Aguilera et al. (2011) reported that in raw Pinto beans 4% of the identified phenolic compounds were flavonols, which included quercetin glucoside (7.15 mg/kg) and kaempferol glucoside (3.13 mg/kg) as key components. In crude extract of adzuki bean Amarowicz et al. (2008) identified quercetin (36.2 mg/kg), quercetin glucoside (181 mg/kg), quercetin galactoside (46.9 mg/kg), quercetin arabinoglucoside (42.8 mg/kg), quercetin rutinoside (38.2 mg/kg), dihydroquercetin (1.15 mg/kg), kaempferol rutinoside (38.2 mg/kg) and myricetin rhamnoside (212 mg/kg).

In the acetone extract of red lentil, kaempferol derivative was 37.6 mg/kg and acylated quercetin hexose was 21.4 mg/kg (Amarowicz et al. 2009). Similarly, quercetin diglycoside (114 mg/kg) was the leading antioxidant in the crude extract of green lentil. Whereas other flavonols identified were acylated quercetin hexose (37.2 mg/kg) and kaempferol glucoside (19.4 mg/kg) (Amarowicz et al. 2010). Xu and Chang (2010) quantified luteolin in 11 cultivars of lentils which were in the range from 21.8 to 77.1 mg/kg. Aguilera et al. (2010) identified luteolin-3'-7-diglucoside as having a concentration of 4.6 mg/kg as the main flavone in lentils. Kaempferol (6.09 mg/kg), quercetin (4.02 mg/kg), and myricetin (2.3 mg/kg) were the free flavonols present in black cowpea seed coats (Gutiérrez-Uribe et al. 2011). Mirali et al. (2014) utilized HPLC-MS for the identification of phenolic compounds present in the seed coats (black, green, and grey) of three lentil genotypes. The contents of oligomeric flavan-3-ols were almost similar in all three genotypes, whereas luteolin including its glycosylated forms was more abundant in the genotype containing black seed coat.

In ten faba bean varieties, total flavonols, and flavones detected in immature seeds ranged from 101 to 377 mg/kg on a fresh weight basis (Baginsky et al. 2013). Myricetin (18.8–119.5 mg/3 kg), myricetin glucoside (9.9–41.9 mg/kg), quercetin galactoside (8.3–92.4 mg/kg), quercetin rutinoside (7.4–56.9 mg/kg), quercetin glucoside (5.1–25.9 mg/kg) and apigenin galactoside (8.9–168 mg/kg) were identified in ten faba bean varieties. Quercetin and its derivatives are the flavonols present only in colored (round purple) beans and they were absent in white kidney beans (García-Lafuente et al. 2014).

Most of the flavonols and flavones are concentrated in insoluble dietary fiber fractions of dark beans and lentils (Dueñas et al. 2016). Dark beans contain kaempferol dihexoside-rhamnoside (1.06 mg/kg) and lentils contain kaempferol (1.64 mg/kg), kaempferol dihexoside (1.37 mg/kg), kaempferol rutinoside (0.30 mg/kg), kaempferol-3-glucoside (0.61 mg/kg), quercetin-3-glucoside (1.0 mg/kg) and luteolin rhamnoside-hexoside (0.36 mg/kg) in insoluble dietary fiber fractions. Myricetin-3-rhamnoside (7.4 mg/kg), luteolin-8-C-glucoside (1.2 mg/kg), quercetin-3-rhamnoside (5.0 mg/kg) and quercetin-3-galactoside (7.2 mg/kg) were reported in Elmo chickpea variety (Magalhães et al. 2017). Gan et al. (2016) reported apigenin-8-C-glucoside and apigenin-6-C-glucoside in mung beans coats in the range from 3.31 to 3.53 and 4.83 to 7.13 mg/g of dry weight basis, respectively. In field pea varieties, the major flavonoids were apigenin-8-C-glucoside (8.5–21.2 mg/kg), apigenin-6-C-glucoside (2.7–24.6 mg/kg), and luteolin-6-C-glucoside (1.4–11.3 mg/kg) (Magalhães et al. 2017). Apigenin-7-β-apiofuranosyl-6,8-di-C–β-glucopyranoside, vicenin 2, apigenin-7-β-glucopyranoside, genistein, and a dihydroflavonol were the major flavonoids reported in Australian sweet lupin (*Lupinus angustifolius*) (Zhong et al. 2019).

In legume seeds, the flavanones reported by Aguilera et al. (2011) were pinocembrin, naringenin, eriodictyol, hesperetin, and sakuranetin. The total flavanones present in Pinto and Cannellini beans were 10.87 and 7.06 mg/kg, respectively. Sakuranetin (4.49 mg/kg), eriodictyol (3.83 mg/kg), naringenin derivative (1.29 mg/kg) and pinocembrin (1.26 mg/kg) were reported in Pinto beans. Eriodictyol derivative (2.01 mg/kg), sakuranetin (1.59 mg/kg), pinocembrin derivative 2 (1.28 mg/kg) and eriodictyol (1.20 mg/kg) were the identified flavanones in Cannellini beans. García-Lafuente et al. (2014) reported naringenin (11.3 mg/kg) and hesperetin (0.14 mg/kg) are the flavanone-glycoside in white kidney bean.

Madrera and Valles (2020) studied 17 accessions of dry beans (*Phaseolus vulgaris* L.) for phenolic composition. The identified flavonoids were myricetin-3-xyloglucoside, myricetin-3-glucoside, quercetin-3-xylorutinoside, myricetin-3-malonylglucoside, quercetin-3-xyloglucoside, quercetin-3-glucoside, quercetin-3-rutininoside, kaempferol-3-xyloglucoside, quercetin-3-acetylglucoside, myricetin, kaempferol-3-glucoside, kaempferol-3-rutininoside, kaempferol-3-acetylglucoside I, kaempferol-3-acetylglucoside II, kaempferol-3-acetylglucoside III, quercetin, and kaempferol. Quercetin-3-glucoside, kaempferol-3-glucoside, and quercetin-3-rutinoside have been reported in black bean (Damián-Medina et al. 2020). Quercetin-3-rutinoside, kaempferol-3-glucoside, Quercetin-3-glucoside, quercetin, and kaempferol have been identified in coffee beans (Król et al. 2020).

6.2.6 Catechins and Procyanidins

Catechins contain a hydroxyl group at C-3 and are called flavan-3-ols. Procyanidins are the oligomers of catechin and epicatechin molecules. These flavonols are mainly present in legumes having colored seed coats. Aguilera et al. (2011) reported that catechins and procyanidins had a concentration of 256.9 mg/kg and the most abundant compounds were (+)-catechin (142.6 mg/kg), (+)-catechin glucoside (45.8 mg/kg), and procyanidin dimer 1 (41.2 mg/kg) in colored Pinto beans. Similarly, catechins, procyanidins, and prodelphinidins are common in raw lentil flour samples (Dueñas et al. 2007). They represent 69% of total phenolic compounds (74.4 mg/kg) in raw lentils as reported by Aguilera et al. (2010). These authors also showed that the main identified compounds in lentils were (+)-catechin-3-glucoside (39.9 mg/kg), procyanidin B2 (8.92 mg/kg), procyanidin trimer (9.3 mg/kg), and procyanidin dimer (4.45 mg/kg). Similarly, Alshikh et al. (2015) showed that catechin was at the concentration range of 267–1899 mg/kg and epicatechin 2535–4946 mg/kg as major compounds present in 11 cultivars of lentil. The dietary insoluble fractions of raw lentils comprise of catechins (1.37 mg/kg) and procyanidins (1.03 mg/kg), whereas they were not reported in dietary soluble fractions (Dueñas et al. 2016). In the crude extract of adzuki bean, catechin glucoside was 688 mg/kg, epicatechin (25.7 mg/kg), epicatechin glucoside (159 mg/kg), and epigallocatechin gallate (0.14 mg/kg) were reported (Amarowicz et al. 2008). Similarly, epicatechin (98.2 mg/kg), catechin (36.0 mg/kg), catechin glucoside (51.9 mg/kg), and epicatechin glucoside (6.7 mg/kg) had been reported in red lentil crude extracts (Amarowicz et al. 2009).

The hull of colored beans i.e., red, brown, and black bean comprise procyanidin B2, C1, and C2 (Amarowicz and Pegg 2008). In a dark-colored seed coat of cranberry beans, catechin was the most predominant flavonoid in the level of 2.5–6.5-fold higher than proanthocyanidins. In free phenolic and alkaline hydrolyzable phenolic fractions of regular-darkening cranberry beans seed coats the level of catechin was 370 and 250 mg/kg and procyanidin dimers were 1800 and 140 mg/kg, respectively (Chen et al. 2015a). These compounds were identified in the free phenol fraction of regular darkening cranberry bean lines. Catechin and proanthocyanidins concentration in whole raw cranberry beans (regular darkening lines) ranged from 142.2 to 203 and 15.3 to 35.29 mg/kg, respectively (Chen et al. 2015b). Mojica et al. (2015) showed that catechin (1.8–5.4%) and epicatechin (3.8–12.4%) were present in seed coats of 15 cultivars of common beans (*Phaseolus vulgaris*). In the crude extract of broad bean, Amarowicz and Shahidi (2017) reported catechin gallate (2060 mg/kg), digallate procyanidin dimer (864 mg/kg), epicatechin (864 mg/kg), and gallate procyanidin dimer (676 mg/kg) as the major flavanols. In a previous study by Ojwang et al. (2013), catechin glucoside was the main dominant compound present in cowpeas. The authors reported that catechin (16.6–297 mg/kg), catechin-7-glucoside (770–2553 mg/kg), epicatechin (2.4–11.3 mg/kg) and (epi)afzelechin glycosides (10.7–243 mg/kg) were also present in six cowpea phenotypes.

The content of procyanidin gallate had been reported to be 12.4 mg/kg, procyanidin dimers (16.0–213 mg/kg), and procyanidin trimers (41.8–42.4 mg/kg) in the crude extract of adzuki bean (Amarowicz et al. 2008). López-Amorós et al. (2006) reported the contents of procyanidins i.e., procyanidin B2, B3, and procyanidin tetramer were in the range of 1–5 mg/kg in lentils on a dry weight basis. Prodelphinidin dimers (161.6 mg/kg), procyanidin dimers (65.9 mg/kg), procyanidin trimers (87.5 mg/kg), digallate procyanidin dimer (83.2 mg/kg), procyanidin gallate (32.8 mg/kg) were reported in the acetone extract of red lentil (Amarowicz et al. 2009). In a later study, Amarowicz et al. (2010) reported prodelphinidin dimer (5.3 mg/kg), digallate procyanidin dimer (7.3 mg/kg), procyanidin dimer 1 (14.1 mg/kg) and procyanidin dimer 2 (100 mg/kg), procyanidin dimer gallate (31.9 mg/kg), digallate procyanidin dimer (14.1 mg/kg), procyanidin trimer (18.6 mg/kg), catechin glucoside (289 mg/kg), catechin gallate (18.7 mg/kg), epicatechin glucoside (59.5 mg/kg) in green lentil. In immature seeds of ten faba bean varieties, the amounts of proanthocyanidins ranged from 543 to 36.6 mg/kg on fresh weight (Baginsky et al. 2013). The amount of catechin (151–978 mg/kg), epicatechin (148–773 mg/kg), procyanidin dimer (71–1352 mg/kg), procyanidin trimer (64–782 mg/kg), and prodelphinidin dimer (81–1150 mg/kg) were also reported to be the key compounds present in those ten faba bean seeds. López et al. (2013) reported a high amount of procyanidin dimers (97.29 mg/kg) in raw dark beans. Among three anthocyanins found in the black soybean extracts, cyanidin-3-O-glucoside was the major one (65–73% of the total), followed by petunidin-3-glucoside (17–23%) and delphinidin-3-O-glucoside (10–12%) (Ryu and Koh 2018).

In a recent study, Orita et al. (2019) reported that black and red cowpeas were rich in proanthocyanidin oligomers such as monomers, dimers, trimers, tetramers, pentamers, hexamers, and heptamers as compared with other grain legumes as shown in Table 6.2. The amount of total proanthocyanidin oligomers were 2.66 mg/g in black cowpea, 1.67 mg/g in red cowpea, 0.68 mg/g in black soybean, 0.61 mg/g in azuki bean, and the lowest values of 0.17 mg/g in black kidney beans.

Similarly, Kawahara et al. (2019) characterize the proanthocyanidins fractions of adzuki beans. The proanthocyanidins enriched fractions were characterized as epicatechin hexamer, heptamer, and octamer, epigallocatechin-epicatechin pentamer, and epigallocatechin-epicatechin hexamer. The authors also identified catechin glucopyranoside, epicatechin dimer, procyanidin B4, arecatannin A1, arecatannin A2, epicatechin nonamer, piceid, and dehydroquercetin rhamnoside using reversed-phase liquid chromatography (Fig. 6.2). These compounds were identified using ESI-TOFMS, NMR, and by comparison with spectra of known compounds.

Table 6.2 Proanthocyanidin contents (mg/gDM) in cowpeas and other black or red legumes grown in Japan. Adapted from Orita et al. (2019)

Sample	Black cowpea	Red cowpea	Black kidney bean	Black soybean	Azuki bean
Seed coat color	Black	Red	Black	Black	Red
Monomers	0.35 ± 0.06[a]	0.23 ± 0.02[b]	0.02 ± 0.00[d]	0.13 ± 0.02[c]	0.06 ± 0.00[c,d]
Dimers	0.38 ± 0.06[a]	0.38 ± 0.03[a]	0.04 ± 0.01[c]	0.09 ± 0.01[c]	0.18 ± 0.01[b]
Trimers	0.30 ± 0.05[a]	0.22 ± 0.02[b]	0.02 ± 0.00[c]	0.06 ± 0.01[c]	0.08 ± 0.01[c]
Tetramers	0.38 ± 0.07[a]	0.26 ± 0.01[b]	0.02 ± 0.00[c]	0.08 ± 0.01[c]	0.09 ± 0.01[c]
Pentamers	0.49 ± 0.09[a]	0.26 ± 0.01[b]	0.03 ± 0.00[c]	0.11 ± 0.01[c]	0.08 ± 0.01[c]
Hexamers	0.39 ± 0.07[a]	0.18 ± 0.01[b]	0.02 ± 0.00[c]	0.10 ± 0.01[b,c]	0.07 ± 0.01[c]
Heptamers	0.37 ± 0.07[a]	0.14 ± 0.01[b]	0.02 ± 0.00[c]	0.10 ± 0.01[b,c]	0.05 ± 0.01[c]
Total proanthocyanidin oligomers	2.66 ± 0.46[a]	1.67 ± 0.08[b]	0.17 ± 0.02[c]	0.68 ± 0.08[c]	0.61 ± 0.02[c]

Results are represented as the mean ± standard deviation (n = 3). Contents are expressed as per gram of dry matter after freeze–drying. Within each row, means followed by different lowercase letters are statistically different ($P < 0.05$)

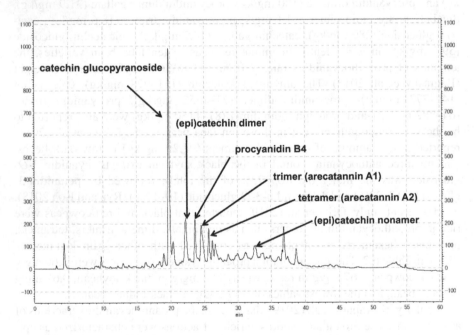

Fig. 6.2 Reversed-phase HPLC chromatographic profile of ethyl acetate eluted fraction using Amberlite XAD-1180N resin. The mobile phase was made of solvent A (0.2% acetic acid) and solvent B (acetonitrile) with gradient elution for 60 min. (Adapted from Kawahara et al. 2019)

6.2.7 Anthocyanidins and Anthocyanins

Anthocyanidins are the main aglycone structures of anthocyanins and consist of an aromatic ring A that is bonded to a heterocyclic ring C, which comprises oxygen forms a carbon-carbon bond with the third aromatic ring B. Anthocyanidins are present in their glycoside form i.e. bonded to a sugar moiety, they are then referred to as anthocyanins as discussed in Chap. 2. Anthocyanins are mostly present in the colored seed coat of the legumes. In red and green lentils, anthocyanin contents reported is three to fourfolds higher in hulls than those in whole seeds (Oomah et al. 2011). Xu and Chang (2010) reported the anthocyanin contents of 157.3 and 665.6 mg/kg in French Green and Pardina lentils, respectively. The authors also showed that sixfolds higher anthocyanin contents were present in yellow pea hull than those in whole seeds. In raw dark beans, López et al. (2013) showed that the most abundant anthocyanins were cyanidin-3-glucoside (88.44 mg/kg) and pelargonidin-3-glucoside (50.72 mg/kg).

In black seed coat, the amounts of monomeric anthocyanin contents (73.7 mg cyanidin-3-glucoside equivalents (CyE)/g) were higher as compared to red seed coat (9.06 mg CyE/g) of peanuts (Attree et al. 2015). In the extract of the round purple bean, the anthocyanins such as malvidin, pelargonidin, and cyanidin glucosides were responsible for the color of this bean (García-Lafuente et al. 2014). Pelargonidin-3-glucoside (1.90 mg/kg), cyanidin-3-glucoside (0.31 mg/kg) and malvidin-3-glucoside (0.06 mg/kg) were the main anthocyanins reported by Dueñas et al. (2015) in kidney bean. The glucosides of delphinidin (9–1290 mg/kg), petunidin (7–1150 mg/kg) and malvidin (1.4–520 mg/kg) had been reported by Mojica et al. (2015) in seed coats of 15 cultivars of bean from Mexico and Brazil. The authors also reported that seed coats of two black bean cultivars had the highest amount of total anthocyanins (2060 and 2500 mg/kg). The contents of cyanidin, pelargonidin, petunidin, malvidin, delphinidin, and the total anthocyanidin in 26 kidney bean varieties ranged from 0 to 1.44, 0 to 0.71, 0 to 0.41, 0 to 0.27, 0 to 4.45, and 0 to 5.84 mg/g of bean coat DW, respectively (Kan et al. 2016). Han et al. (2015) reported that anthocyanins were responsible for the pigmentation of the adzuki bean extracts. They identified malvidin-3-glucoside and peonidin-3-rutinoside as two new anthocyanin compounds in the adzuki beans.

In black soybean (*Glycine max*), delphinidin-galactoside, delphinidin-glucoside, cyanidin-glucoside, petunidin-glucoside, and pelargonidin-glucoside were identified (Kan et al. 2018). This study also reported delphinidin-glucoside, cyanidin-galactoside, cyanidin-glucoside, and pelargonidin-glucoside in red kidney beans, whereas delphinidin-glucoside and cyanidin-glucoside in mung beans, red beans, and lentils. Delphinidin-glucoside has only detected anthocyanin in pea.

The grain legumes (black kidney beans, black soybeans, and azuki beans) commonly consumed in Japan have been studied for their anthocyanins contents by Orita et al. (2019). The authors showed that black cowpeas contained seven anthocyanins, i.e., cyanidin-3-galactoside, cyanidin-3-glucoside, delphinidin-3-galactoside, delphinidin-3-glucoside, malvidin-3-glucoside, peonidin-3-glucoside, and petunidin-3-glucoside at higher levels than other grain legumes. Similarly, cyanidin-3-glucoside, delphinidin-3-glucoside, and petunidin-3-glucoside were higher in black pigmented soybean genotypes than yellow and green genotypes (Krishnan et al. 2020).

In dry beans (*Phaseolus vulgaris*), several anthocyanins have been reported by Madrera and Valles (2020). These include delphinidin-3,5-diglucoside, cyanidin-3,5-diglucoside, petunidin-3,5-diglucoside, pelargonidin-3,5-diglucoside, malvidin-3,5-diglucoside, delphinidin-3-glucoside, cyanidin-3-glucoside, petunidin-3-glucoside, pelargonidin-3-glucoside, malvidin-3-glucoside, pelargonidin-pentoside hexoside, cyanidin-3-(6″ malonylglucoside) and pelargonidin-3-(6″ malonylglucoside).

6.2.8 Tannins

Tannins are important polyphenols with a molecular weight ranging from 500 to 3000 D and are found in complexes with polysaccharides, alkaloids, and proteins. Tannins are mainly present in the testa of legumes, playing important roles in the defense system of these seeds which normally get exposed to oxidative damage from various environmental factors (Shahidi and Ambigaipalan 2015). Legumes with colored seed coats such as lentils, red beans, and black beans contain high amounts of condensed tannins (Amarowicz and Pegg 2008). Condensed tannins (proanthocyanidins) are oligomeric and polymeric flavonoids that release anthocyanidins and catechins by heat treatment in alcoholic solutions. Condensed tannins of lentil, pea, and faba beans consist of catechin, gallocatechin, epigallocatechin, and epicatechin subunits (Jin et al. 2012). The proanthocyanidin subunit composition in pea was primarily prodelphinidin (epigallocatechin and gallocatechin), whereas it was prodelphinidin and procyanidin-type flavan-3-ol subunits in lentils and faba beans. Glycosides of catechin and (epi)afzelechin are the major subunits in proanthocyanidins of cowpea with catechin-7-glucoside as predominant flavan-3-ols accounted for 88% of the monomers (Ojwang et al. 2013). Mirali et al. (2014) showed that catechin, epicatechin, gallocatechin and epigallocatechin were the subunits in oligomers of proanthocyanidins present in the lentil seed coat.

There is a significant variation in the level of condensed tannin content (CTC) among commonly consumed legumes. The CTC of lentil cultivars ranged from 3.73 to 10.20 mg CE/g; common beans varieties from 0.47 to 5.73 mg CE/g; soybean cultivars from 1.06 to 4.04 mg CE/g; yellow pea cultivars ranged from 0.22 to 0.59 mg CE/g and green pea cultivars from 0.23 to 0.61 mg CE/g (Xu and Chang 2007). Legumes having colored or pigmented seed coats are a good natural source of condensed tannins. Seed coats of peanuts have a high level of condensed tannins compared to kernel and cotyledons (Attree et al. 2015). CTC in the seed coat, raw kernel, and cotyledons of six peanut varieties was reported in the range of 29.7 to 84.7, 2.88 to 4.73, and 1.74 to 3.75 mg CE/g, respectively. Alshikh et al. (2015) reported the highest levels of condensed tannins were found in lentils. The CTC in free, esterified, and insoluble bound forms in lentils were in the range of 0.40–2.67, 0.64–4.20, and 0.03–5.57 mg CE/g, respectively.

Tannins content determined using vanillin method in faba bean, broad bean, adzuki bean, red bean, pea, red lentil, and green lentil was 0.416, 0.093, 0.483, 0.434, 0.071, 0.577, and 0.545 absorption units per mg, respectively (Amarowicz et al. 2004). The tannins content in crude extract, low-molecular-weight phenolic fraction, and tannin fraction of adzuki bean was reported to be 136, 45, and 213 absorbance units per mg (Amarowicz et al. 2008), whereas, values of 70, 1.52, and 1.29 absorbance units in red lentil, respectively (Amarowicz et al. 2009). Similarly, green lentils had found to possess more tannin content as the values for crude extract and two fractions were obtained as 93, 1.92, and 252 unit/g (Amarowicz et al. 2010).

In immature seeds of ten bean varieties, the CTC ranged from 309 to 958 mg CE/kg (Baginsky et al. 2013). Condensed tannins are the main polyphenols reported in pigmented bean coats (Gan et al. 2016). Similarly, the mean proanthocyanidin (PA) content reported in faba bean, pea, and lentil were 6543, 336, and 3294 mg/kg on a fresh weight basis, respectively (Jin et al. 2012). In another study, Sreerama et al. (2012) reported that chickpea flour had the highest PA content (5.1 mg CE/g) than horse gram (3.8 mg CE/g) and cowpea (2.1 mg CE/g) flours. Six cowpea phenotypes with different seed coat color (black, red, green, white, light-brown, and golden-brown) contain PA content in the range from 2.2 to 6.3 mg/g DW (Ojwang et al. 2013). Sixty-fold higher PA contents were reported in regular raw dark-colored cranberry beans as compared to raw light-colored cranberry beans (Chen et al. 2015b). The PA content varied widely in 26 kidney bean cultivars due to diversity in color of seed coat and it was observed in the range of 10.52–57.13 mg procyanidin B-2 equivalents/g (Kan et al. 2016). Moreover, proanthocyanidins were not detected in kidney bean cultivars with white color seed coats. Tannins are higher in fresh-cut faba beans as compared to stored samples (Collado et al. 2019). The tannin contents in common beans were 6.13 CE/g and have been found to be declined with high-pressure processing (Belmiro et al. 2020).

6.3 Phenolic Antioxidants in Nuts

Nuts may be defined as hard-shelled dry fruits or seeds with a separable rind or shell and interior kernel. From a botanical perspective, they may be from different taxonomic groups. The common usage of the term "nut" often refers to any hard-walled, edible kernel also called tree nuts. The common tree nuts are almonds (*Prunus amygdalus*), walnuts (*Juglans regia* L.), pistachios (*Pistacia vera* L.), hazelnut (*Corylus avellana* L.), pecan (*Carya illinoinensis*), macadamias (*Macadamia integrifolia*), and Brazil nuts (*Bertholletia excels*). Nuts are nutrient-rich foods. The nuts also contain important metabolites such as proteins, unsaturated fatty acids, dietary fiber, plant sterols, phytochemicals, and micronutrients like tocopherols (Bodoira and Maestri 2020). The consumption of nuts more often has been linked to a lowered risk of cardiovascular disease, especially in patients with diabetes mellitus (Liu

Table 6.3 Total phenolic contents of nine tree nuts and peanuts. Values are mean with SD ($n = 3$). Reproduced with kind permission of Elsevier (Yang et al. 2009)

Edible nut seeds	Total phenolic contents (mg/100 g)		
	Free form	Bound form	Total
Almonds	83.0 ± 1.3	129.9 ± 13	212.9 ± 12.3
Brazil nuts	46.2 ± 5.7	123.1 ± 18.4	169.2 ± 14.6
Cashews	86.7 ± 8.1	229.7 ± 15.1	316.4 ± 7.0
Hazelnuts	22.5 ± 1.1	292.2 ± 48.4	314.8 ± 47.3
Macadamia nuts	36.2 ± 2.6	461.7 ± 51.2	497.8 ± 52.6
Peanuts	352.8 ± 22.2	293.1 ± 25.0	645.9 ± 47.0
Pecans	1227.3 ± 8.4	236.6 ± 28.1	1463.9 ± 32.3
Pine nuts	39.1 ± 0.6	113.8 ± 14.3	152.9 ± 14.1
Pistachios	339.6 ± 15.1	232.2 ± 13.3	571.8 ± 12.5
Walnuts	1325.1 ± 37.4	255.4 ± 25.0	1580.5 ± 58.0

et al. 2019). Phenolic antioxidants, present in nuts may be considered as the major bioactive compounds for health benefits.

The phenolic compounds in nuts are usually present at the highest concentration in the seed coat i.e., the skin or pellicle surrounding the kernel, and may be associated with other plant components, such as carbohydrates and proteins (Jakobek 2015). As shown in Table 6.3, walnuts, pistachios, peanuts, and pecans contain the highest total phenolic contents amongst all types of nuts such as almonds, Brazil nuts, cashews, Hazelnuts, Macadamia nuts, Pecans, and Pine nuts (Yang et al. 2009). The bound phenolic compounds were higher in all nuts except peanuts, pecans, pistachios, and walnuts, where free phenolic contents were significantly higher than their corresponding bound forms.

The total antioxidant activity of nuts was expressed as micromoles of vitamin C equivalents per gram of sample reported by Yang et al. (2009). Among the ten samples tested, walnuts had the greatest antioxidant activity (458.1 µmol/g) followed by pecans (427.0 µmol/g), peanuts (81.3 µmol/g), pistachios (75.9 µmol/g), cashews (29.5 µmol/g), almonds (25.4 µmol/g), Brazil nuts (16.0 µmol/g), pine nuts (14.6 µmol/g), macadamia nuts (13.4 µmol/g), and hazelnuts (7.1 µmol/g). The authors reported a statistically significant difference ($p < 0.05$) in antioxidant activity between walnuts and pecans, pecans and peanuts, and pistachios and cashews as shown in Fig. 6.3. Using both hydrophilic and lipophilic ORAC assays, the total antioxidant capacities (µmol of TE) of the ten nuts reported by Wu et al. (2004) were in the order of pecans > walnuts > hazelnuts > pistachios > almonds > peanuts > cashews > macadamias > Brazil nuts > pine nuts. Based on FRAP, TRAP, and TEAC assays, the contribution of bound phenolics from six nuts, extracted using methanol and alkaline to the total antioxidant capacity (TAC) was evaluated (Pellegrini et al. 2006). In all three assays, walnuts had the highest TAC values followed by pistachios. Pine nuts ranked the lowest of the TAC values in the FRAP and TRAP assays.

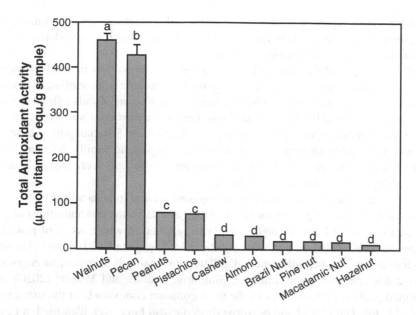

Fig. 6.3 Total antioxidant activity of phytochemical extracts of nine tree nuts and peanuts (mean ± SD, $n = 3$). Bars with on letters in common are significantly different ($p < 0.05$). (Reproduced with kind permission of Elsevier, Yang et al. 2009)

The previous study of Kornsteiner et al. (2006) reported that the antioxidants were found to be lower in cashews and pines. Macadamias also contained the lowest level of antioxidative constituents (polyphenols, tocopherols). Alasalvar et al. (2020) reviewed the bioactivities and health benefits of nuts. There has been an increasing interest in the functional characteristics of nuts as they are an important source of bioactive constituents in our daily diet. The focus of research is now more concentrated on phenolic compounds.

6.3.1 Hydroxybenzoic Acids

Phenolic acids such as hydroxybenzoic acids had been identified in the pericarp of walnut by Luczak et al. (1989) include *p*-hydroxybenzoic, vanillic, gentisic, protocatechuic, syringic, gallic acids. Syringic acid was known to be a principal phenolic acid. In almonds, Sang et al. (2002) isolated protocatechuic acid, vanillic acid, and *p*-hydroxybenzoic acid. Gallic acid had been reported as a major simple phenolic in Mexican pecan nuts (de la Rosa et al. 2011). Other phenolic compounds were protocatechuic, *p*-hydroxybenzoic acids, and catechin. In Brazilian nuts, gallic acid was the major phenolic acid (John and Shahidi 2010). Gallic acid and protocatechuic acids are the common hydroxybenzoic acids in hazelnut, pistachio, and peanut skins (Taş and Gökmen 2017). Gallic acid, *p*-hydroxybenzoic acid, and vanillic

acid have been reported in Maya nuts (Ozer 2017). Protocatechuic, vanillic, and
p-hydroxybenzoic acids were the predominant hydroxybenzoic acid derivatives in
almonds grown in Serbia (Čolić et al. 2017).

Zhang et al. (2009) identified several phenolic compounds in English walnut
seed by spectroscopic methods, namely, p-hydroxybenzoic acid, vanillic acid, ethyl
gallate, protocatechuic acid, and gallic acid. Phenolic compounds extracted from
different fractions of English walnut showed marked antioxidant activities in different *in vitro* model systems (Samaranayaka et al. 2008). In Siberian pine nuts seeds,
Lantto et al. (2009) identified protocatechuic, syringic, and vanillic acids amongst
other phenolic compounds. The least abundant components were *trans*-cinnamic
(122 mg/kg) and *m*-coumaric acids (trace).

Gallic acid was the most abundant compound present in pistachio skin at a concentration of 1.45 mg/g (Tomaino et al. 2010). Similarly, the concentrations of gallic acid between 13 and 36 mg/kg had been reported in whole seeds of pistachio
(Martínez et al. 2016). In addition to gallic acid, Erşan et al. (2016) had identified
several derivatives, including methyl gallate, monogalloyl glucoside, monogalloyl
quinic acid, and hexagalloyl hexose. Similarly, Bodoira and Maestri (2020) also
reported gallic acid derivatives as the most common compounds in the extracts of
pistachio nut. Gallic acid and p-hydroxybenzoic acid have been identified in pecan
nuts (Jia et al. 2018). Similarly, phenolic acids from hazelnut consist primarily of
hydroxybenzoic acid derivatives i.e., gallic, protocatechuic acid, p-hydroxybenzoic
acid, vanillic acid, and syringic acid, as either free or esterified compounds (Pelvan
et al. 2018). In microparticles produced from the Brazilian nut, gallic acid (15.9 mg/
kg), protocatechuic acid (56.4 mg/kg), catechin (190 mg/kg), p-hydroxybenzoic
acid (89.4 mg/kg), 2,4-dihydroxybenzoic acid (14.6 mg/kg) and sinapic acid
(97.3 mg/kg) were reported to be contributed by the nut (Gomes et al. 2019). Gallic
acid, catechin, vanillic acid, and sinapic acid have been found in Ramon nut
(*Brosimum alicastrum*) (Moo-Huchin et al. 2019). In Chilean hazelnuts, benzoic
acid, hydroxybenzoic acid, and sinapic acid hexoside have been identified using
HPLC-MS-MS in the negative ion mode (Pino Ramos et al. 2019).

6.3.2 Hydroxycinnamic Acids

Several hydroxycinnamic acids, such as ferulic acid, vanillic acid, syringic acid,
sinapic acid, coumaric acid, and caffeic acid are abundantly present in nuts.
However, Bodoira and Maestri (2020) stated that the different ways in which concentrations are expressed and the use of different standards for quantification make
a systematic survey of their contents difficult. On the other hand, as occurs with
other phenolic compounds, phenolic acids in nuts may be associated with the cell
wall material. This makes that, depending upon the solvent used, a variable portion
may be inaccessible for extraction. For example, total phenolic contents in Algerian
nuts were significantly higher in 80% methanol, than ethanol and acetone of the

same concentration (Chaalal et al. 2019). Finally, the possibility of overestimation should be considered if the quantification has been made after hydrolysis treatments.

In the pericarp of walnut, *p*-coumaric, ferulic, caffeic, and sinapic acids were identified (Luczak et al. 1989). Amarowicz et al. (2004) reported the presence of phenolic compounds such as vanillic, caffeic, *p*-coumaric, ferulic acids in faba bean, broad bean, adzuki bean, red bean, pea, red lentil. Similarly, chlorogenic, vanillic, and *trans-p*-coumaric acids were identified in the almond skin samples (Mandalari et al. 2010).

Colaric et al. (2005) identified phenolic acids such as chlorogenic, caffeic, *p*-coumaric, ferulic, and sinapic acids amongst other phenolic compounds in ripe fruits of ten walnut cultivars. The predominant phenolic acid present in peanuts with antioxidant potential was *p*-coumaric (Talcott et al. 2005). It was also reported that free *p*-coumaric acid, three esterified derivatives of *p*-coumaric acids were the predominant phenolic antioxidants present in peanuts. Fajardo et al. (1995) confirmed stress-induced the synthesis of free and bound phenolic compounds in peanuts, with *p*-coumaric and ferulic acids as the key compounds. In peanut, phenolic acids have been reported only in the skin. Yu et al. (2005) identified ferulic acid and coumaric acid as the main hydroxycinnamic acids in peanut skins, together with caffeic and chlorogenic acids.

The presence of caffeic and ferulic acids in extracts of hazelnut was reported by Alasalvar et al. (2006). Amongst the identified phenolic acids, *p*-coumaric acid was most abundant in hazelnut kernel, hazelnut green leafy cover, and hazelnut tree leaf, whereas gallic acid was most abundant in hazelnut skin and hazelnut hard shell. Phenolic acids from hazelnut also consist of coumaric acids, in either free or esterified compounds (Pelvan et al. 2018). Phenolic acids account for a minor fraction of the phenolic compounds present in almond kernels. Bolling (2017) review showed the total phenolic acids and aldehydes ranging between 5.2 and 12.0 mg/kg of almonds. In Serbian almonds, chlorogenic acid was the largest hydroxycinnamic acid derivative having a concentration ranging as high as 13.0 mg/kg (Čolić et al. 2017).

Gomes et al. (2019) reported coumaric acid as major hydroxycinnamic acid derivatives in microparticles prepared from Brazilian nuts along with other phenolic compounds. Chlorogenic acids and ferulic acid have been identified as Ramon nut (*Brosimum alicastrum*) by Moo-Huchin et al. (2019). Several derivatives of hydroxycinnamic acids were identified in Chilean hazelnuts by Pino Ramos et al. (2019). These were caffeic acid glucuronate, caffeic acid hexoside, caffeic acid dihexoside, phenyl caffeate, caffeic acid hexoside pentoside, caffeic acid acetate hexoside, caffeic acid hexoside isomer, caffeic acid rhamnoside hexoside, phenethyl caffeate hexoside, ferulic/isoferulic acid rhamnoside hexoside, phenethyl caffeate pentoside, ferulic/isoferulic acid glutarate, and ferulic/isoferulic acid.

6.3.3 Esters of Cinnamic Acids

Cinnamoyl ester made of quinic acid and cinnamic acids has less frequently reported in nuts. Amaral et al. (2005) identified and quantified four phenolic acids, namely, 3-caffeoylquinic acid, 5-caffeoylquinic acid, caffeoyl tartaric acid, and *p*-coumaroyltartaric acid, in hazelnut leaves from ten different cultivars grown in Portugal. The compounds 3-caffeoylquinic acid and 5-caffeoylquinic acid, 3-*p*-coumaroylquinic acid, and 4-*p*-coumaroylquinic acids were identified in different cultivars of walnut grown in Portugal (Pereira et al. 2007). In another study, Ma et al. (2014) reported the presence of esters and conjugates of hydroxycinnamic acids, such as di-*p*-coumaroyltartaric acid, *p*-coumaroyl feruloyl tartaric acid, and *p*-coumaroyl sinapoyl tartaric acid.

6.3.4 Stilbenes

Stilbenes have been reported in several tree nuts and have been found to ameliorate several diseases induced by oxidative stress (El Khawand et al. 2018). *Trans*-resveratrol (3,5,4′-trihydroxystilbene) is the most common phenolic stilbene in nuts. For example, in Sicilian pistachio, the concentration of *trans*-resveratrol ranged from 0.07 to 0.18 mg/kg, for *trans*-piceid from 6.20 to 8.15 mg/kg, and from 6.38 to 8.27 mg/kg for total resveratrol (Grippi et al. 2008).

Peanuts are good sources of resveratrol (Xu et al. 2020). Yu et al. (2005) reported resveratrol to be present in peanut skins. It was also detected in the aqueous extracts from the edible nut of five Turkish cultivars of *Pistacia vera* (Tokuşoğlu et al. 2005). Gentile et al. (2007) found *trans*-resveratrol in the hydrophilic extract of *Pistacia vera*. Halvorsen et al. (2006) ranked the edible pistachio nut amongst the first 50 food products having the highest antioxidant potential.

Ballistreri et al. (2009) reported an average concentration of 12.0 mg/kg of ripe pistachio kernels. Its presence had also been reported in peanut skin (3.6 mg/kg), together with trans-piceatannol (Ma et al. 2014). In wide genotypic variations in trans-resveratrol concentrations may be found among genotypes. On a whole nut weight basis, Tokuşoğlu et al. (2005) had found values from 0.09 to 1.7 mg/kg and from 0.03 to 1.9 mg/kg for different pistachio and peanut cultivars, respectively. Interestingly, almonds also contain polydatin (**6.1**) which is a glucoside (resveratrol-3-glucoside, Fig. 6.4) as the main phenolic stilbene (7.2–8.5 mg/kg). It is concentrated primarily in the skins together with minor amounts of oxyresveratrol (**6.2**), piceatannol (**6.3**), and pterostilbene (**6.4**) (Xie and Bolling 2014).

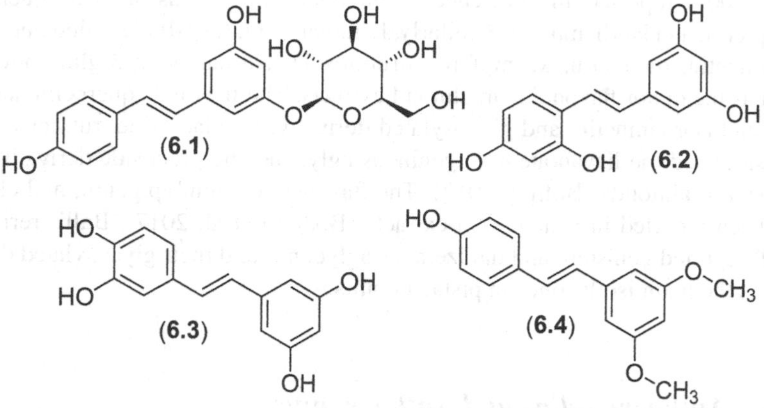

Fig. 6.4 Structures of some stilbenes found in nuts. (**6.1**) Polydatin, (**6.2**) oxyresvertrol, (**6.3**) piceatannol, and (**6.4**) pterostilbene

6.3.5 Flavonoids

Flavonoids are widely distributed in nuts. In a very comprehensive study about flavonoid composition from diverse foods, Harnly et al. (2006) reported significant differences in concentrations among different types of nuts. The compounds quercetin-3-galactoside, quercetin-3-pentoside derivative, quercetin-3-arabinoside, quercetin-3-xyloside, and quercetin-3-rhamnoside were identified in different cultivars of walnut grown in Portugal (Pereira et al. 2007).

Three compounds possessing potential antioxidant activities in the methanolic extract of peanut shells were identified as eriodictyol and luteolin using the HPLC–DAD–TOF/MS technique (Qiu et al. 2012). Gentile et al. (2007) found a remarkable amount of the isoflavones daidzein and genistein (36.8 and 34.0 mg/kg of edible nut, respectively) in the hydrophilic extract of *Pistacia vera*. Seeram et al. (2006) identified quercetin (14.9 mg/kg), luteolin (10.0 mg/kg), eriodictyol (10.2 mg/kg), rutin (1.6 mg/kg), naringenin (1.2 mg/kg), and apigenin (0.2 mg/kg) in pistachio skin. Nine flavonol oligosides of quercetin and kaempferol with glucose, xylose, and rhamnose as sugars were isolated from the chestnut (Hübner et al. 1999). Kapusta et al. (2007) identified a new glycoside, tamarixetin-3-[glucopyranosyl(1 → 3)]-xylopyranosyl-(1 → 2)-glucopyranoside in horse chestnut.

In almond skin samples, (+)-catechin and (−)-epicatechin, isorhamnetin, quercetin, kaempferol and their glycosides such as quercetin-3-rutinoside, quercetin-3-galactoside, quercetin-3-glucoside, kaempferol-3-rutinoside, isorhamnetin-3-rutinoside, kaempferol-3-glucoside, isorhamnetin-3-glucoside, and flavanones naringenin and eriodictyol and their glycosides (eriodictyol-7-glucoside and naringenin-7-glucoside) had been identified (Mandalari et al. 2010). Sang et al. (2002) isolated naringenin glucoside, as well as galactoside, glucoside, and rhamnoglucoside of 3′-methyl quercetin and rhamnoglucoside of kaempferol. Amarowicz et al. (2005) reported the presence of phenolic compounds such as quercetin, kaempferol, and isorhamnetin. Similarly, Esfahlan et al. (2010) identified quercetin, isorhamnetin, quercetin, kaempferol-3-rutinoside, isorhamnetin-3-glucoside, and morin as the major flavonoids in almond extracts. The flavonols quercetin, kaempferol, and isorhamnetin, and glycosylated derivatives (galactoside, rutinoside, and glucoside) and the flavanone naringenin, as aglycone and glucoside derivative, are abundant in almonds (Bolling 2017). The flavones luteolin, apigenin, and chrysin have been reported in peanut skin extracts (Bodoira et al. 2017). Ballistreri et al. (2009) reported genistein and daidzein, as aglycones and their glycosylated derivatives, as the main isoflavones in pistachio nuts.

6.3.6 Anthocyanidins and Anthocyanins

In nuts, Pecans were found to have the highest contents of cyanidin, delphinidin, catechin, epicatechin, epicatechin gallate, and epigallocatechin gallate. Catechin and epicatechin and their gallic acid esters are also abundant in pistachio (Fabani et al. 2013), hazelnut (Del Rio et al. 2011), walnut (Labuckas et al. 2016), cashew nut (Chandrasekara and Shahidi 2011), and peanut (Bodoira et al. 2017; Sarnoski et al. 2012). Various procyanidin oligomers, consisting exclusively of the flavanol monomers and differing in stereochemistry and the points of attachment (linkage) of catechin and epicatechin units with each other, are major compounds in peanut skin (Bodoira et al. 2019). Glycosylated anthocyanidins (cyanidin-3-galactoside and cyanidin-3-glucoside) have been mainly found in pistachios (Ballistreri et al. 2009). Biochanin A and genistein were found in natural and roasted peanuts (Bodoira et al. 2017). Amarowicz et al. (2005) reported the presence of phenolic compounds such as quercetin, kaempferol, and isorhamnetin. Delphinidin and cyanidin as well as procyanidins B2 and B3 in almond extract using an HPLC method and found that procyanidins B2 and B3 were the dominant phenolic compounds in almond extract. Prodelphinidins (3-gallates) including epigallocatechin, epicatechin-3-gallate, and the more common flavan-3-ols, catechin, and epicatechin were identified in pecan kernel (Villarreal-Lozoya et al. 2007).

Proanthocyanidins are also relatively abundant in pistachios (2.40 mg/g), almonds (1.84 mg/g) (Gu et al. 2004), and peanut skin (Bodoira et al. 2017). In this latter case, the types and concentrations of proanthocyanidins may vary depending upon the starting material (raw, roasted, and blanched peanuts). It is an important aspect because the majority of peanut production is destined for snack and oil industries, which utilize roasting and blanching processes to obtain the final products. Peanut skins are removed before processing into products, such as peanut oil and butter. It has been reported that blanching processes deplete skin polyphenols (Bolling 2017). Thus, fewer proanthocyanidin quantities may be obtained when hot water blanched peanut skins are used as source material. Beyond these process-related considerations, peanut skins are certainly high in proanthocyanidins, some of them with a high degree of polymerization (up to 12 interflavan linkages).

Proanthocyanidins present in peanut skins were found to be primarily linked by C4 → C8 bonds, i.e., B- and A-type (this latter has an additional C2 → O → C7 linkage) polymers (Sarnoski et al. 2012). The polymeric proanthocyanidins account for 40% of total testa phenolic compounds; other components of testa phenolics include leucocyanidins and leucopelargonidins (Bodoira and Maestri 2020). Recently, Chandrasekara and Shahidi (2011) reported that whole cashew and nuts testa were better sources of antioxidants compared to the kernel as assessed in different food and biological model systems.

Hazelnut contained only cyanidin and epicatechin, (−)-epigallocatechin, (−)-epigallocatechin 3-gallate, and (+)-catechin. Cyanidin was the most abundant flavonoid and no flavonones and flavonols were found in hazelnut (Alasalvar et al. 2009). Hoffman and Shahidi (2009) found small amounts (8.61–68.22 mg/kg) of paclitaxel in tombul hazelnut hard shells, green shell covers, and leaves. In addition to paclitaxel, 10-diacetyl baccatin III, and baccatin III were also identified in tombul hazelnut extracts. Besides, hazelnut was found to contain the highest amount of condensed tannins amongst seven tree nuts i.e., hazelnut, almond, cashew, chestnut, pecan, pistachio, and walnut.

Identification of phenolic compounds from dry-blanched peanut skins by liquid chromatography-electrospray ionization mass spectrometry was also reported by Ma et al. (2014). Similarly, de Camargo et al. (2015) identified proanthocyanidins as the major phenolic compounds present in peanut skin. Recently, Noguera-Artiaga et al. (2019) showed that the total polymeric procyanidin concentrations fluctuated between 3480 and 59,190 mg/kg on a dry weight basis, (−)-epicatechin being the major monomer contributor.

6.3.7 Tannins

Tannins are also the most abundant compounds in walnuts, hazelnuts, pecans, and almonds. Walnuts are characterized by the majority of hydrolyzable tannins and related compounds (Ballistreri et al. 2009; Labuckas et al. 2016). The structural analysis of the major components leads to their allocation into the groups of gallotannins and ellagitannins. Gallotannins are esters of gallic acid with a polyol carbohydrate, usually glucose. Ellagitannins are esters of hexahydroxydiphenic acid (HHDP) and a polyol (usually glucose) and may also have a galloyl group. Important ellagitannins such as pedunculagin, casuarinin, casuarictin, strictinin, tellimagrandin, praecoxin, rugosin, and isostrictinin have been found in walnuts (Bodoira and Maestri 2020). Ellagic acid, gallotannins, ellagitannins, and derivatives are partially responsible for the astringent taste of walnut kernels. Hazelnuts and pecans have the highest concentrations of total proanthocyanidins, i.e., condensed tannins, (about 5 mg/g, on a wet basis) displaying the highest levels of polymerization (Gu et al. 2004).

The most commonly identified phenolic compounds in walnut are condensed tannins. Walnut phenolic antioxidants are found in the highest concentration in the hull fraction (Labuckas et al. 2016). Anderson et al. (2001) reported that ellagitannins were the most abundant phenolic compounds in walnuts. These include ellagic acid, valoneic acid dilactone, and pedunculagin, which were identified in phenolic extracts from shelled walnuts. The skins of chestnut are rich in tannin. Tannins of Korean chestnut fruit consist mainly of 3,6-digalloylglucose, pyrogallol, and resorcinol (Hwang et al. 2001). Ogawa et al. (2008) reported procyanidin trimers, polymeric proanthocyanidins having a series of heteropolyflavan-3-ols, (+)-catechin, (−)-epicatechin in seed shells of the Japanese horse chestnut.

Tannins present in hazelnuts, pecans, and almonds are from both groups, i.e., hydrolyzable and condensed tannins. In Turkish hazelnut samples, Pelvan et al. (2018) reported the presence of five hydrolyzable tannins and related compounds such as flavogallonic acid dilactone, valoneic acid dilactone, sanguisorbic acid dilactone, bis(hexahydroxydiphenoyl)-glucose (HHDP-glucose), ellagic acid, and derivatives. Bolling (2017) reports compiled data from the literature indicating values ranging between 0.73 and 0.91 mg/g of hydrolyzable tannins including gallotannins, ellagitannins, and phlorotannins of the almond kernel. In Baru (*Dipteryx alata* Vog.) nuts, Oliveira-Alves et al. (2020) showed gallotannins have been one of the major phenolic classes of compounds. The derivatives of tannins were methyl gallate, monogalloylglucose, digalloylglucose, trigalloylglucose, tetragalloylglucose, and pentagalloylglucose.

6.3.8 Phenolic Lignans

Phenolic lignans have been reported to be the minor components of several nut species, such as almonds, hazelnuts, peanuts, cashews, pecans, walnuts, pistachios, and Brazil nuts. These two latter materials could be the major sources of total phenolic lignans with values of 2.0 and 7.8 mg/kg of nut, respectively. Matairesinol, lariciresinol, pinoresinol, and secoisolariciresinol had been frequently found in nuts. It is not certain if they are present as glycosides or in conjugated form because they have been identified after hydrolysis. Secoisolariciresinol seems to be the major compound in most cases, with Brazil nuts being the richest source (7.8 mg/kg). Lariciresinol is found at higher concentrations in almonds (up to 2.4 mg/g) (Bolling 2017).

6.4 Study Questions

- Describe in detail the hydroxybenzoic acid derivatives in legumes and nuts?
- What do you know about the cinnamic acid derivatives present in legumes and nuts?
- What are the esters of cinnamic acids present in legumes or nuts?
- Discuss in detail the composition of flavonoids in legumes.

- What are the different anthocyanins present in legumes?
- Are tannins present in legumes?
- Have you heard about the stilbenes in nuts?
- Describe in detail the flavonoids and anthocyanins present in nuts.
- Write a note on tannins in nuts.

References

Adams MA, Buckley TN, Salter WT, Buchmann N, Blessing CH, Turnbull TL (2018) Contrasting responses of crop legumes and cereals to nitrogen availability. New Phytol 217(4):1475–1483

Aguilera Y, Dueñas M, Estrella I, Hernández T, Benitez V, Esteban RM, Martín-Cabrejas MA (2010) Evaluation of phenolic profile and antioxidant properties of pardina lentil as affected by industrial dehydration. J Agric Food Chem 58(18):10101–10108

Aguilera Y, Estrella I, Benitez V, Esteban RM, Martín-Cabrejas MA (2011) Bioactive phenolic compounds and functional properties of dehydrated bean flours. Food Res Int 44(3):774–780

Alasalvar C, Karamać M, Amarowicz R, Shahidi F (2006) Antioxidant and antiradical activities in extracts of hazelnut kernel (Corylus avellana L.) and hazelnut green leafy cover. J Agric Food Chem 54(13):4826–4832

Alasalvar C, Karamać M, Kosinska A, Rybarczyk A, Shahidi F, Amarowicz R (2009) Antioxidant activity of hazelnut skin phenolics. J Agric Food Chem 57(11):4645–4650

Alasalvar C, Salvadó J-S, Ros E (2020) Bioactives and health benefits of nuts and dried fruits. Food Chem 314:126192

Alshikh N, de Camargo AC, Shahidi F (2015) Phenolics of selected lentil cultivars: antioxidant activities and inhibition of low-density lipoprotein and DNA damage. J Med Food 18:1022–1038

Amaral JS, Ferreres F, Andrade PB, Valentao P, Pinheiro C, Santos A, Seabra R (2005) Phenolic profile of hazelnut (Corylus avellana L.) leaves cultivars grown in Portugal. Nat Prod Res 19(2):157–163

Amarowicz R, Pegg RB (2008) Legumes as a source of natural antioxidants. Eur J Lipid Sci Technol 110(10):865–878

Amarowicz R, Pegg RB (2019) Leguminous seeds as source of phenolic acids, condensed tannins and lignans. In: Martín-Cabrejas MÁ (ed) Legumes: nutritional quality, processing and potential health benefits, vol 8. Royal Society of Chemistry, London, pp 21–48

Amarowicz R, Shahidi F (2017) Antioxidant activity of broad bean seed extract and its phenolic composition. J Funct Foods 38:656–662

Amarowicz R, Troszyńska A, Baryłko-Pikielna N, Shahidi F (2004) Polyphenolics extracts from legume seeds: correlations between total antioxidant activity, total phenolics content, tannins content and astringency. J Food Lipids 11(4):278–286

Amarowicz R, Troszyńska A, Shahidi F (2005) Antioxidant activity of almond seed extract and its fractions. J Food Lipids 12(4):344–358

Amarowicz R, Estrella I, Hernandez T, Troszyńska A (2008) Antioxidant activity of extract of adzuki bean and its fractions. J Food Lipids 15(1):119–136

Amarowicz R, Estrella I, Hernández T, Dueñas M, Troszyńska A, Kosińska A, Pegg RB (2009) Antioxidant activity of a red lentil extract and its fractions. Int J Mol Sci 10(12):5513–5527

Amarowicz R, Estrella I, Hernández T, Robredo S, Troszyńska A, Kosińska A, Pegg RB (2010) Free radical-scavenging capacity, antioxidant activity, and phenolic composition of green lentil (Lens culinaris). Food Chem 121(3):705–711

Anderson KJ, Teuber SS, Gobeille A, Cremin P, Waterhouse AL, Steinberg FM (2001) Walnut polyphenolics inhibit in vitro human plasma and LDL oxidation. J Nutr 131(11):2837–2842

Attree R, Du B, Xu B (2015) Distribution of phenolic compounds in seed coat and cotyledon, and their contribution to antioxidant capacities of red and black seed coat peanuts (*Arachis hypogaea* L.). Ind Crop Prod 67:448–456

Baginsky C, Peña-Neira Á, Cáceres A, Hernández T, Estrella I, Morales H, Pertuzé R (2013) Phenolic compound composition in immature seeds of fava bean (Vicia faba L.) varieties cultivated in Chile. J Food Compos Anal 31(1):1–6

Ballistreri G, Arena E, Fallico B (2009) Influence of ripeness and drying process on the polyphenols and tocopherols of *Pistacia vera* L. Molecules 14(11):4358–4369

Belmiro RH, Tribst AAL, Cristianini M (2020) Effects of high pressure processing on common beans (*Phaseolus Vulgaris* L.): cotyledon structure, starch characteristics, and phytates and tannins contents. Starch 72(3–4):1900212

Bodoira R, Maestri D (2020) Phenolic compounds from nuts: extraction, chemical profiles, and bioactivity. J Agric Food Chem 68(4):927–942

Bodoira R, Rossi Y, Montenegro M, Maestri D, Velez A (2017) Extraction of antioxidant polyphenolic compounds from peanut skin using water-ethanol at high pressure and temperature conditions. J Supercrit Fluids 128:57–65

Bodoira R, Velez A, Rovetto L, Ribotta P, Maestri D, Martínez M (2019) Subcritical fluid extraction of antioxidant phenolic compounds from Pistachio (*Pistacia vera* L.) nuts: experiments, modeling, and optimization. J Food Sci 84(5):963–970

Bolling BW (2017) Almond polyphenols: methods of analysis, contribution to food quality, and health promotion. Compr Rev Food Sci Food Saf 16(3):346–368

Broughton WJ, Hernández G, Blair M, Beebe S, Gepts P, Vanderleyden J (2003) Beans (Phaseolus spp.)—model food legumes. Plant Soil 252(1):55–128

Chaalal M, Ouchemoukh S, Mehenni C, Salhi N, Soufi O, Ydjedd S, Louaileche H (2019) Phenolic contents and in vitro antioxidant activity of four commonly consumed nuts in algeria. Acta Aliment 48(1):125–131

Chandrasekara N, Shahidi F (2011) Effect of roasting on phenolic content and antioxidant activities of whole cashew nuts, kernels, and testa. J Agric Food Chem 59(9):5006–5014

Chen PX, Bozzo GG, Freixas-Coutin JA, Marcone MF, Pauls PK, Tang Y, Zhang B, Liu R, Tsao R (2015a) Free and conjugated phenolic compounds and their antioxidant activities in regular and non-darkening cranberry bean (*Phaseolus Vulgaris* L.) seed coats. J Funct Foods 18:1047–1056

Chen PX, Tang Y, Marcone MF, Pauls PK, Zhang B, Liu R, Tsao R (2015b) Characterization of free, conjugated and bound phenolics and lipophilic antioxidants in regular-and non-darkening cranberry beans (*Phaseolus Vulgaris* L.). Food Chem 185:298–308

Chung I-M, Seo S-H, Ahn J-K, Kim S-H (2011) Effect of processing, fermentation, and aging treatment to content and profile of phenolic compounds in soybean seed, soy curd and soy paste. Food Chem 127(3):960–967

Clifford MN (1999) Chlorogenic acids and other cinnamates—nature, occurrence and dietary burden. J Sci Food Agric 79(3):362–372

Clifford MN, Wight J (1976) The measurement of feruloylquinic acids and caffeoylquinic acids in coffee beans. Development of the technique and its preliminary application to green coffee beans. J Sci Food Agric 27(1):73–84

Clifford MN, Kirkpatrick J, Kuhnert N, Roozendaal H, Salgado PR (2008) LC–MSn analysis of the cis isomers of chlorogenic acids. Food Chem 106(1):379–385

Colaric M, Veberic R, Solar A, Hudina M, Stampar F (2005) Phenolic acids, syringaldehyde, and juglone in fruits of different cultivars of *Juglans regia* L. J Agric Food Chem 53(16):6390–6396

Čolić SD, Akšić MMF, Lazarević KB, Zec GN, Gašić UM, Zagorac DČD, Natić MM (2017) Fatty acid and phenolic profiles of almond grown in Serbia. Food Chem 234:455–463

Collado E, Venzke Klug T, Martínez-Hernández GB, Artés-Hernández F, Martínez-Sánchez A, Aguayo E, Artés F, Fernández JA, Gómez PA (2019) Nutritional and quality changes of minimally processed faba (*Vicia faba* L.) beans during storage: effects of domestic microwaving. Postharvest Biol Technol 151:10–18

Damián-Medina K, Salinas-Moreno Y, Milenkovic D, Figueroa-Yáñez L, Marino-Marmolejo E, Higuera-Ciapara I, Vallejo-Cardona A, Lugo-Cervantes E (2020) In silico analysis of antidiabetic potential of phenolic compounds from blue corn (*Zea mays* L.) and black bean (*Phaseolus Vulgaris* L.). Heliyon 6(3):e03632

de Camargo AC, Regitano-d'Arce MAB, Gallo CR, Shahidi F (2015) Gamma-irradiation induced changes in microbiological status, phenolic profile and antioxidant activity of peanut skin. J Funct Foods 12:129–143

de la Rosa LA, Alvarez-Parrilla E, Shahidi F (2011) Phenolic compounds and antioxidant activity of kernels and shells of Mexican pecan (*Carya illinoinensis*). J Agric Food Chem 59(1):152–162

Del Rio D, Calani L, Dall'Asta M, Brighenti F (2011) Polyphenolic composition of hazelnut skin. J Agric Food Chem 59(18):9935–9941

Drumm TD, Gray JI, Hosfield GL (1990) Variability in the saccharide, protein, phenolic acid and saponin contents of four market classes of edible dry beans. J Sci Food Agric 51(3):285–297

Dueñas M, Hernández T, Estrella I (2007) Changes in the content of bioactive polyphenolic compounds of lentils by the action of exogenous enzymes. Effect on their antioxidant activity. Food Chem 101(1):90–97

Dueñas M, Martínez-Villaluenga C, Limón RI, Peñas E, Frias J (2015) Effect of germination and elicitation on phenolic composition and bioactivity of kidney beans. Food Res Int 70:55–63

Dueñas M, Sarmento T, Aguilera Y, Benitez V, Mollá E, Esteban RM, Martín-Cabrejas MA (2016) Impact of cooking and germination on phenolic composition and dietary fibre fractions in dark beans (*Phaseolus vulgaris* L.) and lentils (*Lens culinaris* L.). LWT Food Sci Technol 66:72–78

El Khawand T, Courtois A, Valls J, Richard T, Krisa S (2018) A review of dietary stilbenes: sources and bioavailability. Phytochem Rev 17(5):1007–1029

Erşan S, Güçlü Üstündağ O, Carle R, Schweiggert RM (2016) Identification of phenolic compounds in red and green pistachio (*Pistacia vera* L.) hulls (exo-and mesocarp) by HPLC-DAD-ESI-(HR)-MSn. J Agric Food Chem 64(26):5334–5344

Esfahlan AJ, Jamei R, Esfahlan RJ (2010) The importance of almond (*Prunus amygdalus* L.) and its by-products. Food Chem 120(2):349–360

Fabani MP, Luna L, Baroni MV, Monferran MV, Ighani M, Tapia A, Wunderlin DA, Feresin GE (2013) Pistachio (*Pistacia vera* var Kerman) from Argentinean cultivars. A natural product with potential to improve human health. J Funct Foods 5(3):1347–1356

Fajardo J, Waniska R, Cuero R, Pettit R (1995) Phenolic compounds in peanut seeds: enhanced elicitation by chitosan and effects on growth and aflatoxin B1 production by *Aspergillus flavus*. Food Biotechnol 9(1–2):59–78

Gan R-Y, Deng Z-Q, Yan A-X, Shah NP, Lui W-Y, Chan C-L, Corke H (2016) Pigmented edible bean coats as natural sources of polyphenols with antioxidant and antibacterial effects. LWT Food Sci Technol 73:168–177

Ganesan K, Xu B (2018) A critical review on phytochemical profile and health promoting effects of mung bean (*Vigna radiata*). Food Sci Human Wellness 7(1):11–33

García-Lafuente A, Moro C, Manchón N, Gonzalo-Ruiz A, Villares A, Guillamón E, Rostagno M, Mateo-Vivaracho L (2014) In vitro anti-inflammatory activity of phenolic rich extracts from white and red common beans. Food Chem 161:216–223

Garcia-Mora P, Frias J, Peñas E, Zieliński H, Giménez-Bastida JA, Wiczkowski W, Zielińska D, Martínez-Villaluenga C (2015) Simultaneous release of peptides and phenolics with antioxidant, ACE-inhibitory and anti-inflammatory activities from pinto bean (*Phaseolus Vulgaris* L. var. pinto) proteins by subtilisins. J Funct Foods 18:319–332

Gentile C, Tesoriere L, Butera D, Fazzari M, Monastero M, Allegra M, Livrea MA (2007) Antioxidant activity of Sicilian pistachio (*Pistacia vera* L. var. Bronte) nut extract and its bioactive components. J Agric Food Chem 55(3):643–648

Giusti F, Capuano E, Sagratini G, Pellegrini N (2019) A comprehensive investigation of the behaviour of phenolic compounds in legumes during domestic cooking and in vitro digestion. Food Chem 285:458–467

Gomes S, Finotelli PV, Sardela VF, Pereira HMG, Santelli RE, Freire AS, Torres AG (2019) Microencapsulated Brazil nut (*Bertholletia excelsa*) cake extract powder as an added-value functional food ingredient. LWT Food Sci Technol 116:108495

Grippi F, Crosta L, Aiello G, Tolomeo M, Oliveri F, Gebbia N, Curione A (2008) Determination of stilbenes in Sicilian pistachio by high-performance liquid chromatographic diode array (HPLC-DAD/FLD) and evaluation of eventually mycotoxin contamination. Food Chem 107(1):483–488

Gu L, Kelm MA, Hammerstone JF, Beecher G, Holden J, Haytowitz D, Gebhardt S, Prior RL (2004) Concentrations of proanthocyanidins in common foods and estimations of normal consumption. J Nutr 134(3):613–617

Gutiérrez-Uribe JA, Romo-Lopez I, Serna-Saldívar SO (2011) Phenolic composition and mammary cancer cell inhibition of extracts of whole cowpeas (*Vigna unguiculata*) and its anatomical parts. J Funct Foods 3(4):290–297

Halvorsen BL, Carlsen MH, Phillips KM, Bøhn SK, Holte K, Jacobs DR Jr, Blomhoff R (2006) Content of redox-active compounds (ie, antioxidants) in foods consumed in the United States. Am J Clin Nutr 84(1):95–135

Han K-H, Kitano-Okada T, Seo J-M, Kim S-J, Sasaki K, Shimada K-i, Fukushima M (2015) Characterisation of anthocyanins and proanthocyanidins of adzuki bean extracts and their antioxidant activity. J Funct Foods 14:692–701

Harnly JM, Doherty RF, Beecher GR, Holden JM, Haytowitz DB, Bhagwat S, Gebhardt S (2006) Flavonoid content of U.S. fruits, vegetables, and nuts. J Agric Food Chem 54(26):9966–9977

Hoffman A, Shahidi F (2009) Paclitaxel and other taxanes in hazelnut. J Funct Foods 1(1):33–37

Hübner G, Wray V, Nahrstedt A (1999) Flavonol oligosaccharides from the seeds of *Aesculus hippocastanum*. Planta Med 65(7):636–642

Hwang J-Y, Hwang I-K, Park J-B (2001) Analysis of physicochemical factors related to the automatic pellicle removal in Korean chestnut (*Castanea crenata*). J Agric Food Chem 49(12):6045–6049

Jakobek L (2015) Interactions of polyphenols with carbohydrates, lipids and proteins. Food Chem 175:556–567

Jia X, Luo H, Xu M, Zhai M, Guo Z, Qiao Y, Wang L (2018) Dynamic changes in phenolics and antioxidant capacity during pecan (*Carya illinoinensis*) kernel ripening and its phenolics profiles. Molecules 23(2):435

Jin AL, Ozga JA, Lopes-Lutz D, Schieber A, Reinecke DM (2012) Characterization of proanthocyanidins in pea (*Pisum sativum* L.), lentil (*Lens culinaris* L.), and faba bean (*Vicia faba* L.) seeds. Food Res Int 46(2):528–535

John JA, Shahidi F (2010) Phenolic compounds and antioxidant activity of Brazil nut (*Bertholletia excelsa*). J Funct Foods 2(3):196–209

Kalogeropoulos N, Chiou A, Ioannou M, Karathanos VT, Hassapidou M, Andrikopoulos NK (2010) Nutritional evaluation and bioactive microconstituents (phytosterols, tocopherols, polyphenols, triterpenic acids) in cooked dry legumes usually consumed in the Mediterranean countries. Food Chem 121(3):682–690

Kan L, Nie S, Hu J, Liu Z, Xie M (2016) Antioxidant activities and anthocyanins composition of seed coats from twenty-six kidney bean cultivars. J Funct Foods 26:622–631

Kan L, Nie S, Hu J, Wang S, Bai Z, Wang J, Zhou Y, Jiang J, Zeng Q, Song K (2018) Comparative study on the chemical composition, anthocyanins, tocopherols and carotenoids of selected legumes. Food Chem 260:317–326

Kapusta I, Janda B, Szajwaj B, Stochmal A, Piacente S, Pizza C, Franceschi F, Franz C, Oleszek W (2007) Flavonoids in horse chestnut (*Aesculus hippocastanum*) seeds and powdered waste water byproducts. J Agric Food Chem 55(21):8485–8490

Kawahara S-i, Ishihara C, Matsumoto K, Senga S, Kawaguchi K, Yamamoto A, Suwannachot J, Hamauzu Y, Makabe H, Fujii H (2019) Identification and characterization of oligomeric proanthocyanidins with significant anti-cancer activity in adzuki beans (*Vigna angularis*). Heliyon 5(10):e02610

Khang DT, Dung TN, Elzaawely AA, Xuan TD (2016) Phenolic profiles and antioxidant activity of germinated legumes. Foods 5(2):27

Kornsteiner M, Wagner K-H, Elmadfa I (2006) Tocopherols and total phenolics in 10 different nut types. Food Chem 98(2):381–387

Krishnan V, Rani R, Pushkar S, Lal SK, Srivastava S, Kumari S, Vinutha T, Dahuja A, Praveen S, Sachdev A (2020) Anthocyanin fingerprinting and dynamics in differentially pigmented exotic soybean genotypes using modified HPLC–DAD method. J Food Meas Charact 14:1966–1975

Król K, Gantner M, Tatarak A, Hallmann E (2020) The content of polyphenols in coffee beans as roasting, origin and storage effect. Eur Food Res Technol 246(1):33–39

Labuckas D, Maestri D, Lamarque A (2016) Molecular characterization, antioxidant and protein solubility-related properties of polyphenolic compounds from walnut (*Juglans regia*). Nat Prod Commun 11(5):1934578X1601100521

Lantto TA, Dorman HD, Shikov AN, Pozharitskaya ON, Makarov VG, Tikhonov VP, Hiltunen R, Raasmaja A (2009) Chemical composition, antioxidative activity and cell viability effects of a Siberian pine (*Pinus sibirica* Du Tour) extract. Food Chem 112(4):936–943

Liu G, Guasch-Ferré M, Hu Y, Li Y, Hu FB, Rimm EB, Manson JE, Rexrode KM, Sun Q (2019) Nut consumption in relation to cardiovascular disease incidence and mortality among patients with diabetes mellitus. Circ Res 124(6):920–929

Llorach R, Favari C, Alonso D, Garcia-Aloy M, Andres-Lacueva C, Urpi-Sarda M (2019) Comparative metabolite fingerprinting of legumes using LC-MS-based untargeted metabolomics. Food Res Int 126:108666

López A, El-Naggar T, Dueñas M, Ortega T, Estrella I, Hernández T, Gómez-Serranillos MP, Palomino OM, Carretero ME (2013) Effect of cooking and germination on phenolic composition and biological properties of dark beans (*Phaseolus vulgaris* L.). Food Chem 138(1):547–555

López-Amorós M, Hernández T, Estrella I (2006) Effect of germination on legume phenolic compounds and their antioxidant activity. J Food Compos Anal 19(4):277–283

Luczak S, Swiatek L, Zadernowski R (1989) Phenolic acids in leaves and pericarp of walnut, *Juglans regia*. Acta Pol Pharm 46:494–499

Luthria DL, Pastor-Corrales MA (2006) Phenolic acids content of fifteen dry edible bean (*Phaseolus vulgaris* L.) varieties. J Food Compos Anal 19(2–3):205–211

Ma Y, Kosińska-Cagnazzo A, Kerr WL, Amarowicz R, Swanson RB, Pegg RB (2014) Separation and characterization of phenolic compounds from dry-blanched peanut skins by liquid chromatography–electrospray ionization mass spectrometry. J Chromatogr A 1356:64–81

Madrera RR, Valles BS (2020) Development and validation of ultrasound assisted extraction (UAE) and HPLC-DAD method for determination of polyphenols in dry beans (*Phaseolus vulgaris*). J Food Compos Anal 85:103334

Magalhães SCQ, Taveira M, Cabrita ARJ, Fonseca AJM, Valentão P, Andrade PB (2017) European marketable grain legume seeds: further insight into phenolic compounds profiles. Food Chem 215:177–184

Malenčić Đ, Kiprovski B, Bursić V, Vuković G, Ćupina B, Mikić A (2018) Dietary phenolics and antioxidant capacity of selected legumes seeds from the Central Balkans. Acta Aliment 47(3):340–349

Mandalari G, Tomaino A, Arcoraci T, Martorana M, Turco VL, Cacciola F, Rich G, Bisignano C, Saija A, Dugo P (2010) Characterization of polyphenols, lipids and dietary fibre from almond skins (*Amygdalus communis* L.). J Food Compos Anal 23(2):166–174

Martínez ML, Fabani MP, Baroni MV, Huaman RNM, Ighani M, Maestri DM, Wunderlin D, Tapia A, Feresin GE (2016) Argentinian pistachio oil and flour: a potential novel approach of pistachio nut utilization. J Food Sci Technol 53(5):2260–2269

Mecha E, Leitão ST, Carbas B, Serra AT, Moreira PM, Veloso MM, Gomes R, Figueira ME, Brites C, Vaz Patto MC, Bronze MR (2019) Characterization of soaking process' impact in common beans phenolic composition: contribute from the unexplored Portuguese germplasm. Foods 8(8):296

Mecha E, Feliciano RP, Rodriguez-Mateos A, Silva SD, Figueira ME, Vaz Patto MC, Bronze MR (2020) Human bioavailability of phenolic compounds found in common beans: the use of high-resolution MS to evaluate inter-individual variability. Br J Nutr 123(3):273–292

Mirali M, Ambrose SJ, Wood SA, Vandenberg A, Purves RW (2014) Development of a fast extraction method and optimization of liquid chromatography-mass spectrometry for the analysis of phenolic compounds in lentil seed coats. J Chromatogr B Anal Technol Biomed Life Sci 969:149–161

Mojica L, Meyer A, Berhow MA, de Mejía EG (2015) Bean cultivars (*Phaseolus vulgaris* L.) have similar high antioxidant capacity, in vitro inhibition of α-amylase and α-glucosidase while diverse phenolic composition and concentration. Food Res Int 69:38–48

Moo-Huchin VM, Canto-Pinto JC, Cuevas-Glory LF, Sauri-Duch E, Pérez-Pacheco E, Betancur-Ancona D (2019) Effect of extraction solvent on the phenolic compounds content and antioxidant activity of Ramon nut (*Brosimum alicastrum*). Chem Pap 73(7):1647–1657

Noguera-Artiaga L, Salvador MD, Fregapane G, Collado-González J, Wojdyło A, López-Lluch D, Carbonell-Barrachina ÁA (2019) Functional and sensory properties of pistachio nuts as affected by cultivar. J Sci Food Agric 99(15):6696–6705

Ogawa S, Kimura H, Niimi A, Katsube T, Jisaka M, Yokota K (2008) Fractionation and structural characterization of polyphenolic antioxidants from seed shells of Japanese horse chestnut (*Aesculus turbinata* BLUME). J Agric Food Chem 56(24):12046–12051

Ojwang LO, Yang L, Dykes L, Awika J (2013) Proanthocyanidin profile of cowpea (*Vigna unguiculata*) reveals catechin-O-glucoside as the dominant compound. Food Chem 139(1-4):35–43

Oliveira-Alves SC, Pereira RS, Pereira AB, Ferreira A, Mecha E, Silva AB, Serra AT, Bronze MR (2020) Identification of functional compounds in baru (*Dipteryx alata* Vog.) nuts: nutritional value, volatile and phenolic composition, antioxidant activity and antiproliferative effect. Food Res Int 131:109026

Oomah BD, Caspar F, Malcolmson LJ, Bellido A-S (2011) Phenolics and antioxidant activity of lentil and pea hulls. Food Res Int 44(1):436–441

Orita A, Musou-Yahada A, Shoji T, Oki T, Ohta H (2019) Comparison of anthocyanins, proanthocyanidin oligomers and antioxidant capacity between cowpea and grain legumes with colored seed coat. Food Sci Technol Res 25(2):287–294

Ozer HK (2017) Phenolic compositions and antioxidant activities of Maya nut (*Brosimum alicastrum*): comparison with commercial nuts. Int J Food Prop 20(11):2772–2781

Pellegrini N, Serafini M, Salvatore S, Del Rio D, Bianchi M, Brighenti F (2006) Total antioxidant capacity of spices, dried fruits, nuts, pulses, cereals and sweets consumed in Italy assessed by three different in vitro assays. Mol Nutr Food Res 50(11):1030–1038

Pelvan E, Olgun EÖ, Karadağ A, Alasalvar C (2018) Phenolic profiles and antioxidant activity of Turkish Tombul hazelnut samples (natural, roasted, and roasted hazelnut skin). Food Chem 244:102–108

Pereira JA, Oliveira I, Sousa A, Valentao P, Andrade PB, Ferreira ICFR, Ferreres F, Bento A, Seabra R, Estevinho L (2007) Walnut (*Juglans regia* L.) leaves: phenolic compounds, antibacterial activity and antioxidant potential of different cultivars. Food Chem Toxicol 45(11):2287–2295

Pino Ramos LL, Jiménez-Aspee F, Theoduloz C, Burgos-Edwards A, Domínguez-Perles R, Oger C, Durand T, Gil-Izquierdo Á, Bustamante L, Mardones C, Márquez K, Contreras D, Schmeda-Hirschmann G (2019) Phenolic, oxylipin and fatty acid profiles of the Chilean hazelnut (*Gevuina avellana*): antioxidant activity and inhibition of pro-inflammatory and metabolic syndrome-associated enzymes. Food Chem 298:125026

Qiu J, Chen L, Zhu Q, Wang D, Wang W, Sun X, Liu X, Du F (2012) Screening natural antioxidants in peanut shell using DPPH–HPLC–DAD–TOF/MS methods. Food Chem 135(4):2366–2371

Ryu D, Koh E (2018) Application of response surface methodology to acidified water extraction of black soybeans for improving anthocyanin content, total phenols content and antioxidant activity. Food Chem 261:260–266

Salawu SO, Bester MJ, Duodu KG (2014) Phenolic composition and bioactive properties of cell wall preparations and whole grains of selected cereals and legumes. J Food Biochem 38(1):62–72

Samaranayaka AG, John JA, Shahidi F (2008) Antioxidant activity of English walnut (*Juglans regia* L.). J Food Lipids 15(3):384–397

Sang S, Lapsley K, Jeong W-S, Lachance PA, Ho C-T, Rosen RT (2002) Antioxidative pheno-lic compounds isolated from almond skins (*Prunus amygdalus* Batsch). J Agric Food Chem 50(8):2459–2463

Sarnoski PJ, Johnson JV, Reed KA, Tanko JM, O'Keefe SF (2012) Separation and characterisation of proanthocyanidins in Virginia type peanut skins by LC–MSn. Food Chem 131(3):927–939

Seeram NP, Zhang Y, Henning SM, Lee R, Niu Y, Lin G, Heber D (2006) Pistachio skin phenolics are destroyed by bleaching resulting in reduced antioxidative capacities. J Agric Food Chem 54(19):7036–7040

Shahidi F, Ambigaipalan P (2015) Phenolics and polyphenolics in foods, beverages and spices: antioxidant activity and health effects—a review. J Funct Foods 18:820–897

Shelembe JS, Cromarty D, Bester MJ, Minnaar A, Duodu KG (2012) Characterisation of phenolic acids, flavonoids, proanthocyanidins and antioxidant activity of water extracts from seed coats of marama bean [*Tylosema esculentum*]—an underutilised food legume. Int J Food Sci Technol 47(3):648–655

Singh B, Singh JP, Kaur A, Singh N (2017a) Phenolic composition and antioxidant potential of grain legume seeds: a review. Food Res Int 101:1–16

Singh B, Singh N, Thakur S, Kaur A (2017b) Ultrasound assisted extraction of polyphenols and their distribution in whole mung bean, hull and cotyledon. J Food Sci Technol 54(4):921–932

Song H, Zhang L, Wu L, Huang W, Wang M, Zhang L, Shao Y, Wang M, Zhang F, Zhao Z, Mei X, Li T, Wang D, Liang Y, Li J, Xu T, Zhao Y, Zhong Y, Chen Q, Lu B (2020) Phenolic acid profiles of common food and estimated natural intake with different structures and forms in five regions of China. Food Chem 321:126675

Sreerama YN, Sashikala VB, Pratape VM (2012) Phenolic compounds in cowpea and horse gram flours in comparison to chickpea flour: evaluation of their antioxidant and enzyme inhibitory properties associated with hyperglycemia and hypertension. Food Chem 133(1):156–162

Talcott ST, Duncan CE, Del Pozo-Insfran D, Gorbet DW (2005) Polyphenolic and antioxidant changes during storage of normal, mid, and high oleic acid peanuts. Food Chem 89(1):77–84

Taş NG, Gökmen V (2017) Phenolic compounds in natural and roasted nuts and their skins: a brief review. Curr Opin Food Sci 14:103–109

Tokuşoğlu Ö, Ünal MK, Yemiş F (2005) Determination of the phytoalexin resveratrol (3,5,4'-tri-hydroxystilbene) in peanuts and pistachios by high-performance liquid chromatographic diode array (HPLC-DAD) and gas chromatography−mass spectrometry (GC-MS). J Agric Food Chem 53(12):5003–5009

Tomaino A, Martorana M, Arcoraci T, Monteleone D, Giovinazzo C, Saija A (2010) Antioxidant activity and phenolic profile of pistachio (*Pistacia vera* L., variety Bronte) seeds and skins. Biochimie 92(9):1115–1122

Upadhyay R, Mohan Rao LJ (2013) An outlook on chlorogenic acids—occurrence, chemistry, technology, and biological activities. Crit Rev Food Sci Nutr 53(9):968–984

Villarreal-Lozoya JE, Lombardini L, Cisneros-Zevallos L (2007) Phytochemical constituents and antioxidant capacity of different pecan [*Carya illinoinensis* (Wangenh.) K. Koch] cultivars. Food Chem 102(4):1241–1249

Wang Y-K, Zhang X, Chen G-L, Yu J, Yang L-Q, Gao Y-Q (2016) Antioxidant property and their free, soluble conjugate and insoluble-bound phenolic contents in selected beans. J Funct Foods 24:359–372

Wu X, Beecher GR, Holden JM, Haytowitz DB, Gebhardt SE, Prior RL (2004) Lipophilic and hydrophilic antioxidant capacities of common foods in the United States. J Agric Food Chem 52(12):4026–4037

Xie L, Bolling BW (2014) Characterisation of stilbenes in California almonds (*Prunus dulcis*) by UHPLC–MS. Food Chem 148:300–306

Xu BJ, Chang SK (2007) A comparative study on phenolic profiles and antioxidant activities of legumes as affected by extraction solvents. J Food Sci 72(2):S159–S166

Xu B, Chang SKC (2010) Phenolic substance characterization and chemical and cell-based antioxidant activities of 11 lentils grown in the northern United States. J Agric Food Chem 58(3):1509–1517

Xu S, Luo H, Chen H, Guo J, Yu B, Zhang H, Li W, Chen W, Zhou X, Huang L, Liu N, Lei Y, Liao B, Jiang H (2020) Optimization of extraction of total trans-resveratrol from peanut seeds and its determination by HPLC. J Sep Sci 43(6):1024–1031

Yang J, Liu RH, Halim L (2009) Antioxidant and antiproliferative activities of common edible nut seeds. LWT Food Sci Technol 42(1):1–8

Yu J, Ahmedna M, Goktepe I (2005) Effects of processing methods and extraction solvents on concentration and antioxidant activity of peanut skin phenolics. Food Chem 90(1-2):199–206

Zhang Y, Chang SKC (2019) Comparative studies on ACE inhibition, degree of hydrolysis, antioxidant property and phenolic acid composition of hydrolysates derived from simulated in vitro gastrointestinal proteolysis of three thermally treated legumes. Food Chem 281:154–162

Zhang Z, Liao L, Moore J, Wu T, Wang Z (2009) Antioxidant phenolic compounds from walnut kernels (*Juglans regia* L.). Food Chem 113(1):160–165

Zhang Y, Pechan T, Chang SKC (2018) Antioxidant and angiotensin-I converting enzyme inhibitory activities of phenolic extracts and fractions derived from three phenolic-rich legume varieties. J Funct Foods 42:289–297

Zhong L, Wu G, Fang Z, Wahlqvist ML, Hodgson JM, Clarke MW, Junaldi E, Johnson SK (2019) Characterization of polyphenols in Australian sweet lupin (*Lupinus angustifolius*) seed coat by HPLC-DAD-ESI-MS/MS. Food Res Int 116:1153–1162

Chapter 7
Phenolic Antioxidants in Beverages

Learning Objectives

In this chapter, the reader will be able to:

- Understand the concept of beverages and their different classes.
- Know about the composition of hydroxybenzoic acid in different beverages.
- Describe the presence of hydroxycinnamic acids in beverages.
- Write a note on the presence of stilbenes in beverages.
- Explain the presence of flavonoids and anthocyanins in beverages.

7.1 Introduction

Human needs water or liquid for maintenance of the proper metabolism and survival. Water is the world's most important consumed drink, obtained from freshwater resources like rivers, lakes, and groundwater. With time, the human taste evolved starting with juice and alcohol production. A large set of drinks had been produced for human consumption by the modern food industry. Thus, the term "beverage" or "drink" has been coined. By definition, *a beverage is a liquid drink intended for human consumption*. However, due to the significant evolution and cultural differences, the beverages varied significantly. Similarly, single or more beverages were restricted by either religion or cultures at complete or restricted levels. For example, the preference for Children is drinking milk whereas adolescents preferred soft drinks/beverages. The adult may prefer tea, coffee, or an alcoholic beverage. According to the European Food Safety Authority (EFSA), children under 2–3 years old must consume 1.3 L/day of water or milk or required beverages. In the case of children aged between 4 and 8 years, the adequate intake (AI) of water needs to be 1.6 L/day. A drink of 2.1 L/day for boys aged between 9 and 13. For girls in the

© Springer Nature Switzerland AG 2021
A. Zeb, *Phenolic Antioxidants in Foods: Chemistry, Biochemistry and Analysis*,
https://doi.org/10.1007/978-3-030-74768-8_7

same age group, the AI has to be 1.9 L/day. In adolescents or adult females (14 years and above) need to consume 2.0 L/day, whereas adolescent or adult males (14 years and above) have to keep the AI at 2.5 L/day (Agostoni et al. 2010).

7.2 Types of Beverages

The intake of beverages other than water warrants several factors, including nutrients composition, water content, mineral contents, and safety considerations. Phenolic compounds are one of the principal components of several plant-based or plant additive based beverages. The taste, culture, and sources thus let us classify beverages. There are several classifications of beverages. A generalized classification is based on the presence of alcohol as below:

- *Alcoholic beverages*: Those beverages which include alcohol at various levels. These include beer, cider, wine, and spirits. Different safety and required levels of alcohol are usually maintained.
- *Non-Alcoholic beverages*: These include all those beverages which lack alcohol but does not require any heat treatment or its consumption in hot conditions. This much diverse class with examples including fruit juices, carbonated beverages, and milk.
- *Hot beverages*: These include those drinks which are prepared using heat or consumed in hot conditions. Examples include tea, coffee, and hot chocolate drinks.

Another classification is based on different factors, such as the composition of the beverages into five classes as given below:

1. *Conventional beverages.* Conventional soft drinks are also known as sugar-sweetened beverages. This includes all types of carbonated beverages, fruit juices, or fruit-flavored juices.
2. *Nutrients based beverages.* This class includes those beverages which are rich in several nutrients. Pure milk is a famous example, which has enough nutrients required for the human body. Fruit juices (100%) made of one or more fruits are the best examples.
3. *Alcoholic beverages.* This includes all alcohol-based beverages such as liquors, spirits, cocktails, wine, and beer. Tremendous research has been conducted on the processing, composition, flavor, and types of these beverages (Jamali et al. 2020). A high intake of alcohol has been one of the main abuse of these beverages.
4. *None Caloric beverages.* This class includes all those hot beverages such as tea, or coffee, and diet beverages with little or no sugar or artificial sweeteners. Tea and coffee consumption have been prevalent among adults for centuries and caffeine is an ingredient of these beverages (Rajeswari and Kalpana 2019).
5. *Functional beverages.* This class of beverages is those which includes prebiotic, probiotic, and symbiotic organisms helpful for maintaining physiological functions. This includes fermented milk and other similar beverages such as herbal beverages. These beverages are quite famous in the European market.

Beverages have one of the biggest markets of billions of dollars worldwide. For example, fermented milk and similar products are currently worth 46 billion euros with 77% of the market reported in Europe, North America, and Asia (Nazhand et al. 2020). Therefore researchers are continuously studying the new aspects of these beverages. Research on composition studies is still the main focus of the food industry. Phenolic compounds are important natural antioxidants present in a variety of beverages. Phenolic compounds present in alcohol-based beverages contributed to the bitterness, flavor, and perception (Luo et al. 2020). In this chapter, the focus is on the types of phenolic compounds present irrespective of beverage nature.

7.3 Phenolic Antioxidants in Beverages

7.3.1 Simple phenols

Simple phenols play an important role in the formation of coffee flavor. More than 800 volatile compounds had been reported in roasted coffee aroma. Among these only 42 had been identified as phenolic compounds (Shahidi and Ambigaipalan 2015). Catechol is the predominant volatile phenolic compound found in the coffee aroma, followed by 4-ethyl guaiacol, 4-ethylcatechol, pyrogallol, quinol, and 4-vinylcatechol (Ongo et al. 2020). In red wine samples, the deduced concentration of catechol was 170 µM using an electrochemical detector (Salvo-Comino et al. 2020). The Italian Robusta coffee was found rich in 4-ethyl guaiacol and phenol. Similarly, guaiacol was detected in all coffee samples with a concentration range of 210.0–3010.0 µg/L and classified as a key odorant with a spicy and phenolic note (Lolli et al. 2020). Guaiacol has been reported in several fruit juices in the Asian Pacific region by Van Luong et al. (2019). The authors showed that guaiacol production of *Alicyclobacillus acidoterrestris* genotypes ranged between 2.4 and 16.1 mg/L. The spores of *A. acidoterrestris* have been found to survive pasteurization during juice processing, and then germinate into vegetative cells leading to spoilage due to the production of off-flavor compounds including guaiacol (Molva and Baysal 2016). In fresh samples of purple sweet potato fermented alcoholic beverage, 2-methoxy-4-vinyl phenol (0.7 mg/L), and 2-methyl phenol (0.3 mg/L), as major simple phenols (Li et al. 2017).

7.3.2 Hydroxybenzoic Acids

Phenolic antioxidants in tea are responsible for its antioxidant activity. Zhao et al. (2010) identified nine phenolic compounds, including gallic acid, protocatechuic acid, (+)-catechin, vanillic acid, and syringic in different beer samples. The major free hydroxybenzoic acid in beers was vanillic in Spanish, German, and Danish brands (Bartolomé et al. 2000). Other phenolic compounds such as sinapic, vanillic, homovanillic, p-hydroxybenzoic, 2,6- and 3,5-dihydroxybenzoic, syringic, gallic, protocatechuic, and caffeic acids were also identified in beer (Bartolomé et al. 2000; Montanari et al. 1999).

In coffee, o-dihydroxybenzene, gallic acid, and protocatechuic acid were identified using capillary electrophoresis with amperometric detection (Chu et al. 2008). Recently Yılmaz and Kolak (2017) reported sinapic, protocatechuic, p-hydroxybenzoic, and vanillic acids in coffee.

Polyphenolic compounds and caffeine are major constituents of tea. The phenolic acids found in tea are more powerful than the antioxidant vitamins C and E as well as β-carotene in an *in vitro* lipoprotein oxidation model (Vinson and Dabbagh 1998). The derivatives of hydroxybenzoic acids had been detected in several herbal beverages (Chandrasekara and Shahidi 2018). In wine samples, Five aromatic acids (gallic, 4-HB, 3-HB, vanillic, and syringic) were detected by Fasciano and Danielson (2016). In the cascara beverage, prepared from the coffee cherry pulp, the detected hydroxybenzoic acids were gallic acid and protocatechuic acid (Heeger et al. 2017). In traditional Borsch beverages obtained from the beetroot and wheat bran, 4-hydroxybenzoic acid, vanillic acid, and syringic acids were present contributing to the antioxidant activity of the beverage (Pasqualone et al. 2018).

In fresh samples of purple sweet potato fermented alcoholic beverage, benzoic acid (0.2 mg/L), benzene acetic acid (0.5 mg/L), 4-hydroxy-3-methoxy-benzene acetic acid (0.5 mg/L), 4-hydroxy-3-methoxy benzoic acid (1.1 mg/L), and 2-hydroxy-6-methyl benzaldehyde (1.6 mg/L) (Li et al. 2017). Gamboa-Gómez et al. (2017) reported gallic acid, 3,4-dihydroxybenzoic acid, 4-hydroxybenzoic acid, 2,4,6 trihydroxybenzaldehyde, and salicylic acid in fermented beverages. Shalash et al. (2017) reported several derivatives of hydroxybenzoic acids such as syringic, vanillic, p-hydroxybenzoic, 2,4-dihydroxybenzoic, sinapic, gentisic and gallic acids in honey, iced tea and canned coffee drink samples.

7.3.3 Hydroxycinnamic Acids

In different beer samples, Zhao et al. (2010) reported several phenolic compounds, including caffeic acid, p-coumaric acid, and ferulic acid. The major free phenolic acids in beers were *m-, p-* and *o*-coumaric and ferulic acids (Montanari et al. 1999), whilst ferulic and p-coumaric acids were the dominant free phenolic acids in Spanish, German and Danish brands (Bartolomé et al. 2000). In fermented beverages, Gamboa-Gómez et al. (2017) reported caffeic acid, coumaric acid, and ferulic acid amongst other phenolic compounds. Shalash et al. (2017) reported several phenolic acids (cinnamic, *m*-coumaric, chlorogenic, syringic, ferulic, *o*-coumaric, p-coumaric, and ferulic acids) in honey, iced tea, and canned coffee drink samples.

Brezová et al. (2009) described that the antioxidant activity of instant coffee was approximately three to four times higher than ground coffee. Further investigations of two coffee components, such as caffeic acid and caffeine, using ABTS, DPPH, and TEMPOL (4-hydroxy-2,2,6,6-tetramethylpiperidine N-oxyl) radicals, showed higher antioxidant activity of caffeic acid but no antioxidant action for caffeine. However, using H_2O_2 hydroxyl radical, both compounds showed a remarkable scavenging activity. In fresh samples of purple sweet potato fermented alcoholic beverage, ethyl 4′-hydroxy-3′-methoxycinnamate (2.4 mg/L), 3-hydroxy-4-methoxy cinnamic acid (0.2 mg/L), and 1.4 mg/L of 4-hydroxy-3-methoxy cinnamic acid (Li et al. 2017).

7.3.4   Esters of Cinnamic acids

Green coffee beans had been found to contain at least five major types of chlorogenic acid isomers, i.e., caffeoylquinic acids, dicaffeoylquinic acids, feruoylquinic acids, p-coumaroylquinic acids, and caffeoyl feruloyl quinic acids (Clifford and Ramirez-Martinez 1991). Chu et al. (2008) identified chlorogenic acid in coffee using capillary electrophoresis with amperometric detection. Monagas et al. (2005) showed that phenolic acids are present in their free form or as glycosylated derivatives and esters of tartaric, quinic, and shikimic acid in both red and white wines.

In fermented beverages, Gamboa-Gómez et al. (2017) reported derivatives of chlorogenic acids which include 4-caffeoylquinic acid, 4,5 dicaffeoylquinic acid, and 3,4 dicaffeoylquinic acid amongst other phenolic compounds. In Cascara beverage prepared from the coffee cherry pulp, only 3-caffeoylquinic acid was major cinnamoyl ester (Heeger et al. 2017). Recently Khamitova et al. (2020) reported three esters of cinnamic acids i.e., 3-caffeoylquinic acid, 5-caffeoylquinic acid, and 3,5-dicaffeoylquinic acid in Arabica and Robusta coffee samples.

7.3.5   Stilbenes

Stilbenes such as *trans*-resveratrol had been identified in red wine (Anli et al. 2006). Vian et al. (2005) reported that *trans-* and *cis*-resveratrol have been shown to exist in wine as both the aglycone and the bound glucoside piceid. Similarly, *trans*-resveratrol and *trans*-piceid in dark chocolate and cocoa liquor extracts and found that chocolate products have higher antioxidant activity than concentrated commercial stilbene extracts (Counet et al. 2006). Stilbenes had been reported in wine along with other phenolic compounds (Li et al. 2009).

Recently, Guerrero et al. (2020) reported that E-piceid was the main stilbene in white wine (mean 155 μg/L). In red wine, Z- and E-piceid (mean 3.73 and 3.16 mg/L, respectively) were predominant. Besides, a large amount of other stilbenes including oligomers such as hopeaphenol (mean 1.55 mg/L) were found in red wines. Similarly, resveratrol (330 μg/L), and δ-viniferin (20 μg/L) have been reported in red wine (El Khawand et al. 2020). The concentration of these stilbenes has been significantly reduced with heat treatment.

7.3.6 *Flavonoids*

The phenolic composition of cocoa has been characterized and quantified (Sánchez-Rabaneda et al. 2003). The compounds identified include epicatechin (**7.1**), gallo-catechin and epigallocatechin, and other phenolic compounds such as flavone and flavonol glycosides such as luteolin-7-glucoside and quercetin-3-arabinoside. Reports on the polyphenolic content of cocoa products vary greatly in the literature, with values ranging from 3.3 to 65 mg/g in cocoa powder or 1.7 to 36.5 mg/g total polyphenols in dark chocolate (Adamson et al. 1999).

Vinson and Dabbagh (1998) indicated that both catechins and theaflavins contribute to the antioxidant characteristics of tea. Green tea is a gently processed tea where the catechin profile closely resembles that originally present in the leaves at harvest (Astill et al. 2001). Black tea, unlike green tea, is a more processed product. Fresh tea leaves are fermented in a process by which oxidative enzymes such as polyphenol oxidase, oxidize catechin monomers and generate a complex mixture of polyphenol derived products including theaflavins, theasinensins, and other poorly characterized complex oxidation polymers known as thearubigins (Ferruzzi 2010). Thearubigin is a polymeric polyphenol formed by condensation of epigallocatechin and epigallocatechin gallate (**7.2**). Thearubigins have been reported as the most abundant pigment in black tea, which comprises polyphenolic oxidation products (Yassin et al. 2015). It was also reported that oxidation is mainly taking place on the B-ring and the galloyl group, where the oxidized components subsequently undergo oxidative coupling for the formation of theaflavins, theasinensins, and polyhydrox-ylated flavan-3-ols, all precursors for thearubigin formation (Yassin et al. 2015). The total flavonol content was reduced from 50 to 53% in green tea to 10% in black tea (Sun et al. 2008). The decreasing order of antioxidant activity of various teas has been reported in the following order: green > oolong > black > puerh tea (Yashin et al. 2011). Catechins are found in particularly high concentrations in teas, especially the unfermented green and white varieties (Freeman and Niemeyer 2008). Relatively high amounts (approximately 30%) of (−)-epicatechin (EC), (−)-epicatechin gallate (ECG), (−)-epigallocatechin (EGC), and (−)-epigallocatechin gallate (EGCG) had been reported in green tea leaves by Amarowicz and Shahidi (1996). Theaflavins, which include theaflavin (**7.3**), theaflavin-3-gallate (**7.4**), theaflavin-3′ gallate (**7.5**), and theaflavin-3,3′-digallate (**7.6**) as shown in Fig. 7.1. are key to the characteristic color and taste of black tea, and account for 2–6% of the solids in brewed black tea (Khan and Mukhtar 2010). The total amount of epicatechin derivatives in Chinese black teas was between 2.4 and 5.1 g/kg; 41.4 and 46.3 g/kg in Chinese oolong tea and in Chinese green tea between 80 and 144.4 g/kg (Chen et al. 2001). The EGCG was the most abundant epicatechin derivative in teas.

In coffee, Chu et al. (2008) reported catechin, and quercetin-3-rutinoside using capillary electrophoresis with amperometric detection. In Cascara beverage prepared from the coffee cherry pulp, only quercetin-3-rutinoside was a major flavonoid reported by Heeger et al. (2017).

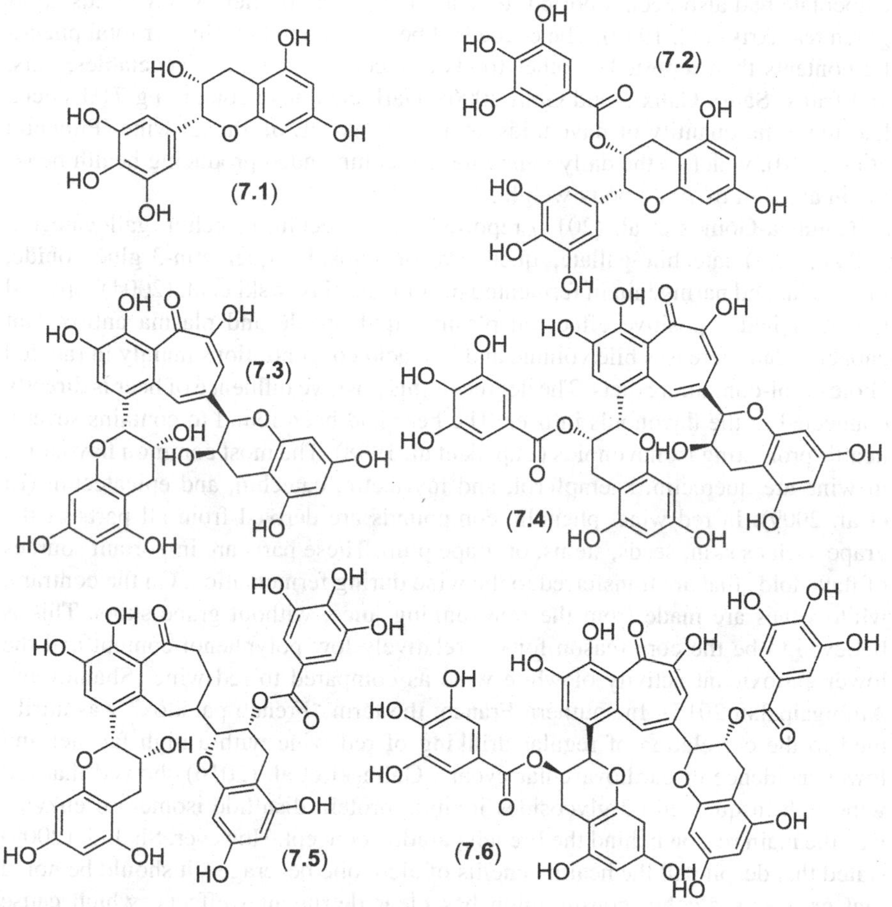

Fig. 7.1 Structures of catechins and theaflavins found in tea. (**7.1**) Epicatechin, (**7.2**) epigallocat-echin gallate, (**7.3**) theaflavin, (**7.4**) theaflavin 3-gallate, (**7.5**) theaflavin 3′-gallate, and (**7.6**) theaflavin 3,3′-digallate

Cocoa bean and its products such as cocoa liquor, cocoa powder, and dark chocolate are a rich source of phenolic compounds (Othman et al. 2007). Indeed cocoa products contain greater antioxidant capacity and higher amounts of flavo-noids per serving than either tea or red wine (Steinberg et al. 2003). In unfer-mented Cocoa beans, a higher amount of phenolic content (12–18%, dry weight) had been reported (Afoakwa et al. 2013). Dreosti (2000) showed that 60% of the total phenolics in raw cocoa beans are epicatechin, catechin, and procyanidin oligomers. These compounds with small amounts of cyanidin glycosides and quercetin glycosides were reported in cocoa (Adamson et al. 1999). In chocolate, epicatechin was the major monomeric polyphenol and the extract of chocolate liquor stimulates *in vitro* cellular immune response (Sanbongi et al. 1997).

Chocolate had also been reported to be a good source of dietary flavonoids, after green tea (Arts et al. 1999). The cocoa had been found to have higher total phenolic contents than reported in other foods like cereals, legumes, vegetables, nuts, and fruits (Saura-Calixto and Goñi 2006). Dark chocolate containing 71% cocoa had the same quantity of flavonoids as that of 196 mL of Tannat wine (Pimentel et al. 2010), which is the daily wine intake recommended producing health benefits in an adult of 70 kg body weight.

Gamboa-Gómez et al. (2017) reported gallocatechin, catechin, gallocatechin gallate, (epi)-catechin gallate, quercetin-3-rutinoside, quercetin-3-glucuronide, myricetin, and naringenin in fermented beverages. Gasowski et al. (2004) reported that beer had a positive effect on plasma lipid profile and plasma antioxidant capacity, and increases bile volume and bile acid concentrations mainly in rats fed cholesterol-containing diets. The degree of this positive influence of beer is directly connected to the flavonoids in beer. The beer had been found to contains several health-promoting isoflavonoids (Lapcík et al. 1998). The most common flavonoids in wine are quercetin, kaempferol, and myricetin, catechin, and epicatechin (Li et al. 2009). In red wine, phenolic compounds are derived from all parts of the grape such as skin, seeds, stems, or grape pulp. These parts are important sources of flavonoids that are transferred to the wine during fermentation. On the contrary, white wines are made from the free-running juice, without grape skins. This is believed to be the core reason for the relatively low polyphenol content and the lower antioxidant activity of white wine as compared to red wine (Shahidi and Ambigaipalan 2015). In southern France, the term "French paradox" was attributed to the correlation of regular drinking of red wine with a high-fat diet and lower incidence of cardiovascular events. Galinski et al. (2016) showed that red wine rich in quercetin-3-glycosides inhibits protein disulfide isomerase enzyme was the main reason behind the French paradox concept. However, Shahidi (2009) stated that despite all the health benefits of alcoholic beverages, it should be noted that excessive alcohol consumption has clear detrimental effects, which cause mortality.

Li et al. (2017) reported 12 flavonoids and its glycosides such as cyanidin-3-sophoroside-5-glucoside (55.5 mg/L), peonidin-3-sophoroside-5-glucoside (106.7 mg/L), cyanidin 3-p-hydroxybenzoyl sophoroside-5-glucoside (18.8 mg/L), peonidin 3-p-hydroxybenzoyl sophoroside-5-glucoside (59.5 mg/L), cyanidin 3-(6″-feruloyl sophoroside)-5-glucoside (17.5 mg/L), peonidin 3-(6″-feruloyl sophoroside)-5-glucoside (22.8 mg/L), cyanidin 3-caffeoyl-p-hydroxybenzoyl sophoroside-5-glucoside (26.0 mg/L), cyanidin 3-(6″-caffeoyl sophoroside)-5-glucoside (62.2 mg/L), peonidin 3-(6″,6‴-dicaffeoyl sophoroside)-5-glucoside (32.1 mg/L), peonidin 3-caffeoyl-p-hydroxybenzoyl sophoroside-5-glucoside (96.6 mg/L), peonidin 3-caffeoyl sophoroside-5-glucoside (175.2 mg/L), and peonidin 3-(6″-caffeoyl-6‴-feruloylsophoroside)-5-glucoside (38.3 mg/L) in fresh samples of purple sweet potato fermented alcoholic beverage.

7.3.7   *Anthocyanidins and Anthocyanins*

Anthocyanidins and anthocyanins have been reported in beverages. Morata et al. (2019) applications of anthocyanins in beverages (Wan et al. 2008). The preliminary work on anthocyanins involves the use of UV-visible spectroscopy for their presence. For example, anthocyanins have been studied using UV–vis absorption spectra in black carrot (BC), purple sweet potato (SP), and grape juice (GJ) extracts in citrate buffer solution at pH 3 as shown in Fig. 7.2 (Gérard et al. 2019). The spectra are characteristic of anthocyanin pigments with absorbance in the range 450–600 nm. The differences in terms of light-absorption properties i.e., absorbance and maximum absorption wavelength indicate that the three extracts are composed of mixtures of different anthocyanin molecules. Most of the studies in this regard are usually based on the total anthocyanin contents in beverages or drinks using UV-visible spectrometry (Purohit et al. 2019). In pomegranate fresh juice, the amount of total anthocyanins were 4.61 mg of cyanidin-3-glucoside/100 mL and pomegranate fermented beverage 1.96 mg of cyanidin-3-glucoside/100 mL (Rios-Corripio and Guerrero-Beltrán 2019). Anthocyanins from grape have been used as a food colorant in kefir and carbonated water (Montibeller et al. 2018).

Individual anthocyanin has also been reported in beverages. Anthocyanins and anthocyanidins such as cyanidin, pelargonin, and pelargonidin had been found in beer at less than 1 mg/L and anthocyanidin and leucoanthocyanidin have been reported in fresh tea (Shahidi and Ambigaipalan 2015). In sports beverages, the addition of black bean coat extract containing petunidin-2-glucoside, delphinidin-glucoside, petunidin-glucoside, and malvidin-glucoside showed a significant contribution to higher antioxidant activity (Aguilera et al. 2016).

Stein-Chisholm et al. (2017) identified ten anthocyanins including five arabinoside and five pyranoside anthocyanins in rabbiteye blueberries. Three minor anthocyanins were also identified for the first time in rabbiteye blueberries were delphinidin-3-(*p*-coumaroyl-glucoside), cyanidin-3-(*p*-coumaroyl-glucoside), and

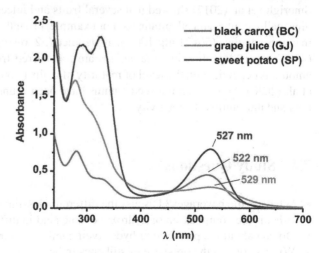

Fig. 7.2 Absorbance spectra of black carrot (BC), grape juice (GJ), and purple sweet potato (SP) extract at C = 500 mg/L in citrate buffer at pH 3. (Reproduced with kind permission of the American Chemical Society, Gérard et al. 2019)

petunidin-3-(*p*-coumaroyl-glucoside). The authors observed a significantly increased (50%) in delphinidin-3-(*p*-coumaroyl-glucoside) after pressing. Zhang et al. (2018) reported the presence of delphinidin-3-sambubioside and cyanidin-3-sambubioside in the *Hibiscus sabdariffa* model beverage. Blood orange-based juices have been found to contain a limited quantity of total anthocyanins as reported by Licciardello et al. (2018). The authors suggested that blood orange-based juices may be called water-based instead of fruit-based juices, as the amount of bioactive content was very low.

Anthocyanins are present in wine and have been one of the hot topics in the wine industry (Nel 2018). However, it is not possible to accommodate all those information in this chapter. Laitila et al. (2019) reported glucoside, acetylglucoside, coumaroylglucoside, carboxypyrano, methylpyrano, vinyl phenol, vinylcatechol, vinyl guaiacol, dimers, and trimers of anthocyanins present in red wine. Recently Alvarez Gaona et al. (2019) studied the anthocyanins in Ancellotta red wine (*Vitis vinifera* L.) and its encapsulated spray-dried powder with low water activity. The retention of total monomeric anthocyanins in the wine powder was found to be greater than 80%. Anthocyanin profiles of Ancellotta liquid wine and wine powder were characterized by using HPLC–DAD and 33 compounds were identified as shown in Table 7.1. Five non-acylated anthocyanins, five acetylated, six cinnamoylated, six vitisin-like pyranoanthocyanins, four hydroxyphenyl-pyranoanthocyanins, and two flavanol–anthocyanin adducts were reported. Microencapsulation of anthocyanins has significant applications in food industries and is found to possess potential health benefits (Yousuf et al. 2016).

7.3.8 Tannins

Tannins have been reported in several fruits and berries as discussed in Chap. 3. Beverages prepared from fruits and berries are therefore found rich in tannins. Smeriglio et al. (2017) showed that several fruits and juices contain proanthocyanidins, gallotannins, and ellagitannins. For example, proanthocyanidins were reported in cranberry juice (20–21 mg/100 g), apple juice (52–69 mg/100 g), and grape juice (45–47 mg/100 g). In wine, the tannins are transferred from the grapes, and their amount is dependent on the level of maturity of fruits (Rousserie et al. 2020). Motta et al. (2020) showed that the seed tannins have been found different than skin tannins and thus antioxidant activity.

7.4 Study Questions

- What are the beverages? Discuss the different classification of beverages.
- What is the composition of hydroxybenzoic acid in different beverages?
- Do you about the presence of hydroxycinnamic acids in beverages?
- Write a note on the presence of stilbenes in beverages.
- What do you know about the flavonoids and anthocyanins in beverages?

Table 7.1 Anthocyanins profile of Ancellotta liquid wine (mg/100 g) and wine powder (mg/100 g) using HPLC–DAD. Reproduced with kind permission of Springer Nature (Alvarez Gaona et al. 2019)

Compounds	Ancellotta 2015 liquid wine	Ancellotta 2015 wine powder
Delphinidin-3-glucoside	7.55 ± 0.7a	35.97 ± 2.3b
Cyanidin-3-glucoside	1.24 ± 0.1a	6.96 ± 0.4b
Petunidin-3-glucoside	8.06 ± 0.7a	39.10 ± 2.5b
Peonidin-3-glucoside	3.94 ± 0.3a	19.06 ± 1.2b
Malvidin-3-glucoside	20.64 ± 2.0a	108.21 ± 7.0b
Delphinidin-3-(6″-acetyl)-glucoside	2.62 ± 0.2a	13.13 ± 0.8b
Cyanidin-3-(6″-acetyl)-glucoside	0.91 ± 0.09a	5.55 ± 0.3b
Petunidin-3-(6″-acetyl)-glucoside	2.50 ± 0.2a	12.44 ± 0.8b
Peonidin-3-(6″-acetyl)-glucoside	0.74 ± 0.07a	4.57 ± 0.3b
Malvidin-3-(6″-acetyl)-glucoside	4.61 ± 0.4a	22.78 ± 1.4b
Delphinidin-3-(6″-p-coumaroyl)-glucoside	0.85 ± 0.08a	5.05 ± 0.3b
Cyanidin-3-(6″-p-coumaroyl)-glucoside	0.13 ± 0.01a	2.07 ± 0.1b
Petunidin-3-(6″-p-coumaroyl)-glucoside	0.69 ± 0.07a	3.44 ± 0.2b
Peonidin-3-(6″-p-coumaroyl)-glucoside	0.82 ± 0.08a	3.20 ± 0.2b
Malvidin-3-(6″-caffeoyl)-glucoside	0.16 ± 0.02a	2.02 ± 0.1b
Malvidin-3-(6″-p-coumaroyl)-glucoside	2.64 ± 0.2a	9.94 ± 0.6b
10-H-pyranomalvidin-3-(6″-acetyl)-glucoside	0.82 ± 0.08a	5.52 ± 0.3b
10-Carboxy-pyranodelphinidin-3-glucoside	0.46 ± 0.04a	3.26 ± 0.2b
10-Carboxy-pyranopetunidin-3-glucoside	0.74 ± 0.07a	5.04 ± 0.3b
10-Carboxy-pyranopeonidin-3-glucoside	0.60 ± 0.06a	6.45 ± 0.4b
10-Carboxy-pyranomalvidin-3-glucoside	1.11 ± 0.1a	6.45 ± 0.4b
10-Carboxy-pyranomalvidin-3-(6″-acetyl) glucoside	1.19 ± 0.1a	6.90 ± 0.4b
10-Hydroxyphenyl-pyranomalvidin-3-glucoside	0.36 ± 0.04a	2.20 ± 0.1b
10-Hydroxyphenyl-pyranomalvidin-3-(6″-acetyl)-glucoside	0.17 ± 0.02	ND
10-Hydroxyphenyl-pyranomalvidin-3-(6″-p-coumaroyl)-glucoside	0.11 ± 0.01	ND
10-Methoxy-hydroxyphenyl-pyranomalvidin-3--glucoside	0.23 ± 0.02	ND
Malvidin-3-glucoside-catechin	0.36 ± 0.03a	3.58 ± 0.2b
Malvidin-3-glucoside-ethyl-catechin	1.34 ± 1.3a	7.13 ± 0.4
Total anthocyanins	65.5 ± 6.3a	340.02 ± 22.1b

ND non-detected; Data is mean with SD (n = 3). Different lowercase letters in the same row indicate significant differences between matrices (Tukey HSD test, $p < 0.05$)

References

Adamson GE, Lazarus SA, Mitchell AE, Prior RL, Cao G, Jacobs PH, Kremers BG, Hammerstone JF, Rucker RB, Ritter KA (1999) HPLC method for the quantification of procyanidins in cocoa and chocolate samples and correlation to total antioxidant capacity. J Agric Food Chem 47(10):4184–4188

Afoakwa EO, Quao J, Takrama J, Budu AS, Saalia FK (2013) Chemical composition and physical quality characteristics of Ghanaian cocoa beans as affected by pulp pre-conditioning and fermentation. J Food Sci Technol 50(6):1097–1105

Agostoni C, Bresson J-L, Fairweather Tait S, Flynn A, Golly I, Korhonen H, Lagiou P, Løvik M, Marchelli R, Martin A (2010) Scientific opinion on dietary reference values for water. EFSA J 8(3):1459

Aguilera Y, Mojica L, Rebollo-Hernanz M, Berhow M, de Mejía EG, Martín-Cabrejas MA (2016) Black bean coats: new source of anthocyanins stabilized by β-cyclodextrin copigmentation in a sport beverage. Food Chem 212:561–570

Alvarez Gaona IJ, Fanzone M, Sari S, Assof M, Pérez D, Chirife J, Zamora MC (2019) Spray-dried Ancellotta red wine: natural colorant with potential for food applications. Eur Food Res Technol 245(12):2621–2630

Amarowicz R, Shahidi F (1996) A rapid chromatographic method for separation of individual catechins from green tea. Food Res Int 29(1):71–76

Anli E, Vural N, Demiray S, Özkan M (2006) Trans-resveratrol and other phenolic compounds in Turkish red wines with HPLC. J Wine Res 17(2):117–125

Arts IC, Hollman PC, Kromhout D (1999) Chocolate as a source of tea flavonoids. Lancet 354(9177):488

Astill C, Birch MR, Dacombe C, Humphrey PG, Martin PT (2001) Factors affecting the caffeine and polyphenol contents of black and green tea infusions. J Agric Food Chem 49(11):5340–5347

Bartolomé B, Pena-Neira A, Gómez-Cordovés C (2000) Phenolics and related substances in alcohol-free beers. Eur Food Res Technol 210(6):419–423

Brezová V, Šlebodová A, Staško A (2009) Coffee as a source of antioxidants: an EPR study. Food Chem 114(3):859–868

Chandrasekara A, Shahidi F (2018) Herbal beverages: bioactive compounds and their role in disease risk reduction—a review. J Tradit Complement Med 8(4):451–458

Chen Z-Y, Zhu QY, Tsang D, Huang Y (2001) Degradation of green tea catechins in tea drinks. J Agric Food Chem 49(1):477–482

Chu Q, Lin M, Yu X, Ye J (2008) Study on extraction efficiency of natural antioxidant in coffee by capillary electrophoresis with amperometric detection. Eur Food Res Technol 226(6):1373–1378

Clifford M, Ramirez-Martinez J (1991) Phenols and caffeine in wet-processed coffee beans and coffee pulp. Food Chem 40(1):35–42

Counet C, Callemien D, Collin S (2006) Chocolate and cocoa: new sources of trans-resveratrol and trans-piceid. Food Chem 98(4):649–657

Dreosti IE (2000) Antioxidant polyphenols in tea, cocoa, and wine. Nutrition 16(7–8):692–694

El Khawand T, Valls Fonayet J, Da Costa G, Hornedo-Ortega R, Jourdes M, Franc C, de Revel G, Decendit A, Krisa S, Richard T (2020) Resveratrol transformation in red wine after heat treatment. Food Res Int 132:109068

Fasciano JM, Danielson ND (2016) Micellar and sub-micellar ultra-high performance liquid chromatography of hydroxybenzoic acid and phthalic acid positional isomers. J Chromatogr A 1438:150–159

Ferruzzi MG (2010) The influence of beverage composition on delivery of phenolic compounds from coffee and tea. Physiol Behav 100(1):33–41

Freeman JD, Niemeyer ED (2008) Quantification of tea flavonoids by high performance liquid chromatography. J Chem Educ 85(7):951

Galinski CN, Zwicker JI, Kennedy DR (2016) Revisiting the mechanistic basis of the French Paradox: red wine inhibits the activity of protein disulfide isomerase in vitro. Thromb Res 137:169–173

Gamboa-Gómez CI, Simental-Mendía LE, González-Laredo RF, Alcantar-Orozco EJ, Monserrat-Juarez VH, Ramírez-España JC, Gallegos-Infante JA, Moreno-Jiménez MR, Rocha-Guzmán NE (2017) In vitro and in vivo assessment of anti-hyperglycemic and antioxidant effects of Oak leaves (*Quercus convallata and Quercus arizonica*) infusions and fermented beverages. Food Res Int 102:690–699

Gasowski B, Leontowicz M, Leontowicz H, Katrich E, Lojek A, Číž M, Trakhtenberg S, Gorinstein S (2004) The influence of beer with different antioxidant potential on plasma lipids, plasma antioxidant capacity, and bile excretion of rats fed cholesterol-containing and cholesterol-free diets. J Nutr Biochem 15(9):527–533

Gérard V, Ay E, Morlet-Savary F, Graff B, Galopin C, Ogren T, Mutilangi W, Lalevée J (2019) Thermal and photochemical stability of anthocyanins from black carrot, grape juice, and purple sweet potato in model beverages in the presence of ascorbic acid. J Agric Food Chem 67(19):5647–5660

Guerrero RF, Valls-Fonayet J, Richard T, Cantos-Villar E (2020) A rapid quantification of stilbene content in wine by ultra-high pressure liquid chromatography–mass spectrometry. Food Control 108:106821

Heeger A, Kosińska-Cagnazzo A, Cantergiani E, Andlauer W (2017) Bioactives of coffee cherry pulp and its utilisation for production of Cascara beverage. Food Chem 221:969–975

Jamali HR, Steel CC, Mohammadi E (2020) Wine research and its relationship with wine production: a scientometric analysis of global trends. Aust J Grape Wine R 26(2):130–138

Khamitova G, Angeloni S, Borsetta G, Xiao J, Maggi F, Sagratini G, Vittori S, Caprioli G (2020) Optimization of espresso coffee extraction through variation of particle sizes, perforated disk height and filter basket aimed at lowering the amount of ground coffee used. Food Chem 314:126220

Khan N, Mukhtar H (2010) Green tea catechins: anticancer effects and molecular targets. In: Fraga CG (ed) Plant phenolics and human health: biochemistry, nutrition, and pharmacology. Wiley, Hoboken, pp 1–49

Köseoğlu Yılmaz P, Kolak U (2017) SPE-HPLC determination of chlorogenic and phenolic acids in coffee. J Chromatogr Sci 55(7):712–718

Laitila JE, Suvanto J, Salminen J-P (2019) Liquid chromatography–tandem mass spectrometry reveals detailed chromatographic fingerprints of anthocyanins and anthocyanin adducts in red wine. Food Chem 294:138–151

Lapcík O, Hill M, Hampl R, Wähälä K, Adlercreutz H (1998) Identification of isoflavonoids in beer. Steroids 63(1):14–20

Li S, Pan M-H, Lo C-Y, Tan D, Wang Y, Shahidi F, Ho C-T (2009) Chemistry and health effects of polymethoxyflavones and hydroxylated polymethoxyflavones. J Funct Foods 1(1):2–12

Li S, An Y, Fu W, Sun X, Li W, Li T (2017) Changes in anthocyanins and volatile components of purple sweet potato fermented alcoholic beverage during aging. Food Res Int 100:235–240

Licciardello F, Arena E, Rizzo V, Fallico B (2018) Contribution of blood orange-based beverages to bioactive compounds intake. Front Chem 6:374

Lolli V, Acharjee A, Angelino D, Tassotti M, Del Rio D, Mena P, Caligiani A (2020) Chemical characterization of capsule-brewed espresso coffee aroma from the most widespread Italian brands by HS-SPME/GC-MS. Molecules 25(5):1166

Luo Y, Kong L, Xue R, Wang W, Xia X (2020) Bitterness in alcoholic beverages: the profiles of perception, constituents, and contributors. Trends Food Sci Technol 96:222–232

Molva C, Baysal AH (2016) The effect of sporulation medium on Alicyclobacillus acidoterrestris guaiacol production in apple juice. LWT Food Sci Technol 69:454–457

Monagas M, Suarez R, Gomez-Cordoves C, Bartolome B (2005) Simultaneous determination of nonanthocyanin phenolic compounds in red wines by HPLC-DAD/ESI-MS. Am J Enol Viticult 56(2):139–147

Montanari L, Perretti G, Natella F, Guidi A, Fantozzi P (1999) Organic and phenolic acids in beer. LWT Food Sci Technol 32(8):535–539

Montibeller MJ, de Lima MP, Tupuna-Yerovi DS, Rios AO, Manfroi V (2018) Stability assessment of anthocyanins obtained from skin grape applied in kefir and carbonated water as a natural colorant. J Food Process Preserv 42(8):e13698

Morata A, López C, Tesfaye W, González C, Escott C (2019) Anthocyanins as natural pigments in beverages. In: Grumezescu AM, Holban AM (eds) Value-added ingredients and enrichments of beverages, vol 14. Academic, New York, pp 383–428

Motta S, Guaita M, Cassino C, Bosso A (2020) Relationship between polyphenolic content, antioxidant properties and oxygen consumption rate of different tannins in a model wine solution. Food Chem 313:126045

Nazhand A, Souto EB, Lucarini M, Souto SB, Durazzo A, Santini A (2020) Ready to use therapeutical beverages: focus on functional beverages containing probiotics, prebiotics and synbiotics. Beverages 6(2):26

Nel A (2018) Tannins and anthocyanins: from their origin to wine analysis—a review. South Afr J Enol Viticult 39(1):1–20

Ongo EA, Montevecchi G, Antonelli A, Sberveglieri V, Sevilla Iii F (2020) Metabolomics fingerprint of Philippine coffee by SPME-GC-MS for geographical and varietal classification. Food Res Int 134:109227

Othman A, Ismail A, Ghani NA, Adenan I (2007) Antioxidant capacity and phenolic content of cocoa beans. Food Chem 100(4):1523–1530

Pasqualone A, Summo C, Laddomada B, Mudura E, Coldea TE (2018) Effect of processing variables on the physico-chemical characteristics and aroma of borş, a traditional beverage derived from wheat bran. Food Chem 265:242–252

Pimentel FA, Nitzke JA, Klipel CB, de Jong EV (2010) Chocolate and red wine—a comparison between flavonoids content. Food Chem 120(1):109–112

Purohit P, Palamthodi S, Lele S (2019) Effect of karwanda (Carissa congesta Wight) and sugar addition on physicochemical characteristics of ash gourd (Benincasa hispida) and bottle gourd (Langenaria siceraria) based beverages. J Food Sci Technol 56(2):1037–1045

Rajeswari VD, Kalpana VN (2019) Preservatives in beverages: perception and needs. In: Grumezescu AM, Holban AM (eds) Preservatives for the beverage industry: the science of beverages, vol 15. Woodhead Publishing, London, pp 1–30

Rios-Corripio G, Guerrero-Beltrán JÁ (2019) Antioxidant and physicochemical characteristics of unfermented and fermented pomegranate (Punica granatum L.) beverages. J Food Sci Technol 56(1):132–139

Rousserie P, Lacampagne S, Vanbrabant S, Rabot A, Geny-Denis L (2020) Influence of berry ripeness on seed tannins extraction in wine. Food Chem 315:126307

Salvo-Comino C, Rassas I, Minot S, Bessueille F, Rodriguez-Mendez ML, Errachid A, Jaffrezic-Renault N (2020) Voltammetric sensor based on electrodeposited molecularly imprinted chitosan film on BDD electrodes for catechol detection in buffer and in wine samples. Mater Sci Eng C 110:110667

Sanbongi C, Suzuki N, Sakane T (1997) Polyphenols in chocolate, which have antioxidant activity, modulate immune functions in humansin vitro. Cell Immunol 177(2):129–136

Sánchez-Rabaneda F, Jáuregui O, Casals I, Andrés-Lacueva C, Izquierdo-Pulido M, Lamuela-Raventós RM (2003) Liquid chromatographic/electrospray ionization tandem mass spectrometric study of the phenolic composition of cocoa (Theobroma cacao). J Mass Spectrom 38(1):35–42

Saura-Calixto F, Goñi I (2006) Antioxidant capacity of the Spanish Mediterranean diet. Food Chem 94(3):442–447

Shahidi F (2009) Nutraceuticals and functional foods: whole versus processed foods. Trends Food Sci Technol 20(9):376–387

Shahidi F, Ambigaipalan P (2015) Phenolics and polyphenolics in foods, beverages and spices: antioxidant activity and health effects—a review. J Funct Foods 18:820–897

Shalash M, Makahleh A, Salhimi SM, Saad B (2017) Vortex-assisted liquid-liquid-liquid microextraction followed by high performance liquid chromatography for the simultaneous determination of fourteen phenolic acids in honey, iced tea and canned coffee drinks. Talanta 174:428–435

Smeriglio A, Barreca D, Bellocco E, Trombetta D (2017) Proanthocyanidins and hydrolysable tannins: occurrence, dietary intake and pharmacological effects. Br J Pharmacol 174(11):1244–1262

Steinberg FM, Bearden MM, Keen CL (2003) Cocoa and chocolate flavonoids: implications for cardiovascular health. J Am Diet Assoc 103(2):215–223

Stein-Chisholm RE, Beaulieu JC, Grimm CC, Lloyd SW (2017) LC–MS/MS and UPLC–UV evaluation of anthocyanins and anthocyanidins during rabbiteye blueberry juice processing. Beverages 3(4):56

Sun T, Ho C-T, Shahidi F (2008) Bioavailability and metabolism of tea catechins in human subjects. In: Ho C-T, Lin JK, Shahidi F (eds) Tea and tea products. CRC Press, Boca Raton, pp 120–138

Van Luong TS, Moir CJ, Kaur M, Frank D, Bowman JP, Bradbury MI (2019) Diversity and guaiacol production of Alicyclobacillus spp. from fruit juice and fruit-based beverages. Int J Food Microbiol 311:108314

Vian MA, Tomao V, Gallet S, Coulomb P, Lacombe J (2005) Simple and rapid method for cis-and trans-resveratrol and piceid isomers determination in wine by high-performance liquid chromatography using Chromolith columns. J Chromatogr A 1085(2):224–229

Vinson JA, Dabbagh YA (1998) Tea phenols: antioxidant effectiveness of teas, tea components, tea fractions and their binding with lipoproteins. Nutr Res 18(6):1067–1075

Wan X, Li D, Zhang Z (2008) Antioxidant properties and mechanisms of tea polyphenols. In: Ho CT, Lin JK, Shahidi F (eds) Tea and tea products. CRC Press, Boca Raton, pp 139–167

Yashin A, Yashin Y, Nemzer B (2011) Determination of antioxidant activity in tea extracts, and their total antioxidant content. Am J Biomed Sci 3(4):322–335

Yassin GH, Koek JH, Kuhnert N (2015) Model system-based mechanistic studies of black tea thearubigin formation. Food Chem 180:272–279

Yousuf B, Gul K, Wani AA, Singh P (2016) Health benefits of anthocyanins and their encapsulation for potential use in food systems: a review. Crit Rev Food Sci Nutr 56(13):2223–2230

Zhang Y, Sang J, F-f C, Sang J, C-q L (2018) β-Cyclodextrin-assisted extraction and green chromatographic analysis of *Hibiscus sabdariffa* L. anthocyanins and the effects of gallic/ferulic/caffeic acids on their stability in beverages. J Food Meas Character 12(4):2475–2483

Zhao H, Chen W, Lu J, Zhao M (2010) Phenolic profiles and antioxidant activities of commercial beers. Food Chem 119(3):1150–1158

Chapter 8
Phenolic Antioxidants in Herbs and Spices

Learning Objectives

In this chapter, the reader will be able to:

- Understand what are herbs and spices and how they are classified.
- Describe the composition of simple phenols in herbs and spices.
- Discuss the occurrence of hydroxybenzoic acids in herbs and spices.
- Describe the presence of hydroxycinnamic acids and their derivatives in spices and herbs.
- Write a detailed note on the occurrence of flavonoids in herbs and spices.

8.1 Introduction

Herbs and spices are terms used to describe those plants or plant parts having savory or aromatic properties that are used for flavoring foods. Herbs usually refer to the plant leaves and flowers, whereas spices consist of seeds, berries, roots, buds, and bark (Table 8.1). Human is using small plants or their parts since antiquity. For example, they were used by the ancient Egyptians and for centuries have been used in South Asia (El-Sayed and Youssef 2019). Herbs and spices may be classified into numerous groups depending on their color or flavor characteristics i.e., hot like cayenne pepper, black and white peppers, mustard, and chilies; slight flavor i.e., coriander, and paprika; aromatic spices i.e., clove, cumin, dill fennel, nutmeg, mace, and cinnamon; and aromatic herbs such as thyme, marjoram, shallot, basil, bay leaf, onion, and garlic (El-Sayed and Youssef 2019). Based on color such as that of turmeric and herbaceous like sage, rosemary, or based on their taste such as sweet, bitter, spicy, sour, and sharp (Embuscado 2019).

© Springer Nature Switzerland AG 2021
A. Zeb, *Phenolic Antioxidants in Foods: Chemistry, Biochemistry and Analysis*,
https://doi.org/10.1007/978-3-030-74768-8_8

Table 8.1 Sources of herbs or spices

Part used	Herbs and spices
Leaves	Basil, oregano, bay leaf, thyme, tarragon, mint, marjoram, sage, curry leaf
Bark	Cinnamon, cassia
Flower/bud, pistil	Clove, saffron
Fruits/berries	Clove, chili, black pepper, allspice
Bulbs	Onion, shallot, garlic, leek
Roots	Ginger, turmeric, carrot
Aril	Mace
Seeds	Ajowan, aniseed, caraway, celery, coriander, dill, fennel, fenugreek, mustard

The increasing curiosity for spices and herbs by the food industry and scientists is due to their strong antioxidant and antimicrobial properties, which surpass many currently used natural and synthetic antioxidants (Suhaj 2006). These properties had been attributed to several compounds including vitamins, flavonoids, terpenoids, carotenoids, phytoestrogens, and minerals, amongst others. These compounds present in herbs and spices are used as preservative agents in foods due to their antioxidant functions (Martínez-Graciá et al. 2015). Some of the herbs and spices are usually used to flavor dishes, were excellent sources of phenolic compounds with good antioxidant activity (Carlsen et al. 2010; Embuscado 2015).

The antioxidant-rich marinades comprising beer and white wine alone or mixed with herbs commonly used as meat flavoring (garlic, ginger, thyme, rosemary, and red chili pepper) had shown inhibitory effects on the formation of heterocyclic aromatic amines (HAs) in pan-fried beef (Viegas et al. 2012). It was also found that all selected marinades exhibit a reduction in total HAs formation in pan-fried meat. Shahidi et al. (1995) showed that in comminuted pork systems, the antioxidant potential of ground clove, ginger, oregano, rosemary, sage, and thyme were significant. Spices at a concentration of 200–2000 ppm inhibited the formulation of the 2-thiobarbituric acid reactive substances (TBARS) by 12–96% over 21-days of storage at 4 °C. The antioxidant potential of spices was reported as clove > sage > rosemary = oregano > thyme = ginger. Besides, some herbs such as rosemary and sage are used to produce drugs classified as phytopharmaceuticals, which represent a significant part of the world pharmaceutical market (Kähkönen et al. 1999).

Spices such as garlic, pepper, and bird chili have been extensively used in Southeast Asian cuisines. Besides their culinary values, these spices can also act as preservatives especially when used in fermented foods (Kittisakulnam et al. 2017). Herbs and spices are rich in phenolic compounds, which are responsible for antioxidant activity (Zheng and Wang 2001). Wojdyło et al. (2007) reported the phenolic contents and antioxidant (DPPH, ABTS and FRAP) activities of 32 herbs as shown in Table 8.2. The authors tested 19 families such as Labiatae (6 tested spices), Compositae (5 tested spices), Umbelliferae (4 tested spices), and Asteraceae (2 tested spices). The samples from Asteraceae and Compositae had shown high amounts of polyphenols (124 and 530 mg GAE/kg DW, respectively). Thirty-nine spices have been studied for phenolic contents and antioxidant capacities by Assefa

Table 8.2 Antioxidant capacity and total phenolic content in selected herbs. Redrawn with kind permission of Elsevier (Wojdyło et al. 2007)

Name	Family	Parts used	Total phenolic content (mg/kg)[*]	TEAC (μM trolox/100 g dw)[*]		
				ABTS	DPPH	FRAP
Salvia officinalis	Labiatae	Herbal	82.5	17.0	41.2	167
Origanum vulgare	Labiatae	Herbal	1.5	19.9	79.6	405
Marrubium vulgare	Labiatae	Herbal	38.6	11.8	22.5	138
Rosmarinus officinalis	Labiatae	Herbal	17.1	38.7	513	662
Melisa officinalis	Labiatae	Herbal	132	10.6	36.1	61.8
Artemisia vulgaris	Compositae	Herbal	38.3	7.42	74.7	51.7
Inula helenium	Compositae	Root	36.5	8.75	144	60.1
Silybum marianum	Compositae	Seed	47.7	12.3	34.3	65.7
Taraxacum officinale	Compositae	Root	126	1.76	213	15.9
Tanacetum vulgare	Compositae	Leaf	16.8	37.3	469	455
Petroselinum sativum	Umbelliferae	Root	20.2	11.8	39.9	40.9
Carum carvi	Umbelliferae	Fruit	0.7	13.1	153	75.6
Levisticum officinale	Umbelliferae	Herbal	7.2	18.9	232	123
Archangelica officinalis	Umbelliferae	Leaf	2.9	0.45	7.34	13.8
Achillea millefolium	Asteraceae	Herbal	95.5	11.2	200	191
Echinacea purpurea	Asteraceae	Leaf	151.5	12.3	75.0	94.6
Acorus calamus	Araceae	Rhizome	124.5	8.66	79.9	78.9
Humulus lupulus	Cannabaceae	Cone	71.4	10.8	83.2	50.3
Herniara glebra	Caryophyllaceae	Herbal	0.00	48.1	50.5	66.4
Glycyrrhiza glabra	Fabaceae	Herbal	11.5	30.8	177	67.3
Hypericum perforatum	Hypericaceae	Herbal	5.5	57.8	82.3	420
Juglans regia	Juglandaceae	Leaf	2.4	27.3	119	128
Thymus vulgaris	Lamiaceae	Herbal	5.8	35.4	295	693
Cynamonum zeylanicum	Lauraceae	Seed	1.3	140	253	233
Trigonella foenum-graecum	Leguminosae	Seed	76.0	6.74	364	21.6
Myristica fragrans	Myristicaceae	Fruit	89.5	33.3	182	218
Syzygium aromaticum	Myrtaceae	Fruit	89.6	346	884	2133
Epilobium hirsutum	Onagraceae	Herbal	40.3	69.5	2021	275
Polygonum aviculare	Polygonaceae	Herbal	112	19.2	141	161
Valeriana officinalis	Valerianaceae	Herbal	111	10.7	25.8	59.3
Chelidonium majus	Papaveraceae	Herbal	20.9	9.56	300	62.2
Curcuma longa	Zingiberaceae	Rhizome	17.2	19.5	100	62.6

[*]Values are means of triplicate measurements

et al. (2018). The authors showed that cloves had the most distinct and potent anti-oxidant capacity, followed by all spice and cinnamon. The polyphenols were found responsible for their fascinating antioxidant capacities of spices. Similarly, several other studies (Konczak et al. 2010; Lu et al. 2011) also shown total phenolic contents and different antioxidant activities of the herbs and spices. Herbs and spices are rich sources of several volatile aroma compounds, which are not phenolic compounds, therefore, those will not be focused here. Individual classes of phenolic compounds have also been reported.

8.2   Phenolic Antioxidants in Herbs and Spices

8.2.1   Simple Phenols

Simple phenols have characteristic aroma and flavor. The main aroma constituents of *Syzygium aromaticum* buds are 2-methoxy-4-allylphenol, and eugenyl acetate reported by Lee and Shibamoto (2001). The authors also found that eugenol had a good antioxidant activity by inhibits malondialdehyde formation from cod liver oil and also the formation of hexanal. The key components of essential oil of the clove leaf are eugenol (76.8%), followed by β-caryophyllene (17.4%), α-humulene (2.1%), and eugenyl acetate (1.2%) (Jirovetz et al. 2006). Suhaj (2006) reported that according to the USDA report, ascorbic acid, beta-carotene, camphene, carvacrol, eugenol, methyl eugenol, terpinen-4-ol were the antioxidants present in black pepper. Similarly, eugenol was reported in turmeric by Cousins et al. (2007).

In grape marc distillate used as herb liquor, simple phenolic compounds like 4-hydroxy-3-methoxybenzyl alcohol, benzyl alcohol, 2-phenyl ethanol, guaiacol, and isoeugenol were reported (Rodríguez-Solana et al. 2016). Carvacrol and euge-nol were identified in spices like oregano, Padang cassia, ness, and clove by Chan et al. (2018). Recently, eugenol has been reported in spices from Cameron, such as *Xylopia aethiopica* (1326.2 mg/kg), *Aframomum citratum* (493.9 mg/kg), *Ricinodendron heudelotii* (125.3 mg/kg), *Piper capense* (566.8 mg/kg), and *Tetrapleura tetraptera* (683.8 mg/kg) by Sokamte et al. (2019).

8.2.2   Hydroxybenzoic Acids

Oregano (*Origanum vulgare* L.), a widely used herb, usually used in the preparation of Western dishes. The compound 4-(3,4-dihydroxybenzoyloxymethyl)phenyl-*O*-β-D-glucopyranoside was a major constituent of oregano and might contribute to its antioxidant activity (Nakatani and Kikuzaki 1987). In the methanolic extract of oregano, protocatechuic acid was identified by Kikuzaki and Nakatani (1993). In turmeric, protocatechuic acid, syringic acid, and vanillic acid had been reported amongst other phenolic compounds (Cousins et al. 2007).

Wojdyło et al. (2007) studied 32 herbs and spices for phenolic compounds. The authors found that in *Salvia officinalis,* caffeic acid (2960 mg/kg), *p*-coumaric acid (103 mg/kg), and ferulic acid (135 mg/kg) was present. Caffeic acid (6490 mg/kg) was only present in *Origanum vulgare.* Caffeic acid and ferulic acid were found only in *Marrubium vulgare, Rosmarinus officinalis, Melisa officinalis, Artemisia vulgaris, Inula helenium, Silybum marianum, Tanacetum vulgare, Petroselinum sativum, Carum carvi,* and other selected herbs.

Hydroxybenzoic acid was only reported in several Australian native herbs (Konczak et al. 2010). Similarly, Lu et al. (2011), reported protocatechuic acid as the main hydroxybenzoic acid in Tsaoko amomum fruit (19.9 mg/kg) consumed in China. Protocatechuic acid, gallic acid, syringic acid, *p*- and *m*-hydroxybenzoic acids were detected in culinary herbs (rosemary, thyme, oregano, and bay) and spices such as cinnamon and cumin (Vallverdú-Queralt et al. 2014). Another study reported gallic acid, syringic acid, protocatechuic acid, homovanillic acid glucoside, *p*-hydroxybenzoic acid, m-hydroxybenzoic acid, and homovanillic acid in culinary herbs and spices (Vallverdú-Queralt et al. 2015).

In chamomile (*Matricaria chamomilla* L.) and St. John's wort (*Hypericum perforatum*), *p*-hydroxybenzoic acid, and gallic acid were reported by Sentkowska et al. (2016). Both free and bound form of gallic acid, *p*-hydroxybenzoic acid, *p*-hydroxybenzaldehyde, protocatechuic acid, 3,4-dihydroxybenzaldehyde, syringic acid, vanillic acid, and vanillin were detected in Ceylon cinnamon and Cassia cinnamon (Klejdus and Kováčik 2016).

Protocatechuic acid and gallic acid were present both in free as well as bound form in spices such as Ness, Blume, oregano, clove, Chinese cinnamon, Houtt, and pomegranate peel (Chan et al. 2016). In wild thyme extract, gallic acid (0.6 mg/g), protocatechuic acid (1.03 mg/g), and syringic acid (1.56 mg/g) have been found (Janiak et al. 2017). The main phenolic compound in the extract from black pepper was 3,4-dihydroxybenzaldehyde (55 mg/kg) as reported by Lackova et al. (2017). They also reported protocatechuic acid, gallic acid, *p*-hydroxybenzoic acid, syringic acid, vanillic acid, vanillin, and cinnamic acid. In the extract of paprika, protocatechuic aldehyde was in the range of 0.1–0.6 mg/L, homovanillic acid (2.8–8.9 mg/L), vanillin (1.9–11.2 mg/L), syringaldehyde (4.5–10.8 mg/L) (Barbosa et al. 2020).

8.2.3   *Hydroxycinnamic Acids*

Kikuzaki and Nakatani (1993) isolated important hydroxycinnamic acid i.e., caffeic acid, and a phenyl glucoside from the methanol extract of leaves of oregano. In turmeric, *p*-coumaric acid, and protocatechuic acid had been reported amongst other phenolic compounds (Cousins et al. 2007). Ferulic acid, caffeic acid, and *p*-coumaric acid were reported in several Australian native herbs (Konczak et al. 2010). Similarly, Lu et al. (2011), reported caffeic acid in *Angelica dahurica* root (259.3 mg/kg), ferulic acid (6.5 mg/kg) in dried ginger consumed in China.

In culinary herbs such as rosemary, thyme, oregano, and bay and spices like cinnamon and cumin, the compound detected were caffeic acid, caffeic-hexoside, ferulic-hexoside, ferulic acid, *p*-coumaric acid, homovanillic-hexoside as reported by Vallverdú-Queralt et al. (2014). Similarly, caffeic acid, caffeic acid hexoside, coumaric acid, ferulic acid, and ferulic acid hexoside were reported in Ground dried dill (*Anethum graveolens*), marjoram (*Origanum majorana*), turmeric (*Curcuma longa*), caraway (*Carum carvi*), and nutmeg (*Myristica fragans*) (Vallverdú-Queralt et al. 2015). In Romanian dried dill, sinapic and vanillic acid were the major phenolic acids (Violeta et al. 2017).

In chamomile and St. John's wort, *p*-coumaric acid and caffeic acid were reported by Sentkowska et al. (2016). Similarly, in Australian native herb *Prostanthera rotundifolia,* 1-glucopyranosyl sinapate (14.6 mg/g), 4-methoxy cinnamic acid (94.7 mg/g), and coumaric acid glucoside (24.0 mg/g). In black pepper, caffeic acid, ferulic acid, coumaric acid, and sinapic acid had been identified (Lackova et al. 2017). Another study on black pepper oleoresin showed decaffeoylacteoside, and N-trans-Coumaroyltaramine (Olalere et al. 2019).

In two cultivars of cinnamon i.e., Cassia cinnamon and Ceylon cinnamon, Klejdus and Kováčik (2016) reported caffeic acid, ferulic acid, *p*-coumaric acid, sinapic acid, and cinnamic acid. In spices like Ness, Blume, oregano, clove, Chinese cinnamon, Houtt, and pomegranate peel, cinnamaldehyde, cinnamic acid, cinnamoyl acetate, caffeic acid were present both in free and bound forms (Chan et al. 2016). In caffeic acid hexoside, caffeic acid, and caffeic acid derivatives have been reported in thyme, while caffeic acid and its trimer reported in peppermint (Pereira et al. 2016). Similarly, in wild thyme, Janiak et al. (2017) reported caffeic acid (1.27 mg/g), and *p*-coumaric acid (1.17 mg/g).

Caffeic acid and *p*-coumaric acid were reported in lemongrass (Sepahpour et al. 2018). Coumaric acid and trans-cinnamaldehyde were identified in spices like oregano, Padang cassia, ness, and clove by Chan et al. (2018). Hayat et al. (2019) reported caffeic acid (0.525 mg/g), ferulic acid (0.855 mg/g) and coumaric acid (0.785 mg/g) in fennel seeds. In two thymus cultivars i.e., *thymus vulgaris* and *thymus daenensis,* Bistgani et al. (2019) reported coumaric acid, trans-2-hydroxycinnamic acid, and caffeic acid.

In spices from Cameron, such as *Xylopia aethiopica, Aframomum citratum, Ricinodendron heudelotii, Piper capense*, and *Tetrapleura tetraptera*, showed the presence of caffeic acid, *p*-coumaric acid, ferulic acid, sinapic acid, and trans-cinnamic acid (Sokamte et al. 2019). In spice *Ammodaucus leucotrichus*, sinapic, ferulic and *p*-coumaric were reported in higher amount in butanol and ethylacetate and showed high antioxidant activities (Mouderas et al. 2019). In the extract of paprika, chlorogenic acid was in the range of 5–10.1 mg/L (Barbosa et al. 2020).

8.2.4   Esters of Hydroxycinnamic Acids

In the methanolic extract of oregano, a new compound i.e., 2-caffeoyloxy-3-[2-(4-hydroxy benzyl)-4,5-dihydroxy]phenylpropionic acid (**8.2**) along with rosmarinic acid (**8.1**) were identified by Kikuzaki and Nakatani (1993). Rosmarinic acid (Fig. 8.1) is a caffeoyl ester that has been shown to be an important antioxidant and

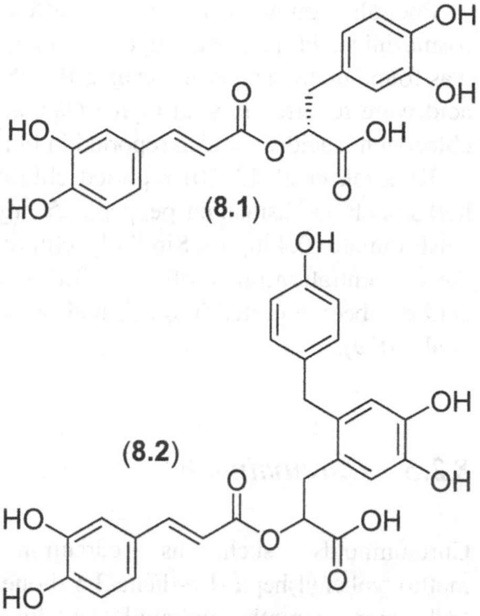

Fig. 8.1 Structures of rosmarinic acid (**8.1**) and 2-caffeoyloxy-3-[2-(4-hydroxy benzyl)-4,5-dihydroxy]phenylpropionic acid (**8.2**)

anti-inflammatory compound (Chun et al. 2005). Amarowicz et al. (2009) reported the antioxidative effects of ethanolic extract of oregano on butter.

Caffeic acid, caffeic-hexoside, protocatechuic acid, 3-caffeoylquinic, 4-caffeoylquinic, 5-caffeoylquinic, and coumaroylquinic acids were detected in culinary herbs (rosemary, thyme, oregano, and bay) and spices (cinnamon and cumin) (Vallverdú-Queralt et al. 2014). Neochlorogenic acid, 3-coumaroylquinic acid, chlorogenic acid, cryptochlorogenic acid, and 4-coumaroylquinic acid were reported in ground dried dill, marjoram, turmeric, caraway, and nutmeg (Vallverdú-Queralt et al. 2015).

Amongst herbs and spices, the highest level of attention has been received by rosemary as sources of phenolic antioxidants (Berdahl and McKeague 2015; Shahidi and Ambigaipalan 2015). In previous studies, sage (*Salvia officinalis* L.) and rosemary had shown a similar pattern of phenolic compounds, and their antioxidant activity was ascribed mainly to carnosic acid, carnosol, and rosmarinic acid (Zhang et al. 2012). Shahidi and Ambigaipalan (2015) proposed that the polyphenols of rosemary may greatly increase the functionality of food for health and wellness. For example, the rosemary extracts are effective antioxidants but measured to be less effective than BHA and BHT (Berdahl and McKeague 2015). Ahn et al. (2002) reported that rosemary was significantly less effective than synthetic BHA/BHT for suppression of oxidative changes in cooked ground beef. Other studies had shown that rosemary extracts are effective oxidative stabilizers of various meat and poultry products as well as vegetable and fish oils (Shahidi and Ambigaipalan 2015). In addition, rosemary extracts are effective antioxidants in snack foods, nuts, baked goods and pet foods (Grüner-Richter et al. 2012). In chamomile and St. John's wort, only chlorogenic acid was reported by Sentkowska et al. (2016).

Phenolic compounds such as 3-caffeoylquinic acid, 5-caffeoylquinic acid, and rosmarinic acid have been reported in peppermint, while rosmarinic acid hexoside was found in thyme (Pereira et al. 2016). Similarly, rosmarinic acid and chlorogenic acid were reported in wild thyme (Janiak et al. 2017). Chlorogenic acid and neo-chlorogenic acid were also reported in black pepper (Lackova et al. 2017).

Konczak et al. (2010) reported chlorogenic acid in several Australian native herbs such as Tasmannia pepper (1.5 mg/g, DW), Anise Myrtle (7.8 mg/g), and Bush tomato (0.4 mg/g). Similarly, chlorogenic acid was present in torch ginger at the concentration range of 11.5–21.8 mg/g (Sepahpour et al. 2018). Chlorogenic acid has been reported in the fennel seeds at a concentration of 0.52 mg/g (Hayat et al. 2019).

8.2.5   Curcuminoids

Curcuminoids such as curcumin (catechin(1E,6E)-1,7-bis(4-hydroxy-3-methoxyphenyl)hepta-1,6-diene-3,5-dione, **8.3**), demethoxy-curcumin ((1E,6E)-1-(4-hydroxy-3-methoxyphenyl)-7-(4-hydroxyphenyl)hepta-1,6-diene-3,5-dione, **8.4**), and bisdemethoxy-curcumin ((1E,6E)-1,7-bis(4-hydroxyphenyl)hepta-1,6-diene-3,5-dione, **8.5**) as shown in Fig. 8.2 are the major antioxidative compounds of

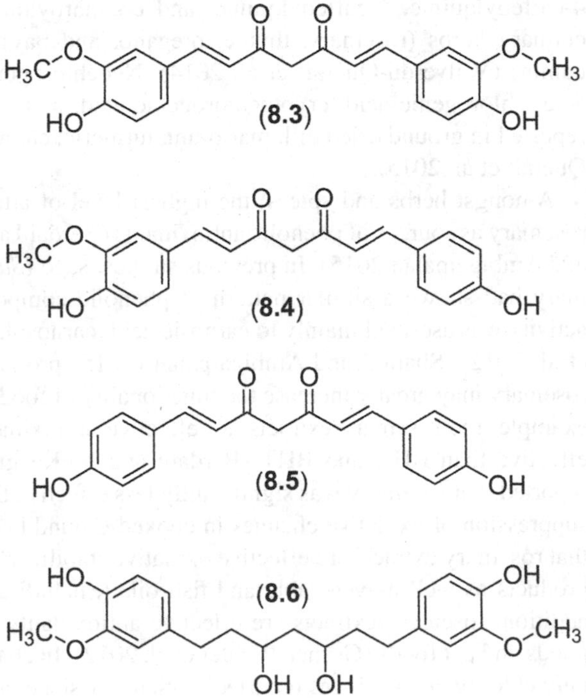

Fig. 8.2 Structures of curcumin (**8.3**), demethoxy-curcumin (**8.4**), bisdemethoxy-curcumin (**8.5**), and octahydro-curcumin (**8.6**)

turmeric (Schieffer 2002). In addition, Cousins et al. (2007) reported also reported turmerin and turmeronol in turmeric. Recently, curcumin, demethoxy-curcumin and Bisdesmethoxycurcumin were reported in turmeric by Sepahpour et al. (2018). The highest amount of these compounds were reported in acetone, followed by ethanol and lowest in water. In black pepper oleoresin, octahydrocurcumin (1,7-bis(4-hydroxy-3-methoxyphenyl)heptane-3,5-diol, **8.6**) has been detected (Olalere et al. 2019). Recently Sabir et al. (2021) reported bisdemethoxycurcumin, curcuminol, demethoxycurcumin, curcumin, and curcumin-O-glucuronide in turmeric rhizome.

8.2.6 Flavonoids

Flavonoids have been reported to be linked with the antioxidant abilities of rosemary (Suhaj 2006). Kaempferol, chrysin, pinocembrin, and naringenin had been found in rosemary honey (Escriche et al. 2014). Wang et al. (1998) identified seven flavonoids, luteolin-7-glucopyranoside, as a major compound in the ethanolic extract of sage. In oregano extracts, flavonoids such as luteolin, hispidulin, apigenin, acacetin, diosmetin, herbacetin, quercetin, and naringin had also been found in oregano extracts (Cavero et al. 2006).

Fenugreek seeds are a rich source of polyphenols (Kenny et al. 2013). Several flavonoids were identified such as apigenin and many kaempferol and quercetin glycosides (Chatterjee et al. 2009) and vitexin, tricin, naringenin, quercetin, and tricin-7-glucopyranoside had been found in Fenugreek seed (Shahidi and Ambigaipalan 2015). Wojdyło et al. (2007) studied 32 herbs and spices, most of which contain quercetin, kaempferol, luteolin, apigenin, isorhamnetin, and myricetin. Konczak et al. (2010) reported Quercetin, Quercetin hexoside, Quercetin pentoside, Rutin, Rutin hexoside, Kaempferol/luteolin hexoside, Myricetin, Hesperetin rhamnoside, Hesperetin pentoside, and Hesperetin hexoside in several Australian native herbs.

Similarly, Lu et al. (2011), reported apigenin and luteolin in cumin consumed in China at a concentration of 56.1 and 79.7 mg/kg, respectively. The authors also found quercetin only in green prickle yash (6.5 mg/kg), naringenin only in dried tangerine peel (1994.5 mg/kg), while quercetin-3-rutinoside in green prickle yash, Sichuan pepper, dried tangerine peel, Bay leaf, and Fennel at a concentration of 2214.7, 1724.7, 126.7, 929.4 and 12,431 mg/kg, respectively.

Kaempferol-3-glucoside, kaempferol, and quercetin were detected in culinary herbs such as rosemary, thyme, oregano, and bay and spices (cinnamon and cumin) by Vallverdú-Queralt et al. (2014). In ground dried dill, marjoram, turmeric, caraway, and nutmeg, epicatechin, apigenin-dihexoside, apigenin hexoside pentoside, rutin, quercetin-3-glucoside, kaempferol-3-glucoside, apigenin-7-glucoside, naringenin hexoside, eriodictyol, kaempferol, quercetin, naringenin, and apigenin were found (Vallverdú-Queralt et al. 2015). Violeta et al. (2017) reported that myricetin was a major flavonoid in parsley and celery leaves while quercetin-3-rutinoside was the main flavonoid in dill and lovage. In curry leaf, quercetin-3-rutinoside, quercetin-3-glucoside, myricetin, and quercetin were reported higher in ethanol extract as compared to acetone, methanol, and water by Sepahpour et al. (2018). The authors also reported luteolin-7-glucoside in lemon grass.

In the thyme, several flavonoids had been reported (Martins et al. 2015). This includes apigenin 6,8-di-C-glucoside, luteolin-7-glucuronide, luteolin-7-glucoside, quercetin-glucuronide, eriodictyol, and its derivatives, such as methyleriodictyol-pentosylhexoside reported by Pereira et al. (2016). The authors also found luteolin-7-rutinoside, naringenin-rutinoside, luteolin-7-glucuronide, and luteolin-diglucuronide in peppermint. Luteolin-7-glucoside (4.09 mg/g), luteolin (1.55 mg/g), eriodictyol (0.65 mg/g), apigenin (0.99 mg/g), and naringin (0.74 mg/g), were reported in wild thyme (Janiak et al. 2017).

In chamomile and St. John's wort, luteolin, rutin, hesperidin, and apigenin were reported by Sentkowska et al. (2016). Apigenin, dihydro-kaempferol, eriodictyol, homo-eriodictyol, naringenin, chalcones, quercetin, and rutin were reported in black pepper (Lackova et al. 2017). Recently, Sokamte et al. (2019) showed the presence of catechin, epicatechin, rutin, and quercetin in spices from Cameron, such as *Xylopia aethiopica, Aframomum citratum, Ricinodendron heudelotii, Piper capense,* and *Tetrapleura tetraptera.* Naringenin and quercetin were reported in spice (*Ammodaucus leucotrichus*), in higher amounts in butanol and ethylacetate, and showed high antioxidant activities (Mouderas et al. 2019).

Quercetin was present in dried ground red pepper fruits in different forms: not only as dihydrocapsaicin but also as quercetin-3-deoxyhexoside-glucuronide and quercetin-3-deoxyhexoside, whereas luteolin occurred as luteolin-7-dihexoside, luteolin-6-C-hexoside-8-C-pentoside, and luteolin-7-malonyl-dihexosyl-pentoside (Durak et al. 2018). The amount of apigenin-6-C-hexoside-8-C-pentoside was 11.78 mg/g DW. The total concentration of chili phenolic compounds (966.2 mg/g) was much higher than cinnamon (42.6 mg/g) (Durak et al. 2014) and ginger (5.4 mg/g) (Durak et al. 2015). In fennel seeds, the amount of rutin and quercetin reported were 0.955 and 1.145 mg/g, respectively (Hayat et al. 2019). The highest amount of flavonoids was reported in methanol extracts (Rezaei and Ghasemi Pirbalouti 2019). Bistgani et al. (2019) reported naringenin and apigenin in two thymus cultivars i.e., *thymus vulgaris* and *thymus daenensis.*

In the extract of paprika, quercetin-3-rutinoside was in the range of 1.3–6.1 mg/L, nepetin-7-glucoside (none-0.2 mg/L), along with trace amounts of other flavonoids (Barbosa et al. 2020). Recently, Slimestad et al. (2020) studied different spices and herbs from the commercial Norwegian kitchens for phenolic composition. These spices and herbs include basil, chive, coriander, dill, mint, oregano, parsley, rosemary, tarragon, and thyme. As shown in Fig. 8.3, the eight flavonoids were reported. These include as isorhamnetin-3,7-diglucoside, isorhamnetin-3-malonylglucoside-7-glucoside, hesperetin-7-glucoside, apigenin-7-apiosylglucoside, isorhamnetin-3-glucoside, diosmetin-7-apiosylglucoside, apigenin-7-malonylapiosylglucoside and diosmetin-7-malonylapiosylglucoside.

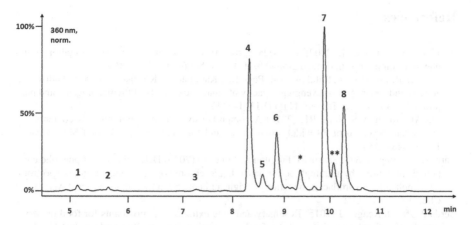

Fig. 8.3 UHPLC chromatogram at 360 nm of a crude extract of parsley, *Petroselinum crispum*. Peaks were identified as isorhamnetin-3,7-diglucoside (**1**), isorhamnetin-3-malonylglucoside-7-glucoside (**2**), hesperetin-7-glucoside (**3**), apigenin-7-apiosylglucoside (Apiin) (**4**), isorhamnetin-3-glucoside (**5**), diosmetin-7-apiosylglucoside (**6**), apigenin-7-malonylapiosylglucoside (**7**), and diosmetin-7-malonylapiosylglucoside (**8**). A compound (*) was detected which exhibited similar spectral characteristics to **7**, whereas (**) was similar to **8**. (Reproduced with kind permission of Elsevier, Slimestad et al. 2020)

8.2.7   Anthocyanidins and Anthocyanins

Konczak et al. (2010) reported cyanidin-3-glucoside (23.9 mg/g) and cyanidin-3-ru-tinoside (55.3 mg/g, DW) were only present in the berry of Tasmannia pepper, among the selected native Australian herbs. In ground dried dill, marjoram, turmeric, caraway, and nutmeg, Vallverdú-Queralt et al. (2015) reported proanthocyanidin trimers and proanthocyanidin hexamer. Sabir et al. (2021) showed the presence of qurecetin-3-D-galactoside, casuarinin, and isorhamnetin in Turmeric rhizome.

8.3   Study Questions

- What are herbs and spices?
- How herbs and spices are classified?
- What are the simple phenols in herbs and spices?
- Discuss the occurrence of hydroxybenzoic acids in herbs and spices.
- What do you know about the presence of hydroxycinnamic acids and their derivatives in spices and herbs?
- Write a detailed note on the occurrence of flavonoids in herbs and spices.

References

Ahn J, Grün I, Fernando L (2002) Antioxidant properties of natural plant extracts containing polyphenolic compounds in cooked ground beef. J Food Sci 67(4):1364–1369

Amarowicz R, Żegarska Z, Rafałowski R, Pegg RB, Karamać M, Kosińska A (2009) Antioxidant activity and free radical-scavenging capacity of ethanolic extracts of thyme, oregano, and marjoram. Eur J Lipid Sci Technol 111(11):1111–1117

Assefa AD, Keum Y-S, Saini RK (2018) A comprehensive study of polyphenols contents and antioxidant potential of 39 widely used spices and food condiments. J Food Meas Charact 12(3):1548–1555

Barbosa S, Campmajó G, Saurina J, Puignou L, Núñez O (2020) Determination of phenolic compounds in paprika by ultrahigh performance liquid chromatography–tandem mass spectrometry: application to product designation of origin authentication by chemometrics. J Agric Food Chem 68(2):591–602

Berdahl DR, McKeague J (2015) Rosemary and sage extracts as antioxidants for food preservation. In: Handbook of antioxidants for food preservation. Elsevier, Amsterdam, pp 177–217

Bistgani ZE, Hashemi M, DaCosta M, Craker L, Maggi F, Morshedloo MR (2019) Effect of salinity stress on the physiological characteristics, phenolic compounds and antioxidant activity of *Thymus vulgaris* L. and *Thymus daenensis* Celak. Ind Crop Prod 135:311–320

Carlsen MH, Halvorsen BL, Holte K, Bøhn SK, Dragland S, Sampson L, Willey C, Senoo H, Umezono Y, Sanada C (2010) The total antioxidant content of more than 3100 foods, beverages, spices, herbs and supplements used worldwide. Nutr J 9(1):3

Cavero S, García-Risco MR, Marín FR, Jaime L, Santoyo S, Señoráns FJ, Reglero G, Ibanez E (2006) Supercritical fluid extraction of antioxidant compounds from oregano: chemical and functional characterization via LC–MS and in vitro assays. J Supercrit Fluids 38(1):62–69

Chan C-L, Gan R-Y, Corke H (2016) The phenolic composition and antioxidant capacity of soluble and bound extracts in selected dietary spices and medicinal herbs. Int J Food Sci Technol 51(3):565–573

Chan C-L, Gan R-Y, Shah NP, Corke H (2018) Polyphenols from selected dietary spices and medicinal herbs differentially affect common food-borne pathogenic bacteria and lactic acid bacteria. Food Control 92:437–443

Chatterjee S, Variyar PS, Sharma A (2009) Stability of lipid constituents in radiation processed fenugreek seeds and turmeric: role of phenolic antioxidants. J Agric Food Chem 57(19):9226–9233

Chun S-S, Vattem DA, Lin Y-T, Shetty K (2005) Phenolic antioxidants from clonal oregano (*Origanum vulgare*) with antimicrobial activity against Helicobacter pylori. Process Biochem 40(2):809–816

Cousins M, Adelberg J, Chen F, Rieck J (2007) Antioxidant capacity of fresh and dried rhizomes from four clones of turmeric (*Curcuma longa* L.) grown in vitro. Ind Crop Prod 25(2):129–135

Durak A, Gawlik-Dziki U, Pecio Ł (2014) Coffee with cinnamon—impact of phytochemicals interactions on antioxidant and anti-inflammatory in vitro activity. Food Chem 162:81–88

Durak A, Gawlik-Dziki U, Kowlska I (2015) Coffee with ginger—interactions of biologically active phytochemicals in the model system. Food Chem 166:261–269

Durak A, Kowalska I, Gawlik-Dziki U (2018) UPLC–MS method for determination of phenolic compounds in chili as a coffee supplement and their impact of phytochemicals interactions on antioxidant activity in vitro. Acta Chromatogr 30(1):66–71

El-Sayed SM, Youssef AM (2019) Potential application of herbs and spices and their effects in functional dairy products. Heliyon 5(6):e01989

Embuscado ME (2015) Herbs and spices as antioxidants for food preservation. In: Shahidi F (ed) Handbook of antioxidants for food preservation. Woodhead Publishing, Cambridge, pp 251–283

Embuscado ME (2019) Bioactives from culinary spices and herbs: a review. J Food Bioact 6:68–99

Escriche I, Kadar M, Juan-Borras M, Domenech E (2014) Suitability of antioxidant capacity, flavonoids and phenolic acids for floral authentication of honey. Impact of industrial thermal treatment. Food Chem 142:135–143

Grüner-Richter S, Otto F, Weidner E (2012) Impregnation of oil containing fruits. J Supercrit Fluids 66:321–327

Hayat K, Abbas S, Hussain S, Shahzad SA, Tahir MU (2019) Effect of microwave and conventional oven heating on phenolic constituents, fatty acids, minerals and antioxidant potential of fennel seed. Ind Crop Prod 140:111610

Janiak MA, Slavova-Kazakova A, Kancheva VD, Ivanova M, Tsrunchev T, Karamać M (2017) Effects of γ-irradiation of wild thyme (Thymus serpyllum L.) on the phenolic compounds profile of its ethanolic extract. Polish J Food Nutr Sci 67(4):309–316

Jirovetz L, Buchbauer G, Stoilova I, Stoyanova A, Krastanov A, Schmidt E (2006) Chemical composition and antioxidant properties of clove leaf essential oil. J Agric Food Chem 54(17):6303–6307

Kähkönen MP, Hopia AI, Vuorela HJ, Rauha J-P, Pihlaja K, Kujala TS, Heinonen M (1999) Antioxidant activity of plant extracts containing phenolic compounds. J Agric Food Chem 47(10):3954–3962

Kenny O, Smyth T, Hewage C, Brunton N (2013) Antioxidant properties and quantitative UPLC-MS analysis of phenolic compounds from extracts of fenugreek (Trigonella foenumgraecum) seeds and bitter melon (Momordica charantia) fruit. Food Chem 141(4):4295–4302

Kikuzaki H, Nakatani N (1993) Antioxidant effects of some ginger constituents. J Food Sci 58(6):1407–1410

Kittisakulnam S, Saetae D, Suntornsuk W (2017) Antioxidant and antibacterial activities of spices traditionally used in fermented meat products. J Food Process Preserv 41(4):e13004

Klejdus B, Kováčik J (2016) Quantification of phenols in cinnamon: a special focus on "total phenols" and phenolic acids including DESI-Orbitrap MS detection. Ind Crop Prod 83:774–780

Konczak I, Zabaras D, Dunstan M, Aguas P (2010) Antioxidant capacity and phenolic compounds in commercially grown native Australian herbs and spices. Food Chem 122(1):260–266

Lackova Z, Buchtelova H, Buchtova Z, Klejdus B, Heger Z, Brtnicky M, Kynicky J, Zitka O, Adam V (2017) Anticarcinogenic effect of spices due to phenolic and flavonoid compounds—in vitro evaluation on prostate cells. Molecules 22(10):1626

Lee K-G, Shibamoto T (2001) Antioxidant property of aroma extract isolated from clove buds [Syzygium aromaticum (L.) Merr. et Perry]. Food Chem 74(4):443–448

Lu M, Yuan B, Zeng M, Chen J (2011) Antioxidant capacity and major phenolic compounds of spices commonly consumed in China. Food Res Int 44(2):530–536

Martínez-Graciá C, González-Bermúdez CA, Cabellero-Valcárcel AM, Santaella-Pascual M, Frontela-Saseta C (2015) Use of herbs and spices for food preservation: advantages and limitations. Curr Opin Food Sci 6:38–43

Martins N, Barros L, Santos-Buelga C, Silva S, Henriques M, Ferreira IC (2015) Decoction, infusion and hydroalcoholic extract of cultivated thyme: antioxidant and antibacterial activities, and phenolic characterisation. Food Chem 167:131–137

Mouderas F, Lahfa FB, Mezouar D, Benahmed NEH (2019) Valorization and identification of bioactive compounds of a spice Ammodaucus leucotrichus. Adv Traditional Med 20:159–168

Nakatani N, Kikuzaki H (1987) A new antioxidative glucoside isolated from oregano (Origanum vulgare L.). Agric Biol Chem 51(10):2727–2732

Olalere OA, Abdurahman NH, Yunus RM, Alara OR, Ahmad MM (2019) Mineral element determination and phenolic compounds profiling of oleoresin extracts using an accurate mass LC-MS-QTOF and ICP-MS. J King Saud Univ Science 31(4):859–863

Pereira E, Pimenta AI, Calhelha RC, Antonio AL, Verde SC, Barros L, Santos-Buelga C, Ferreira ICFR (2016) Effects of gamma irradiation on cytotoxicity and phenolic compounds of Thymus vulgaris L. and Mentha x piperita L. LWT Food Sci Technol 71:370–377

Rezaei M, Ghasemi Pirbalouti A (2019) Phytochemical, antioxidant and antibacterial properties of extracts from two spice herbs under different extraction solvents. J Food Meas Charact 13(3):2470–2480

Rodríguez-Solana R, Salgado JM, Domínguez JM, Cortés-Diéguez S (2016) Phenolic compounds and aroma-impact odorants in herb liqueurs elaborated by maceration of aromatic and medicinal plants in grape marc distillates. J I Brew 122(4):653–660

Sabir SM, Zeb A, Mahmood M, Abbas SR, Ahmad Z, Iqbal N (2021) Phytochemical analysis and biological activities of ethanolic extract of (Curcuma longa) rhizome. Braz J Biol 81(3):737–740

Schieffer GW (2002) Pressurized liquid extraction of curcuminoids and curcuminoid degradation products from turmeric (Curcuma longa) with subsequent HPLC assays. J Liq Chromatogr Relat Technol 25(19):3033–3044

Sentkowska A, Biesaga M, Pyrzynska K (2016) Effects of brewing process on phenolic compounds and antioxidant activity of herbs. Food Sci Biotechnol 25(4):965–970

Sepahpour S, Selamat J, Abdul Manap MY, Khatib A, Abdull Razis AF (2018) Comparative analysis of chemical composition, antioxidant activity and quantitative characterization of some phenolic compounds in selected herbs and spices in different solvent extraction systems. Molecules 23(2):402

Shahidi F, Ambigaipalan P (2015) Phenolics and polyphenolics in foods, beverages and spices: antioxidant activity and health effects—a review. J Funct Foods 18:820–897

Shahidi F, Pegg RB, Saleemi ZO (1995) Stabilization of meat lipids with ground spices. J Food Lipids 2(3):145–153

Slimestad R, Fossen T, Brede C (2020) Flavonoids and other phenolics in herbs commonly used in Norwegian commercial kitchens. Food Chem 309:125678

Sokamte TA, Mbougueng PD, Tatsadjieu NL, Sachindra NM (2019) Phenolic compounds characterization and antioxidant activities of selected spices from Cameroon. S Afr J Bot 121:7–15

Suhaj M (2006) Spice antioxidants isolation and their antiradical activity: a review. J Food Compos Anal 19(6–7):531–537

Vallverdú-Queralt A, Regueiro J, Martínez-Huélamo M, Rinaldi Alvarenga JF, Leal LN, Lamuela-Raventos RM (2014) A comprehensive study on the phenolic profile of widely used culinary herbs and spices: rosemary, thyme, oregano, cinnamon, cumin and bay. Food Chem 154:299–307

Vallverdú-Queralt A, Regueiro J, Alvarenga JFR, Martinez-Huelamo M, Leal LN, Lamuela-Raventos RM (2015) Characterization of the phenolic and antioxidant profiles of selected culinary herbs and spices: caraway, turmeric, dill, marjoram and nutmeg. Food Sci Technol 35(1):189–195

Viegas O, Amaro LF, Ferreira IM, Pinho O (2012) Inhibitory effect of antioxidant-rich marinades on the formation of heterocyclic aromatic amines in pan-fried beef. J Agric Food Chem 60(24):6235–6240

Violeta N, Trandafir I, Cosmulescu S (2017) Bioactive compounds, antioxidant activity and nutritional quality of different culinary aromatic herbs. Not Bot Horti Agrobo 45(1):179–184

Wang M, Li J, Rangarajan M, Shao Y, LaVoie EJ, Huang T-C, Ho C-T (1998) Antioxidative phenolic compounds from sage (Salvia officinalis). J Agric Food Chem 46(12):4869–4873

Wojdyło A, Oszmiański J, Czemerys R (2007) Antioxidant activity and phenolic compounds in 32 selected herbs. Food Chem 105(3):940–949

Zhang Y, Smuts JP, Dodbiba E, Rangarajan R, Lang JC, Armstrong DW (2012) Degradation study of carnosic acid, carnosol, rosmarinic acid, and rosemary extract (Rosmarinus officinalis L.) assessed using HPLC. J Agric Food Chem 60(36):9305–9314

Zheng W, Wang SY (2001) Antioxidant activity and phenolic compounds in selected herbs. J Agric Food Chem 49(11):5165–5170

Chapter 9
Phenolic Antioxidants in Edible Oils

Learning Objectives
In this chapter, the reader will be able to:

- Describe the different classes of polyphenols in soybean oil.
- Learn about the composition of phenolic compounds in olive oils.
- Describe the phenolic profile of sunflower oil of different edible oils.

9.1 Introduction

Edible oils are obtained from plant sources, while fats are obtained from animals. Both are important components of our daily diet. Edible oils are obtained from the oilseeds, which are the source of several types of antioxidants. Edible oils are also rich in phenolic compounds. They are used as a frying medium for foods (Zeb 2019). Sunflower, soybean, olive, rapeseed, sesame, canola, corn, cottonseed, palm, and coconut oils are major frying oils. In the year 2019–2020, the consumption of palm oil was 74.6, soybean 56.8, rapeseed 27.8, sunflower 19, palm kernel 8.56, peanut 5.17, coconut 3.56, and olive oils 3.23 million metric tons (Shahbandeh 2020). The annual consumption has been surpassing 200 million metric tons. The major producers of edible oils are Canada, the USA, China, and Malaysia.

The increasing trends in consumption may be due to several factors including significant growth in world population, production of new foods and frying industries, and change from animal fats to edible oils during frying. In recent years, edible oils such as soybean, canola, flax, borage, and evening primrose have received great attention due to their known useful health effects. These effects were attributed to the antioxidants present in edible oils. Oil bearing seeds are the major source of phenolic antioxidants.

© Springer Nature Switzerland AG 2021

A. Zeb, *Phenolic Antioxidants in Foods: Chemistry, Biochemistry and Analysis*,
https://doi.org/10.1007/978-3-030-74768-8_9

9.2 Phenolic Antioxidants in Edible Oils

The composition of edible oils is categorized into two components, i.e., major components, and minor components. The former components of edible oils are lipids constituting free fatty acids, monoacylglycerols, diacylglycerols, triacylglycerols, phospholipids, sterols, fat-soluble vitamins, oxidized lipids, and squalene (Zeb and Murkovic 2013). The major component is largely comprised of triacylglycerols (95–98%). The minor components include phenolic compounds, volatile and aromatic substances, and impurities. Phenolic compounds are also present as a major player of quality characteristics and antioxidant capacity. The presence of polyphenolic compounds in flaxseed, soybean, and olive oils showed a large percentage had been identified as free form, as compared to bound forms as shown in Fig. 9.1. About 90% of phenolic compounds were in free forms in full-fat oil-bearing plants (Alu'datt et al. 2013). The flaxseeds contain high amounts of phenolic content than soybean.

Phenolic compounds had been ascribed to the antioxidant activity of the edible oils. Higher antioxidant activity had been shown by the full-fat meals in comparison to defatted meals (Alu'datt et al. 2013) as given in Fig. 9.2. The bound phenolic

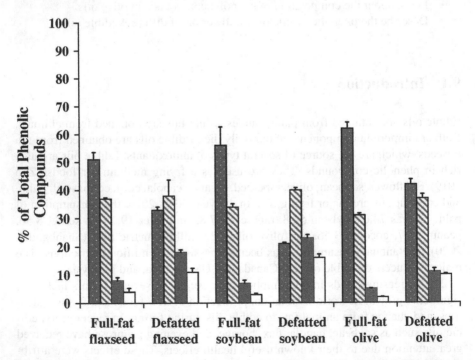

Fig. 9.1 Phenolic compounds are distributed in flaxseed, soybean, and olive meal. The leftmost column in every sample showed free phenolic compounds (methanol/23 °C extraction); followed by free phenolic compounds (methanol/60 °C extraction); the third column from the left side is bound phenolic compounds (NaOH hydrolysis); the fourth column is bound phenolic compounds (HCl hydrolysis). (Reproduced with kind permission of Elsevier, Alu'datt et al. 2013)

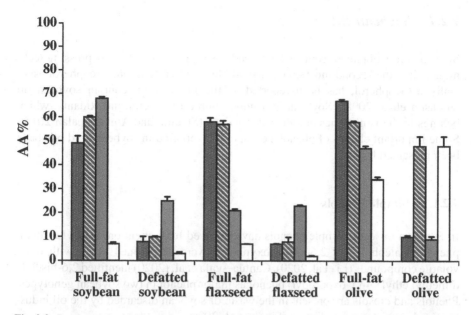

Fig. 9.2 Percent antioxidant activity of flaxseed, soybean, and olive meal. The leftmost column in every sample showed free phenolic compounds (methanol/23 °C extraction); followed by free phenolic compounds (methanol/60 °C extraction); the third column from the left side is bound phenolic compounds (NaOH hydrolysis); the fourth column is bound phenolic compounds (HCl hydrolysis). (Reproduced with kind permission of Elsevier, Alu'datt et al. 2013)

compounds exhibited higher antioxidant capacity (68%) as compared to free phenolic compounds. Full-fat olive meals having free phenolics exhibited greater antioxidant capacity when compared to the defatted meal. Different edible oils showed different antioxidant activities in different solvent systems. The amounts of total phenolics also depend on extraction techniques of oils. For example, the amount of total phenolic contents were higher in cold pressed rapeseed, soybean, and sunflower oils than their corresponding refined oils (Siger et al. 2005). The contents may vary with cultivation, time of harvest, type of soil (Alu'datt et al. 2017), the difference in the extraction procedure, and analytical methods of determination. The size of oilseeds also showed to affect the contents of phenolic compounds. Lee et al. (2008) showed that large soybean seeds contain lower total phenolic content than medium and small size seeds. Phenolic compounds provide stability to edible oils (Wu et al. 2019). The types of phenolic compounds may also vary with the type of edible oils. Details of these edible oils containing phenolic antioxidants have been reviewed recently (Zeb 2021). The author showed that simple phenols, hydroxycinnamic acids, tyrosol, and hydroxytyrosol were higher in olive oil, hydroxybenzoic acids in soybean oil, esters of hydroxycinnamic acids and flavonoids in sunflower oil, and lignans in sesame oil and flaxseed oil. Furthermore, hydroxybenzoic acids, hydroxycinnamic acid, and flavonoids were the major phenolic compounds present in all studied edible oils.

9.2.1 Soybean Oil

Soybean oil is obtained from soybean seeds using predominately by pressing techniques. It is the second most-consumed edible oil after palm oil. Tocopherol especially α-tocopherol, has been reported as the main antioxidant in soybean oil (Johnson et al. 2008). Soybean is a good source of several antioxidants, which belongs to the isoflavones and their derivatives (Shahidi and Ambigaipalan 2015). Some important classes of phenolic compounds reported in soybean seed oils have been presented here.

9.2.1.1 Simple Phenols

In Soybean oilseeds, simple phenols have received little attention. Benzyl alcohol, phenethyl alcohol, and 2,6-dimethoxyphenol were determined along with other volatile compounds (Li et al. 2020). Cantúa Ayala et al. (2020) identified 2,6-bis(1,1-dimethyl ethyl)-4-(1-oxopropyl) phenol as major phenol in two soybean genotypes. Phenols and cresols are present in the waste of soybean discarded by the oil industries and caused water pollution (Singh et al. 2020).

9.2.1.2 Hydroxybenzoic Acids

Syringic acid, vanillic acid, and 4-hydroxybenzoic acid having strong antioxidant capacity had been reported in soybeans (Arai et al. 1966; Hammerschmidt and Pratt 1978). Soybean oil contains several derivatives of hydroxybenzoic acids. 4-hydroxybenzoic acid (130 mg/kg), and syringic acid (264 mg/kg) had been reported as insoluble esters of oilseed flour, whereas 4-hydroxybenzoic acid (9.0 mg/kg), and syringic acid (22.0 mg/kg) had been liberated from the insoluble residues (Dabrowski and Sosulski 1984). Soybean sprouts of different cultivars grown in Korea had been reported to contain p-hydroxybenzoic acid (Kim et al. 2006).

Protocatechuic acid, 4-hydroxybenzoic acid, syringic acid, and salicylic acid had been reported in raw soybean flour (Prabakaran et al. 2018). Supplementation of gallic acid and methyl gallate into soybean oil has shown promising results in terms of thermal stability (Farhoosh and Nystrom 2018). Alu'datt et al. (2013) reported that in full fat and defatted soybean meals, the amount of gallic acid was 1.2 and 1.09%, respectively. Similarly, protocatechuic acid, hydroxybenzoic acid, vanillic acid, and syringic acids were also reported. Vanillin and syringic acid have been found dominant compounds present in soybean oil (Silva et al. 2020).

9.2.1.3 Hydroxycinnamic Acids

Caffeic, ferulic, and p-coumaric which are strong antioxidants had been reported in soybeans (Arai et al. 1966; Hammerschmidt and Pratt 1978). Chlorogenic acid, isochlorogenic acid, caffeic acid, ferulic acids had been reported to the present in the extracts of methanol or water of dried, and fresh soybean (Hammerschmidt and Pratt 1978). The order of increasing antioxidant capacity was *p*-coumaric, ferulic, chlorogenic, and caffeic acids. Soybean contains caffeic acid, sinapic acid, ferulic acid, and *p*-hydroxybenzoic *p*-coumaric acid both in full-fat and defatted soybeans (Alu'datt et al. 2013).

9.2.1.4   Flavonoids

In soybean, isoflavonoids existed largely (99%) in the form of its glycosides. Among them, genistein (64%), daidzein (23%), and glycetin-7-glycosides were 13% (Naim et al. 1973). In the β-carotene/linoleate model system, the order of increasing antioxidant activity of the isoflavones in soybean was glycetin, daidzein, genistein, quercetin, and 6,7,4′-trihydroxyisoflavones, respectively (Pratt and Birac 1979). Soybean also contains daidzein, daidzin, malonyl daidzin, genistein, genistin, acetyl-genistin, malonyl genistin, glycitin, and malonyl glycitin as reported by Riedl et al. (2007). The authors found that total isoflavones varied significantly in the range of 1573–7710 nmol/g in seeds obtained from several locations and cultivar combinations. In Maryland grown soybean lines, the predominant flavonoids were daidzein, genistein, and glycetin (Slavin et al. 2009). Soybean from Central Europe also showed a higher composition of daidzein, genistein, and glycitin (Malenčić et al. 2012).

Cho et al. (2013) reported several isoflavones in soybean seeds using HPLC. These include 12 isoflavones (1) daidzin, (2) glycetin, (3) genistin, (4) malonyl daidzin, (5) malonyl glycitin, (6) acetyl daidzin, (7) acetyl glycetin, (8) malonyl genistin, (9) daidzein, (10) glycitein, (11) acetyl genistin, and (12) genistein as shown in Fig. 9.3. Furthermore, the mean contents of isoflavone were 3079.4, 2393.4, 2373.9, and 1821.8 mg/kg in green, yellow, black, and brown soybeans, respectively.

Quercetin-3-rutinoside (6.57 and 43.31%), hesperidin (10.5 and 17.2%), and quercetin (0.52 and 0.71%) were reported in full-fat and defatted soybean meals. The soybean coat contain diadzein, genistein, glycitin and their acetyl derivatives (Chen et al. 2020b).

9.2.1.5   Anthocyanins

In different colored skins soybean cultivars, nine anthocyanins have been reported such as catechin-cyanidin-3-glucoside, delphinidin-3-galactoside, delphinidin-3-glucoside, cyanidin-3-galactoside, cyanidin-3-glucoside, petunidin-3-glucoside, pelargonidin-3-glucoside, peonidin-3-glucoside, and cyanidin (Cho et al. 2013). In addition, cyanidin-3-glucoside, delphinidin-3-glucoside, and petunidin-3-glucoside were present at a concentration of 11.0, 1.97, and 0.557 mg/g, respectively only in black soybeans. Krishnan et al. (2020) reported cyanidin-3-glucoside (C3G), delphinidin-3-glucoside (D3G), and petunidin-3-glucoside in different soybean genotypes. The maximum amount of C3G was 4.9 mg/g, while the least values were 3.56 mg/g.

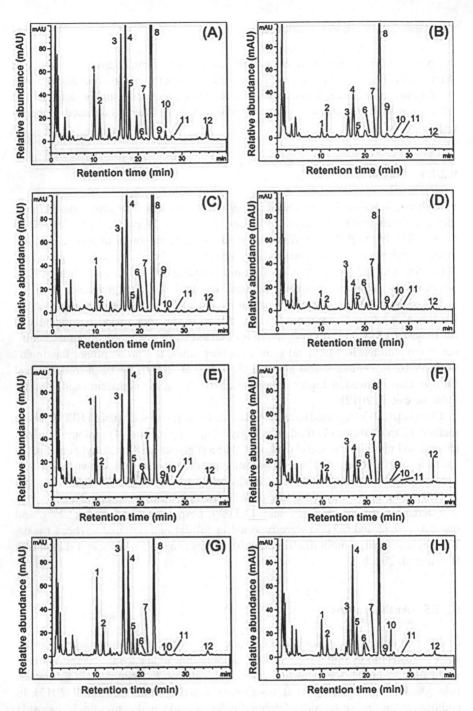

Fig. 9.3 Comparison of HPLC chromatograms concerning the highest and the lowest isoflavone contents from different colored seed coat soybeans. (**A**) Daepung (2010, yellow soybean), (**B**) Hwangkeumkong (2010, yellow soybean), (**C**) Cheongja 2 (2009, black soybean), (**D**) Seonheukkong (2009, black soybean), (**E**) Galmikong (2010, brown soybean), (**F**) Jinyoulkong (2009, brown soybean), (**G**) Nogchae (2010, Green soybean). (**H**) Cheongdu 1 (2009, Green soybean). Peak numbers represent individual phenolic compounds, (1) daidzin, (2) glycetin, (3) genistin, (4) malonyl daidzin, (5) malonyl glycitin, (6) acetyl daidzin, (7) acetyl glycetin, (8) malonyl genistin, (9) daidzein, (10) glycitein, (11), acetyl genistin, and (12) genistein. (Reproduced with kind permission of Elsevier, Cho et al. 2013)

9.2.2   Olive Oil

Olive oil is an important edible oil obtained from olive seeds. The olive plant is traditionally grown in the Mediterranean regions, for food uses. The world's top producing countries are Spain, Italy, Greece, Turkey, Morocco, Tunisia, Portugal, and Syria. Olive oil is a rich source of triacylglycerols comprising of 98–99%. Olive oil is rich in oleic acid in triacylglycerol forms (Zeb and Murkovic 2011; Boskou et al. 2006). Oleic acid may also be present as the free acid. The international olive oil council has recommended a few standards for olive oil types. These are extra-virgin olive oil (having 0.8% of free acidity), virgin olive oil (2.0% of free acidity), refine olive oil (0.3% of free acidity), olive pomace oil (1% of free acidity), and ordinary virgin oil (3.3% of free acidity). Olive oil is an extensively studied edible oil. Several books on the oil may be referred for details (Boskou 2015b; Harwood and Aparicio 2000). The details of the minor constituents can also be found (Boskou 2009, 2015a). Among the minor constituents, phenolic compounds are a major player in imparting different properties like antioxidants and health benefits. Several review articles have been focused on the minor constituents of olive oil (Uncu and Ozen 2020; Cicerale et al. 2008; Mateos et al. 2019; Ray et al. 2019; Guo et al. 2018; Wani et al. 2018). These review papers were focused on all antioxidants or minor constituents of olive oils and not on the individual class of phenolic compounds. It is not possible to accommodate all information available on olive oil phenolic compounds in this chapter, however, summary of latest findings on the individual class of polyphenolic compounds are presented.

During the growth of the olive fruit, the phenolic compounds increased significantly, whereas decreased when fruits reaching maturation (Franco et al. 2014). The amount of total phenolic contents (TPC) in olive oils usually ranging from 190 to 500 mg/kg (Khalatbary 2013). Similarly, in extra-virgin olive oil, the TPC typically ranging from 250 to 925 mg/kg (Kotsiou and Tasioula-Margari 2016). Other factors such as variety, climate, soil conditions, storage, extraction, and analysis of polyphenolic compounds may affect the composition. Several classes of polyphenols are present in olive oils, these are presented as a separate class, to better understand the phenolic antioxidant chemistry of olive oil.

9.2.2.1   Simple Phenols

In olive oils, simple phenols are usually reported as part of volatile fractions using GC-MS (Kalua et al. 2007). Guaiacol was reported as important simple phenol as odorants in virgin olive oils having different profiles of flavors (Reiners and Grosch 1998). A large number of simple phenols had been reported to form during the pyrolysis of olive waste, showing a possibility of the presence of simple phenols in olive oils during thermal treatment (Petrov et al. 2008). These include syringol, eugenol, guaiacol, cresol, phenol, catechol, and their derivatives. Aromatisation of olive oil with the ingredient to boost its sensory and nutritional characteristics. The final product is flavored olive oil, which may have high acceptability and thermal stability (Issaouia et al. 2019). When olive oil was aromatized with basil, eugenol was the major component produced (Veillet et al. 2010). Hachicha Hbaieb et al. (2016) reported that storage, volatile phenols such as phenol, guaiacol, 4-ethylphenol, 4-ethylguaiacol, 4-vinyl phenol, and 4-vinylguaiacol were present in virgin olive oils. The authors showed that 4-ethylphenol, 4-ethylguaiacol, 4-vinylphenol, and 4-vinylguaiacol had higher amounts in Arbequina oils at the end week of storage (Fig. 9.4). The formation of volatile phenols had a positive correlation with the period of storage and temperature. Recently, Lukić et al. (2019) reported p-cresol, benzyl alcohol, acetophenone, 2,3,6-trimethylphenol, 2-methoxy phenol, phenol, 3-ethylphenol, 3-ethyl benzaldehyde, and 2-phenyl ethanol in virgin olive oils of different geographical origins.

9.2.2.2 Hydroxybenzoic Acids

Olive oils contain several hydroxybenzoic acids, which include protocatechuic acid, vanillic acid, syringic acid, 3,4-dihydroxyphenylacetic acid, 4-hydroxyphenyl acetic acid, and 4-hydroxybenzoic acid (Cabrini et al. 2001). In three Spanish extra-virgin olive oils obtained from the mixtures of Morisca, Verdial de Badajoz, Picual, Arbequina, and Manzanilla de Sevilla cultivars, the only reported vanillic acid was in the range of 0.30–0.34 mg/kg (Reboredo-Rodríguez et al. 2014). Similarly, the virgin olive oils of four Turkish cultivars showed 4-hydroxybenzoic acid from 0.7 to 0.57 mg/kg, 2,3-dihydroxybenzoic acid from 0.19 to 0.41 mg/kg, vanillic acid (0.3–0.6 mg/kg), vanillin (0.02 mg/kg), and syringic acid (0.46–1.40 mg/kg) (Kesen et al. 2014). Similarly, another study of the three Turkish virgin olive oils obtained from Ayvalik, Gemlik, Memecik cultivars contain p-hydroxybenzoic acid, 2,3-dihydroxybenzoic acid, vanillic acid, and vanillin (Kelebek et al. 2015).

Ramos-Escudero et al. (2015) reported vanillic acid (0.6–2.7 mg/kg), and vanillin (0.3–1.1 mg/kg) in olive oils of nine varieties (Cuquillo, Empeltre, Manzanilla, Cornicabra, Picual, Arbequina, Lechin, Picudo, and Hojiblanca). Vanillin, 3-hydroxyphenyl acetic acid, and 4-phenylacetic acid had been reported in 36 extra virgin olive oils from 2 Turkish cultivars (Jolayemi et al. 2016). In multi-varietal Spanish extra virgin olive oils, vanillic acid (0.06–0.506 mg/kg), vanillin (none-0.642 mg/kg), and homovanillic acid (0.363–7.9 mg/kg) were present (Becerra-Herrera et al. 2018).

Vanillin, benzaldehyde, 4-methyl benzaldehyde, and benzoic acid were reported in virgin olive oils (Lukić et al. 2019). In olive oils of hundred different cultivars and their blends, 4-hydroxybenzoic acid, and vanillic acid were the main hydroxybenzoic acids (Pedan et al. 2019). In another study, Ramírez-Anaya et al. (2019) reported gallic acid (0.012 mg/kg), syringic acid (0.019 mg/kg), and vanillin (0.368 mg/kg) in extra virgin olive oils.

Hydroxybenzoic acid added to olive oils showed strong thermal stability than other oils (Castada et al. 2020). The amount of hydroxybenzoic acids has been enhanced by biostimulation of the olive plant with beneficial microbes as reported recently by Dini et al. (2020). The amount of 4-hydroxybenzoic acid was enhanced

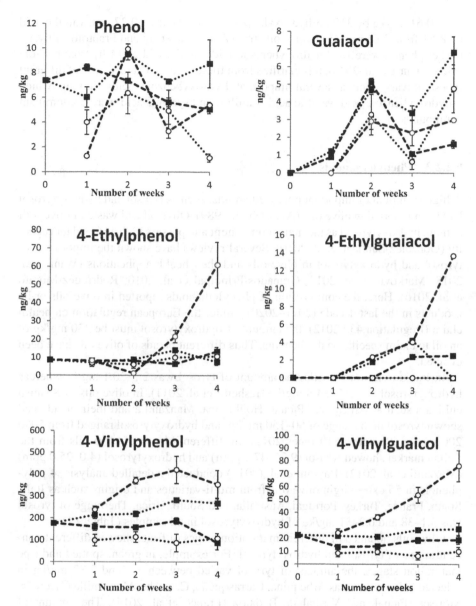

Fig. 9.4 Changes of volatile phenols concentrations (mg/kg olive oil, means of two replicates) in virgin olive oil obtained from Arbequina (open circle) and Chétoui (filled square) fruits stored at 4 °C (filled circle) and 25 °C (dashed lines) during 4 weeks. Vertical bars mean (standard deviation (SD)). The point-0 corresponds to the initial harvest date. (Reproduced with kind permission of Elsevier, Hachicha Hbaieb et al. 2016)

from 0.61 mg/kg by 31%, 3-hydroxybenzoic acid 77% from 0.27, and vanillic acid by 80% from 2.66 mg/kg. In olive oils from Ascolana tenera and Frantoio cultivars, the key phenols were 3,4-dihydroxybenzoic acid, vanillic acid, and 4-hydroxybenzoic acid (Peršurić et al. 2020). It is clarified from these studies that olive oils of different types/cultivars contain several important hydroxybenzoic acids, among which 4-hydroxybenzoic acid, vanillic acid, vanillin, and syringic acid were predominant compounds.

9.2.2.3   Phenylethanoids

A higher amount of important phenylethanoids, such as tyrosol and hydroxytyrosol had been reported in olive oils (Amiot et al. 1989). Olive oil and waste of olive mill contain hydroxytyrosol as the major component and possess high antioxidant capacity (González-Santiago et al. 2010). Several reviews have shown the importance of tyrosol and hydroxytyrosol in olive oils and their health applications (Wani et al. 2018; Marković et al. 2019; Granados-Principal et al. 2010; Rodríguez-Morató et al. 2016). Here, the composition of phenylethanoids reported in olive oils or its products in the last decade (2011–2020). Under the European regulation on health claim (Regulation 432/2012), the amount of hydroxytyrosol must be 250 mg/kg of an oil to claim specific health benefits. Thus different blends of olive oils have been evolved.

In Iranian virgin olive oils, the amount of tyrosol was 2.1–20.1 mg/L, whereas hydroxytyrosol was 0.7–84.8 mg/L (Hashemi et al. 2011). In olive oils of Spanish cultivars such as Arbequina, Picual, Hojiblanca, Manzanilla and their blends had shown tyrosol in the range of 60–150 mg/kg, and hydroxytyrosol ranged from 50 to 200 mg/kg (Romero and Brenes 2012). In different blends of olive oils from the Italian market showed tyrosol (9.75–47.1 ppm) and hydroxytyrosol (4.0–25.0 ppm) (Mazzotti et al. 2012). Bayram et al. (2012) conducted a detailed analysis of polyphenols in 55 extra-virgin olive oils from multi-varieties and origins such as Italy, Spain, France, Turkey, Portugal, Australia, and South Africa. The range of tyrosol was 3.1–38 and 0.2–31 mg/kg of hydroxytyrosol in all samples (Table 9.1).

Olive oils obtained in different maturation stages of fruits showed different concentrations of tyrosol and hydroxytyrosol. For example, in green, spotted and ripe maturation stages, the amount of tyrosol varied between 1.3 and 7.82 mg/kg in selected cultivars such as Arbequina, Carrasqueña, Corniche, Manzanilla Cacereña, Morisca, Picual, and Verdial de Badajoz (Franco et al. 2014). The amount of hydroxytyrosol varied between 0.39 and 3.77 mg/kg in all samples at the selected maturation stages. In another study of three extra-virgin olive oils of Spanish origin obtained from the mixtures of Morisca, Verdial de Badajoz, Picual, Arbequina, and Manzanilla de Sevilla cultivars, tyrosol was in the range of 8.4–51.9 mg/kg, while hydroxytyrosol was 9.8–14.9 mg/kg (Reboredo-Rodríguez et al. 2014).

The virgin olive oils of four Turkish cultivars showed tyrosol in the range of 12.4–18.7 mg/kg, while hydroxytyrosol from 1.56 to 5.68 mg/kg (Kesen et al. 2014). Another study of the three Turkish virgin olive oils obtained from Ayvalik,

Table 9.1 Composition of tyrosol and hydroxytyrosol in different types of olive oils

Type of olive oil	Cultivar/ samples numbers	Country	Analytical technique	Tyrosol (unit)	Hydroxytyrosol (unit)	Reference
Virgin	–	Iran	HPLC-UV	2.1–20.1 mg/L	0.7–84.8 mg/L	Hashemi et al. (2011)
Extra virgin and refined	Arbequina, Picual, Hojiblanca, Manzanilla	Spain	HPLC-UV	50–200 mg/kg	60–150	Romero and Brenes (2012)
Virgin	Olio bari, Olio bio, Olio 41, Olio gabro 3, Olio gabro 4, Olio carolea	Italy	HPLC-MS	9.75–47.1 ppm	4.02–25.08 ppm	Mazzotti et al. (2012)
Virgin	–	Italy	HPLC-DAD & GC-FID	360–1295 mg/kg	565–1630 mg/kg	Purcaro et al. (2014)
Extra virgin	Multi-varietal, 55 different olive oils	Italy, Spain, France, Turkey, Portugal, Australia and South Africa	HPLC-CAD	3.1–38 mg/kg	0.2–31 mg/kg	Bayram et al. (2012)
Virgin	Arbequina, Carrasqueña, Corniche, Manzanilla Cacereña, Morisca, Picual, and Verdial de Badajoz	Spain	HPLC-UV	Green: 1.95–6.25 mg/kg Spotted: 1.3–7.26 mg/kg Ripe: 0.64–7.82 mg/kg	Green: 1.28–3.31 mg/kg Spotted: 0.75–3.77 mg/kg Ripe: 0.39–1.6 mg/kg	Franco et al. (2014)
Extra virgin	3 oils mixture of Morisca, Verdial de Badajoz, Picual, Arbequina, and Manzanilla de Sevilla	Spain	HPLC-DAD	8.4–51.9 mg/kg	9.8–14.9 mg/kg	Reboredo-Rodríguez et al. (2014)
Virgin	Nizip yaglik, Kilis yaglik, Nizip yaglik-Bornova, Kilis yaglik-Bornova	Turkey	HPLC-MS	12.4–18.7 mg/kg	1.56–5.68 mg/kg	Kesen et al. (2014)

(continued)

Table 9.1 (continued)

Type of olive oil	Cultivar/ samples numbers	Country	Analytical technique	Tyrosol (unit)	Hydroxytyrosol (unit)	Reference
Virgin	Ayvalik, Gemlik, Memecik	Turkey	LC-MS/MS	6.93–14.9 mg/kg	0.8–17.3 mg/kg	Kelebek et al. (2015)
Virgin	Cuquillo, Empeltre, Manzanilla, Cornicabra, Picual, Arbequina, Lechin, Picudo, and Hojiblanca	Spain	HPLC	1.9–8.7 mg/kg	ND-8.1 mg/kg	Ramos-Escudero et al. (2015)
Extra virgin	Ayvalik and Memecik (36 oils)	Turkey	HPLC-DAD	Ayvalik: 1.19–9.26 mg/kg Memecik: 2.49–18 mg/kg	Ayvalik: 0.08–2.29 mg/kg Memecik: 0.24–2.65 mg/kg	Jolayemi et al. (2016)
Extra virgin	Carpinetana (3 samples), Dritta (6 samples), Gentile di Chieti (3 samples), Grignano (4 samples), Intosso (6 samples), Leccino (5 samples), Rustica (6 samples)	Italy	HPLC-DAD	3.5–6.3 mg/kg	2.2–7.1 mg/kg	Ambra et al. (2017)
Extra virgin	NR, n = 10	Italy	LC-MS/MS	23.8–86.1 mg/kg	1.72–113.8 mg/kg	Bartella et al. (2018)
Extra virgin	Multi-varietal (=50)	Spain	UHPLC-ESI-MS/MS	ND-29.9 mg/kg	6.9–72.7 mg/kg	Becerra-Herrera et al. (2018)
Extra virgin	different varieties, n = 163	Italy	HPLC-DAD	0.7–52.0 mg/kg	0.0–34.0 mg/kg	Paradiso et al. (2019)
Extra virgin	Tuscan (63 samples) and Apulian (45 samples): 100 oils	Italy	HPLC	0.29–8.9 mg/20 g	1.38–7.24 mg/20 g	Bellumori et al. (2019)

(continued)

Table 9.1 (continued)

Type of olive oil	Cultivar/ samples numbers	Country	Analytical technique	Tyrosol (unit)	Hydroxytyrosol (unit)	Reference
Virgin	NR, n = 30	Greece, Italy, Slovenia, Germany, Spain	UHPLC	1.26– 5.70 mg/20 g	1.14– 5.12 mg/20 g	Tsimidou et al. (2019)
Extra virgin	Arbequina, Picual, Madural, Cordovil de Serpa, Cobrançosa, Verdeal Alentejana, Carrasquenha, Blanqueta and Galega Vulgar	Portugal	HPLC-DAD	1.135– 5.715 mg/20 g	0.651– 5.259 mg/20 g	Pereira et al. (2020)
Extra virgin	NR, n = 10	Italy	HPLC-UV, PS-MS/MS	ND–285 mg/kg	ND–220 mg/kg	Bartella et al. (2020)
Extra virgin	Arbequina, Arbosana, Coratina, Koroneiki, Frantoio, and Picual	Brazil	UHPLC-ToF-MS	27.7– 330.5 mg/kg	0.37–1.89 mg/kg	Crizel et al. (2020)
Virgin	Mono-varietal, at different stages	Italy	UHPLC-ESI-MS	2.9–85.6 mg/kg	9.3–91.2 mg/kg	Di Stefano and Melilli (2020)

Gemlik, Memecik cultivars showed tyrosol (6.93–14.9 mg/kg), and hydroxytyrosol (0.8–17.3 mg/kg) as main compounds (Kelebek et al. 2015). Ramos-Escudero et al. (2015) studied olive oils of nine varieties (Cuquillo, Empeltre, Manzanilla, Cornicabra, Picual, Arbequina, Lechin, Picudo, and Hojiblanca) for phenolic composition. The amount of tyrosol was in the range of 1.9-8.7 mg/kg, while hydroxytyrosol was ND-8.1 mg/kg.

The amount of tyrosol was in the range of 3.5–6.3 mg/kg while that of hydroxytyrosol was 2.2–7.1 mg/kg in mono-varietal Italian extra virgin olive oils reported by Ambra et al. (2017). The authors also reported the four compounds of secoiridoids (derivatives of hydroxytyrosol). Several phenolic compounds including tyrosol and hydroxytyrosol have been reported in Spanish extra virgin olive oils (n = 50) (Becerra-Herrera et al. 2018). Tyrosol ranged from none to 29.9 mg/kg, whereas hydroxytyrosol was 6.9–72.7 mg/kg. The microwave-assisted extraction of tyrosol, hydroxytyrosol, and their derivatives showed a significantly high amount of these compounds in Italian olive oils of extra virgin quality (Bartella et al. 2018). The amount of tyrosol was in the range of 23.8–86.1 mg/kg, whereas that of hydroxytyrosol was 1.72–113.8 mg/kg.

Similarly, malaxation temperature and harvest timing also affected the composition of these phenylethanoids. In 36 extra-virgin olive oils of 2 Turkish cultivars (Ayvalik and Memecik), the tyrosol and hydroxytyrosol were affected by malaxation temperatures of 27, 37, and 47 °C as reported by Jolayemi et al. (2016). The tyrosol was in the range of 1.19–9.26 mg/kg in Ayvalik olive oils, whereas in Memecik, tyrosol was 2.49–18 mg/kg. The amount of hydroxytyrosol was ranging from 0.08 to 2.29 mg/kg in Ayvalik olive oils, whereas in Memecik, hydroxytyrosol was 0.24–2.65 mg/kg. Refining of olive oils has also been found to affect the tyrosol and hydroxytyrosol contents in olive oils (Lucci et al. 2020).

Bellumori et al. (2019) studied 100 samples of Italian extra virgin oils, which were comprised of Tuscan (63 samples) and Apulian (45 samples) for the phenolic profile. The amount of tyrosol was 0.29–8.9 mg/20 g, whereas hydroxytyrosol was 1.38–7.24 mg/20 g. The authors also showed that hydrolysis of the olive oils for the analysis of tyrosol, and hydroxytyrosol can give better results to suits the European Food Safety Authority level of these compounds. In another study by Paradiso et al. (2019), the amount of tyrosol was ranged from 0.7 to 52.0 mg/kg, whereas hydroxytyrosol was 0.0–34.0 mg/kg in several extra virgin olive oils (n = 163) from Italy. The authors found that hydroxytyrosol eluted as peak 1, followed by tyrosol. Other derivatives were 3,4-DHPEA-EDA (dialdehydic form of decarboxymethyl elenolic acid linked to 3,4-DHPEA), p-DHPEA-EDA (oxidized dialdehydic form of decarboxymethyl elenolic acid linked to p-HPEA), 3,4-DHPEA-EA (oleuropein aglycone), and p-DHPEA-EA (ligstroside aglycone) as shown in Fig. 9.5. These compounds were previously reported in olive oils (Bartella et al. 2018).

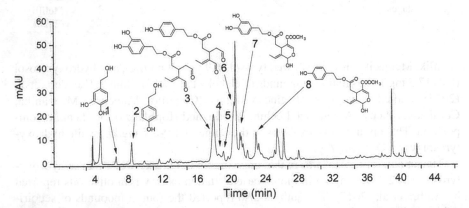

Fig. 9.5 Sample chromatogram with the phenolic compounds included in the hydroxytyrosol derivatives class. (1) Hydroxytyrosol, (2) tyrosol, (3), dialdehydic form of decarboxymethyl elenolic acid linked to 3,4-DHPEA (3,4-DHPEA-EDA), (4), oxidized dialdehydic form of decarboxymethyl elenolic acid linked to p-HPEA (p-DHPEA-EDA), (7), oleuropein aglycone (3,4-DHPEA-EA), and (8) Ligstroside aglycone (p-DHPEA-EA). (Reproduced with kind permission of Elsevier, Paradiso et al. 2019)

In mono-varietal extra virgin olive oils from Portugal showed tyrosol (1.13–5.71 mg/20 g) and hydroxytyrosol (0.651–5.25 mg/20 g) in nine cultivars namely Arbequina, Picual, Madural, Cordovil de Serpa, Cobrançosa, Verdeal Alentejana, Carrasquenha, Blanqueta and Galega Vulgar (Pereira et al. 2020). The authors also reported several secoiridoids derivatives of hydroxyelenolic acid, hydroxyoleuropein aglycone, elenolic acid, oleacin, oleocanthal, oleuropein aglycone, methyloleuropein aglycone, and ligstroside aglycone. The structures of some of these compounds. Bartella et al. (2020) quantified the amount of tyrosol and hydroxytyrosol ranging from none-285 mg/kg, and none-220 mg/kg, respectively. Similarly, Tsimidou et al. (2019) showed the utilization of UHPLC for the analyses of tyrosol and hydroxytyrosol in olive oils from Greece, Italy, Spain, Germany, and Slovenia. The authors showed that tyrosol ranged from 1.26 to 5.70 mg/20 g, whereas hydroxytyrosol ranged from 1.14 to 5.12 mg/20 g in the selected samples. Brazilian extra virgin olive oils also showed tyrosol in the range of 27.7–330.5 mg/kg, and hydroxytyrosol in smaller amounts (0.37–1.89 mg/kg) reported by Crizel et al. (2020). These studies revealed that olive oils, from different origins, have enough amount of tyrosol, hydroxytyrosol, and their derivatives, which are beneficial for human consumption.

9.2.2.4   Hydroxycinnamic Acids

A comprehensive study of polyphenols composition in 55 olive oil (extra-virgin) samples from different varieties and origins (Italy, Spain, France, Turkey, Portugal, Australia, and South Africa) had shown caffeic acid (0.07–0.70 mg/kg), ferulic acid (0.08–0.25 mg/kg), and p-coumaric acid (not detected-1.74 mg/kg) (Bayram et al. 2012). In three Spanish extra-virgin olive oils obtained from the mixtures of Morisca, Verdial de Badajoz, Picual, Arbequina, & Manzanilla de Sevilla cultivars, the only reported phenol was p-coumaric acid (0.32–1.01 mg/kg) (Reboredo-Rodríguez et al. 2014). Similarly, the virgin olive oils of four Turkish cultivars, Kesen et al. (2014) reported p-coumaric acid (0.63–2.01 mg/kg), ferulic acid (0.06–0.38 mg/kg) and cinnamic acid (0.51–1.86 mg/kg). Another study of the three Turkish olive oils (virgin) obtained from Ayvalik, Gemlik, Memecik cultivars reported p-coumaric acid, ferulic acid, and cinnamic acid (Kelebek et al. 2015).

Ramos-Escudero et al. (2015) reported p-coumaric acid (0.1–1.5 mg/kg), and cinnamic acid (1.3–15.5 mg/kg) in olive oils of nine Spanish varieties such as Cuquillo, Empeltre, Manzanilla, Cornicabra, Picual, Arbequina, Lechin, Picudo, and Hojiblanca. In virgin olive oils of different crop seasons and varieties, caffeic acid, vanillic acid, vanillin, p-coumaric acid, and ferulic acid had been reported by Franco et al. (2014). The authors showed that there was a strong correlation between crop seasons and varieties with the composition of ferulic acid. Similarly, in 36 extra virgin olive oils from 2 Turkish cultivars, the phenolic acids were p-coumaric acid, and ferulic acid was reported (Jolayemi et al. 2016).

Becerra-Herrera et al. (2018) reported p-coumaric acid (none-5.77 mg/kg), o-coumaric acid (0.802–0.107 mg/kg), and cinnamic acid (none-0.263 mg/kg) in

Spanish multi-varietal extra virgin olive oils. Caffeic acid, ferulic acid, coumaric acid were reported in olive oil of extra-virgin quality (Melguizo-Rodríguez et al. 2019). In hundred different cultivars of olive oils and their blends, coumaric acid and ferulic acid have been reported by Pedan et al. (2019). In extra virgin olive oil, caffeic acid (0.011 mg/g), p-coumaric acid (0.05 mg/g), and o-coumaric acid (0.022 mg/g) were present (Ramírez-Anaya et al. 2019).

Biostimulation has enhanced the amount of p-coumaric acid (1.42 mg/kg by 69%), cinnamic acid (0.43 mg/kg by 9%), and ferulic acid (0.064 mg/kg by 73%) (Dini et al. 2020). Di Stefano and Melilli (2020) reported the changes in the p-coumaric acid, caffeic acid, and ferulic acid in mono-varietal Italian olive oil under fruit maturation stages. In olive oils from Ascolana tenera and Frantoio cultivars, the amount of caffeic acid, ferulic acid, and p-coumaric acid were reported as major cinnamic acid derivatives (Peršurić et al. 2020). In olive oils from different Italian cultivars such as Frantoio, Leccino, Picholine, Kalamon, and Picual, only ferulic acid was reported by Rocchetti et al. (2020). These studies revealed that olive oils contain caffeic acid, ferulic acid, and p-coumaric acid as predominant phenolic acids.

9.2.2.5 Coumarins and Chromans

Coumarins and chromans have been reported in limited publications. For example, chroman such as hydroxy-isochromans, 1-phenyl-6,7-dihydroxy-isochroman, and 1-(3′-methoxy-4′-hydroxy) phenyl-6,7-dihydroxy-isochroman had been reported in diverse samples of olive oil of extra virgin quality (Bianco et al. 2002). Similarly, 7-isopentenyloxycoumarin (1.82 µg/mL) and 7-hydroxycoumarin also known as Umbelliferone (6.6 µg/mL) have been reported recently in olive oil (Ferrone et al. 2018; Alu'datt et al. 2018).

9.2.2.6 Flavonoids

In olive fruits the two main flavonoids present in high amounts are quercetin-3-rutinoside and luteolin-7-glucoside (Romani et al. 1999). The leaves of olive are rich in a wide range of flavonoids, which help the plant during growth and fruit maturation and protect it from negative UV radiation (Dias et al. 2020). Olive oil obtained from a ripe fruit has been reported to contain flavonoids, the predominant of which was quercetin-3-rutinoside (Huang et al. 2020).

Reboredo-Rodríguez et al. (2014) reported luteolin (2.3–6.2 mg/kg), and apigenin (0.99–2.2 mg/kg) in three extra-virgin Spanish olive oils obtained from the mixtures of Morisca, Verdial de Badajoz, Picual, Arbequina, and Manzanilla de Sevilla cultivars. Similarly, in the olive oils of four Turkish cultivars with quality of virgin, Kesen et al. (2014) reported apigenin (1.17–4.96 mg/kg), and luteolin (1.51–7.57 mg/kg) as key flavonoids. Similarly, three Turkish virgin olive oils obtained from Ayvalik, Gemlik, Memecik cultivars exhibited the occurrence of apigenin, and luteolin (Kelebek et al. 2015).

Jolayemi et al. (2016) reported luteolin and luteolin glucoside in 36 extra virgin olive oils from two Turkish cultivars. Becerra-Herrera et al. (2018) reported apigenin in the range of 0.202–2.46 mg/kg and luteolin in the range of 0.854–3.6 mg/kg in several Spanish multi-varietal extra virgin olive oils. Apigenin and luteolin were recently reported in extra-virgin olive oil by Franco et al. (2014) and Melguizo-Rodríguez et al. (2019).

Similarly, in a hundred samples of olive oils of different cultivars and their blends, luteolin-7-glucoside, apigenin-7-glucoside, apigenin, and luteolin were reported by Pedan et al. (2019). In extra virgin olive oil, luteolin was 0.101 mg/kg and apigenin was 0.142 mg/kg reported by Ramírez-Anaya et al. (2019). The amount of luteolin was enhanced by a maximum of 70% from 3.1 mg/kg and apigenin by a maximum of 10% from 0.228 mg/kg using biostimulation (Dini et al. 2020). Apigenin, apigenin-7-glucoside, diosmetin, luteolin, luteolin-7-glucoside, naringenin, quercetin, and rutin have been identified in olive oils from Ascolana tenera and Frantoio cultivars (Peršurić et al. 2020). Similarly, Di Stefano and Melilli (2020) also reported apigenin (none-4.15 mg/kg), apigenin-7-glucoside (0.64–6.13 mg/kg) and diosmetin (0.34–9.19 mg/kg) in olive oils affected by the maturation of olive fruits. Luteolin and apigenin were also reported in Brazilian extra virgin olive oils (Crizel et al. 2020). These results indicated that luteolin and apigenin are the major components of flavonoids present in olive oils.

9.2.2.7 Anthocyanins

Romani et al. (1999) showed that olive fruits were rich in important anthocyanins such as cyanidin-3-glucoside and cyanidin-3-rutinoside. In olive oils from different Italian cultivars such as Frantoio, Leccino, Picholine, Kalamon, and Picual, cyanidin was found (Rocchetti et al. 2020).

9.2.2.8 Lignans

Lignans such as pinoresinol and acetoxypinoresinol have been reported in olive oils. The pinoresinol (1.1–14.0 mg/kg) had been found as a major phenolic compound in 55 olive oil samples of extra virgin quality from different countries such as Italy, Spain, France, Turkey, Portugal, Australia, and South Africa (Bayram et al. 2012). Similarly, Reboredo-Rodríguez et al. (2014) reported pinoresinol in the range of 2.9–5.8 mg/kg in three Spanish extra-virgin olive oils obtained from the mixtures of Morisca, Verdial de Badajoz, Picual, Arbequina, and Manzanilla de Sevilla cultivars. Ramos-Escudero et al. (2015) reported pinoresinol in the range of 3.9–7.8 mg/kg in olive oils of nine varieties from Spain. Recently, Becerra-Herrera et al. (2018) reported pinoresinol in the range of none to 7.0 mg/kg in several Spanish multivarietal extra virgin olive oils.

Pinoresinol and 1-acetoxypinoresinol were recently identified in a hundred olive oils of different cultivars and their blends (Pedan et al. 2019). The amount of pinoresinol was 0.454 mg/kg in extra-virgin olive oil (Ramírez-Anaya et al. 2019). The amount of these lignans was enhanced by biostimulation. Dini et al. (2020) showed that pinoresinol enhanced by a maximum of 57% from 0.095 mg/kg, and acetoxypinoresinol enhanced from 4.3 mg/kg by a maximum of 62% in various treatments. Peršurić et al. (2020) reported pinoresinol at a concentration of 0.73 and 0.51 mg/kg in olive oils of Ascolana tenera and Frantoio cultivars. Matairesinol has been reported in olive oils from different cultivars such as Frantoio, Leccino, Picholine, Kalamon, and Picual from Italy (Rocchetti et al. 2020). In Brazilian extra virgin olive oils, acetoxypinoresinol was the main lignans (Crizel et al. 2020).

9.2.3   Brassica Oils

Canola oil is an important edible oil obtained from the oilseeds of brassica plants (*Brassica Rapa, Brassica napus*, and *brassica juncea*). It is different from the rapeseed oil-based on the amount of erucic acid. Erucic acid is a 22-carbons fatty acid with one double bond, it is chemically known as Z-docosa-13-enoic acid. Rapeseed oil received second placed among the edible oils in the world market. The rapeseed cultivation is mainly carried out for edible purposes, animal feeds, and the production of biodiesel (Chew 2020). Mustard oil is obtained from seeds of *Brassica juncea*. The terms canola, rapeseed, and mustard oils are used interchangeably and usually refers to the oils from brassica seeds. Canola oil has a low amount of erucic acid. Canola or rapeseed oil is rich in polyphenolic compounds and possess strong antioxidant activity. The ethanolic extracts of canola oil were similar to that of tert-butylhydroquinone (TBHQ) and stronger than butylated hydroxyanaline (BHA), butylated hydroxytoluene (BHT), and the mixture of these antioxidants (Wanasundara and Shahidi 1994). Supplementation of canola flour to meat showed higher inhibition of lipid oxidation (73–97%) measured as thiobarbituric acid reactive substances (Shahidi and Naczk 2003). The phenolic portion of canola and rapeseed oils contain several types of polyphenols such as phenolic acids, flavones, and flavonols. The details of some of these phenolic compounds had been summarized by Shahidi (1990) and Naczk et al. (1998). However, due to the continuous scientific research on the subjects, updated knowledge is presented here.

9.2.3.1   Hydroxybenzoic Acids

Canola and rapeseed oils contain several hydroxybenzoic acids. For example, 4-hydroxybenzoic acid, vanillic acid, gentisic acid, protocatechuic acid, and syringic acid had been reported both in free as well as in bound forms in rapeseeds flour and hulls (Krygier et al. 1982). Gentisic acid, hydroxybenzoic acid, salicylic acid, and syringic acid had been reported in canola seeds, canola meal, and their by-products (Shahidi and Ambigaipalan 2015). The rapeseed hulls contain protocatechuic acid, *p*-hydroxybenzoic acid, and syringic acid as principle hydroxybenzoic acids (Liu et al. 2012).

Jun et al. (2014) described that 80% of methanol extract of canola seeds contains gallic acid and protocatechuic acid, whereas ethyl acetate extract contains gallic acid, protocatechuic acid, salicylic acid, and 4-hydroxybenzoic acid. Thus ethyl acetate was found more suitable for the extraction of hydroxybenzoic acid in canola oils. In mustard oil from Pakistan, Zeb and Rahman (2017) reported gallic acid (7.14 mg/kg), and vanillic acid hexoside (6.03 mg/kg) obtained through hot pressing techniques.

Song et al. (2019) reported that an LC-MS/MS-based method for the analyses of phenolic compounds. It was found that *p*-hydroxybenzoic acid (0.05 mg/kg) was only present in hot-pressed rapeseed oils, whereas syringic acid was 0.22 and 0.43 mg/kg in cold pressed and hot pressed oils. In ten defatted rapeseed meal from China, gallic acid, 4-hydroxybenzoic acid, and syringic acid were present, whereas 4-hydroxybenzoic acid has a major contribution in terms of quantity of polyphenols (Zhang et al. 2019).

9.2.3.2 Hydroxycinnamic Acids

Caffeic acid, cinnamic acid, *p*-coumaric acid, ferulic acid, sinapic, and their derivatives have been found in canola seeds, canola meal, and their byproducts (Shahidi and Ambigaipalan 2015). Krygier et al. (1982) reported *p*-coumaric acid, *cis*-ferulic acid, *trans*-ferulic acid, caffeic acid, *cis*-sinapic acid, and *trans*-sinapic acid in rapeseed hulls and flour. These phenolic compounds showed high radical scavenging potentials. Rapeseed was reported rich in sinapic acid, and its derivatives like sinapine, and esters, which constitute a major part of phenolic compounds (Nowak et al. 1992). Sinapine is chemically sinapoylcholine found commonly in brassica seeds and vegetables (Fig. 9.6). In rapeseed, sinapic acid has also been reported in the

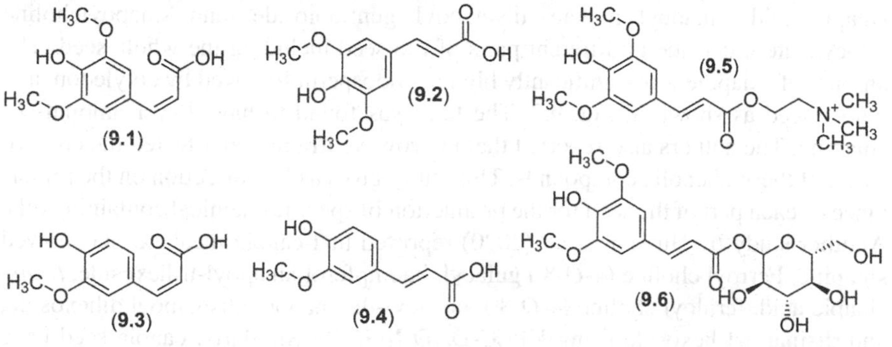

Fig. 9.6 Structures of hydroxycinnamic acids and derivatives. (**9.1**) cis-sinapic acid, (**9.2**) trans-sinapic acid, (**9.3**) cis-ferulic acid, (**9.4**) trans-ferulic acid, (**9.5**) sinapine, and (**9.6**) glucopyranosyl sinapate

form of a glucosidic ester, glucopyranosyl sinapate (Amarowicz and Shahidi 1994). Similarly, rapeseed and canola hulls were found rich in sinapic acid (Amarowicz et al. 2000). Thiyam et al. (2006) showed that press cakes obtained from rapeseed and mustard oils contain sinapine, sinapic acid, and their derivatives.

In Chinese rapeseed, sinapine was the main compound in hulls of defatted rapeseed as well as in dehulled flours, and their amount was ranged from 0.93 to 1.76 and 15.6 to 21.8 mg/g, respectively as reported by Liu et al. (2012). The authors also showed the presence of sinapoyl glucoside isomers, *cis*-sinapic acid, *trans*-sinapic acid, disinapoyl gentiobioside, and disinapoyl glucoside in the dehulled rapeseed. The hulls had been found to contain caffeic acid, coumaric acid, trans-ferulic acid, and trans-sinapic acid.

The methanolic extracts of canola seeds contain caffeic acid, *trans*-sinapic acid, and chlorogenic acid, while ethyl acetate fraction contains caffeic acid, ferulic acid, *trans*-sinapic acid, and chlorogenic acid (Jun et al. 2014). Siger and Józefiak (2016) reported sinapine, sinapic acid, *p*-coumaric acid, ferulic acid, and sinapic acid methyl ester Polish rapeseed oils. The authors also showed that the roasting of the rapeseeds significantly affected the composition of phenolic acids.

Trans-sinapic acid, ferulic acid, *p*-coumaric acid, sinapine, and sinapic acid methyl ester have been reported in Polish rapeseed oils (Rękas et al. 2017). Similarly, in mustard oil from Pakistan, sinapine (3.15 mg/kg), sinapic acid (4.02 mg/kg), sinapic acid glucoside (1.865 mg/kg), and sinapoyl dihexoside (0.89 mg/kg) obtained through hot pressing techniques (Zeb and Rahman 2017).

The hydroxycinnamic acids composition of rapeseed oils obtained using cold-pressed and hot pressing showed *p*-coumaric acid, sinapic acid, ferulic acid, cinnamic acid, and sinapine (Song et al. 2019). Rapeseed cultivars from China showed sinapic, caffeic, ferulic, and *p*-coumaric acids in free forms, whereas sinapine, sinapoyl glucoside, and disinapoyl gentiobiose were present as bound or esterified forms (Zhang et al. 2019). Sinapic acid was the major compound. In another study, Mikołajczak et al. (2019) also reported *p*-coumaric acid and sinapic acid in rapeseed oils.

Cong et al. (2019) studied rapeseeds for phenolic composition. The identified compounds were sinapine, sinapoyl glucoside, quercetin sinapoyl di-hexosepentose, sinapic acid, sinapoyl malate, disinapoyl gentiobioside, and sinapoylcholine thiocyanate-glucoside in different parts of the seed including the whole seed. The amount of sinapine was significantly high in endosperm followed by cotyledon, and whole seed as shown in Fig. 9.7. The hull was found to have lower amounts of sinapine. The authors also reported that microwave irradiation affected the composition of these phenolic compounds. This study gives a clear direction on the importance of each part of the seed for the production of specific chemical containing oils. Another study by Hussain et al. (2020) reported that canola meal extract showed sinapine, Ferroyl choline (4-*O*-8′) guiacyl, kaempferol-sinapoyl-trihexoside, *trans*-sinapic acid, feruloyl choline (4-*O*-8′) guaiacyl-di-sinapoyl, disinapoyl dihexoside, and disinapoyl hexoside using HPLC-DAD-MS/MS. Similarly, canola seed cake was found to contain cis-, trans-sinapic acids, its nine glycosylated derivatives, and eight sinapoyl choline derivatives (Zardo et al. 2020). These compounds were also

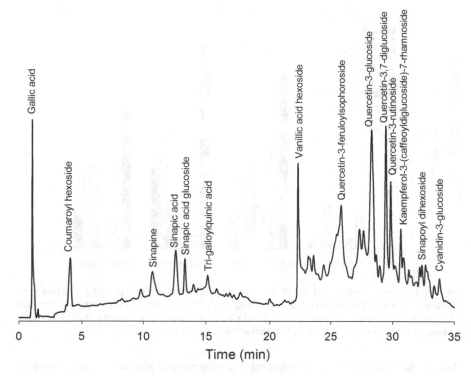

Fig. 9.7 Representative HPLC-DAD chromatograms of mustard oil at 320 nm showing flavonoids and other phenolic compounds. (Source: Zeb and Rahman 2017)

identified with help of HPLC-DAD-MS/MS. These studies concluded that sinapine, sinapic acid, coumaric acid, and ferulic acid are major fingerprinting compounds in brassica edible oils.

9.2.3.3 Flavonoids

In rapeseed meal, flavonoid glucosides such as kaempferol-3-(sinapoyl sophoroside)-7-glucoside and kaempferol-3-(sinapoyl glucoside)-7-sophoroside were reported (Shahidi and Ambigaipalan 2015). Figure 9.8 showed the separation of polyphenols in mustard oil. The mustard oil obtained from hot pressing had been found to contain quercetin-3-feruloylsophoroside (7.98 mg/kg), quercetin-3-glucoside (17.6 mg/kg), quercetin-3,7-diglucoside (5.01 mg/kg), quercetin-3-rutinoside (2.73 mg/kg), and kaempferol-3-(caffeoyldiglucoside)-7-rhamnoside (3.76 mg/kg) (Zeb and Rahman 2017). These compounds had shown to decline with thermal oxidation.

The baby leaf white and yellow cultivars of rapeseeds contain kaempferol-3-hydroxyferuoyl-sophoroside-7-glucosides, isorhamnetin-3-glucoside-7-glucoside, kaempferol-3-sinapoyl-sophoroside-7-diglucoside, and kaempferol-3-sinapoyl-sophoroside-7-glucoside (Groenbaek et al. 2019). Similarly, Zardo et al. (2020) studied canola seed cake using HPLC-DAD-MS/MS and identified several

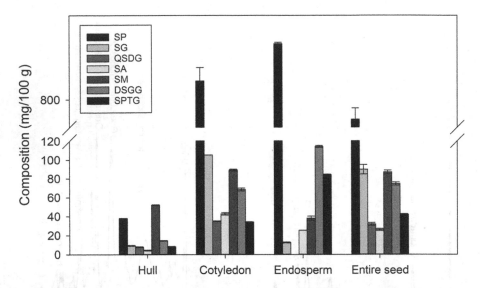

Fig. 9.8 Content of sinapic acid and its derivatives in different parts of rapeseed. (Redrawn from the source with kind permission of Elsevier, Cong et al. 2019). Sinapine (SP), sinapoyl-glucoside (SG), quercetin-sinapoyl-di-hexosepentose (QSDG), sinapic acid (SA), sinapoyl-malic acid (SM), disinapoyl gentiobioside (DSGG), and sinapoylcholine thiocyanate-glucoside (SPTG)

flavonoids. These include kaempferol-3-sophoroside-7-glucoside, kaempferol-3-sinapoylsophoroside-7-glucoside, kaempferol-6-sinapoylglucoside-3,7-diglucoside, kaempferol-3-sophoroside, and kaempferol-3-sinapoylsophoroside. This study thus confirmed that those flavonoids which are present in the seed, may possibly be leached to oil during extraction.

9.2.3.4 Tannins

Rapeseed and canola oils contained soluble as well as insoluble tannins (Shahidi and Naczk 1995; Amarowicz et al. 2000). The amount of tannins in canola and rapeseed hulls had been reported up to 6% tannins (Naczk et al. 2000). Thus hulls can be used as a source of these antioxidants (Amarowicz et al. 2000).

9.2.4 Sunflower Oil

Sunflower oil is a highly important edible oil obtained from seeds of sunflower (*Helianthus annuus*) and is widely used as a frying medium. It is rich in oleic acid and linoleic acid (Zeb 2019). The amount of linoleic acid (53.9 mg/kg) is a usually higher component of triacylglycerols than oleic acid having an amount of 23.9 mg/kg (Zeb and Murkovic 2013; Zeb 2011). Sunflower oil is also rich in several types of natural antioxidants, which include phenolic compounds. Sunflower seeds have been found to contain several phenolic compounds, with significant antioxidant activity (De Leonardis et al. 2003). The defatted meal and seed shells of sunflower had shown higher antioxidant activities, which was attributed to phenolic compounds in the samples (Schmidt and Pokorný 2005; De Leonardis et al. 2005). The majority of the phenolic compounds had been reported in the kernel, whereas less amount had been reported in the seed shells (Pedrosa et al. 2000). The byproduct of oil extract such as sunflower cake contains 2–4% of the polyphenolic compounds (Weisz et al. 2009). The amount of these compounds can be enhanced in sunflower seeds using different concentration of calcium, zinc and gibberellic acid (Jan et al. 2018). Sunflower seeds or its products especially oil contain hydroxybenzoic acids, phenylethanoids, hydroxycinnamic acids, esters of hydroxycinnamic acids, and flavonoids.

9.2.4.1 Hydroxybenzoic Acids

Hydroxybenzoic acids such as 4-hydroxybenzoic acid, and sinapic acid had been found in sunflower oilseeds flour (Kozlowska et al. 1983). Similarly, both free and bound hydroxybenzoic acids had been reported in sunflower flour. These include *p*-hydroxybenzoic acid (60 mg/kg), vanillic acid (8.0 mg/kg), syringic acid (22.0 mg/kg), and *cis*-sinapic acid (9609 mg/kg) were present as soluble esters (Dabrowski and Sosulski 1984).

Shi et al. (2015) reported gallic acid (229 mg/L), vanillic acid (59 mg/L) in sunflower oil. Gallic acid, protocatechuic acid, 4-hydroxybenzoic acid, vanillic acid, syringic acid, and salicylic acid have been reported in kernels, hulls, and press residues of the sunflower oilseeds (Zoumpoulakis et al. 2017). Benzoic acid was present at the level of 380 mg/L in sunflower oil from Japan (Xuan et al. 2018).

Sonmezdag et al. (2019) reported 4-hydroxybenzoic acid (0.5 mg/kg), vanillic acid (1.35 mg/kg) in high oleic sunflower oil from the commercial market in Turkey. Similarly, Zeb and Ullah (2019) reported 4-hydroxybenzoic acid (6.6 mg/kg), gallic acid (1.23 mg/kg), vanillic acid hexoside (0.38 mg/kg), and syringic acid (0.11 mg/kg) in sunflower oil from Pakistan.

9.2.4.2 Phenylethanoids

Hydroxytyrosol had been found in sunflower oil as well as high-oleic sunflower oil (Vázquez-Velasco et al. 2011). Hydroxytyrosol has been reported in the press residue of the sunflower oilseeds and was absent in kernels and hulls (Zoumpoulakis et al. 2017).

9.2.4.3 Hydroxycinnamic Acids

Caffeic acid ferulic acid had been reported in kernels and shells of oilseed sunflowers of different origins (Italy, France, and Germany) (Weisz et al. 2009). *Trans*-ferulic acid and *cis*-ferulic acid have been reported in sunflower seeds (Karamać et al.

2012). Shi et al. (2015) reported chlorogenic acid (2.0 mg/L), caffeic acid (12 mg/L), and ferulic acid (17 mg/L) in sunflower oil. In sunflower oilseed products such as kernels, hulls, and press residues, caffeic acid, p-coumaric acid, ferulic acid o-coumaric acid, and cinnamic acid had been reported (Zoumpoulakis et al. 2017).

Italian sunflower oil has shown ferulic acid having an amount of 18,120 mg/L as reported by Ferrone et al. (2018). In sunflower oil marketed in Japan, Xuan et al. (2018) reported p-coumaric acid at a concentration of 42 mg/L. Similarly, sunflower oil from Pakistan contains p-coumaric acid (1.12 mg/kg), and caffeic acid (0.21 mg/kg) (Zeb and Ullah 2019). In Turkish high oleic sunflower oil, caffeic acid (0.95 mg/kg), p-coumaric acid (0.51 mg/kg), ferulic acid (0.38 mg/kg), and sinapic acid (0.26 mg/kg) were identified and quantified using LC-DAD-ESI-MS by Sonmezdag et al. (2019). The authors found that the amount of these phenolic compounds decrease with successive frying.

9.2.4.4   Esters of Hydroxycinnamic Acids

In sunflower kernels and shells of different origins, several esters of hydroxycinnamic acids have been reported by Weisz et al. (2009). The authors found that 5-caffeoylquinic acid was the major compound having concentration up to 591 mg/kg in the sunflower seed shells, whereas it was 30,505 mg/kg in kernels. Other compounds were 5-coumaroylquinic acid, 5-feruloylquinic acid, 3-caffeoylquinic acid, 4-caffeoylquinic acid, 5-caffeoylquinic acid, caffeoylquinic acid, 3,4-dicaffeoylquinic acid, 3,5-di-caffeoylquinic acid, and 4,5-di-caffeoylquinic acid as shown in Table 9.2.

In sunflower seeds, Karamać et al. (2012) reported several derivatives of esters of hydroxycinnamic acids. These include 3-caffeoylquinic acid, 4-caffeoylquinic acid, 5-caffeoylquinic acid, 3-coumaroylquinic acid, 4-coumaroylquinic acid, and 5-coumaroylquinic acid, six derivatives of dicaffeoylquinic acid, five derivatives of caffeoyl-dimethoxy-cinnamoyl quinic acids, and oligomers of ferulic acid. Shi et al. (2015) reported chlorogenic acid (2.0 mg/L) in sunflower oil. In several sunflower oilseeds from Pakistan, 3,4-di-caffeoylquinic acid, 3,5-di-caffeoylquinic acid was the major phenolic compounds (Jan et al. 2018). Ultrasound extraction of sunflower oilseeds cake also showed high amounts of chlorogenic acid (Zardo et al. 2019).

In Italian sunflower oilseed and oil, 3-caffeoylquinic acid, 4-caffeoylquinic acid, p-coumaroylquinic acid, and their derivatives have been reported (Romani et al. 2017). It was also reported that different agricultural farming such as old organic, young organic, and conventional farming affected the concentration of these phenolic compounds. Sonmezdag et al. (2019) reported chlorogenic acid at a concentration of 1.66 mg/kg in Turkish high oleic acid sunflower oil. The amount of chlorogenic acid significantly declined during the frying process. In sunflower oil of Pakistani origin, 5-caffeoylquinic acid was reported at a concentration of 4.49 mg/kg (Zeb and Ullah 2019).

Table 9.2 Contents of esters of hydroxycinnamic acids in oilseed sunflower kernels and shells (mg/kg of DM). Reproduced from the original data with kind permission of Elsevier (Weisz et al. 2009). Data has been converted to mg/kg from the original data represented in mg/100 g

Identity	Oilseed kernels and shells								
	Schilfer	Geiger	Geiger	Italy	Italy	France	France	Germany	Germany
	Oil extraction residue	Kernels	Shells	Kernels	Shells	Kernels	Shells	Kernels	shells
5-Coumaroylquinic acid	890	113	21	52	12	112	19	100	17
5-Feruloylquinic acid	170	113	5.0	62	12	190	5.0	38	8.0
3-Caffeoylquinic acid	3162	4399	29	4237	42	3941	34	2475	48
4-Caffeoylquinic acid	531	875	21	575	24	862	30	863	47
5-Caffeoylqinic acid	20,864	24,670	266	19,457	475	30,505	450	22,714	591
Caffeoylquinic acid	156	365	15	378	37	514	48	387	33
3,4-Di-caffeoylquinic acid	23.5 ± 1.0	288	12	266	11	274	10	290	14
3,5-Di-caffeoylquinic acid	2601	2112	48	2747	81	3325	44	2128	46
4,5-Di-caffeoylquinic acid	1249	1209	26	1350	37	1703	26	1314	43

9.2.4.5 Flavonoids

Quercetin derivatives, quercetin diglycoside, quercetin rutinoside, quercetin gluc-uronide, and flavanone were reported in sunflower seeds (Karamać et al. 2012). Similarly, Zoumpoulakis et al. (2017) reported quercetin-3-rutinoside, luteolin, and eriodictyol in kernels, hulls, and press residues of sunflower oilseeds.

Xuan et al. (2018) reported esculetin (3553 mg/L), isoquercetin (2257 mg/L), fisetin (2900 mg/L), luteolin (862 mg/L), and kaempferol (731 mg/L) in sun-flower oil from the Japanese market. Quercetin-3-galactoside (0.09 mg/kg), kaempferol-3-glucoside (0.13 mg/kg) and quercetin-3-rutinoside (2.7 mg/kg) was reported in high oleic sunflower oil of Turkish origin (Sonmezdag et al. 2019). In sunflower oil of Pakistan, Zeb and Ullah (2019) reported five flavonoids in the form of glycosides. These include spinacetine glucuronide (6.31 mg/kg), isorhamnetin-3-(hydroxyferuloylglucoside)-7-glucoside (0.21 mg/kg), apigenin-2-pentoxide-8-hexoside (1.13 mg/kg), quercetin-3-(sinapoyldiglucoside)-7-glucoside (1.22 mg/kg) and quercetin-3-sinapolysophoroside-7-glucoside (0.22 mg/kg).

9.2.5 Flaxseed Oil

Flaxseed oil is obtained from seeds of the plant (*Linum usitatissimum*). It is also known as linseed oil. The main countries where flaxseed is grown are China, India, Ethiopia, and the USA with a total world production of 2.65 million tonnes (Zeb 2021). The oil is rich in α-linolenic acid (50–60%), and oleic acid (15–25%). Due to the high unsaturation, the oil oxidizes rapidly by forming polymeric structures (Zeb 2019). The oil is used for food processing but the uses have been limited due to the rapid rancidity and low smoke point (107 °C). The oil is also rich in several types of phenolic compounds (Herchi et al. 2014). The amount of phenolic compounds may vary with cultivars, analysis, and extracts such as a defatted extract or non-defatted extract (Bekhit et al. 2018). The polyphenolic compounds may be present in free or bounded with proteins or lipids (Alu'datt et al. 2019).

9.2.5.1 Simple phenols

Flaxseeds from Tunisia showed vanillin as the principal simple phenol (Herchi et al. 2011a). Among the simple phenols, vanillin is present in flaxseed oil (Herchi et al. 2011b). The phenolic extract of flaxseed oil was studied using UPLC, showed vanil-lin in the range of 2–6 µg/100 g of oil (Hasiewicz-Derkacz et al. 2015).

9.2.5.2 Hydroxybenzoic Acids

Hydroxybenzoic acid, gentisic acid, vanillic acid, and sinapic acid had been reported in free as well as bound forms in defatted flour of flaxseed (Dabrowski and Sosulski 1984). Flaxseeds and oils from Tunisian varieties contain *p*-hydroxybenzoic acid and vanillic acid as main hydroxybenzoic acid derivatives (Herchi et al. 2011a, b). Hasiewicz-Derkacz et al. (2015) reported syringic aldehyde in the range of 0.3–0.5 µg/100 g of oil. In Italian flaxseeds oil, the amount of vanillic acid was 0.54 mg/kg and *p*-hydroxybenzoic acid was 1.08 mg/kg of the phenolic dry extract obtained from the oil (Sorice et al. 2016). Tańska et al. (2018) reported that cold-pressed flaxseed oil contains *p*-hydroxybenzoic acid (1.17 µg/100 g) and protocatechuic acid (1.48 µg/100 g).

9.2.5.3 Hydroxycinnamic Acids

Flaxseed contains several derivatives of hydroxycinnamic acids such as *p*-coumaric acid 4-glucoside and ferulic acid 4-glucoside (Johnsson et al. 2002). Flaxseeds were found to contain ferulic acid in higher amounts (Herchi et al. 2011a). In oil of Tunisian cultivars of flaxseed, coumaroyl methyl ester, ferulic acid methyl ester, and ferulic acid had been reported (Herchi et al. 2011b). Ferulic acid was the main fingerprinting phenolic compound in the plants of the Linaceae family (Siger et al. 2008).

In flaxseeds, oil from Poland showed the presence of caffeic acid, coumaric acid, chlorogenic acid, and ferulic acid (Hasiewicz-Derkacz et al. 2015). Similarly, Corbin et al. (2015) reported glucosides of caffeic acid, coumaric acid, and ferulic acid in flaxseeds from France. In Italian flaxseeds oil, the amount of ferulic acid was 1.64 mg/kg of the phenolic dry extract (Sorice et al. 2016). Similarly, Waszkowiak and Gliszczyńska-Świgło (2016) reported *p*-coumaric acid glucoside, caffeic acid glucoside, ferulic acid glucoside, *p*-coumaric acid, ferulic acid, *p*-hydroxybenzoic acid, caffeic acid, *p*-coumaric acid ester, and ferulic acid ester in alkaline and alkaline-acid hydrolysates of flaxseeds.

Caffeic acid, *p*-coumaric acid, and ferulic acid had been reported both in free as well as in bound form in six flaxseed cultivars from China by Wang et al. (2017). The authors reported the range of caffeic acid from 5.18 to 8.7 mg/kg, *p*-coumaric acid from 9.2 to 17.8 mg/kg, ferulic acid from 3.99 to 46.4 mg/kg. In cold-pressed flaxseed oil, coumaric acid (0.23 µg/100 g) was the only reported hydroxycinnamic acid (Tańska et al. 2018). Socrier et al. (2019) reported coumaric acid glucoside in flaxseeds. Recently, caffeic acid, *p*-coumaric acid, and ferulic acid were found in flaxseeds extract (de Magalhães et al. 2020).

9.2.5.4 Flavonoids

Flaxseed and its oil contain several important flavonoids. Catechin, kaempferol, epicatechin, and quercetin-3-rutinoside had been reported in several cultivars of flaxseeds (Kaur et al. 2017). The total bound flavonoid contents in flaxseed oils of six cultivars from China were ranged from 1317 to 2506 mg/kg, whereas the total free flavonoid contents were in the range of 1026–2492 mg/kg as catechin equivalents (Wang et al. 2017).

9.2.5.5 Lignans

Secoisolariciresinol, Pinoresinol, Matairesinol, and Diphyllin were identified in flaxseed oil of Tunisian origin using HPLC–ESI–TOF MS by Herchi et al. (2011b). Coniferyl aldehyde was present in the range of 0.5–3.0 μg/100 g of oil (Hasiewicz-Derkacz et al. 2015). In another study, secoisolariciresinol di-glucoside was reported in French flaxseeds cultivars (Corbin et al. 2015). Secoisolariciresinol, anhydrosecolariciresinol, and pinoresinol were recently reported in flaxseeds by Socrier et al. (2019). Flaxseed oil of different varieties contains several lignans, which provide stability to the oil-in-water nano-emulsion (Cheng et al. 2019).

9.2.6 Sesame Oil

Sesame oil is obtained from sesame (*Sesamum indicum*) seeds. Sesame oil contains high amounts of linoleic acid (40–50%), and oleic acid (30–43%). Historically sesame seed oil was used for cooking and lightening by the Indus Valley Civilization (modern-day Pakistan) (Saeed et al. 2015). The major producer in the world market is now South Asian countries (Latif and Anwar 2011). It is one of the important frying oil in China, South Korea, and India. Sesame oil is rich in tocopherols, while phenolic compounds are also present in sufficient concentration (Konsoula and Liakopoulou-Kyriakides 2010). The seeds cake contains several phenolic compounds and possesses significant antioxidant activity (Mohdaly et al. 2011; Şahin and Elhussein 2018). The presence of sesamol, a phenolic compound has been reported to provide sesame oil with superior oxidative stability (Shahidi et al. 1992). The thermal stability of sesame oil was attributed to the occurrence of lignans and tocopherols (Abou-Gharbia et al. 2000).

9.2.6.1 Simple Phenols

Among the derivatives of simple phenols, sesamol had been widely reported as a fingerprinting phenolic compound in sesame seeds or their oil. Sesamol was present at 2.51 mg/g in sesame oil of Pakistani origin (Zeb et al. 2017). Sesamol is soluble in water and oils and thus can be used as potential additives for foods and edible oils. It has been reported to be present in the range of 2.84–131.38 mg/kg in Chinese varieties and oil blends (Liu et al. 2019).

9.2.6.2 Hydroxybenzoic Acids

The sesame cake from Egypt contains 4-hydroxybenzoic acid, and vanillic acid (Mohdaly et al. 2013). Sesame oil from China contains vanillic acid, gallic acid, and syringic acid (Wu et al. 2016). In sesame oil from Pakistan, the amount of protocatechuic acid was 3.6 mg/kg and syringic acid was 4.31 mg/kg (Zeb et al. 2017). Vanillyl alcohol, vanillic acid, and syringic acid have been reported in sesame oil from the USA (Deme et al. 2018). Similarly, 2-(4-Hydroxyphenyl) ethanol, vanillic acid, and syringic acid have been identified in sesame oil of Chinese origin (Lang et al. 2019). The Egyptian "Giza 32" variety of sesame cake contains gallic acid, protocatechuic acid, sesamol, syringic acid, syringic acid hexoside, vanillic acid, and several derivatives (Mekky et al. 2019). Gallic acid, protocatechuic acid, and 4-hydroxybenzoic acid have been reported in Chinese sesame seeds (Chen et al. 2020a). These results indicated that vanillic acid, syringic acid, and gallic acid are principal hydroxybenzoic acids present in sesame oils of different origins.

9.2.6.3 Hydroxycinnamic Acids

Sesame cake contains cinnamic acid and *p*-coumaric acid as observed by Mohdaly et al. (2013). Coumaric, 2-hydroxycinnamic, caffeic, ferulic, sinapoyl alcohol, and sinapic acids had been reported in Chinese sesame oil (Wu et al. 2016). Similarly, ferulic acid, cinnamic acid, and caffeic acid were reported in Iranian sesame oil (Khezeli et al. 2016). Sesame oil from the USA showed dihydro sinapic acid, *p*-coumaric acid, and ferulic acid (Deme et al. 2018). Caffeic acid, ferulic acid, sinapoyl alcohol, and sinapic acid were present in sesame oil obtained from white and black seeds (Lang et al. 2019). In sesame seed flour, *p*-coumaric acid, ferulic acid, and caffeic acid have been reported (de Magalhães et al. 2020). Mekky et al. (2019) studied sesame seeds using RP-HPLC–DAD–QTOF-MS. The authors reported the presence of coumaric acid, ferulic acid, sinapic acid, and their glycosides among the 86 phenolic compounds. In Chinese sesame seeds, caffeic acid, *p*-coumaric acid, and ferulic acids were the main identified compounds (Chen et al. 2020a). In conclusion, *p*-coumaric acid, sinapic acid, caffeic acid, and ferulic acid are the fingerprinting compounds in sesame oils of different origins.

9.2.6.4 Stilbenes

Wu et al. (2016) reported trans-resveratrol in Chinese sesame oil. Another study from China also showed that sesame oil obtained from the seeds of black and white cultivars of sesame contains trans-resveratrol (Lang et al. 2019).

9.2.6.5 Flavonoids

Apigenin, luteolin, catechin, epicatechin, quercetin, daidzein, genistein, daidzin, genistin, sesamin (Wu et al. 2016). Kaemferol-3-feruloylsophoroside-7-glucoside (2.5 mg/kg), kaempferol-3-(p-coumaroyldiglucoside)-7-glucoside (3.82 mg/kg), quercetin-3-(sinapoyldiglucoside)-7-glucoside (2.93 mg/kg), quercetin-3-galactopyranoside (1.9 mg/kg), quercetin-3-triglucoside (3.24 mg/kg), epicatechin (1.39 mg/kg) and quercetin-3,4-diglucoside-3-(6-feruloy-glucoside) (5.25 mg/kg) was present in sesame oil from Pakistan (Zeb et al. 2017).

Daidzein, apigenin, genistein, catechin, epicatechin, daidzin, and genistin have been found in sesame oils of black and white cultivars (Lang et al. 2019). Luteolin, apigenin, quercetin, and kaempferol were reported both in free as well in glycosides form in sesame seeds (Mekky et al. 2019) suggesting the possible presence of these compounds in the oil. Quercetin was the only flavonoid identified in Chinese sesame seeds (Chen et al. 2020a). Similarly, de Magalhães et al. (2020) have reported quercetin and kaempferol. These results indicated that quercetin, catechin, daidzein, genistein, and apigenin are the principal flavonoids present in sesame oil.

9.2.6.6 Lignans

Sesame oil is rich in important lignans such as sesamin, sesamol, and sesamolin. In Chinese sesame, the amount of these lignans increased significantly with roasting time (Ji et al. 2019). These lignans have reported both in free or bound forms. However, their concentration is higher in free forms (Ha et al. 2017; Zhou et al. 2016). 7-Hydroxymatairesinol (13.9 mg/kg), 7-Hidroxysecoisolariciresinol (0.23 mg/kg), Conidendrin (1.75 mg/kg), Dimethylmatairesinol (1.48 mg/kg), Lariciresinol (2.74 mg/kg), Medioresinol (7.6 mg/kg), Pinoresinol/Matairesinol (17.4 mg/kg), Todolactol A (45.3 mg/kg), sesamin (536.2 mg/kg), sesaminol (26.7 mg/kg), Sesamolinol (14.8 mg/kg) and Syringaresinol (4.73 mg/kg) had been reported in sesame seeds (Ghisoni et al. 2017). Lin et al. (2017) reported that sesamol, sesamin, and sesamolin were present both in free as well as in bound form. Sesaminol triglucoside was reported as a bound form.

Sesaminol, Sesaminol trihexoside, Hydroxysesamolin trihexoside, Sesaminol dipentosides, Pinoresinol dihexosides, and Pinoresinol malonyl dihexosides have been reported recently in sesame seeds (Mekky et al. 2019). Similarly, only sesamin was reported in oils of both black and white cultivars of sesame (Lang et al. 2019). Sesaminol diglucoside has been reported in sesame seed cake with potential antioxidant, anti-collagenase, and anti-hyaluronidase activities (Nantarat et al. 2020). It is concluded that sesame oil is rich in sesamin, sesamolin, and sesaminol as major lignans.

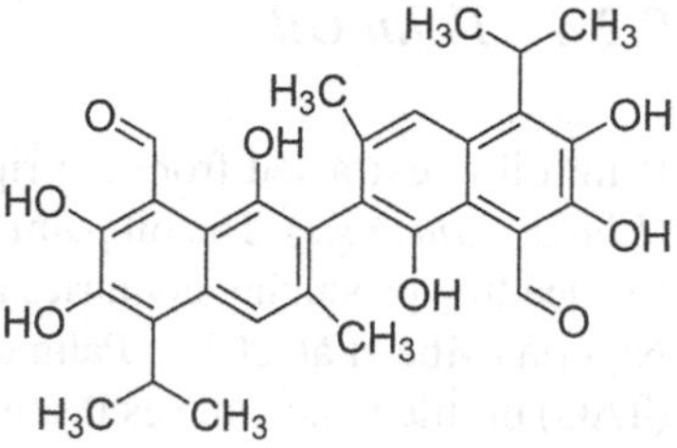

Fig. 9.9 Chemical
structure of Gossypol, a
toxic phenolic compound
present in cottonseed oil

9.2.7 Cottonseed Oil

Cottonseed oil is obtained from seeds of cotton (*Gossypium hirsutum* and *Gossypium herbaceum*). The seed kernel is used for the extraction of oil. It is commonly used for cooking, salad dressing, and the production of mayonnaise. The oil is rich in unsaturated fatty acids (more than 70%). Due to high unsaturation, cottonseed oil has a high smoke point and is thus usually not preferred as a frying medium. Another disadvantage is the presence of a toxic phenolic compound called Gossypol (Fig. 9.9). Thus, genetic modification and technologies are used for reducing the level of gossypol in cottonseed. The commercial benefits of cotton are, however, more than its oil. Most of the studies are based on genetic engineering for the production of cotton (Gao et al. 2020). As for the frying medium, the cooking process such as microwave and thermal heating had enhanced the extraction of total phenolic compounds and decrease total gossypol contents (Taghvaei et al. 2014, 2015). The addition of phenolic compounds such as gallic acid to cottonseed oil has enhanced shelf life, thermal stability, and antioxidant activity (Kurtulbaş et al. 2018). In addition to high unsaturation, cottonseed oil also contain other phenolic compounds but received little attention.

In earlier studies, vanillic acid and *p*-hydroxybenzoic acid had been reported in cottonseed oil (Jones 1979). Besides, gallic acid and 3,4-dihydroxybenzoic acid were identified in whole cottonseed by-products. Glandless cottonseed had been reported to contain kaempferol-3-apiosyl-rhamnosyl-glucoside, quercetin-3-apiosyl rhamnosyl-glucoside, quercetin-3-apiosyl-glucoside, rutin, and quercetin-3-glucoside (Zhang et al. 2001). Another study reported quercetin-3-apiofuranosyl-rhamnopyranosyl-glucopyranoside, kaempferol-3-apiofuranosyl-rhamnopyranosyl-glucopyranoside, quercetin-3-apiofuranosyl-glucopyranoside, quercetin-3-glucopyranoside, kaempferol-3-rhamnopyranosyl-glucopyranoside, quercetin-3-rhamnopyranosyl-glucopyranoside, kaempferol-3-rhamnopyranoside (Piccinelli et al. 2007).

9.2.8 Palm Oil

Palm oil is extracted from the ripened mesocarp of the fruits of the oil palm tree (*Elaeis guineensis*). The oil palm fruit is a drupe formed in spiky tight bunches. The five leading producing countries are Indonesia, Malaysia, Thailand, Colombia, and Nigeria (Mba et al. 2015). Palm oil has a unique fatty acid (FA) and triacylglycerol (TAG) profile which makes it suitable for numerous food applications. It is the only vegetable oil with almost 50–50 composition of saturated and unsaturated fatty acids. Palm oil is the top prime among frying oils. In addition to its unique fatty acid composition, it has a high smoke point of about 230 °C. Palm oil contains minor components that demonstrate major nutritional and health benefits. These micronutrients include carotenoids, tocopherols, tocotrienols, sterols, phospholipids, glycerolipids, and squalene. Palms, in general, are rich in oils, terpenoids, and phenolic compounds.

Agostini-Costa (2018) reviewed the bioactive composition of palm fruits. Phenolic compounds were associated with the total antioxidant capacity of fruits. *p*-Hydroxybenzoic and cinnamic acid derivatives were detected in several species of palms. The predominance of *p*-hydroxybenzoic acid, along with ferulic and traces of *p*-coumaric acid, was found in the fruits of 22 genera of palms (Chakraborty et al. 2006).

Other phenolics derived from benzoic acids, such as catechuic, vanillic, gallic, and syringic acids have been detected in palms, together with their aldehyde forms such as vanillin, and protocatechuic aldehyde, *p*-hydroxybenzoic aldehyde, and syringaldehyde, which were found in the leaves and other tissues of *E. guineensis* (Kawamura et al. 2014). Along with ferulic, other cinnamic acids, such as caffeic, sinapic, and chlorogenic (3-caffeoylquinic acid) and their isomers have been found in the mesocarp of many species (Chakraborty et al. 2006).

Simple patterns of anthocyanins have been observed in palms, in which fruit pigmentation appears to be mainly based on cyanidin-3-*O*-glucoside and cyanidin-3-*O*-rutinoside. Additionally, pelargonidin-3-*O*-rutinoside, peonidin-3-*O*-glucoside, and peonidin-3-*O*-rutinoside were described in the fruits of *E. edulis* (Bicudo et al. 2014), along with cyanidin-3-*O*-sambubioside and pelargonidin 3-*O*-glucoside in the fruits of *E. oleracea*. Peonidin-3-(6″-malonylglucoside) and delphinidin 3-(6″-acetyl) glucoside were reported in *E. oleracea* fruits (Chakraborty et al. 2006). Syringic acid, ferulic acid, *p*-coumaric acid, and ferulic acid were quantified in palm oil using HPLC (Lamarca et al. 2018). Recently, Abdullah and Ramli (2020) studied the phenolic compounds in different palm oil products. Benzoic acid derivatives such as gallic acid, *p*-hydroxybenzoic acid, vanillic acid, and syringic acid in high amounts as compared to cinnamic acids/hydroxycinnamic acid derivatives which were present in lower concentrations were caffeic acid, *p*-coumaric acid, and ferulic acid.

9.3 Study Questions

- Discuss the different classes of polyphenols in soybean oil.
- Write a detailed note on the composition of phenolic compounds in olive oils.
- What do you know about the phenolic profile of sunflower oil?
- Write a detailed note on the flaxseed oil.
- Discuss various classes of phenolic compounds present in sesame oil.

References

Abdullah F, Ramli NAS (2020) Identification of hydrophilic phenolic compounds derived from palm oil products. J Oil Palm Res 32(2):258–270

Abou-Gharbia HA, Shehata AAY, Shahidi F (2000) Effect of processing on oxidative stability and lipid classes of sesame oil. Food Res Int 33(5):331–340

Agostini-Costa TS (2018) Bioactive compounds and health benefits of some palm species traditionally used in Africa and the Americas—a review. J Ethnopharmacol 224:202–229

Alu'datt MH, Rababah T, Ereifej K, Alli I (2013) Distribution, antioxidant and characterisation of phenolic compounds in soybeans, flaxseed and olives. Food Chem 139(1):93–99

Alu'datt MH, Rababah T, Alhamad MN, Al-Mahasneh MA, Almajwal A, Gammoh S, Ereifej K, Johargy A, Alli I (2017) A review of phenolic compounds in oil-bearing plants: distribution, identification and occurrence of phenolic compounds. Food Chem 218:99–106

Alu'datt MH, Rababah T, Alhamad MN, Al-Rabadi GJ, Tranchant CC, Almajwal A, Kubow S, Alli I (2018) Occurrence, types, properties and interactions of phenolic compounds with other food constituents in oil-bearing plants. Crit Rev Food Sci Nutr 58(18):3209–3218

Alu'datt MH, Rababah T, Kubow S, Alli I (2019) Molecular changes of phenolic–protein interactions in isolated proteins from flaxseed and soybean using Native-PAGE, SDS-PAGE, RP-HPLC, and ESI-MS analysis. J Food Biochem 43(5):e12849

Amarowicz R, Shahidi F (1994) Chromatographic separation of glucopyranosyl sinapate from canola meal. J Am Oil Chem Soc 71(5):551–552

Amarowicz R, Naczk M, Shahidi F (2000) Antioxidant activity of crude tannins of canola and rapeseed hulls. J Am Oil Chem Soc 77(9):957

Ambra R, Natella F, Lucchetti S, Forte V, Pastore G (2017) α-Tocopherol, β-carotene, lutein, squalene and secoiridoids in seven monocultivar Italian extra-virgin olive oils. Int J Food Sci Nutr 68(5):538–545

Amiot M-J, Fleuriet A, Macheix J-J (1989) Accumulation of oleuropein derivatives during olive maturation. Phytochemistry 28(1):67–69

Arai S, Suzuki H, Fujimaki M, Sakurai Y (1966) Studies on flavor components in soybean: part II. Phenolic acids in defatted soybean flour. Agric Biol Chem 30(4):364–369

Bartella L, Mazzotti F, Napoli A, Sindona G, Di Donna L (2018) A comprehensive evaluation of tyrosol and hydroxytyrosol derivatives in extra virgin olive oil by microwave-assisted hydrolysis and HPLC-MS/MS. Anal Bioanal Chem 410(8):2193–2201

Bartella L, Mazzotti F, Sindona G, Napoli A, Di Donna L (2020) Rapid determination of the free and total hydroxytyrosol and tyrosol content in extra virgin olive oil by stable isotope dilution analysis and paper spray tandem mass spectrometry. Food Chem Toxicol 136:111110

Bayram B, Esatbeyoglu T, Schulze N, Ozcelik B, Frank J, Rimbach G (2012) Comprehensive analysis of polyphenols in 55 extra virgin olive oils by HPLC-ECD and their correlation with antioxidant activities. Plant Foods Hum Nutr 67(4):326–336

Becerra-Herrera M, Vélez-Martín A, Ramos-Merchante A, Richter P, Beltrán R, Sayago A (2018) Characterization and evaluation of phenolic profiles and color as potential discriminating features among Spanish extra virgin olive oils with protected designation of origin. Food Chem 241:328–337

Bekhit AE-DA, Shavandi A, Jodjaja T, Birch J, Teh S, Mohamed Ahmed IA, Al-Juhaimi FY, Saeedi P, Bekhit AA (2018) Flaxseed: composition, detoxification, utilization, and opportunities. Biocatal Agric Biotechnol 13:129–152

Bellumori M, Cecchi L, Innocenti M, Clodoveo ML, Corbo F, Mulinacci N (2019) The EFSA health claim on olive oil polyphenols: acid hydrolysis validation and total hydroxytyrosol and tyrosol determination in italian virgin olive oils. Molecules 24(11):2179

Bianco A, Coccioli F, Guiso M, Marra C (2002) The occurrence in olive oil of a new class of phenolic compounds: hydroxy-isochromans. Food Chem 77(4):405–411

Bicudo MOP, Ribani RH, Beta T (2014) Anthocyanins, phenolic acids and antioxidant properties of Juçara fruits (*Euterpe edulis* M.) along the on-tree ripening process. Plant Foods Hum Nutr 69(2):142–147

Boskou D (2009) Olive oil: minor constituents and health. CRC Press, Boca Raton

Boskou D (2015a) Olive and olive oil bioactive constituents. Elsevier, Amsterdam

Boskou D (2015b) Olive oil: properties and processing for use in food. In: Talbot G (ed) Specialty oils and fats in food and nutrition: properties, processing and applications, vol 1. Woodhead Publishing, London

Boskou D, Blekas G, Tsimidou M (2006) Olive oil composition. In: Boskou D (ed) Olive oil, 2nd edn. AOCS Press, Champaign, pp 41–72

Cabrini L, Barzanti V, Cipollone M, Fiorentini D, Grossi G, Tolomelli B, Zambonin L, Landi L (2001) Antioxidants and total peroxyl radical-trapping ability of olive and seed oils. J Agric Food Chem 49(12):6026–6032

Cantúa Ayala JA, Olivas AF, Valenzuela Soto JH, Pagaza YR, Hernández Castillo FD, López PF, Torres JM, Chávez ER (2020) Preference for oviposition by sweetpotato whitefly, *bemisia tabaci* (gennadius) in two soybean genotypes, and volatile release. Southwest Entomol 45(1):99–108, 110

Castada HZ, Sun Z, Barringer SA, Huang X (2020) Thermal degradation of p-hydroxybenzoic acid in Macadamia nut oil, olive oil, and corn oil. J Am Oil Chem Soc 97(3):289–300

Chakraborty M, Das K, Dey G, Mitra A (2006) Unusually high quantity of 4-hydroxybenzoic acid accumulation in cell wall of palm mesocarps. Biochem Syst Ecol 34(6):509–513

Chen Y, Lin H, Lin M, Zheng Y, Chen J (2020a) Effect of roasting and in vitro digestion on phenolic profiles and antioxidant activity of water-soluble extracts from sesame. Food Chem Toxicol 139:111239

Chen Y, Shan S, Cao D, Tang D (2020b) Steam flash explosion pretreatment enhances soybean seed coat phenolic profiles and antioxidant activity. Food Chem 319:126552

Cheng C, Yu X, McClements DJ, Huang Q, Tang H, Yu K, Xiang X, Chen P, Wang X, Deng Q (2019) Effect of flaxseed polyphenols on physical stability and oxidative stability of flaxseed oil-in-water nanoemulsions. Food Chem 301:125207

Chew SC (2020) Cold-pressed rapeseed (*Brassica napus*) oil: chemistry and functionality. Food Res Int 131:108997

Cho KM, Ha TJ, Lee YB, Seo WD, Kim JY, Ryu HW, Jeong SH, Kang YM, Lee JH (2013) Soluble phenolics and antioxidant properties of soybean (*Glycine max* L.) cultivars with varying seed coat colours. J Funct Foods 5(3):1065–1076

Cicerale S, Conlan XA, Sinclair AJ, Keast RSJ (2008) Chemistry and health of olive oil phenolics. Crit Rev Food Sci Nutr 49(3):218–236

Cong Y, Cheong L-Z, Huang F, Zheng C, Wan C, Zheng M (2019) Effects of microwave irradiation on the distribution of sinapic acid and its derivatives in rapeseed and the antioxidant evaluation. LWT Food Sci Technol 108:310–318

Corbin C, Fidel T, Leclerc EA, Barakzoy E, Sagot N, Falguiéres A, Renouard S, Blondeau J-P, Ferroud C, Doussot J, Lainé E, Hano C (2015) Development and validation of an efficient ultrasound assisted extraction of phenolic compounds from flax (*Linum usitatissimum* L.) seeds. Ultrason Sonochem 26:176–185

Crizel RL, Hoffmann JF, Zandoná GP, Lobo PMS, Jorge RO, Chaves FC (2020) Characterization of extra virgin olive oil from Southern Brazil. Eur J Lipid Sci Technol 122(4):1900347

Dabrowski KJ, Sosulski FW (1984) Composition of free and hydrolyzable phenolic acids in defatted flours of ten oilseeds. J Agric Food Chem 32(1):128–130

De Leonardis A, Macciola V, Di Rocco A (2003) Oxidative stabilization of cold-pressed sunflower oil using phenolic compounds of the same seeds. J Sci Food Agric 83(6):523–528

De Leonardis A, Macciola V, Di Domenico N (2005) A first pilot study to produce a food antioxidant from sunflower seed shells (*Helianthus annuus*). Eur J Lipid Sci Technol 107(4):220–227

de Magalhães BEA, Santana DA, Silva IMJ, Minho LAC, Gomes MA, Almeida JRGS, Lopes dos Santos WN (2020) Determination of phenolic composition of oilseed whole flours by HPLC-DAD with evaluation using chemometric analyses. Microchem J 155:104683

Deme P, Narasimhulu CA, Parthasarathy S (2018) Identification and evaluation of anti-inflammatory properties of aqueous components extracted from sesame (*Sesamum indicum*) oil. J Chromatogr B 1087–1088:61–69

Di Stefano V, Melilli MG (2020) Effect of storage on quality parameters and phenolic content of Italian extra-virgin olive oils. Nat Prod Res 34(1):78–86

Dias MC, Pinto DCGA, Freitas H, Santos C, Silva AMS (2020) The antioxidant system in *Olea europaea* to enhanced UV-B radiation also depends on flavonoids and secoiridoids. Phytochemistry 170:112199

Dini I, Graziani G, Fedele FL, Sicari A, Vinale F, Castaldo L, Ritieni A (2020) Effects of trichoderma biostimulation on the phenolic profile of extra-virgin olive oil and olive oil by-products. Antioxidants 9(4):284

Farhoosh R, Nystrom L (2018) Antioxidant potency of gallic acid, methyl gallate and their combinations in sunflower oil triacylglycerols at high temperature. Food Chem 244:29–35

Ferrone V, Genovese S, Carlucci M, Tiecco M, Germani R, Preziuso F, Epifano F, Carlucci G, Taddeo VA (2018) A green deep eutectic solvent dispersive liquid-liquid micro-extraction (DES-DLLME) for the UHPLC-PDA determination of oxyprenylated phenylpropanoids in olive, soy, peanuts, corn, and sunflower oil. Food Chem 245:578–585

Franco MN, Galeano-Díaz T, López Ó, Fernández-Bolaños JG, Sánchez J, De Miguel C, Gil MV, Martín-Vertedor D (2014) Phenolic compounds and antioxidant capacity of virgin olive oil. Food Chem 163:289–298

Gao L, Chen W, Xu X, Zhang J, Singh TK, Liu S, Zhang D, Tian L, White A, Shrestha P, Zhou XR, Llewellyn D, Green A, Singh SP, Liu Q (2020) Engineering trienoic fatty acids into cottonseed oil improves low-temperature seed germination, plant photosynthesis and cotton fibre quality. Plant Cell Physiol 61(7):1335–1347

Ghisoni S, Chiodelli G, Rocchetti G, Kane D, Lucini L (2017) UHPLC-ESI-QTOF-MS screening of lignans and other phenolics in dry seeds for human consumption. J Funct Foods 34:229–236

González-Santiago M, Fonollá J, Lopez-Huertas E (2010) Human absorption of a supplement containing purified hydroxytyrosol, a natural antioxidant from olive oil, and evidence for its transient association with low-density lipoproteins. Pharmacol Res 61(4):364–370

Granados-Principal S, Quiles JL, Ramirez-Tortosa CL, Sanchez-Rovira P, Ramirez-Tortosa MC (2010) Hydroxytyrosol: from laboratory investigations to future clinical trials. Nutr Rev 68(4):191–206

Groenbaek M, Tybirk E, Neugart S, Sundekilde UK, Schreiner M, Kristensen HL (2019) Flavonoid glycosides and hydroxycinnamic acid derivatives in baby leaf rapeseed from white and yellow flowering cultivars with repeated harvest in a 2-years field study. Front Plant Sci 10:355

Guo Z, Jia X, Zheng Z, Lu X, Zheng Y, Zheng B, Xiao J (2018) Chemical composition and nutritional function of olive (*Olea europaea* L.): a review. Phytochem Rev 17(5):1091–1110

Ha TJ, Lee MH, Seo WD, Kang JE, Lee JH (2017) Changes occurring in nutritional components (phytochemicals and free amino acid) of raw and sprouted seeds of white and black sesame (*Sesamum indicum* L.) and screening of their antioxidant activities. Food Sci Biotechnol 26(1):71–78

Hachicha Hbaieb R, Kotti F, Gargouri M, Msallem M, Vichi S (2016) Ripening and storage conditions of Chétoui and Arbequina olives: part I. Effect on olive oils volatiles profile. Food Chem 203:548–558

Hammerschmidt PA, Pratt DE (1978) Phenolic antioxidants of dried soybeans. J Food Sci 43(2):556–559

Harwood J, Aparicio R (2000) Handbook of olive oil: analysis and properties. Springer, Berlin

Hashemi P, Nazari Serenjeh F, Ghiasvand AR (2011) Reversed-phase dispersive liquid-liquid microextraction with multivariate optimization for sensitive HPLC determination of tyrosol and hydroxytyrosol in olive oil. Anal Sci 27(9):943–943

Hasiewicz-Derkacz K, Kulma A, Czuj T, Prescha A, Żuk M, Grajzer M, Łukaszewicz M, Szopa J (2015) Natural phenolics greatly increase flax (Linum usitatissimum) oil stability. BMC Biotechnol 15(1):62

Herchi W, Sakouhi F, Boukhchina S, Kallel H, Pepe C (2011a) Changes in fatty acids, tocochromanols, carotenoids and chlorophylls content during flaxseed development. J Am Oil Chem Soc 88(7):1011–1017

Herchi W, Sawalha S, Arráez-Román D, Boukhchina S, Segura-Carretero A, Kallel H, Fernández-Gutierrez A (2011b) Determination of phenolic and other polar compounds in flaxseed oil using liquid chromatography coupled with time-of-flight mass spectrometry. Food Chem 126(1):332–338

Herchi W, Arráez-Román D, Trabelsi H, Bouali I, Boukhchina S, Kallel H, Segura-Carretero A, Fernández-Gutierrez A (2014) Phenolic compounds in flaxseed: a review of their properties and analytical methods. An overview of the last decade. J Oleo Sci 63(1):7–14

Huang S, Wang Q, Wang Y, Ying R, Fan G, Huang M, Agyemang M (2020) Physicochemical characterization and antioxidant activities of Chongqing virgin olive oil: effects of variety and ripening stage. J Food Meas Charact 14(4):2010–2020

Hussain S, Rehman AU, Luckett DJ, Blanchard CL, Obied HK, Strappe P (2020) Phenolic compounds with antioxidant properties from canola meal extracts inhibit adipogenesis. Int J Mol Sci 21(1):1

Issaouia M, Bendini A, Souid S, Flamini G, Barbieri S, Toschi TG, Hammami M (2019) Flavored olive oils: focus on their acceptability and thermal stability. Grasas Aceites 70(1):293

Jan AU, Hadi F, Zeb A, Islam Z (2018) Identification and quantification of phenolic compounds through reversed phase HPLC-DAD method in sunflower seeds under various treatments of potassium nitrate, zinc sulphate and gibberellic acid. J Food Meas Charact 12(1):269–277

Ji J, Liu Y, Shi L, Wang N, Wang X (2019) Effect of roasting treatment on the chemical composition of sesame oil. LWT Food Sci Technol 101:191–200

Johnson LA, White PJ, Galloway R (2008) Soybeans: chemistry, production, processing, and utilization. AOCS Press, Urbana

Johnsson P, Peerlkamp N, Kamal-Eldin A, Andersson RE, Andersson R, Lundgren LN, Åman P (2002) Polymeric fractions containing phenol glucosides in flaxseed. Food Chem 76(2):207–212

Jolayemi OS, Tokatli F, Ozen B (2016) Effects of malaxation temperature and harvest time on the chemical characteristics of olive oils. Food Chem 211:776–783

Jones LA (1979) Gossypol and some other terpenoids, flavonoids, and phenols that affect quality of cottonseed protein. J Am Oil Chem Soc 56(8):727–730

Jun H-I, Wiesenborn DP, Kim Y-S (2014) Antioxidant activity of phenolic compounds from canola (Brassica napus) seed. Food Sci Biotechnol 23(6):1753–1760

Kalua CM, Allen MS, Bedgood DR, Bishop AG, Prenzler PD, Robards K (2007) Olive oil volatile compounds, flavour development and quality: a critical review. Food Chem 100(1):273–286

Karamać M, Kosińska A, Estrella I, Hernández T, Dueñas M (2012) Antioxidant activity of phenolic compounds identified in sunflower seeds. Eur Food Res Technol 235(2):221–230

Kaur R, Kaur M, Gill BS (2017) Phenolic acid composition of flaxseed cultivars by ultra-performance liquid chromatography (UPLC) and their antioxidant activities: effect of sand roasting and microwave heating. J Food Proc Preserv 41(5):e13181

Kawamura F, Saary NS, Hashim R, Sulaiman O, Hashida K, Otsuka Y, Nakamura M, Ohara S (2014) Subcritical water extraction of low-molecular-weight phenolic compounds from oil palm biomass. Japan Agric Res Quart 48(3):355–362

Kelebek H, Kesen S, Selli S (2015) Comparative study of bioactive constituents in turkish olive oils by LC-ESI/MS/MS. Int J Food Prop 18(10):2231–2245

Kesen S, Kelebek H, Selli S (2014) LC–ESI–MS characterization of phenolic profiles Turkish olive oils as influenced by geographic origin and harvest year. J Am Oil Chem Soc 91(3):385–394

Khalatbary AR (2013) Olive oil phenols and neuroprotection. Nutr Neurosci 16(6):243–249

Khezeli T, Daneshfar A, Sahraei R (2016) A green ultrasonic-assisted liquid-liquid microextraction based on deep eutectic solvent for the HPLC-UV determination of ferulic, caffeic and cinnamic acid from olive, almond, sesame and cinnamon oil. Talanta 150:577–585

Kim EH, Kim SH, Chung JI, Chi HY, Kim JA, Chung IM (2006) Analysis of phenolic compounds and isoflavones in soybean seeds (Glycine max (L.) Merill) and sprouts grown under different conditions. Eur Food Res Technol 222(1):201

Konsoula Z, Liakopoulou-Kyriakides M (2010) Effect of endogenous antioxidants of sesame seeds and sesame oil to the thermal stability of edible vegetable oils. LWT Food Sci Technol 43(9):1379–1386

Kotsiou K, Tasioula-Margari M (2016) Monitoring the phenolic compounds of Greek extra-virgin olive oils during storage. Food Chem 200:255–262

Kozlowska H, Zadernowski R, Sosulski FW (1983) Phenolic acids in oilseed flours. Food Nahrung 27(5):449–453

Krishnan V, Rani R, Pushkar S, Lal SK, Srivastava S, Kumari S, Vinutha T, Dahuja A, Praveen S, Sachdev A (2020) Anthocyanin fingerprinting and dynamics in differentially pigmented exotic soybean genotypes using modified HPLC–DAD method. J Food Meas Charact 14:1966–1975

Krygier K, Sosulski F, Hogge L (1982) Free, esterified, and insoluble-bound phenolic acids. 2. Composition of phenolic acids in rapeseed flour and hulls. J Agric Food Chem 30(2):334–336

Kurtulbaş E, Bilgin M, Şahin S (2018) Assessment of lipid oxidation in cottonseed oil treated with phytonutrients: kinetic and thermodynamic studies. Ind Crop Prod 124:593–599

Lamarca RS, Matos RC, Costa Matos MA (2018) Determination of phenolic acids in palm oil samples by HPLC-UV-AD using homemade flow cell. Anal Methods 10(37):4535–4542

Lang H, Yang R, Dou X, Wang D, Zhang L, Li J, Li P (2019) Simultaneous determination of 19 phenolic compounds in oilseeds using magnetic solid phase extraction and LC-MS/MS. LWT Food Sci Technol 107:221–227

Latif S, Anwar F (2011) Aqueous enzymatic sesame oil and protein extraction. Food Chem 125(2):679–684

Lee S-J, Kim J-J, Moon H-I, Ahn J-K, Chun S-C, Jung W-S, Lee O-K, Chung I-M (2008) Analysis of isoflavones and phenolic compounds in Korean soybean [Glycine max (L.) Merrill] seeds of different seed weights. J Agric Food Chem 56(8):2751–2758

Li X, Song L, Xu X, Wen C, Ma Y, Yu C, Du M, Zhu B (2020) One-step coextraction method for flavouring soybean oil with the dried stipe of Lentinus edodes (Berk.) sing by supercritical CO_2 fluid extraction. LWT Food Sci Technol 120:108853

Lin X, Zhou L, Li T, Brennan C, Fu X, Liu RH (2017) Phenolic content, antioxidant and antiproliferative activities of six varieties of white sesame seeds (Sesamum indicum L.). RSC Adv 7(10):5751–5758

Liu Q, Wu L, Pu H, Li C, Hu Q (2012) Profile and distribution of soluble and insoluble phenolics in Chinese rapeseed (Brassica napus). Food Chem 135(2):616–622

Liu W, Zhang K, Yang G, Yu J (2019) A highly efficient microextraction technique based on deep eutectic solvent formed by choline chloride and p-cresol for simultaneous determination of lignans in sesame oils. Food Chem 281:140–146

Lucci P, Bertoz V, Pacetti D, Moret S, Conte L (2020) Effect of the refining process on total hydroxytyrosol, tyrosol, and tocopherol contents of olive oil. Foods 9(3):292

Lukić I, Carlin S, Horvat I, Vrhovsek U (2019) Combined targeted and untargeted profiling of volatile aroma compounds with comprehensive two-dimensional gas chromatography for differentiation of virgin olive oils according to variety and geographical origin. Food Chem 270:403–414

Malenčić D, Cvejić J, Miladinović J (2012) Polyphenol content and antioxidant properties of colored soybean seeds from Central Europe. J Med Food 15(1):89–95

Marković AK, Torić J, Barbarić M, Brala CJ (2019) Hydroxytyrosol, tyrosol and derivatives and their potential effects on human health. Molecules 24(10):2001

Mateos R, Sarria B, Bravo L (2019) Nutritional and other health properties of olive pomace oil. Crit Rev Food Sci Nutr 60(20):3506–3521

Mazzotti F, Benabdelkamel H, Di Donna L, Maiuolo L, Napoli A, Sindona G (2012) Assay of tyrosol and hydroxytyrosol in olive oil by tandem mass spectrometry and isotope dilution method. Food Chem 135(3):1006–1010

Mba OI, Dumont M-J, Ngadi M (2015) Palm oil: processing, characterization and utilization in the food industry—a review. Food Biosci 10:26–41

Mekky RH, Abdel-Sattar E, Segura-Carretero A, Contreras MM (2019) Phenolic compounds from sesame cake and antioxidant activity: a new insight for agri-food residues' significance for sustainable development. Foods 8(10):432

Melguizo-Rodríguez L, Manzano-Moreno FJ, Illescas-Montes R, Ramos-Torrecillas J, de Luna-Bertos E, Ruiz C, García-Martínez O (2019) Bone protective effect of extra-virgin olive oil phenolic compounds by modulating osteoblast gene expression. Nutrients 11(8):1722

Mikołajczak N, Tańska M, Konopka I (2019) Impact of the addition of 4-vinyl-derivatives of ferulic and sinapic acids on retention of fatty acids and terpenoids in cold-pressed rapeseed and flaxseed oils during the induction period of oxidation. Food Chem 278:119–126

Mohdaly AAA, Smetanska I, Ramadan MF, Sarhan MA, Mahmoud A (2011) Antioxidant potential of sesame (Sesamum indicum) cake extract in stabilization of sunflower and soybean oils. Ind Crop Prod 34(1):952–959

Mohdaly AAA, Hassanien MFR, Mahmoud A, Sarhan MA, Smetanska I (2013) Phenolics extracted from potato, sugar beet, and sesame processing by-products. Int J Food Prop 16(5):1148–1168

Naczk M, Amarowicz R, Sullivan A, Shahidi F (1998) Current research developments on polyphenolics of rapeseed/canola: a review. Food Chem 62(4):489–502

Naczk M, Amarowicz R, Pink D, Shahidi F (2000) Insoluble condensed tannins of canola/rapeseed. J Agric Food Chem 48(5):1758–1762

Naim M, Gestetner B, Kirson I, Birk Y, Bondi A (1973) A new isoflavone from soya beans. Phytochemistry 12(1):169–170

Nantarat N, Mueller M, Lin W-C, Lue S-C, Viernstein H, Chansakaow S, Sirithunyalug J, Leelapornpisid P (2020) Sesaminol diglucoside isolated from black sesame seed cake and its antioxidant, anti-collagenase and anti-hyaluronidase activities. Food Biosci 36:100628

Nowak H, Kujawa K, Zadernowski R, Roczniak B, Kozłowska H (1992) Antioxidative and bactericidal properties of phenolic compounds in rapeseeds. Lipid/Fett 94(4):149–152

Paradiso VM, Squeo G, Pasqualone A, Caponio F, Summo C (2019) An easy and green tool for olive oils labelling according to the contents of hydroxytyrosol and tyrosol derivatives: extraction with a natural deep eutectic solvent and direct spectrophotometric analysis. Food Chem 291:1–6

Pedan V, Popp M, Rohn S, Nyfeler M, Bongartz A (2019) Characterization of phenolic compounds and their contribution to sensory properties of olive oil. Molecules 24(11):2041

Pedrosa MM, Muzquiz M, García-Vallejo C, Burbano C, Cuadrado C, Ayet G, Robredo LM (2000) Determination of caffeic and chlorogenic acids and their derivatives in different sunflower seeds. J Sci Food Agric 80(4):459–464

Pereira C, Costa Freitas AM, Cabrita MJ, Garcia R (2020) Assessing tyrosol and hydroxytyrosol in Portuguese monovarietal olive oils: revealing the nutraceutical potential by a combined spectroscopic and chromatographic techniques-based approach. LWT Food Sci Technol 118:108797

Peršurić Ž, Saftić Martinović L, Zengin G, Šarolić M, Kraljević Pavelić S (2020) Characterization of phenolic and triacylglycerol compounds in the olive oil by-product pâté and assay of its antioxidant and enzyme inhibition activity. LWT Food Sci Technol 125:109225

Petrov N, Budinova T, Razvigorova M, Parra J, Galiatsatou P (2008) Conversion of olive wastes to volatiles and carbon adsorbents. Biomass Bioenergy 32(12):1303–1310

Piccinelli AL, Veneziano A, Passi S, Simone FD, Rastrelli L (2007) Flavonol glycosides from whole cottonseed by-product. Food Chem 100(1):344–349

Prabakaran M, Lee J-H, Ahmad A, Kim S-H, Woo K-S, Kim M-J, Chung I-M (2018) Effect of storage time and temperature on phenolic compounds of soybean (*Glycine max* L.) flour. Molecules 23(9):2269

Pratt DE, Birac PM (1979) Source of antioxidant activity of soybeans and soy products. J Food Sci 44(6):1720–1722

Purcaro G, Cordero C, Liberto E, Bicchi C, Conte LS (2014) Toward a definition of blueprint of virgin olive oil by comprehensive two-dimensional gas chromatography. J Chromatogr A, 1334:101–111

Ramírez-Anaya JP, Castañeda-Saucedo MC, Olalla-Herrera M, Villalón-Mir M, de la Serrana HL, Samaniego-Sánchez C (2019) Changes in the antioxidant properties of extra virgin olive oil after cooking typical Mediterranean vegetables. Antioxidants 8(8):246

Ramos-Escudero F, Morales MT, Asuero AG (2015) Characterization of bioactive compounds from monovarietal virgin olive oils: relationship between phenolic compounds-antioxidant capacities. Int J Food Prop 18(2):348–358

Ray NB, Hilsabeck KD, Karagiannis TC, McCord DE (2019) Chapter 36—Bioactive olive oil polyphenols in the promotion of health. In: Singh RB, Watson RR, Takahashi T (eds) The role of functional food security in global health. Academic, New York, pp 623–637

Reboredo-Rodríguez P, Rey-Salgueiro L, Regueiro J, González-Barreiro C, Cancho-Grande B, Simal-Gándara J (2014) Ultrasound-assisted emulsification–microextraction for the determination of phenolic compounds in olive oils. Food Chem 150:128–136

Reiners J, Grosch W (1998) Odorants of virgin olive oils with different flavor profiles. J Agric Food Chem 46(7):2754–2763

Rękas A, Ścibisz I, Siger A, Wroniak M (2017) The effect of microwave pretreatment of seeds on the stability and degradation kinetics of phenolic compounds in rapeseed oil during long-term storage. Food Chem 222:43–52

Riedl KM, Lee JH, Renita M, St Martin SK, Schwartz SJ, Vodovotz Y (2007) Isoflavone profiles, phenol content, and antioxidant activity of soybean seeds as influenced by cultivar and growing location in Ohio. J Sci Food Agric 87(7):1197–1206

Rocchetti G, Senizza B, Giuberti G, Montesano D, Trevisan M, Lucini L (2020) Metabolomic study to evaluate the transformations of extra-virgin olive oil's antioxidant phytochemicals during in vitro gastrointestinal digestion. Antioxidants 9(4):302

Rodríguez-Morató J, Boronat A, Kotronoulas A, Pujadas M, Pastor A, Olesti E, Pérez-Mañá C, Khymenets O, Fitó M, Farré M, de la Torre R (2016) Metabolic disposition and biological significance of simple phenols of dietary origin: hydroxytyrosol and tyrosol. Drug Metab Rev 48(2):218–236

Romani A, Mulinacci N, Pinelli P, Vincieri FF, Cimato A (1999) Polyphenolic content in five tuscany cultivars of *Olea europaea* L. J Agric Food Chem 47(3):964–967

Romani A, Pinelli P, Moschini V, Heimler D (2017) Seeds and oil polyphenol content of sunflower (*Helianthus annuus* l.) grown with different agricultural management. Adv Hortic Sci 31(2):85–88

Romero C, Brenes M (2012) Analysis of total contents of hydroxytyrosol and tyrosol in olive oils. J Agric Food Chem 60(36):9017–9022

Saeed F, Qamar A, Nadeem MT, Ahmed RS, Arshad MS, Afzaal M (2015) Nutritional composition and fatty acid profile of some promising sesame cultivars. Pakistan J Food Sci 25(2):98–103

Şahin S, Elhussein EAA (2018) Assessment of sesame (*Sesamum indicum* L.) cake as a source of high-added value substances: from waste to health. Phytochem Rev 17(4):691–700

Schmidt S, Pokorný J (2005) Potential application of oilseeds as sources of antioxidants for food lipids–a review. Czech J Food Sci 23(3):93–102

Shahbandeh M (2020) Vegetable oils: global consumption by oil type 2013/14 to 2019/2020. Statistica. https://www.statista.com/statistics/263937/vegetable-oils-global-consumption/. Accessed 5 May 2020

Shahidi F (1990) Canola and rapeseed: production, chemistry, nutrition, and processing technology. Springer Science & Business Media, Berlin

Shahidi F, Ambigaipalan P (2015) Phenolics and polyphenolics in foods, beverages and spices: antioxidant activity and health effects—a review. J Funct Foods 18:820–897

Shahidi F, Naczk M (1995) Food phenolics. Technomic Publishing, Lancaster

Shahidi F, Naczk M (2003) Phenolics in food and nutraceuticals. CRC Press, Boca Raton

Shahidi F, Janitha PK, Wanasundara PD (1992) Phenolic antioxidants. Crit Rev Food Sci Nutr 32(1):67–103

Shi Z, Qiu L, Zhang D, Sun M, Zhang H (2015) Dispersive liquid–liquid microextraction based on amine-functionalized Fe_3O_4 nanoparticles for the determination of phenolic acids in vegetable oils by high-performance liquid chromatography with UV detection. J Sep Sci 38(16):2865–2872

Siger A, Józefiak M (2016) The effects of roasting and seed moisture on the phenolic compound levels in cold-pressed and hot-pressed rapeseed oil. Eur J Lipid Sci Technol 118(12):1952–1958

Siger A, Nogala-Kalucka M, Lampart-Szczapa E, Hoffman A (2005) Antioxidant activity of phenolic compounds of selected cold-pressed and refined plant oils. Rośliny Oleiste Oilseed Crops 26(2):549–559

Siger A, Nogala-Kalucka M, Lampart-Szczapa E (2008) The content and antioxidant activity of phenolic compounds in cold-pressed plant oils. J Food Lipids 15(2):137–149

Silva B, Souza MM, Badiale-Furlong E (2020) Antioxidant and antifungal activity of phenolic compounds and their relation to aflatoxin B1 occurrence in soybeans (*Glycine max* L.). J Sci Food Agric 100(3):1256–1264

Singh T, Bhatiya AK, Mishra PK, Srivastava N (2020) An effective approach for the degradation of phenolic waste: phenols and cresols. In: Singh P, Kumar A, Borthakur A (eds) Abatement of environmental pollutants. Elsevier, Amsterdam, pp 203–243

Slavin M, Cheng ZH, Luther M, Kenworthy W, Yu LL (2009) Antioxidant properties and phenolic, isoflavone, tocopherol and carotenoid composition of Maryland-grown soybean lines with altered fatty acid profiles. Food Chem 114(1):20–27

Socrier L, Quéro A, Verdu M, Song Y, Molinié R, Mathiron D, Pilard S, Mesnard F, Morandat S (2019) Flax phenolic compounds as inhibitors of lipid oxidation: elucidation of their mechanisms of action. Food Chem 274:651–658

Song JG, Cao C, Li J, Xu YJ, Liu Y (2019) Development and validation of a QuEChERS-LC-MS/MS method for the analysis of phenolic compounds in rapeseed oil. J Agric Food Chem 67(14):4105–4112

Sonmezdag AS, Kesen S, Amanpour A, Guclu G, Kelebek H, Selli S (2019) LC-DAD-ESI-MS/MS and GC-MS profiling of phenolic and aroma compounds of high oleic sunflower oil during deep-fat frying. J Food Process Preserv 43(3):e13879

Sorice A, Guerriero E, Volpe MG, Capone F, La Cara F, Ciliberto G, Colonna G, Costantini S (2016) Differential response of two human breast cancer cell lines to the phenolic extract from flaxseed oil. Molecules 21(3):319

Taghvaei M, Jafari SM, Assadpoor E, Nowrouzieh S, Alishah O (2014) Optimization of microwave-assisted extraction of cottonseed oil and evaluation of its oxidative stability and physicochemical properties. Food Chem 160:90–97

Taghvaei M, Jafari SM, Nowrouzieh S, Alishah O (2015) The influence of cooking process on the microwave-assisted extraction of cottonseed oil. J Food Sci Technol 52(2):1138–1144

Tańska M, Mikołajczak N, Konopka I (2018) Comparison of the effect of sinapic and ferulic acids derivatives (4-vinylsyringol vs. 4-vinylguaiacol) as antioxidants of rapeseed, flaxseed, and extra virgin olive oils. Food Chem 240:679–685

Thiyam U, Stockmann H, Schwarz K (2006) Antioxidant activity of rapeseed phenolics and their interactions with tocopherols during lipid oxidation. J Am Oil Chem Soc 83(6):523–528

Tsimidou MZ, Nenadis N, Mastralexi A, Servili M, Butinar B, Vichi S, Winkelmann O, García-González DL, Toschi TG (2019) Toward a harmonized and standardized protocol for the determination of total hydroxytyrosol and tyrosol content in virgin olive oil (VOO). The pros of a fit for the purpose ultra high performance liquid chromatography (UHPLC) procedure. Molecules 24(13):2429

Uncu O, Ozen B (2020) Importance of some minor compounds in olive oil authenticity and quality. Trends Food Sci Technol 100:164–176

Vázquez-Velasco M, Esperanza Díaz L, Lucas R, Gómez-Martínez S, Bastida S, Marcos A, Sánchez-Muniz FJ (2011) Effects of hydroxytyrosol-enriched sunflower oil consumption on CVD risk factors. Br J Nutr 105(10):1448–1452

Veillet S, Tomao V, Chemat F (2010) Ultrasound assisted maceration: an original procedure for direct aromatisation of olive oil with basil. Food Chem 123(3):905–911

Wanasundara UN, Shahidi F (1994) Canola extract as an alternative natural antioxidant for canola oil. J Am Oil Chem Soc 71(8):817–822

Wang H, Wang J, Qiu C, Ye Y, Guo X, Chen G, Li T, Wang Y, Fu X, Liu RH (2017) Comparison of phytochemical profiles and health benefits in fiber and oil flaxseeds (*Linum usitatissimum* L.). Food Chem 214:227–233

Wani TA, Masoodi FA, Gani A, Baba WN, Rahmanian N, Akhter R, Wani IA, Ahmad M (2018) Olive oil and its principal bioactive compound: hydroxytyrosol—a review of the recent literature. Trends Food Sci Technol 77:77–90

Waszkowiak K, Gliszczyńska-Świgło A (2016) Binary ethanol–water solvents affect phenolic profile and antioxidant capacity of flaxseed extracts. Eur Food Res Technol 242(5):777–786

Weisz GM, Kammerer DR, Carle R (2009) Identification and quantification of phenolic compounds from sunflower (*Helianthus annuus* L.) kernels and shells by HPLC-DAD/ESI-MSn. Food Chem 115(2):758–765

Wu R, Ma F, Zhang LX, Li PW, Li GM, Zhang Q, Zhang W, Wang XP (2016) Simultaneous determination of phenolic compounds in sesame oil using LC-MS/MS combined with magnetic carboxylated multi-walled carbon nanotubes. Food Chem 204:334–342

Wu G, Chang C, Hong C, Zhang H, Huang J, Jin Q, Wang X (2019) Phenolic compounds as stabilizers of oils and antioxidative mechanisms under frying conditions: a comprehensive review. Trends Food Sci Technol 92:33–45

Xuan TD, Gangqiang G, Minh TN, Quy TN, Khanh TD (2018) An overview of chemical profiles, antioxidant and antimicrobial activities of commercial vegetable edible oils marketed in Japan. Foods 7(2):21

Zardo I, de Espíndola SA, Marczak LDF, Sarkis J (2019) Optimization of ultrasound assisted extraction of phenolic compounds from sunflower seed cake using response surface methodology. Waste Biomass Valori 10(1):33–44

Zardo I, Rodrigues NP, Sarkis JR, Marczak LD (2020) Extraction and identification by mass spectrometry of phenolic compounds from canola seed cake. J Sci Food Agric 100(2):578–586

Zeb A (2011) Effects of β-carotene on the thermal oxidation of fatty acids. Afr J Biotechnol 10(68):15346–15352

Zeb A (2019) Chemistry of the frying medium. In: Food frying: chemistry, biochemistry and safety, 1st edn. Wiley, New York, pp 71–113

Zeb A, Murkovic M (2011) Carotenoids and triacylglycerols interactions during thermal oxidation of refined olive oil. Food Chem 127(4):1584–1593

Zeb A, Murkovic M (2013) Pro-oxidant effects of β-carotene during thermal oxidation of edible oils. J Am Oil Chem Soc 90(6):881–889

Zeb A, Rahman SU (2017) Protective effects of dietary glycine and glutamic acid toward the toxic effects of oxidized mustard oil in rabbits. Food Funct 8(1):429–436

Zeb A, Ullah F (2019) Effects of spinach leaf extracts on quality characteristics and phenolic profile of sunflower oil. Eur J Lipid Sci Technol 121(1):1800325

Zeb A, Muhammad B, Ullah F (2017) Characterization of sesame (*Sesamum indicum* L.) seed oil from Pakistan for phenolic composition, quality characteristics and potential beneficial properties. J Food Meas Charact 11(3):1362–1369

Zeb A (2021) A comprehensive review on different classes of polyphenolic compounds present in edible oils. Food Res Int 143(1):110312

Zhang QJ, Yang M, Zhao Y, Luan X, Ke YG (2001) Isolation and structure identification of flavonol glycosides from glandless cotton seeds. Acta Pharm Sin 36(11):827–831

Zhang M, Zheng C, Yang M, Zhou Q, Li W, Liu C, Huang F (2019) Primary metabolites and polyphenols in rapeseed (*Brassica napus* L.) cultivars in China. J Am Oil Chem Soc 96(3):303–317

Zhou L, Lin X, Abbasi AM, Zheng B (2016) Phytochemical contents and antioxidant and antiproliferative activities of selected black and white sesame seeds. Biomed Res Int 2016:8495630

Zoumpoulakis P, Sinanoglou VJ, Siapi E, Heropoulos G, Proestos C (2017) Evaluating modern techniques for the extraction and characterisation of sunflower (*Hellianthus annus* L.) seeds phenolics. Antioxidants 6(3):46

Chapter 10
Phenolic Antioxidants in Dairy Products

Learning Objectives

In this chapter, the reader will be able to:

- Learn about the phenolic compounds in dairy products.
- Describe in detail the occurrence of simple phenols and their significance in dairy products.
- Understand the applications of phenolic antioxidants in dairy products.

10.1 Introduction

Milk and dairy products are obtained from animals like goats, cows, buffaloes, and camels. Milk is considered a complete diet of having all the required nutrients. Fermented milk, yogurt, butter, cheeses, and creams are the main dairy products produced from the milk. Dairy industries have been contributing a big share to the world market. The total revenue for the year from dairy products was 771,303 million worldwide (Statista 2020). The huge revenue represents the significant consumption of dairy products. Thus, it warrants continuous research on the different aspects to maintain a healthy human life.

Phenolic compounds can be found in dairy milk. The sources of these compounds are plant feeds. O'Connell and Fox (2001) reviewed the presence of phenolic compounds in milk and dairy products and their effects on their production and quality. The authors revealed that certain properties of dairy products such as astringency, beer hazes, specific (dis)coloration, and off-flavors are due to phenolic compounds. The occurrence may be due to the result of several factors, for example, the feeding of particular fodder crops, the catabolism of proteins by bacteria, contamination with sanitizing agents, process induced assimilation, or their deliberate addition as a specific flavoring or functional ingredients. The latter has been reviewed by Alenisan et al. (2017) to possess antioxidant activities. Table 10.1 shows the effects

© Springer Nature Switzerland AG 2021

A. Zeb, *Phenolic Antioxidants in Foods: Chemistry, Biochemistry and Analysis*,
https://doi.org/10.1007/978-3-030-74768-8_10

Table 10.1 Antioxidant activity in dairy products supplemented with plant-based natural additives. Updated with the latest research and part of data are redrawn with kind permission of Elsevier (Alenisan et al. 2017)

Samples	Additives	Method	Antioxidant activity	Reference
Cow milk, camel milk	Glycine max L. (Soybean).	(1,1-Diphenyl-2-picrylhydrazyl radical) (DPPH)	26.4% 61.8% 15.4% 53.1%	Shori (2013)
Yogurt	*Azadirachta indica*	(1,1-Diphenyl-2-picrylhydrazyl radical) (DPPH)	35.9% 53.1%	Shori and Baba (2013)
Cow's milk	Grape residue silage	Reducing power (mg GAE L^{-1})	44.6 mg GAE L^{-1}	Santos et al. (2014)
Yogurt	Pomegranate peel extracts (PPE)	Radical scavenging activity (RSA %) 2,2'-azino-bis (3-ethylbenzothiazoline-6-sulfonic acid) (ABTS) (before and after inoculation with the starter)	19.1% (before) 17.9% (after) 3.18% (before) 1.91% (after) 53.4% (before) 43.8% (after) 14.0% (before) 12.2% (after)	El-Said et al. (2014)
Cow milk yogurt, camel milk yogurt	*Allium sativum*	(1,1-Diphenyl-2-picrylhydrazyl radical) (DPPH)	26.4% 37.9% 15.4% 26.1%	Shori and Baba (2014)
Low-fat cheese	Catechin Control with polyethylene glycol Control (neither catechin nor polyethylene glycol)	FRAP ORAC	14.5 mmol kg^{-1} 215 µmol TE g^{-1} 5 mmol kg^{-1} 110 µmol TE g^{-1} 4.8 mmol kg^{-1} 105 µmol TE g^{-1}	Rashidinejad et al. (2014)
Semi-hard cheese	Coated with dehydrated rosemary leaves Uncoated	Kinetic study (Ozawa's non-isothermal method)	85.9 kJ mol^{-1} 59.0 kJ mol^{-1}	Marinho et al. (2015)
Yogurt	Potassium sorbate (E202) *Foeniculum vulgare* Mill. (fennel) *Matricaria recutita* L. (chamomile)	Reducing power(1,1-diphenyl-2-picrylhydrazyl radical) (DPPH)	32.4 mg mL^{-1} (EC_{50}) 195 mg mL^{-1} (EC_{50}) 29 mg mL^{-1} (EC_{50}) 111 mg mL^{-1} (EC_{50}) 27 mg mL^{-1} (EC_{50}) 94 mg mL^{-1} (EC_{50}) 16.4 mg mL^{-1} (EC_{50}) 45 mg mL^{-1} (EC_{50})	Caleja et al. (2016)
Yogurt	*Ascophyllum nodosum* (0.5%) *Fucus vesiculosus* (0.5%)	(1,1-Diphenyl-2-picrylhydrazyl radical) (DPPH)	20% 32% 47%	O'Sullivan et al. (2016)
Cheese	Green pink pepper Mature pink pepper (*Schinus terebinthifolius Raddi*)	(2,2-Diphenyl-1-picrylhydrazyl) DPPH	21.7 µg mL^{-1} (EC_{50}) 23.5 µg mL^{-1} (EC_{50})	da Silva et al. (2016)

(continued)

Table 10.1　(continued)

Samples	Additives	Method	Antioxidant activity	Reference
Full-fat cheese	Green tea extract (GTE) Without GTE	Ferric reducing antioxidant power (FRAP) Oxygen radical absorbance capacity (ORAC)	13 mmol kg^{-1}(1000 ppm) 240 Teq, mmol g^{-1} (1000 ppm) 8.2 mmol kg^{-1}(1000 ppm) 190 Teq, mmol g^{-1} (1000 ppm)	Rashidinejad et al. (2016)
Raw milk Whole UHT milk Skimmed UHT milk Whole pasteurized milk Skimmed pasteurized milk	Bronopol (as preservatives)	2,2′-Azino-Bis(3-Ethilbenzotiazolina-6 Sulfonic Acid) (ABTS assay)	22.6–38.3 mol L^{-1} TE 24.1–43.0 mol L^{-1} TE 29.3–44.7 mol L^{-1} TE 19.4–39.8 mol L^{-1} TE 22.6–41.4 mol L^{-1} TE	Niero et al. (2017)
Yogurt	Phytomix-3+ mangosteen	(1,1-Diphenyl-2-picrylhydrazyl radical) (DPPH)	40%	Shori et al. (2018)
Yogurt	Apple peel polyphenol extract	DPPH free radical scavenging activity	39.2–67.1%	Ahmad et al. (2020)

of various plant-based extracts on the antioxidant activities of dairy products. The addition of bioactive phenolic compounds helps enhance the shelf life of dairy products (Fazilah et al. 2018). Han et al. (2011) showed that Cheese curds with polyphenolic compounds at a concentration of 0.5 mg/mL showed effective free radical-scavenging activity.

In camel and cow's milk, the addition of soybean extract revealed 15.4–61.7% of DPPH antioxidant activity reported by Shori (2013). *Azadirachta indica* added to yogurt showed an increase in the DPPH radical scavenging activity (Shori and Baba 2013). Grape residue silage was added to cow's milk and showed a reducing power of 44.6 mg GAE L^{-1} (Santos et al. 2014). The pomegranate extract peel extract showed a significant increase in the antioxidant activities of stirred yogurt up to 25% (El-Said et al. 2014). In yogurts made from cow and camel milk, the plain *Allium sativum* extract showed a DPPH antioxidant activity in the range of 15.4–37.9% (Shori and Baba 2014). It was found that the addition of *A. sativum* caused higher antioxidant activities, proteolysis, and inhibition enzymes such as α-amylase and α-glucosidase in camel milk yogurt than in cow milk yogurt. In semi-hard cheese, the addition of rosemary leaves enhances the antioxidant activity (Marinho et al. 2015).

Caleja et al. (2016) reported the decoctions of *Matricaria recutita* L. (chamomile) and *Foeniculum vulgare* Mill. (fennel) in yogurt. The range of DPPH antioxidant activity ranged from 16.4 to 195 mg mL^{-1} as EC$_{50}$. Similarly, the addition of plain *Ascophyllum nodosum* (0.5%) and *Fucus vesiculosus* (0.5%) in yogurt showed

20–47% DPPH activity (O'Sullivan et al. 2016). However, extract enriched yogurt did not protect against DNA damage. In another study, da Silva et al. (2016) showed that the addition of green pink pepper and mature pink pepper enhanced DPPH activity. Similarly, green tea catechins enhanced the antioxidant activity of cheese (Rashidinejad et al. 2014, 2016).

The ABTS antioxidant assay of the raw milk, whole UHT milk, skimmed UHT milk, whole pasteurized milk, and skimmed pasteurized milk having Bronopol were in the range of 19.4–29.3 mol/L of Trolox Equivalent (TE) (Niero et al. 2017). Shori et al. (2018) showed that phytomix-3+ mangosteen added to yogurt significantly enhanced (40%) antioxidant activity than its counterpart plain yogurt. Ahmad et al. (2020) showed that the survival of probiotics and antioxidant activity of the yogurts were enhanced significantly with the addition of apple peel polyphenol extract.

Phenolic compounds are present in milk and its subsequent products. O'Connell and Fox (2001) reviewed the significance and application of PCs on milk and dairy products. It is well-known that animal bodies do not synthesize PCs, but they come from plant-based feeds. *Trifolium repens* and *Medicago sativa* are very effective feed in terms of milk production, quality, and health of the animals (Aerts et al. 1999). PCs present in milk and dairy products which are derived from pasture, animal metabolism, and amino acid catabolism or microbial activity in dairy products are referred to as indigenous PCs while those added directly (intentionally or not) or added during specific processing steps are referred to as exogenous PCs.

Though the significance of indigenous PCs on the sensory attributes of milk and dairy products has not yet been revealed fully, it is proposed that at low levels, indigenous PCs may impart desirable sweet, smoky, or caramel flavor notes to a range of dairy products but at higher levels cause undesirable sensory traits, e.g., sharp, medicinal and sheep yard (O'Connell and Fox 2001). It should be noted that the impact of indigenous PCs on the sensory attributes of dairy products is, on most occasions, likely to be very subtle and when they are present at sub-threshold concentrations they contribute to a delicate balance of flavor components. Also, the impact of indigenous PCs is dependent on the medium, i.e., the contribution of a certain phenolic sensory undertone may be desirable in the piquant taste of Roquefort cheese but undesirable in the milder taste of Cheddar cheese. It has been proposed that the distinct taste of Romano cheese prepared from sheep milk and the desirable background taste of Parmesan cheese are due partly to PCs (Ha and Lindsay 1991). Likely, differences in the PC profile and concentration between milk from different species are responsible for species-related differences in cheeses. Species-related differences in the profile or concentrations of PCs may be a consequence of differences in amino acid catabolism or feeding patterns between ruminants. Differences must be found in the concentration of PCs in herbage and cheeses as a function of pasture altitude or botanical diversity (O'Connell and Fox 2001). Indigenous PCs are also thought to be key determinants in the flavor of smear-ripened cheeses such as Livarot, Maroilles, and Pont-l'Evêque (Urbach 1997). These results indicated that phenolic compounds from plants serve as a functional ingredient in dairy products.

10.2 Phenolic Antioxidants in Dairy Products

10.2.1 Simple Phenols

Thiophenol, phenol, cresols, ethyl phenols, thymol, carvacrol had been reported to be present in bovine, caprine, ovine milk (Lopez and Lindsay 1993). Some of these may also be formed from amino acids. Kilic and Lindsay (2005) showed that alkyl phenol and their conjugates are present in cow, sheep, goat milk as derivatives such as sulfate, phosphate, and glucuronides. Other PCs have been found at lower concentrations, including 4-allylphenol, catechol, 4-(methyl or ethyl) catechol, guaiacol, 4-(methyl, ethyl, or vinyl) guaiacol, and vanillin. Phenol and cresols strongly contributed to the flavor of smoked Provolone cheese (Ha and Lindsay 1991), whereas 3,4-dimethylphenol and cresols were responsible for the specific flavor of sheep milk.

It has also been shown that indigenous PCs, particularly phenol, o- or p-cresol, and guaiacol contribute to the desirable taste of butter and that they are present in butter above their flavor threshold (Stark et al. 1976). The development of a phenolic off-flavor in in-bottle sterilized milk is due to the presence of p-cresol, produced by *Bacillus circulans* (Badings and Neeter 1980), while *Lactobacillus casei* subsp. *alactosus* has been implicated in phenolic flavor defects in Cheddar cheese. It has also been reported that the development of phenolic flavor defects in Gouda cheese is due to p-cresol, produced by an atypical salt-resistant strain of *Lactobacillus*, which originated from rennet that had not been filtered adequately (O'Connell and Fox 2001). The addition of phenol to cheese milk, at 0.1 mg kg^{-1} for Cheddar cheese, or at 1–10 mg kg^{-1} for Camembert, Roquefort, or blue cheeses, as flavoring agents have been patented (Dunn and Lindsay 1985). The sensory properties of a range of simple phenols were listed by Maga (1978). The addition of phenolic preparations at 100 mg kg^{-1} to cheese milk resulted in approximately 60–80% being retained in the curd and imparted a desirable smoky taste to the cheese. The principal PCs in smoked cheese are guaiacol, phenol, eugenol, and syringol (Kornreich and Issenberg 1972).

The presence of Maillard reaction derived PCs (cresol, phenol) in whey powder, concentrated milk powder, lactic casein, and skim milk powder has been reported (Ferretti and Flanagan 1972); their presence in the latter three systems is thought to contribute to off-flavors. Heat treatment of milk (146 °C for 4 s) has been reported to result in the production of vanillin from the thermal oxidation of coniferyl alcohol, which in turn was derived from the microbial degradation of plant lignin by the intestinal microflora of ruminants (Schutte 1999). It has been proposed that vanillin and maltol contribute to the distinct taste of heated milk. A relationship between the presence of PCs in ripe Camembert and heat treatment of cheese milk has been reported (O'Connell and Fox 2001).

The role of vanillin in the flavor of ice cream and yogurt is well established. The interaction of vanillin with milk proteins and the effect thereof on the flavor attributes of these products have not been well investigated. The ability of milk proteins to interact with vanillin and other simple PCs and to reduce their flavor perception has been established (Reiners et al. 2000; Guichard and Langourieux 2000). Hansen and Heinis (1991) proposed that vanillin interacts via a cysteine–aldehyde condensation reaction and/or Schiff base formation, while Reiners et al. (2000) proposed that hydrophobic interactions are principally responsible. Recently, Anantharamkrishnan et al. (2020) showed that several flavor compounds interact with β-lactoglobulin using covalent bonds includes Schiff base, Michael addition, and disulfide linkages.

Another way in which PCs may be incorporated into milk and dairy products is through contamination with disinfectants, insecticides, or herbicides, in which case the PCs are invariably chlorinated. For example, 6-chloro-o-cresol may occasionally be responsible for undesirable flavor taints in milk. 6-Chloro-o-cresol, which is formed on chlorinating water that contains PCs, has a sensory threshold of 0.5 μg kg^{-1} in milk, meaning that 1 mg can taint 20 tonnes of milk (O'Connell and Fox 2001). Other chlorophenols, e.g., 2,4,6- or 2,4,5-trichlorophenol, 2,3,4,5-tetrachlorophenol, and pentachlorophenol, have far higher sensory thresholds, usually approaching 0.1 mg kg^{-1} but under some circumstances may be as low as 0.001 mg kg^{-1}. The principal sources of chlorophenols in milk are udder creams, insecticides, chlorination of water containing phenols (Rêgo et al. 2019; Özdemir et al. 2019). Other routes through which milk and dairy products may become contaminated with PCs include contamination of ingredients such as sugar or phosphates, contamination with boiler water to which algicides have been added, air-borne contamination, and the use of pentachlorophenol as a wood preservative. Chlorophenols impart a phenolic or medicinal taste to milk but they can also be metabolized by microbes to chloroanisoles which impart a musty taste to milk.

10.2.2 Hydroxybenzoic Acids

Hydroxybenzoic acids and their esters have been reported in daily products. In strawberry yogurt, 4-hydroxybenzoic acid was present at a concentration of 201, 1222 ng/mL in cherry yogurt, and 112 ng/mL in strawberry ice-cream (Redeuil et al. 2009). Similarly, gallic acid was present in strawberry ice-cream (434 ng/mL), strawberry yogurt (296 ng/mL), and raspberry yogurt (296 ng/mL). Protocatechuic acid was only reported in cherry yogurt (2501 ng/mL), while vanillic acid was present in plum yogurt (254 ng/mL) and cherry yogurt (510 ng/mL) (Redeuil et al. 2009). Yuan et al. (2013) showed the presence of methyl p-hydroxybenzoic acid ester, ethyl p-hydroxybenzoic acid ester, propyl p-hydroxybenzoic acid ester, and butyl p-hydroxybenzoic acid ester in acidic milk beverage and yogurt using high-pressure liquid chromatography.

10.2.3 Hydroxycinnamic Acids

Hydroxycinnamic acids have been reported in fruit-based yogurts. Redeuil et al. (2009) showed that caffeic acid was present in white grape yogurt (70 ng/mL), strawberry yogurt (80 ng/mL), plum yogurt (162 ng/mL), apricot yogurt (61 ng/mL), raspberry yogurt (338 ng/mL), cherry yogurt (10,464 ng/mL), strawberry ice-cream (270 ng/mL), and apricot ice-cream (284 ng/mL). The amount of 5-caffeoylquinic acid was significantly higher in all samples except grape and strawberry yogurts. Ferulic acid was only found in strawberry ice-cream, whereas *p*-coumaric acid, sinapic acid, and caftaric acid were also present in some yogurts.

The coumaric acid and chlorogenic acids were reported in peach yogurts and were found to provide stability to the yogurts (Oliveira et al. 2017). Phenolic acids have been found to bind to the milk protein (β-casein). Binding of phenolic acids leads to loss of random coil structure of β-casein and weakens the activity of phenolic acids (Li et al. 2020). The strength of the binding constant decreased in the following order: caffeic acid > chlorogenic acid > ferulic acid > syringic acid > gallic acid > 3,4-dihydroxybenzoic acid.

Ianni and Martino (2020) reviewed the efficacy of grape pomace to the dairy cows and their relation with milk quality. The author suggested that "introduction in the diet of lactating cows made it possible to obtain dairy products characterized by improved nutritional properties and high health functionality" (Ianni et al. 2019).

10.2.4 Flavonoids

Flavonoids are major phenolic compounds with several health benefits. They are present in nearly all plant foods (Zeb 2020). Flavonoids are present in dairy products. They are directly sourced to the food supplement in dairy products. Catechin, epicatechin, and epigallocatechin have been reported in different fruits yogurts (Redeuil et al. 2009). In peach yogurt, catechin and quercetin-3-rutinoside were important phenolic compounds playing a significant role in the stability of the product (Oliveira and Pintado 2015; Oliveira et al. 2017). Flavonoids present in the ruminant feed are protective against diseases and also contributed to the quality of milk (Olagaray and Bradford 2019). The addition of naringin to β-casein in milk or dairy product resulted in the formation of naringin-β-casein binding that can serve as a good carrier of naringin (Pacheco et al. 2020).

10.2.5 Proanthocyanidin and Anthocyanidins

The use of anthocyanins as colorants in dairy products such as yogurt and ice cream has been proposed (O'Connell and Fox 2001). Anthocyanins have a red color at acidic pH, little color at neutral pH, and blue color at alkaline pH values. However, at alkaline-neutral pH values, anthocyanins are inherently unstable and are converted to their corresponding carbinol bases, which are colorless. Consequently, the use of most anthocyanins as colorants in dairy products with an alkaline or neutral pH is possible only if the product is stored at a low temperature, e.g., ice cream. However, anthocyanins that have been acetylated with caffeic acid are stable at neutral pH values and thus are suitable colorants in an array of dairy products. A commercially available anthocyanin-rich extract from grape skin, Enocianina, has been suggested as a suitable colorant for yogurt (O'Connell and Fox 2001).

Proanthocyanidins and anthocyanidins have been reported in dairy products. For example, Redeuil et al. (2009) reported a high amount of procyanidin B1 in strawberry ice-cream (3486 ng/mL) and in strawberry yogurt (1798 ng/mL), whereas and procyanidin B2 was reported only in cherry yogurt. Similarly, Oliveira and Pintado (2015) showed the presence of pelargonidin-3-glucoside and pelargonidin-3-rutinoside in peach yogurt. Recently Simitzis et al. (2019) showed that supplementation of hesperidin or naringin in the diet of dairy ewes at doses of 6000 mg/kg showed an increase in the oxidative stability of milk without affecting milk quality.

10.3 Applications of Phenolic Antioxidants in Dairy Products

Phenolic compounds are present in milk and its subsequent products, which are either metabolite of the feeds or added in the products. Phenolic antioxidants have several applications in foods including dairy products (Zeb 2020). O'Connell and Fox (2001) reviewed the significance and application of PCs on milk and dairy products. Similarly, Cutrim and Cortez (2018) also reviewed the applications of phenolic compounds in dairy products. It is well-known that animal bodies do not synthesize PCs, but they come from plant-based feeds. *Trifolium repens* and *Medicago sativa* are very effective feeds in terms of milk production, quality, and health of the animals (Aerts et al. 1999). Thiophenol, phenol, cresols, ethyl phenols, thymol, carvacrol had been reported to be present in bovine, caprine, ovine milk (Lopez and Lindsay 1993). Some of these may also be formed from amino acids. Dairy milk is thus a good source of several PCs including equol and phenolic antioxidants (Tsen et al. 2014). These secondary metabolites are formed by the bovine's gut bacterial flora from the PCs present in the feed. The lipid concentration of the milk was positively correlated with PCs. Kilic and Lindsay (2005) showed that alkyl phenol and their conjugates are present in cow, sheep, goat milk as derivatives such as sulfate, phosphate, and glucuronides.

The effects of dietary PCs on milk quality and composition have been reported to be significantly affected. Propolis extracts have been shown to improve milk quality when supplemented with the diet of dairy cows (Aguiar et al. 2014). However, different PC have different effects on milk quality. The differences in the phenolic profile are perhaps produced by variations in the feed (Lopez and Lindsay 1993). For examples, when cows were feed high levels of particular crops, other PCs may also be identified in the milk, i.e., ptaquiloside from *Pteridium aquilinum* at up to 50 mg/L (Alonso-Amelot et al. 1996) or genistein up to 30 μg/L, equol up to 300 μg/L and daidzein up to 5 μg/L (King et al. 1998). The latter compounds were found to be all derived from clover. Some feed or plant such as grape pomace may not produce any effects on the milk composition, but an induced modification in the fatty acid profile of the milk (Ianni et al. 2019). Similarly, in the diet of dairy goats, pistachio by-products had been used as forage, which had beneficial effects on the fatty acid profile of milk (Sedighi-Vesagh et al. 2015; Ghaffari et al. 2014). Hazelnut as an ingredient in sheep dietary feed showed significant improvement in milk quality and sensory attributes (Caccamo et al. 2019).

The extracts of four grape varieties and grape callus rich in PCs were added into yogurt as functional ingredients (Karaaslan et al. 2011). The results had shown that the callus culture of grape has the potential to be used as a food supplement. Chouchouli et al. (2013) showed that fortification of grape seed extract to yogurt had enhanced antioxidant activity. A slow decline of simple and total phenolic contents occurred with storage time both in full fat and non-fat yogurts. However, polyphenols were present in all yogurts after 32 days of storage. A similar study had reported grape peel extract was effective at 20–25% in stirred yogurt (El-Said et al. 2014).

Due to the antioxidant nature of PCs, they are used as food additives. Milk supplemented with polyphenols from grape seeds and apple extracts had been shown to suppress flavor characteristics (Axten et al. 2008). It was also reported that the apple extracts have large flavor profile implications for product development due to their high level of bitterness. O'Connell and Fox (1999) studied the effects of different PCs on the thermal stability of milk. They found that some PCs did not disturb the heat stability of milk such as guaiacol, thymol, chlorogenic acid, vanillin, BHA, propyl gallate, and BHT. A significant reduction in the stability of skim milk was shown by quinic acid. Caffeic acid, vanillic acid and ferulic acid were found to enhanced thermal stability.

PCs when added to cheese curd, showed high retention coefficient values (Han et al. 2011). The retention coefficient values were highest in flavone, followed by hesperetin, and lowest in catechin added cheese as compared to control. The selected PCs showed three times higher values of antiradical activity than control with PCs suggesting beneficial effects. Similarly, PCs present in the extract of *Matricaria recutita* L. (chamomile) were incorporated into cottage cheese (Caleja et al. 2015). It was found that these PCs enhanced the antioxidant activity and storage stability of cheese. Different compounds or plant extracts have been utilized for the enrichment of dairy products (Table 10.2). For example, the essential oil of *Mintha spicata* in white cheese (Fadavi and Beglaryan 2015), phenolic compounds from tea in cheese (Han et al. 2007, 2011), extract of grape in yogurt (Karaaslan et al. 2011), olive oil waste phenols in yogurt (Petrotos et al. 2012), grape seed extract in yogurt (Chouchouli et al. 2013), green tea extract in cheese (Giroux et al. 2013), catechin in cheese (Rashidinejad et al. 2014), blackcurrant extract and cyanidin 3-glucopyranoside chloride in yogurt (Sun-Waterhouse et al. 2013), wine grape pomace in yogurt (Tseng and Zhao 2013), white grape polyphenols in Chihuahua cheese (Pimentel-González et al. 2015), catechin and epigallocatechin gallate

Table 10.2 Overview of recent applications of polyphenols in dairy products. Updated and reproduced with kind permission of John Wiley & Sons (Cutrim and Cortez 2018)

Dairy product	Compounds and utilization	Reference
White cheese	Essential oil of *Mintha spicata* L (0.5, 0.75, 1.0, 1.5, 2.0 and 2.5 mL/kg)	Fadavi and Beglaryan (2015)
Cheese	Catechin, tannic acid, homovanillic acid, hesperetin, epigallocatechin gallate, 2-phenylchromone, extracts of whole grape and green tea, and dehydrated cranberry juice powder (0.5 mg/mL)	Han et al. (2007)
Yogurt	Extracts of grape varieties (Cabernet Sauvignon, Chardonnay, Syrah, and Merlot) and grape callus (1% v/v)	Karaaslan et al. (2011)
Traditional Greek-type and European-type sheep's yogurts	Olive mill wastewater polyphenols (500 ppm), nonencapsulated and starch encapsulated forms	Petrotos et al. (2012)
Full-fat and nonfat yogurt	Seed extracts of grapes Agiorgitiko and Moschofilero (0.27, 0.53 or 0.67 mg/mL)	Chouchouli et al. (2013)
Cheddar-type cheese	Green tea extract (1 and 2 g/kg)	Giroux et al. (2013)
Low-fat hard cheese	Catechin in polyethylene glycol (125, 250 and 500 ppm)	Rashidinejad et al. (2014)
Drinking yogurt	Blackcurrant extract and cyanidin 3-o-b-glucopyranoside chloride (2% BPE and 0.15% Cyanidin)	Sun-Waterhouse et al. (2013)
Low-fat yogurt	Wine grape pomace (*Vitis vinifera* L. cv Pinot Noir; 1%, 2% or 3% w/w) and liquid extract or freeze-dried extract (2%)	Tseng and Zhao (2013)
Chihuahua cheese	White grape polyphenols free and encapsulated by multiple emulsions	Pimentel-González et al. (2015)
Low-fat hard cheese	Catechin and epigallocatechin gallate encapsulated within soy lecithin liposomes (250 ppm each)	Rashidinejad et al. (2014)
Yogurt	Skins of hazelnut (*Corylus avellana* L.) var. ('Tonda Gentile Trilobata', 'San Giovanni' and 'Georgia)' (3% and 6%)	Bertolino et al. (2015)
Serra da Estrela cheese	Dried chestnut (*Castanea sativa* Mill.) flowers (248 mg/cheese) and its decoction extracts (799 mg/cheese) and lemon balm (*Melissa officinalis*; 368 mg/cheese) and its decoction extracts (380 mg/cheese)	Carocho et al. (2015)
Ultrafiltered Feta cheese	Peppermint extract (220, 440, and 660 µg/g cheese)	Fadavi and Beglaryan (2015)
Cheese	Whole grape, grape seed and grape skin extracts (0.1%, 0.2% and 0.3% w/v)	Felix da Silva et al. (2015)
Low-fat yoghurt	*Pleurotus ostreatus* aqueous extract (0.25%, 0.50%, 0.75% and 1%)	Vital et al. (2015)

(continued)

Table 10.2 (continued)

Dairy product	Compounds and utilization	Reference
Fermented goat's milk	Grape pomace extract (2%)	Dos Santos et al. (2017)
Cheddar cheese	Green tea extract (0.1% w/w)	Lamothe et al. (2016)
Cheddar-type cheese	*Inula britannica* extract (0.25%, 0.5%, 0.75% and 1% w/v)	Lee et al. (2016)
Probiotic yogurt	Green, white and black tea extract (2% w/w)	Muniandy et al. (2016)
Cottage cheese	Rosemary extract (0.1 g of free and 0.9 g of alginate-base microencapsulated in 200 g of cheese)	Ribeiro et al. (2016)
Robiola-type cheese	Barbera and Chardonnay skin powders (0.8%, 1.6% and 2.4% w/w)	Torri et al. (2016)
Yogurt	Freeze-dried stevia extract (0.25% and 0.5%)	de Carvalho et al. (2019)
Cheese	Different dates aqueous extract	Qureshi et al. (2019)
Cheese	Solid or molten chocolate 0%, 5%, 10%, and 15% (w/w)	Ashkezary et al. (2020)
Milk, yogurt, and Kefir	Propolis extract (0%, 0.5%, 1.0%, 1.5%, and 2.0%).	Chon et al. (2020)

encapsulated within soy lecithin liposomes in cheese (Rashidinejad et al. 2014), Skins of hazelnut in yogurt (Bertolino et al. 2015), dried chestnut flowers and its decoction extracts and lemon balm and its decoction extracts in cheese (Carocho et al. 2015), Peppermint extract in cheese (Fadavi and Beglaryan 2015), whole grape, grape seed and grape skin extracts in cheese (Felix da Silva et al. 2015), *Pleurotus ostreatus* aqueous extract in yogurt (Vital et al. 2015), grape pomace extract in fermented goat's milk (Dos Santos et al. 2017), green tea extract in Cheddar cheese (Lamothe et al. 2016), *Inula britannica* extract in cheese (Lee et al. 2016), green, white and black tea extract in yogurt (Muniandy et al. 2016), rosemary extract in cottage cheese (Ribeiro et al. 2016), grape skin extracts in cheese (Torri et al. 2016), freeze-dried stevia extract in yogurt (de Carvalho et al. 2019), different dates aqueous extract in soft cheese (Qureshi et al. 2019), solid or molten chocolate in cheese (Ashkezary et al. 2020), and propolis extract in dairy products (Chon et al. 2020). In conclusion, the use of PCs in different dairy products showed beneficial effects.

The use of PCs as antimicrobial agents began with Lister in 1867 and today, PCs are used as antimicrobial agents in an array of products, including food, paint, leather, metalworking fluids, textiles, and petroleum (O'Connell and Fox 2001). The ability of an array of PCs, e.g., ferulic acid tea catechins, oleuropein, ellagic acid, and *p*-coumaric acid, to inhibit the growth of bacteria (*Salmonella enteritidis*, *Staphylococcus aureus*, *Listeria monocytogenes*) and fungi in milk have been reported (Van het Hof et al. 1998). The antimicrobial effect of PCs, which is not exclusive to milk systems is probably related to the inhibition of bacterial enzymes, alterations in cell wall permeability, an increase in the hydrogen ion activity of the microbial environment, a reduction in the surface and/or interfacial tension and perhaps most importantly, chelation of essential minerals, particularly iron with concomitant impairment of the microbial oxidative metabolic system (Chung et al. 1998). Interestingly, Rosenthal et al. (1997) reported that tea catechins and ferulic acid inhibit the growth of pathogenic bacteria (coliforms and *Salmonella*) with little effect on lactic acid bacteria, presumably because of the fermentative metabolism of the latter.

10.4   Study Questions

- What do you know about the phenolic compounds in dairy products?
- Describe in detail the occurrence of simple phenols and their significance in dairy products.
- Write a note on the application of phenolic antioxidants in dairy products.

References

Aerts RJ, Barry TN, McNabb WC (1999) Polyphenols and agriculture: beneficial effects of proanthocyanidins in forages. Agric Ecosyst Environ 75(1):1–12

Aguiar SC, Cottica SM, Boeing JS, Samensari RB, Santos GT, Visentainer JV, Zeoula LM (2014) Effect of feeding phenolic compounds from propolis extracts to dairy cows on milk production, milk fatty acid composition, and the antioxidant capacity of milk. Anim Feed Sci Technol 193:148–154

Ahmad I, Khalique A, Shahid MQ, Ahid Rashid A, Faiz F, Ikram MA, Ahmed S, Imran M, Khan MA, Nadeem M, Afzal MI, Umer M, Kaleem I, Shahbaz M, Rasool B (2020) Studying the influence of apple peel polyphenol extract fortification on the characteristics of probiotic yoghurt. Plants (Basel) 9(1):77

Alenisan MA, Alqattan HH, Tolbah LS, Shori AB (2017) Antioxidant properties of dairy products fortified with natural additives: a review. J Assoc Arab Univ Basic Appl Sci 24:101–106

Alonso-Amelot ME, Castillo U, Smith BL, Lauren DR (1996) Bracken ptaquiloside in milk. Nature 382(6592):587–587

Anantharamkrishnan V, Hoye T, Reineccius GA (2020) Covalent adduct formation between flavor compounds of various functional group classes and the model protein β-lactoglobulin. J Agric Food Chem 68(23):6395–6402

Ashkezary MR, Bonanno A, Todaro M, Settanni L, Gaglio R, Todaro A, Alabiso M, Maniaci G, Mazza F, Grigoli AD (2020) Effects of adding solid and molten chocolate on the physicochemical, antioxidant, microbiological, and sensory properties of ewe's milk cheese. J Food Sci 85(3):556–566

Axten L, Wohlers M, Wegrzyn T (2008) Using phytochemicals to enhance health benefits of milk: impact of polyphenols on flavor profile. J Food Sci 73(6):H122–H126

Badings H, Neeter R (1980) Recent advances in the study of aroma compounds of milk and dairy products. Neth Milk Dairy J 34(1):9–30

Bertolino M, Belviso S, Dal Bello B, Ghirardello D, Giordano M, Rolle L, Gerbi V, Zeppa G (2015) Influence of the addition of different hazelnut skins on the physicochemical, antioxidant, polyphenol and sensory properties of yogurt. LWT Food Sci Technol 63(2):1145–1154

Caccamo M, Valenti B, Luciano G, Priolo A, Rapisarda T, Belvedere G, Marino VM, Esposto S, Taticchi A, Servili M, Pauselli M (2019) Hazelnut as ingredient in dairy sheep diet: effect on sensory and volatile profile of cheese. Front Nutr 6:125

Caleja C, Barros L, Antonio AL, Ciric A, Barreira JCM, Sokovic M, Oliveira MBPP, Santos-Buelga C, Ferreira ICFR (2015) Development of a functional dairy food: exploring bioactive and preservation effects of chamomile (*Matricaria recutita* L.). J Funct Foods 16:114–124

Caleja C, Barros L, Antonio AL, Carocho M, Oliveira MBPP, Ferreira ICFR (2016) Fortification of yogurts with different antioxidant preservatives: a comparative study between natural and synthetic additives. Food Chem 210:262–268

Carocho M, Barreira JC, Antonio AL, Bento A, Morales P, Ferreira IC (2015) The incorporation of plant materials in "Serra da Estrela" cheese improves antioxidant activity without changing the fatty acid profile and visual appearance. Eur J Lipid Sci Technol 117(10):1607–1614

Chon J-W, Seo K-H, Oh H, Jeong D, Song K-Y (2020) Chemical and organoleptic properties of some dairy products supplemented with various concentration of propolis: a preliminary study. J Dairy Sci Biotechnol 38(2):59–69

Chouchouli V, Kalogeropoulos N, Konteles SJ, Karvela E, Makris DP, Karathanos VT (2013) Fortification of yoghurts with grape (*Vitis vinifera*) seed extracts. LWT Food Sci Technol 53(2):522–529

Chung K-T, Wei C-I, Johnson MG (1998) Are tannins a double-edged sword in biology and health? Trends Food Sci Technol 9(4):168–175

Cutrim CS, Cortez MAS (2018) A review on polyphenols: classification, beneficial effects and their application in dairy products. Int J Dairy Technol 71(3):564–578

da Silva DG, Funck GD, Mattei FJ, da Silva WP, Fiorentini ÂM (2016) Antimicrobial and anti-oxidant activity of essential oil from pink pepper tree (*Schinus terebinthifolius* Raddi) in vitro and in cheese experimentally contaminated with *Listeria monocytogenes*. Innovative Food Sci Emerg Technol 36:120–127

de Carvalho MW, Arriola NDA, Pinto SS, Verruck S, Fritzen-Freire CB, Prudêncio ES, Amboni RDMC (2019) Stevia-fortified yoghurt: stability, antioxidant activity and in vitro digestion behaviour. Int J Dairy Technol 72(1):57–64

Dos Santos KM, de Oliveira IC, Lopes MA, Cruz APG, Buriti FC, Cabral LM (2017) Addition of grape pomace extract to probiotic fermented goat milk: the effect on phenolic content, probiotic viability and sensory acceptability. J Sci Food Agric 97(4):1108–1115

Dunn HC, Lindsay RC (1985) Evaluation of the role of microbial strecker-derived aroma compounds in unclean-type flavors of cheddar cheese. J Dairy Sci 68(11):2859–2874

El-Said MM, Haggag HF, Fakhr El-Din HM, Gad AS, Farahat AM (2014) Antioxidant activities and physical properties of stirred yoghurt fortified with pomegranate peel extracts. Ann Agric Sci 59(2):207–212

Fadavi A, Beglaryan R (2015) Optimization of UF-Feta cheese preparation, enriched by peppermint extract. J Food Sci Technol 52(2):952–959

Fazilah NF, Ariff AB, Khayat ME, Rios-Solis L, Halim M (2018) Influence of probiotics, prebiotics, synbiotics and bioactive phytochemicals on the formulation of functional yogurt. J Funct Foods 48:387–399

Felix da Silva D, Matumoto-Pintro PT, Bazinet L, Couillard C, Britten M (2015) Effect of commercial grape extracts on the cheese-making properties of milk. J Dairy Sci 98(3):1552–1562

Ferretti A, Flanagan VP (1972) Steam volatile constituents of stale nonfat dry milk. Role of the Maillard reaction in staling. J Agric Food Chem 20(3):695–698

Ghaffari MH, Tahmasbi AM, Khorvash M, Naserian AA, Vakili AR (2014) Effects of pistachio by-products in replacement of alfalfa hay on ruminal fermentation, blood metabolites, and milk fatty acid composition in Saanen dairy goats fed a diet containing fish oil. J Appl Anim Res 42(2):186–193

Giroux HJ, De Grandpré G, Fustier P, Champagne CP, St-Gelais D, Lacroix M, Britten MJDS (2013) Production and characterization of Cheddar-type cheese enriched with green tea extract. Dairy Sci Technol 93(3):241–254

Guichard E, Langourieux S (2000) Interactions between β-lactoglobulin and flavour compounds. Food Chem 71(3):301–308

Ha JK, Lindsay RC (1991) Volatile branched-chain fatty acids and phenolic compounds in aged italian cheese flavors. J Food Sci 56(5):1241–1247

Han X, Shen T, Lou H (2007) Dietary polyphenols and their biological significance. Int J Mol Sci 8(9):950–988

Han J, Britten M, St-Gelais D, Champagne CP, Fustier P, Salmieri S, Lacroix M (2011) Polyphenolic compounds as functional ingredients in cheese. Food Chem 124(4):1589–1594

Hansen AP, Heinis JJ (1991) Decrease of vanillin flavor perception in the presence of casein and whey proteins. J Dairy Sci 74(9):2936–2940

Ianni A, Martino G (2020) Dietary grape pomace supplementation in dairy cows: effect on nutritional quality of milk and its derived dairy products. Foods 9(2):168

Ianni A, Di Maio G, Pittia P, Grotta L, Perpetuini G, Tofalo R, Cichelli A, Martino G (2019) Chemical–nutritional quality and oxidative stability of milk and dairy products obtained from Friesian cows fed with a dietary supplementation of dried grape pomace. J Sci Food Agric 99(7):3635–3643

Karaaslan M, Ozden M, Vardin H, Turkoglu H (2011) Phenolic fortification of yogurt using grape and callus extracts. LWT Food Sci Technol 44(4):1065–1072

Kilic M, Lindsay RC (2005) Distribution of conjugates of alkylphenols in milk from different ruminant species. J Dairy Sci 88(1):7–12

King RA, Mano MM, Head RJ (1998) Assessment of isoflavonoid concentrations in Australian bovine milk samples. J Dairy Res 65(3):479–489

Kornreich MR, Issenberg P (1972) Determination of phenolic wood smoke components as trimethylsilyl ethers. J Agric Food Chem 20(6):1109–1113

Lamothe S, Langlois A, Bazinet L, Couillard C, Britten M (2016) Antioxidant activity and nutrient release from polyphenol-enriched cheese in a simulated gastrointestinal environment. Food Funct 7(3):1634–1644

Lee NK, Jeewanthi RKC, Park EH, Paik HD (2016) Physicochemical and antioxidant properties of Cheddar-type cheese fortified with *Inula britannica* extract. J Dairy Sci 99(1):83–88

Li T, Li X, Dai T, Hu P, Niu X, Liu C, Chen J (2020) Binding mechanism and antioxidant capacity of selected phenolic acid—β-casein complexes. Food Res Int 129:108802

Lopez V, Lindsay RC (1993) Metabolic conjugates as precursors for characterizing flavor compounds in ruminant milks. J Agric Food Chem 41(3):446–454

Maga JA (1978) Simple phenol and phenolic compounds in food flavor. CRC Crit Rev Food Sci Nutr 10(4):323–372

Marinho MT, Bersot LS, Nogueira A, Colman TAD, Schnitzler E (2015) Antioxidant effect of dehydrated rosemary leaves in ripened semi-hard cheese: a study using coupled TG–DSC–FTIR (EGA). LWT Food Sci Technol 63(2):1023–1028

Muniandy P, Shori AB, Baba AS (2016) Influence of green, white and black tea addition on the antioxidant activity of probiotic yogurt during refrigerated storage. Food Packag Shelf Life 8:1–8

Niero G, Penasa M, Currò S, Masi A, Trentin AR, Cassandro M, De Marchi M (2017) Development and validation of a near infrared spectrophotometric method to determine total antioxidant activity of milk. Food Chem 220:371–376

O'Connell JE, Fox PF (1999) Effects of phenolic compounds on the heat stability of milk and concentrated milk. J Dairy Res 66(3):399–407

O'Connell JE, Fox PF (2001) Significance and applications of phenolic compounds in the production and quality of milk and dairy products: a review. Int Dairy J 11(3):103–120

O'Sullivan AM, O'Grady MN, O'Callaghan YC, Smyth TJ, O'Brien NM, Kerry JP (2016) Seaweed extracts as potential functional ingredients in yogurt. Innovative Food Sci Emerg Technol 37:293–299

Olagaray KE, Bradford BJ (2019) Plant flavonoids to improve productivity of ruminants—a review. Anim Feed Sci Technol 251:21–36

Oliveira A, Pintado M (2015) Stability of polyphenols and carotenoids in strawberry and peach yoghurt throughout in vitro gastrointestinal digestion. Food Funct 6(5):1611–1619

Oliveira A, Coelho M, Alexandre EMC, Pintado M (2017) Impact of storage on phytochemicals and milk proteins in peach yoghurt. J Food Meas Charact 11(4):1804–1814

Özdemir C, Özdemir S, Oz E, Oz F (2019) Determination of organochlorine pesticide residues in pasteurized and sterilized milk using QuEChERS sample preparation followed by gas chromatography–mass spectrometry. J Food Process Preserv 43(11):e14173

Pacheco AFC, Nunes NM, de Paula HMC, Coelho YL, da Silva LHM, Pinto MS, Pires ACS (2020) β-casein monomers as potential flavonoids nanocarriers: thermodynamics and kinetics of β-casein-naringin binding by fluorescence spectroscopy and surface plasmon resonance. Int Dairy J 108:104728

Petrotos KB, Karkanta FK, Gkoutsidis PE, Giavasis I, Papatheodorou KN, Ntontos AC (2012) Production of novel bioactive yogurt enriched with olive fruit polyphenols. World Acad Sci Eng Technol 64:867–872

Pimentel-González D, Aguilar-García M, Aguirre-Álvarez G, Salcedo-Hernández R, Guevara-Arauza J, Campos-Montiel R (2015) The process and maturation stability of chihuahua cheese with antioxidants in multiple emulsions. J Food Process Preserv 39(6):1027–1035

Qureshi TM, Amjad A, Nadeem M, Murtaza MA, Munir M (2019) Antioxidant potential of a soft cheese (paneer) supplemented with the extracts of date (*Phoenix dactylifera* L.) cultivars and its whey. Asian Austral J Anim 32(10):1591–1602

Rashidinejad A, Birch EJ, Sun-Waterhouse D, Everett DW (2014) Delivery of green tea catechin and epigallocatechin gallate in liposomes incorporated into low-fat hard cheese. Food Chem 156:176–183

Rashidinejad A, Birch EJ, Everett DW (2016) Antioxidant activity and recovery of green tea catechins in full-fat cheese following gastrointestinal simulated digestion. J Food Compos Anal 48:13–24

Redeuil K, Bertholet R, Kussmann M, Steiling H, Rezzi S, Nagy K (2009) Quantification of flavan-3-ols and phenolic acids in milk-based food products by reversed-phase liquid chromatography-tandem mass spectrometry. J Chromatogr A 1216(47):8362–8370

Rêgo ICV, Santos GNV, Santos GNV, Ribeiro JS, Lopes RB, Santos SB, Sousa A, Mendes RA, Taketomi ATF, Vasconcelos AA (2019) Organochlorine pesticides residues in commercial milk: a systematic review. Acta Agron 68(2):99–107

Reiners J, Nicklaus S, Guichard E (2000) Interactions between β-lactoglobulin and flavour compounds of different chemical classes. Impact of the protein on the odour perception of vanillin and eugenol. Lait 80(3):347–360

Ribeiro A, Caleja C, Barros L, Santos-Buelga C, Barreiro MF, Ferreira IC (2016) Rosemary extracts in functional foods: extraction, chemical characterization and incorporation of free and microencapsulated forms in cottage cheese. Food Funct 7(5):2185–2196

Rosenthal I, Rosen B, Bernstein S (1997) Phenols in milk. Evaluation of ferulic acid and other phenols as antifungal agents. Milchwissenschaft 52:134–138

Santos NW, Santos GTD, Silva-Kazama DC, Grande PA, Pintro PM, de Marchi FE, Jobim CC, Petit HV (2014) Production, composition and antioxidants in milk of dairy cows fed diets containing soybean oil and grape residue silage. Livest Sci 159:37–45

Schutte L (1999) Development and application of dairy flavors. In: Flavor chemistry: thirty years of progress. Springer, Boston, pp 155–165

Sedighi-Vesagh R, Naserian AA, Ghaffari MH, Petit HV (2015) Effects of pistachio by-products on digestibility, milk production, milk fatty acid profile and blood metabolites in Saanen dairy goats. J Anim Physiol Anim Nutr 99(4):777–787

Shori AB (2013) Antioxidant activity and viability of lactic acid bacteria in soybean-yogurt made from cow and camel milk. J Taibah Univ Sci 7(4):202–208

Shori AB, Baba AS (2013) Antioxidant activity and inhibition of key enzymes linked to type-2 diabetes and hypertension by *Azadirachta indica*-yogurt. J Saudi Chem Soc 17(3):295–301

Shori AB, Baba AS (2014) Comparative antioxidant activity, proteolysis and in vitro α-amylase and α-glucosidase inhibition of Allium sativum-yogurts made from cow and camel milk. J Saudi Chem Soc 18(5):456–463

Shori AB, Rashid F, Baba AS (2018) Effect of the addition of phytomix-3+ mangosteen on antioxidant activity, viability of lactic acid bacteria, type 2 diabetes key-enzymes, and sensory evaluation of yogurt. LWT Food Sci Technol 94:33–39

Simitzis P, Massouras T, Goliomytis M, Charismiadou M, Moschou K, Economou C, Papadedes V, Lepesioti S, Deligeorgis S (2019) The effects of hesperidin or naringin dietary supplementation on the milk properties of dairy ewes. J Sci Food Agric 99(14):6515–6521

Stark W, Urbach G, Hamilton JS (1976) Volatile compounds in butter oil: V. The quantitative estimation of phenol, o-methoxyphenol, m- and p-cresol, indole and skatole by cold-finger molecular distillation. J Dairy Res 43(3):479–489

Statista (2020) Dairy products and eggs. Statistica. https://www.statista.com/outlook/40010000/100/dairy-products-eggs/worldwide. Accessed 5 May 2020

Sun-Waterhouse D, Zhou J, Wadhwa SS (2013) Drinking yoghurts with berry polyphenols added before and after fermentation. Food Control 32(2):450–460

Torri L, Piochi M, Marchiani R, Zeppa G, Dinnella C, Monteleone E (2016) A sensory- and consumer-based approach to optimize cheese enrichment with grape skin powders. J Dairy Sci 99(1):194–204

Tsen SY, Siew J, Lau EKL, Roslee FA, Chan HM, Loke W (2014) Cow's milk as a dietary source of equol and phenolic antioxidants: differential distribution in the milk aqueous and lipid fractions. Dairy Sci Technol 94(6):625–632

Tseng A, Zhao Y (2013) Wine grape pomace as antioxidant dietary fibre for enhancing nutritional value and improving storability of yogurt and salad dressing. Food Chem 138(1):356–365

Urbach G (1997) The flavour of milk and dairy products: II. Cheese 50(3):79–89

Van het Hof K, Kivits G, Weststrate J, Tijburg L (1998) Bioavailability of catechins from tea: the effect of milk. Eur J Clin Nutr 52(5):356–359

Vital ACP, Goto PA, Hanai LN, Gomes-da-Costa SM, de Abreu Filho BA, Nakamura CV, Matumoto-Pintro PT (2015) Microbiological, functional and rheological properties of low fat yogurt supplemented with Pleurotus ostreatus aqueous extract. LWT Food Sci Technol 64(2):1028–1035

Yuan FQ, Wang J, Sun JM, Cui X, Chen J, Lin LM, Chang JJ, Song XD (2013) Determination of p-hydroxybenzoic acid esters in milk and dairy products by high pressure liquid chromatography. J Food Saf Qual 4(3):890–894

Zeb A (2020) Concept, mechanism, and applications of phenolic antioxidants in foods. J Food Biochem 44(9):e13394

Part II
Biochemistry of Phenolic Antioxidants

Chapter 11
Biosynthesis of Phenolic Antioxidants

Learning Objectives

In this chapter, the reader will be able to:

- Learn the shikimate pathway for the biosynthesis of phenolic acids.
- Describe how chlorogenic acids are biosynthesized.
- Understand the biosynthesis of stilbenes and coumarins.
- Describe how flavonoids are biosynthesized in foods.
- Know about the biosynthesis of anthocyanins.
- Understand how phenolic glycosides are formed.

11.1 Introduction

The metabolism in plant foods resulted in two classes of metabolites. The first one is called primary metabolites i.e., carbohydrates, lipids, proteins, enzymes, whereas phenolic compounds are classified into secondary metabolites. In the former case, the metabolic pathways of proteins, carbohydrates, lipids, and nucleic acids are interconnected resulting in diverse metabolic reactions. The metabolic pathways of primary metabolites in plant foods are also linked with metabolic pathways of secondary metabolites. In the latter case, the substrates for the biosynthesis are usually primary metabolites such as sugars or amino acids. The biosynthesis of phenolic compounds in foods has been studied with aimed to explore the effects of different stresses (Trivellini et al. 2016; Xu et al. 2020). The advent of modern analytical and biochemical techniques clarifies the biosynthetic pathways of phenolic compounds in foods (Cunha et al. 2017; Emir et al. 2020). Thus, the knowledge of fundamental biochemistry, biochemical, and analytical techniques is important for a true understanding of biosynthesis. In this chapter, the biosynthesis of phenolic compounds

© Springer Nature Switzerland AG 2021
A. Zeb, *Phenolic Antioxidants in Foods: Chemistry, Biochemistry and Analysis*,
https://doi.org/10.1007/978-3-030-74768-8_11

present in foods has been discussed, based on each class. The biosynthesis of secondary metabolites in foods may differ between plant species, due to the difference in the nature of the plant. Thus the profile of secondary metabolites differs between plant species. Plant growing under different environments and conditions also showed differences in the metabolite profile (Zeb and Hussain 2020; Zeb and Imran 2019).

11.2   Biosynthesis of Hydroxybenzoic Acids

Benzoic acids are the building blocks of most of the phenolic compounds in foods. Among the benzoic acid derivatives, 2-hydroxy benzoic acid (salicylic acid), 4-hydroxyl benzoic acid, 2,3-dihydroxybenzoic acid, 3,4-dihydroxybenzoic acid (protocatechuic acid), 3,4,5-trihydroxybenzoic acid (gallic acid) are widely studied phenolic compounds in foods. The biosynthesis of hydroxybenzoic acid occurred using several metabolic pathways (Marchiosi et al. 2020; Widhalm and Dudareva 2015). However, there is still much to learn, over the past several years comparative transcriptomic and traditional genetic and biochemical approaches have led to significant advances in the identification of the molecular players involved in producing plant benzoic acids and their products. Benzoic acid is predominately biosynthesized using shikimate and the core phenylpropanoid pathways. Several authors reviewed these pathways in detail (Tzin and Galili 2010; Maeda and Dudareva 2012; Widhalm and Dudareva 2015; Marchiosi et al. 2020). Some of these have not yet been fully elucidated, although significant progress has been made in recent years, which has been covered here along with previously reported proposed pathways.

11.2.1   The Shikimate/Chorismate Pathway

The shikimate pathway is widely known as metabolic reactions occurring in the plastid. The pathway is not present in animals and humans, and thus the end products of these pathways are obtained from plant foods. The pathway involves seven or more chemical reactions and is a major link between primary and secondary metabolism. In microorganisms, the shikimate pathway produces aromatic amino acids phenylalanine, tyrosine, and tryptophan, which are the molecular building blocks for protein biosynthesis (Weaver and Herrmann 1997). Recently the pathway has been widely studied in a variety of plants (Santos-Sánchez et al. 2019). The shikimate can also be produced in the laboratory using several precursors and their reactions have been reported (Candeias et al. 2018). In plant foods, the pathway has been studied at molecular as well as chemical levels. For example, Zhu et al. (2016) reported the four genes involved in the shikimate pathway in the tea plant.

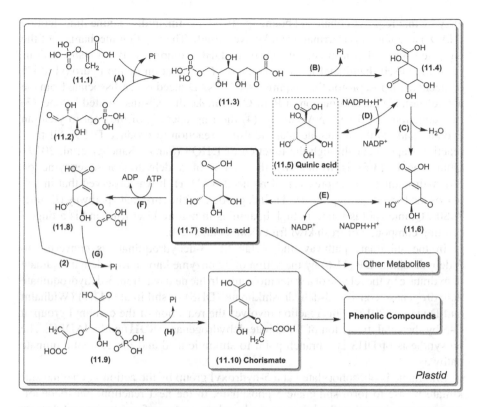

Fig. 11.1 The Shikimate pathway. Enzymes involved are represented by alphabets, (**A**) 3-Deoxy-D-arabino-heptulosonate-7-phosphate synthase, (**B**) 3-Dehydroquinate synthase, (**C**) 3-Dehydroquinate dehydratase, (**D**) quinate dehydrogenase, (**E**) shikimate dehydrogenase, (**F**) shikimate kinase, (**G**) 5-enolpyruvylshikimate 3-phosphate synthase, and (H) chorismate synthase. Metabolites produced are represented by numerals, (**11.1**) phophoenolpyruvate, (**11.2**) erythrose-4-phosphate, (**11.3**) 3-deoxy-D-arabino-heptulosonate 7-phosphate (**DAHP**); (**11.4**) 3-dehydroquinate (**DHQ**); (**11.5**) quinic acid; (**11.6**) 3-dehydroshikimate; (**11.7**) shikimate; (**11.8**) shikimate 3-phosphate; (**11.9**) 5-enolpyruvylshikimate 3-phosphate (**EPSP**) and (**11.10**) Chorismate. The highlighted encircled compounds are key to the biosynthesis of phenolic compounds. NAPH is reduced nicotinamide adenine dinucleotide, whereas NADP is its oxidized form

In the first reaction, two metabolites from primary metabolism are precursors. These are phosphoenolpyruvate (**11.1**) from glycolysis and erythrose-4-phosphate (**11.2**) from the pentose phosphate pathway. This reaction is catalyzed by enzyme 3-deoxy-D-arabino-heptulosonate-7-phosphate synthase with the release of inorganic phosphate (Pi) and 3-deoxy-D-arabino-heptulosonate 7-phosphate (**DAHP**) (**11.3**) to the medium in the plastid (Fig. 11.1). Two isoenzymes of DAHPS have been found for the catalysis of this first reaction step. One isozyme needs only Mn^{2+}, and the other, either Co^{2+}, Mg^{2+}, or Mn^{2+} for the catalysis (Schmid and Amrhein 1995). In the second reaction, DAHP releases an inorganic phosphate by the action of an enzyme known as 3-dehydroquinate synthase (**DHQS**) forming 3-dehydroquinate (DHQ, **11.4**). The enzyme DHQS is a carbon-oxygen lyase enzyme that requires cobalt and bound oxidized nicotinamide adenine dinucleotide (NAD$^+$) as cofactors (Herrmann and Weaver 1999). The reaction mechanism of the enzyme-catalyzed conversion of DAHP to DHQ comprises five transformations from the DAHP hemiketal form, a pyranose: (1) oxidation of the hydroxyl at C5 adjacent to the lost proton that requires NAD$^+$ (NAD$^+$ need never dissociate from the active site), (2) the elimination of Pi of C7 to make the α,β-unsaturated ketone, (3) the reduction of C5 with NADH$^+$ H$^+$, (4) the ring-opening of the enol to yield an enolate, and (5) the intramolecular aldol-like reaction to produce DHQ. All five-reaction steps occur through the function of DHQS (Santos-Sánchez et al. 2019). Quinic acid (**11.5**) is formed by the action of a dehydrogenase enzyme on 3-dehydroquinate in the presence of reduced NADPH. It was observed that in oak leaves the formation of quinic acid is independent of the shikimate pathway (Boudet 1980). Quinic acid is found in high quantities in mature kiwi fruit and is a finger-printing compound of fresh kiwi fruit (Kim et al. 2020).

In the shikimate pathway, the metabolite 3-dehydroquinate is converted to 3-dehydroshikimate (**11.6**) by the action of an enzyme known as 3-dehydroquinate dehydratase by the release of a water molecule. In the next reaction, 3-dehydroquinate dehydrogenase converts 3-dehydroshikimate (DHS) to shikimate (**11.7**) (Widhalm and Dudareva 2015). This reaction involves the reduction of the carbonyl group at C-5 by the catalytic action of Shikimate dehydrogenase (**SDH**) with NADPH. The biosynthesis of DHS is a branch point to shikimic acid and the catabolic quinate pathway.

Shikimate is phosphorylated at a 5-hydroxyl group by the action of enzyme shi-kimate kinase to form shikimate 3-phosphate. In the next reaction, phosphoenol-pyruvate condensed with shikimate 3-phosphate to form 5-enolpyruvylshikimate 3-phosphate (**EPSP**) catalyzed by 5-enolpyruvylshikimate 3-phosphate synthase (**EPSPS**). The reaction mechanism involves the protonation of PEP to the subse-quent nucleophilic attack of the hydroxyl at C-5 of S3P to form an intermediate that loses Pi to form EPSP (Santos-Sánchez et al. 2019). EPSPS is the most studied enzyme of the shikimate pathway because it plays a crucial role in the penultimate step. If this enzyme is inhibited, there is an accumulation of shikimic acid. The synthesis of aromatic amino acid is thus disabled, leading to the death of the plant (Cao et al. 2012). This enzyme is a good target for pesticides. Pesticide like Glyphosate inhibits EPSPS and is a potent nonselective herbicide that mimics the carbocation of PEP and binds EPEPS competitively and thus destroys foods and crops (Sellin et al. 1992; Cao et al. 2012; Zabalza et al. 2017).

In the last reaction, the enzyme chorismate synthase converts 5-enolpyruvylshikimate 3-phosphate to chorismate. This reaction involves the 1,4-*trans* elimination of the inorganic phosphate group at C-3. It is believed that choris-mate synthase needs reduced Flavin mononucleotide (FMNH$_2$) as a cofactor. Kinetic isotope effect study showed that the FMNH$_2$ transfers an electron to the substrate reversibly (Bornemann et al. 2000). Chorismate is the final molecule of the shikimate/chorismate pathway. Chorismate is a key branch point for the synthesis of phenylalanine, tyrosine, and tryptophan, and also other phenolic compounds.

11.2.2   Biosynthesis of Gallic Acid

The biosynthetic pathway of gallic acid has been proposed and confirmed using several biochemical techniques. Isotopic tracer techniques have been found to results in significant contribution in the biosynthetic pathway of gallic acid (Werner et al. 1999; Haslam 1992; Haslam and Cai 1994). Several biosynthetic pathways have been reported. These include shikimate, phenylalanine, cinnamic acid, coumaric acid, and other phenolic acids (Vermerris and Nicholson 2007). The first and foremost important one is formation from shikimate. There is a possibility that different pathways co-existed in different species or even within one species. For example, Saijo (1983) reported that gallic acid is produced in young tea shoots from the dehydrogenation of shikimate, as well as the phenylpropanoid pathway. This was confirmed by isotope tracer studies of shikimic acid-G-^{14}C, phenylalanine-U-^{14}C, and trans-cinnamic acid-3-^{14}C. Similarly, a retrobiosynthetic NMR study with labeled glucose supplied to the young leaves of Sumac highlighted that the carboxylic group of gallic acid comes from an intermediate of the shikimate pathway (Werner et al. 1997). From crude extracts of birch leaves, gallic acid biosynthesis has been reported from 3-dehydroshikimate and NADP$^+$ as shown in Fig. 11.2 by 5-dehydroshikimate dehydrogenase (Ossipov et al. 2003). It has also been suggested that, besides its already known catalytic properties, a shikimate

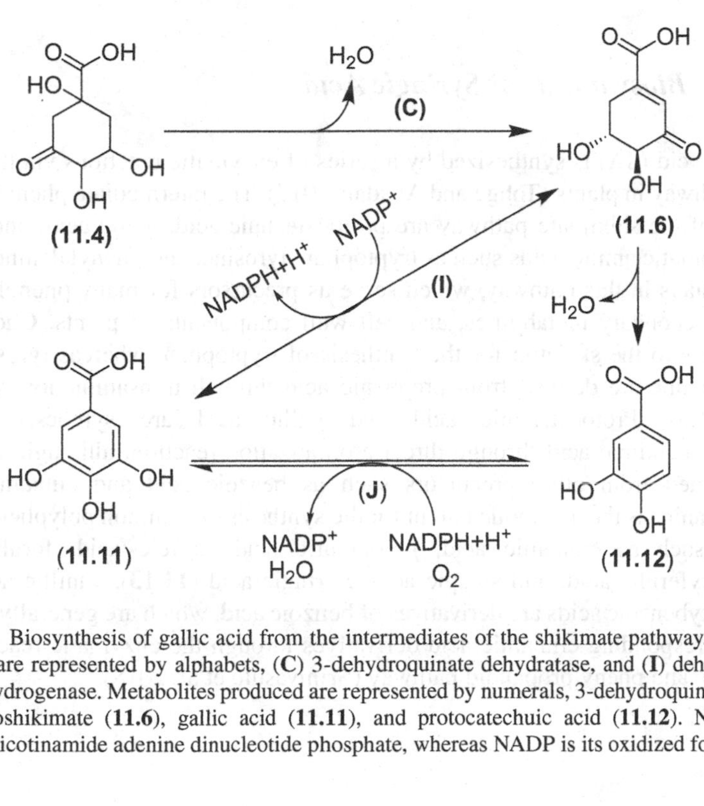

Fig. 11.2 Biosynthesis of gallic acid from the intermediates of the shikimate pathway. Enzymes involved are represented by alphabets, (**C**) 3-dehydroquinate dehydratase, and (**I**) dehydroshikimate dehydrogenase. Metabolites produced are represented by numerals, 3-dehydroquinate (**11.4**), 3-dehydroshikimate (**11.6**), gallic acid (**11.11**), and protocatechuic acid (**11.12**). NADPH is reduced nicotinamide adenine dinucleotide phosphate, whereas NADP is its oxidized form

dehydrogenase (SDH) could be responsible for the biosynthesis of gallic acid in walnut (*Juglans regia*) (Muir et al. 2011). In grapevine, the enzyme dehydroquinate dehydratase/shikimate dehydrogenase was found to utilize $NADP^+$ to produce gallic acid (Bontpart et al. 2016). It was also postulated that the C-terminal SDH domain of the bifunctional enzyme dehydroquinate dehydratase/shikimate dehydrogenase in tea leaves is involved in the biosynthesis of gallic acid (Huang et al. 2019). In the shikimate pathway, the quinic acid can also be converted to gallic acid (Srinivasulu et al. 2018). Different elicitors can be used to enhance the production of gallic acid in foods (Osman et al. 2018).

11.2.3 Biosynthesis of Protocatechuic Acid

Protocatechuic acid is synthesized from the intermediates of the shikimate pathway. In this case, 3-dehydroshikimate is converted to protocatechuic acid as shown in Fig. 11.2. Activity measurements from mung bean seedlings, on the other hand, indicate that 3-dehydroshikimate can be converted to protocatechuic acid (Widhalm and Dudareva 2015). Protocatechuic acid (**11.12**) upon hydroxylation produces gallic acid. The enzyme catalyzing this reaction is not well understood (Muir et al. 2011).

11.2.4 Biosynthesis of Syringic Acid

Syringic acid (SA) is synthesized by a series of enzymatic reactions via the shikimate pathway in plants (Tohge and Alisdair 2017). The intermediate phenolic compounds of the shikimate pathway are protocatechuic acid, gallic acid, and quinic acid. Aromatic amino acids such as tryptophan, tyrosine, and phenylalanine are the end products in this pathway, which serve as precursors for many phenolic compounds, secondary metabolites, and cell wall components of plants. Chorismate contributes to the skeleton for the synthesis of tryptophan, whereas tyrosine and phenylalanine are derived from prephenic acid through transamination reactions (Fig. 11.3). Protocatechuic acid and gallic acid are synthesized from 3-dehydroshikimic acid through direct aromatization reaction, although these are also formed from other precursors such as benzoic acid and cinnamic acid. Phenylalanine is the metabolic parent for the synthesis of common polyphenol compounds such as cinnamic acid, *p*-coumaric acid, caffeic acid, ferulic acid, 5-hydroxyferulic acid, and sinapic acid. Syringic acid (**11.13**), vanillic acid, and 4-hydroxybenzoic acids are derivatives of benzoic acid, which are generally derived from corresponding cinnamic acid derivatives through the enzymatic reactions of shikimate and phenylpropanoid pathway (Srinivasulu et al. 2018).

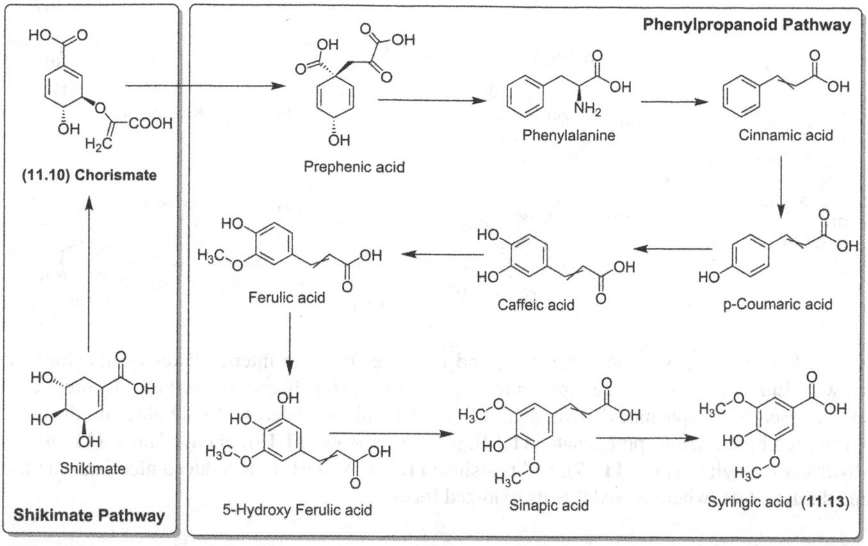

Fig. 11.3 Biosynthesis of syringic acid (**11.13**) from the intermediates of the shikimate pathway and phenylpropanoid pathway. The enzymes and complete sequence of reactions are given in Fig. 11.4

11.3 Biosynthesis of Hydroxycinnamic Acids

In foods, hydroxycinnamic acids are widely reported phenolic acids as discussed in Chaps. 3–10. Hydroxycinnamic acid composition varies with climate, water resources, and cultivars of the food plants (Zeb 2019). They play a significant role in the medicinal as well as nutritional properties of the foods (Zeb 2020). Thus, it is important to know about how they are biosynthesized in plant foods. Hydroxycinnamic acids are biosynthesized by the phenylpropanoid pathway as discussed below.

11.3.1 Phenylpropanoid Pathway

Phenylpropanoids constitute a large group of phenolic compounds produced from amino acids such as phenylalanine and tyrosine. Phenylpropanoid contains a phenyl group and a propene side chain. Phenylpropanoids are either synthesized from the Shikimate pathway or directly contributed by these amino acids. The phenylpropanoid pathway is common to several classes of phenolic compounds, including hydroxybenzoic acids. To understand, this pathway, first we need to know about the biosynthesis of phenylalanine and tyrosine. As shown in Fig. 11.4, chorismate produced from the shikimate pathway is converted to prephenate (**11.14**) by the action of the enzyme chorismate mutase, an isomerase enzyme. Prephenate forms either

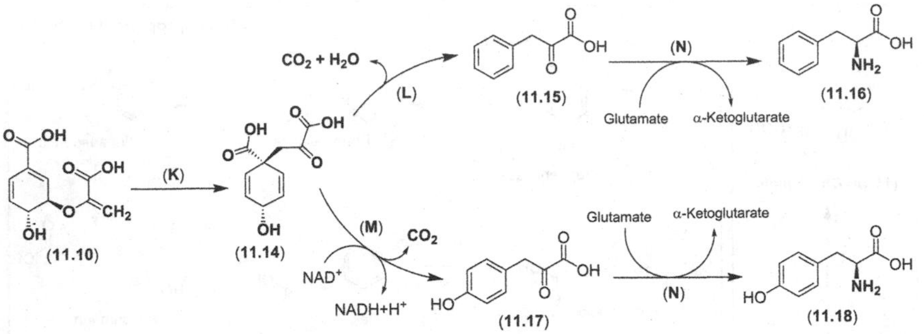

Fig. 11.4 Biosynthesis of phenylalanine, and tyrosine from the intermediates of the shikimate pathway. Enzymes involved are represented by alphabets, (**K**) chorismate mutase, (**L**) prephenate dehydratase, (**M**) prephenate dehydrogenase, and (**N**) aminotransferase. Metabolites produced are represented by numerals, prephenate (**11.14**), phenylpyruvate (**11.15**), phenylalanine (**11.16**), and 4-hydroxy phenylpyruvate (**11.17**), and tyrosine (**11.18**). NADH+H is reduced nicotinamide adenine dinucleotide, whereas NAD⁺ is its oxidized form

phenylalanine or tyrosine. In the first case, prephenate dehydratase converts prephenate to phenylpyruvate by the release of carbon dioxide and water. In a transamination reaction, phenylpyruvate is converted to phenylalanine by the action of the aminotransferase enzyme. Glutamate is required for this reaction. In the second case, prephenate is converted to 4-hydroxy phenylpyruvate by the action of prephenate dehydrogenase in the presence of oxidized NAD. Tyrosine is finally formed by the action of aminotransferase enzyme from 4-hydroxy phenylpyruvate in the presence of glutamate.

The phenylpropanoid pathway starts from the phenylalanine as a precursor and is converted to cinnamic acid (**11.19**). This reaction is catalyzed by phenylalanine ammonia-lyase (**PAL**) releasing the NH_2 group as shown in Fig. 11.5. PAL has been purified and characterized from several plants. It is a large protein (240,000–330,000 kDa) composed of four subunits. PAL genes from several plants have been sequenced including parsley, beans, lemon, rice, and wheat (El-Seedi et al. 2012). Upon hydroxylation, cinnamic acid forms *p*-coumaric acid (**11.20**) by the action of the enzyme cinnamate 4-Hydroxylase (**C4H**). Tyrosine is also converted to *p*-coumaric acid by isozyme PAL known as tyrosine ammonia-lyase (**TAL**) as reported by Rosler et al. (1997). These authors also showed that PAL and TAL activities reside in the same polypeptide, and *p*-coumaric acid was produced both from phenylalanine and tyrosine. The addition of coenzyme A (CoASH) with *p*-coumaric acid forms *p*-coumaroyl-CoA (**11.21**) by the action of enzyme 4-coumaroyl-CoA ligase (**4CL**). Caffeoyl-CoA (**11.22**) is formed by the hydroxylation of *p*-coumaroyl-CoA by the action of enzyme coumarate 3-hydroxylase (**C3H**). Release of coenzyme A forms caffeic acid (**11.23**) by the enzyme 4CL. Caffeic acid is converted to ferulic acid (**11.24**) by the action of enzyme known as caffeic acid *O*-methyltransferase (**COMT**).

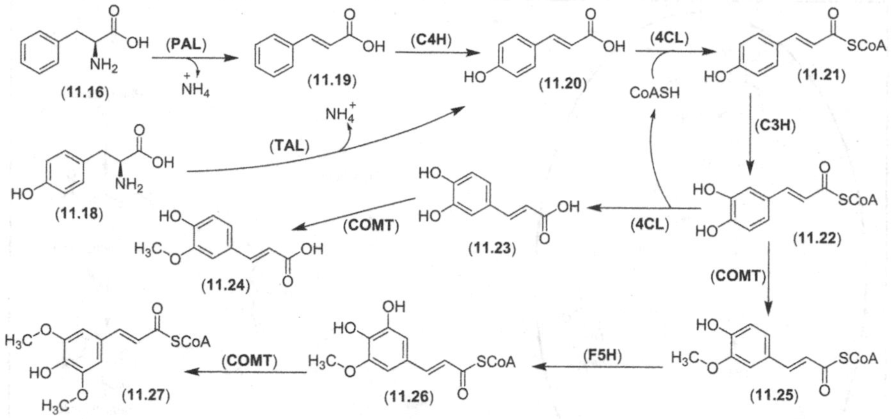

Fig. 11.5 Biosynthesis of cinnamic acid (**11.19**), *p*-coumaric acid (**11.20**), caffeic acid (**11.23**), ferulic acid (**11.24**), feruloyl-CoA (**11.25**), 5-hydroxyferuloyl-CoA (**11.26**), and sinapoyl-CoA (**11.27**). Enzymes involved are represented by alphabets, (**PAL**) phenylalanine ammonia-lyase, (**C4H**) cinnamate 4-hydroxylase, (**4CL**) 4-coumaroyl-CoA ligase, (**TAL**) tyrosine ammonia-lyase, (**C4H**) coumarate 3-hydroxylase, and (**4CL**) 4-hydroxycinnamoyl-CoA ligase

Caffeoyl-CoA upon the action of caffeoyl-CoA methyltransferase (**COMT**) forming feruloyl CoA (**11.25**), which may release free ferulic acid upon hydrolysis. 5-Hydroxyferuloyl-CoA (**11.26**) is formed by hydroxylation of feruloyl-CoA catalyzed by ferulate 5-hydroxylase (**F5H**) enzyme. Sinapoyl-CoA (**11.27**) is formed from the 5-hydroxyferuloyl-CoA. This reaction is catalyzed by a 5-hydroxyferulic acid methyltransferase (**COMT**) enzyme.

11.3.2   Biosynthesis of Salicylic Acid

Biochemical studies using isotope feeding experiments have suggested that plants PAL catalyzed biosynthesize hydroxybenzoic acids such as salicylic acid (SA) from cinnamate (Chen et al. 2009). PAL is induced under a variety of conditions of abiotic and biotic stresses. The SA can be formed from cinnamate through coumarate or benzoate depending on whether the hydroxylation of the aromatic ring takes place before or after the chain-shortening reactions. In sunflower, potato, and pea, isotope-labeled experiments indicated that SA was formed from benzoate, which is synthesized by cinnamate chain shortening reactions most likely through a β-oxidation process analogous to fatty acid β-oxidation (Klämbt 1962). Feeding of C^{14}-labeled phenylalanine and cinnamate to young *Primula acaulis* and *Gaultheria procumbens* leaf segments indicated that SA was formed through coumarate (el-Basyouni et al. 1964). In the same plants, the labeled SA was also formed after treatment with C^{14}-labeled benzoate, which suggests that these plants may use both pathways for SA biosynthesis. Similarly, in young tomato seedlings, SA appeared

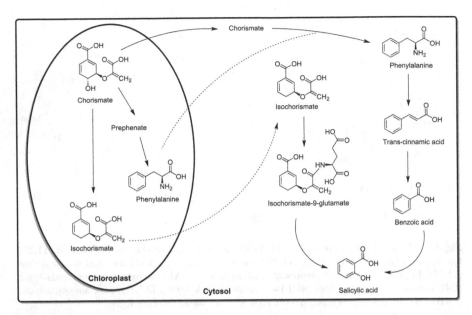

Fig. 11.6 Biosynthesis of Salicylic acid from the intermediates of the shikimate pathway. The pathway occurs both in chloroplast and cytosol

to be formed mostly from cinnamate through benzoate, but after infection with *Agrobacterium tumefaciens*, 2-hydroxylation of cinnamate to *o*-coumarate was favored (Chadha and Brown 1974).

In several lines of rice, evidence suggests that SA was synthesized from cinnamate through benzoate (Silverman et al. 1995). More label was incorporated into SA when C[14]-labeled benzoate was fed than when ^{14}C-labeled cinnamate was used, consistent with benzoic acid being the immediate precursor of SA as shown in Fig. 11.6. Similar results were also obtained from the labeling experiments in potato and cucumber (Coquoz et al. 1998; Lefevere et al. 2020). Furthermore, a benzoic acid 2-hydroxylase (**BA2H**) activity was detected in rice. Despite the extensive biochemical and molecular evidence, none of the enzymes required for the conversion of SA from cinnamate in the PAL pathway has been isolated from plants (Chen et al. 2009).

11.4 Biosynthesis of Phenolic Aldehydes and Alcohols

The biosynthetic pathway of phenolic aldehydes and alcohols have been widely studied in microorganisms such as bacteria. In foods, these compounds have been reported rarely. Among these compounds, vanillin, vanillic acid, and their derivatives are the most important flavor phenolic compounds, widely reported in some foods. Kundu (2017) reviewed the biosynthesis of vanillin in detail. The biosynthesis of vanillin has been studied in vanilla beans using mass spectrometry-based metabolomics (Gu et al. 2017). The vanillin is an important target for biotechnological interventions and applications in foods (Banerjee and Chattopadhyay 2019).

Vanillin and its derivatives are synthesized by the shikimate and phenylpropanoid pathway as the center of metabolism. Studies showed that vanillin is synthesized from L-phenylalanine through cinnamic acid, 4-coumaric acid, caffeic acid, and ferulic acid (Gallage and Møller 2015). These reactions involve phenylalanine ammonia-lyase (PAL), hydroxylation, O-methylation, and lastly a chain-shortening reaction (Gallage et al. 2014) as discussed above. Ferulic acid is directly converted to vanillin by the action of the enzyme vanillin synthase (**VAN**) as shown in Fig. 11.7. The conversion of ferulic acid to vanillin is catalyzed by **VAN** and is intended to proceed sequentially by two partial reactions composed of an initial hydration addition reaction followed by a retro-aldol elimination reaction. In *vanilla planifolia*, this reaction is catalyzed by a single enzyme namely *V. planifolia* vanillin synthase (***Vp*VAN**) (Gallage et al. 2014). It was also found that VpVAN localizes to the inner part of the vanilla pod and high transcript levels are found in single cells located a few cell layers from the inner epidermis.

Another pathway of vanillin biosynthesis starts from 4-coumaroyl-CoA (**11.21**) forming 4-benzaldehyde (**11.30**) by the action of 4-hydroxy cinnamoyl-CoA hydratase/lyase (**4HCH**). Further hydroxylation occurred forming 3,4-dihydroxybenzadehyde (**11.31**) by hydroxybenzaldehyde synthase (**HBS**) from 4-hydroxybenzaldehyde. Vanillin (**11.28**) is formed from 3,4-dihydroxybenzadehyde by O-methyltransferase (**OMT**) (Kundu 2017). Vanillin upon hydroxylation produces vanillic acid (**11.29**) as shown in Fig. 11.7. Wickramasinghe and Munafo (2020) studied the biosynthetic pathways of edible mushrooms using isotopic labeling. The authors reported that isotopically labeled vanillic acid was converted to both vanillin and vanillyl alcohol. Similarly, the conversion of ferulic acid in pineapple peel and crown leaves were converted to vanillin and vanillic acid by *Aspergillus niger* (Tang and Hassan 2020).

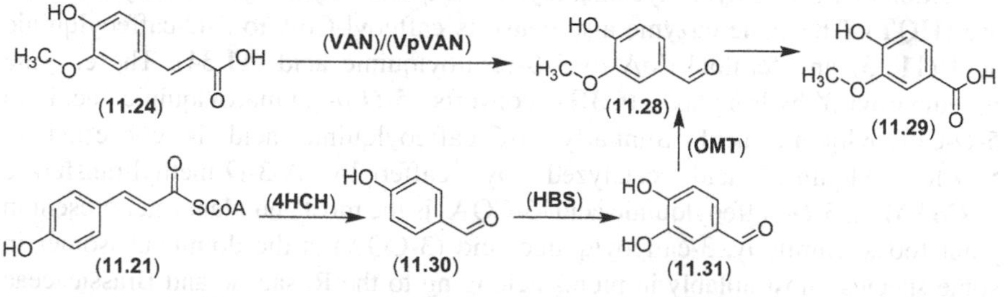

Fig. 11.7 Biosynthesis of vanillin, and vanillic acid. Enzymes involved are represented by alphabets, 4-hydroxy cinnamoyl-CoA hydratase/lyase (**4HCH**), hydroxybenzaldehyde synthase (**HBS**), and O-methyltransferase (**OMT**). Intermediates and products of the reactions are 4-coumaroyl-CoA (**11.21**), 4-benzaldehyde (**11.30**), 3,4-dihydroxybenzadehyde (**11.31**), vanillin (**11.28**), and vanillic acid (**11.29**)

11.5   Biosynthesis of Hydroxycinnamic Acid Esters

The esters of hydroxycinnamic acids are produced from the intermediates of the shikimate pathway as discussed in the phenylpropanoid pathway. Strack and Sharma (1985) examined protoplasts from leaves of radish for the subcellular localization of p-coumaric, caffeic, ferulic, and sinapic acid esters of malic acid and the enzyme(s) involved in their syntheses. The author isolated vacuoles from leaf protoplasts that contained all the hydroxycinnamic acid esters as well as all the dependent enzyme activities. The acyltransferases which catalyze the formation of hydroxycinnamic acid esters of sugar acids (gluconate, glucarate, galactarate, glucuronolactone) and hydro-aromatic acids (quinate, shikimate) involving hydroxycinnamoyl-CoAs as acyl donors have been partially purified from primary leaves of rye (Strack et al. 1987). Biosynthesis of esters of hydroxycinnamic acids such as chlorogenic acids, or their derivatives is given below.

11.5.1   Biosynthesis of Chlorogenic acids

Esters of hydroxycinnamic acids include chlorogenic acids (CGAs), which are a large group of phenolic compounds made of hydroxycinnamic acids and quinic acids. A wider definition of CGAs includes compounds formed with various quinic acid epimers, quinic acid methyl ethers, alkyl quinates, deoxyquinic acid, 2-hydroxyquinic acid, and shikimic acid and its epimers, plus the analogous compounds that are esterified with a hydroxybenzoic acid, hydroxyphenyl acetic acid or 3-(4'-hydroxyphenyl)propionic acid. Aliphatic acid substituents may also be present, occasionally in the absence of an aromatic residue (Clifford et al. 2017).

Chlorogenic acids are biosynthesized from the phenylpropanoid pathway, with the precursor of phenylalanine. The coumaroyl-CoA formed in this pathway is converted to caffeoyl-CoA and then to feruloyl-CoA as shown in Fig. 11.8. Coumaroyl-CoA upon addition of quinic acid produced 5-O-p-coumaroylquinic acid (**11.32**) by the action of enzyme hydroxycinnamoyl CoA: quinate hydroxycinnamoyl transferase (**HQT**). The same enzyme also converts caffeoyl-CoA to 5-O-caffeoylquinic acid (**11.33**) and feruloyl-CoA to 5-O-feruloylquinic acid (**11.34**). The enzyme p-coumaroyl-3'-hydroxylase (**C3H**) converts 5-O-p-coumaroylquinic acid to 5-O-caffeoylquinic acid. Similarly, 5-O-caffeoylquinic acid is converted to 5-O-feruloylquinic acid catalyzed by caffeoyl-CoA-3-O-methyltransferase (**CCoAMT**). 5-O-caffeoylquinic acid (5-CQA) is the most abundant ester present in plant foods. Similarly, 3-caffeoylquinic acid (3-CQA) is the dominant isomer in some species, most notably in plums belonging to the Rosaceae and Brassicaceae families (Clifford et al. 2017). From an enzymatic perspective, information is available only on the biosynthesis of 5-CQA and the perceived understanding is that other CQAs are derived from 5-CQA although there are no data on the seemingly specific isomerases involved in such conversions. Little is also known about the

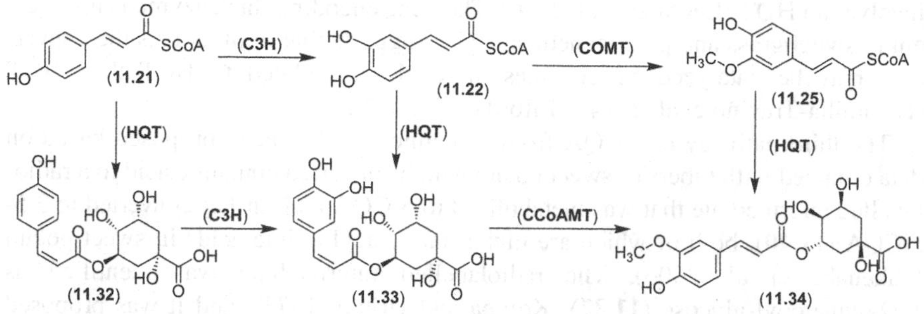

Fig. 11.8 Biosynthesis of quinic acid esters such as 5-*O*-*p*-coumaroylquinic acid (**11.32**), 5-*O*-caffeoylquinic acid (**11.33**), and 5-*O*-feruloylquinic acid (**11.34**). Enzymes involved are represented by alphabets, hydroxycinnamoyl CoA: quinate hydroxycinnamoyl transferase (**HQT**), *p*-coumaroyl-3′-hydroxylase (**C3H**), and caffeoyl-CoA-3-*O*-methyltransferase (**CCoAMT**). Intermediates and products of the reactions are 4-coumaroyl-CoA (**11.21**), caffeoyl-CoA (**11.22**), and feruloyl-CoA (**11.25**)

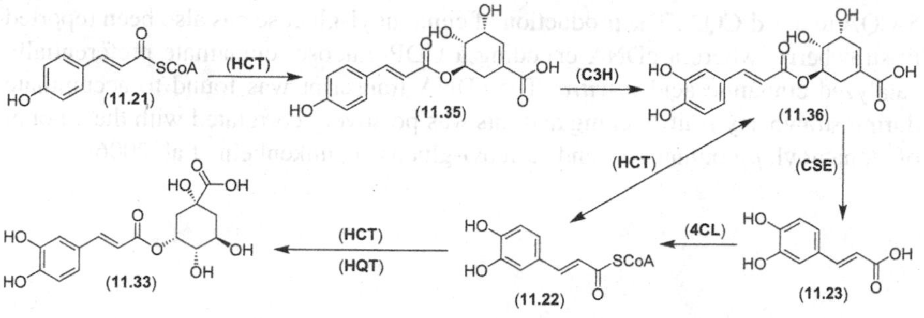

Fig. 11.9 Biosynthesis of 5-Caffeoylquinic acid. Enzymes involved are hydroxycinnamoyl CoA: quinate hydroxycinnamoyl transferase (**HQT**), *p*-coumaroyl-3′-hydroxylase (**C3H**), caffeoyl shikimate esterase (**CSE**), and 4-cinnamoyl CoA ligase (**4CL**). Intermediates and products of the reactions are 5-*O*-*p*-coumaroylshikimic acid (**11.35**), and 5-*O*-caffeoylshikimic acid (**11.36**)

biosynthesis of di- and triCQAs or acyl-quinic acids containing substituents other than caffeic acid.

Research with globe artichoke, switchgrass (*Panicum virgatum*), and chicory has shown a further route to acyl-quinic acids that contains a shikimic acid shunt (Moglia et al. 2014). It involves a hydroxycinnamoyl-CoA: shikimate hydroxycinnamoyl transferase (**HCT**), or possibly two such enzymes in switchgrass (Escamilla-Treviño et al. 2014) that catalyze the conversion of *p*-coumaroyl-CoA to 5-*O*-*p*-coumaroylshikimic acid (**11.35**), which is further converted to 5-*O*-caffeoylshikimic acid (**11.36**) by *p*-coumaroylshikimate-3′-hydroxylase (**C3H**) (Fig. 11.9). The caffeoyl shikimate generated could be converted to caffeoyl-CoA by an HCT acting in the reverse direction. Alternatively, a caffeoyl shikimate esterase could release caffeic acid, which is converted to caffeoyl-CoA by a ligase. Conversion of caffeoyl-CoA to 5-CQA in globe artichoke and other dicots probably involves an HQT (Legrand et al. 2016). The gene encoding this enzyme is not present in switchgrass and phylogenetic analysis suggests that in monocots the conversion may be catalyzed by enzymes more closely related to HCT than HQT (Escamilla-Treviño et al. 2014; Clifford et al. 2017).

The third pathway to 5-CQA from cinnamic acid has been proposed based on data obtained with tubers of sweet potato which converted cinnamic acid to a radiolabelled intermediate that was metabolized to 5-CQA and further converted to 3,5-diCQA (**11.40**), both of which are major native acyl-quinic acids in sweet potato (Takenaka et al. 2006). The radiolabelled intermediate was identified as 1-*O*-cinnamoyl-glucose (**11.37**) (Kojima and Uritani 1973), and it was proposed that it was converted to 5-CQA via 1-*O*-*p*-coumaroyl-glucose (**11.38**) and 1-*O*-caffeoyl-glucose (**11.39**) as shown in Fig. 11.10. Enzymes that catalyze the first three steps in this postulated pathway, a UDP-glucose: cinnamate glucosyltransferase (Shimizu and Kojima 1984), a *p*-coumaroyl-glucose hydroxylase, and hydroxycinnamoyl glucose: quinate hydroxycinnamoyltransferase has been detected in sweet potato roots (Clifford et al. 2017). Besides, a partially purified enzyme has been isolated from sweet potato which catalyzes the single-step conversion of 5-CQA to 3,5-diCQA. The production of cinnamoyl-glucose has also been reported in strawberry, where a cDNA encoding a UDP-glucose: cinnamate preferentially catalyzed cinnamic acid *in-vitro*. The cDNA transcript was found to accumulate during strawberry fruit ripening and this was positively correlated with the amount of cinnamoyl, *p*-coumaroyl-, and caffeoyl-glucose (Lunkenbein et al. 2006).

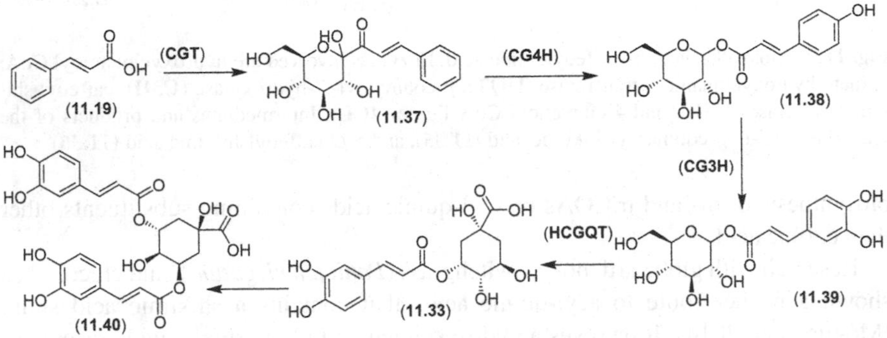

Fig. 11.10 Proposed pathway for the biosynthesis of 5-*O*-caffeoylquinic acid in sweet potato. Enzymes are UDP-glucose: cinnamate glucosyltransferase (**CGT**), cinnamate-glucose 4′-hydroxylase (**CG4H**), *p*-coumaroyl-glucose 3′-hydroxylase (**CG3H**), and hydroxycinnamoyl glucose: quinate hydroxycinnamoyltransferase (**HCGQT**). Intermediates are 1-*O*-cinnamoyl-glucose (**11.37**), 1-*O*-*p*-coumaroyl-glucose (**11.38**), 1-*O*-caffeoyl glucose (**11.39**), and 3,5-dicaffeoylquinic acid (**11.40**)

11.6 Biosynthesis of Coumarins

Coumarins, also known as 1,2-benzopyrones, include a large class of secondary metabolites that are universally found throughout the plant kingdom. In *Citrus sinensis* and *Citrus limonia* seedlings, coumarins have been reported to be formed by the shikimate pathway (Amaral et al. 2020). Vanholme et al. (2019) showed that coumarin synthase (**COSY**) is a key enzyme in the biosynthesis of coumarins. The biosynthesis of coumarins also involves the phenylpropanoid pathway. Enzyme 2-hydroxylase converts *trans*-cinnamic acid (**11.19**) to coumaric acid (**11.20**). Uridine diphosphate-glucose (**UDP-Glc**) is used as a carrier of glucose to coumaric acid, which is carried out by glucosyltransferase (**GT**), to form trans-coumaric acid-2-*O*-glucoside (**11.41**). Coumarin synthase (**COSY**) releases glucose and forms coumarin (1,2-benzopyrones, **11.42**) as shown in Fig. 11.11.

Coumaric acid is converted to coumaroyl-CoA, which upon hydroxylation produces 2-hydroxy-coumaroyl-CoA (**11.43**). This reaction is carried out by the enzyme *p*-coumaroyl CoA 2′-hydroxylase (**C2H**) (Yao et al. 2017). Shimizu (2014)

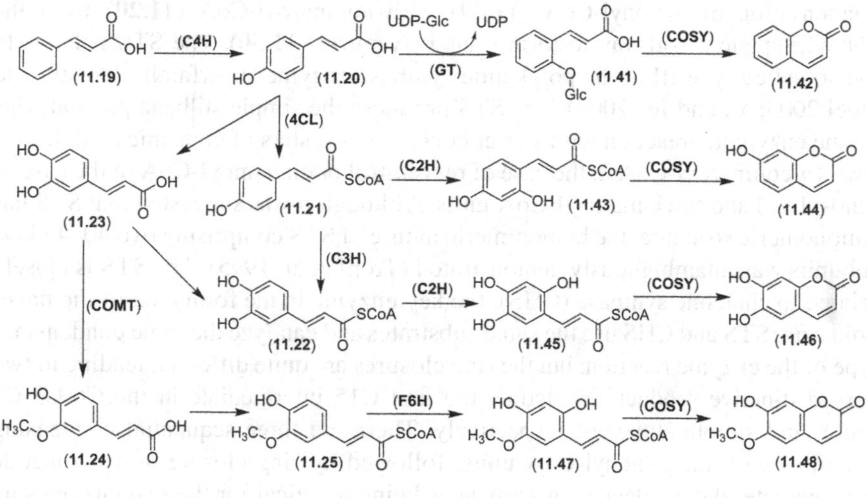

Fig. 11.11 Biosynthetic pathway of coumarins. Intermediate compounds are *trans*-cinnamic acid (**11.19**), coumaric acid (**11.20**), coumaric acid 2-*O*-glucoside (**11.41**), coumarin (**11.42**), coumaroyl-CoA (**11.21**), 2-hydroxyl coumaroyl-CoA (**11.43**), umbelliferone (**11.44**), caffeic acid (**11.23**), caffeoyl-CoA (**11.22**), 2-hydroxyl caffeoyl-CoA (**11.45**), esculetin (**11.46**), ferulic acid (**11.24**), feruloyl-CoA (**11.25**), 6-hydroxyl feruloyl-CoA (**11.47**), and scopoletin (**11.48**). The enzymes involved are glucosyltransferase (**GT**), 4-cinnamoyl CoA ligase (**4CL**), coumarin synthase (**COSY**), *p*-coumaroyl-3′-hydroxylase (**C3H**), *p*-coumaroyl CoA 2′-hydroxylase (**C2H**), *p*-coumaroyl CoA 2′-hydroxylase (**C2H**), and caffeoyl-CoA methyltransferase (**COMT**)

showed that a spontaneous reaction converts *p*-coumaroyl-CoA to umbelliferone (**11.44**) with the release of coenzyme A. Zhao et al. (2019) showed that this reaction is enzymatic, not spontaneous. Vanholme et al. (2019) showed that this reaction involves trans–cis isomerization followed by a lactonization which is carried out by Coumarin synthase (**COSY**).

Similarly, caffeoyl-CoA formed through cinnamic acid is hydroxylated to 2-hydroxy-caffeoyl-CoA (**11.45**), which release esculetin (**11.46**). An essential enzyme for the biosynthesis of the major scopoletin (**11.48**) and its derivatives in Arabidopsis, is *p*-coumaroyl CoA 2′-hydroxylase (**C2H**), which belongs to a large enzyme family of the 2-oxoglutarate and Fe^{2+}-dependent dioxygenases (Siwinska et al. 2018). Scopoletin is further hydroxylated by scopoletin 8-hydroxylase to form fraxetin.

11.7 Biosynthesis of Stilbenes

Stilbene is formed on the phenylalanine/polymalonate route, the last step of this biosynthesis pathway being catalyzed by stilbene synthase (**STS**) (Jeandet et al. 2010; Chong et al. 2009; Dubrovina and Kiselev 2017). The STS catalyzes the condensation of malonyl-CoA (**11.49**) with coumaroyl-CoA (**11.20**) from the phenylpropanoid pathway to form trans-resveratrol (**11.50**). The STS belongs to the so-called type III of the polyketide synthase enzyme superfamily (Austin and Noel 2003; Yu and Jez 2008). The STS produces the simple stilbene phytoalexins in one enzymatic reaction with starter coenzyme A-esters of cinnamic acid derivatives (*p*-coumaroyl-CoA in the case of resveratrol or cinnamoyl-CoA in the case of pinosylvin) and three malonyl-CoA units. Although it was suggested that STS has a monomeric structure, the homodimeric nature of STS comprising two 40–45 kDa subunits was unambiguously demonstrated (Tropf et al. 1995). The STS is closely related to chalcone synthase (**CHS**), the key enzyme in the formation of the flavonoid ring. STS and CHS use the same substrates and catalyze the same condensing-type of the enzyme reaction, but the ring closures are quite different, leading to two very distinctive products, chalcone, the first C15 intermediate in the C6-C3-C6 route, and simple stilbenes, respectively. There are three sequential condensing reactions with the malonyl-CoA units, followed by ring closure of a tetraketide intermediate, the condensation sequences being identical for the two enzymes up to the tetraketide stage. These condensation mechanisms have extensively been studied. In the STS-catalysed pathway leading to stilbenes, the discovery of an aldol switch was shown to determine the type of ring-folding (Austin et al. 2004). Finally, it has been confirmed that electronic effects rather than steric factors balance competing cyclization specificities in CHS and STS. Peroxidase (**PER**) forms trans-ε-viniferin (**11.52**), whereas UDP-glucose addition is catalyzed by 3-glucosyl-*O*-transferase (**3GT**) as shown in Fig. 11.12.

In *Vitis vinifera*, one of the plant species in which the highest amounts of resveratrol are naturally found, genome sequencing has revealed a large expansion of

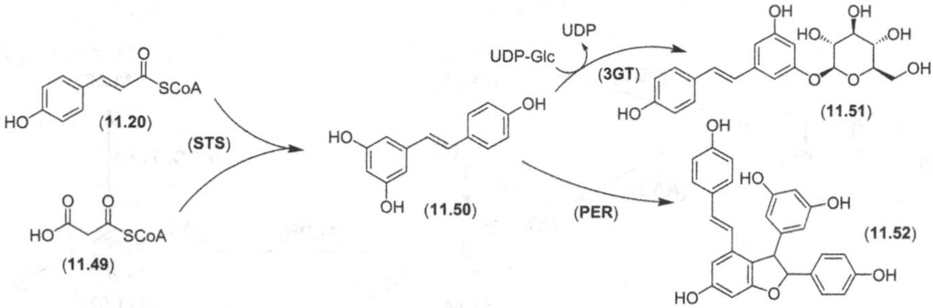

Fig. 11.12 Biosynthesis of Stilbenes in foods from the precursor of the phenylpropanoid pathway. Intermediates are coumaroyl-CoA (**11.20**), malonyl-CoA (**11.49**), trans-resveratrol (**11.50**), piceid (**11.51**) and trans-ε-viniferin (**11.52**). Enzymes are stilbene synthase (**STS**), peroxidases (**PER**), and 3-glucosyl-*O*-transferase (**3GT**)

STS genes (43 genes identified with 20 of these previously being shown to be expressed), suggesting the great importance of stilbene metabolism for this species (Jaillon et al. 2007). The STS cDNA and genomic clones have been described from groundnut (Lanz et al. 1990), and grapevine (Sparvoli et al. 1994). The STS and CHS contain, at the same position, a single essential cysteine residue (Cys 164), which most likely represents the active site (Suh et al. 2000). A single change of histidine residue (close to the active site) to glutamine is responsible for the substrate specificity in STS: in the STS that produce pinosylvin using cinnamoyl-CoA as a starter CoA ester, and resveratrol using *p*-coumaroyl-CoA, these positions are occupied by Glutamine-Histidine, and Histidine-Glutamine, respectively (Jeandet et al. 2010). Zhu et al. (2020) showed that light irradiation significantly prompted stilbene accumulation in peanuts by upregulating the expression of genes and enzymes in the stilbene biosynthesis-related pathway, and UV-C was more effective to promote stilbene accumulation.

11.8 Biosynthesis of Phenylethanoids

Tyrosol is synthesized from tyrosine. There are two possible biosynthesis pathways for tyrosol synthesis in plants (Chung et al. 2017). In the first proposed pathway, tyrosine is converted into tyramine (**11.53**) by tyrosine decarboxylase (**TDC**). Subsequent oxidation and reduction of tyramine result in the formation of 4-hydroxyphenylacetaldehyde (**11.54**) by the action of enzyme tyramine oxidase (**TYO**) as shown in Fig. 11.13. Tyrosine is also converted to 4-hydroxyphenylacetaldehyde (**11.54**) by aromatic acetaldehyde synthase (**AAS**). Tyrosol is synthesized by several known enzymes in different plants, including alcohol dehydrogenase (**ALDH**). Tyrosol is hydroxylated to hydroxytyrosol by either polyphenol oxidase (**PPO**) or tyrosol hydroxylase (**TLH**). However, growing evidence indicates that tyrosol is synthesized via tyramine, as TDC was identified

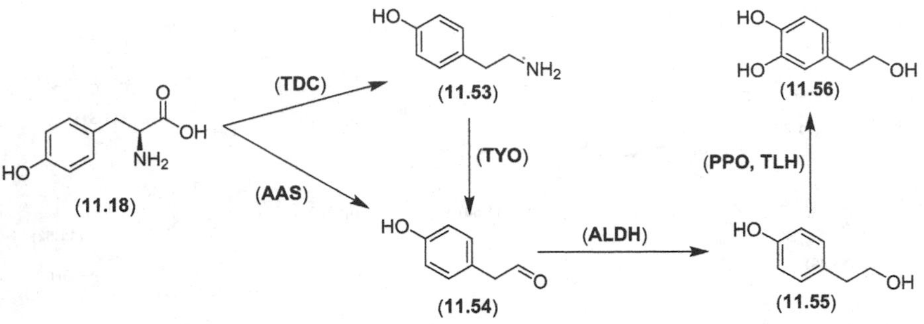

Fig. 11.13 Biosynthesis of Tyrosol and hydroxytyrosol in foods. Enzymes are tyrosine decarboxylase (**TDC**), tyramine oxidase (**TYO**), aromatic acetaldehyde synthase (**AAS**), alcohol dehydrogenase (ALDH), polyphenol oxidase (**PPO**), and tyrosol hydroxylase (**TLH**). Intermediates are tyrosine (**11.18**), tyramine (**11.53**), 4-hydroxyphenylacetaldehyde (**11.54**), tyrosol (**11.55**), and hydroxytyrosol (**11.56**)

in *Rhodiola sachalinensis* (Zhou et al. 2020). The carbon backbone of tyrosol is C6-C2. Plant phenolic compounds have been synthesized via cinnamic acid, which has a C6-C3 carbon backbone and is derived from phenylalanine. To synthesize other phenolic compounds such as flavonoids and stilbenes, malonyl-CoA serves as a carbon donor to transfer two carbons to hydroxycinnamic acid. The synthesis of C6-C1 phenolic compounds, such as benzoic acid relies on the coenzyme A-dependent β-oxidation of cinnamoyl-CoA (Vogt 2010; Alagna et al. 2012). These results support that tyrosol is synthesized through tyrosine decarboxylation.

Alagna et al. (2012) reported the content of these intermediate compounds varied significantly among the olive cultivars and decreased during fruit development and maturation, with some compounds showing specificity for certain cultivars. The authors observed a strong correlation between phenolic compound concentrations and transcripts putatively involved in their biosynthesis, suggesting a transcriptional regulation of the corresponding pathways. A recent study showed that hydroxytyrosol biosynthetic genes are highly expressed in young olive fruit (Mougiou et al. 2018). A recent study of Guodong et al. (2019) using the third generation of full-length transcriptome technology together with metabolomics confirmed these biosynthetic metabolites and pathways.

11.9 Biosynthesis of Flavonoids

The flavonoid pathway is the main process by which various flavonoids are produced in plant foods. The genes involving the biosynthetic pathway have been studied in maize (Andersen et al. 2008), grape (Petrussa et al. 2013), buckwheat (Matsui and Walker 2020), kiwifruit (Liang et al. 2018), safflower (Chen et al. 2018), lettuce (Gurdon et al. 2019) and raspberry (Lebedev et al. 2019). The precursors include *p*-coumaroyl-CoA, and malonyl-CoA, which originates from carbohydrate

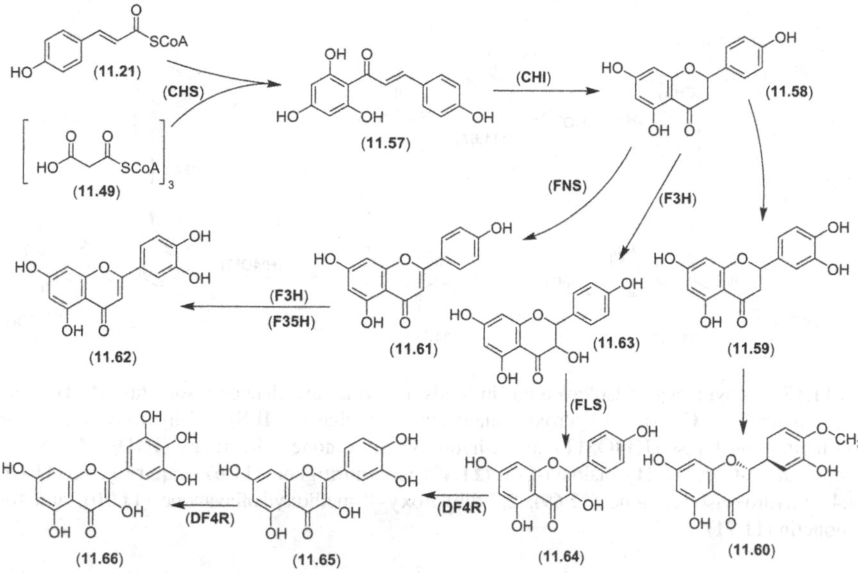

Fig. 11.14 Biosynthesis of Flavonoids in foods. Enzymes are chalcone synthase (**CHS**), chalcone isomerase (**CHI**), flavone synthase (**FNS**), flavonoid 3-hydroxylase (**F3H**), flavonoid 3′,5′-hydroxylase (**F35H**), dihydroflavonol 4-reductase (**DF4R**), and Flavonol synthase (**FLS**). Intermediates are chalcone (**11.57**), naringenin (**11.58**), eriodictyol (**11.59**), hesperetin (**11.60**), apigenin (**11.61**), luteolin (**11.62**), dihydrokaempferol (**11.63**), kaempferol (**11.64**), quercetin (**11.65**), and myricetin (**11.66**)

metabolism. Consequently, chalcone (2′,4,4′,6′-tetrahydroxychalcone, **11.57**) is formed by the condensation of one *p*-coumaroyl-CoA and three malonyl-CoA under the catalysis of chalcone synthase (**CHS**) as shown in Fig. 11.14. Chalcone synthase was found to be a rate-limiting enzyme after **4CL** (Zang et al. 2019). Zhao et al. (2020a) showed that in citrus fruits, chalcone then cyclizes to form naringenin (**11.58**) through the addition of 5-hydroxyl on A ring and alkene with the catalysis of chalcone isomerase (**CHI**). A ring originates from three molecules of malonyl-CoA and the C ring originates from one molecule of *p*-coumaroyl-CoA, which has been verified through isotope labeling experiments (Nabavi et al. 2020). The general chiral flavanone (naringenin) is a central intermediate, which can be further transferred into different classes of flavonoids (flavanones, flavones, and flavonols) by a well-characterized enzymatically derived process in citrus plants (Zhao et al. 2020a).

Flavanone is biosynthesized from naringenin using hydroxylase enzymes to eriodictyol (**11.59**) and hesperetin (**11.60**). Flavone biosynthesis starts with the dehydrogenation of naringenin at C2-C3 to form apigenin (**11.61**) by flavone synthase (**FNS**). Then, luteolin (**11.62**) and diosmetin are produced with the help of F3H/ F35H and methylase enzymes, respectively. The hydroxylation of naringenin at C3 under the catalysis of flavanone 3-hydroxylase (F3H) produces dihydrokaempferol

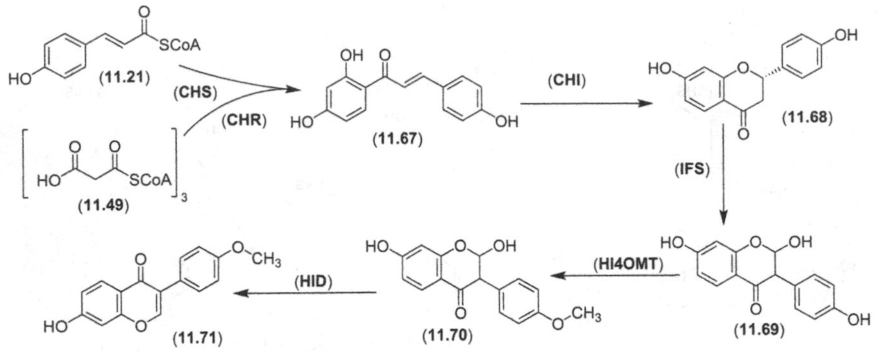

Fig. 11.15 Biosynthesis of Isoflavonoids in foods. Enzymes are chalcone isomerase (**CHI**), chalcone synthase (**CHS**), 2-hydroxyisoflavanone synthase (**IFS**), 2-hydroxyisoflavanone 4′-O-methyltransferase (**HI4OMT**), and 2-hydroxyisoflavanone dehydratase (**HID**). Metabolites are coumaroyl-CoA (**11.21**), malonyl-CoA (**11.49**), isoliquiritigenin (**11.67**), liquiritigenin (**11.68**), 2,7,4′-trihydroxyisoflavanone (**11.69**), 2,7-dihydroxy-4′-methoxyisoflavanone (**11.70**), and formononetin (**11.71**)

(**11.63**). It can be further converted to dihydroquercetin and dihydromyricetin by flavonoid 3-hydroxylase (**F3H**) and flavonoid 3′,5′-hydroxylase (**F35H**). Flavonol synthase (**FLS**) is responsible for the synthesis of kaempferol (**11.64**), which is converted to quercetin (**11.65**) by the action of the enzyme dihydroflavonol 4-reductase (**DF4R**). Quercetin is also converted to myricetin (**11.66**) by the DF4R enzyme.

The second pathway is the biosynthesis of isoflavonoids from coumaroyl-CoA as shown in Fig. 11.15. Isoliquiritigenin (**11.67**) is produced by the coupled catalytic action of CHS and chalcone reductase (**CHR**). The latter enzyme is only identified in papilionoid legumes (like *Glycine max, Medicago sativa, Glycyrrhiza echinata, Glycyrrhiza glabra*) (García-Calderón et al. 2020). The CHR acts in a coupled catalytic action with CHS (Dao et al. 2011). Chalcone isomerase (**CHI**) converts isoliquiritigenin to liquiritigenin (**11.68**). Isoflavone synthase (**IFS**) is a membrane-associated enzyme belonging to the CYP93C subfamily of cytochrome P450 mono-oxygenases that builds the isoflavonoid skeleton from 4′,7-dihydroxyflavanone substrate (liquiritigenin) by an unusual aryl migration reaction to form 2,7,4′-trihydroxyisoflavanone (**11.69**). The IFS has been reported in nearly all legumes (García-Calderón et al. 2020). The enzyme 2-hydroxyisoflavanone 4′-O-methyltransferase (**HI4OMT**) forms 2,7-dihydroxy-4′-methoxy isoflavanone (**11.70**). Formononetin (**11.71**) is formed by 2-hydroxyisoflavanone dehydratase (**HID**). The substrate specificity of HID may differ among species. In soybean, it accepts 2,5,7,4′-tetrahydroxyisoflavanone or 2,7,4′-trihydroxyisoflavanone as substrate, which is then dehydrated to produce a double bond between C-2 and C-3, yielding genistein, or daidzein (Veitch 2007; Latif et al. 2020). Further hydroxylation and reduction may produce other isoflavonoids.

11.10 Biosynthesis of Anthocyanidins

The biosynthesis of anthocyanidins is linked with the biosynthesis of flavonoids. Generally, by the action of chalcone synthase (**CHS**), three malonyl-CoA molecules and one *p*-coumaroyl-CoA molecule can condense to produce a naringenin chalcone. The B ring of the naringenin flavanone can be further hydroxylated by flavonoid 3′-hydroxylase (**F3′H**) or flavonoid 3′5′-hydroxylase (**F3′5′H**) to produce eriodictyol (**11.59**) or pentahydroxyflavanone (**11.74**), respectively (Bogs et al. 2006). All of the three (2*S*)-flavanones can be modified by the catalysis of flavanone 3β-hydroxylase (**F3H**) to produce the corresponding dihydroflavonols (Fig. 11.16). Furthermore, in grapevine, the direct enzymatically oxidized product of naringenin, dihydrokaempferol (**11.63**), is also the potential substrate for F3′H and F3′5′H, to produce the corresponding dihydroflavonols, dihydro-quercetin (**11.78**) and

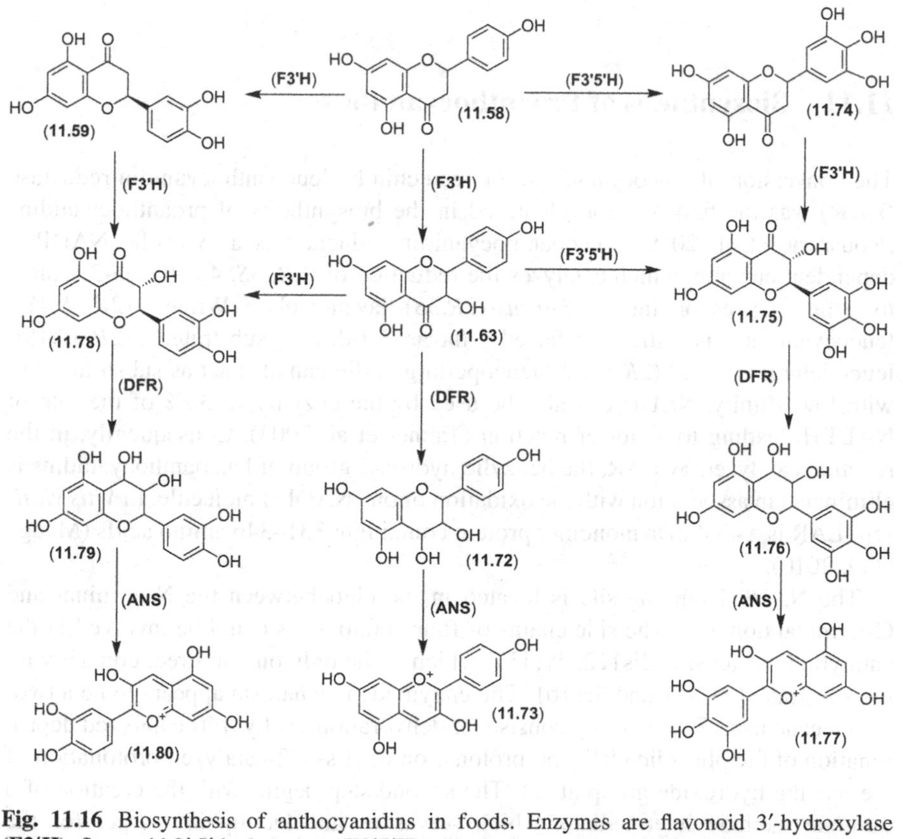

Fig. 11.16 Biosynthesis of anthocyanidins in foods. Enzymes are flavonoid 3′-hydroxylase (**F3′H**), flavonoid 3′,5′-hydroxylase (**F3′5′H**), dihydroflavonol 4-reductase (**DFR**), and anthocyanidin synthase (**ANS**). Metabolites are naringenin (**11.58**), eriodictyol (**11.59**), dihydrokaempferol (**11.63**), leucopelargonidin (**11.72**), pelargonidin (**11.73**), pentahydroxy-flavanone (**11.74**), dihydromyricetin (**11.75**), leucodelphinidin (**11.76**), delphinidin (**11.77**), dihydroquercetin (**11.78**), leucocyanidin (**11.79**), and cyanidin (**11.80**)

dihydromyricetin (**11.75**), respectively (Sparvoli et al. 1994). The enzyme dihydro-flavonol 4-reductase (**DFR**) convert dihydroflavonols into leucoanthocyanidins, such as leucocyanidin (**11.79**), leucopelargonidin (**11.72**), and leucodelphinidin (**11.76**). The crystal structure of the DFR in *Vitis vinifera* was determined and the crucial region of substrate binding and recognition was also confirmed (Petit et al. 2007). Enzyme anthocyanidin synthase (**ANS**) convert leucoanthocyanidins to their corresponding anthocyanidins such as pelargonidin (**11.73**), delphinidin (**11.77**), and cyanidin (**11.80**) (He et al. 2010). ANS is a 2-oxoglutarate-dependent oxygen-ase that is believed to abstract hydrogen radical from C2 of leucoanthocyanidin to yield radical. Several modifications have been reported to form other anthocyani-dins and anthocyanins. The transcriptional complexes that regulate the biosynthetic pathway in tomato fruits and their efficiency in inducing anthocyanin pigmentation have been reported (Colanero et al. 2020). The basic helix-loop-helix transcription factor has been inferred to play an important role in blue and purple grain traits in common wheat (Zhao et al. 2020b).

11.11 Biosynthesis of Proanthocyanidins

The conversion of leucocyanidin to (+)-catechin by leucoanthocyanidin reductase (**LAR**) was the first reaction identified in the biosynthesis of proanthocyanidins (Rousserie et al. 2019). Leucoanthocyanidin reductase is a cytosolic NADPH-dependent enzyme, which catalyzes the reduction of (2*R*, 3*S*, 4*S*)-flavan-3,4-diols to the corresponding 2,3-*trans*-(2*R*,3*S*)-flavan-3-ols. While (2*R*,3*S*,4*S*)-leucocyanidin is the preferred flavan-3,4-diols substrate, (2*R*,3*S*,4*S*)-leucodelphinidin and (2*R*,3*S*,4*S*)-leucoperlargonidin can also act as substrates, but with low affinity. NADH can also be used by the enzyme, at 30% of the rate of NADPH, leading to a slower reaction (Tanner et al. 2003). Consequently, in the reaction catalyzed by LAR, the benzylic hydroxyl group of leucoanthocyanidins is eliminated in association with the oxidation of one NADPH molecule. In *Vitis vinif-era*, LAR is a 45.4 kDa monomer protein containing 331–346 amino acids (Maugé et al. 2010).

The NADPH binding site is located in the cleft between the N-terminal and C-terminal domains. The side chains of four amino acids could be involved in the catalytic mechanism: His122, Tyr137, which is the only one in direct contact with the substrate, Lys140, and Ser161. The enzymatic mechanism appears to be a two-step mechanism. The first step consists of dehydration via Lys140-catalyzed depro-tonation of the phenolic OH7 and protonation of His-122-catalyzed protonation of the leaving hydroxide group at C4. The second step begins with the creation of a quinone methide intermediate, which serves as an electrophilic target for the NADPH with the help of the ammonium form of the Lys-140. Then, hydrogen from NADPH is transferred to the C4 of the phenolic structure, which leads to the forma-tion of the (+)-catechin (Rousserie et al. 2019). Anthocyanidin reductase (**ANR**) is known to be a cytosolic NADPH-dependent enzyme. This enzyme catalyzes the

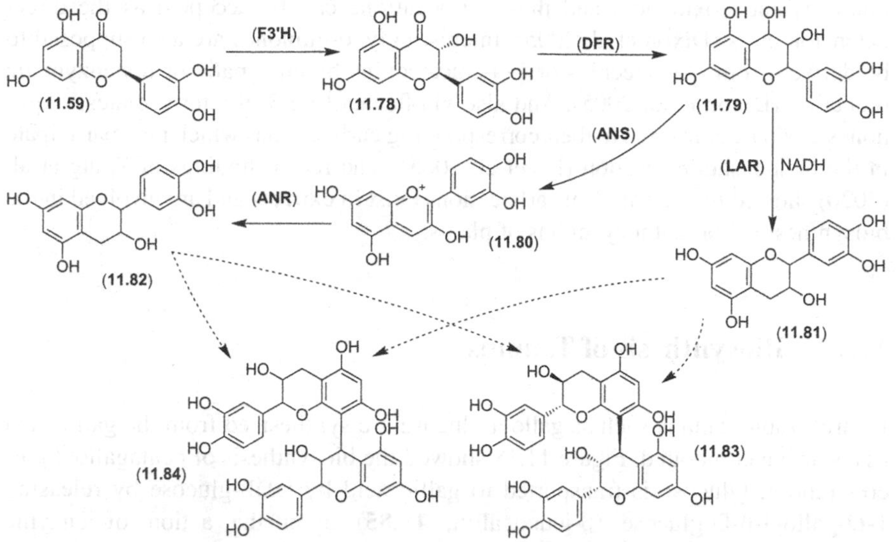

Fig. 11.17 Biosynthesis of proanthocyanidins in foods. Enzymes are flavonoid 3'-hydroxylase (**F3'H**), dihydroflavonol 4-reductase (**DFR**), anthocyanidin synthase (**ANS**), and anthocyanidin reductase (**ANR**). Metabolites are eriodictyol (**11.59**), dihydroquercetin (**11.78**), leucocyanidin (**11.79**), cyanidin (**11.80**), catechin (**11.81**), epicatechin (**11.82**), proanthocyanidin B1 (**11.83**), and proanthocyanidin B2 (**11.84**)

double NADPH reduction of anthocyanidins, which leads to the production of C3 epimers (2*S*, 3*R*) and (2*S*, 3*S*)-flavan-3-ols, i.e., the naturally rare (+)-epicatechin and (−)-catechin. Polymerization of catechin (**11.81**), or epicatechin (**11.82**), resulted in the formation of proanthocyanidins such as proanthocyanidin B1 (**11.83**), and proanthocyanidin B2 (**11.84**) as shown in Fig. 11.17. However, the identity of the final "enzyme" which catalyzes the polymerization reaction of these precursors to form PAs and the exact mechanism whereby the elementary units of flavanols are assembled in vivo remain unknown (He et al. 2008). Most of the models state that the electrophilic C4 position of the extension unit (flavan-3,4-diol) condenses with the nucleophilic C8 or C6 position of the start/terminal unit (flavan-3-ol) to produce PAs. In one previously proposed route, 2*R*,3*S*,4*S*-leucoanthocyanidins are first converted into 2*R*,3*S*-quinone methides, which can be used as the 2*R*,3*S*-extension units directly. Meanwhile, they can also be converted into 2*R*,3*R*-quinone methides via the flavan-3-en-3-ol intermediates, which can be condensed as the 2*R*,3*R*-extension units directly (Hemingway and Laks 1985). Furthermore, these trans- or cis-quinone methides could be converted to their corresponding carbocations which could be attacked by (+)-catechins or (−)-epicatechins directly to produce the PAs.

In another model, flavan-3-ols [(+)-catechins or (−)-epicatechins] are converted to their corresponding quinone methides by the catalysis of polyphenol oxidase (**PPO**). These quinone methides can be further converted to carbocations via flavan-3-en-3-ol intermediates or be reduced to carbocations through coupled non-enzymatic oxidation, and these carbocations can be accepted as the direct extension units (Dixon et al. 2005). Interestingly, o-quinones are also supposed to be derived from (+)-catechins or (−)-epicatechins by enzymatic or nonenzymatic reduction (Dixon et al. 2005). And also via flavan-3-en-3-ol intermediates, o-quinones could be converted to their corresponding carbocations which may participate in the condensation reaction (He et al. 2008). The recent findings of Wang et al. (2020) showed that flavan-3-ol carbocations exist in extracts and are involved in the biosynthesis of proanthocyanidins of plants.

11.12   Biosynthesis of Tannins

Hydrolyzable tannins such as galloyl glucose are synthesized from the gallic acid as a starting compound. Figure 11.18 showed the biosynthesis of pentagalloyl glucopyranose. Glucose is transported to gallic acid by UDP-glucose by releasing 1-O-galloyl-β-D-glucose (β-glucogallin, **11.85**) upon the action of enzyme

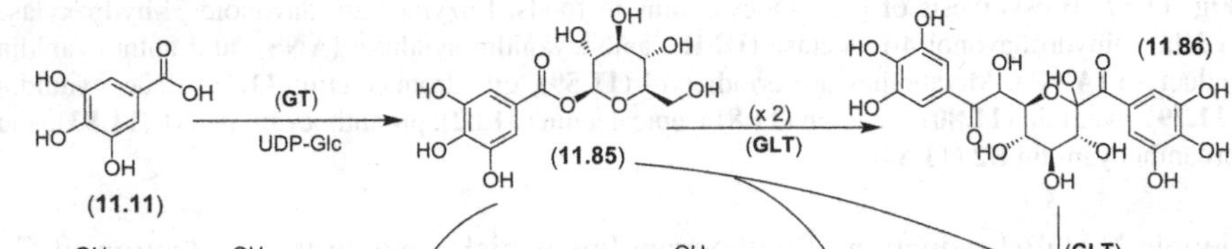

Fig. 11.18 Biosynthesis of hydrolysable tannins in foods. Enzymes are glucosyltransferase (**GT**) and galloyltransferase (**GLT**). Metabolites are gallic acid (**11.11**), 1-O-galloyl-β-D-glucose (**11.85**), 1,6-digalloyl glucopyranose (**11.86**), 1,2,6-trigalloyl glucopyranose (**11.87**), 1,2,3,6-tetragalloyl glucopyranose (**11.88**), 1,2,3,4,6-pentagalloyl glucopyranose (**11.89**), tellima-grandin II (**11.90**) and 3,4,5,3′,4′,5′-hexahydroxydiphenic acid (**11.91**)

glucosyltransferase (**GT**). The GT enzyme contributes to the production of ellagic acid/ellagitannins in strawberries and raspberries (Schulenburg et al. 2016). The enzyme galloyltransferase (**GLT**) has been reported to carry out transfer or condensation reactions. Two molecules of 1-*O*-galloyl-β-D-glucose form 1,6-digalloyl glucopyranose (**11.86**). Upon the addition of subsequent addition of 1-*O*-galloyl-β-D-glucose, 1,2,6-trigalloyl glucopyranose (**11.87**), 1,2,3,6-tetragalloyl glucopyranose (**11.88**), and 1,2,3,4,6-pentagalloyl glucopyranose (**11.89**) are. formed (Grundhöfer et al. 2001). The enzyme producing pentagalloyl glucopyranose was named β-glucogallin: 1,2,6-tri-*O*-galloyl-β-d-glucose 3-*O*-galloyltransferase (Hagenah and Gross 1993). Hexagalloyl glucopyranose and heptagalloyl glucopyranose have been formed in a similar manner.

Ellagitannins are originated from the dehydrogenation of neighboring galloyl groups of 1,2,3,4,6- pentagalloylglucose forming yielding tellimagrandin II (**11.90**). Further hydrolysis yield 3,4,5,3′,4′,5′-hexahydroxydiphenic acid (**11.91**). With the loss of water molecule and lactonization, ellagic acid (**11.92**) is formed. This reaction is catalyzed by a laccase/polyphenol oxidase type enzyme (Niemetz and Gross 2005). Several ellagitannins such as vescalagin, punicalagin, castalagin present in some fruits, nuts, and seeds, such as pomegranates, black raspberries, raspberries, strawberries, walnuts, and almonds (Landete 2011) are also releasing ellagic acid.

11.13 Biosynthesis of Lignans

The biosynthesis of lignans consists of complex pathways based on the diverse nature of these phenolic compounds in foods. Suzuki and Umezawa (2007) reviewed the biosynthesis of lignans and norlignans. Similarly, Touré and Xueming (2010) also reviewed the biosynthesis of lignans in flaxseed. The biosynthesis of sesame lignans up to (+)-sesamin is well characterized. The pathway branches out from a common cascade of phenylpropanoids metabolite feruloyl-CoA to form coniferaldehyde (**11.93**), and then by stereoselective radical coupling of two molecules of a monolignol, coniferyl alcohol (**11.94**), guided by a dirigent protein (**DP**) (Davin et al. 1997; Davin and Lewis 2003). This coupling reaction results in the formation of the (+)-enantiomer of pinoresinol (**11.95**), a central precursor of lignans. Pinoresinol is then converted to (+)-sesamin via sequential oxygenation resulting in the formation of two methylenedioxy bridges (Fig. 11.19). This reaction is catalyzed by a single cytochrome P450 monooxygenase CYP81Q1, also known as piperitol/sesamin synthase to form piperitol (**11.98**), and sesamin (**11.99**) (Murata et al. 2017). Sesamin forms sesamolin (**11.100**), and sesaminol (**11.101**). Pinoresinol is also converted to lariciresinol (**11.96**) by the enzyme pinoresinol reductase (**PR**). Secoisolariciresinol (**11.97**) is formed by the action of another reductase enzyme known as lariciresinol reductase (**LR**). Secoisolariciresinol or lariciresinol may act as precursors of several other lignans.

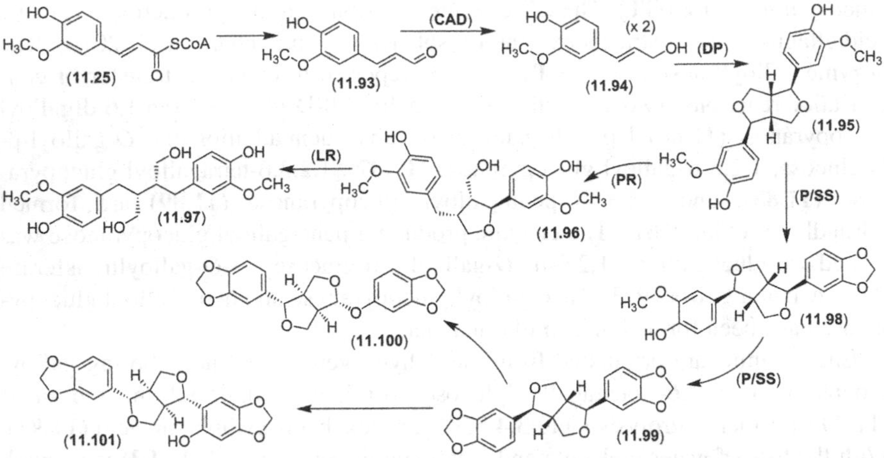

Fig. 11.19 Biosynthesis of lignans in foods. Enzymes include enzymes of shikimate pathway and coniferyl alcohol dehydrogenase (**CAD**), dirigent protein (**DP**), pinoresinol reductase (**PR**), lariciresinol reductase (**LR**), piperitol/sesamin synthase (**P/SS**). Metabolites are feruloyl-CoA (**11.25**), coniferaldehyde (**11.93**), coniferyl alcohol (**11.94**), pinoresinol (**11.95**), lariciresinol (**11.96**), secoisolariciresinol (**11.97**), piperitol (**11.98**), sesamin (**11.99**), sesamolin (**11.100**), and sesaminol (**11.101**)

11.14 Biosynthesis of Glycosides

Glycosides are a diverse group of phenolic compounds with antioxidant properties. The sugar residue may be a monosaccharide or oligosaccharide. The glucose residue is obtained from photosynthesis and is converted to form uridine diphosphate glucose UDP-glucose (**11.102**). The UDP-glucose is used as a carrier of glucose to a phenolic compound to form glycoside and release uridine diphosphate (**11.103**) as shown in Fig. 11.20. The enzyme was found to be specific for phenol β-glucosides and substituted phenol β-glucosides (such as arbutin, salicin, p- and m-methoxyphenol glucoside, resorcinol glucoside, and mandelonitrile glucoside) and is known as glycosyltransferase (**GT**) (Yang et al. 2018). This enzyme is central in poplar salicinoid phenolic glycoside biosynthesis (Fellenberg et al. 2020). The enzyme β-glucosidase is responsible for providing the glucose or sugar moiety to the phenolic compound (Bassanini et al. 2019).

The biosynthesis of flavonoid glycosides has been well reported in the literature (Yang et al. 2018). For example in citrus, hesperetin (**11.60**) is converted to hesperetin-7-glucoside (**11.104**) by UDP-glucose: flavanone-7-O-glucosyl-transferase (**UFGT**) as shown in Fig. 11.21. UDP-rhamnose: flavanone-glucoside rhamnosyl-transferase (**UFGRT**) adds another sugar residue to form neohesperedin (**11.105**) as reported by Lewinsohn et al. (1989).

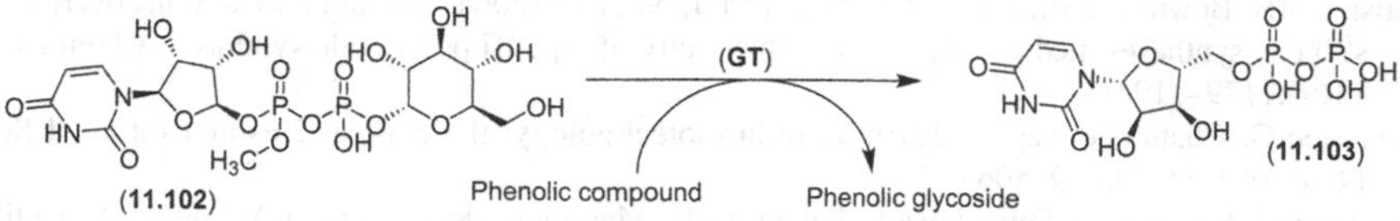

Fig. 11.20 Representative Glycosylation reaction in foods. The enzyme is glycosyltransferase (GT) and metabolites are uridine diphosphate glucose (**11.102**) and uridine diphosphate (**11.103**)

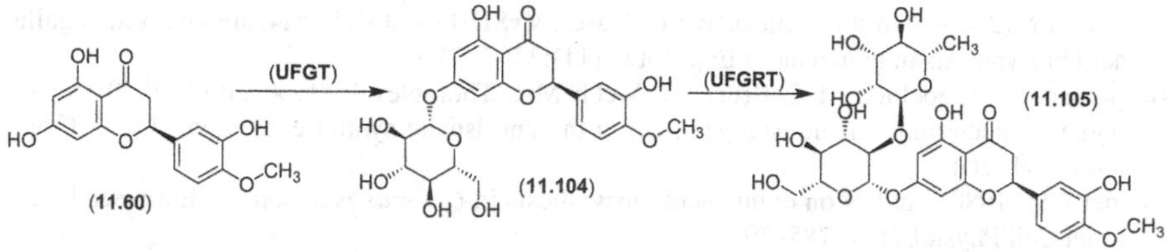

Fig. 11.21 Glycosylation reaction of flavonoids in foods. Enzymes are UDP-glucose: flavanone-7-o-glucosyl-transferase (**UFGT**) and UDP-rhamnose: flavanone-glucoside rhamnosyl-transferase (**UFGRT**). Metabolites are hesperetin (**11.60**), hesperetin-7-glucoside (**11.104**), and neohesperedin (**11.105**)

11.15 Study Questions

- Discuss the shikimate pathway for the biosynthesis of phenolic acids.
- How chlorogenic acids are biosynthesized?
- Write a detailed note on the biosynthesis of stilbenes and coumarins.
- How flavonoids are biosynthesized in foods?
- What do you know about the biosynthesis of anthocyanins?
- How phenolic glycosides are formed?

References

Alagna F, Mariotti R, Panara F, Caporali S, Urbani S, Veneziani G, Esposto S, Taticchi A, Rosati A, Rao R, Perrotta G, Servili M, Baldoni L (2012) Olive phenolic compounds: metabolic and transcriptional profiling during fruit development. BMC Plant Biol 12(1):162

Amaral JC, da Silva MM, da Silva MFGF, Alves TC, Ferreira AG, Forim MR, Fernandes JB, Pina ES, Lopes AA, Pereira AMS, Novelli VM (2020) Advances in the biosynthesis of pyranocoumarins: isolation and 13C-incorporation analysis by high-performance liquid chromatography–ultraviolet–solid-phase extraction–nuclear magnetic resonance data. J Nat Prod 83(5):1409–1415

Andersen JR, Zein I, Wenzel G, Darnhofer B, Eder J, Ouzunova M, Lübberstedt T (2008) Characterization of phenylpropanoid pathway genes within European maize (*Zea mays* L.) inbreds. BMC Plant Biol 8(1):2

Austin MB, Noel JP (2003) The chalcone synthase superfamily of type III polyketide synthases. Nat Prod Rep 20(1):79–110

Austin MB, Bowman ME, Ferrer J-L, Schröder J, Noel JP (2004) An aldol switch discovered in stilbene synthases mediates cyclization specificity of type III polyketide synthases. Chem Biol 11(9):1179–1194

Banerjee G, Chattopadhyay P (2019) Vanillin biotechnology: the perspectives and future. J Sci Food Agric 99(2):499–506

Bassanini I, Kapešová J, Petrásková L, Pelantová H, Markošová K, Rebroš M, Valentová K, Kotik M, Káňová K, Bojarová P (2019) Glycosidase-catalyzed synthesis of glycosyl esters and phenolic glycosides of aromatic acids. Adv Syn Catal 361(11):2627–2637

Bogs J, Ebadi A, McDavid D, Robinson SP (2006) Identification of the flavonoid hydroxylases from grapevine and their regulation during fruit development. Plant Physiol 140(1):279–291

Bontpart T, Marlin T, Vialet S, Guiraud J-L, Pinasseau L, Meudec E, Sommerer N, Cheynier V, Terrier N (2016) Two shikimate dehydrogenases, VvSDH3 and VvSDH4, are involved in gallic acid biosynthesis in grapevine. J Exp Bot 67(11):3537–3550

Bornemann S, Theoclitou M-E, Brune M, Webb MR, Thorneley RNF, Abell C (2000) A secondary β deuterium kinetic isotope effect in the chorismate synthase reaction. Bioorg Chem 28(4):191–204

Boudet AM (1980) Studies on quinic acid biosynthesis in Quercus pedunculata Ehrh. seedlings. Plant Cell Physiol 21(5):785–792

Candeias NR, Assoah B, Simeonov SP (2018) Production and synthetic modifications of Shikimic acid. Chem Rev 118(20):10458–10550

Cao G, Liu Y, Zhang S, Yang X, Chen R, Zhang Y, Lu W, Liu Y, Wang J, Lin M (2012) A novel 5-enolpyruvylshikimate-3-phosphate synthase shows high glyphosate tolerance in Escherichia coli and tobacco plants. PLoS One 7(6):e38718

Chadha KC, Brown SA (1974) Biosynthesis of phenolic acids in tomato plants infected with Agrobacterium tumefaciens. Can J Bot 52(9):2041–2047

Chen Z, Zheng Z, Huang J, Lai Z, Fan B (2009) Biosynthesis of salicylic acid in plants. Plant Signal Behav 4(6):493–496

Chen J, Tang X, Ren C, Wei B, Wu Y, Wu Q, Pei J (2018) Full-length transcriptome sequences and the identification of putative genes for flavonoid biosynthesis in safflower. BMC Genomics 19(1):548

Chong J, Poutaraud A, Hugueney P (2009) Metabolism and roles of stilbenes in plants. Plant Sci 177(3):143–155

Chung D, Kim SY, Ahn J-H (2017) Production of three phenylethanoids, tyrosol, hydroxytyrosol, and salidroside, using plant genes expressing in Escherichia coli. Sci Rep 7(1):2578

Clifford MN, Jaganath IB, Ludwig IA, Crozier A (2017) Chlorogenic acids and the acyl-quinic acids: discovery, biosynthesis, bioavailability and bioactivity. Nat Prod Rep 34(12):1391–1421

Colanero S, Tagliani A, Perata P, Gonzali S (2020) Alternative splicing in the anthocyanin fruit gene encoding an R2R3 MYB transcription factor affects anthocyanin biosynthesis in tomato fruits. Plant Commun 1(1):100006

Coquoz J-L, Buchala A, Métraux J-P (1998) The biosynthesis of salicylic acid in potato plants. Plant Physiol 117(3):1095–1101

Cunha AG, Brito ES, Moura CF, Ribeiro PR, Miranda MR (2017) UPLC-qTOF-MS/MS-based phenolic profile and their biosynthetic enzyme activity used to discriminate between cashew apple (Anacardium occidentale L.) maturation stages. J Chromatogr B Anal Technol Biomed Life Sci 1051:24–32

Dao TTH, Linthorst HJM, Verpoorte R (2011) Chalcone synthase and its functions in plant resistance. Phytochem Rev 10(3):397

Davin LB, Lewis NG (2003) An historical perspective on lignan biosynthesis: monolignol, allylphenol and hydroxycinnamic acid coupling and downstream metabolism. Phytochem Rev 2(3):257

Davin LB, Wang H-B, Crowell AL, Bedgar DL, Martin DM, Sarkanen S, Lewis NG (1997) Stereoselective bimolecular phenoxy radical coupling by an auxiliary (dirigent) protein without an active center. Science 275(5298):362–367

Dixon RA, Xie D-Y, Sharma SB (2005) Proanthocyanidins—a final frontier in flavonoid research? New Phytol 165(1):9–28

Dubrovina AS, Kiselev KV (2017) Regulation of stilbene biosynthesis in plants. Planta 246(4):597–623

el-Basyouni SZ, Chen D, Ibrahim RK, Neish AC, Towers GHN (1964) The biosynthesis of hydroxybenzoic acids in higher plants. Phytochemistry 3(4):485–492

El-Seedi HR, El-Said AMA, Khalifa SAM, Göransson U, Bohlin L, Borg-Karlson A-K, Verpoorte R (2012) Biosynthesis, natural sources, dietary intake, pharmacokinetic properties, and biological activities of hydroxycinnamic acids. J Agric Food Chem 60(44):10877–10895

Emir A, Emir C, Yildirim H (2020) Characterization of phenolic profile by LC-ESI-MS/MS and enzyme inhibitory activities of two wild edible garlic: *Allium nigrum* L. and *Allium subhirsutum* L. J Food Biochem 44(4):e13165

Escamilla-Treviño LL, Shen H, Hernandez T, Yin Y, Xu Y, Dixon RA (2014) Early lignin pathway enzymes and routes to chlorogenic acid in switchgrass (*Panicum virgatum* L.). Plant Mol Biol 84(4):565–576

Fellenberg C, Corea O, Yan L-H, Archinuk F, Piirtola E-M, Gordon H, Reichelt M, Brandt W, Wulff J, Ehlting J, Peter Constabel C (2020) Discovery of salicyl benzoate UDP-glycosyltransferase, a central enzyme in poplar salicinoid phenolic glycoside biosynthesis. Plant J 102(1):99–115

Gallage NJ, Møller BL (2015) Vanillin–bioconversion and bioengineering of the most popular plant flavor and its de novo biosynthesis in the vanilla orchid. Mol Plant 8(1):40–57

Gallage NJ, Hansen EH, Kannangara R, Olsen CE, Motawia MS, Jørgensen K, Holme I, Hebelstrup K, Grisoni M, Møller BL (2014) Vanillin formation from ferulic acid in *Vanilla planifolia* is catalysed by a single enzyme. Nat Commun 5(1):4037

García-Calderón M, Pérez-Delgado CM, Palove-Balang P, Betti M, Márquez AJ (2020) Flavonoids and isoflavonoids biosynthesis in the model legume lotus japonicus; connections to nitrogen metabolism and photorespiration. Plants (Basel) 9(6):774

Grundhöfer P, Niemetz R, Schilling G, Gross GG (2001) Biosynthesis and subcellular distribution of hydrolyzable tannins. Phytochemistry 57(6):915–927

Gu F, Chen Y, Hong Y, Fang Y, Tan L (2017) Comparative metabolomics in vanilla pod and vanilla bean revealing the biosynthesis of vanillin during the curing process of vanilla. AMB Express 7(1):116

Guodong R, Jianguo Z, Xiaoxia L, Ying L (2019) Identification of putative genes for polyphenol biosynthesis in olive fruits and leaves using full-length transcriptome sequencing. Food Chem 300:125246

Gurdon C, Poulev A, Armas I, Satorov S, Tsai M, Raskin I (2019) Genetic and phytochemical characterization of lettuce flavonoid biosynthesis mutants. Sci Rep 9(1):3305

Hagenah S, Gross GG (1993) Biosynthesis of 1,2,3,6-tetra-*O*-galloyl-β-d-glucose. Phytochemistry 32(3):637–641

Haslam E (1992) Gallic acid and its metabolites. In: Hemingway RW, Laks PE (eds) Plant polyphenols: synthesis, properties, significance. Springer US, Boston, pp 169–194

Haslam E, Cai Y (1994) Plant polyphenols (vegetable tannins): gallic acid metabolism. Nat Prod Rep 11:41–66

He F, Pan QH, Shi Y, Duan CQ (2008) Biosynthesis and genetic regulation of proanthocyanidins in plants. Molecules 13(10):2674–2703

He F, Mu L, Yan G-L, Liang N-N, Pan Q-H, Wang J, Reeves MJ, Duan C-Q (2010) Biosynthesis of anthocyanins and their regulation in colored grapes. Molecules 15(12):9057–9091

Hemingway RW, Laks PE (1985) Condensed tannins: a proposed route to 2 R, 3 R-(2, 3-cis)-proanthocyanidins. J Chem Soc Chem Commun 11:746–747

Herrmann KM, Weaver LM (1999) The shikimate pathway. Annu Rev Plant Biol 50(1):473–503

Huang K, Li M, Liu Y, Zhu M, Zhao G, Zhou Y, Zhang L, Wu Y, Dai X, Xia T, Gao L (2019) Functional analysis of 3-dehydroquinate dehydratase/shikimate dehydrogenases involved in shikimate pathway in *Camellia sinensis*. Front Plant Sci 10:1268

Jaillon O, Aury J-M, Noel B, Policriti A, Clepet C, Casagrande A, Choisne N, Aubourg S, Vitulo N, Jubin C, Vezzi A, Legeai F, Hugueney P, Dasilva C, Horner D, Mica E, Jublot D, Poulain J, Bruyère C, Billault A, Segurens B, Gouyvenoux M, Ugarte E, Cattonaro F, Anthouard V, Vico V, Del Fabbro C, Alaux M, Di Gaspero G, Dumas V, Felice N, Paillard S, Juman I, Moroldo M, Scalabrin S, Canaguier A, Le Clainche I, Malacrida G, Durand E, Pesole G, Laucou V, Chatelet P, Merdinoglu D, Delledonne M, Pezzotti M, Lecharny A, Scarpelli C, Artiguenave F, Pè ME, Valle G, Morgante M, Caboche M, Adam-Blondon A-F, Weissenbach J, Quétier F, Wincker P, The French–Italian Public Consortium for Grapevine Genome C (2007) The grapevine genome sequence suggests ancestral hexaploidization in major angiosperm phyla. Nature 449(7161):463–467

Jeandet P, Delaunois B, Conreux A, Donnez D, Nuzzo V, Cordelier S, Clément C, Courot E (2010) Biosynthesis, metabolism, molecular engineering, and biological functions of stilbene phytoalexins in plants. Biofactors 36(5):331–341

Kim YM, Park YS, Park Y-K, Ham K-S, Kang S-G, Shafreen RMB, Lakshmi SA, Gorinstein S (2020) Characterization of bioactive ligands with antioxidant properties of kiwifruit and persimmon cultivars using in vitro and in silico studies. Appl Sci Basel 10(12):4218

Klämbt HD (1962) Conversion in plants of benzoic acid to salicylic acid and its β-D-glucosid. Nature 196(4853):491–491

Kojima M, Uritani I (1973) Studies on chlorogenic acid biosynthesis in sweet potato root tissue in special reference to the isolation of a chlorogenic acid intermediate. Plant Physiol 51(4):768–771

Kundu A (2017) Vanillin biosynthetic pathways in plants. Planta 245(6):1069–1078

Landete JM (2011) Ellagitannins, ellagic acid and their derived metabolites: a review about source, metabolism, functions and health. Food Res Int 44(5):1150–1160

Lanz T, Schröder G, Schröder J (1990) Differential regulation of genes for resveratrol synthase in cell cultures of *Arachis hypogaea* L. Planta 181(2):169–175

Latif S, Weston PA, Barrow RA, Gurusinghe S, Piltz JW, Weston LA (2020) Metabolic profiling provides unique insights to accumulation and biosynthesis of key secondary metabolites in annual pasture legumes of mediterranean origin. Meta 10(7):267

Lebedev VG, Subbotina NM, Maluchenko OP, Krutovsky KV, Shestibratov KA (2019) Assessment of genetic diversity in differently colored raspberry cultivars using SSR markers located in flavonoid biosynthesis genes. Agronomy 9(9):518

Lefevere H, Bauters L, Gheysen G (2020) Salicylic acid biosynthesis in plants. Front Plant Sci 11:338

Legrand G, Delporte M, Khelifi C, Harant A, Vuylsteker C, Mörchen M, Hance P, Hilbert J-L, Gagneul D (2016) Identification and characterization of five BAHD acyltransferases involved in hydroxycinnamoyl ester metabolism in chicory. Front Plant Sci 7:741

Lewinsohn E, Britsch L, Mazur Y, Gressel J (1989) Flavanone glycoside biosynthesis in citrus. Plant Physiol 91(4):1323–1328

Liang D, Shen Y, Ni Z, Wang Q, Lei Z, Xu N, Deng Q, Lin L, Wang J, Lv X, Xia H (2018) Exogenous melatonin application delays senescence of kiwifruit leaves by regulating the antioxidant capacity and biosynthesis of flavonoids. Front Plant Sci 9:426

Lunkenbein S, Bellido M, Aharoni A, Salentijn EMJ, Kaldenhoff R, Coiner HA, Muñoz-Blanco J, Schwab W (2006) Cinnamate metabolism in ripening fruit. Characterization of a UDP-glucose:cinnamate glucosyltransferase from strawberry. Plant Physiol 140(3):1047–1058

Maeda H, Dudareva N (2012) The shikimate pathway and aromatic amino acid biosynthesis in plants. Annu Rev Plant Biol 63(1):73–105

Marchiosi R, dos Santos WD, Constantin RP, de Lima RB, Soares AR, Finger-Teixeira A, Mota TR, de Oliveira DM, Foletto-Felipe MdP, Abrahão J, Ferrarese-Filho O (2020) Biosynthesis and metabolic actions of simple phenolic acids in plants. Phytochem Rev.

Matsui K, Walker AR (2020) Biosynthesis and regulation of flavonoids in buckwheat. Breed Sci 70(1):74–84

Maugé C, Granier T, d'Estaintot BL, Gargouri M, Manigand C, Schmitter J-M, Chaudière J, Gallois B (2010) Crystal structure and catalytic mechanism of leucoanthocyanidin reductase from *Vitis vinifera*. J Mol Biol 397(4):1079–1091

Moglia A, Lanteri S, Comino C, Hill L, Knevitt D, Cagliero C, Rubiolo P, Bornemann S, Martin C (2014) Dual catalytic activity of hydroxycinnamoyl-coenzyme a quinate transferase from tomato allows it to moonlight in the synthesis of both mono- and dicaffeoylquinic acids. Plant Physiol 166(4):1777–1787

Mougiou N, Trikka F, Trantas E, Ververidis F, Makris A, Argiriou A, Vlachonasios KE (2018) Expression of hydroxytyrosol and oleuropein biosynthetic genes are correlated with metabolite accumulation during fruit development in olive, Olea europaea, cv. Koroneiki. Plant Physiol Biochem 128:41–49

Muir RM, Ibáñez AM, Uratsu SL, Ingham ES, Leslie CA, McGranahan GH, Batra N, Goyal S, Joseph J, Jemmis ED (2011) Mechanism of gallic acid biosynthesis in bacteria (*Escherichia coli*) and walnut (*Juglans regia*). Plant Mol Biol 75(6):555–565

Murata J, Ono E, Yoroizuka S, Toyonaga H, Shiraishi A, Mori S, Tera M, Azuma T, Nagano AJ, Nakayasu M, Mizutani M, Wakasugi T, Yamamoto MP, Horikawa M (2017) Oxidative rearrangement of (+)-sesamin by CYP92B14 co-generates twin dietary lignans in sesame. Nat Commun 8(1):2155

Nabavi SM, Šamec D, Tomczyk M, Milella L, Russo D, Habtemariam S, Suntar I, Rastrelli L, Daglia M, Xiao J, Giampieri F, Battino M, Sobarzo-Sanchez E, Nabavi SF, Yousefi B, Jeandet P, Xu S, Shirooie S (2020) Flavonoid biosynthetic pathways in plants: versatile targets for metabolic engineering. Biotechnol Adv 38:107316

Niemetz R, Gross GG (2005) Enzymology of gallotannin and ellagitannin biosynthesis. Phytochemistry 66(17):2001–2011

Osman NI, Jaafar Sidik N, Awal A (2018) Efficient enhancement of gallic acid accumulation in cell suspension cultures of *Barringtonia racemosa* L. by elicitation. Plant Cell Tissue Organ Cult 135(2):203–212

Ossipov V, Salminen J-P, Ossipova S, Haukioja E, Pihlaja K (2003) Gallic acid and hydrolysable tannins are formed in birch leaves from an intermediate compound of the shikimate pathway. Biochem Syst Ecol 31(1):3–16

Petit P, Granier T, d'Estaintot BL, Manigand C, Bathany K, Schmitter J-M, Lauvergeat V, Hamdi S, Gallois B (2007) Crystal structure of grape dihydroflavonol 4-reductase, a key enzyme in flavonoid biosynthesis. J Mol Biol 368(5):1345–1357

Petrussa E, Braidot E, Zancani M, Peresson C, Bertolini A, Patui S, Vianello A (2013) Plant flavonoids—biosynthesis, transport and involvement in stress responses. Int J Mol Sci 14(7):14950–14973

Rosler J, Krekel F, Amrhein N, Schmid J (1997) Maize phenylalanine ammonia-lyase has tyrosine ammonia-lyase activity. Plant Physiol 113(1):175–179

Rousserie P, Rabot A, Geny-Denis L (2019) From flavanols biosynthesis to wine tannins: what place for grape seeds? J Agric Food Chem 67(5):1325–1343

Saijo R (1983) Pathway of gallic acid biosynthesis and its esterification with catechins in young tea shoots. Agric Biol Chem 47(3):455–460

Santos-Sánchez NF, Salas-Coronado R, Hernández-Carlos B, Villanueva-Cañongo C (2019) Shikimic acid pathway in biosynthesis of phenolic compounds. In: Soto-Hernández M, García-Mateos R, Palma-Tenango M (eds) Plant physiological aspects of phenolic compounds. IntechOpen, London

Schmid J, Amrhein N (1995) Molecular organization of the shikimate pathway in higher plants. Phytochemistry 39(4):737–749

Schulenburg K, Feller A, Hoffmann T, Schecker JH, Martens S, Schwab W (2016) Formation of β-glucogallin, the precursor of ellagic acid in strawberry and raspberry. J Exp Bot 67(8):2299–2308

Sellin C, Forlani G, Dubois J, Nielsen E, Vasseur J (1992) Glyphosate tolerance in *Cichorium-Intybus* L Var Magdebourg. Plant Sci 85(2):223–231

Shimizu B-I (2014) 2-Oxoglutarate-dependent dioxygenases in the biosynthesis of simple couma-
    rins. Front Plant Sci 5:549
Shimizu T, Kojima M (1984) Partial purification and characterization of UDPG: t-cinnamate
    glucosyltransferase in the root of sweet potato, *Ipomoea batatas* Lam. J Biochem Tokyo
    95(1):205–212
Silverman P, Seskar M, Kanter D, Schweizer P, Metraux J-P, Raskin I (1995) Salicylic acid in rice
    (biosynthesis, conjugation, and possible role). Plant Physiol 108(2):633–639
Siwinska J, Siatkowska K, Olry A, Grosjean J, Hehn A, Bourgaud F, Meharg AA, Carey M,
    Lojkowska E, Ihnatowicz A (2018) Scopoletin 8-hydroxylase: a novel enzyme involved in cou-
    marin biosynthesis and iron-deficiency responses in Arabidopsis. J Exp Bot 69(7):1735–1748
Sparvoli F, Martin C, Scienza A, Gavazzi G, Tonelli C (1994) Cloning and molecular analysis of
    structural genes involved in flavonoid and stilbene biosynthesis in grape (*Vitis vinifera* L.).
    Plant Mol Biol 24(5):743–755
Srinivasulu C, Ramgopal M, Ramanjaneyulu G, Anuradha CM, Suresh Kumar C (2018) Syringic
    acid (SA)—a review of its occurrence, biosynthesis, pharmacological and industrial impor-
    tance. Biomed Pharmacother 108:547–557
Strack D, Sharma V (1985) Vacuolar localization of the enzymatic synthesis of hydroxycinnamic
    acid esters of malic acid in protoplasts from *Raphanus sativus* leaves. Physiol Plantarum
    65(1):45–50
Strack D, Keller H, Weissenbock G (1987) Enzymatic synthesis of hydroxycinnamic acid esters of
    sugar acids and hydroaromatic acids by protein preparations from rye (*Secale cereale*) primary
    leaves. J Plant Physiol 131(1):61–73
Suh D-Y, Kagami J, Fukuma K, Sankawa U (2000) Evidence for catalytic cysteine–histidine dyad
    in chalcone synthase. Biochem Biophys Res Commun 275(3):725–730
Suzuki S, Umezawa T (2007) Biosynthesis of lignans and norlignans. J Wood Sci 53(4):273–284
Takenaka M, Nanayama K, Isobe S, Murata M (2006) Changes in caffeic acid derivatives in
    sweet potato (*Ipomoea batatas* L.) during cooking and processing. Biosci Biotechnol Biochem
    70(1):172–177
Tang PL, Hassan O (2020) Bioconversion of ferulic acid attained from pineapple peels and pineap-
    ple crown leaves into vanillic acid and vanillin by *Aspergillus niger* I-1472. BMC Chem 14(1):7
Tanner GJ, Francki KT, Abrahams S, Watson JM, Larkin PJ, Ashton AR (2003) Proanthocyanidin
    biosynthesis in plants purification of legume leucoanthocyanidin reductase and molecular clon-
    ing of its cDNA. J Biol Chem 278(34):31647–31656
Tohge T, Alisdair RF (2017) An Overview of compounds derived from the shikimate and phenyl-
    propanoid pathways and their medicinal importance. Mini Rev Med Chem 17(12):1013–1027
Touré A, Xueming X (2010) Flaxseed lignans: source, biosynthesis, metabolism, antioxidant activ-
    ity, bio-active components, and health benefits. Compr Rev Food Sci Food Saf 9(3):261–269
Trivellini A, Lucchesini M, Maggini R, Mosadegh H, Villamarin TSS, Vernieri P, Mensuali-Sodi
    A, Pardossi A (2016) Lamiaceae phenols as multifaceted compounds: bioactivity, industrial
    prospects and role of "positive-stress". Ind Crop Prod 83:241–254
Tropf S, Kärcher B, Schröder G, Schröder J (1995) Reaction mechanisms of homodimeric plant
    polyketide synthases (stilbene and chalcone synthase): a single active site for the condensing
    reaction is sufficient for synthesis of stilbenes, chalcones, and 6′-deoxychalcones. J Biol Chem
    270(14):7922–7928
Tzin V, Galili G (2010) The biosynthetic pathways for shikimate and aromatic amino acids in
    Arabidopsis thaliana. Arabidopsis Book 8:e0132
Vanholme R, Sundin L, Seetso KC, Kim H, Liu X, Li J, De Meester B, Hoengenaert L, Goeminne
    G, Morreel K, Haustraete J, Tsai H-H, Schmidt W, Vanholme B, Ralph J, Boerjan W (2019)
    COSY catalyses trans–cis isomerization and lactonization in the biosynthesis of coumarins.
    Nat Plants 5(10):1066–1075
Veitch NC (2007) Isoflavonoids of the leguminosae. Nat Prod Rep 24(2):417–464
Vermerris W, Nicholson R (2007) Phenolic compound biochemistry. Springer Science & Business
    Media, Berlin

Vogt T (2010) Phenylpropanoid biosynthesis. Mol Plant 3(1):2–20

Wang P, Liu Y, Zhang L, Wang W, Hou H, Zhao Y, Jiang X, Yu J, Tan H, Wang Y, Xie D-Y, Gao L, Xia T (2020) Functional demonstration of plant flavonoid carbocations proposed to be involved in the biosynthesis of proanthocyanidins. Plant J 101(1):18–36

Weaver LM, Herrmann KM (1997) Dynamics of the shikimate pathway in plants. Trends Plant Sci 2(9):346–351

Werner I, Bacher A, Eisenreich W (1997) Retrobiosynthetic NMR studies with 13C-labeled glucose: formation of gallic acid in plants and fungi. J Biol Chem 272(41):25474–25482

Werner I, Bacher A, Eisenreich W (1999) Analysis of gallic acid biosynthesis via quantitative prediction of isotope labeling patterns. In: Gross GG, Hemingway RW, Yoshida T, Branham SJ (eds) Plant polyphenols 2: chemistry, biology, pharmacology, ecology. Springer, Boston, pp 43–61

Wickramasinghe PCK, Munafo JP (2020) Biosynthesis of benzylic derivatives in the fermentation broth of the edible mushroom, *Ischnoderma resinosum*. J Agric Food Chem 68(8):2485–2492

Widhalm Joshua R, Dudareva N (2015) A familiar ring to it: biosynthesis of plant benzoic acids. Mol Plant 8(1):83–97

Xu M, Rao J, Chen B (2020) Phenolic compounds in germinated cereal and pulse seeds: classification, transformation, and metabolic process. Crit Rev Food Sci Nutr 60(5):740–759

Yang B, Liu H, Yang J, Gupta VK, Jiang Y (2018) New insights on bioactivities and biosynthesis of flavonoid glycosides. Trends Food Sci Technol 79:116–124

Yao R, Zhao Y, Liu T, Huang C, Xu S, Sui Z, Luo J, Kong L (2017) Identification and functional characterization of a p-coumaroyl CoA 2′-hydroxylase involved in the biosynthesis of coumarin skeleton from *Peucedanum praeruptorum* Dunn. Plant Mol Biol 95(1):199–213

Yu O, Jez JM (2008) Nature's assembly line: biosynthesis of simple phenylpropanoids and polyketides. Plant J 54(4):750–762

Zabalza A, Órcaray L, Fernández-Escalada M, Zulet-González A, Royuela M (2017) The pattern of shikimate pathway and phenylpropanoids after inhibition by glyphosate or quinate feeding in pea roots. Pest Biochem Phys 141:96–102

Zang Y, Zha J, Wu X, Zheng Z, Ouyang J, Koffas MAG (2019) In vitro naringenin biosynthesis from p-coumaric acid using recombinant enzymes. J Agric Food Chem 67(49):13430–13436

Zeb A (2019) Chemo-metric analysis of the polyphenolic profile of *Cichorium intybus* L. leaves grown on different water resources of Pakistan. J Food Meas Charact 13(1):728–734

Zeb A (2020) Concept, mechanism, and applications of phenolic antioxidants in foods. J Food Biochem 44(9):e13394

Zeb A, Hussain A (2020) Chemo-metric analysis of carotenoids, chlorophylls, and antioxidant activity of *Trifolium hybridum*. Heliyon 6(1):e03195

Zeb A, Imran M (2019) Carotenoids, pigments, phenolic composition and antioxidant activity of *Oxalis corniculata* leaves. Food Biosci 32:100472

Zhao Y, Jian X, Wu J, Huang W, Huang C, Luo J, Kong L (2019) Elucidation of the biosynthesis pathway and heterologous construction of a sustainable route for producing umbelliferone. J Biol Eng 13(1):44

Zhao C, Wang F, Lian Y, Xiao H, Zheng J (2020a) Biosynthesis of citrus flavonoids and their health effects. Crit Rev Food Sci Nutr 60(4):566–583

Zhao S, Xi X, Zong Y, Li S, Li Y, Cao D, Liu B (2020b) Overexpression of ThMYC4E enhances anthocyanin biosynthesis in common wheat. Int J Mol Sci 21(1):137

Zhou Y, Zhu J, Shao L, Guo M (2020) Current advances in acteoside biosynthesis pathway elucidation and biosynthesis. Fitoterapia 142:104495

Zhu X-J, Zhao Z, Xin H-H, Wang M-L, Wang W-D, Chen X, Li X-H (2016) Isolation and dynamic expression of four genes involving in shikimic acid pathway in *Camellia sinensis* 'Baicha 1' during periodic albinism. Mol Biol Rep 43(10):1119–1127

Zhu T, Yang J, Zhang D, Cai Q, Zhou D, Tu S, Liu Q, Tu K (2020) Effects of white LED light and UV-C radiation on stilbene biosynthesis and phytochemicals accumulation identified by UHPLC–MS/MS during peanut (*Arachis hypogaea* L.) germination. J Agric Food Chem 68(21):5900–5909

Chapter 12
Metabolism of Phenolic Antioxidants

Learning Objectives

In this chapter, the reader will be able to:
- Learn the *in-vitro* metabolism of phenolic compounds.
- Understand the bioaccessibility and bioavailability of phenolic compounds.
- Know how phenolic compounds are made bioavailable and metabolized in humans.

12.1 Introduction

Phenolic compounds are consumed in the form of foods. Consumption of phenolic-rich foods has shown protective and beneficial properties in humans. Upon consumption, phenolic compounds are modified to form new compounds with potential health benefits or interact with other primary metabolites. In this chapter, the metabolism of the phenolic compound with antioxidant properties include the catabolic or anabolic reactions. However, it worth mentioning that the metabolism of phenolic compounds is first studied in the in-vitro system followed by in-vivo evaluation. The studies on the *in-vitro* model system can offer valuable insights that are relevant to human health only if they are carefully designed and critically interpreted. It is obvious that to use any biological action, phenolic compounds and/or their *in-vivo* metabolites must reach the appropriate internal compartment at concentrations that are sufficient and maintained for an adequate period of time. Despite several drawbacks of the *in-vitro* studies on phenolic compounds, it is still the only way to migrate to the *in-vivo* system.

© Springer Nature Switzerland AG 2021   333
A. Zeb, *Phenolic Antioxidants in Foods: Chemistry, Biochemistry and Analysis*,
https://doi.org/10.1007/978-3-030-74768-8_12

12.2 *In-Vitro* Metabolism of Phenolic Compounds

In-vitro studies have been carried out using several types of cultured cells. For example, the uptake and metabolism of the phenolic compounds present in the diet have been studied using intestinal and hepatic cells (Aragonès et al. 2017). These studies have demonstrated the biochemical changes occurring during the metabolism of phenolic compounds, which help identify the specific generated metabolites (Table 12.1). Phenolic compounds undergo several reactions such as glucuronidation, sulfonation, and methylation. Several different isoenzymes catalyze glucuronidation. These include isoforms of uridine diphosphate (UDP)-glucuronosyltransferase, depending on the type of cells assayed (Wong et al. 2009). The bioavailability and metabolism of phenolic compounds are significantly affected by conjugation and/or deconjugation reactions. These reactions can modify their ability to be transported through membranes and result in metabolic changes, thus mod their biological potentials (Warner et al. 2016). It has been suggested that the deconjugation of phenolic conjugated metabolites may be necessary before they can exert beneficial effects (Rodriguez-Mateos et al. 2014b). Metabolism of phenolic compounds is also organ-dependent and starts in the gut. Several studies have been dedicated to colonic cells because they are exposed to phenolic compounds in-vivo. In particular, Caco-2 cells can perform glucuronidation, sulfonation, and methylation reactions on several phenolic compounds (Kern et al. 2003). Several phenolic compounds and their conjugates have been detected and identified in Caco-2 cell lysates, confirming their active uptake and metabolism (Gonzalez-Sarrias et al. 2009). The liver is the most important organ responsible for the biotransformation of phenolic compounds. The HepG2 cells were found to mimic liver metabolism and perform glucuronidation, sulfonation, and methylation reactions (Mateos et al. 2006).

12.2.1 *Hydroxycinnamic Acids and Derivatives*

The principal metabolites formed after incubation of Caco-2 cells with different hydroxycinnamic acids such as caffeic, ferulic, *p*-coumaric, and sinapic acids are their glucuronidated and sulfated conjugates (Kern et al. 2003). A moderate uptake of caffeic and ferulic acids has been observed in human hepatic cells, where the former gives rise to a wide range of metabolites, including methylated, methylglucuronidated, glucuronidated, and sulfated conjugates, whereas the latter results only in sulfated or glucuronidated forms. Chlorogenic acid can be absorbed by human hepatoma HepG2 cells, but it does not undergo further intracellular metabolism (Mateos et al. 2006). On the other hand, primary rat hepatocytes can transform chlorogenic acid into its methyl esters (Kahle et al. 2011). This could be due to a significantly higher level of phase I and II enzyme expression in primary cells, as compared with hepatoma cells (Wilkening et al. 2003). In colonic adenocarcinoma cell models (Caco-2 and SW480), hydroxycinnamic acids with a catechol-type

Table 12.1 Some studies where the metabolism of phenolic compounds has been studied in intestinal, hepatic cells and other types of cell lines

Cell model	Cell line	Parent compound	Newly-formed metabolites	Cell lysates/ Cell media	Reference
Human colon cancer cells	Caco-2	Quercetin	Quercetin glucuronides; quercetin sulphates; methyl-quercetin	Cells and media	del Mar et al. 2016)
		Quercetin 3-O-glucoside	Methylquercetin-3-glucoside		
		Quercetin 3-O-glucuronide	Methylquercetin-3-glucuronide		
	HT-29	Quercetin	Quercetin 3-glucuronide; quercetin 3′-glucuronide; quercetin 4′-glucuronide; quercetin 3′-methyl-3-glucuronide; quercetin 3′-methyl-4′-glucuronide; quercetin 4′methyl-3′-glucuronide	Media	van der Woude et al. 2004)
			3′-Methylquercetin; 4′-methylquercetin; 3′-methylquercetin glucuronide; 4′-methylquercetin glucuronide; 3′-methylquercetin sulphate; 4′-methylquercetin sulphate	Cells and media	de Boer et al. 2006)
	Caco-2/TC7 cells	Kaempferol	Kaempferol-3-glucuronide; kaempferol-4′-glucuronide; kaempferol-7-glucuronide; kaempferol-sulphate	Cells and media	Barrington et al. 2009)
		Galangin	Galangin-3-glucuronide; galangin-5-glucuronide; galangin-7-glucuronide; galangin-sulphate	Cells and media	
	Caco-2	(−)-Epicatechin	(−)-Epicatechin-3′-β-D-glucuronide; 3′-O-Methyl-(−)-epicatechin-5-β-D-glucuronide; 3′-O-methyl-(−)-epicatechin-7-β-D-glucuronide; 4′-O-methyl-(−)-epicatechin-5-β-D-glucuronide; 4′-O-methyl-(−)-epicatechin-7-β-D-glucuronide; 3′-O-methyl-(−)-epicatechin-7-sulphate	Media (apical side)	Rodriguez-Mateos et al. 2014a)

(continued)

Table 12.1 (continued)

Cell model	Cell line	Parent compound	Newly-formed metabolites	Cell lysates/ Cell media	Reference
			3'-O-Methyl-(−)-epicatechin-7-β-D-glucuronide; 3'-O-methyl-(−)-epicatechin-7-sulphate	Cells	
			(−)-Epicatechin-3'-β-D-glucuronide; 3'-O-methyl-(−)-epicatechin-7-β-D-glucuronide; 4'-O-methyl-(−)-epicatechin-5-β-D-glucuronide; 3'-O-methyl-(−)-epicatechin-7-sulphate	Media (basolateral side)	
	Caco-2	(−)-Epicatechin-3'-β-d-glucuronide, (−)-epicatechin-3'-sulphate, 3'-O-methyl-(−)-epicatechin-5-sulphate, and 3'-O-methyl-(−)-epicatechin-7-sulphate	(no newly-formed metabolites)	Media (apical side)	Rodriguez-Mateos et al. 2014b)
			(no newly-formed metabolites)	Cells	
			(no newly-formed metabolites)	Media (basolateral side)	
	HT-29	(−)-Epigallocate-chin-3-gallate	(−)-Epigallocatechin-3-gallate-4''-O-glucuronide; 4''-O-methyl-(−)-epigallocatechin-3-gallate	Cells	Hong et al. (2002)

Caco-2	Apigenin	Apigenin glucuronide	Media (apical and basolateral sides)	Ng et al. (2005)
	Baicalein	Baicalein glucuronide		
	Chrysin	Chrysin glucuronide		
	Luteolin	Luteolin glucuronide		
Caco-2/TC7 cells	Genistein	Genistein glucuronide; genistein sulphate	Media (apical and basolateral sides)	Chen et al. (2005)
	Daidzein	Daidzein glucuronide; daidzein sulphate		
	Glycitein	Glycitein glucuronide; glycitein sulphate		
	Formononetin	Formononetin glucuronide; formononetin sulphate		
	Biochanin A	Biochanin A glucuronide; biochanin A sulphate		
	Prunetin	Prunetin glucuronide; prunetin sulphate		
Caco-2	Hesperetin	Hesperetin-7-O-glucuronide; hesperetin-7-O-sulphate	Media (apical and basolateral sides)	Brand et al. (2008)
Caco-2	Caffeic acid	Ferulic acid; isoferulic acid	Media (apical side)	Kern et al. (2003)
	Ferulic acid	Sulphate caffeic acid		
	p-Coumaric acid	Methyl-p-coumarate-sulphate; p-coumaric acid-sulphate; methyl-p-coumarate-glucuronide (tentatively identified)		
	Sinapic acid	Sinapic acid-sulphate		

(continued)

Table 12.1 (continued)

Cell model	Cell line	Parent compound	Newly-formed metabolites	Cell lysates/ Cell media	Reference
	Caco-2	Punicalagin	Ellagic acid; dimethyl ellagic acid; dimethyl ellagic acid-glucuronide; isomer of dimethyl ellagic acid-sulphate	Media	Larrosa et al. (2006)
			Dimethyl ellagic acid-glucuronide	Cells	
		Ellagic acid	Dimethyl ellagic acid; dimethyl ellagic acid-glucuronide; isomer of dimethyl ellagic acid-sulphate	Media	
			Dimethyl ellagic acid; dimethyl ellagic acid-glucuronide; isomers of dimethyl ellagic acid-sulphate; ellagic acid -derived metabolites (unidentified)	Cells	
		Urolithin A	Urolithin A glucuronide; urolithin A sulphate; urolithin A disulphate	Cells and media	González-Sarrías et al. (2009)
			Urolithin A glucuronide	Media	González-Sarrías et al. (2014)
			Urolithin A-3 glucuronide; urolithin A-8 glucuronide; urolithin A sulphate	Media	González-Sarrías et al. (2017)
		Isourolithin A	Isourolithin A-3-glucuronide; isourolithin A-9-glucuronide; isourolithin A-sulphate	Media	González-Sarrías et al. (2017)
		Urolithin B	Urolithin B glucuronide; urolithin B sulphate; urolithin A disulphate; urolithin A; urolithin A glucuronide	Cells and media	González-Sarrías et al. (2009)

Cell type	Substrate	Metabolites	Media	Reference
		Urolithin A glucuronide	Media	Gonzalez-Sarrias et al. (2014)
	Urolithin C	Glucuronide conjugates	Media	Gonzalez-Sarrias et al. (2014)
	Urolithin D	(no newly-formed metabolites)	Media	
Caco-2 adherent cells	Urolithin A, urolithin C, and ellagic acid	Ellagic acid; urolithin A; urolithin A-3 glucuronide; urolithin A-8 glucuronide; urolithin A sulphate; urolithin C glucuronide; urolithin C methyl-sulphate	Media	Núñez-Sánchez et al. (2016)
	Urolithin A, isourolithin A, urolithin B, urolithin C, and ellagic acid	Ellagic acid; urolithin A; urolithin A sulphate; Isourolithin A; isourolithin A-9 glucuronide; isourolithin A sulphate; urolithin B; urolithin B glucuronide; urolithin C glucuronide; urolithin C methyl-sulphate		
Caco-2 derived spheroids cultures	Urolithin A, urolithin C, and ellagic acid	Ellagic acid; urolithin A; urolithin A-3 glucuronide; urolithin A-8 glucuronide; urolithin C	Media	Núñez-Sánchez et al. (2016)
	Urolithin A, isourolithin A, urolithin B, urolithin C, and ellagic acid	Ellagic acid; urolithin A; urolithin A-3 glucuronide; urolithin A-8 glucuronide; isourolithin A; urolithin B; urolithin C		

(continued)

Table 12.1 (continued)

Cell model	Cell line	Parent compound	Newly-formed metabolites	Cell lysates/ Cell media	Reference
	HT-29	Urolithin A	Urolithin A-8-glucuronide; urolithin A-3-glucuronide	Media	González-Sarrías et al. (2017)
		Isourolithin A	Isourolithin A-9-glucuronide; isourolithin A-3-glucuronide	Media	González-Sarrías et al. (2017)
		Urolithin B	Urolithin B-glucuronide	Media	González-Sarrías et al. (2014)
		Urolithin C	Urolithin C-glucuronide; urolithin C-methyl-glucuronide	Media	González-Sarrías et al. (2014)
		Urolithin D	(no newly-formed metabolites)	Media	González-Sarrías et al. (2014)
	SW480	Urolithin A	(no newly-formed metabolites)	Media	González-Sarrías et al. 2014; González-Sarrías et al. (2017)
		Isourolithin A	(no newly-formed metabolites)	Media	González-Sarrías et al. (2017)
		Urolithin B	(no newly-formed metabolites)	Media	González-Sarrías et al. (2014)
		Urolithin C			
		Urolithin D			

CCD18-co	Ellagic acid	(no newly-formed metabolites)	Cells and media	Gonzalez-Sarrías et al. (2009)
	Urolithin A	(no newly-formed metabolites)	Cells and media	Gonzalez-Sarrías et al. (2009)
	Isourolithin A	(no newly-formed metabolites)	Media	González-Sarrías et al. (2017)
	Urolithin B	(no newly-formed metabolites)	Media	González-Sarrías et al. (2017)
		(no newly-formed metabolites)	Cells and media	Gonzalez-Sarrías et al. (2009)
Primary tumour cells from a patient with colorectal cancer	Urolithin A, urolithin C, and ellagic acid	Ellagic acid; urolithin A; urolithin A-3 glucuronide; urolithin A-8 glucuronide; urolithin C	Media	Núñez-Sánchez et al. (2016)
	Urolithin A, isourolithin A, urolithin B, urolithin C, and ellagic acid	Ellagic acid; urolithin A; urolithin A-3 glucuronide; urolithin A-8 glucuronide; isourolithin A; isourolithin A-3 glucuronide; urolithin B; urolithin B glucuronide; urolithin C		
Caco-2	Resveratrol	Resveratrol glucuronide; resveratrol sulphate (tentatively identified)	Media (apical and basolateral sides)	Kaldas et al. (2003)

(continued)

Table 12.1 (continued)

Cell model	Cell line	Parent compound	Newly-formed metabolites	Cell lysates/ Cell media	Reference
	HT-29	Resveratrol	Resveratrol-3-glucuronide; resveratrol-4'-O-glucuronide; resveratrol-3-O-sulphate	Media	Patel et al. (2013)
			Resveratrol-3-O-sulphate	Cells	
		Resveratrol-3-O-sulphate and resveratrol-4'-O-sulphate (3:2)	Resveratrol-4'-O-glucuronide; resveratrol sulphate glucuronides	Media	
			Resveratrol; resveratrol-4'-O-glucuronide; resveratrol sulphate glucuronide	Cells	
	HCA-7	Resveratrol	Resveratrol-3-glucuronide; resveratrol-4'-O-glucuronide; resveratrol-3-O-sulphate	Media	
			Resveratrol-3-O-sulphate	Cells	
		Resveratrol-3-O-sulphate and resveratrol-4'-O-sulphate (3:2)	Resveratrol-4'-O-glucuronide; resveratrol sulphate glucuronides	Media	
			Resveratrol	Cells	
	HCEC	Resveratrol	Resveratrol-4'-O-sulphate; resveratrol disulphate	Media	
			Resveratrol-3-O-sulphate	Cells	
		Resveratrol-3-O-sulphate and resveratrol-4'-O-sulphate (3:2)	Resveratrol-4'-O-glucuronide; resveratrol disulphate	Media	
			(no newly-formed metabolites)	Cells	

Cell model	Cell line	Parent compound	Newly-formed metabolites	Cell lysates/ Cell media	Reference
Human hepatoma cells	HepG2	Quercetin	Quercetin 3-glucuronide; quercetin 3′-gluc-uronide; quercetin 4′-glucuronide; quercetin 7-glucuronide, 3′-methylquercetin; 3′-methyl-quercetin-3-glucuronide; 3′-methylquercetin-4′-glucuronide; 3′-methylquercetin-7-glucuronide; quercetin-3′-sulphate	Media	O'Leary et al. (2003)
		Quercetin 3-glucuronide	Quercetin 3-glucuronide; quercetin 3′-gluc-uronide; 3′-methylquercetin; 4′-methylquerce-tin; quercetin 3′-methyl-3-glucuronide; quercetin 4′-methyl-3′-glucuronide; quercetin 3′-sulphate; quercetin 7-sulphate	Media	van der Woude et al. (2004)
		Quercetin 3-glucuronide	3′-Methylquercetin-3-glucuronide; 4′-methyl-quercetin-3-glucuronide; quercetin-3′-sulphate	Media	O'Leary et al. (2003)
		Quercetin 4′-glucuronide	(no newly-formed metabolites)		
		Quercetin 7-glucuronide	3′-Methylquercetin-7-glucuronide; 4′-methyl-quercetin-7-glucuronide; quercetin-3′-sulphate		

(continued)

Table 12.1 (continued)

Cell model	Cell line	Parent compound	Newly-formed metabolites	Cell lysates/ Cell media	Reference
	HepG2	(−)-Epicatechin	3'-O-Methyl-(−)-epicatechin-7-β-D-glucuronide; 4'-O-methyl-(−)-epicatechin-5-β-D-glucuronide; 3'-O-Methyl-(−)-epicatechin-7-sulphate	Media	Rodriguez-Mateos et al. 2014a)
			3'-O-methyl-(−)-epicatechin-7-β-D-glucuronide; 4'-O-methyl-(−)-epicatechin-5-β-D-glucuronide; 3'-O-Methyl-(−)-epicatechin-7-sulphate	Cells	
		(−)-epicatechin-3'-β-D-glucuronide, (−)-epicatechin-3'-sulphate, 3'-O-methyl-(−)-epicatechin-5-sulphate, and 3'-O-methyl-(−)-epicatechin-7-sulphate	(no newly-formed metabolites)	Cells and media	

Cell model	Cell line	Parent compound	Newly-formed metabolites	Cell lysates/Cell media	Reference
Human hepatic cells	HepG2	Caffeic acid	Ferulic acid; caffeic acid-glucuronide; caffeic acid-methylglucuronide; caffeic acid-sulphate	Media	Mateos et al. (2006)
			(no newly-formed metabolites)	Cells	Wong and Williamson (2013)
			Ferulic acid; caffeic acid-sulphate	Media	Wong and Williamson (2013)
		Ferulic acid	Ferulic acid-glucuronide	Media	Mateos et al. (2006)
			(no newly-formed metabolites)	Cells	Wong and Williamson (2013)
			Ferulic acid-sulphate	Media	Wong and Williamson (2013)
		Chlorogenic acid	Chlorogenic acid isomer	Media	Mateos et al. (2006)
			(no newly-formed metabolites)	Cells	Mateos et al. (2006)
	HepG2	Resveratrol	Resveratrol glucuronide; resveratrol sulphate	Media	Lançon et al. (2007)
	Primary hepatocytes	Quercetin	Quercetin 3-glucuronide; quercetin 3′-glucuronide; quercetin 7-glucuronide; 3′-methylquercetin; quercetin 3′-methyl-3-glucuronide; quercetin 3′-methyl-4′-glucuronide; quercetin 3′-methyl-7-glucuronide; quercetin 4′-methyl-3′-glucuronide; quercetin 4′-methyl-7-glucuronide	Media	van der Woude et al. (2004)

(continued)

Table 12.1 (continued)

Cell model	Cell line	Parent compound	Newly-formed metabolites	Cell lysates/ Cell media	Reference
Rat hepatic cells	Primary hepatocytes	Quercetin	Quercetin 3-O-glucuronide; quercetin 3'-O-glucuronide; quercetin 4'-O-glucuronide; quercetin 7-O-glucuronide; 3'-O-methylquercetin; 3'-O-methylquercetin 3-O-glucuronide; 3'-O-methylquercetin 7-O-glucuronide	Cells and media	Kahle et al. (2011)
	Primary hepatocytes	(+)-catechin	(no newly-formed metabolites)	Cells and media	Kahle et al. (2011)
		(−)-epicatechin			
		Procyanidin B$_2$			
	Primary hepatocytes	Caffeic acid	Ferulic acid; isoferulic acid	Cells and media	Kahle et al. (2011)
		p-coumaric acid	No derived metabolites		
		4-p-coumaroylquinic acid	3-p-Coumaroylquinic acid; 5-p-coumaroylquinic acid; D-(−)-quinic acid; methyl-p-coumarate		
		Phloretin	Phloretin 2'-O-glucuronide		
		Chlorogenic acid	1-Caffeoylquinic acid; 3-caffeoylquinic acid; 4-caffeoylquinic acid; D-(−)-quinic acid; methyl caffeate		

Vascular endothelial cells	HAEC	Hesperetin and hesperetin-glucuronide	Hesperetin 3'-O-sulphate; hesperetin 7-O-sulphate	Media	Giménez-Bastida et al. (2016)
		Hesperetin-O-glucuronide	Hesperetin	Media	Gonzalez-Sarrias et al. (2014)
		Urolithin B	Urolithin B-sulphate	Media	Spigoni et al. (2016)
		Urolithin B-glucuronide	Urolithin B; urolithin B-sulphate		
		Urolithin A	Urolithin A-sulphate; urolithin A-disulphate		
	HUVEC	Epicatechin	3'-O-Methyl-epicatechin-7-sulphate; 3'-O-methyl-epicatechin-7-β-d-glucuronide	Cells and media	Rodriguez-Mateos et al. (2014a)
		Epicatechin-3'-glucuronide	(no newly-formed metabolites)		
		Epicatechin-3'-sulphate			
		3'-O-methyl-epicatechin-5-sulphate			
		3'-O-methyl-epicatechin-7-sulphate			
		Quercetin-3-O-glucuronide	Quercetin	Unspecified	Tribolo et al. (2013)

(continued)

Table 12.1 (continued)

Cell model	Cell line	Parent compound	Newly-formed metabolites	Cell lysates/ Cell media	Reference
Vascular endothelial cells and immune cells	HUVEC and THP-1-derived macrophages	Urolithin B	Urolithin B-sulphate	Media	Mele et al. (2016)
		Urolithin A	Urolithin A-sulphate		
		Urolithin C	Urolithin C-sulphate; methyl-O-urolithin C; dimethyl-O-urolithin C; methyl-O-urolithin C-sulphate		
		Urolithin D	Methyl-O-urolithin D; dimethyl-O-urolithin D		
		Ellagic acid	Methyl-O-ellagic acid; dimethyl-O-ellagic acid		
Immune cells	RAW264 Differentiated THP-1 J774-1	Quercetin-3-O-glucuronide	Quercetin; 3′-O-methyl-quercetin; 4′-O-methyl-quercetin	Cells	Kawai et al. 2008; Ishisaka et al. (2013)
	Primary human macrophages	Naringenin-7-O-glucuronide	(no newly-formed metabolites)	Cells	Dall' Asta et al. (2013)
		Narigenin-4′-O-glucuronide			

Human dermal fibroblasts				
FEK4	Quercetin	2'-Glutathionyl-quercetin; quercetin quinone/quinone methide	Cells	Spencer et al. (2003)
	4'-O-methyl-quercetin	Quercetin		
	3'-O-methyl-quercetin	Quercetin; quercetin quinone/quinone methide		
	Quercetin-7-O-β-D-glucuronide	(no newly-formed metabolites)		
	Hesperetin	Hesperetin; hesperetin-7-O-glucuronide; hesperetin-5-O-glucuronide	Cells and media	Protegente et al. (2003)
	Hesperetin-7-O-glucuronide	(no newly-formed metabolites)		
	Hesperetin-5-O-glucuronide			

(continued)

Table 12.1 (continued)

Cell model	Cell line	Parent compound	Newly-formed metabolites	Cell lysates/ Cell media	Reference
Human breast cancer cells	JIMT-1	Hydroxytyrosol and tyrosol	Methyl hydroxytyrosol	Cells and media	García-Villalba et al. (2012)
		Oleuropein aglycon (Ol Agl)	Dihydrogenated Ol Agl; methyl Ol Agl; dihydrogenated methyl Ol Agl		
		Deacetoxy Ol Agl (DOA)	Dihydrogenated methyl DOA; methyl DOA		
		Ligstroside aglycon (Lig Agl)	Dihydrogenated ligstroside aglycon		
		Deacetoxy ligstroside aglycon	(no newly-formed metabolites)		
		Luteolin	Dihydroxy luteolin; luteolin hydrated; hydroxy luteolin; methyl luteolin		
		Apigenin	(no newly-formed metabolites)		
		Pinoresinol	Demethoxy pinoresinol; pinoresinol sulphate; acetoxy-syringaresinol		
	MCF-7	Urolithin A	Urolithin A-sulphate	Cells	Larrosa et al. (2006)
			Urolithin A-glucuronide Urolithin A-sulphate Urolithin A-disulphate	Media	
		Urolithin B	Urolithin B-sulphate Urolithin A Urolithin A-sulphate	Media	
Blood cells	Human primary blood cells	δ-(3,4-dihydroxy-phenyl)-γ-valerolactone (M1)	M1-glutathione; M1-cysteine conjugate; open-chained ester form of M1; M1-sulphated; hydroxybenzoic acid; M1-methylated; M1-acetylated	Cells	Mülek et al. (2015)

			Cells and media	
Central nervous system cells	Mouse primary cortical neurons	Quercetin	3'-O-Methyl-quercetin; 4'-O-methylquercetin; 2'-glutathionyl-quercetin	Cells and media	Vafeiadou et al. (2008)
		Epicatechin	(no newly-formed metabolites)	Cells	Spencer et al. (2001)
		3'-O-methyl-epicatechin			
Cardiac cells	Neonatal rat ventricular myocytes and fibroblasts	Urolithin A	Urolithin A-glucuronide	Media	Sala et al. (2015)
		Urolithin B	Urolithin B-sulphate; Urolithin B-glucuronide		
		Urolithin B-glucuronide	Urolithin B; Urolithin B-sulphate		
		Urolithin C	Methyl-O-Urolithin C; Methyl-O-Urolithin C-sulphate; Methyl-O-Urolithin C-glucuronide		

(continued)

structure were degraded by intracellular Caco-2 UDP-glucuronosyltransferases and catechol-O-methyltransferases to form glucuronide and methyl conjugates (Martini et al. 2019). Intestinal metabolism revealed that hydroxycinnamates were preferentially hydrolyzed and subsequently methylated and better absorbed (Gómez-Juaristi et al. 2020).

12.2.2   Stilbenes

Resveratrol can easily cross cell membranes and this was confirmed in some intestinal cell lines (Caco-2, HT-29, HCA-7, and HCEC) when these were incubated with physiologically relevant concentrations of the compound (Patel et al. 2013). In Caco-2 enterocytes, it seems that sulphation is more important than glucuronidation (Kaldas et al. 2003). Resveratrol is also rapidly conjugated and metabolized into glucuronide and sulfate derivatives in HepG2 cells. The two main resveratrol metabolites were monosulfate and disulfate (Lançon et al., 2007). In plant culture cells, resveratrol formed glycosides as reported by Shimoda et al. (2020). In leukemia cell lines, methoxy derivatives of stilbenes showed the formation of conjugated compounds which are responsible for apoptosis. The metabolic profiles of resveratrol (RV) and its analogs (polydatin [PD], oxyresveratrol [ORV], acetylresveratrol [ARV]) in human bladder cancer T24 cells were studied by Yang et al. (2019). The authors showed that the inhibitory potencies to T24 cells in the order of ORV > ARV > RV > PD were related to the structure and metabolism of RV and its analogs.

12.2.3   Flavonoids

In Caco-2 cells, the aglycone of quercetin is metabolized to glucuronidated, sulfated, and methylated forms, whereas quercetin-3-glucoside and quercetin-3-glucuronide are only methylated (del Mar et al. 2016). In human colon adenocarcinoma cells HT-29, quercetin forms isomers of quercetin glucuronides and methyl-quercetin glucuronides and is directly transported out of the cell. Also, both unconjugated forms and as glucuronides of methoxylated quercetin such as isorhamnetin and tamarixetin had been reported in the culture medium of HT-29 cells treated with quercetin (de Boer et al. 2006). Another study showed that quercetin-3-glucuronide and quercetin-4'-glucuronide were the major metabolites (van der Woude et al. 2004).

In HepG2 cells, quercetin undergoes extensive methylation, sulphation, and glucuronidation (van der Woude et al. 2004). Generally, flavonol glucuronides have been found to follow two metabolic paths in hepatic cells: methylation in the 4'- or 3'-position, or deglucuronidation followed by sulphation. De-glucuronidation and sulphation of quercetin appear to be the favored pathways, followed by re-glucuronidation at some different positions of the parent compounds (O'Leary et al. 2003).

Barrington et al. (2009) had proved that kaempferol and galangin are taken up and rapidly conjugated by Caco-2/TC7 cells to yield several glucuronide and sulfate conjugates. In cultured human blood cells, the metabolism of the microbial flavan-3-ol metabolite δ-(3,4-dihydroxy-phenyl)-γ-valerolactone had been reported by Mülek et al. (2015). The major metabolites were glutathione and methylated conjugates.

Glucuronide metabolites were not observed in any of the vascular endothelial cell models, despite being reported as the most relevant urolithin circulating metabolites in-vivo in humans (Mena et al. 2015). This suggests that both sulphation and methylation are the common metabolic process undergone by urolithins in human primary vascular cells, whereas glucuronidation might be limited to specific cell types including enterocytes and hepatocytes.

The uptake and metabolism of epicatechin in different cell models were investigated by Rodriguez-Mateos et al. (2014a). They evaluated the nature of various metabolites of epicatechin in Caco-2 cells and identified mainly glucuronidated, methyl-glucuronidated and methyl-sulfated conjugates. In HepG2 cells, the major metabolites of epicatechin are methyl-epicatechin-glucuronides and methyl-epicatechin-sulfates. Similar results were obtained after incubating HepG2 cells with a mixture of related epicatechin metabolites. Epigallocatechin-3-gallate (EGCG) is absorbed by HT-29 cells and metabolized to glucuronidated and methylated products (Hong et al. 2002). Moreover, under typical cell culture conditions at pH 7.2–7.4, the EGCG was rapidly auto-oxidized, leading to the formation of products with dimeric structures.

Some flavones, such as apigenin, luteolin, chrysin, and baicalein, are susceptible to glucuronidation in a Caco-2 cell model (Ng et al. 2005). The hydroxyl group at position-7 was found to be the preferred site of conjugation. Therefore, the glucuronidation of chrysin and luteolin is favored in comparison to that of apigenin and baicalein. Similarly, in Caco-2/TC7 cells, the glucuronidation and sulphation of isoflavones (biochanin A, daidzein, formononetin, genistein, glycitein, and prunetin) were reported (Chen et al. 2005).

Hesperetin was absorbed by Caco-2 cells and that it is then conjugated to produce glucuronide and sulfate metabolites (Brand et al. 2008). The metabolism of hesperetin and its conjugates in vascular endothelial cells (HAECs) treated with normal and inflammatory conditions i.e., TNFα-stimulated and unstimulated cells showed a time-dependent decrease in concentrations (Giménez-Bastida et al. 2016). This decline was associated with the accumulation of two sulfate conjugates, namely hesperetin-3′-sulphate and hesperetin-7-sulphate. On the other hand, the uptake of epicatechin by HUVECs led to the appearance of distinct metabolites conjugated with methyl and glucuronide groups to yield 3′-methyl-epicatechin-7-sulphate and 3′-methyl-epicatechin-7-glucuronide. This fact indicates that human endothelial cells could contain glucuronosyl-transferases specifically capable of glucuronidation of epicatechin, whereas they do not seem able to conjugate with glucuronic acid other scaffolds, such as hesperetin and urolithins (Aragonès et al. 2017). In human dermal fibroblasts, hesperetin glucuronides were present derived from foreskin (FEK4) treated with 30 µM of the aglycone hesperetin (Proteggente et al. 2003).

However, hesperetin glucuronides were not able to enter fibroblasts and be subjected to de-glucuronidation or any other metabolic reaction. Similarly, in these cells, 4-methylumbelliferone was also glucuronidated endorsing the idea that human skin fibroblasts contain glucuronosyl-transferases capable of glucuronidation of some phenolic structures. In the case of quercetin, treated with cultured FEK4 fibroblasts, a time-dependent appearance of different products, such as 2'-glutathionyl quercetin and a quercetin quinone/quinone methide were reported (Spencer et al. 2003). The metabolites such as 3'-methyl-quercetin and 4'-methyl-quercetin also metabolized when in contact with dermal fibroblasts, with the latter efficiently demethylated and the former oxidized to form quinone/quinone methide products. Contrarily to these quercetin phase II metabolites, quercetin-7-β-D-glucuronide was not metabolized, likely due to its inability to enter the fibroblast. In conclusion, skin cells are able to metabolize flavonoid metabolites following glucouronidation (hesperetin) suggesting that glucuronosyl-transferases are only specific for flavanones.

The de-conjugation of phenolic compounds in different peripheral cell models had been reported (Aragonès et al. 2017). The deglucuronidation of quercetin in HUVECs was considered to be essential in the vascular functionality of these cells, via nitric oxide expression regulation, when they were exposed to quercetin-3-glucuronide (Tribolo et al. 2013). De-glucuronidation and then sulphation had also been described after incubation of physiological concentrations of urolithin B-glucuronide in HAECs (Spigoni et al. 2016). On the other hand, when the major in vivo metabolites of epicatechin i.e., epicatechin-3'-glucuronide, epicatechin-3'-sulphate, 3'-methyl-epicatechin-5-sulphate, and 3'-methyl-epicatechin-7-sulphate were incubated with HUVEC cells, neither further phase II metabolism nor deconjugation was detected, suggesting that these cells do not metabolize these compounds (Rodriguez-Mateos et al. 2014a). Significantly different metabolism had been reported by endothelial cells from different vascular zones. This was because human endothelial cells could present phenotypical and physiological variations among cells located in different parts of the circulatory system (Aragonès et al. 2017).

Quercetin-3-glucuronide is metabolized into the more active aglycone and partially into methylated forms i.e., 3'- and 4'-methyl-quercetin in RAW264 murine macrophage-like cell line (Kawai et al. 2008). Similarly, Ishisaka et al. (2013) showed that quercetin-3-glucuronide was adsorbed to the cell surface of human-derived THP-1 monocytic cells and was readily deconjugated into the free form. Remarkably, this macrophage-mediated deconjugation was significantly greater in cultured lipopolysaccharide-stimulated macrophages. However, no de-conjugated forms were identified in primary human polarized macrophages treated with naringenin conjugates i.e., naringenin-7-glucuronide and naringenin-4'-glucuronide (Dall'Asta et al. 2013). Similarly, urolithins A, B, C, and D undergo extensive metabolism in THP-1-derived macrophages which were treated with human hypercholesterolemic serum or acetylated low-density lipoproteins to induce foam cell formation (Mele et al. 2016). The metabolic reactions undergone by differently hydroxylated urolithins in contact with THP-1-derived macrophages were the same previously described for vascular endothelial cells, i.e. sulphation and methylation.

The metabolism of phenolic compounds has also been studied in primary cultures of mouse cortical neurons. Vafeiadou et al. (2008) reported that quercetin was rapidly conjugated with glutathione in glial cells and produce 2'-glutathionyl-quercetin. Besides, it also undergoes intracellular methylation to produce 3'-methylquercetin and 4'-methylquercetin. In comparison with quercetin, epicatechin, and 3'-methyl epicatechin were not further transformed into other metabolites by primary cortical neurons (Spencer et al. 2001). This may be due to the selective activity of phase I and II enzymes present in neurons.

In neonatal rat ventricular myocytes and cardiac fibroblasts under hyperglucidic conditions, the main changes carried out by these cultured cells included sulphation, glucuronidation, and, in the case of urolithin B-glucuronide, de-glucuronidation (Sala et al. 2015). Besides, three different methylated metabolites were observed in these cells when exposed to physiologically reliable concentrations of urolithin C.

12.2.4 · Anthocyanins

The metabolites formed by the *in-vitro* cells from the metabolism of anthocyanins have been the topic of recent research. In Caco-2 cells, Kay et al. (2009) had demonstrated that cyanidin-3-glucoside rapidly and spontaneously degrades to protocatechuic acid and phloroglucinaldehyde before metabolism. Similarly, gallic acid, 3-methyl gallic acid, and 2,4,6-trihydroxybenzaldehyde were the metabolites of anthocyanin extract containing delphinidin-3-glucoside, petunidin-3-glucoside, peonidin-3-glucoside, and malvidin-3-glucoside (73%) (Forester and Waterhouse 2010). In HT-29 Colon Cancer Cells, malvidin glucoside forms vanillic acid, gallic acid, and hydroxyphenyl acetic acid (López de las Hazas et al. 2017). Similarly, pelargonidin glucoside forms *p*-hydroxybenzoic acid directly or through hydroxyphenylpropionic acid and ultimately metabolized to tyrosol as shown in Fig. 12.1. Recently uptake of cyanidin-3-glucoside was greater than cyanidin-3-galactoside in human gastric epithelial (NCI-N87) cells (Sigurdson et al. 2018).

12.2.5 Ellagitannins

Punicalagin (**12.6**), the main pomegranate ellagitannin, when incubated with Caco-2 cells, is hydrolyzed to ellagic acid (**12.7**), which then enters the cells (Larrosa et al. 2006). Ellagic acid appears to be rapidly methylated by the action of the enzyme catechol *O*-methyl transferase (**COMT**), which was present in these intestinal cells, to form dimethyl-ellagic acid (**12.8**) or its monomethyl derivatives. The most abundant metabolite detected under these conditions was dimethyl-ellagic acid-glucuronide (**12.10**), with other metabolites such as dimethyl ellagic acid sulfate (**12.9**) being produced in smaller amounts (Fig. 12.2).

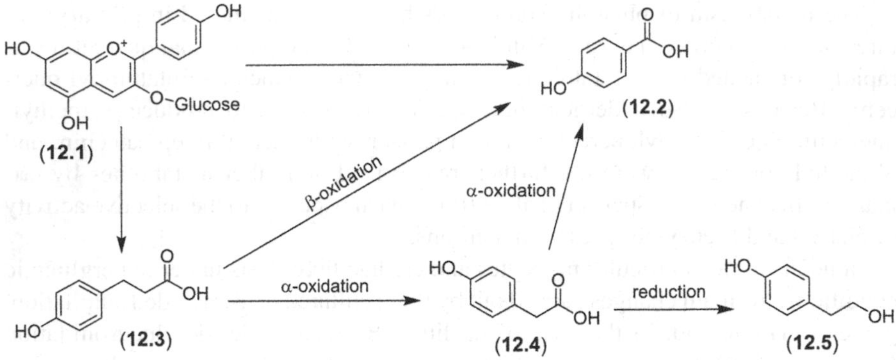

Fig. 12.1 Metabolism of anthocyanins. Metabolites are pelargonidin glucose (**12.1**), *p*-hydroxybenzoic acid (**12.2**), *p*-hydroxyphenylpropionic acid (**12.3**), *p*-hydroxyphenyl acetic acid (**12.4**), and tyrosol (**12.5**).

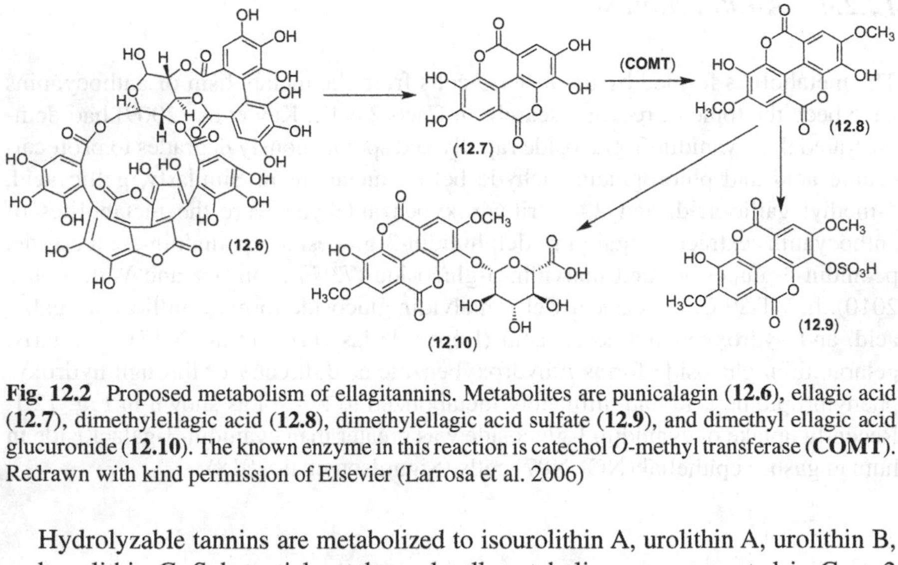

Fig. 12.2 Proposed metabolism of ellagitannins. Metabolites are punicalagin (**12.6**), ellagic acid (**12.7**), dimethylellagic acid (**12.8**), dimethylellagic acid sulfate (**12.9**), and dimethyl ellagic acid glucuronide (**12.10**). The known enzyme in this reaction is catechol *O*-methyl transferase (**COMT**). Redrawn with kind permission of Elsevier (Larrosa et al. 2006)

Hydrolyzable tannins are metabolized to isourolithin A, urolithin A, urolithin B, and urolithin C. Substantial uptake and cell metabolism were reported in Caco-2 cells following treatment with isourolithin A, urolithin A, urolithin B, and urolithin C (González-Sarrías et al. 2017; Gonzalez-Sarrias et al. 2014; Núñez-Sánchez et al. 2016; Gonzalez-Sarrias et al. 2009). These microbial metabolites were reported to be glucuronidated by HT-29 cells, whereas no apparent urolithin transformation occurs when SW480 human colon cancer cells or CCD18-Co human normal colon cells are used (Gonzalez-Sarrias et al. 2014; González-Sarrías et al. 2017). Moreover, conjugated metabolites were not detected following the incubation of Caco-2, HT-29, and SW480 cells in the presence of urolithin D, which suggests the absence of the formation of these conjugates in cultured intestinal cells. Urolithins have been metabolized by both HUVECs (Mele et al. 2016), and HAECs (Spigoni et al. 2016). The metabolizing activity in both cell models was limited to methylation and sulphation.

12.3 *In-Vitro* Metabolism of Phenolic Antioxidants

The metabolism of phenolic compounds in animal models has been a widely studied topic in food science, nutrition, and biochemistry. Phenolic compounds are
administered either alone, as a mixture, or in the form of food or plant food extracts.
These studies provide insight into the metabolism of phenolic compounds and its
extrapolation to the human. These studies revealed that metabolism involves, bioaccessibility, bioavailability, digestion, absorption, and consequent metabolism. In
this section, the *in-vivo* metabolism is limited to animal models.

12.3.1 *Bioaccessibility*

Bioaccessibility is referred to the fraction of the total amount of phenolic compound that is potentially available for absorption into the bloodstream. This term
includes digestive transformations of foods into material ready for assimilation,
the absorption/assimilation into intestinal epithelium cells as well as the pre-sys-
temic, intestinal, and hepatic metabolism. Shahidi et al. (2018) stated that most of
the mechanisms of action and related pharmacokinetic parameters that depend on
phenolic compound uptake are still poorly demonstrated and clarified. Thus bioaccessibility of phenolic compounds depends on the *in-vitro* methods (Carbonell
et al. 2014).

12.3.2 *Bioavailability, Release, and Absorption*

The bioavailability, release, and absorption of phenolic compounds have been
extensively reviewed (Shahidi and Peng 2018; Teng and Chen 2019). Genistein
absorption was also investigated by using a rat small intestine perfusion model in
which genistein solution was perfused into isolated rat small intestine; genistein
was recovered from vascular perfusion media, blood vessel, intestine tissue, and
non-absorbable effluent. The results showed 99.8% recovery, and 46% of genistein could be absorbed. Among absorbed genistein, about 40% of it came from
vascular perfusion media, 6% was from blood vessels, and the intestinal tissue
(Andlauer et al. 2000). The study of absorption of catechin and tannic acid in rat
small intestine segments showed that both tannic acid (50%) and catechin (30%)
could enter into the small intestinal cells, but only catechin (10%) passed through
the gut wall and arrived into the incubation buffers whilst no tannic acid was
detected (Carbonaro et al. 2001). The bioavailability of phenolic compounds has
been extensively reviewed (Karakaya 2004; Barros and Maróstica Jr. 2019).

Rats were fed with 14C-labeled hydroxycinnamates obtained from cultured spinach cell walls, approximately 25% label was found in body tissues after 2 h suggesting that hydroxycinnamates were absorbed from the foregut (Chesson et al. 1999). Similarly, ferulic acid given to rats was absorbed by passive diffusion or by facilitated transport that appears not to be saturated, even at a luminal concentration of 50 μmol/L (Adam et al. 2002). In rat jejunal segments, ferulic acid absorption was controlled by the Na+/dependent carrier-mediated transport process (Chesson et al. 1999). When Wistar rats were fed a diet enriched with ferulic acid (10–50 μmol/L), ferulic acid (56%) was found to be absorbed through the small intestine, and 5–7% of the perfused dose was determined in bile conjugated with glucuronic acid or sulfate. In another study, Sprague-Dawley rats were fed with ferulic acid at a dose of 5.15 mg/kg body weight, showed a maximum amount in plasma in 30 min (Rondini et al. 2002).

The absorption of quercetin, rutin, and quercetin-3-glucose in the stomach of Wistar rats showed that gastric juice has no effects on glycosides (Crespy et al. 2002). It was found that quercetin was rapidly absorbed by the stomach, and was recovered in the bile at 20 min after infusion. The absorption mechanisms of luteolin and luteolin-7-glucoside, using Sprague Dawley rats, were studied by Shimoi et al. (1998). The authors concluded that luteolin-7-glucoside was first hydrolyzed to luteolin and then absorbed in the form of luteolin and glucuronides.

The absorption of anthocyanins in animal models has been reviewed by Sandoval-Ramírez et al. (2018). Table 12.2 shows some important in-vivo studies of anthocyanins bioavailability. Anthocyanins from grape upon supplementation to rats showed absorption in the liver and kidneys (Vanzo et al. 2008). Bilberry anthocyanins were absorbed by the liver and kidneys (Sakakibara et al. 2009). The oral supplementation of anthocyanins from wild blueberry was not reported in the heart, kidneys, liver, brain, and bladder (Del Bò et al. 2010). Ichiyanagi et al. (2006) showed that bilberry extract containing anthocyanins was absorbed and present in the liver and kidneys. An intravenously administered dose of 668 nmol of cyanidin 3-glucoside (C3G) in anesthetized Wistar rats revealed the rapid distribution of C3G in the brain (Fornasaro et al. 2016). The amount of C3G was also reported in the liver and kidneys. Vanzo et al. (2011) treated rats with cyanidin 3-glucoside and found it to be absorbed in the liver and kidneys. In another study by Marczylo et al. (2009), mice received C3G by either gavage at 500 mg/kg or tail vein injection at 1 mg/kg and were found to be absorbed and detected in the heart, liver, and kidneys. Systemic bioavailabilies for parent C3G and total anthocyanins were 1.7% and 3.3%, respectively. The anthocyanins in tart cherry had been reported to contain cyanidin-3-glucosylrutinoside, cyanidin-3-rutinoside, cyanidin-3-rutinoside 5-glucoside, and peonidin-3-rutinoside. Upon oral supplementation, the highest concentration of cyanidin-3-glucosylrutinoside (2339 pg/g) was detected in the bladder, followed by cyanidin-3-rutinoside 5-glucoside (916 pg/g) in the liver of rats (Kirakosyan et al. 2015). It was found that organs not involved in metabolism and elimination had less tissue anthocyanin content compared to the other organs studied.

Table 12.2 Anthocyanin bioavailability in animals and total anthocyanin concentrations detected in analyzed tissues

Anthocyanin source	Animal model	Administration routes and dosage	Duration	Total anthocyanin (pmol/g) in different tissues				References
				Heart	Brain	Liver	Kidney	
Bilberry extract	Wistar rat	Oral, 400 mg/kg, IV 5 mg/kg	15 min			206.1	673.2	Ichiyanagi et al. (2006)
Grape extract	Wistar rat	Oral, 8 mg/kg	10 min			2.65×10^3	3.03×10^3	Vanzo et al. (2008)
Bilberry extract	Mice	Oral, 617.6 mg/kg	2 week	ND	ND	1.73×10^5	1.14×10^5	Sakakibara et al. (2009)
C3G extract	Mice	Oral, 500 mg/kg	5, 10, 20, 30, 60, 90, 120 min	3.6×10^3		$3.53 \times z10^4$	2.17×10^5	Marczylo et al. (2009)
		IV, 1 mg/kg	5, 10, 15, 20, 30, 60, 90, 120 min	720.0		80.00	970.0	
C3G extract	Wistar rat	IV, 670 nmol	0.25, 1, 5, 15 min			1011.0	138.0	Vanzo et al. (2011)
Tart cherry powder	Wistar rat	Oral, 200 mg/kg/day	3 week	1.11	0.27	0.43	2.07	Kirakosyan et al. (2015)
		Oral, 2000 mg/kg/day	3 week	0.51	0.43	1.52	1.65	
C3G extract	Wistar rat	IV, 668 nmol	0, 25, 5, 10, 15, 20 min	44.98		599.0	3.68×10^3	Fornasaro et al. (2016)

IV intravenous, *ND* not detected, *R. Fat* retroperitoneal fat, *SD* Sprague–Dawley

12.3.3   Metabolism

Phase I and II metabolic pathways have been reported in animal models. In rats fed with ferulic acid, the main forms in plasma were sulfated metabolites (58% of the total ferulic acid), followed by free ferulic acid (24%) and glucuronidated metabolites (18%) (Rondini et al. 2002). It was concluded that the correlation between the decrease of glucuronidated conjugates with time, and the increase of sulfoglucuronidated or sulfated forms in bladder urine could be accepted as an indicator of the conjugation in the kidney by the action of phase II enzymes, such as sulfatase or β-glucuronidase. Transportation of intact quercetin-3-glucoside across a perfused rat gut and its preferential uptake from rat intestine in-vivo were determined in two different studies (Manach et al. 1999; Gee et al. 2000). However, only conjugated metabolites of quercetin were found in the serum, whereas deglycosylation was found to occur rapidly. Bariexca et al. (2019) provided rats with quercetin, catechin, and hesperetin for 3 days. Sulphation and methylation were found the major reactions forming quercetin-3-sulfate and methyl catechin.

Anthocyanin metabolism *in-vivo* affects molecular size and polarity and, as a result, alters metabolite solubility among biometrics (Kalt 2019). The major metabolites of C3G, when supplemented to rats, were products of methylation and glucuronidation (Marczylo et al. 2009). Cyanidin was a minor metabolite in the gut. C3G and its metabolites were recovered from murine tissues. Cyanidin 3-glucoside fed rats showed the formation of methylation product peonidin-3-glucoside in plasma, liver, and kidneys as reported by Vanzo et al. (2011). The authors also detected three additional methylated metabolites in traces, namely, delphinidin-3-glucoside, petunidin-3-glucoside, and malvidin 3-glucoside.

12.4   Bioavailability of Phenolic Compounds in Humans

The metabolism of phenolic compounds in humans involves ingestion, digestion, absorption, interaction with other metabolites, and consequent metabolism and excretion. Either nutrients or non-nutrient, bioaccessibility, and bioavailability could be regarded as bio-efficiency. Specifically, bioaccessibility in the digestion and absorption efficiency (or digestibility and absorptivity) of a certain food constituent or drug ingested by oral administration, normally expressed as a percentage of the actual amount released and absorbed constituent to its total content (Shahidi and Peng 2018). However, for bioavailability, there are significant differences between that of nutrients and non-nutriments. In the nutrition area, bioavailability is crudely defined as the utilized or stored proportion of the total administered quantity. As a food component, phenolic compounds are not fully released and those released are poorly absorbed. Furthermore, the absorbed phenolic molecules cannot be completely transported to the action site in order to exert their bioactive effects. In this connection, their physicochemical properties including the degree of polymerization or glycosylation and molecular properties, polarity, and interaction status with nutrients, as well as individual physiological conditions such as the expression of transport protein and status of tissues are important factors to consider.

12.4.1 Ingestion of Phenolic Compounds

Phenolic compounds are ingested by a human through plant foods. Plant foods are taken by mouth and are masticated to a smaller size. The different tastes of phenolic compounds are perceived by the taste buds which define the acceptability of the foods. Once the food is in the mouth, the teeth, saliva, and tongue play significant roles in mastication by preparing the food into a bolus. Saliva helps in the processing of foods to the stomach. Mastication, or chewing, is a particularly important part of the digestive process, especially for fruits and vegetables, as these have indigestible cellulose coats, which must be physically broken down. Similarly, digestive enzymes such as the Amylase enzyme broken-down carbohydrates into smaller forms by acting on the surfaces of food particles. When the food is mechanically broken down, the saliva enzymes also contribute to their chemical process of the food as well. The combined action of these processes modifies the food from large particles to a soft mass that can be swallowed and can travel the length of the esophagus.

12.4.2 Digestion and Absorption

When food containing phenolic compounds are ingested, they can be modified in the oral cavity by the hydrolyzing activity of saliva, although after passing through the stomach most reach the small intestine and thereafter the colon (Rodriguez-Mateos et al. 2014b). Within the GI tract, their absorption is associated with the hydrolyzing activity of a cascade of enzymes. In the GIT tract, simple phenols like gallic acid and isoflavones which have small molecular weights are easily absorbed (Carbonell et al. 2014). In the small intestine, cleavage of the sugar unit in phenolic glycosides is mediated through the action of lactase phloridzin hydrolase (LPH), located in the brush border of epithelial cells. LPH exhibits broad specificity for flavonoid-O-β-D-glucosides, and the released aglycones can enter the epithelial cells by passive diffusion due to their increased lipophilicity (Day et al. 2000). An alternative hydrolytic step is mediated by a cytosolic β-glucosidase (CBG). In this case, the active sodium-dependent glucose transporter, SGLT-1, is thought to be involved in the transport of flavonoid glycosides into epithelial cells, a necessary step to facilitate the action of CBG (Gee et al. 2000). Once absorbed, flavonoids and related phenolics follow the common metabolic pathway of exogenous organic substances and, like drugs and most xenobiotics, undergo phase II enzymatic metabolism. They can be conjugated with glucuronic acid, sulfate, and methyl groups, in reactions catalyzed by UDP-glucuronosyltransferases (UGTs), sulphotransferases (SULTs), and catechol-O-methyltransferases (COMT), respectively (Del Rio et al. 2013). Phase II metabolism first occurs in the wall of the small intestine, after which metabolites pass through the portal vein to the liver, where they may undergo further conversions before entering the systemic circulation and eventually undergoing renal excretion. It is possible that some recycling of phase II metabolites from the liver back to the small intestine may occur via enterohepatic recirculation in the bile.

12.4.3   Phenolic Compounds in the Bloodstream

Several interactions take place in the GI lumen as reviewed by Dominguez-Avila et al. (2017). The orally ingested phenolic compounds (PCs) pass through the gut epithelia before reaching the bloodstream and/or lymphatic system. Bioaccessibility of PCs is very important for their enteric and systemic bioavailability, which is determined by many factors including PC type, the amount in the edible sources, the nature of its matrix, and several host-related factors. Once released, both hydrophilic e.g., phenolic acids or glycosylated, and lipophilic e.g., procyanidin or prenylated PCs are transported by the various mechanisms (Shahidi and Peng 2018) that have been previously characterized for other xenobiotics and nutrients. Transcellular mechanisms such as passive diffusion (PD), carrier-mediated active (AT), and facilitated (FT) transport, and paracellular transport in tight junctions (TJ) have been reported and all of them make a particular contribution to PC transport and homeostasis, as shown in Fig. 12.3.

Transport between the cells is facilitated by transmembrane pumps, channels, and carriers expressed in a polarized manner, while para-cellular transport is driven

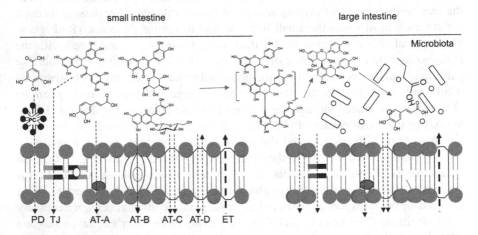

Fig. 12.3 Proposed transport mechanisms for phenolic compounds (PCs). Passive diffusion (**PD**), paracellular (**TJ**), single-solute (**AT-A**), translocase-type (**AT-B**) and co-solute symporter (**AT-C**) and antiporter (**AT-D**) active transporters, efflux transport (**ET**). Reproduced with kind permission of the American Chemical Society (Dominguez-Avila et al. 2017)

by solute gradients. Transport out of epithelial cells is further facilitated by basolateral membrane transporters that deliver both free and vesicle-contained PCs into the bloodstream and lymphatic system or efflux them back to the lumen. However, the novel transporters and mechanisms for the intestinal handling of PCs and the inhibitory action of PCs on well-known transporters have been elucidated in recent years, which has helped to complete our understanding of the first-pass metabolism of PCs.

In passive diffusion (PD), hydrophobic and smaller molecular weight PCs are easily transported. The luminal processing of hydrophobic PCs such as alkylated-PCs or procyanidin heterodimers involves bile salt-emulsification, micellar formation, and apical membrane translocation (Dominguez-Avila et al. 2017). Highly complex PCs are hydrolyzed to smaller molecules and then transported by PD. In para-cellular transport (PT), Hydrophobic and neutral PCs and other nanoparticles can easily pass between the narrow gaps (i.e., TJ) that separate the neighboring enterocytes by certain proteins called claudins. Small molecules that are charged, such as chlorogenic, gallic, and rosmarinic acids are passed by PT, although their permeation rate is low (5 mM), and the non-absorbed molecules travel directly to the large bowel. The active (AT) involves different energy-coupled mechanisms that create ion/solute gradients across membranes. The facilitated transport (FT), passes solutes across membranes (apical/basolateral) down their electrochemical gradient and does not use energy at all. Intracellular homeostasis and the deleterious accumulation of PCs are controlled by their influx (AT, FT) and efflux transport (AT). The most investigated efflux transporters in the intestine are P-glycoprotein. This protein transports hydrophobic, amphipathic, and cationic molecules in the basal-to-apical and blood-to-lumen directions, and its secretion varies across different segments of the intestine. It transports flavonoids back to the lumen, but other fruit flavonoids such as those found in grapefruit juice its pharmacokinetics.

12.5 Metabolism of Phenolic Compounds in Humans

The metabolism of phenolic compounds in the human GI system involves complex processes including stages as discussed above. The metabolic pathways of most of the phenolic compounds are still not clear. Therefore, some proposed pathways have been reported in the literature based on the end products obtained either in excretion or in the bloodstream. Also, the diverse and complex nature of edible plant foods presents the metabolism more complicated and still needs to be explored. Granato et al. (2020) concluded that any functional effects in the *in-vitro* model system should be avoided as these results do not represent the real biological effect in humans and a high intake of polyphenol in a specific population may have adverse effects. However, several studies showed the metabolism and consequent beneficial effects of phenolic compounds. There is a significant role of gastrointestinal (GI) microbiota in the metabolism of phenolic compounds in humans (Vetrani et al. 2016). However, it is not possible to accommodate all available literature regarding each phenolic compound and its metabolic fate in humans. Therefore, this section will present the concept of metabolism of each class of phenolic compounds with the support of specific examples.

12.5.1   Metabolism of Hydroxybenzoic Acids

Hydroxybenzoic acids (HBAs) consumed in foods undergo substantial metabolic processes. Free phenolic acids can be released from food and beverage matrices in the stomach, the muscles of which reduce food particle size and further enhance the release of phenolic compounds and their absorption. The majority of the hydroxybenzoic acids are conjugated or bound to other components of foods, therefore, hydrolysis is required to release them. Intestinal or microbial esterases carried out hydrolysis and are considered as the major route of the hydrolytic pathway. The free HBAs are absorbed through a specific transport system into the bloodstream. Microbial degradation of vanillin (**12.11**) formed vanillic acid (**12.12**), which upon demethylation produces protocatechuic acid (**12.13**) (Bento-Silva et al. 2020). Catechol (**12.14**) is formed from protocatechuic acid using a decarboxylation reaction. Catechol forms 2,6-dihydroxycyclohexane carboxylic acid (**12.15**), followed by 2-hydroxycyclohexane carboxylic acid (**12.16**) and then 3-hydrophilic acid (**12.17**). The final product is either acetoacetate (**12.18**) which is excreted through urine or carbon dioxide and water (Fig. 12.4).

The metabolic transformations include conjugation with sulfate, glucuronate, and glycine. Methylation may also occur, as may demethylation, dehydroxylation, and decarboxylation (Tomás-Barberán and Clifford 2000). Some of the dietary HBAs can be interconverted during metabolism; for example, salicylic to gentisic, or protocatechuic to vanillic acid. In humans, the metabolism of benzoic acid per se and its significance have been reviewed (Tremblay and Qureshi 1993). It is excreted mainly as hippuric acid with a much smaller amount as the glucuronide, except at very high doses when the glucuronide becomes more important.

Benzoic acid (**12.20**) is converted to hippuric acid (**12.21**) by its conjugation with glycine (Fig. 12.5). Benzoic acid may also be formed endogenously from β-oxidation of 3-phenyl propionic acid, a gut flora metabolite of tyrosine in man (Tomás-Barberán and Clifford 2000). In a study on the metabolism of 50 mg of gallic acid (**12.19**) in a single volunteer, the metabolite 4-methyl gallic acid (**12.22**) appeared rapidly in the plasma and urine (Shahrzad and Bitsch 1998). It has been reported that gallic acid, ellagic acid, and related acids may be released from gallate esters by gut flora metabolism in the colon, and possibly at the low pH in the stomach, and this would tend to increase circulating levels, although the pharmacokinetics would be very different (Heleno et al. 2015).

Salicylic acid is excreted largely unchanged, but also as ester and ethereal conjugates with glucuronic acid, and as the glycine conjugate i.e., salicyluric acid. The isomeric 3-hydroxybenzoic acid and 4-hydroxybenzoic acid have been less studied. Both are known as gut flora metabolites of many dietary phenols, whereas 4-HBA is also a natural constituent of foods. The early data suggest that 4-HBA is metabolized like 2-HBA, but is subject also to ethereal conjugation with sulfate and decarboxylation by the gut microflora.

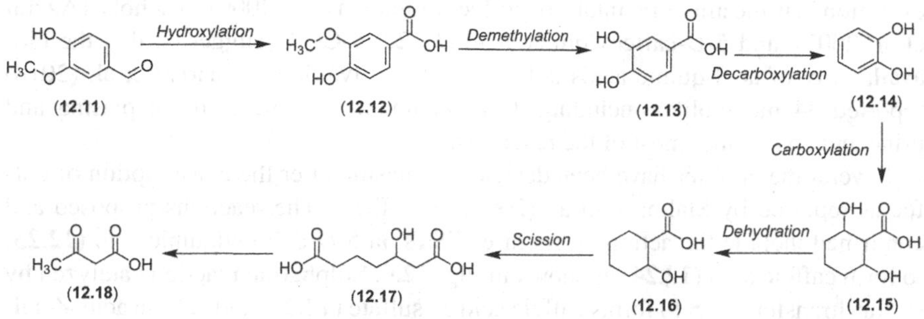

Fig. 12.4 Metabolism of hydroxybenzoic acids in humans. Metabolites are vanillin (**12.11**), vanillic acid (**12.12**), protocatechuic acid (**12.13**), Catechol (**12.14**), 2,6-dihydroxycyclohexane carboxylic acid (**12.15**), 2-hydroxycyclohexane carboxylic acid (**12.16**), 3-hydrophilic acid (**12.17**), and acetoacetate (**12.18**)

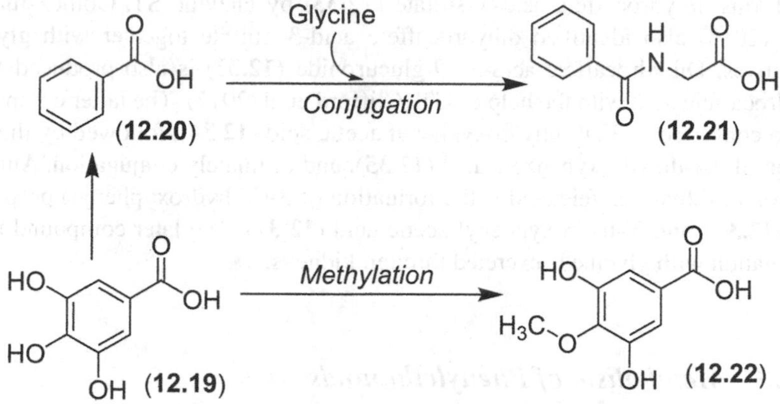

Fig. 12.5 Metabolism of gallic acid (**12.19**) in humans. Metabolites are benzoic acid (**12.20**), hippuric acid (**12.21**), and 4-methyl gallic acid (**12.22**)

The literature on the metabolism of gentisic acid shows that some are excreted unchanged, and some as conjugates at position 5 as sulfate and glucuronide (Scheline 1978). The isomer protocatechuic acid, under its 3,4-hydroxylation pattern, can be methylated and dehydroxylated to yield vanillic acid and 3-HBA respectively, and some may be excreted unchanged.

12.5.2 Metabolism of Hydroxycinnamic Acids

Hydroxycinnamic acids are present abundantly in tea and coffee. Clifford et al. (2017) reviewed the metabolism of caffeoylquinic acids in detail. Dicaffeoylquinic acids have been found to cross the membrane because they have greater hydrophobicity. The presence of low concentrations of CQAs in plasma and their low-level excretion in urine after oral intake of coffee (Stalmach et al. 2009), artichoke (Azzini et al. 2007), and 5-O-caffeoylquinic acid (**12.23**, 5-CQA) suggested that the bio-availability of acyl-quinic acids is limited. Recently, Gómez-Juaristi et al. (2018) reported 34 metabolites including dihydrocinnamoylquinic acids in plasma and urine and confirming most of the reactions.

Several metabolites have been detected in plasma after the consumption of coffee as reported by Stalmach et al. (2009) and others. The reactions proposed and confirmed include the action of esterase (EST) on 5-O-caffeoylquinic acid (**12.23**) to form caffeic acid (**12.24**) as shown in Fig. 12.6. Sulphation reaction catalyzed by sulfuryltransferase (**ST**) forms caffeic acid 3'-sulfate (**12.25**) and caffeic acid 4'-sulfate (**12.26**). Caffeic acid is converted to isoferulic acid (**12.27**) by catechol-O-methyltransferase (**COMT**). Isoferulic acid either form isoferulic acid-3'-sulfate (**12.27**) by the action of enzyme **ST** or isoferulic acid-3-O-glucuronide (**12.29**) by UDP-glucuronyltransferase (**GT**). Caffeic acid crosses the colon and is converted to dihydrocaffeic acid (**12.31**) by the action of the reductase enzyme (**RA**). Dihydrocaffeic acid forms dihydrocaffeic acid-3'-sulfate (**12.33**) by enzyme ST. Gómez-Juaristi et al. (2018) also identified dihydrocaffeic acid-3'-sulfate together with glycine derivatives. Dihydrocaffeic acid-3'-O-glucuronide (**12.32**) is also produced from dihydrocaffeic acid with the help of GT (Clifford et al. 2017). The latter compound is also converted to 3',4'-dihydroxyphenyl acetic acid (**12.34**), followed by the formation of 3,4-dihydroxybenzoic acid (**12.35**), and ultimately conjugation. Another reaction of dihydrocaffeic acid is the formation of 3-(3'-hydroxyphenyl) propionic acid (**12.36**), and 3'-hydroxyphenyl acetic acid (**12.37**). The later compound upon conjugation with glycine is excreted through kidneys.

12.5.3 Metabolism of Phenylethanoids

Tyrosol and hydroxytyrosol are important phenylethanoids present in olive oil and its products. Tuck and Hayball (2002) postulated the metabolic pathway of hydroxytyrosol metabolism as shown in Fig. 12.7. Hydroxytyrosol (**12.38**), produce hydroxytyrosol-4-sulfate (**12.39**) by the action of sulfuryltransferase (**ST**), or hydroxytyrosol-3-glucuronide (**12.40**) using UDP-glucuronyltransferase (**GT**). The action of alcohol dehydrogenase (**AD**) and methyltransferase (**MT**), produces 3,4-dihydroxyphenyl acetaldehyde (**12.41**), 3,4-dihydroxyphenyl acetic acid (**12.42**), and 4-hydroxy-3-methoxyphenylacetic acid (**12.43**), respectively. Hydroxytyrosol may also form 4-hydroxy-3-methoxyphenyl ethanol (**12.44**), and 4-hydroxy-3-methoxyphenylacetaldehyde (**12.45**) by the action of enzymes catechol-O-methyltransferase (**COMT**), and AD, respectively.

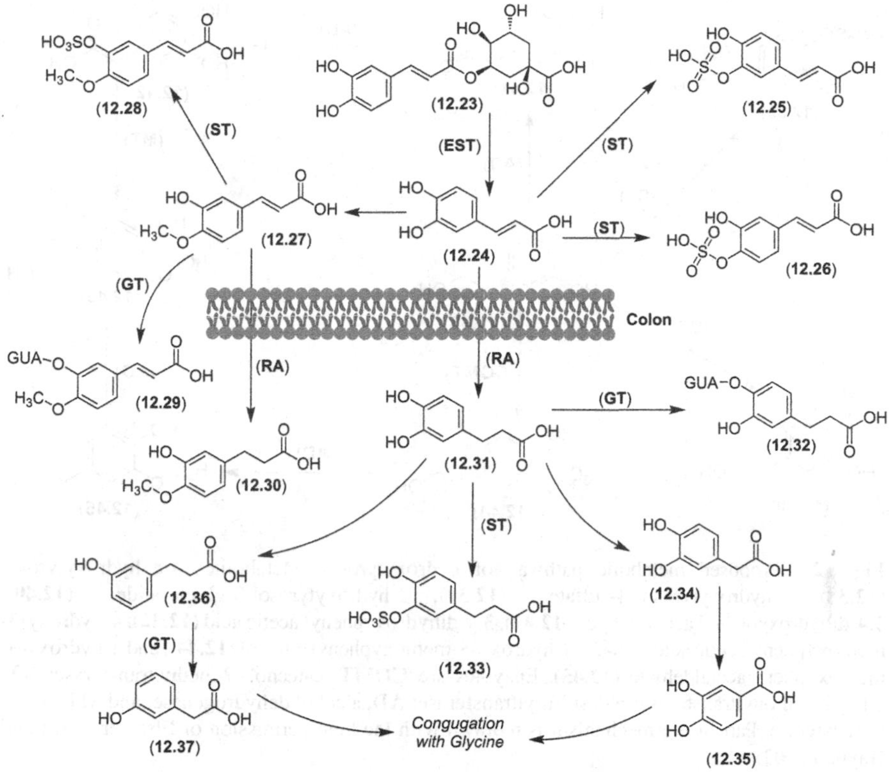

Fig. 12.6 Proposed metabolism of caffeoylquinic acids. Metabolites are 5-*O*-caffeoylquinic acid (**12.23**), caffeic acid (**12.24**), caffeic acid 3′-sulfate (**12.25**), caffeic acid 4′-sulfate (**12.26**), isoferulic acid (**12.27**), isoferulic acid-3′-sulfate (**12.27**), isoferulic acid-3-*O*-glucuronide (**12.29**), dihydroisoferulic acid (**12.30**), dihydrocaffeic acid (**12.31**), dihydrocaffeic acid-3′-*O*-glucuronide (**12.32**), dihydrocaffeic acid-3′-sulfate (**12.33**), 3′,4′-dihydroxyphenyl acetic acid (**12.34**), 3,4-dihydroxybenzoic acid (**12.35**), 3-(3′-hydroxyphenyl) propionic acid (**12.36**), and 3′-hydroxyphenyl acetic acid (**12.37**), Enzymes are **COMT**, catechol-*O*-methyltransferase; **EST**, esterase; **RA**, reductase; **GT**, UDP-glucuronyltransferase; **ST**, sulfuryltransferase; **GUA**, glucuronide moiety. Data are redrawn from Ludwig et al. (2013); Clifford et al. (2017); Stalmach et al. (2009); Gómez-Juaristi et al. (2018) and Bento-Silva et al. (2020)

12.5.4   Metabolism of Flavonoids

Gut bacteria hydrolyses glycosides, glucuronides, sulfates, amides, esters, and lactones. The flavonoid skeleton undergoes ring fission forming products that are subjected to reduction, decarboxylation, demethylation, and dehydroxylation reactions (Selma et al. 2009). These complex modifications generate low molecular weight catabolites that can be efficiently absorbed in situ. Some of these products undergo further phase II metabolism locally and/or in the liver before entering the circulation and being excreted in the urine in substantial amounts that typically exceed the excretion of phenolic metabolites absorbed in the upper GI tract (Crozier et al.

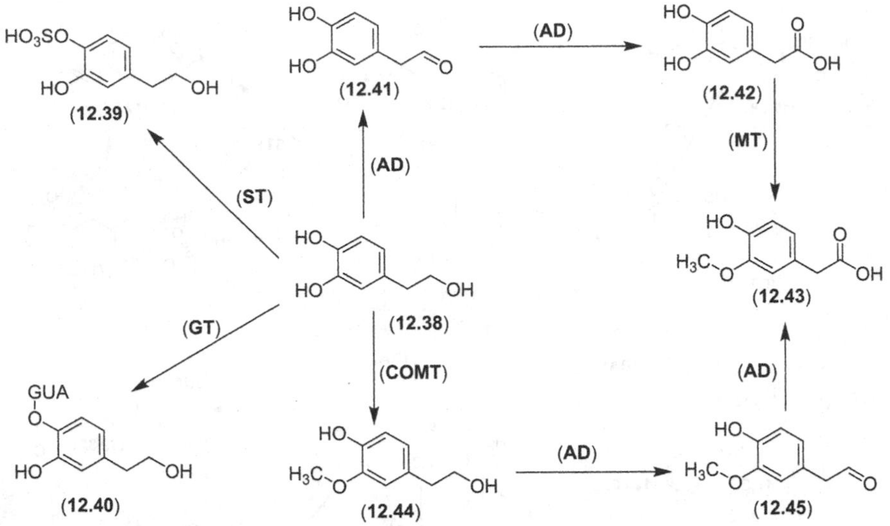

Fig. 12.7 Proposed metabolic pathway of hydroxytyrosol. Metabolites are hydroxytyrosol (**12.38**), hydroxytyrosol-4-sulfate (**12.39**), hydroxytyrosol-3-glucuronide (**12.40**), 3,4-dihydroxyphenyl acetaldehyde (**12.41**), 3,4-dihydroxyphenyl acetic acid (**12.42**), 4-hydroxy-3-methoxyphenylacetic acid (**12.43**), 4-hydroxy-3-methoxyphenyl ethanol (**12.44**), and 4-hydroxy-3-methoxyphenylacetaldehyde (**12.45**). Enzymes are **COMT**, catechol-O-methyltransferase; **GT**, UDP-glucuronyltransferase; **ST**, sulfuryltransferase; **AD**, alcohol dehydrogenase; and **MT**, methyltransferase. Part of the mechanism is redrawn with the kind permission of Elsevier (Tuck and Hayball 2002)

2010). Besides, more than half the carbon of glycosides and the flavonoid A ring may be metabolized to short-chain fatty acids and, therefore, be available for energy metabolism by the host and ultimately be exhaled as CO_2 (Rodriguez-Mateos et al. 2014b).

In the intestine, and liver, the enzymes of biotransformation act upon flavonoids and their colonic metabolites (Hollman 2004). The kidney also contains enzymes capable of biotransformation of flavonoids. Conjugation of the polar hydroxyl groups with glucuronic acid, sulfate, or glycine has been reported for flavonoids and their colonic metabolites. Besides, O-methylation by the enzyme catechol-O-methyltransferase (COMT) plays an important role in the inactivation of the catechol moiety, that is, the two adjacent (ortho) aromatic hydroxyl groups, of flavonoids and their colonic metabolites (Fig. 12.8). De-glycosylation of glycosides in the brush border membrane of the small intestine was reported by Day et al. (2000). Similarly, kaempferol-O-glycosides and apigenin-C-glycoside are phenylacetic acids using deglucosylation, reduction, ring fission, and dehydroxylation reactions (Vollmer et al. 2018).

The conjugation reactions are very efficient in humans, evidenced by the fact that flavonoids predominantly occur in plasma and urine as conjugates, and that it is difficult to detect flavonoid aglycones in the plasma because they are mostly below the limit of detection of the analytical methods used. The presence of flavonoid conjugates in humans, including O-methylated conjugates, is apparent from differential

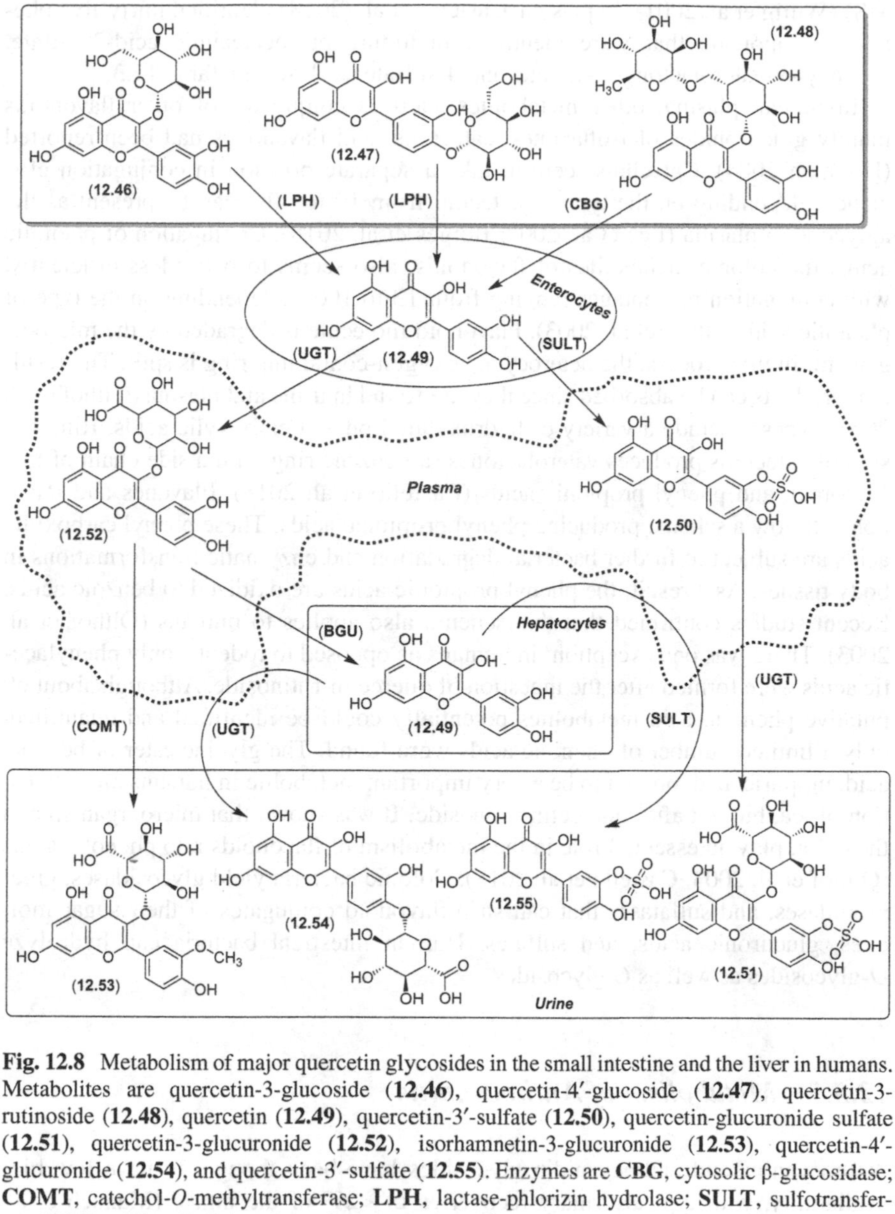

Fig. 12.8 Metabolism of major quercetin glycosides in the small intestine and the liver in humans. Metabolites are quercetin-3-glucoside (**12.46**), quercetin-4′-glucoside (**12.47**), quercetin-3-rutinoside (**12.48**), quercetin (**12.49**), quercetin-3′-sulfate (**12.50**), quercetin-glucuronide sulfate (**12.51**), quercetin-3-glucuronide (**12.52**), isorhamnetin-3-glucuronide (**12.53**), quercetin-4′-glucuronide (**12.54**), and quercetin-3′-sulfate (**12.55**). Enzymes are **CBG**, cytosolic β-glucosidase; **COMT**, catechol-*O*-methyltransferase; **LPH**, lactase-phlorizin hydrolase; **SULT**, sulfotransferase; **UGT**, uridine-5′-diphosphate glucuronosyltransferase

High Performance Liquid Chromatography (HPLC) analyses with and without hydrolysis of the sample with a mixture of β-glucuronidases and sulfatases: flavonols, flavones, catechins, flavanones, and anthocyanins. In several human intervention studies, the nature of the conjugates has been identified. Major conjugates of quercetin after onions supplementation were the 3′-sulfate, the 3′-methoxy-3-glucuronide, and the 3-glucuronide (Day et al. 2001). The 3′-sulfate could not be confirmed with liquid chromatography-mass spectrometry/mass spectrometry (LC-MS/

MS) (Wittig et al. 2001). In plasma, Castello et al. (2018) identified thirty-five phenolic compounds that were identified including protocatechuic acid-3-sulfate, 5-phenyl-γ-valerolactone-3′-sulfate, and 4-sulfate as shown in Table 12.3.

In human plasma, other metabolites such as conjugates of other flavonoids mainly glucuronides of isoflavones, catechins, and flavanones had been reported (Hollman 2004). Catechins seem to take a separate position in conjugation efficiency: depending on the type of catechin from 10 to 80% can be present as the aglycone in plasma (Lee et al. 2002; Borges et al. 2018). Conjugation of phenolic acids, the colonic metabolites of flavonoids, also seems to occur less efficiently, with conjugation percentages ranging from 13 to 100%, depending on the type of phenolic acid (Olthof et al. 2003). Flavonoid molecule is degraded by the microorganism. In this process, the heterocyclic oxygen-containing ring is split. The resultant products can be absorbed since they are found in urine and plasma (Olthof et al. 2003). These include a variety of hydroxylated phenyl carboxylic acids. Ring fission of catechins produces valerolactones (a benzene ring with a side chain of five C-atoms), and phenyl propionic acids (Castello et al. 2018). Flavones and flavanones follow a scheme producing phenyl propionic acids. These phenyl carboxylic acids are subject to further bacterial degradation and enzymatic transformations in body tissues. As a result, the phenyl propionic acids are oxidized to benzoic acids. Recent studies confirmed that this scheme also applies to humans (Olthof et al. 2003). There was one exception; in humans as opposed to rodents, only phenylacetic acids were formed after the ingestion of quercetin-rutinoside. Although about 60 putative phenolic acid metabolites potentially could be identified and quantified, only a limited number of phenolic acids were found. The glycine ester of benzoic acid, hippuric acid, proved to be a very important metabolite in humans after ingestion of tea, but not after quercetin rutinoside. It was shown that microorganisms in the colon play an essential role in the metabolism of flavonoids into phenolic acids (Olthof et al. 2003; Castello et al. 2018). Colonic bacteria yield glycosidases, glucuronidases, and sulfatases that can strip flavonoid conjugates of their sugar moieties, glucuronic acids, and sulfates. Human intestinal bacteria can hydrolyze O-glycosides as well as C-glycosides.

12.5.5 Metabolism of Anthocyanins

Anthocyanins and anthocyanidin glycosides have been thought to have low bioavailability, with typical urinary recoveries of <2% of the intake (Manach et al. 2004; Rodriguez-Mateos et al. 2014b). Dietary anthocyanins that reach the microbiota are those not absorbed in the upper GI level together with their metabolites excreted in the bile and/or from the enterohepatic circulation. There are two potential mechanisms of anthocyanin glycosides absorption: one may contain SGLT1 and GLUT2 as specific glucose transporters, another is likely to involve the extracellular hydrolysis of anthocyanins by brush border enzymes such as lactase phloridzin hydrolase before passive diffusion of the aglycone (Fig. 12.9). The complex

Table 12.3 Pharmacokinetic parameters of phenolic metabolites detected in human plasma (values are mean with SEM of n = 10) (reproduced with kind permission of Elsevier (Castello et al. 2018))

Id.#	Phenolic compounds	Cmax (nM)	Tmax (h)	$t_{1/2}$ (h)	AUC_{0-24} (nmol h L^{-1})
Simple phenols and hydroxybenzoic acids					
1	Gallic acid	124.3 ± 31.9	3.8 ± 0.7	20.5 ± 9.4	607.2 ± 211.3
2	Vanillic acid-4-glucuronide	61.3 ± 9.0	5.7 ± 0.4	6.2 ± 1.4	410.9 ± 82.2
4	Protocatechuic acid	5.9 ± 1.2	4.1 ± 0.8	40.4 ± 13.7	40.9 ± 7.5
3	Protocatechuic acid-3-glucuronide	3.1 ± 0.6	3.0 ± 0.8	6.7 ± 2.3	20.5 ± 3.8
9	Protocatechuic acid-3-sulphate	408.5 ± 68.7	2.1 ± 0.3	3.8 ± 0.4	1088.6 ± 126.3
10	Benzoic acid-4-sulphate	56.7 ± 9.4	3.0 ± 0.7	37.2 ± 14.6	521.3 ± 76.3
13	Vanillic acid-4-sulphate	117.0 ± 30.2	4.0 ± 0.5	4.8 ± 0.6	381.9 ± 121.7
6	Catechol-sulphate	2.9 ± 1.0	11.2 ± 2.9	6.3 ± 1.2	20.8 ± 7.5
7	Methylpyrogallol-sulphate	512.4 ± 117.5	5.9 ± 0.5	3.6 ± 0.5	2724.2 ± 584.2
8	Methylcatechol-sulphate	47.1 ± 7.6	2.4 ± 0.5	3.3 ± 0.6	184.6 ± 23.6
5	4-Hydroxyhippuric acid	98.7 ± 15.1	4.6 ± 0.5	14.8 ± 4.2	874.0 ± 173.4
Hydroxyphenylpropionic and hydroxycinnamic acids					
12	Ferulic acid 4-glucuronide	72.8 ± 19.9	7.0 ± 2.0	5.7 ± 1.4	567.5 ± 208.1
24	Feruloylglycine	26.0 ± 3.9	9.3 ± 3.3	26.3 ± 4.9	175.1 ± 28.0
23	Dihydrocaffeic acid-sulphate	8.3 ± 1.5	7.0 ± 2.0	40.1 ± 17.0	50.8 ± 19.9
26	Dihydroferulic acid-sulphate	7.7 ± 1.3	8.5 ± 2.7	17.5 ± 6.8	56.4 ± 16.9
28	Ferulic acid-4-sulphate	10.0 ± 1.8	5.2 ± 0.6	17.9 ± 7.0	63.4 ± 16.2
(Epi)catechin derivatives					
11	(Epi)catechin-glucuronide-sulphate	6.8 ± 1.5	4.8 ± 0.6	0.9 ± 0.3	20.0 ± 4.5
20	(Epi)catechin-glucuronide	135.5 ± 14.2	1.7 ± 0.3	2.3 ± 0.4	459.9 ± 44.5
27	(Epi)catechin-sulphate, isomer 1	87.0 ± 22.8	1.6 ± 0.2	1.9 ± 0.4	166.2 ± 31.1
29	(Epi)catechin-sulphate, isomer 2	94.9 ± 22.8	2.5 ± 0.5	2.9 ± 0.5	290.2 ± 50.1
32	Methyl(epi)catechin-sulphate, isomer 1	12.6 ± 1.9	2.7 ± 0.8	8.1 ± 1.3	53.1 ± 10.2
Phenyl-γ-valerolactones and phenyl-valeric acids					
31	5-Phenyl-valeric acid-sulphate-glucuronide	11.0 ± 1.1	7.9 ± 2.0	36.0 ± 14.1	106.2 ± 29.1
17	5-(3'-Hydroxyphenyl)-γ-valerolactone-4'-glucuronide	268.4 ± 68.2	5.3 ± 0.6	2.1 ± 0.5	1098.2 ± 219.2
19	5-(4'-Hydroxyphenyl)-γ-valerolactone-3'-glucuronide	1171.2 ± 242.7	5.2 ± 0.6	12.4 ± 8.3	6224.8 ± 1175.4

(continued)

Table 12.3 (continued)

Id.#	Phenolic compounds	Cmax (nM)	Tmax (h)	$t_{1/2}$ (h)	AUC_{0-24} (nmol h L^{-1})
30	5-(Hydroxyphenyl)-γ-valerolactone-sulphate isomers	893.7 ± 201.3	6.3 ± 2.0	4.8 ± 1.1	5196.3 ± 1457.9
25	5-Phenyl-γ-valerolactone-3'-glucuronide	88.4 ± 44.7	9.1 ± 2.5	11.0 ± 2.7	617.0 ± 178.2
35	5-Phenyl-γ-valerolactone-3'-sulphate	69.2 ± 25.2	11.0 ± 2.9	8.1 ± 1.9	441.0 ± 135.7
21	5-(3',4'-Dihydroxyphenyl)-γ-valerolactone	14.3 ± 3.2	7.0 ± 2.0	39.5 ± 26.4	66.1 ± 16.5

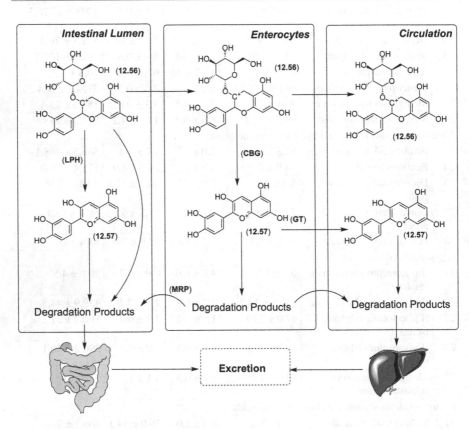

Fig. 12.9 Simplified metabolic scheme of the transport mechanism for cyanidin-3-glucoside (**12.56**) and cyanidin (**12.57**) in intestinal absorption and circulation. Enzymes are CBG, cytosolic β-glucosidase; **LPH**, lactase-phlorizin hydrolase; **MRP**, multidrug resistant protein; and GT, uridine-5'-diphosphate glucuronosyltransferase

metabolism of anthocyanins has been obtained in a stable isotope-labeled feeding study with human volunteers. Subjects ingested 0.5 g (1,114 μmoL) of $^{13}C5$-labelled cyanidin-3-glucoside and after 48 h, relative bioavailability was assessed to be as 12% of the 13-C dose with 5.4% excreted in urine and 6.9% in-breath, while feces accounted for a further 32% of the administered 13C-label (Czank et al. 2013). The C max of the anthocyanin-derived circulating metabolites varied between 11 nmol/L for 4-hydroxybenzoic acid-3-glucuronide (protocatechuic acid-3-glucuronide) and 1962 nmol/L for hippuric acid, with Tmax ranging from 1.8 h for cyanidin-3-glucuronide to 30.1 h for sulfate of vanillic acid (De Ferrars et al. 2014). Anthocyanins are rapidly absorbed in the stomach and detected in the blood circulation and urine in the forms of intact, methylated, glucuronidated, and sulfo-conjugated (Fernandes et al. 2015). The degradation products of anthocyanins from red grape and red wine include phenolic acids, phenol aldehydes, and phenols (Han et al. 2019a).

12.5.6 Metabolism of Tannins

Urolithins are the key colonic metabolites of ellagitannins (ETs). Seeram et al. (2004) studied the ingestion of pomegranate juice containing ellagic acid (**12.5**) and ellagitannin (**12.59**) in humans. Ellagic acid (EA) was detected in human plasma at a maximum concentration after 1 h post-ingestion but was rapidly eliminated by 4 h. Therefore, the presence of free EA in human plasma could be due to its release from the hydrolysis of ETs, facilitated by physiological pH and/or gut microbiota as reported by García-Villalba et al. (2020). A plausible mechanism for ellagitannin degradation by human microbiota, via hydrolysis into ellagic acid and its microbial transformation into urolithin B, has been proposed. A study performed on human volunteers fed with a single dose of ellagitannin-rich dietary sources demonstrated a large inter-individual variation in the metabolite profiles among subjects within each group, suggesting the involvement of variable colonic microbiota (Espín et al. 2013) (Fig. 12.10). Following the metabolism of ETs, urolithins A and B are formed and conjugated in the liver before their excretion in urine over 12–56 h after a single administration of food containing ETs and EA. Therefore, these urolithins circulate in the blood and must also reach the target organs, where the effects of ETs are noticeable (Landete 2011).

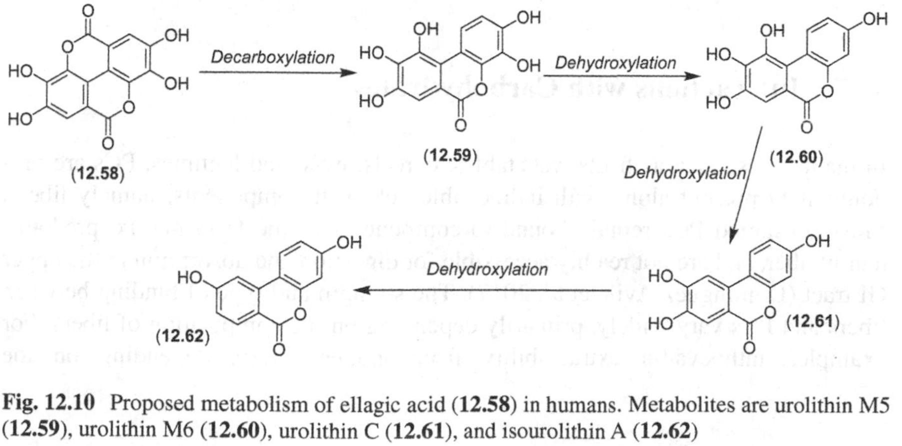

Fig. 12.10 Proposed metabolism of ellagic acid (**12.58**) in humans. Metabolites are urolithin M5 (**12.59**), urolithin M6 (**12.60**), urolithin C (**12.61**), and isourolithin A (**12.62**)

12.6 Interactions with Dietary Proteins

Dietary proteins can form hydrogen bonds and hydrophobic interactions with phenolic compounds (PCs) recently reviewed by Dominguez-Avila et al. (2017) and. The binding of PCs formed varied molecular structures and can reach colloidal-size aggregate complexes (Brudzynski and Maldonado-Alvarez 2015). One of the most commonly known effects of PC–protein interactions is protein precipitation by tannins that results in an astringent sensation (Victor de and Nuno 2012). Astringency due to PC–protein interactions are mediated by the combination of hydrogen bonding and hydrophobic interactions, but it is disrupted when certain carbohydrates (xanthan gum > pectin > gum arabic) strongly bind the PCs first (Mateus et al. 2004). PC interactions with milk proteins have been thoroughly studied (Han et al. 2019b). For example, resveratrol, genistein, and curcumin interact with α and β-caseins through hydrophobic interactions and hydrogen bonds (Bourassa et al. 2013). Catechins of green and black tea with low molecular weights have a strong affinity for casein micelles and alter the secondary structures of the proteins.

PC–protein interactions are not random because the physicochemical characteristics of PCs and proteins can be used to predict their non-covalent interactions. Binding is stronger when PCs have a higher logP (a higher logP indicates hydrophobic nature), a double bond at position 2–3, and a keto group at position 4. Furthermore, the isoelectric point, number of acidic/basic amino acid residues, number of proline residues, and secondary structures allow for similar predictions to be made for proteins (Nagy et al. 2012). For catechins, a gallate group increases protein affinity, whereas a pyrogallol group decreases it. In the case of PCs binding to milk β-lactoglobulin, affinity increases with the number of hydroxyl groups in the PC (Qie et al. 2020). However, this is not a general rule because the position of the hydroxyl group is more important than the number in most cases.

The PC–protein interactions confer protection to the PC. Therefore, dairy products such as yogurt and cheese had been used for these purposes (Dominguez-Avila et al. 2017). For example, Cheddar, Chihuahua, Gouda, Pecorino, and other cheese types have been used as delivery vehicles for PCs. However, such incorporation also modifies the cheese-making ability of milk and alters sensory parameters.

12.7 Interactions with Carbohydrates

In plant foods, such as fruits, vegetables, cereals, nuts, and legumes, PCs are predominately present along with indigestible cell wall components, namely fibers. Most consumed PCs remain bound to components of the food matrix, predominantly fiber, and are not readily accessible for digestion and absorption in the upper GI tract (Dominguez-Avila et al. 2017). The strength and type of binding between fibers and PCs vary widely, primarily depending on the composition of fibers. For example, anthocyanin extractability from grapes differs depending on the composition of cell wall polysaccharides, e.g., cellulose, lignin, etc., which suggests a varied degree of conjugation of PCs with fibers. Some fibers, particularly pectin, can form secondary structures, i.e., hydrophobic pockets, to encapsulate PCs. Thus, PCs interact with fibers through a combination of hydrogen bonds formed between the hydroxyl groups of PCs and ether bonds of fibers and the hydrophobic interactions that occur within pockets. However, due to the varied nature of PCs and fibers, their interactions are not universal and vary according to the chemical structures of both. An example of this has been described by Wang et al. (2013), who showed that the in-vitro adsorption of flavonoids to oat β-glucan differs depending on the type of flavonoid (flavonol > flavone > flavanone > isoflavone), while structural modifications (hydroxylation, glycosylation, methylation, methoxylation, esterification or galloylation) can widely increase or decrease adsorption. Furthermore, PCs can bind to fibers via non-covalent bonds when they are from different food sources (Fig. 12.11). For example, cellulose can rapidly (<1 min) bind with catechin, ferulic

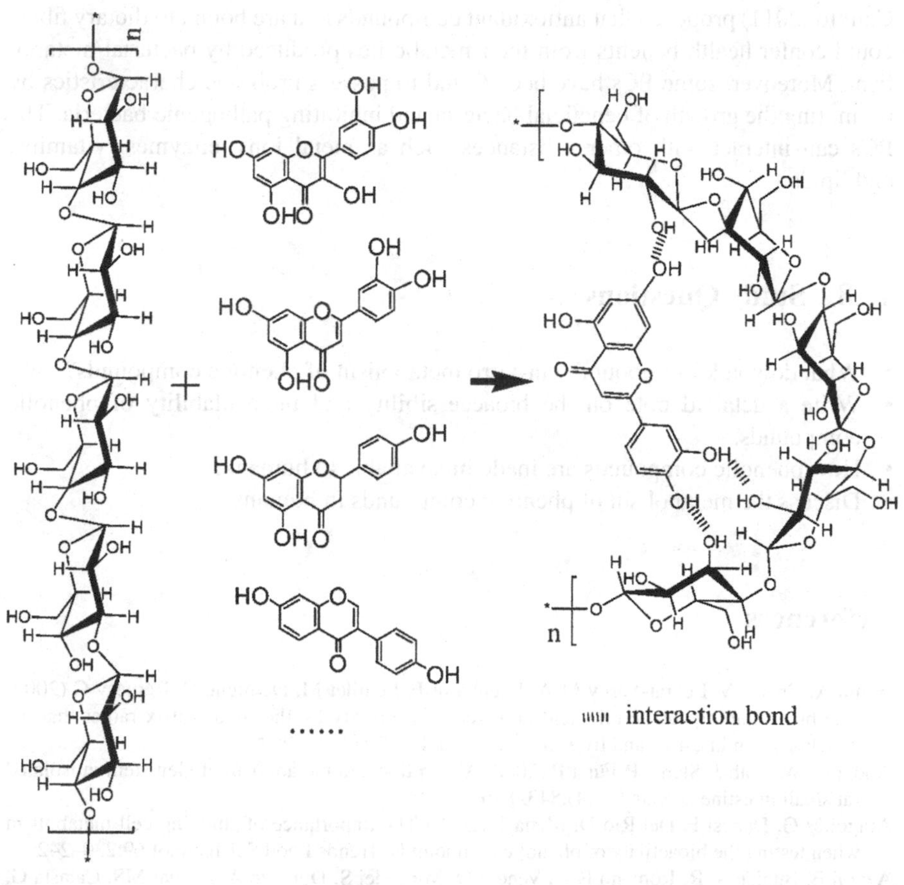

Fig. 12.11 Interactions of phenolic compounds with carbohydrates. Reproduced with kind permission of the American Chemical Society (Wang et al. 2013)

acid, chlorogenic acid, gallic acid, and cyanidin-3-glycoside in in-vitro studies (Phan et al. 2015). Because these interactions are largely inevitable in plant foods consumed by individuals, approaches such as industrial food processing and cooking have been identified to make PCs more accessible for absorption in the upper GI tract by weakening these interactions or binding. The *in-vitro* cellulase treatment of oat can liberate PCs, significantly enhancing the PC content and antioxidant capacity. In a study of healthy men, ferulic acid was found to be more bioavailable in bread made with flour that was pretreated with yeast, xylanase, cellulase, β-glucanase, and feruloyl esterase, showing the interactions of ferulic acid with selected carbohydrates (Mateo Anson et al. 2010).

The majority of PCs contained in non-processed plant foods do not reach the circulation because of the previously mentioned interactions with the fibers that transport them to the large intestine. However, the reduced bioavailability of PCs caused by fibers is compensated for by the health benefits of the breakdown products derived from bacterial metabolism in the lower GI tract. For example, Saura-Calixto (2011) proposed that antioxidant compounds that are bound to dietary fibers could confer health benefits from their metabolites produced by bacterial metabolism. Moreover, some PCs have been found to possess probiotic characteristics by promoting the growth of beneficial bacteria and inhibiting pathogenic bacteria. The PCs can interact with other substances such as metal ions, enzymes, vitamins, and lipids.

12.8   Study Questions

- What do you know about the in-vitro metabolism of phenolic compounds?
- Write a detailed note on the bioaccessibility and bioavailability of phenolic compounds.
- How phenolic compounds are made bioavailable to humans?
- Discuss the metabolism of phenolic compounds in humans.

References

Adam A, Crespy V, Levrat-Verny M-A, Leenhardt F, Leuillet M, Demigné C, Rémésy C (2002) The bioavailability of Ferulic acid is governed primarily by the food matrix rather than its metabolism in intestine and liver in rats. J Nutr 132(7):1962–1968

Andlauer W, Kolb J, Stehle P, Fürst P (2000) Absorption and metabolism of Genistein in isolated rat small intestine. J Nutr 130(4):843–846

Aragonès G, Danesi F, Del Rio D, Mena P (2017) The importance of studying cell metabolism when testing the bioactivity of phenolic compounds. Trends Food Sci Technol 69:230–242

Azzini E, Bugianesi R, Romano F, Di Venere D, Miccadei S, Durazzo A, Foddai MS, Catasta G, Linsalata V, Maiani G (2007) Absorption and metabolism of bioactive molecules after oral consumption of cooked edible heads of *Cynara scolymus* L. (cultivar Violetto di Provenza) in human subjects: a pilot study. Br J Nutr 97(5):963–969

Bariexca T, Ezdebski J, Redan BW, Vinson J (2019) Pure polyphenols and cranberry juice high in anthocyanins increase antioxidant capacity in animal organs. Foods 8(8):340

Barrington R, Williamson G, Bennett RN, Davis BD, Brodbelt JS, Kroon PA (2009) Absorption, conjugation and efflux of the flavonoids, kaempferol and galangin, using the intestinal CaCo-2/TC7 cell model. J Funct Foods 1(1):74–87

Bento-Silva A, Koistinen VM, Mena P, Bronze MR, Hanhineva K, Sahlstrøm S, Kitrytė V, Moco S, Aura A-M (2020) Factors affecting intake, metabolism and health benefits of phenolic acids: do we understand individual variability? Eur J Nutr 59(4):1275–1293

de Boer VCJ, de Goffau MC, Arts ICW, Hollman PCH, Keijer J (2006) SIRT1 stimulation by polyphenols is affected by their stability and metabolism. Mech Ageing Dev 127(7):618–627

Borges G, Ottaviani JI, van der Hooft JJJ, Schroeter H, Crozier A (2018) Absorption, metabolism, distribution and excretion of (−)-epicatechin: a review of recent findings. Mol Asp Med 61:18–30

Bourassa P, Bariyanga J, Tajmir-Riahi HA (2013) Binding sites of resveratrol, Genistein, and curcumin with Milk α- and β-caseins. J Phys Chem B 117(5):1287–1295

Brand W, Van Der Wel PA, Rein MJ, Barron D, Williamson G, Van Bladeren PJ, Rietjens IM (2008) Metabolism and transport of the citrus flavonoid hesperetin in Caco-2 cell monolayers. Drug Metab Dispos 36(9):1794–1802

Brudzynski K, Maldonado-Alvarez L (2015) Polyphenol-protein complexes and their consequences for the redox activity, structure and function of honey. A current view and new hypothesis–a review. Polish J Food Nutr Sci 65(2):71–80

Carbonaro M, Grant G, Pusztai A (2001) Evaluation of polyphenol bioavailability in rat small intestine. Eur J Nutr 40(2):84–90

Carbonell-Capella JM, Buniowska M, Barba FJ, Esteve MJ, Frígola A (2014) Analytical methods for determining bioavailability and bioaccessibility of bioactive compounds from fruits and vegetables: a review. Compr Rev Food Sci Food Saf 13(2):155–171

Castello F, Costabile G, Bresciani L, Tassotti M, Naviglio D, Luongo D, Ciciola P, Vitale M, Vetrani C, Galaverna G, Brighenti F, Giacco R, Del Rio D, Mena P (2018) Bioavailability and pharmacokinetic profile of grape pomace phenolic compounds in humans. Arch Biochem Biophys 646:1–9

Chen J, Lin H, Hu M (2005) Absorption and metabolism of genistein and its five isoflavone analogs in the human intestinal Caco-2 model. Cancer Chemother Pharmacol 55(2):159–169

Chesson A, Provan GJ, Russell WR, Scobbie L, Richardson AJ, Stewart C (1999) Hydroxycinnamic acids in the digestive tract of livestock and humans. J Sci Food Agric 79(3):373–378

Clifford MN, Jaganath IB, Ludwig IA, Crozier A (2017) Chlorogenic acids and the acyl-quinic acids: discovery, biosynthesis, bioavailability and bioactivity. Nat Prod Rep 34(12):1391–1421

Crespy V, Morand C, Besson C, Manach C, Demigne C, Remesy C (2002) Quercetin, but not its glycosides, is absorbed from the rat stomach. J Agric Food Chem 50(3):618–621

Crozier A, Del Rio D, Clifford MN (2010) Bioavailability of dietary flavonoids and phenolic compounds. Mol Asp Med 31(6):446–467

Czank C, Cassidy A, Zhang Q, Morrison DJ, Preston T, Kroon PA, Botting NP, Kay CD (2013) Human metabolism and elimination of the anthocyanin, cyanidin-3-glucoside: a 13C-tracer study. Am J Clin Nutr 97(5):995–1003

Dall'Asta M, Derlindati E, Curella V, Mena P, Calani L, Ray S, Zavaroni I, Brighenti F, Del Rio D (2013) Effects of naringenin and its phase II metabolites on in vitro human macrophage gene expression. Int J Food Sci Nutr 64(7):843–849

Day AJ, Cañada FJ, Dìaz JC, Kroon PA, Mclauchlan R, Faulds CB, Plumb GW, Morgan MR, Williamson G (2000) Dietary flavonoid and isoflavone glycosides are hydrolysed by the lactase site of lactase phlorizin hydrolase. FEBS Lett 468(2–3):166–170

Day AJ, Mellon F, Barron D, Sarrazin G, Morgan MRA, Williamson G (2001) Human metabolism of dietary flavonoids: identification of plasma metabolites of quercetin. Free Radic Res 35(6):941–952

De Ferrars R, Czank C, Zhang Q, Botting N, Kroon P, Cassidy A, Kay C (2014) The pharmacokinetics of anthocyanins and their metabolites in humans. Brit J Pharmacol 171(13):3268–3282

Del Bò C, Ciappellano S, Klimis-Zacas D, Martini D, Gardana C, Riso P, Porrini M (2010) Anthocyanin absorption, metabolism, and distribution from a wild blueberry-enriched diet (*vaccinium angustifolium*) is affected by diet duration in the Sprague–dawley rat. J Agric Food Chem 58(4):2491–2497

Del Rio D, Rodriguez-Mateos A, Spencer JP, Tognolini M, Borges G, Crozier A (2013) Dietary (poly) phenolics in human health: structures, bioavailability, and evidence of protective effects against chronic diseases. Antioxid Redox Signal 18(14):1818–1892

Dominguez-Avila JA, Wall-Medrano A, Velderrain-Rodriguez GR, Chen CYO, Salazar-Lopez NJ, Robles-Sanchez M, Gonzalez-Aguilar GA (2017) Gastrointestinal interactions, absorption, splanchnic metabolism and pharmacokinetics of orally ingested phenolic compounds. Food and Function 8(1):15–38

Espín JC, Larrosa M, García-Conesa MT, Tomás-Barberán F (2013) Biological significance of Urolithins, the gut microbial Ellagic acid-derived metabolites: the evidence so far. Evid Based Complement Alternat Med 2013:270418

Fernandes I, Faria A, de Freitas V, Calhau C, Mateus N (2015) Multiple-approach studies to assess anthocyanin bioavailability. Phytochem Rev 14(6):899–919

Forester SC, Waterhouse AL (2010) Gut metabolites of anthocyanins, gallic acid, 3-o-methylgallic acid, and 2,4,6-trihydroxybenzaldehyde, inhibit cell proliferation of Caco-2 cells. J Agric Food Chem 58(9):5320–5327

Fornasaro S, Ziberna L, Gasperotti M, Tramer F, Vrhovšek U, Mattivi F, Passamonti S (2016) Determination of cyanidin 3-glucoside in rat brain, liver and kidneys by UPLC/MS-MS and its application to a short-term pharmacokinetic study. Sci Rep 6(1):22815

de Freitas Queiroz Barros HD, Maróstica Jr. MR (2019) Phenolic compound bioavailability using in vitro and in vivo models. In: Campos MRS (ed) Bioactive Compounds. Woodhead Publishing, pp 113–126

García-Villalba R, Carrasco-Pancorbo A, Oliveras-Ferraros C, Menéndez JA, Segura-Carretero A, Fernández-Gutiérrez A (2012) Uptake and metabolism of olive oil polyphenols in human breast cancer cells using nano-liquid chromatography coupled to electrospray ionization–time of flight-mass spectrometry. J Chromatogr B 898:69-77

García-Villalba R, Beltrán D, Frutos MD, Selma MV, Espín JC, Tomás-Barberán FA (2020) Metabolism of different dietary phenolic compounds by the urolithin-producing human-gut bacteria *Gordonibacter urolithinfaciens* and *Ellagibacter isourolithinifaciens*. Food Funct 11(8):7012–7022

Gee JM, DuPont MS, Day AJ, Plumb GW, Williamson G, Johnson IT (2000) Intestinal transport of quercetin glycosides in rats involves both Deglycosylation and interaction with the hexose transport pathway. J Nutr 130(11):2765–2771

Giménez-Bastida JA, González-Sarrías A, Vallejo F, Espín JC, Tomás-Barberán FA (2016) Hesperetin and its sulfate and glucuronide metabolites inhibit TNF-α induced human aortic endothelial cell migration and decrease plasminogen activator inhibitor-1 (PAI-1) levels. Food Funct 7(1):118–126

Gómez-Juaristi M, Martínez-López S, Sarria B, Bravo L, Mateos R (2018) Absorption and metabolism of yerba mate phenolic compounds in humans. Food Chem 240:1028–1038

Gómez-Juaristi M, Sarria B, Goya L, Bravo-Clemente L, Mateos R (2020) Experimental confounding factors affecting stability, transport and metabolism of flavanols and hydroxycinnamic acids in Caco-2 cells. Food Res Int 129:108797

Gonzalez-Sarrias A, Azorin-Ortuno M, Yanez-Gascon M-J, Tomas-Barberan FA, Garcia-Conesa M-T, Espin J-C (2009) Dissimilar in vitro and in vivo effects of ellagic acid and its microbiota-derived metabolites, urolithins, on the cytochrome P450 1A1. J Agric Food Chem 57(12):5623–5632

Gonzalez-Sarrias A, Giménez-Bastida JA, Núñez-Sánchez MÁ, Larrosa M, García-Conesa MT, Tomás-Barberán FA, Espín JCJEjon (2014) Phase-II metabolism limits the antiproliferative activity of urolithins in human colon cancer cells. Eur J Nutr. 53 (3):853–864

González-Sarrías A, Núñez-Sánchez MÁ, García-Villalba R, Tomás-Barberán FA, Espín JC (2017) Antiproliferative activity of the ellagic acid-derived gut microbiota isourolithin a and comparison with its urolithin a isomer: the role of cell metabolism. Eur J Nutr 56(2):831–841

Granato D, Mocan A, Câmara JS (2020) Is a higher ingestion of phenolic compounds the best dietary strategy? A scientific opinion on the deleterious effects of polyphenols in vivo. Trends Food Sci Technol 98:162–166

Han F, Yang P, Wang H, Fernandes I, Mateus N, Liu Y (2019a) Digestion and absorption of red grape and wine anthocyanins through the gastrointestinal tract. Trends Food Sci Technol 83:211–224

Han J, Chang Y, Britten M, St-Gelais D, Champagne CP, Fustier P, Lacroix M (2019b) Interactions of phenolic compounds with milk proteins. Eur Food Res Technol 245(9):1881–1888

Heleno SA, Martins A, Queiroz MJRP, Ferreira ICFR (2015) Bioactivity of phenolic acids: metabolites versus parent compounds: a review. Food Chem 173:501–513

Hollman PCH (2004) Absorption, bioavailability, and metabolism of flavonoids. Pharm Biol. 42(suppl1):74–83

Hong J, Lu H, Meng X, Ryu J-H, Hara Y, Yang CS (2002) Stability, cellular uptake, biotransformation, and efflux of tea polyphenol (−)-epigallocatechin-3-gallate in HT-29 human colon adenocarcinoma cells. Cancer Res 62(24):7241–7246

Ichiyanagi T, Shida Y, Rahman MM, Hatano Y, Konishi T (2006) Bioavailability and tissue distribution of anthocyanins in bilberry (Vaccinium myrtillus L.) extract in rats. J Agric Food Chem 54(18):6578–6587

Ishisaka A, Kawabata K, Miki S, Shiba Y, Minekawa S, Nishikawa T, Mukai R, Terao J, Kawai Y (2013) Mitochondrial dysfunction leads to deconjugation of quercetin glucuronides in inflammatory macrophages. PLoS One 8(11):e80843

Kahle K, Kempf M, Schreier P, Scheppach W, Schrenk D, Kautenburger T, Hecker D, Huemmer W, Ackermann M, Richling E (2011) Intestinal transit and systemic metabolism of apple polyphenols. Eur J Nutr 50(7):507–522

Kaldas MI, Walle UK, Walle T (2003) Resveratrol transport and metabolism by human intestinal Caco-2 cells. J Pharm Pharmacol 55(3):307–312

Kalt W (2019) Anthocyanins and their C6-C3-C6 metabolites in humans and animals. Molecules 24(22):4024

Karakaya S (2004) Bioavailability of phenolic compounds. Crit Rev Food Sci Nutr 44(6):453–464

Kawai Y, Nishikawa T, Shiba Y, Saito S, Murota K, Shibata N, Kobayashi M, Kanayama M, Uchida K, Terao J (2008) Macrophage as a target of quercetin glucuronides in human atherosclerotic arteries: implication in the anti-atherosclerotic mechanism of dietary flavonoids. J Biol Chem 283(14):9424–9434

Kay CD, Kroon PA, AJM C (2009) The bioactivity of dietary anthocyanins is likely to be mediated by their degradation products. Mol Nutr Food Res 53(S1):S92–S101

Kern SM, Bennett RN, Needs PW, Mellon FA, Kroon PA, Garcia-Conesa M-T (2003) Characterization of metabolites of hydroxycinnamates in the in vitro model of human small intestinal epithelium Caco-2 cells. J Agric Food Chem 51(27):7884–7891

Kirakosyan A, Seymour EM, Wolforth J, McNish R, Kaufman PB, Bolling SF (2015) Tissue bioavailability of anthocyanins from whole tart cherry in healthy rats. Food Chem 171:26–31

Lançon A, Hanet N, Jannin B, Delmas D, Heydel J-M, Lizard G, Chagnon M-C., Artur Y, Latruffe N (2007) Resveratrol in human hepatoma HepG2 cells: Metabolism and inducibility of detoxifying enzymes. Drug Metab Dispos 35(5):699–703

Landete JM (2011) Ellagitannins, ellagic acid and their derived metabolites: a review about source, metabolism, functions and health. Food Res Int 44(5):1150–1160

Larrosa M, Tomás-Barberán FA, Espín JC (2006) The dietary hydrolysable tannin punicalagin releases ellagic acid that induces apoptosis in human colon adenocarcinoma Caco-2 cells by using the mitochondrial pathway. J Nutr Biochem 17(9):611–625

Lee M-J, Maliakal P, Chen L, Meng X, Bondoc FY, Prabhu S, Lambert G, Mohr S, Yang CS (2002) Pharmacokinetics of tea catechins after ingestion of green tea and (−)-epigallocatechin-3-gallate by humans. Cancer Epidem Biomar 11(10):1025–1032

López de las Hazas M-C, Mosele JI, Macià A, Ludwig IA, Motilva M-J (2017) Exploring the colonic metabolism of grape and strawberry anthocyanins and their in vitro apoptotic effects in HT-29 Colon Cancer cells. J Agric Food Chem 65(31):6477–6487

Ludwig IA, Paz de Peña M, Concepción C, Alan C (2013) Catabolism of coffee chlorogenic acids by human colonic microbiota. Biofactors 39(6):623–632

Manach C, Texier O, Morand C, Crespy V, Régérat F, Demigné C, Rémésy C (1999) Comparison of the bioavailability of quercetin and catechin in rats. Free Radic Biol Med 27(11):1259–1266

Manach C, Scalbert A, Morand C, Rémésy C, Jiménez L (2004) Polyphenols: food sources and bioavailability. Am J Clin Nutr 79(5):727–747

del Mar Contreras M, Borrás-Linares I, Herranz-López M, Micol V, Segura-Carretero A (2016) Further exploring the absorption and enterocyte metabolism of quercetin forms in the Caco-2 model using nano-LC-TOF-MS. Electrophoresis 37(7–8):998–1006

Marczylo TH, Cooke D, Brown K, Steward WP, Gescher AJ (2009) Pharmacokinetics and metabolism of the putative cancer chemopreventive agent cyanidin-3-glucoside in mice. Cancer Chemother Pharmacol 64(6):1261–1268

Martini S, Conte A, Tagliazucchi D (2019) Antiproliferative activity and cell metabolism of Hydroxycinnamic acids in human Colon adenocarcinoma cell lines. J Agric Food Chem 67(14):3919–3931

Mateo Anson N, Aura AM, Selinheimo E, Mattila I, Poutanen K, van den Berg R, Havenaar R, Bast A, Haenen GRMM (2010) Bioprocessing of wheat bran in whole wheat bread increases the bioavailability of phenolic acids in men and exerts Antiinflammatory effects ex vivo. J Nutr 141(1):137–143

Mateos R, Goya L, Bravo L (2006) Uptake and metabolism of Hydroxycinnamic acids (Chlorogenic, Caffeic, and Ferulic acids) by HepG2 cells as a model of the human liver. J Agric Food Chem 54(23):8724–8732

Mateus N, Carvalho E, Luìs C, de Freitas V (2004) Influence of the tannin structure on the disruption effect of carbohydrates on protein–tannin aggregates. Anal Chim Acta 513(1):135–140

Mele L, Mena P, Piemontese A, Marino V, López-Gutiérrez N, Bernini F, Brighenti F, Zanotti I, Del Rio D (2016) Antiatherogenic effects of ellagic acid and urolithins in vitro. Arch Biochem Biophys 599:42–50

Mena P, Calani L, Bruni R, Del Rio D (2015) Bioactivation of high-molecular-weight polyphenols by the gut microbiome. In: Tuohy K, Del Rio D (eds) Diet-microbe interactions in the gut. Academic, San Diego, pp 73–101

Mülek M, Fekete A, Wiest J, Holzgrabe U, Mueller MJ, Högger P (2015) Profiling a gut microbiota-generated catechin metabolite's fate in human blood cells using a metabolomic approach. J Pharmaceut Biomed 114:71–81

Nagy K, Courtet-Compondu M-C, Williamson G, Rezzi S, Kussmann M, Rytz A (2012) Non-covalent binding of proteins to polyphenols correlates with their amino acid sequence. Food Chem 132(3):1333–1339

Ng SP, Wong KY, Zhang L, Zuo Z, Lin G (2005) Evaluation of the first-pass glucuronidation of selected flavones in gut by Caco-2 monolayer model. Pharm Pharmaceut Sci 8(1):1–9

Núñez-Sánchez MÁ, Karmokar A, González-Sarrías A, García-Villalba R, Tomás-Barberán FA, García-Conesa MT, Brown K, Espín JC (2016) In vivo relevant mixed urolithins and ellagic acid inhibit phenotypic and molecular colon cancer stem cell features: a new potentiality for ellagitannin metabolites against cancer. Food Chem Toxicol 92:8–16

O'Leary KA, Day AJ, Needs PW, Mellon FA, O'Brien NM, Williamson G (2003) Metabolism of quercetin-7- and quercetin-3-glucuronides by an in vitro hepatic model: the role of human β-glucuronidase, sulfotransferase, catechol-O-methyltransferase and multi-resistant protein 2 (MRP2) in flavonoid metabolism. Biochem Pharmacol 65(3):479–491

Olthof MR, Hollman PCH, Buijsman MNCP, van Amelsvoort JMM, Katan MB (2003) Chlorogenic acid, Quercetin-3-Rutinoside and black tea phenols are extensively metabolized in humans. J Nutr 133(6):1806–1814

Patel KR, Andreadi C, Britton RG, Horner-Glister E, Karmokar A, Sale S, Brown VA, Brenner DE, Singh R, Steward WP, Gescher AJ, Brown K (2013) Sulfate metabolites provide an intracellular Pool for resveratrol generation and induce autophagy with senescence. 5 (205):205ra133-205ra133

Phan ADT, Netzel G, Wang D, Flanagan BM, D'Arcy BR, Gidley MJ (2015) Binding of dietary polyphenols to cellulose: structural and nutritional aspects. Food Chem 171:388–396

Proteggente AR, Basu-Modak S, Kuhnle G, Gordon MJ, Youdim K, Tyrrell R, Rice-Evans CA (2003) Hesperetin glucuronide, a photoprotective agent arising from flavonoid metabolism in human skin fibroblasts. Photochem Photobiol 78(3):256–261

Qie X, Chen Y, Quan W, Wang Z, Zeng M, Qin F, Chen J, He Z (2020) Analysis of β-lactoglobulin–epigallocatechin gallate interactions: the antioxidant capacity and effects of polyphenols under different heating conditions in polyphenolic–protein interactions. Food Funct 11(5):3867–3878

Rodriguez-Mateos A, Toro-Funes N, Cifuentes-Gomez T, Cortese-Krott M, Heiss C, Spencer JPE (2014a) Uptake and metabolism of (−)-epicatechin in endothelial cells. Arch Biochem Biophys 559:17–23

Rodriguez-Mateos A, Vauzour D, Krueger CG, Shanmuganayagam D, Reed J, Calani L, Mena P, Del Rio D, Crozier A (2014b) Bioavailability, bioactivity and impact on health of dietary flavonoids and related compounds: an update. Arch Toxicol 88(10):1803–1853

Rondini L, Peyrat-Maillard M-N, Marsset-Baglieri A, Berset C (2002) Sulfated Ferulic acid is the Main in vivo metabolite found after short-term ingestion of free Ferulic acid in rats. J Agric Food Chem 50(10):3037–3041

Sakakibara H, Ogawa T, Koyanagi A, Kobayashi S, Goda T, Kumazawa S, Kobayashi H, Shimoi K (2009) Distribution and excretion of bilberry anthocyanins in mice. J Agric Food Chem 57(17):7681–7686

Sala R, Mena P, Savi M, Brighenti F, Crozier A, Miragoli M, Stilli D, Del Rio D (2015) Urolithins at physiological concentrations affect the levels of pro-inflammatory cytokines and growth factor in cultured cardiac cells in hyperglucidic conditions. J Funct Foods 15:97–105

Sandoval-Ramírez BA, Catalán Ú, Fernández-Castillejo S, Rubió L, Macià A, Solà R (2018) Anthocyanin tissue bioavailability in animals: possible implications for human health. A systematic review. J Agric Food Chem 66(44):11531–11543

Saura-Calixto F (2011) Dietary Fiber as a carrier of dietary Antioxidants: an essential physiological function. J Agric Food Chem 59(1):43–49

Scheline RR (1978) Mammalian metabolism of plant xenobiotics. Academic Press, London

Seeram NP, Lee R, Heber D (2004) Bioavailability of ellagic acid in human plasma after consumption of ellagitannins from pomegranate (Punica granatum L.) juice. Clin Chim Acta 348(1):63–68

Selma MV, Espín JC, Tomás-Barberán FA (2009) Interaction between Phenolics and gut microbiota: role in human health. J Agric Food Chem 57(15):6485–6501

Shahidi F, Peng H (2018) Bioaccessibility and bioavailability of phenolic compounds. J Food Bioact 4:11–68

Shahidi F, Chandrasekara A, Zhong Y (2018) Bioactive phytochemicals in vegetables. In: Siddiq M, Uebersax MA (eds) Handbook of vegetables vegetable processing, vol 2. Wiley, London, pp 181–222

Shahrzad S, Bitsch I (1998) Determination of gallic acid and its metabolites in human plasma and urine by high-performance liquid chromatography. J Chromatogr B Biomed Sci Appl 705(1):87–95

Shimoda K, Kubota N, Uesugi D, Kobayashi Y, Hamada H, Hamada H (2020) Glycosylation of stilbene compounds by cultured plant cells. Molecules 25(6):1437

Shimoi K, Okada H, Furugori M, Goda T, Takase S, Suzuki M, Hara Y, Yamamoto H, Kinae N (1998) Intestinal absorption of luteolin and luteolin 7-O-β-glucoside in rats and humans. FEBS Lett 438(3):220–224

Sigurdson GT, Atnip A, Bomser J, Giusti MM (2018) Aglycone structures and glycosylations affect anthocyanin transport and uptake in human gastric epithelial (NCI-N87) cells. J Food Compos Anal 65:33–39

Spencer JPE, Schroeter H, Crossthwaithe AJ, Kuhnle G, Williams RJ, Rice-Evans C (2001) Contrasting influences of glucuronidation and O-methylation of epicatechin on hydrogen per-oxide-induced cell death in neurons and fibroblasts. Free Radic Biol Med 31(9):1139–1146

Spencer JP, Kuhnle GG, Williams RJ, Rice-Evans CJBJ (2003) Intracellular metabolism and bio-activity of quercetin and its in vivo metabolites. Biochem J 372 (1):173–181

Spigoni V, Mena P, Cito M, Fantuzzi F, Bonadonna RC, Brighenti F, Cas AD, Rio DD (2016) Effects on nitric oxide production of Urolithins, gut-derived Ellagitannin metabolites, in human aortic endothelial cells. Molecules 21(8):1009

Stalmach A, Mullen W, Barron D, Uchida K, Yokota T, Cavin C, Steiling H, Williamson G, Crozier A (2009) Metabolite profiling of hydroxycinnamate derivatives in plasma and urine after the ingestion of coffee by humans: identification of biomarkers of coffee consumption. Drug Metab Dispos 37(8):1749–1758

Teng H, Chen L (2019) Polyphenols and bioavailability: an update. Crit Rev Food Sci Nutr 59(13):2040–2051

Tomás-Barberán FA, Clifford MN (2000) Dietary hydroxybenzoic acid derivatives – nature, occur-rence and dietary burden. J Sci Food Agric 80(7):1024–1032

Tremblay GC, Qureshi IA (1993) The biochemistry and toxicology of benzoic acid metabolism and its relationship to the elimination of waste nitrogen. Pharmacol Therapeut 60(1):63–90

Tribolo S, Lodi F, Winterbone MS, Saha S, Needs PW, Hughes DA, Kroon PA (2013) Human metabolic transformation of quercetin blocks its capacity to decrease endothelial nitric oxide synthase (eNOS) expression and Endothelin-1 secretion by human endothelial cells. J Agric Food Chem 61(36):8589–8596

Tuck KL, Hayball PJ (2002) Major phenolic compounds in olive oil: metabolism and health effects. J Nutr Biochem 13(11):636–644

Vafeiadou K, Vauzour D, Rodriguez-Mateos A, Whiteman M, Williams RJ, Spencer JPE (2008) Glial metabolism of quercetin reduces its neurotoxic potential. Arch Biochem Biophys 478(2):195–200

Vanzo A, Terdoslavich M, Brandoni A, Torres AM, Vrhovsek U, Passamonti S (2008) Uptake of grape anthocyanins into the rat kidney and the involvement of bilitranslocase. Mol Nutr Food Res 52(10):1106–1116

Vanzo A, Vrhovsek U, Tramer F, Mattivi F, Passamonti S (2011) Exceptionally fast uptake and metabolism of Cyanidin 3-glucoside by rat kidneys and liver. J Nat Prod 74(5):1049–1054

Vetrani C, Rivellese AA, Annuzzi G, Adiels M, Borén J, Mattila I, Orešič M, Aura A-M (2016) Metabolic transformations of dietary polyphenols: comparison between in vitro colonic and hepatic models and in vivo urinary metabolites. J Nutr Biochem 33:111–118

Victor de F, Nuno M (2012) Protein/polyphenol interactions: past and present contributions. Mechanisms of astringency perception. Curr Org Chem 16(6):724–746

Vollmer M, Esders S, Farquharson FM, Neugart S, Duncan SH, Schreiner M, Louis P, Maul R, Rohn S (2018) Mutual interaction of phenolic compounds and microbiota: metabolism of com-plex phenolic Apigenin-C- and Kaempferol-O-derivatives by human fecal samples. J Agric Food Chem 66(2):485–497

Wang Y, Liu J, Chen F, Zhao G (2013) Effects of molecular structure of polyphenols on their non-covalent interactions with oat β-glucan. J Agric Food Chem 61(19):4533–4538

Warner EF, Zhang Q, Raheem KS, O'Hagan D, O'Connell MA, Kay CD (2016) Common phenolic metabolites of flavonoids, but not their unmetabolized precursors, reduce the secretion of vas-cular cellular adhesion molecules by human endothelial cells. J Nutr 146(3):465–473

Wilkening S, Stahl F, Bader A (2003) Comparison of primary human hepatocytes and hepatoma cell line Hepg2 with regard to their biotransformation properties. Drug Metab Dispos 31(8):1035–1042

Wittig J, Herderich M, Graefe EU, Veit M (2001) Identification of quercetin glucuronides in human plasma by high-performance liquid chromatography–tandem mass spectrometry. J Chromatogr B Biomed Sci Appl 753(2):237–243

Wong CC, Williamson G (2013) Inhibition of hydroxycinnamic acid sulfation by flavonoids and their conjugated metabolites. Biofactors 39(6):644–651

Wong YC, Zhang L, Lin G, Zuo Z (2009) Structure–activity relationships of the glucuronidation of flavonoids by human glucuronosyltransferases. Expert Opinion on Drug Metabolism and Toxicology 5(11):1399–1419

van der Woude H, Boersma MG, Vervoort J, Rietjens IMCM (2004) Identification of 14 quercetin phase II mono- and mixed conjugates and their formation by rat and human phase II in vitro model systems. Chem Res Toxicol 17(11):1520–1530

Yang Y, Zhang G, Li C, Wang S, Zhu M, Wang J, Yue H, Ma X, Zhen Y, Shu X (2019) Metabolic profile and structure-activity relationship of resveratrol and its analogs in human bladder cancer cells. Cancer Manag Res 11:4631–4642

Chapter 13
Applications of Phenolic Antioxidants

13.1 Introduction

Phenolic compounds (PCs) are present in plant foods, which are known for their significant antioxidant role. A large number of PCs have been identified and the analysis of more plant foods is still under consideration. PCs are known to act as an antioxidant by reacting with a variety of free radicals. The mechanism of actions of phenolic antioxidants involved either by hydrogen atom transfer, single electron transfer, sequential proton loss electron transfer, and transition metal chelation. In foods, the PCs had been studied for antioxidant activity using several in-vitro methods. Milk and dairy products are supplemented to enhance color, taste, storage stability, and consequent quality. This chapter critically reviewed the application of PCs as food additives, that is, antimicrobial, and antioxidant agents. The PCs are also used as flavoring agents. Studies also showed that PCs in foods possess several health benefits such as antibacterial, antihyperlipidemic, anticancer, antioxidants, cardioprotective, neuroprotective, and antidiabetic properties (Zeb 2020). Due to the toxicity of synthetic antioxidants, phenolic compounds from foods or plants can be used as functional food ingredients. Several foods that need to be evaluated for the phenolic profile in different parts of the world are still unexplored.

13.2 Applications of Phenolic Antioxidants in Foods

13.2.1 As an Antioxidants in Foods

Foods especially plant foods are rich in PCs. Escarpa and Gonzalez (2001) reviewed the distribution, synthesis, and analytical techniques used for the determination of PCs in foods. Similarly, the antioxidant and bioactivity of phenolic acids have been reviewed (Heleno et al. 2015). However, due to the significance of the topic,

© Springer Nature Switzerland AG 2021
A. Zeb, *Phenolic Antioxidants in Foods: Chemistry, Biochemistry and Analysis*,
https://doi.org/10.1007/978-3-030-74768-8_13

tremendous growth in scientific literature has been observed in the last two decades. Shahidi and Ambigaipalan (2015) also reviewed PCs in foods, beverages, and spices. The authors summarized synthetic as well as phenolic antioxidants from the natural sources, their mode of action, degradation, health effects, and toxicological properties. It was reported that natural phenolic antioxidants are safer than synthetic antioxidants and their sources in foods need to be exploited in detail. The antioxidant functions of foods are usually measured in the *in-vitro* system using different analytical techniques. There are several antioxidant assay techniques such as azino-bis(3-ethylbenzothiazoline-6-sulfonic acid) (ABTS), ferric reducing antioxidant power (FRAP), cupric reducing antioxidant capacity (CUPRAC), oxygen radical absorbance capacity (ORAC), Trolox equivalent antioxidant capacity (TEAC), diphenyl-1-picrylhydrazyl (DPPH), total antioxidant activity (TAA) (Ruiz-Torralba et al. 2018) and photo-chemiluminescence (PoCL) (Table 13.1). Thaipong et al. (2006) and Sridhar and Charles (2019) compared ABTS, FRAP, DPPH, and ORAC assays in guava fruits. The authors reported that these assays were in close agreement with one another. However, a comparison of these methods is not possible due to differences in the mechanism of action, type of substrate, sample matrix, and extraction solvents (Harnly 2017). The DPPH radical scavenging activity has been the most extensively used method in leafy vegetables (Zeb et al. 2018; Zeb 2016; Zeb and Habib 2018; Zeb 2015b). Similarly, CUPRAC, FRAP, DPPH, and TEAC assays had been used to assess the antioxidant activity of different legumes (Koley et al. 2019).

Beans such as red beans and adzuki beans have been studied for different antioxidant activity including peroxyl radical scavenging activity (PRSA) (Amarowicz

Table 13.1 Important representative *in-vitro* antioxidant activity measurement methods used for foods (reproduced with kind permission of John Wiley and Sons (Zeb 2020))

Methods	Foods	References
Oxygen radical absorbance capacity (ORAC)	Guava fruit	Thaipong et al. (2006)
Total radical-trapping antioxidant potential (TRAP)	Fruits	Ruiz-Torralba et al. (2018)
Trolox equivalent antioxidant capacity (TEAC)	Legumes, fruits	Thaipong et al. (2006); Koley et al. (2019)
2,2-Diphenyl-1-picrylhydrazyl (DPPH)	Guava fruit, leafy vegetables, fruits	Ruiz-Torralba et al. (2018); Thaipong et al. (2006); Zeb (2015)
Ferric reducing antioxidant power (FRAP)	Guava fruit, fruits	Ruiz-Torralba et al. (2018); Thaipong et al. (2006); Koley et al. (2019)
ABTS	Guava fruit, grape, bakery products	Ruiz-Torralba et al. (2018); Koley et al. (2019); Sridhar and Charles (2019)
Total antioxidant activity (TAA)	Red beans	Thaipong et al. (2006)
Cupric reducing antioxidant capacity (CUPRAC)	Legumes	Koley et al. (2019)
Peroxyl radical scavenging activity (PRSA)	Beans	Koley et al. (2019); Amarowicz et al. (2017)

et al. 2017; López-Alarcón and Lissi 2005; Bai et al. 2017). Five methods such as TEAC, DPPH, ORAC, red blood cell hemolysis and ESR (electron spin resonance for direct free radical evaluation) have been studied in twenty-five different phenolic compounds from flavonols, anthocyanins, flavanones, flavan-3-ols, phenolic acids, and others and food beverages (Tabart et al. 2009). These methods are used for a quick assessment and as screening tools for *in-vitro* system and thus cannot be ruled out (Granato et al. 2018; Alam et al. 2013; Schaich et al. 2015), however, none of these methods, independently or together are valid to the real physiological antioxidant capacity of the human body (Tsao and Li 2012). These methods can tell us, how much a PC can play antioxidant function. The antioxidant functions in these methods are based on either hydrogen donation or electron transfer mechanisms. The antioxidant activities of foods are strongly dependent on the total PC contents as well as other antioxidants present in the foods. Karakaya et al. (2001) reported that the total phenols of liquid foods varied between 68 to 4162 mg/L and in solid foods, they ranged from 735–3994 mg/kg. The TAA of liquid and solid foods ranged from 0.61 to 6.78 mM and 0.63 to 8.62 mM, respectively. The FRAP, TEAC, DPPH with TPC contents of the different fresh fruits available in the Spanish market were evaluated (Ruiz-Torralba et al. 2018). The authors showed that total antioxidant capacity (TAC) was 31, 17, and 9 times more in berries than for apples, bananas, and oranges, respectively. The commonly consumed fruits in Spain were bananas, grapes, oranges, strawberries, and tangerines, which account for 76% of the TAC of fruits. The TAC and TPC increases by 2.4 times were due to the consumption of unpeeled fruit. The oranges in the Spanish diet had provided a higher portion of TAC and TPC i.e., 33 and 32%, respectively. Singh et al. (2017) reviewed the grain legume for phenolic profile and antioxidant potential. The authors showed that phenolic acids, flavonoids, and condensed tannins were the primary PCs present in grain seeds of legumes. These results indicated that PCs are contributing significantly to the antioxidant activities of the foods.

13.2.2 Applications in Dairy Products

Phenolic compounds are present in milk and its subsequent products, which are either metabolite of the feeds or added in the products. O'Connell and Fox (2001) reviewed the significance and application of PCs on milk and dairy products. It is well-known that animal bodies do not synthesize PCs, but they come from plant-based feeds. *Trifolium repens* and *Medicago sativa* are very effective feeds in terms of milk production, quality, and health of the animals (Aerts et al. 1999). Thiophenol, phenol, cresols, ethyl phenols, thymol, carvacrol had been reported to be present in bovine, caprine, ovine milk (Lopez and Lindsay 1993). Some of these may also be formed from amino acids. Dairy milk is thus a good source of several PCs including equol and phenolic antioxidants (Tsen et al. 2014). These secondary metabolites are formed by the bovine's gut bacterial flora from the PCs present in the feed. The lipid concentration of the milk was positively correlated with PCs. Kilic and Lindsay

(2005) showed that alkyl phenol and their conjugates are present in cow, sheep, goat milk as derivatives such as sulfate, phosphate, and glucuronides.

The effects of dietary PCs on milk quality and composition have been reported to be significantly affected. Propolis extracts have been shown to improve milk quality when supplemented with the diet of dairy cows (Aguiar et al. 2014). However, different PC have different effects on milk quality. The differences in the phenolic profile are perhaps produced by variations in the feed (Lopez and Lindsay 1993). For examples, when cows were feed high levels of particular crops, other PCs may also be identified in the milk, i.e., ptaquiloside from *Pteridium aquilinum* at up to 50 mg/L (Alonso-Amelot et al. 1996) or genistein up to 30µg/L, equol up to 300µg/L and daidzein up to 5µg/L (King et al. 1998). The latter compounds were found to be all derived from clover. Some feed or plant such as grape pomace may not produce any effects on the milk composition, but an induced modification in the fatty acid profile of the milk (Ianni et al. 2019). Similarly, in the diet of dairy goats, pistachio by-products had been used as forage, which had beneficial effects on the fatty acid profile of milk (Sedighi-Vesagh et al. 2015; Ghaffari et al. 2014).

The extracts of four grape varieties and grape callus rich in PCs were added into yogurt as functional ingredients (Karaaslan et al. 2011). The results had shown that the callus culture of grape has the potential to be used as a food supplement. Chouchouli et al. (2013) showed that fortification of grape seed extract to yogurt had enhanced antioxidant activity. A slow decline of simple and total phenolic contents occurred with storage time both in full fat and non-fat yogurts. However, polyphenols were present in all yogurts after 32 days of storage. A similar study had reported grape peel extract was effective at 20–25% in stirred yogurt (El-Said et al. 2014).

Due to the antioxidant nature of PCs, they are used as food additives. Milk supplemented with polyphenols from grape seeds and apple extracts had been shown to suppress flavor characteristics (Axten et al. 2008). It was also reported that the apple extracts have large flavor profile implications for product development due to their high level of bitterness. O'Connell and Fox (1999) studied the effects of different PCs on the thermal stability of milk. They found that some PCs did not disturb the heat stability of milk such as guaiacol, thymol, chlorogenic acid, vanillin, BHA, propyl gallate, and BHT. A significant reduction in the stability of skim milk was shown by quinic acid. Caffeic acid, vanillic acid and ferulic acid were found to enhanced thermal stability.

PCs when added to cheese curd, showed high retention coefficient values (Han et al. 2011). The retention coefficient values were highest in flavone, followed by hesperetin, and lowest in catechin added cheese as compared to control. The selected PCs showed three times higher values of antiradical activity than control with PCs suggesting beneficial effects. Similarly, PCs present in the extract of *Matricaria recutita* L. (chamomile) were incorporated into cottage cheese (Caleja et al. 2015). It was found that these PCs enhanced the antioxidant activity and storage stability of cheese. In conclusion, the use of PCs in different dairy products showed beneficial effects.

13.2.3 As Food Additives

As discussed above PCs possess strong antioxidant properties and thus may be suitable as food additives or bio-preservatives. Due to the toxic properties of most of the synthetic additives, recently the demand for the natural preservative such as PCs has grown tremendously. Natural food additives are gaining interest because of consumer habits and benefits to human health (Carocho et al. 2014). Natural food additives may be classified into three types; antimicrobial, antioxidant additives, and flavoring agents.

13.2.3.1 Antimicrobial Food Additives

Microbial additives help to protect foods against the microorganisms present or their possible contamination in the future. Sulfur-containing molecules such as sulfur dioxide, sodium, and potassium bisulfite, and, sodium and potassium bisulfates are used as antimicrobial agents in foods. The mechanism of action of these antimicrobials involves the uptake of SH groups from sulfites by the cell membrane of microorganisms, where they react with DNA, proteins, and enzymes (Lück and Jager 1997). Sulfites can act freely or be combined with organic acids, are used in winemaking, and in many other foodstuffs that are prone to microbiological decay. PCs have been used as antimicrobial food additives. The experiments usually involve the preliminary antimicrobial evaluation of the PC and then their utilization in foods. In chicken products, the antimicrobial potential of extracts from pomegranate peel and seed is very effective (Kanatt et al. 2010). The concentration of antimicrobials desired to reach a substantial antibacterial potential was found to be in the range of 0.5 to 20μL/g in foods such as milk, dairy products, meat products, fruit, vegetables, and cooked rice (Kanatt et al. 2010; Burt 2004). Similarly, the antimicrobial activity against growth and bacterial byproducts such as gram-positive and negative, spores, spoilage, yeasts, and molds in foods were studied for phenolic antioxidants (Raccach 1984). The amount of phenolic compounds was in the range of 30–10,000 ppm that had antimicrobial activity in food products.

In fish and fish products, the PCs have an emerging antimicrobial role (Maqsood et al. 2014; Maqsood et al. 2013). PC alone or a mixture of two or more can be used as effective antimicrobial agents (Lee and Lee 2010). Cueva et al. (2010) reported the antimicrobial activity of 13 phenolic acids against the commensal, probiotic, and pathogenic bacteria. The authors showed that *E. coli* showed the highest susceptibility toward phenolic acids at a concentration of at least 1000μg/mL, followed by other pathogenic bacteria. The structure-activity relationship study revealed that the number and position of substitutions on the benzene ring of phenolic acids influenced the antimicrobial potential. The PCs have to be beneficial in food packaging for preventing food deterioration (Cooksey 2005).

13.2.3.2 Antioxidant Food Additives

Food deterioration is one of the most significant issues from home to food industries. Food deterioration starts with lipid oxidation. Lipid free radicals produced either by auto-oxidation or thermal oxidation results in the deterioration of the food (Shahidi and Zhong 2010; St Angelo 1996). During thermal stress such as frying, food components such as lipids, proteins, and carbohydrates are oxidized to form different products (Ziaiifar et al. 2008; Zeb 2019; Zhang et al. 2012). Antioxidants are therefore added to prevent oxidation of foods. Natural antioxidants either alone or as a combination or in the form of the extract can be used to prevent oxidation of foods and consequent negative side effects (Zeb et al. 2017; Zeb and Haq 2016; Zeb 2018; Singh et al. 2018; Mancini et al. 2017). PCs have been used to minimize the effects of oxidized lipids in foods and their adverse effects (Zeb 2019; Zeb and Haq 2016; Zeb and Hussain 2014; Zeb and Akbar 2018). Significant struggles have been carried out to reduce the threat associated with heterocyclic amines (HAs) to humans. Supplementations of antioxidants such as PCs have been considered hopeful measures to decline exposure to Has (Vitaglione and Fogliano 2004). The PCs can inhibit the formation of HA or block or suppress HAs toxic biotransformation (Jiang and Xiong 2016). Exposure of phenolic compounds (gallic acid, vanillic acid, catechin) and extracts of vegetables to high temperatures had shown that PCs were comparatively stable (Volf et al. 2014). This showed that phenolic compounds alone or in combination can be used as food additives during the thermal processing of foods.

The uses of different spices, plant derivatives such as leaves, stem, fruit, and root extracts or pulp as food additives have been known to be very effective (Paz et al. 2015; Murcia et al. 2004; Hinneburg et al. 2006; Martillanes et al. 2017). The by-products of exotic fruits have also been reviewed to possess effective antioxidant additives (Ayala-Zavala et al. 2011). Grape seed extract (GSE) rich in phenolic antioxidants was effective in increasing the antioxidant and quality attributes of bread (Peng et al. 2010). The findings showed that GSE was promising as functional food additives. Similarly, tomato powder had been an effective source of phenolic antioxidants together with carotenoids for use as food additives (Lavelli et al. 2001). The multifunctional effects include protection against oxidative stress in the *in-vitro* and *in-vivo* model (Zeb and Haq 2016; Li et al. 2018).

A stevia fraction (SF) has been extracted from the leaves of *Stevia rebaudiana*, which contains caffeic acid, rosmarinic acid, chlorogenic acid, di-caffeoylquinic acid quercetin-3-glycoside, cyanidin-3-glucoside, kaempferol, quercetin, and apigenin (Milani et al. 2017). The SF showed a multi-functional basis for these PCs. These PCs upon fortification to whey-protein isolate had increased 80% of the antioxidant activity. Similarly, extracts of the onion skin fortified bean paste in the range of 5 to 50 mg of PCs per 100 g of beans showed an increase in phenolic and antioxidant activity (Swieca et al. 2014). It was concluded that fortification with PCs improves the health-promoting properties of bean paste. Though, the contacts of PCs with the food matrix play a significant part in the formation of the nutraceutical and nutritional value of foods.

All these studies revealed that fortification of phenolic antioxidants to foods increases the antioxidant activity, However, the study of Pinelo et al. (2004) suggested that in a complex system of food matrix it will be very hard to find the exact antioxidant role of each PC and their interactions with other PC presents on the system. The authors showed that in comparison to the control the lower anti-radical activity values of the mixture during the complete period of storage indicated that interaction among these polyphenols promotes a negative synergistic effect. It is worth to mention that this study was carried out at three different temperatures (22, 37, and 60 °C), which might be different in the thermal system with higher applied temperatures.

13.2.4 As Flavouring Agent in Foods

In addition to the above, PCs have been used as flavoring agents in foods. In this case, the main purpose is to have a specific flavor/taste of the food product. One of the most important PC is Vanillin, which is considered as one of the most widely used flavoring agents for sweet foods such as biscuits, desserts, and ice cream. It had been reported that vanillin or its derivatives improves the pleasant flavor as well as also acts as an antioxidant in foods rich in polyunsaturated fatty acids (Burri et al. 1989; Mahal et al. 2001). It was shown that the addition of 0·01–0·5% of vanillin helped keep intact the quality of cereal flakes that were pre-cooked and dried. In addition to the flavor properties, vanillin has a protective function against oxidative stress (Sefi et al. 2019). Wood smoke is commonly used to smoke flavorings and preservation methods for the curing of foods like sausages, ham, fish, bacon, and cheese (Bortolomeazzi et al. 2007).

The PCs also give bitter or astringent flavors to several foods such as fruit-based products, beer, wine, tree nuts, chocolate, coffee, and tea. Astringency is a perceptible response, whereas bitterness is a taste response. Anthocyanidins and condensed tannins had been shown to cause both astringency and taste responses (Soto-Vaca et al. 2012). The bitter and astringent responses by these compounds had been affected by the differences in composition, polymer size, and degree of galloylation. The specific responses may also be affected by pH, sweetness, viscosity, and the amount of ethanol present (Teissedre and Landrault 2000; Waterhouse 2002).

Several types of berries like blackberry, strawberry, blueberry, mulberry, and raspberry are known to possess PCs with respective flavors and are therefore used in a variety of foods (Zadernowski et al. 2005). Polyphenols from blueberry and cranberry juices were used as a natural source of PC flavors in soybean flour (Roopchand et al. 2012). However, not all of the PCs present in the berries are contributing to flavor. The characteristic flavor of most of the exotic tropical fruits is one of their most attractive attributes to consumers (Lasekan and Abbas 2012). These flavor compounds may or may not be produced from the PCs. In conclusion, PCs contributed to the flavoring agents in foods.

13.3 Pharmacological Significance

Phenolic compounds have been well known for their pharmacological significance since history to date. The toxicity or side effects of synthetic drugs forced scientists to look for drugs having ameliorative properties against several diseases. Plant foods are rich in these bioactive compounds with potential pharmacological significance. Fraga (2010) presented the pharmacological importance of several classes of phenolic compounds. Similarly, Brahmachari (2013) also focused their work on the pharmacological significance of dietary phenolic compounds. The review articles on the subject have been focused on either the significance of plant foods or the class of phenolic compounds (Srinivasulu et al. 2018; Naveed et al. 2018; Rather et al. 2016; Ediriweera et al. 2017; Takooree et al. 2019; Abedi et al. 2020). It is, however, beyond the scope of this chapter to present all available information regarding the pharmacological significance of phenolic antioxidants in foods. However, a summary of the health significance of the phenolic compounds is presented.

13.4 Health Applications of Phenolic Antioxidants

Phenolic compounds are widely known for their health benefits (Scalbert et al. 2005). Since PCs are present in varieties of foods as discussed in Chap. 3-10, therefore it is not possible to accommodate all the health benefits here in this chapter. Bioactivity, functionality, health benefits, and bioavailability of PCs from whole grains were reviewed recently by Călinoiu and Vodnar (2018). Similarly, PCs in honey and their associated health benefits were reviewed by Cianciosi et al. (2018). Another review reported the occurrence, biological activity, and health benefits of flavonoids and phenolic acids from Oregano (Gutiérrez-Grijalva et al. 2018). The prospective health properties of olive oil and plant polyphenols were reviewed by Gorzynik-Debicka et al. (2018). However, these reviews were limited to one or two food items.

    Phenolic compounds in foods are highly reactive. They stop free radicals as well as metal ions chelators that are capable to catalyze oxidative reactions in foods (Ozcan et al. 2014). The health-promoting properties of PCs are influenced by their structure, solubility, conjugation with other PCs or other compounds, and absorption and consequently on their metabolism (Singh et al. 2017). Owing to the antioxidant and antibacterial potential many physiological activities are affected by PCs. They apply the modulatory effect on the cell components having important functions like growth, production, and cell death (Crozier et al. 2009). PCs such as catechins, flavonoids, and their derivatives have been used for medicinal purposes owing to their well-known health beneficial properties (Andarwulan et al. 2012; Soobrattee et al. 2005). PCs from *Trifolium repens* showed beneficial effects against drug induced nephro-toxicity (Ahmad and Zeb 2020b). Some of the main health properties of PCs are given below (Table 13.2).

Table 13.2 Some of the health benefits of PCs with the mechanism of actions or effects (reproduced with kind permission of John Wiley and Sons (Zeb 2020))

Phenolic compounds	Pathological condition	Mechanism of actions	References
Quercetin, Kaempferol, genistein, resveratrol	Colon cancer	Suppresses COX-2 expression by inhibiting tyrosine kinases important for induction of COX-2 gene expression	Lee et al. (1998)
Catechins	Neurodegenerative diseases	Enhance activity of SOD and catalase	Levites et al. (2001)
(−)-Epigallocatechin gallate	Neurodegenerative conditions	Decreases the expression of proapoptotic genes (bax, bad, caspase-1 and -6, cyclin-dependent kinase inhibitor) thus maintaining the integrity of the mitochondrial membrane	Levites et al. (2003)
(−)-Epigallocatechin gallate	Cancer, diabetic retinopathy, chronic inflammation	Suppression of angiogenesis by inhibiting growth factors triggered the activation of receptors and PKC.	Wollin and Jones (2001), Jung et al. (2001), Cao and Cao (1999)
		Downregulation of VEGF production in tumor cells.	
		Repression of AP-1, NF-κB and STAT-1 transcription factor pathways.	
		Inhibits capillary endothelial cell proliferation and blood vessel formation.	
Proanthocyanidin (GSPE)	Cardiovascular disorders	Inhibitory effects on proapoptotic and cardio regulatory genes.	Bagchi et al. (2003)
		Modulating apoptotic regulatory bcl-X_L, p53 and c-myc genes	
Ferulic acid	Diabetes	Decrease lipid peroxidation and enhances the level of glutathione and antioxidant enzymes	Mahesh et al. (2004), Balasubashini et al. (2004)
Phlorotannins	Diabetes, colon cancer	Inhibits cell proliferation	Nwosu et al. (2011)
Flavonoids, Oleuropein Hydroxytyrosol Tyrosol	Hypercholesterolemia	Inhibition of cholesterol synthesis	Fki et al. (2005)

(continued)

Table 13.2 (continued)

Phenolic compounds	Pathological condition	Mechanism of actions	References
Quercetin, Procyanidins	Oxidative stress in human astrocytoma U373 MG cell line	Enhance the activity of antioxidant enzymes and expression of proteins	Martín et al. (2011)
Ellagic acid	Oxidative stress-induced by oxidized lipids	Enhance the activity of GSH and antioxidant enzymes	Zeb and Akbar (2018)
PCs extract	In-vitro and in-vivo diabetes	Normalizes glucose, HbA1c and serum lipids	Naz et al. (2019)
PCs extract	Hepato-toxicity, Nephro-toxicity	Enhance the activity of GSH and antioxidant enzymes	Ahmad and Zeb (2019), Ahmad and Zeb (2020b)

13.4.1 Antibacterial Properties

In legumes seeds, PCs are the most extensively spread compounds with strong antioxidant potential. The growth inhibition of several pathogens and decay of microorganisms are affected by PCs. Moreover, PCs present in edible beans inhibit fungal amylase activity (Telles et al. 2017). Anthocyanins not only impart color to foods but also have adequate antioxidant activities. Anthocyanins have been reported to possess several health benefits including anti-carcinogenic agents, anti-inflammatory, protection against heart diseases (Bagchi et al. 2003), High amounts of anthocyanins (delphinidin and phenolic acids) present in beans might have health benefits and are normally utilized as functional ingredients in foods (Mojica et al. 2015). In conclusion, PCs possess strong antibacterial activity in both in-vitro and in-vivo model system.

13.4.2 Anti-Hyperlipidemic Properties

An epidemiological study (Graf et al. 2005) and *in-vivo* animal model study revealed that PCs exhibited a significant role in suppressing biosynthesis and deposition of triglycerides and declined reactive oxygen species in the body (Zeb and Haq 2016; Fki et al. 2005). Alshikh et al. (2015) reported that PCs present in the lentils were effective during *in-vitro* inhibition of human LDL peroxidation induced by cupric ion. Fki et al. (2005) studied the effects of PCs from olive on hypercholesterolemia. The authors showed a significant reduction in the serum cholesterol contents of rats. Germinated pigeon pea contains more PCs than un-germinated peas and had been found to an effective food source for controlling lipid peroxidation and hyperglycemia (Uchegbu and Ishiwu 2016).

Hyperlipidemia in model animals induced by thermally oxidized lipids was controlled by PCs such as ellagic acid (Zeb 2018; Zeb and Akbar 2018). PCs play significant beneficial remedial functions in diabetes mellitus and the management of obesity (He and Giusti 2010; Wollin and Jones 2001). These reports indicated that PCs present in a variety of foods have strong anti-hyperlipidemic properties.

13.4.3   Cytotoxic, Anti-Tumor and and Anti-Cancer Properties

The role of PCs in the inhibition and treatment of cancers and heart diseases had been reported (Hertog et al. 1995). Adebamowo et al. (2005) reported the link between lower incidences of breast cancer and the consumption of lentils and or beans rich in flavonols. Epigallocatechin gallate inhibits VEGF induction in human colon carcinoma cells, thus inhibiting tumor growth (Jung et al. 2001). PCs rich extract from edible marine algae had the potential to act as antiproliferative and anti-diabetic agents (Nwosu et al. 2011). The phenolic extract (188.1 mg GAE/l) of black cowpea seeds inhibited 65% of the in-vitro proliferation and decline viability of mammary cancer cells (MCF-7) (Gutiérrez-Uribe et al. 2011). Similarly, the PCs present in Trifolium repens revealed protective effects on drug-induced hepatotoxicity in rats (Ahmad and Zeb 2019). The PCs present in dark beans showed neuroprotective and antitumor effects in in-vitro astrocyte cultures from human renal adenocarcinoma, melanoma, breast adenocarcinoma, and glioblastoma (López et al. 2013). Quercetin, kaempferol, genistein, and resveratrol were protective against colon cancer by suppressing the expression of COX-2 by inhibiting tyrosine kinases enzymes (Lee et al. 1998). These results showed that PCs have the potential to act as anti-cancer drugs, however, Wang et al. (2011) showed that food components that show strong antioxidant potency cannot always act as anti-cancer. Rodríguez-García et al. (2019) reviewed the correlation of uptake of dietary flavonoids and the incidence of cancer in different continents. This epidemiological review based on the studies (January 2008 to March 2019) concluded that a high intake of dietary flavonoids is associated with a reduced risk of cancer. Phenolic acids and flavonoids have shown promising results against breast, colon, prostate, and lung cancers (Bonta 2020). Microencapsulation of phenolic compounds has been reported to enhance anti-tumor activity (Rahaiee et al. 2020; Riahi-Chebbi et al. 2019). However, what we know is that PCs influenced several intracellular molecular targets that lead to apoptosis and blocking of tumour cell transformation and proliferation. Most of the studies on anti-cancer effects of PCs in foods are limited to specific cell lines (Teixeira-Guedes et al. 2019), and the exact mechanism is still under research especially in human subjects (do Carmo et al. 2018). Thus comprehensive molecular and metabolites studies especially through metabolomics, followed by clinical trials are therefore needed.

13.4.4 Antioxidants Properties

The health benefits of PCs have been attributed to their antioxidant and protective role against chronic diseases induced by free radicals (Xu and Chang 2010; Ozcan et al. 2014). Consumption of foods rich in PCs may decline the risk of many health issues. Lotito and Frei (2006) reviewed the consumption of flavonoid-rich foods and an increase in the antioxidant activity of human plasma. Olive oil rich in PCs had beneficial effects in healthy human adults by increasing plasma antioxidant capacity (Oliveras-López et al. 2014). PCs rich beverages sold in the US market have strong potency and have been beneficial for health (Seeram et al. 2008). Results also showed that due to the bioavailability of PCs in the juice and their ability to bind with the lipid fraction, the lipid peroxidation in the serum was significantly reduced (García-Alonso et al. 2006) and balancing oxidative status and health (Burton-Freeman 2010). However, the increase in the plasma antioxidant may not be due to PCs but vitamin C by the consumption of certain juices such as blueberry and cranberry.

Grape juice rich in PCs enhanced serum glutathione contents, catalase, superoxide dismutase, glutathione peroxidase, and antioxidant activities in healthy individuals (Toaldo et al. 2016). The grape juice rich in catechin, isoquercetin, and proanthocyanidin B1 had improved antioxidant status and lipid profile in the healthy individual doing the intense physical exercise (Toscano et al. 2017). Rodríguez-Pérez et al. (2019) concluded that PCs are multifunctional and anti-obesity agents. Several recent studies (Sobeh et al. 2018; Magrone et al. 2020; Bodoira and Maestri 2020) showed that PCs in a variety of foods possess in-vivo antioxidant activity.

PCs present in food have potential inhibitory effects on the activity of NADPH oxidase, scavenge free radicals, and produce an ameliorative role in hypertension (Yousefian et al. 2019). Barraza-Garza et al. (2020) showed that a higher polyphenol amount showed better protection against oxidative stress in *in-vitro* cell culture. However, the in-vitro antioxidant potential cannot be extrapolated to the in-vivo model due to the complex nature of the biological organism. The ingestion of higher amounts of PCs in foods may not be the best dietary strategy to overcome oxidative stress as these compounds reversed their role to pro-oxidative (Granato et al. 2020). Thus, nutritional, toxicological, and medical data are required to presume a dietary higher PC ingestion as a safe supplement.

13.4.5 Cardio-Protective Properties

PCs in foods possess significant cardio-protective properties. Several authors (Rodriguez-Mateos et al. 2014; Rangel-Huerta et al. 2015; Tangney and Rasmussen 2013; Scalbert et al. 2005; Quiñones et al. 2013; Geleijnse and Hollman 2008; De Pascual-Teresa et al. 2010; Potì et al. 2019) have reviewed the topic in details and therefore is beyond the limit of this chapter. However, some important foods such as tea and wine have been reported to be very important for cardiovascular health. Tea also inhibits angiogenesis (Cao and Cao 1999; Jung et al. 2001). Wine is also rich in PCs which have a protective role in cardiovascular disease (Abas et al. 2010).

Ahmadi et al. (2019) showed that the cardio-protective effects of green grape juice were due to the presence of PCs. Caffeic acid and chlorogenic acids possess antioxidant and cardio-protective properties (Agunloye et al. 2019). These PCs lower the blood pressure of the hypertensive rats by reducing angiotensin-converting enzyme, cholinesterase, and arginase activity. Hydroxytyrosol has been reported to possess significant cardio-protective properties (Bertelli et al. 2020). Both syringic acid and resveratrol produce cardio-protective effects by lowering total cholesterol, triglycerides, low-density lipoprotein cholesterol, very-low-density lipoprotein cholesterol, and TBARS, and a significant decrease in high-density lipoprotein cholesterol in serum and heart (Manjunatha et al. 2020). The isoproterenol-induced cardiotoxicity in rats was reversed through reducing nuclear factor-kappa B (NF-kB) and tumor necrosis factor-α (TNF-α) pathways. In conclusion, PCs help decrease risk factors for cardiovascular diseases.

13.4.6 Neuroprotective Properties

PCs such as flavonoids possess a variety of neuroprotective properties. These properties include the protection of neurons from the neurotoxins induced damage, a decrease of neuro-inflammation, memory, learning, and cognitive function promotion (Nehlig 2013). Bastianetto et al. (2009) reviewed the data until 2009 and suggested that PCs target multiple enzymes or proteins, which leads to their neuroprotective actions, possibly through specific plasma membrane binding sites. PCs present in Spanish red wine had declined the ROS generation and enhanced the activity and expression of the antioxidant enzymes such as catalase (CAT), superoxide dismutase (SOD), glutathione reductase (GR), and glutathione peroxidase (GPO) (Martín et al. 2011). The neuroprotective properties of wine polyphenols are now widely known to us (Basli et al. 2012). Phenolic antioxidants from *Trifolium repens* leaves extract showed significant antioxidant and anticholinesterase activity (Ahmad and Zeb 2020a). PCs such as catechins present in green tea were protective during neurodegeneration by enhancing SOD and catalase activities (Levites et al. 2001; Levites et al. 2003). Several recent studies showed that PCs present in foods possess neuroprotective properties (Zhao et al. 2020; Khalatbary and Khademi 2020; Silva and Pogačnik 2020; Mohd Sairazi and Sirajudeen 2020). In conclusion, strong pieces of evidence are available to show that PCs in foods have significant neuroprotective properties.

13.4.7 In-Vitro and In-Vivo Anti-Diabetic Properties

The PCs had shown to play important roles in the *in-vitro* inhibition of enzymes like lipases, α-amylase, α-glucosidase, and β-glucosidase activities (Tan et al. 2017; Zhang et al. 2010; Zhang et al. 2015). The α-glucosidase upon inhibition declines the glucose digestion as well as absorption in the intestine. Thus it plays an important role in the management of type-2 diabetes by controlling post-prandial glycaemic response (Balasubramaniam et al. 2013). Similarly, strong inhibitory activity had been shown by myricetin against α-amylase, α-glucosidase, and lipase found in black legumes such as black soybean and black turtle beans (Tan et al. 2017). Recently, phenolic antioxidants from *Sedum adenotrichum* leave extract showed to possess anti-diabetic properties in both *in-vitro* and *in-vivo* models (Naz et al. 2019). Lipid peroxidation in diabetic rats was alleviated by ferulic acid supplementation (Balasubashini et al. 2004). PCs can be used as an ingredient to serve as functional food and nutraceuticals to promote health, weight management, and controlling diabetes (Zhang et al. 2015). Extracts of edible seaweeds rich in PCs inhibit colon cancer proliferation, α-amylase, and α-glucosidase activity (Nwosu et al. 2011). PC-rich extracts from Soybean showed inhibitory effects on *in-vitro* activities of α-amylase, α-glucosidase, and angiotensin-I converting enzyme (Ademiluyi and Oboh 2013). These results established that phenolic compounds should be included in the diet as they possess several important properties for human health.

13.5 Oxidative Stress, Aging and Phenolic Antioxidants

13.5.1 Oxidative Stress and Phenolic Antioxidants

Oxidative stress is a specific condition of a cell or group of cells under which reactive oxygen-containing compounds are a higher number than under normal conditions. It can also be defined as the disturbance of oxidant to antioxidant balance ratio in the favor of oxidant species in the biological system (Zeb 2018). This means that the internal defense system is lacking the potential of neutralizing the oxidant species. These oxidant species may be reactive oxygen species or reactive nitrogen species (RNS) (Pisoschi and Pop 2015) or can be oxidized lipids (Zeb 2015a). Thus, several biological parameters are used to determine oxidative stress. These include some characteristics of cellular substances such as lipid peroxidation as malondialdehyde, glutathione, oxidized glutathione, oxidized DNA, superoxide anion (SOA), cytochrome C, oxidized hemoglobin (Oxy Hb), 8-hydroxy-2′-deoxyguanosine (8-OHdG), carboxylated proteins, nitrogen oxides, and lipid hydroperoxides. Several antioxidant enzymes are also used to determine the oxidative stress such as superoxide dismutase (SOD), catalase (CAT), nitrogen oxides (NOx), glutathione peroxidase (GPx), glutathione reductase (GR), glutathione S-transferase (GST), lactate dehydrogenase (LDH), hem-oxygenase (HO-1), xanthine oxidase (XO), myeloperoxidase (MPO), caspases, protein kinase C, oxidoreductases, and lipoxygenases. Specific gene expression and cellular conditions such as cell viability, hemolysis, apoptosis, and cell survival are also determined to elucidate oxidative stress.

Oxidative stress is responsible for causing several metabolic changes that ultimately resulted in chronic diseases. Oxidative stress is an accountable factor for chronic cardiovascular diseases (Qasim et al. 2016; Kim et al. 2016) and especially in cardiovascular aging (Wu et al. 2014). Oxidative stress severely affects liver structure and function (Pillon Barcelos et al. 2017) and also causes liver cancer (Ivanov et al. 2017). Oxidative stress also contributes to different neurological diseases (Salim 2017). An antioxidant enzyme such as catalase plays an important role in oxidative stress-induced calcium signaling in the nervous system (Naziroglu 2012). Oxidative stress is one of the main causes of various types of cancers in animals and humans (Kruk and Aboul-Enein 2017).

The antioxidant from plants was, therefore, used to ameliorate the oxidative stress. Due to the diverse nature of oxidative stress in the biological system, researchers are continuously working on the role of different phytochemicals from plants. Several phytochemicals have been extensively used to cope with oxidative stress in a variety of biological systems (Forcados et al. 2017; Marotta et al. 2012).

Exposure to harmful stimuli such as physical injury, chemical and mechanical stress, metabolic disorders, redox imbalance, and absence of oxygen or glucose are all known to contribute to compromised immune regulatory functions, which may induce autoimmune diseases (Zhang and Tsao 2016). These stimuli can cause acute or chronic, localized, or systemic inflammations. Acute inflammation involves the release of cell-derived mediators such as cytokines, prostaglandins, and ROS that are produced to protect cells and tissues. Physiologically produced ROS such as nitric oxide (NO), hydroxyl radical (OH) and superoxide anion (O_2) from various types of immune cells or respiratory burst in neutrophils are beneficial to human health because of their role in preventing pathogen invasion, depleting malignant cells and improving wound healing. However, persistent and long term immune responses can cause a homeostatic imbalance of the immune regulatory functions, leading to irreversible damage to the tissues. Overexposure to various stimuli such as pollutants, smoke, drugs, xenobiotics, ionizing radiation, and heavy metal ions, induce excess ROS production, the main exogenous causative factor for oxidative stress. Vital biomolecules including lipids, proteins, and DNA can be irreversibly and permanently damaged by highly reactive intermediates such as ROS (Reuter et al. 2010).

Dietary phenolic compounds are powerful antioxidants *in-vitro*, being able to neutralize free radicals by donating an electron or hydrogen atom to a wide range of reactive oxygen, nitrogen, and chlorine species, including O_2, OH, peroxyl radicals RO_2, hypochlorous acid (HOCl), and peroxynitrous acid (ONOOH). Phenolics interrupt the propagation stage of the lipid autoxidation chain reactions as effective radical scavengers or act as metal chelators to convert hydroperoxides or metal pro-oxidants into stable compounds. Phenolic compounds as metal chelators can directly inhibit Fe^{3+} reduction thereby reducing the production of reactive OH (Zeb 2020).

Both phenolic acids and flavonoids possess effective radical scavenging activity, however, the metal chelating potential and reducing power can vary depending on their structural features (Perron and Brumaghim 2009). Excessive accumulation of ROS or depletion of intermediates with antioxidant capacity alters the redox balance and leads to oxidative stress. An integrated cellular enzymatic redox system consisted of catalase (CAT), superoxide dismutase (SOD), glutathione peroxidase (GPx), glutathione reductase (GR) and peroxiredoxins (PRXs) is a cellular defense mechanism that maintains the oxidative equilibrium (Bhattacharyya et al. 2014). However, this mechanism is deficient under oxidative stress from excess ROS.

Phenolic compounds are produced by plant cells in response to stress such as microbial infection as a defense mechanism. Such protective properties of phenolic compounds may enable themselves to exert similar effects in mammalian systems after consumption of food containing these compounds. This has been recently recognized and developed into a novel theory termed xenohormesis (Howitz and Sinclair 2008). Dietary polyphenols are one of the most important xenobiotics with physiological relevance to human health. Consumption or supplementation of dietary polyphenols can restore the redox homeostasis and prevent systemic or localized inflammation by enhancing the activities of the antioxidant enzymes SOD, CAT, GPx, and GR. Expressions of these detoxifying and antioxidant enzymes are modulated by a key transcription factor nuclear factor erythroid-related factor (Nrf)-2 which can be activated by ROS at the cellular level. Nrf2 translocates into the nucleus and regulates antioxidant-responsive elements (ARE)-mediated transcriptions of various genes encoding the above-mentioned antioxidant enzymes. The Nrf2-Keap1 pathway controls the modulation of the redox homeostasis and detoxification. Polyphenols may induce Nrf2 activation to up-regulate cellular antioxidant enzymes (Kansanen et al. 2013). Dietary polyphenols, especially flavonoids, are also capable of triggering Nrf2 translocation to induce subsequent activation of the endogenous antioxidant actions through ligand interaction with cytosolic aryl hydrocarbon receptor (AhR). Flavonols and isoflavones and their derivatives have shown an agonistic potential to regulate AhR mediated signaling in cells [24]. Several flavonols and flavones, including quercetin, luteolin, apigenin, and chrysin, have been confirmed as AhR agonistic regulators (Köhle and Bock 2006). Polyphenols including flavonoids and their metabolites are not only AhR agonists but also induce dissociation of the Keap1/Nrf2 complex. Effects on the cross-link between AhR and Nrf2 signaling pathways are the key molecular mechanism underlying the ability of polyphenols in promoting endogenous antioxidant defensive systems based on SOD, CAT, GPx, and GR, to restore the cellular redox homeostasis.

Besides, polyphenols can also suppress oxidative stress by reducing inflammatory responses via interfering with nuclear factor kappa B (NFkB) and mitogen-activated protein kinase (MAPK) controlled inflammatory signaling cascades (Chuang and McIntosh 2011). Activation of these cellular processes leads to innate magnification of regulatory immune responses. As a result, pro-inflammatory cytokines, including interleukin (IL)-1b, IL-6, IL-8, tumor necrosis factor (TNF)-α, and interferin (IFN)-g, are released into the circulation system which, if not properly regulated, can trigger irreversible systemic inflammation and disrupted immune homeostasis (Zhang and Tsao 2016). While many polyphenols have shown regulatory activity on reducing pro-inflammatory biomarkers, the underlying mechanisms of these compounds are still not well understood. However, those known mechanisms are presented in Chap. 12.

13.5.2 Aging and Phenolic Antioxidants

Aging is a complex biological process characterized by a gradual loss of physiological integrity, leading to the decline of almost all physiological functions and increased vulnerability to death (Corrêa et al. 2018). Phenolic antioxidants in foods are used to treat or prevent aging. For example, myricetin present in berries and red wine among several food matrices inhibited wrinkle formation in mouse skin induced by chronic UVB irradiation (Jung et al. 2010). Similarly, Kim et al. (2013) found that caffeic acid, S-allyl cysteine, and uracil significantly inhibited the degradation of type I procollagen and the expressions of MMPs in-vivo.

Human studies on the anti-aging potential of phenolic compounds from foods are mostly based on epidemiological studies, which showed that plant foods containing phenolic compounds help in the pathogenesis of age-related diseases. The major epidemiological and clinical studies on green tea consumption and human cancer prevention in different organs, cancer being the current major cause of mortality throughout the world were reviewed by Khan and Mukhtar (2013). The authors also presented evidence for the association between tea drinking and a diminished occurrence of diabetes, arthritis, and disturbances in the neurological system (all age-related diseases) in humans. Epidemiological studies have shown an inverse relationship between nut intake and chronic diseases such as cardiovascular diseases and cancers. Significant research has been focused on the beneficial effects of phenolic compounds in skincare. However, despite the past decade's advances, our knowledge regarding the potential of phenolic compounds as anti-aging agents is still restricted.

13.6 Study Questions

- Describe in detail the applications of phenolic antioxidants in foods.
- What do you kno about the health applications of phenolic compounds?
- How phenolic antioxidants in foods are helpful in oxidative stress and aging?

References

Abas F, Alkan T, Goren B, Taskapilioglu O, Sarandol E, Tolunay S (2010) Neuroprotective effects of postconditioning on lipid peroxidation and apoptosis after focal cerebral ischemia/reperfusion injury in rats. Turk Neurosurg 20(1):1–8

Abedi F, Razavi BM, Hosseinzadeh H (2020) A review on gentisic acid as a plant derived phenolic acid and metabolite of aspirin: comprehensive pharmacology, toxicology, and some pharmaceutical aspects. Phytother Res 34(4):729–741

Adebamowo CA, Cho E, Sampson L, Katan MB, Spiegelman D, Willett WC, Holmes MD (2005) Dietary flavonols and flavonol-rich foods intake and the risk of breast cancer. Int J Cancer 114(4):628–633

Ademiluyi AO, Oboh G (2013) Soybean phenolic-rich extracts inhibit key-enzymes linked to type 2 diabetes (α-amylase and α-glucosidase) and hypertension (angiotensin I converting enzyme) in vitro. Exp Toxicol Pathol 65(3):305–309

Aerts RJ, Barry TN, McNabb WC (1999) Polyphenols and agriculture: beneficial effects of proanthocyanidins in forages. Agric Ecosyst Environ 75(1):1–12

Aguiar SC, Cottica SM, Boeing JS, Samensari RB, Santos GT, Visentainer JV, Zeoula LM (2014) Effect of feeding phenolic compounds from propolis extracts to dairy cows on milk production, milk fatty acid composition, and the antioxidant capacity of milk. Anim Feed Sci Tech 193:148–154

Agunloye OM, Oboh G, Ademiluyi AO, Ademosun AO, Akindahunsi AA, Oyagbemi AA, Omobowale TO, Ajibade TO, Adedapo AA (2019) Cardio-protective and antioxidant properties of caffeic acid and chlorogenic acid: mechanistic role of angiotensin converting enzyme, cholinesterase and arginase activities in cyclosporine induced hypertensive rats. Biomed Pharmacother 109:450–458

Ahmad S, Zeb A (2019) Effects of phenolic compounds from aqueous extract of Trifolium repens against acetaminophen-induced hepatotoxicity in mice. J Food Biochem 43(9):e12963

Ahmad S, Zeb A, Ayaz M, Murkovic M (2020a) Characterization of phenolic compounds using UPLC-HRM and HPLC-DAD and anticholinesterase and antioxidant activities of Trifolium repens L. leaves. Eur Food Res Technol 246:485–496

Ahmad S, Zeb A (2020b) Nephroprotective property of Trifolium repens leaf extract against paracetamol-induced kidney damage in mice. 3 Biotech 10(12):541

Ahmadi L, El-Kubbe A, Roney SK (2019) Potential cardio-protective effects of green grape juice: a review. Curr Nutr Food Sci 15(3):202–207

Alam MN, Bristi NJ, Rafiquzzaman M (2013) Review on in vivo and in vitro methods evaluation of antioxidant activity. Saudi Pharmaceut J 21(2):143–152

Alonso-Amelot ME, Castillo U, Smith BL, Lauren DR (1996) Bracken ptaquiloside in milk. Nature 382(6592):587–587

Alshikh N, de Camargo AC, Shahidi F (2015) Phenolics of selected lentil cultivars: antioxidant activities and inhibition of low-density lipoprotein and DNA damage. J Med Food 18:1022–1038

Amarowicz R, Karamać M, Dueñas M, Pegg RB (2017) Antioxidant activity and phenolic composition of a red bean (Phasoelus vulgaris) extract and its fractions. Nat Prod Commun 12(4):541–544

Andarwulan N, Kurniasih D, Apriady RA, Rahmat H, Roto AV, Bolling BW (2012) Polyphenols, carotenoids, and ascorbic acid in underutilized medicinal vegetables. J Funct Foods 4(1):339–347

Axten L, Wohlers M, Wegrzyn T (2008) Using phytochemicals to enhance health benefits of milk: impact of polyphenols on flavor profile. J Food Sci 73(6):H122–H126

Ayala-Zavala JF, Vega-Vega V, Rosas-Domínguez C, Palafox-Carlos H, Villa-Rodriguez JA, Siddiqui MW, Dávila-Aviña JE, González-Aguilar GA (2011) Agro-industrial potential of exotic fruit byproducts as a source of food additives. Food Res Int 44(7):1866–1874

Bagchi D, Sen CK, Ray SD, Das DK, Bagchi M, Preuss HG, Vinson JA (2003) Molecular mechanisms of cardioprotection by a novel grape seed proanthocyanidin extract. Mutat Res/Fund Mol
Mech Mutagenesis 523-524:87–97

Bai Y, Xu Y, Wang BY, Li SS, Guo F, Hua HM, Zhao YL, Yu ZG (2017) Comparison of phenolic
compounds, antioxidant and antidiabetic activities between selected edible beans and their different growth periods leaves. J Funct Foods 35:694–702

Balasubashini MS, Rukkumani R, Viswanathan P, Menon VP (2004) Ferulic acid alleviates lipid
peroxidation in diabetic rats. Phytother Res 18(4):310–314

Balasubramaniam V, Mustar S, Mustafa Khalid N, Abd Rashed A, Mohd Noh MF, Wilcox MD,
Chater PI, Brownlee IA, Pearson JP (2013) Inhibitory activities of three Malaysian edible seaweeds on lipase and α-amylase. J Appl Phycol 25(5):1405–1412

Barraza-Garza G, Pérez-León JA, Castillo-Michel H, de la Rosa LA, Martinez-Martinez A, Cotte
M, Alvarez-Parrilla E (2020) Antioxidant effect of phenolic compounds (PC) at different concentrations in IEC-6 cells: a spectroscopic analysis. Spectrochim Acta A Mol Biomol Spectrosc
227:117570

Basli A, Soulet S, Chaher N, Mérillon J-M, Chibane M, Monti J-P, Richard T (2012) Wine polyphenols: potential agents in neuroprotection. Oxid Med Cell Long 2012:805762

Bastianetto S, Dumont Y, Han Y, Quirion R (2009) Comparative neuroprotective properties of
stilbene and catechin analogs: action via a plasma membrane receptor site? CNS Neurosci
Ther 15(1):76–83

Bertelli M, Kiani AK, Paolacci S, Manara E, Kurti D, Dhuli K, Bushati V, Miertus J, Pangallo
D, Baglivo M (2020) Hydroxytyrosol: a natural compound with promising pharmacological
activities. J Biotechnol 309:29–33

Bhattacharyya A, Chattopadhyay R, Mitra S, Crowe SE (2014) Oxidative stress: An essential factor in the pathogenesis of gastrointestinal mucosal diseases. Physiol Rev 94(2):329–354

Bodoira R, Maestri D (2020) Phenolic compounds from nuts: extraction, chemical profiles, and
bioactivity. J Agric Food Chem 68(4):927–942

Bonta RK (2020) Dietary phenolic acids and flavonoids as potential anti-cancer agents: current
state of the art and future perspectives. Anti-Cancer Agent 20(1):29–48

Bortolomeazzi R, Sebastianutto N, Toniolo R, Pizzariello A (2007) Comparative evaluation of the
antioxidant capacity of smoke flavouring phenols by crocin bleaching inhibition, DPPH radical
scavenging and oxidation potential. Food Chem 100(4):1481–1489

Brahmachari G (2013) Chemistry and pharmacology of naturally occurring bioactive compounds.
CRC, Boca Ratan

Burri J, Graf M, Lambelet P, Löliger J (1989) Vanillin: more than a flavouring agent—a potent
antioxidant. J Sci Food Agric 48(1):49–56

Burt S (2004) Essential oils: their antibacterial properties and potential applications in foods—a
review. Int J Food Microbiol 94(3):223–253

Burton-Freeman B (2010) Postprandial metabolic events and fruit-derived phenolics: a review of
the science. Br J Nutr 104(S3):S1–S14

Caleja C, Barros L, Antonio AL, Ciric A, Barreira JCM, Sokovic M, Oliveira MBPP, Santos-
Buelga C, Ferreira ICFR (2015) Development of a functional dairy food: exploring bioactive
and preservation effects of chamomile (Matricaria recutita L.). J Funct Foods 16:114–124

Călinoiu LF, Vodnar DC (2018) Whole grains and phenolic acids: a review on bioactivity, functionality, health benefits and bioavailability. Nutrients 10(11):1615

Cao Y, Cao R (1999) Angiogenesis inhibited by drinking tea. Nature 398(6726):381–381

Carocho M, Barreiro MF, Morales P, Ferreira ICFR (2014) Adding molecules to food, pros
and cons: a review on synthetic and natural food additives. Compr Rev Food Sci Food Saf
13(4):377–399

Chouchouli V, Kalogeropoulos N, Konteles SJ, Karvela E, Makris DP, Karathanos VT (2013)
Fortification of yoghurts with grape (Vitis vinifera) seed extracts. LWT Food Sci Technol
53(2):522–529

Chuang C-C, McIntosh MK (2011) Potential mechanisms by which polyphenol-rich grapes prevent obesity-mediated inflammation and metabolic diseases. Annu Rev Nutr 31(1):155–176

Cianciosi D, Forbes-Hernández TY, Afrin S, Gasparrini M, Reboredo-Rodriguez P, Manna PP, Zhang J, Bravo Lamas L, Martínez Flórez S, Agudo Toyos P, Quiles JL, Giampieri F, Battino M (2018) Phenolic compounds in honey and their associated health benefits: a review. Molecules 23(9):2322

Cooksey K (2005) Effectiveness of antimicrobial food packaging materials. Food Addit Contam 22(10):980–987

Corrêa RCG, Peralta RM, Haminiuk CWI, Maciel GM, Bracht A, Ferreira ICFR (2018) New phytochemicals as potential human anti-aging compounds: reality, promise, and challenges. Crit Rev Food Sci Nutr 58(6):942–957

Crozier A, Jaganath IB, Clifford MN (2009) Dietary phenolics: chemistry, bioavailability and effects on health. Nat Prod Rep 26(8):1001–1043

Cueva C, Moreno-Arribas MV, Martín-Álvarez PJ, Bills G, Vicente MF, Basilio A, Rivas CL, Requena T, Rodríguez JM, Bartolomé B (2010) Antimicrobial activity of phenolic acids against commensal, probiotic and pathogenic bacteria. Res Microbiol 161(5):372–382

De Pascual-Teresa S, Moreno DA, García-Viguera C (2010) Flavanols and anthocyanins in cardiovascular health: a review of current evidence. Int J Mol Sci 11(4):1679–1703

do Carmo MAV, Pressete CG, Marques MJ, Granato D, Azevedo L (2018) Polyphenols as potential antiproliferative agents: scientific trends. Curr Opin Food Sci 24:26–35

Ediriweera MK, Tennekoon KH, Samarakoon SR (2017) A review on ethnopharmacological applications, pharmacological activities, and bioactive compounds of *Mangifera indica* (mango). Evid Based Complement Alternat Med 2017:6949835

El-Said MM, Haggag HF, Fakhr El-Din HM, Gad AS, Farahat AM (2014) Antioxidant activities and physical properties of stirred yoghurt fortified with pomegranate peel extracts. Ann Agric Sci 59(2):207–212

Escarpa A, Gonzalez MC (2001) An overview of analytical chemistry of phenolic compounds in foods. Crit Rev Anal Chem 31(2):57–139

Fki I, Bouaziz M, Sahnoun Z, Sayadi S (2005) Hypocholesterolemic effects of phenolic-rich extracts of Chemlali olive cultivar in rats fed a cholesterol-rich diet. Bioorgan Med Chem 13(18):5362–5370

Forcados GE, James DB, Sallau AB, Muhammad A, Mabeta P (2017) Oxidative stress and carcinogenesis: potential of phytochemicals in breast Cancer therapy. Nutr Cancer 69(3):365–374

Fraga CG (2010) Plant phenolics and human health: biochemistry, nutrition, and pharmacology. The Wiley-IUBMB series on biochemistry and molecular biology. Wiley, Hoboken, NJ

García-Alonso J, Ros G, Vidal-Guevara ML, Periago MJ (2006) Acute intake of phenolic-rich juice improves antioxidant status in healthy subjects. Nutr Res 26(7):330–339

Geleijnse JM, Hollman PC (2008) Flavonoids and cardiovascular health: which compounds, what mechanisms? Am J Clin Nutr 88(1):12–13

Ghaffari MH, Tahmasbi AM, Khorvash M, Naserian AA, Vakili AR (2014) Effects of pistachio by-products in replacement of alfalfa hay on ruminal fermentation, blood metabolites, and milk fatty acid composition in Saanen dairy goats fed a diet containing fish oil. J Appl Anim Res 42(2):186–193

Gorzynik-Debicka M, Przychodzen P, Cappello F, Kuban-Jankowska A, Marino Gammazza A, Knap N, Wozniak M, Gorska-Ponikowska M (2018) Potential health benefits of olive oil and plant polyphenols. Int J Mol Sci 19(3):686

Graf BA, Milbury PE, Blumberg JB (2005) Flavonols, flavones, flavanones, and human health: epidemiological evidence. J Med Food 8(3):281–290

Granato D, Shahidi F, Wrolstad R, Kilmartin P, Melton LD, Hidalgo FJ, Miyashita K, Jv C, Alasalvar C, Ismail AB, Elmore S, Birch GG, Charalampopoulos D, Astley SB, Pegg R, Zhou P, Finglas P (2018) Antioxidant activity, total phenolics and flavonoids contents: should we ban in vitro screening methods? Food Chem 264:471–475

Granato D, Mocan A, Câmara JS (2020) Is a higher ingestion of phenolic compounds the best dietary strategy? A scientific opinion on the deleterious effects of polyphenols in vivo. Trends Food Sci Technol 98:162–166

Gutiérrez-Grijalva EP, Picos-Salas MA, Leyva-López N, Criollo-Mendoza MS, Vazquez-Olivo G, Heredia JB (2018) Flavonoids and phenolic acids from oregano: occurrence, biological activity and health benefits. Plants (Basel) 7(1):2

Gutiérrez-Uribe JA, Romo-Lopez I, Serna-Saldívar SO (2011) Phenolic composition and mammary cancer cell inhibition of extracts of whole cowpeas (Vigna unguiculata) and its anatomical parts. J Funct Foods 3(4):290–297

Han J, Britten M, St-Gelais D, Champagne CP, Fustier P, Salmieri S, Lacroix M (2011) Polyphenolic compounds as functional ingredients in cheese. Food Chem 124(4):1589–1594

Harnly J (2017) Antioxidant methods. J Food Compos Anal 64:145–146

He J, Giusti MM (2010) Anthocyanins: natural colorants with health-promoting properties. Annu Rev Food Sci Technol 1(1):163–187

Heleno SA, Martins A, Queiroz MJRP, Ferreira ICFR (2015) Bioactivity of phenolic acids: metabolites versus parent compounds: a review. Food Chem 173:501–513

Hertog MGL, Kromhout D, Aravanis C, Blackburn H, Buzina R, Fidanza F, Giampaoli S, Jansen A, Menotti A, Nedeljkovic S, Pekkarinen M, Simic BS, Toshima H, Feskens EJM, Hollman PCH, Katan MB (1995) Flavonoid intake and long-term risk of coronary heart disease and cancer in the seven countries study. Arch Intern Med 155(4):381–386

Hinneburg I, Damien Dorman HJ, Hiltunen R (2006) Antioxidant activities of extracts from selected culinary herbs and spices. Food Chem 97(1):122–129

Howitz KT, Sinclair DA (2008) Xenohormesis: sensing the chemical cues of other species. Cell 133(3):387–391

Ianni A, Di Maio G, Pittia P, Grotta L, Perpetuini G, Tofalo R, Cichelli A, Martino G (2019) Chemical–nutritional quality and oxidative stability of milk and dairy products obtained from Friesian cows fed with a dietary supplementation of dried grape pomace. J Sci Food Agric 99(7):3635–3643

Ivanov AV, Valuev-Elliston VT, Tyurina DA, Ivanova ON, Kochetkov SN, Bartosch B, Isaguliants MG (2017) Oxidative stress, a trigger of hepatitis C and B virus-induced liver carcinogenesis. Oncotarget 8(3):3895–3932

Jiang J, Xiong YL (2016) Natural antioxidants as food and feed additives to promote health benefits and quality of meat products: a review. Meat Sci 120:107–117

Jung YD, Kim MS, Shin BA, Chay KO, Ahn BW, Liu W, Bucana CD, Gallick GE, Ellis LM (2001) EGCG, a major component of green tea, inhibits tumour growth by inhibiting VEGF induction in human colon carcinoma cells. Br J Cancer 84(6):844–850

Jung SK, Lee KW, Kim HY, Oh MH, Byun S, Lim SH, Heo Y-S, Kang NJ, Bode AM, Dong Z, Lee HJ (2010) Myricetin suppresses UVB-induced wrinkle formation and MMP-9 expression by inhibiting Raf. Biochem Pharmacol 79(10):1455–1461

Kanatt SR, Chander R, Sharma A (2010) Antioxidant and antimicrobial activity of pomegranate peel extract improves the shelf life of chicken products. Int J Food Sci Technol 45(2):216–222

Kansanen E, Kuosmanen SM, Leinonen H, Levonen A-L (2013) The Keap1-Nrf2 pathway: mechanisms of activation and dysregulation in cancer. Redox Biol 1(1):45–49

Karaaslan M, Ozden M, Vardin H, Turkoglu H (2011) Phenolic fortification of yogurt using grape and callus extracts. LWT Food Sci Technol 44(4):1065–1072

Karakaya S, El SN, Taş AA (2001) Antioxidant activity of some foods containing phenolic compounds. Int J Food Sci Nutr 52(6):501–508

Khalatbary AR, Khademi E (2020) The green tea polyphenolic catechin epigallocatechin gallate and neuroprotection. Nutr Neurosci 23(4):281–294

Khan N, Mukhtar H (2013) Tea and health: studies in humans. Curr Pharm Design 19(34):6141–6147

Kilic M, Lindsay RC (2005) Distribution of conjugates of alkylphenols in milk from different ruminant species. J Dairy Sci 88(1):7–12

Kim SR, Jung YR, An HJ, Kim DH, Jang EJ, Choi YJ, Moon KM, Park MH, Park CH, Chung KW
      (2013) Anti-wrinkle and anti-inflammatory effects of active garlic components and the inhibi-
      tion of MMPs via NF-κB signaling. PLoS One 8(9):e73877
Kim H, Yun J, Kwon SM (2016) Therapeutic strategies for oxidative stress-related cardiovascular
      diseases: removal of excess reactive oxygen species in adult stem cells. Oxid Med Cell Long
      2016:2483163
King RA, Mano MM, Head RJ (1998) Assessment of isoflavonoid concentrations in Australian
      bovine milk samples. J Dairy Res 65(3):479–489
Köhle C, Bock KW (2006) Activation of coupled ah receptor and Nrf2 gene batteries by dietary
      phytochemicals in relation to chemoprevention. Biochem Pharmacol 72(7):795–805
Koley TK, Maurya A, Tripathi A, Singh BK, Singh M, Bhutia TL, Tripathi PC, Singh B (2019)
      Antioxidant potential of commonly consumed underutilized leguminous vegetables. Int J Veg
      Sci 25(4):362–372
Kruk J, Aboul-Enein HY (2017) Reactive oxygen and nitrogen species in carcinogenesis: implica-
      tions of oxidative stress on the progression and development of several cancer types. Mini Rev
      Med Chem 17(11):904–919
Lasekan O, Abbas KA (2012) Distinctive exotic flavor and aroma compounds of some exotic tropi-
      cal fruits and berries: a review. Crit Rev Food Sci Nutr 52(8):726–735
Lavelli V, Hippeli S, Dornisch K, Peri C, Elstner EF (2001) Properties of tomato powders as addi-
      tives for food fortification and stabilization. J Agric Food Chem 49(4):2037–2042
Lee O-H, Lee B-Y (2010) Antioxidant and antimicrobial activities of individual and combined
      phenolics in Olea europaea leaf extract. Bioresour Technol 101(10):3751–3754
Lee S, Kuan C, Yang C, Yang S (1998) Bioflavonoids commonly and potently induce tyrosine
      dephosphorylation/inactivation of oncogenic proline-directed protein kinase FA in human
      prostate carcinoma cells. Anticancer Res 18(2A):1117–1121
Levites Y, Weinreb O, Maor G, Youdim MB, Mandel S (2001) Green tea polyphenol
      (−)-epigallocatechin-3-gallate prevents N-methyl-4-phenyl-1, 2, 3, 6-tetrahydropyridine-
      induced dopaminergic neurodegeneration. J Neurochem 78(5):1073–1082
Levites Y, Amit T, Mandel S, Youdim MBH (2003) Neuroprotection and neurorescue against Aβ
      toxicity and PKC-dependent release of nonamyloidogenic soluble precursor protein by green
      tea polyphenol (−)-epigallocatechin-3-gallate. FASEB J 17(8):952–954
Li CC, Liu C, Fu MB, Hu KQ, Aizawa K, Takahashi S, Hiroyuki S, Cheng JR, von Lintig J, Wang
      XD (2018) Tomato powder inhibits hepatic steatosis and inflammation potentially through
      restoring SIRT1 activity and adiponectin function independent of carotenoid cleavage enzymes
      in mice. Mol Nutr Food Res 62(8):1700738
Lopez V, Lindsay RC (1993) Metabolic conjugates as precursors for characterizing flavor com-
      pounds in ruminant milks. J Agric Food Chem 41(3):446–454
López A, El-Naggar T, Dueñas M, Ortega T, Estrella I, Hernández T, Gómez-Serranillos MP,
      Palomino OM, Carretero ME (2013) Effect of cooking and germination on phenolic composition
      and biological properties of dark beans (Phaseolus vulgaris L.). Food Chem 138(1):547–555
López-Alarcón C, Lissi E (2005) Interaction of pyrogallol red with peroxyl radicals. A basis
      for a simple methodology for the evaluation of antioxidant capabilities. Free Radic Res
      39(7):729–736
Lotito SB, Frei B (2006) Consumption of flavonoid-rich foods and increased plasma anti-
      oxidant capacity in humans: cause, consequence, or epiphenomenon? Free Radic Biol Med
      41(12):1727–1746
Lück E, Jager M (1997) Antimicrobial food additives: characteristics, uses, effects, vol 2. Springer
      Science and Business Media, Heidelberg, Germany
Magrone T, Magrone M, Russo MA, Jirillo E (2020) Recent advances on the anti-inflammatory and
      antioxidant properties of red grape polyphenols: in vitro and in vivo studies. Antioxidants 9(1):35
Mahal HS, Badheka LP, Mukherjee T (2001) Radical scavenging properties of a flavouring agent–
      vanillin. Res Chem Intermed 27(6):595–604

Mahesh T, Sri Balasubashini MM, Menon VP (2004) Photo-irradiated curcumin supplementation in streptozotocin-induced diabetic rats: effect on lipid peroxidation. Therapie 59(6):639–644

Mancini S, Paci G, Preziuso G (2017) Effect of dietary *Curcuma longa* L. powder on lipid oxidation of frozen pork. Large Anim Rev 23(3):111–113

Manjunatha S, Shaik AH, Prasad ME, Al Omar SY, Mohammad A, Kodidhela LD (2020) Combined cardio-protective ability of syringic acid and resveratrol against isoproterenol induced cardio-toxicity in rats via attenuating NF-kB and TNF-α pathways. Sci Rep 10(1):3426

Maqsood S, Benjakul S, Shahidi F (2013) Emerging role of phenolic compounds as natural food additives in fish and fish products. Crit Rev Food Sci Nutr 53(2):162–179

Maqsood S, Benjakul S, Abushelaibi A, Alam A (2014) Phenolic compounds and plant phenolic extracts as natural antioxidants in prevention of lipid oxidation in seafood: a detailed review. Compr Rev Food Sci Food Saf 13(6):1125–1140

Marotta F, Kumari A, Catanzaro R, Solimene U, Jain S, Minelli E, Harada M (2012) A phyto-chemical approach to experimental metabolic syndrome-associated renal damage and oxidative stress. Rejuvenation Res 15(2):153–156

Martillanes S, Rocha-Pimienta J, Cabrera-Bañegil M, Martín-Vertedor D, Delgado-Adámez J (2017) Application of phenolic compounds for food preservation: food additive and active packaging. In: Soto-Hernández M, Palma Tenango M, García-Mateos G (eds) Phenolic compounds–biological activity. IntechOpen, London, UK, pp 39–58

Martín S, González-Burgos E, Carretero ME, Gómez-Serranillos MP (2011) Neuroprotective properties of Spanish red wine and its isolated polyphenols on astrocytes. Food Chem 128(1):40–48

Milani PG, Formigoni M, Lima YC, Piovan S, Peixoto GML, Camparsi DM, Rodrigues WDD, da Silva JQP, Avincola AD, Pilau EJ, da Costa CEM, da Costa SC (2017) Fortification of the whey protein isolate antioxidant and antidiabetic activity with fraction rich in phenolic compounds obtained from *Stevia rebaudiana* (Bert.). Bertoni leaves. J Food Sci Tech Mys 54(7):2020–2029

Mohd Sairazi NS, Sirajudeen K (2020) Natural products and their bioactive compounds: neuroprotective potentials against neurodegenerative diseases. Evid-Based Complement Altern Med 2020:6565396

Mojica L, Meyer A, Berhow MA, de Mejía EG (2015) Bean cultivars (*Phaseolus vulgaris* L.) have similar high antioxidant capacity, in vitro inhibition of α-amylase and α-glucosidase while diverse phenolic composition and concentration. Food Res Int 69:38–48

Murcia MA, Egea I, Romojaro F, Parras P, Jiménez AM, Martínez-Tomé M (2004) Antioxidant evaluation in dessert spices compared with common food additives. Influence of irradiation procedure. J Agric Food Chem 52(7):1872–1881

Naveed M, Hejazi V, Abbas M, Kamboh AA, Khan GJ, Shumzaid M, Ahmad F, Babazadeh D, FangFang X, Modarresi-Ghazani F, WenHua L, XiaoHui Z (2018) Chlorogenic acid (CGA): a pharmacological review and call for further research. Biomed Pharmacother 97:67–74

Naz D, Muhamad A, Zeb A, Shah I (2019) In vitro and in vivo antidiabetic properties of phenolic antioxidants from *Sedum adenotrichum*. Front Nutr 6:177

Naziroglu M (2012) Molecular role of catalase on oxidative stress-induced ca(2+) signaling and TRP cation channel activation in nervous system. J Recept Signal Transduct Res 32(3):134–141

Nehlig A (2013) The neuroprotective effects of cocoa flavanol and its influence on cognitive performance. Brit J Clin Pharmacol 75(3):716–727

Nwosu F, Morris J, Lund VA, Stewart D, Ross HA, McDougall GJ (2011) Anti-proliferative and potential anti-diabetic effects of phenolic-rich extracts from edible marine algae. Food Chem 126(3):1006–1012

O'Connell JE, Fox PF (1999) Effects of phenolic compounds on the heat stability of milk and concentrated milk. J Dairy Res 66(3):399–407

O'Connell JE, Fox PF (2001) Significance and applications of phenolic compounds in the production and quality of milk and dairy products: a review. Int Dairy J 11(3):103–120

Oliveras-López M-J, Berná G, Jurado-Ruiz E, López-García de la Serrana H, Martín F (2014) Consumption of extra-virgin olive oil rich in phenolic compounds has beneficial antioxidant effects in healthy human adults. J Funct Foods 10:475–484

Ozcan T, Akpinar-Bayizit A, Yilmaz-Ersan L, Delikanli B (2014) Phenolics in human health. Int J Chem Eng Appl 5(5):393

Paz M, Gúllon P, Barroso MF, Carvalho AP, Domingues VF, Gomes AM, Becker H, Longhinotti E, Delerue-Matos C (2015) Brazilian fruit pulps as functional foods and additives: evaluation of bioactive compounds. Food Chem 172:462–468

Peng X, Ma J, Cheng K-W, Jiang Y, Chen F, Wang M (2010) The effects of grape seed extract fortification on the antioxidant activity and quality attributes of bread. Food Chem 119(1):49–53

Perron NR, Brumaghim JL (2009) A review of the antioxidant mechanisms of polyphenol compounds related to iron binding. Cell Biochem Biophys 53(2):75–100

Pillon Barcelos R, Freire Royes LF, Gonzalez-Gallego J, Bresciani G (2017) Oxidative stress and inflammation: liver responses and adaptations to acute and regular exercise. Free Radic Res 51(2):222–236

Pinelo M, Manzocco L, Nuñez MJ, Nicoli MC (2004) Interaction among phenols in food fortification: negative synergism on antioxidant capacity. J Agric Food Chem 52(5):1177–1180

Pisoschi AM, Pop A (2015) The role of antioxidants in the chemistry of oxidative stress: a review. Eur J Med Chem 97:55–74

Potì F, Santi D, Spaggiari G, Zimetti F, Zanotti I (2019) Polyphenol health effects on cardiovascular and neurodegenerative disorders: a review and meta-analysis. Int J Mol Sci 20(2):351

Qasim M, Bukhari SA, Ghani MJ, Masoud MS, Huma T, Arshad M, Haque A, Ibrahim Z, Javed S, Rajoka MI (2016) Relationship of oxidative stress with elevated level of DNA damage and homocysteine in cardiovascular disease patients. Pak J Pharm Sci 29 (6(Suppl)):2297-2302

Quiñones M, Miguel M, Aleixandre A (2013) Beneficial effects of polyphenols on cardiovascular disease. Pharmacol Res 68(1):125–131

Raccach M (1984) The antimicrobial activity of phenolic antioxidants in foods: a review. J Food Safety 6(3):141–170

Rahaiee S, Assadpour E, Faridi Esfanjani A, Silva AS, Jafari SM (2020) Application of nano/microencapsulated phenolic compounds against cancer. Adv Colloid Interf Sci 279:102153

Rangel-Huerta OD, Pastor-Villaescusa B, Aguilera CM, Gil A (2015) A systematic review of the efficacy of bioactive compounds in cardiovascular disease: phenolic compounds. Nutrients 7(7):5177–5216

Rather MA, Dar BA, Sofi SN, Bhat BA, Qurishi MA (2016) *Foeniculum vulgare*: a comprehensive review of its traditional use, phytochemistry, pharmacology, and safety. Arab J Chem 9:S1574–S1583

Reuter S, Gupta SC, Chaturvedi MM, Aggarwal BB (2010) Oxidative stress, inflammation, and cancer: how are they linked? Free Radic Biol Med 49(11):1603–1616

Riahi-Chebbi I, Souid S, Othman H, Haoues M, Karoui H, Morel A, Srairi-Abid N, Essafi M, Essafi-Benkhadir K (2019) The phenolic compound kaempferol overcomes 5-fluorouracil resistance in human resistant LS174 colon cancer cells. Sci Rep 9(1):1–20

Rodríguez-García C, Sánchez-Quesada C, Gaforio JJ (2019) Dietary flavonoids as cancer chemopreventive agents: An updated review of human studies. Antioxidants 8(5):137

Rodriguez-Mateos A, Heiss C, Borges G, Crozier A (2014) Berry (poly)phenols and cardiovascular health. J Agric Food Chem 62(18):3842–3851

Rodríguez-Pérez C, Segura-Carretero A, del Mar CM (2019) Phenolic compounds as natural and multifunctional anti-obesity agents: a review. Crit Rev Food Sci Nutr 59(8):1212–1229

Roopchand DE, Grace MH, Kuhn P, Cheng DM, Plundrich N, Poulev A, Howell A, Fridlender B, Lila MA, Raskin I (2012) Efficient sorption of polyphenols to soybean flour enables natural fortification of foods. Food Chem 131(4):1193–1200

Ruiz-Torralba A, Guerra-Hernández EJ, García-Villanova B (2018) Antioxidant capacity, polyphenol content and contribution to dietary intake of 52 fruits sold in Spain. CyTA: J Food 16(1):1131–1138

Salim S (2017) Oxidative stress and the central nervous system. J Pharmacol Exp Ther 360(1):201–205

Scalbert A, Manach C, Morand C, Rémésy C, Jiménez L (2005) Dietary polyphenols and the prevention of diseases. Crit Rev Food Sci Nutr 45(4):287–306

Schaich KM, Tian X, Xie J (2015) Hurdles and pitfalls in measuring antioxidant efficacy: a critical evaluation of ABTS, DPPH, and ORAC assays. J Funct Foods 14:111–125

Sedighi-Vesagh R, Naserian AA, Ghaffari MH, Petit HV (2015) Effects of pistachio by-products on digestibility, milk production, milk fatty acid profile and blood metabolites in Saanen dairy goats. J Anim Physiol an N 99(4):777–787

Seeram NP, Aviram M, Zhang Y, Henning SM, Feng L, Dreher M, Heber D (2008) Comparison of antioxidant potency of commonly consumed polyphenol-rich beverages in the United States. J Agric Food Chem 56(4):1415–1422

Sefi M, Elwej A, Chaâbane M, Bejaoui S, Marrekchi R, Jamoussi K, Gouiaa N, Boudawara-Sellemi T, El Cafsi M, Zeghal N, Soudani N (2019) Beneficial role of vanillin, a polyphenolic flavoring agent, on maneb-induced oxidative stress, DNA damage, and liver histological changes in Swiss albino mice. Hum Exp Toxicol 38(6):619–631

Shahidi F, Ambigaipalan P (2015) Phenolics and polyphenolics in foods, beverages and spices: antioxidant activity and health effects–a review. J Funct Foods 18:820–897

Shahidi F, Zhong Y (2010) Lipid oxidation and improving the oxidative stability. Chem Soc Rev 39(11):4067–4079

Silva RF, Pogačnik L (2020) Polyphenols from food and natural products: neuroprotection and safety. Antioxidants 9(1):61

Singh B, Singh JP, Kaur A, Singh N (2017) Phenolic composition and antioxidant potential of grain legume seeds: a review. Food Res Int 101:1–16

Singh P, Singh TP, Gandhi N (2018) Prevention of lipid oxidation in muscle foods by milk proteins and peptides: a review. Food Rev Intl 34(3):226–247

Sobeh M, Esmat A, Petruk G, Abdelfattah MAO, Dmirieh M, Monti DM, Abdel-Naim AB, Wink M (2018) Phenolic compounds from Syzygium jambos (Myrtaceae) exhibit distinct antioxidant and hepatoprotective activities in vivo. J Funct Foods 41:223–231

Soobrattee MA, Neergheen VS, Luximon-Ramma A, Aruoma OI, Bahorun T (2005) Phenolics as potential antioxidant therapeutic agents: mechanism and actions. Mut Res/Fund Mol Mech Mutagen 579(1):200–213

Soto-Vaca A, Gutierrez A, Losso JN, Xu Z, Finley JW (2012) Evolution of phenolic compounds from color and flavor problems to health benefits. J Agric Food Chem 60(27):6658–6677

Sridhar K, Charles AL (2019) In vitro antioxidant activity of Kyoho grape extracts in DPPH and ABTS assays: estimation methods for EC50 using advanced statistical programs. Food Chem 275:41–49

Srinivasulu C, Ramgopal M, Ramanjaneyulu G, Anuradha CM, Suresh Kumar C (2018) Syringic acid (SA) – a review of its occurrence, biosynthesis, pharmacological and industrial importance. Biomed Pharmacother 108:547–557

St Angelo AJ (1996) Lipid oxidation on foods. Crit Rev Food Sci Nutr 36(3):175–224

Swieca M, Seczyk L, Gawlik-Dziki U, Dziki D (2014) Bread enriched with quinoa leaves - the influence of protein-phenolics interactions on the nutritional and antioxidant quality. Food Chem 162:54–62

Tabart J, Kevers C, Pincemail J, Defraigne J-O, Dommes J (2009) Comparative antioxidant capacities of phenolic compounds measured by various tests. Food Chem 113(4):1226–1233

Takooree H, Aumeeruddy MZ, Rengasamy KRR, Venugopala KN, Jeewon R, Zengin G, Mahomoodally MF (2019) A systematic review on black pepper (Piper nigrum L.): from folk uses to pharmacological applications. Crit Rev Food Sci Nutr 59(supll):S210-S243

Tan Y, Chang SKC, Zhang Y (2017) Comparison of α-amylase, α-glucosidase and lipase inhibitory activity of the phenolic substances in two black legumes of different genera. Food Chem 214:259–268

Tangney CC, Rasmussen HE (2013) Polyphenols, inflammation, and cardiovascular disease. Curr Atheroscler Rep 15(5):324

Teissedre P-L, Landrault N (2000) Wine phenolics: contribution to dietary intake and bioavailability. Food Res Int 33(6):461–467

Teixeira-Guedes CI, Oppolzer D, Barros AI, Pereira-Wilson C (2019) Phenolic rich extracts from cowpea sprouts decrease cell proliferation and enhance 5-fluorouracil effect in human colorectal cancer cell lines. J Funct Foods 60:103452

Telles AC, Kupski L, Furlong EB (2017) Phenolic compound in beans as protection against mycotoxins. Food Chem 214:293–299

Thaipong K, Boonprakob U, Crosby K, Cisneros-Zevallos L, Hawkins Byrne D (2006) Comparison of ABTS, DPPH, FRAP, and ORAC assays for estimating antioxidant activity from guava fruit extracts. J Food Compos Anal 19(6):669–675

Toaldo IM, Cruz FA, da Silva EL, Bordignon-Luiz MT (2016) Acute consumption of organic and conventional tropical grape juices (*Vitis labrusca* L.) increases antioxidants in plasma and erythrocytes, but not glucose and uric acid levels, in healthy individuals. Nutr Res 36(8):808–817

Toscano LT, Silva AS, Toscano LT, Tavares RL, Biasoto ACT, de Camargo AC, da Silva CSO, Gonçalves MCR, Shahidi F (2017) Phenolics from purple grape juice increase serum antioxidant status and improve lipid profile and blood pressure in healthy adults under intense physical training. J Funct Foods 33:419–424

Tsao R, Li H (2012) Antioxidant properties in vitro and in vivo: realistic assessments of efficacy of plant extracts. Plant Sci Rev 7(9):11–13

Tsen SY, Siew J, Lau EKL, Afiqah bte Roslee F, Chan HM, Loke WMJDS (2014) Cow's milk as a dietary source of equol and phenolic antioxidants: differential distribution in the milk aqueous and lipid fractions. Dairy Sci Technol 94 (6):625–632

Uchegbu NN, Ishiwu CN (2016) Germinated pigeon pea (*Cajanus cajan*): a novel diet for lowering oxidative stress and hyperglycemia. Food Sci Nutr 4(5):772–777

Vitaglione P, Fogliano V (2004) Use of antioxidants to minimize the human health risk associated to mutagenic/carcinogenic heterocyclic amines in food. J Chromatogr B 802(1):189–199

Volf I, Ignat I, Neamtu M, Popa VI (2014) Thermal stability, antioxidant activity, and photooxidation of natural polyphenols. Chem Pap 68(1):121–129

Wang S, Meckling KA, Marcone MF, Kakuda Y, Tsao R (2011) Can phytochemical antioxidant rich foods act as anti-cancer agents? Food Res Int 44(9):2545–2554

Waterhouse AL (2002) Wine phenolics. Ann N Y Acad Sci 957(1):21–36

Wollin SD, Jones PJH (2001) Alcohol, red wine and cardiovascular disease. J Nutr 131(5):1401–1404

Wu J, Xia S, Kalionis B, Wan W, Sun T (2014) The role of oxidative stress and inflammation in cardiovascular aging. Biomed Res Int 2014:615312

Xu B, Chang SKC (2010) Phenolic substance characterization and chemical and cell-based antioxidant activities of 11 lentils grown in the northern United States. J Agric Food Chem 58(3):1509–1517

Yousefian M, Shakour N, Hosseinzadeh H, Hayes AW, Hadizadeh F, Karimi G (2019) The natural phenolic compounds as modulators of NADPH oxidases in hypertension. Phytomedicine 55:200–213

Zadernowski R, Naczk M, Nesterowicz J (2005) Phenolic acid profiles in some small berries. J Agric Food Chem 53(6):2118–2124

Zeb A (2015a) Chemistry and liquid chromatography methods for the analyses of primary oxidation products of triacylglycerols. Free Radic Res 49(5):549–564

Zeb A (2015b) Phenolic profile and antioxidant potential of wild watercress (*Nasturtium officinale* L.). Springerplus 4:714

Zeb A (2016) Phenolic profile and antioxidant activity of melon (*Cucumis Melo* L.) seeds from Pakistan. Foods 5 (4)

Zeb A (2018) Ellagic acid in suppressing in vivo and in vitro oxidative stresses. Mol Cell Biochem 448(1–2):27–41

Zeb A (2019) Food frying: chemistry, biochemistry and safety, vol 1st. Wiley, London, UK

Zeb A (2020) Concept, mechanism, and applications of phenolic antioxidants in foods. J Food Biochem 44(9):e13394

Zeb A, Akbar A (2018) Ellagic acid suppresses the oxidative stress induced by dietary-oxidized tallow. Oxid Med Cell Long 2018:7408370

Zeb A, Habib A (2018) Lipid oxidation and changes in the phenolic profile of watercress (*Nasturtium officinale* L.) leaves during frying. J Food Measur Charact 12:677–2684

Zeb A, Haq I (2016) The protective role of tomato powder in the toxicity, fatty infiltration and necrosis induced by oxidized tallow in rabbits. J Food Biochem 40:428–435

Zeb A, Haq I (2016) Polyphenolic composition, lipid peroxidation and antioxidant properties of chapli kebab during repeated frying process. J Food Measur Charact 12:555–563

Zeb A, Hussain S (2014) Sea buckthorn seed powder provides protection in the oxidative stress produced by thermally oxidized sunflower oil in rabbits. J Food Biochem 38(5):498–508

Zeb A, Muhammad B, Ullah F (2017) Characterization of sesame (*Sesamum indicum* L.) seed oil from Pakistan for phenolic composition, quality characteristics and potential beneficial properties. J Food Meas Charact 11(3):1362–1369

Zeb A, Haq A, Murkovic M (2018) Effects of microwave cooking on carotenoids, phenolic compounds and antioxidant activity of *Cichorium intybus* L. (chicory) leaves. Eur Food Res Technol 245(2):365–374

Zhang H, Tsao R (2016) Dietary polyphenols, oxidative stress and antioxidant and anti-inflammatory effects. Curr Opin Food Sci 8:33–42

Zhang L, Li J, Hogan S, Chung H, Welbaum GE, Zhou K (2010) Inhibitory effect of raspberries on starch digestive enzyme and their antioxidant properties and phenolic composition. Food Chem 119(2):592–599

Zhang Q, Saleh ASM, Chen J, Shen Q (2012) Chemical alterations taken place during deep-fat frying based on certain reaction products: a review. Chem Phys Lipids 165(6):662–681

Zhang B, Deng Z, Ramdath DD, Tang Y, Chen PX, Liu R, Liu Q, Tsao R (2015) Phenolic profiles of 20 Canadian lentil cultivars and their contribution to antioxidant activity and inhibitory effects on α-glucosidase and pancreatic lipase. Food Chem 172:862–872

Zhao D, Simon JE, Wu Q (2020) A critical review on grape polyphenols for neuroprotection: strategies to enhance bioefficacy. Crit Rev Food Sci Nutr 60(4):597–625

Ziaiifar AM, Achir N, Courtois F, Trezzani I, Trystram G (2008) Review of mechanisms, conditions, and factors involved in the oil uptake phenomenon during the deep-fat frying process. Int J Food Sci Technol 43(8):1410–1423

Chapter 14
Molecular Mechanism of Phenolic Antioxidants

Learning Objectives
In this chapter, the reader will be able to:
- Learn the basic mechanism of antioxidant action.
- Understand the mechanism of phenolic antioxidants in in-vitro studies.
- Discuss the molecular mechanism involved in the in-vivo studies.
- Describe how oxidative stress can be controlled by phenolic antioxidants.

14.1 Introduction

Chemical, biochemical, clinical, and epidemiological evidence proves the chemo-protective effects of phenolic antioxidants against oxidative stress-mediated disorders. The pharmacological actions of phenolic antioxidants stem mainly from their free radical scavenging and metal chelating properties as well as their effects on cell signaling pathways and gene expression (Soobrattee et al. 2005). The *in-vitro* and *in-vivo* studies showed that phenolic compounds from foods are beneficial in maintaining human health. The free radical scavenging and antioxidant activities of phenolic compounds depend on the arrangement of functional groups about the nuclear structure. Both the number and configuration of H-donating hydroxyl groups are the main structural features influencing the antioxidant capacity of phenolic compounds (Zeb 2020) as discussed in Chap. 1. This chapter presented the basic concept of phenolic compound reaction at the molecular level with specific reference to only a few representative compounds.

© Springer Nature Switzerland AG 2021
A. Zeb, *Phenolic Antioxidants in Foods: Chemistry, Biochemistry and Analysis,*
https://doi.org/10.1007/978-3-030-74768-8_14

14.2 Basic Mechanism

Phenolic compounds had been reported to produce beneficial effects in oxidative stress by regulating several pathways. These include activation of antioxidant response using the Nrf2 (nuclear erythroid factor 2) (Baluchnejadmojarad et al. 2017; Ding et al. 2014); inhibition of cyclooxygenase 2 (COX-2) (Chatterjee et al. 2012) and cytokines (Anderson and Teuber 2010) with the help of NF-kB (nuclear factor-kappa B) (Ahad et al. 2014); modulation of cell survival or apoptosis (Ho et al. 2014); enhanced the activity of biological antioxidants and antioxidant enzymes (Anitha et al. 2013). The details of these oxidative stress studies are shown in Fig. 14.1. The figure showed some important sources of phenolic compounds, and oxidative stress induced by different *in-vitro* and *in-vivo* stressors. It has been observed that oxidative stress was studied in antioxidant enzymes, DNA damage, and oxidized DNA. Such studies were classed as general studies. The mechanistic studies involved the determination of cytokines, transcription factors, kinase enzymes, and cell properties. Cano et al. (2019) upregulation of Mitogen-Activated Protein Kinases (MAPK) enzymes in aging is still controversial, increasing evidence shows that dysregulation of those enzymes is associated with biological processes that contribute to aging such as irreversible senescence. The authors suggested

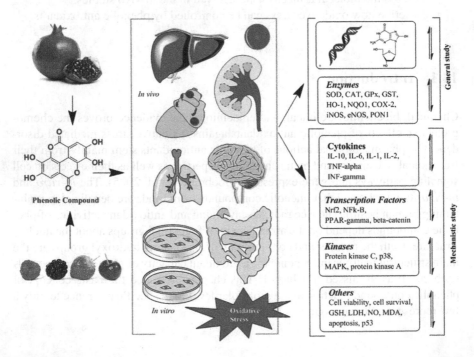

Fig. 14.1 The role of ellagic acid in oxidative stress at different organ site. The parameters studied in the mechanistic studies and general protective studies (*in-vitro* and *in-vivo*) are shown (redrawn with kind permission of Springer Nature (Zeb 2018))

that suggest that phenolic compounds that regulate MAPK enzymes and reduce senescent cells can be potentially used to improve longevity and prevent/treat age-related diseases.

14.3 *In-Vitro* Studies

Hydrogen peroxide-induced oxidative stress was studied in trout erythrocyte cells (Fedeli et al. 2004). A phenolic compound such as ellagic acid (EA) was used (1–30 µM) for 1 h to treat oxidative stress (Table 14.1). The percent hemolysis and inhibitory chemiluminescence signal were studied. Ellagic acid significantly reduced the hydrogen peroxide luminol chemiluminescence and rate of hemolysis in stress trout cells. The authors reported that EA at low concentration was protective against the DNA breakage. In Chinese Hamster lung (V79-4) cells and human osteogenic sarcoma (HOS) cells, hydrogen peroxide was used to induce oxidative stress (Han et al. 2006). The activity of antioxidant enzymes (SOD, CAT, GPx) was significantly elevated by the treatment of EA (4–100 µg/mL) in V79-4 cells and apoptosis in HOS cells through up-regulation of Bax and activation of caspase-3. In human fibroblast (IMR-90) cells, ellagic acid (10 µg/mL) was used to treat the oxidative stress induced by hydrogen peroxide (Chen et al. 2007). DNA damage and lipid peroxidation was found to be significantly reduced by EA, while intracellular GSH was enhanced. The authors suggested that EA and other polyphenols protect cells from the oxidative stress induced by hydrogen peroxide. Khanduja et al. (2006) reported the effect of hydrogen peroxide-induced oxidative stress in normal human peripheral blood mononuclear cells (PBMCs). The authors reported that EA significantly inhibited lipid peroxidation, and protected DNA damage in PBMC cells against oxidative stress. These studies clearly showed that hydrogen peroxide-induced oxidative stress and its consequent negative effects can be reduced by EA.

Different kinds of radiation have also been effective in producing oxidative stress. For example, γ-irradiation of HeLa cells for 4 weeks induced oxidative stress, which was ameliorated by 10–1000 µM/L of EA (Bhosle et al. 2005; Bhosle et al. 2010). These authors reported that EA is protected against radiation-induced oxidative stress in splenic lymphocytes of tumor transplanted mice. The changes in the antioxidant enzymes and tumor burden was significantly reduced. These authors also reported that EA generated ROS in tumor cell lines and was effective in killing tumor cells. However, the generation of ROS in the tumor cells with a high concentration (1000 µM/L) of EA may be due to its pro-oxidant action. Two pairs of isomers of resveratrol dimers trans-ε-viniferin and (+)-cis-ε-viniferin, astilbin, and isoastilbin, and resveratrol tetramer (−)-hopeaphenol were isolated from grape and their cellular antioxidant activities were evaluated in human hepatoma (Hep G2) cells (Xueyan et al. 2018). Results indicated that astilbin has the strongest antioxidant activity (10 µg/mL), while isoastilbin was the weakest one. The five stilbenes could down-regulation of oxidative stress genes and apoptosis genes. Phenolic extracts derived from rice bran (RB) are recognized to have antioxidant and

Table 14.1 Effects of ellagic acid on the oxidative stress induced by different stressor in in-vitro studies (reproduced with kind permission of Springer Nature (Zeb 2018))

Oxidative stressor	Ellagic acid	Duration	Target cells	Effects of EA on Markers	References
H_2O_2	50–200 μM	2 h	PBMC cells	MDA↑	Khanduja et al. (2006)
H_2O_2	4–100 μg/mL	48 h	Chinese hamster lung cells and human osteogenic sarcoma cells	MDA↓, SOD↑, CAT↑, GPx↑, RSA↑, cell survival↓	Han et al. (2006).
UV-A	25-75 μM	24 h	Human keratinocytes	MDA↓, SOD, MDA↓, Nrf2↓, LDH↓, HO-1↓	Hseu et al. (2012)
t-butyl hydroperoxide	0.1–0.25 μM	4 h	Rat hepatocytes	MDA↓, cell viability↑, XO↓	Singh et al. (1999)
Lung Cancer	10–80 μM	96 h	Human lung carcinoma cells	MDA↓, GSH↑, LDH↓, NF-kB↓, HO-1↓	Kim et al. (2013)
Colon Cancer	20–100 μM	48 h	Human Colon adenocarcinoma cells	Cell viability↑, iROS, Caspase-3↑, Bax↑, Bcl-2↓, apoptosis↑	Umesalma et al. (2015)
Lipopolysaccharide	50–200 μg/mL	24 h	RAW 264.7 cells	NO↓, cell viability↑, CoX-2	BenSaad et al. (2017)
H_2O_2	10 μg/mL	3 h	IMR-90 cells	DNA damage↓, iGSH↑, lipid peroxidation↓	Chen et al. (2007)
Iron, H_2O_2 and t-BHP	6.2–26 μM	2 h	PC-12 cells	% survival↑, GSH↑, MDA↓	Pavlica and Gebhardt (2005)
Nicotine	10–300 μM	1 h	Lymphocytes	MDA↓, GSH↑, GPx↑, SOD↑, CAT↑	Sudheer et al. (2007)
Oxidized LDL	50 μM/L	3 h	Rat thoracic aortic smooth muscle cells	Cell viability↑, ox-LDL↓	Chang et al. (2008)
H_2O_2	1–30 μM	1 h	Trout erythrocytes	% hemolysis ↓, inhibitory CL signal (%)↑	Fedeli et al. (2004)
Leukemia	10–80 μM/L	30 m	MOLT cells	DCFH-DA oxidation assay↓	Mertens-Talcott et al. (2005)

(continued)

Table 14.1 (continued)

Oxidative stressor	Ellagic acid	Duration	Target cells	Effects of EA on Markers	References
Oxidized LDL	5–20 µM	24 h	Human umbilical cords cells	ROS↓, cell viability↑, eNOS↓	Ou et al. (2010)
Carbonyl cyanide and rotenone	10 µM	18 h	Chang human liver cells	MDA↓, GSH↑, iROS↓, apoptosis↓, cell viability↑	Hwang et al. (2010)
UV-B	1–100 µM/L	48 h	Human keratinocytes HaCaT cells and human fibroblast	Cell viability↑, gene expression↓	Bae et al. (2010)
Oxidized LDL	5–20 µM	16 h	Human endothelial cells	SOD↑, ROS↓, iNOS↓, eNOS↓	Lee et al. (2010)
Bladder cancer	35.6–43.9 µM	723 h	Human bladder cancer T24 cells	MDA↓, iROS↓, SOD↑, apoptosis↑, p53 expression↑	Qiu et al. (2013)
γ-Radiation	10–1000 µM/L	4 weeks	HeLa cells	SOD↑, GPx↑, GR↑, cell viability↑	Bhosle et al. (2005)
Lipid oxidation	30 µM	24 h	Astrocytes	Cell viability↑, ROS↓	Yang et al. (2008)
Lipid oxidation	300–900 µg/mL	24 h	Meat and Fork	MDA↓, oxy Hb↓	Hayes et al. (2009)

anti-inflammatory potential. The extracts derived from RB down-regulated the expression of four genes, ICAM1, CD39, CD73, and NOX4, and up-regulated the expression of another four genes, nuclear factor erythroid 2-related factor 2 (Nrf2), quinone oxidoreductase 1 (NQO1), heme oxygenase 1 (HO1), and endothelial nitric oxide synthase (eNOS), indicating an antioxidant/anti-inflammatory effect for RB against endothelial dysfunction (Saji et al. 2019).

Similarly, UV-B radiation also caused oxidative stress in human keratinocytes, HaCaT cells, and human fibroblast cell lines, which was ameliorated by the treatment with EA at the dose of 1–100 µM/L for 48 h (Bae et al. 2010). The authors showed that ellagic acid prevented collagen destruction and inflammatory responses (IL-1b and IL-6), which were induced by UV-B radiation. Similarly, Hseu et al. (2012) reported the mechanism of protection of EA in the UV-A-induced oxidative stress in human keratinocytes cells. The oxidative stress markers such as SOD, MDA, LDH, and HO-1 were studied. The authors reported that EA activated the antioxidant system through the upregulation and stabilization of Nrf2 cells. These studies concluded that EA can be used as a drug or chemotherapeutic agent, or as a food supplement for curing radiation-induced oxidative stress in the skin.

Certain chemicals such as t-butyl hydroperoxides (t-BHP) is a very good oxidative stressor in cell cultures. Singh et al. (1999) reported the protective effects of EA on the t-BHP induced oxidative stress in hepatocytes cells from rats. A very small amount (0.1–0.25 μM) of EA was used for 4 h. The authors studied MDA, XO, and cell viability to determine oxidative stress. They concluded that EA inhibited both enzymatically and non-enzymatically, the generation of superoxide anions and hydroxyl radicals in the EA treated cells. Similarly, the oxidative stress induced by iron, hydrogen peroxide, and t-BHP were measured in PC-12 cells, which was ameliorated by EA at the dose of 6.2–26 μM (Pavlica and Gebhardt 2005). The formation of ROS, depletion of GSH, and decrease in the % survival of the cells were inhibited by EA treatment. Nicotine was also found to induce oxidative stress in lymphocyte cells, which was ameliorated by treatment with EA at the dose of 10–300 μM for 1 h (Sudheer et al. 2007). In pc12 cells, ellagic acid and chlorogenic acids protected against oxidative stress as reported by Pavlica and Gebhardt (2005). The protective effects were studied in terms of lipid peroxides, hydroperoxides, SOD, CAT, GPx, and GSH. It was also reported that the protective effect of EA was similar to another antioxidant, i.e. N-acetyl cysteine. Thus EA was beneficial against nicotine-induced genotoxicity. These studies suggested that EA act against the chemical-induced oxidative stress in cell cultures.

In Chang human liver cell lines, carbonyl cyanide, and rotenone-induced oxidative stress (Hwang et al. 2010). The treatment of EA at the concentration of 10 μM for 18 h has resulted in the reduction of oxidative stress. The cell death, elevation of GSH, ALT, and AST were controlled by EA, suggesting cytoprotective effects. In a recent study of RAW 264.7 microphage cells, the oxidative stress was induced by lipopolysaccharide, which was treated with 50–200 μg/mL of EA for 24 h (BenSaad et al. 2017). It was reported that EA inhibited the production of NO, PGE2, IL-6 in oxidatively stressed cells.

Chang et al. (2008) showed that EA significantly suppressed the oxidative stress induced by oxidized LDL in rat thoracic aortic smooth muscle cells (RASMC). The authors reported that EA inhibited the oxidized LDL induced proliferation of RASMC and phosphorylation of extracellular signal-regulated kinase. EA was also found to block the cell cycle progression and reduced the expression of proliferating cell nuclear antigen (PCNA). The authors suggested that EA plays an antioxidant role in the prevention of atherosclerosis. Similarly, Ou et al. (2010) also showed that oxidized LDL produced oxidative stress in human umbilical cord cells by increasing ROS, eNOS, and cell viability. When stress cells were treated with EA (5–20 μM) for 24 h, the oxidative stress markers were reduced to control, suggesting beneficial effects. Lee et al. (2010) studied the mechanism of EA (5–20 μM) for 16 h on the oxidized LDL-induced oxidative stress in human endothelial cell lines. The authors reported that EA deactivated NADPH oxidase, suppressed ROS generation, inhibited LOX-1 induced endothelial dysfunction, and down-regulated intracellular NOS. These studies suggested that EA has strong protective effects against the oxidized LDL-induced oxidative stress and consequently atherosclerosis and atherogenesis.

As discussed above oxidative stress in the cell can trigger a variety of transcription factors such as NF-κB, AP-1, p53, HIF-1α, PPAR-γ, β-catenin, and Nrf2. The activation of these factors can result in the expression of more than 500 different genes, such as genes for growth factors, inflammatory cytokines, cell cycle regulatory molecules, and importantly anti-inflammatory molecules. These changes transform a normal cell into a tumor or cancer cell. Thus, there is a strong link between oxidative stress and cancer (Reuter et al. 2010; Martinez-Useros et al. 2017; Ma-On et al. 2017). Ellagic acid was protective against the oxidative stress in leukemia cells (MOLT cells) when treated at a concentration of 10–80 µM/L for 30 min (Mertens-Talcott et al. 2005). In human bladder cancer T24 cells, the EA was highly effective at the concentration of 35.6–43.9 µM for 72 h reported by Qiu et al. (2013). An increase in mRNA and expression of Phospho-p38 MAPK and a decrease of mRNA and expression of MEKK1 and Phospho-c-Jun in T24 cells treated with EA were observed. The authors also reported the activation of caspase-3 and the expression of PPAR-c protein increased in EA induced apoptosis suggested a reduction in the oxidative stress in bladder cancer. In human lung carcinoma cells, EA (10–80 µM) for 96 h was highly protective by regulating the levels of GSH, LDH, F-kB, HO-1, and GR (Kim et al. 2013). Similarly, in human colon adenocarcinoma cells, the oxidative stress was significantly suppressed by EA at the concentration of 20–100 µM for 48 h (Umesalma et al. 2015). The cell viability, iROS, cell proliferation, apoptosis, NO levels, and CoX-2 were normalized by EA. Thus, EA can be used for oxidative stress in the majority of cancers.

During oxidative stress, lipid peroxidation is an important factor that can be controlled using phenolic compounds (Rong et al. 2017). In-vitro studies showed that EA is a good inhibitor of lipid peroxidation in cellular studies such as astrocytes (Yang et al. 2008), or foods such as meat or fork (Hayes et al. 2009; Hayes et al. 2010). Recent results also illustrated that citrus specie finger limes rich in phenolic compounds inhibited the NO-releasing and the inflammation-related cytokines including IL-1β, IL-6, and TNFα elevation (Wang et al. 2019). The effects of protocatechuic aldehyde (PCA), vanillic acid (VA), and trans-ferulic acids were studied for their protective role in vitro model of oxidative stress-induced neuroblastoma SH-SY5Y cells (Gay et al. 2018). Pre-treated SH-SY5Y cells with phenolic compounds also helped to upregulate H_2O_2-induced depletion of the expressions of sirtuin-1 (SIRT1) and forkhead box O (FoxO) 3a as well as induce the levels of antioxidants i.e., SOD and catalase and anti-apoptotic B-cell lymphoma 2 (Bcl-2) proteins. The role of the Nrf2-signaling pathway in the cellular antioxidant activity of açaí seed extract (ASE) containing gallic acid, 3,4-dihydrobenzoic acid, catechin, syringic acid, epicatechin, epigallocatechin gallate, and quercetin-3-O-rutinoside in H_2O_2-induced oxidative stress in human endothelial cells (HUVEC) was reported by Soares et al. (2017). The authors showed that ASE decreased H_2O_2-induced cytotoxicity, prevented migratory capacity loss and oxidative stress. ASE promoted up-regulation of the master antioxidant transcription factor Nrf2 through activation of ERK, thus leading to an increased expression of antioxidant enzymes. ASE alone or in pre-treatment and treatment groups significantly enhanced protein expression level of Nrf2 protein compared to H_2O_2 and control groups (1.5-fold, $p < 0.001$) without influencing its repressor protein Keap 1 (Fig. 14.2) that it was enhanced only in H_2O_2group ($p < 0.0001$).

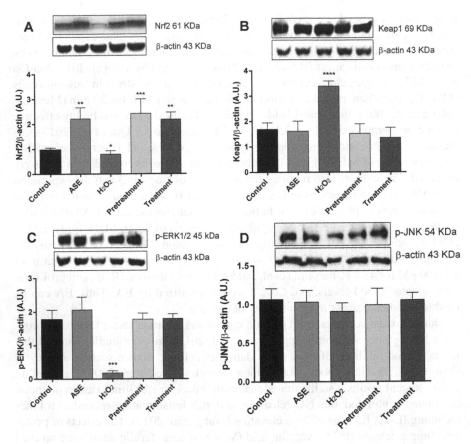

Fig. 14.2 Effect of ASE on protein expression of Keap1-Nrf2 pathway, Nrf2 (A), Keap1 (B), p-ERK ½ (C) and p-JNK (D), in HUVEC pretreated for 4 h with ASE (10 μg/mL) and then incubate it with H_2O_2 (200 μM) for 16 h (Pretreatment), or treated with ASE and H_2O_2 together for 16 h (Treatment). Results are expressed as mean ± standard deviation of one-way ANOVA followed by Tukey *post hoc*. $^{**}p < 0.001$; $^{***}p = 0.0001$; $^{****}p < 0.0001$, compared to control group (reproduced with kind permission of Elsevier (Soares et al. 2017))

In conclusion, the *in vitro* studies revealed the uses of phenolic compounds, especially the amount of EA were in the range of 0.1 to 100 μM and the duration of the study was from 1 h to 720 h. *In-vitro* studies were mostly focused on human cell cultures, which is one of the important points for extending treatment to human subjects. The majority of studies showed the mechanism of phenolic compounds involves the activation of specific genes of the antioxidant defense system. However, the disadvantages of the *in-vitro* studies are incorrectly extrapolated to humans, and the absence of bio-kinetics leads to misinterpretation of the results.

14.4   *In-Vivo* Studies

Anthocyanins, followed by hydroxybenzoic and hydroxycinnamic acids comprise the most studied phenolic classes (Martins et al. 2016). Flavones, flavonols, coumestans, diferuloylmethane, and simple phenols, were also studied. Similar to the phenolic extracts, rats (Wistar, Sprague–Dawley, Rowett Hooded Lister strain) and mice (ICR, Kunming) are the most commonly used animal models, followed by Hartley albino guinea-pigs and the marine fish. The measured biochemical parameters include MDA, CAT, SOD, GSH, GSH-Px, Px, ALT, AST, AAPH, TAOC, ORAC, 4-HNE, heme oxygenase-1 (HO-1), lipid hydroperoxides, 8-OHdG, learning and memory abilities, auditory brain response, immunohistochemical and western blot analysis (Nrf2, NQO1 and HO-1), as well as the recovery levels of ferulic (FA) and *p*-coumaric (PCA) acids, and related metabolites.

Since 2010, a significant improvement of the *in-vitro* studies was observed, in which the antioxidant potential of phenolic compounds has been studied; however, a small portion of studies reported theirs *in-vivo* potential. The most tested phenolic compounds are purchased from commercial sources, while those isolated from natural matrices are considerably scarce. Leaves of *Apium graveolens* L. var. *dulce* (apiin), corn bran, *Alpinia oxyphylla* Miq. (protocatechuic acid), *Abies koreana* E.H. Wilson (cyanidin-3-glucoside, delphinidin-3-glucoside, malvidin 3-glucoside, peonidin-3-glucoside, and petunidin-3-glucoside) and *Ipomoea batatas* (L.) Lam. (anthocyanin) seems to be the main phenolic matrices from which phenolic compounds were isolated, towards the evaluation of *in-vivo* antioxidant effects. It is a fact that numerous phenolic compounds are common in several plant species, but the existence of tenuous differences in their relative abundance is, normally, a predictor of the bioactive potential (Avello et al. 2013).

Numerous and significant advances have been reported, highlighting the antioxidant potential of individual phenolic compounds. Among them, ellagic acid, apigenin, quercetin, kaempferol, myricetin, luteolin, isorhamnetin, neochlorogenic, caffeic, ferulic, and *p*-coumaric acids, and their derivatives are the most reported (Zeb 2018; Sulaiman et al. 2011). So, and despite the pieces of evidence of promising antioxidant potential, mainly through *in-vitro* experiments, it is urgent to validate the *in vivo* activity of these and other phenolic constituents of plant species. In addition, apart from the effect of slight differences on the phenolic composition, geographical and environmental variations, harvesting time, storage, and even extraction procedures are also able to produce considerable variations in phenolic plant contents, and consequently on their total antioxidant potential. Several studies have also reported that different cultural conditions can produce significant changes in the phenolic composition of the same plant (Sulaiman et al. 2011; Avello et al. 2013).

The effects of ellagic acid on the *in-vivo* studies are shown in Table 14.2. Tetrachloro-dibenzo *p*-dioxin (TCDD) has been found to induce oxidative stress in animals. Hassoun et al. (1997) studied the TCDD induced oxidative damage to the embryo and placental tissues in mice. The authors showed that EA provided better

Table 14.2 Effects of ellagic acid on the oxidative stress induced by different stressor in animal models (*in-vivo* studies) (reproduced with kind permission of Springer Nature (Zeb 2018))

Oxidative stressor	Ellagic acid	Duration	Target site	Effects of EA on Markers	References
TCDD	10 mg/kg	21 days	Kidneys	LPO↓, GSH↑, SOD↑, GPx↑, GST↑	Vijayapadma et al. (2014)
TCDD	10 mg/kg	13 days	Brain	MDA↓, SO anion↓	Hassoun et al. (2004)
TCDD	10 μg/mL	3 h	Placenta and Fetal	SO anion↓, Cyt c↓, MDA↓	Hassoun et al. (1997)
Cyclophosphamide	50–100 mg/kg	7 days	Kidneys	GR↑, GPx↑, GST↑, MDA↓, XO↑, CAT↑, LDH↓	Rehman et al. (2012)
Cyclophosphamide	2 mg/kg	8 weeks	Sperm and Blood	MDA↓, GSH↑, GPx↑, CAT↑, SOD↑	Ceribasi et al. (2012)
Cisplatin	2 mg/kg	8 weeks	Testes	MDA↓, SOD↑	Turk et al. (2011)
Cisplatin	10 mg/kg	10 days	Liver and heart	MDA↓, GSH↑, CAT↑, GPx↑	Yuce et al. (2007)
Cisplatin	10 mg/kg	10 days	Kidneys	MDA↓, GSH↑, CAT↑, GPx↑	Atessahin et al. (2006)
Cyclosporin A	15 mg/kg	21 days	Testes and Sperm	MDA↓, GSH↑, GPx↑, CAT↑	Turk et al. (2010)
Cyclosporin A	50 mg/kg	21 days	Liver	MDA↓, Hydroperoxides↓, GSH↑, SOD↑, CAT↑, GST↑	Pari and Sivasankari (2008)
Cyclosporin A	2.5 mg/kg	30 days	Kidneys and Plasma	MDA↓	Sonaje et al. (2007)
Diabetes	20–100 mg/kg	14 days	Kidneys	MDA↓, CAT↑, SOD↑, GSH↑, GR↑, GPx↑	Ahad et al. (2014)
Diabetes	50 mg/kg	21 days	Sciatic nerves	MDA↓, TOS↓, CAT↑, NOx↓, OSI↓	Uzar et al. (2012)
Diabetes	2%	12 weeks	Heart	GSH↑, GSSG↓, ROS↓, GPx↑, CAT↑, MDA↓, SOD↑, TNF-α↓, IL-6↓, IL-β↓	Chao et al. (2009)
Carbon tetrachloride	50–100 mg/kg	7 days	Liver	MDA↓, GSH↑, CAT↑	Girish and Pradhan (2012)
Isoproterenol	7.5–15 mg/kg	12 days	Heart	Apoptosis↓, NADH deH↑, Cyt- c oxidase↑, cell viability↑	Kannan et al. (2012)
Doxorubicin	0.25–1 g /kg	72 h	Heart	MDA↓, GSH↑, GPx↑, SOD↑, XO↓, ROS↓	Lin and Yin (2013)
Galactosamine	20 mg/kg	24 h	Kidneys	GSH↑, GST↑, caspase-3↓, LDH↓	Ayhanci et al. (2016)

(continued)

Table 14.2 (continued)

Oxidative stressor	Ellagic acid	Duration	Target site	Effects of EA on Markers	References
Selenite	200 mg/kg	14 days	Eyes and erythrocytes	CAT↑, GPx↑, SOD↑, GST↑, GR↑	Sakthivel et al. (2008)
Nickel	500 µg/kg and 1 mM/kg	16 h	Liver and Kidneys	GSH↑, GST↓, GSR↓, GPx↑	Ahmed et al. (1999)
Adriamycin	2 mg/kg	8 weeks	Testes and Sperm	CAT↑, SOD↑, GSH↑, MDA↓, GPx↑	Ceribasi et al. (2012)
1,2-dimethyl hydrazine	60 mg/kg	15 weeks	Colon	5'-Nucleotidase↓, γ-GT↓, ALP↓ and LDH↓	Umesalma and Sudhandiran (2010)
Beta-amyloid	10–100 mg/kg	21 days	Hippocampal	MDA↓, CAT↑, GSH↑, NF-kB↓, Nrf2↑	Kiasalari et al. (2017)
Ammonia	1–6 mg/ml	1 h	Stomach	MDA↓, SOD↑	Iino et al. (2001)
Dextran sulfate sodium	10–100 mg/kg	7 days	Colon	MPO↓, TBARS↓	Ogawa et al. (2002)
6-hydroxydopamine	50 mg/kg	1 weeks	Brain	MDA↓, ROS↓, MAO-B↓, HO-1↓, Nrf2↑	Baluchnejadmojarad et al. (2017)
Carrageenan and Indomethacin	10–200 mg/kg	4 h	Serum, hind paws	MDA↓, GSH↑, NF-αB↓, IL-1β↓, IL-10↓, β-actin↓	El-Shitany et al. (2014)
4-hydroxy-17-β-estradiol and CuCl₂	100–1500 ppm	10 days	Liver	Oxidative DNA↓, CAT↓, SOD↑	Aiyer et al. (2008)
Dalton lymphoma	40–80 µM	15 days	Dalton lymphoma tissues	MDA↓, protein kinase C↓, cell proliferation↓, cell viability↓	Mishra and Vinayak (2013)
Dalton lymphoma	40–80 µM	19 days	Dalton lymphoma tissues	LDH↓, caspase-3↑, MDA↓, protein carbonylation and protein kinase C↓	Mishra and Vinayak (2015)
Cancer	60–80 mg/kg	15 days	Liver	MDA↓, Carboxylated proteins↓	Mishra and Vinayak (2011)
Lipid oxidation	0.25–2 mM	8 weeks	Liver and lungs	MDA↓, GSH, GR↑, GPx↑, SOD↑	Majid et al. (1991)
Apo-lipoprotein E deficiency	30 mg/kg	14 weeks	Cardiovascular system	HO-1↓, Nrf2↑, NOS↓	Ding et al. (2014)
Hyperlipidemia	1% in diet	8 weeks	Serum	MDA↓, SO anion↓, 8-OhdG↓, Caspase-8↓ and 9↓	Yu et al. (2005)
Inflammation and Fibrosis	100 mg/kg	10 weeks	Pancreas	ROS↓, MPO↓	Suzuki et al. (2009)
Oxidized lipids	50–150 mg/kg	2 weeks	Liver	MDA, GSH, GST	(Zeb and Akbar 2018)

protection against the TCDD induced oxidative damage to fetal tissues than vitamin E succinate. EA had significantly reduced the superoxide anions, lipid peroxidation, and DNA damage in mice suggesting a beneficial role in fetal development. Similarly, the same set of antioxidant compounds (ellagic acid (EA) and Vitamin E succinate) were used against the TCDD induced oxidative stress in the brain tissue of mice by Hassoun et al. (2004). The authors used 10 μg/mL of EA against the lipid peroxidation and superoxide anions. They showed that vitamin E succinate provided better protection than EA, however, EA can still be used as a beneficial agent for oxidative stress in the nervous system. Vijayapadma et al. (2014) reported the mechanism of EA protection in TCDD induced oxidative stress. The authors showed that EA mediated the suppression of CYP1A1 activity and enhanced the antioxidant mechanism. They suggested that reduction in CYP1A1 activity might be due to either the direct binding of EA to CYP1A1 or due to its aryl hydrocarbon receptor antagonism.

Cyclophosphamide (CP) was another oxidative stressor that had been used for studying the in-vivo effects of EA. Ceribasi et al. (2012) showed that CP caused abnormality in sperm, plasma MDA level, and erythrocyte SOD and decrease in CAT activity, necrosis, and production of immature germ cells as well as congestion and testicular atrophy. These effects were normalized by treating with EA, suggesting protective effects. In another study, Rehman et al. (2012) reported the effects of CP on nephrotoxicity in rats. The authors studied the oxidative stress markers such as GSH, GST, GR, CAT, and MDA in CP and EA treated rats. They observed that CP increased MDA, and depleted GSH, CAT, QR, and induced DNA strand breaks. The EA showed a prominent protective role against CP induced renal injury, DNA damage, and genotoxicity. However, this study does not provide any mechanism behind the EA protective role.

Cisplatin is a chemotherapy agent used in the treatment of a variety of cancers. However, severe toxicity of the cisplatin had resulted in oxidative stress. Atessahin et al. (2006) studied the effects of EA (10 mg/kg) against the cisplatin-induced oxidative stress in rats for 10 days. The authors showed that EA significantly reduced raised plasma creatinine, urea, and calcium levels and neutralized the harmful effects of cisplatin on oxidative stress markers. Their results also indicated that the EA might have a protective role against cisplatin-induced nephrotoxicity and oxidative stress in rats, but the effects were not enough to inhibit renal dysfunction induced by cisplatin. Another study by the same research group (Yuce et al. 2007) reported that EA (10 mg/kg) alone or in combination with cisplatin can be useful for chemotherapy for cisplatin-induced oxidative stress in the liver and heart. Similarly, another study by Turk et al. (2011) showed that EA (2 mg/kg) for 8 weeks was highly protective against the cisplatin-induced oxidative stress in testes. The authors suggested that lycopene and EA can be successfully used in cisplatin-induced toxicity. These studies, however, do not provide any information on the mechanism of EA protection.

Cyclosporine A (CsA) is a drug used for the suppression of the immune system in certain diseases or transplant surgeries. CsA has been found to induce oxidative stress. Sonaje et al. (2007) encapsulated EA (2.5 mg/kg) into nanoparticles to improve oral bioavailability against cyclosporine A induced oxidative stress and nephrotoxicity in rats. Their results revealed that EA-nanoparticle formulations were highly effective in the prevention of CsA induced oxidative stress at a much lower dose level suggesting beneficial properties. Pari and Sivasankari (2008) treated the CsA induced oxidative stress rats with 50 mg/kg of EA for 21 days. The oxidative stress on the liver was determined using MDA, hydroperoxides, GSH, SOD, CAT, and GST, which was found to be normalized by the treatment of EA. Their results indicated that EA played a significant role in protecting CsA induced oxidative damage in the liver. Similarly, Turk et al. (2010) also reported the protective role of EA in CsA induced oxidative stress in reproductive systems. The authors reported that CsA in combination with EA reduced the toxicity and oxidative damage produced by CsA alone on testicular tissues, sperm, and oxidant/antioxidant parameters. These studies, however, do not provide any information about the mechanism of action of EA against the CsA induced oxidative stress.

Oxidative stress is also one of the main secondary factors in diabetes, which was found to be controlled by EA. Chao et al. (2009) studied the mechanism of 2% EA and caffeic acid (CA) action in diabetic mice. They reported that the hypo-lipidemic effects of EA were higher than in CA. EA had significantly declined the cardiac levels of MDA, ROS, IL-1β, IL-6, tumor necrosis factor (TNF)-α, and normalized the cardiac activity of GPx, SOD, and CAT. The authors clearly showed that EA significantly up-regulated cardiac mRNA expression of GPx1, SOD, and CAT; and down-regulated IL-1β, IL-6, TNF-α, and MCP-1 mRNA expression in diabetic mice. It was concluded that EA can be used in oxidative stress induced by diabetes for controlling diabetic cardiomyopathy. In another study, Uzar et al. (2012) treated diabetic rats with EA (50 mg/kg) for 21 days and studied the oxidative stress marker in the nervous system. EA prevented brain and sciatic nerve damage in diabetes-induced oxidative stress. The authors studied MDA, paraoxonase (PON-1), total oxidant status (TOS), CAT, NOx, and oxidative stress index (OSI) of the diabetic and EA-fed rats. However, the mechanism of antioxidant action of EA has not been reported. Ahad et al. (2014) studied the role of EA (20–100 mg/kg) in diabetic neuropathy in rats for 14 days. Their results revealed that EA had significantly inhibited the activation of renal NF-κB and ameliorates renal pathology and suppressed transforming growth factor-beta (TGF-β) in renal tissues. EA also reduced the serum pro-inflammatory cytokines, IL-1β, IL-6, and TNF-α. The authors suggested that EA revealed renal protective effects in diabetic rats partly through anti-hyperglycemia, which was accompanied by attenuation of inflammatory processes via inhibition of the NF-κB pathway. These studies concluded that EA has a protective role in diabetes-induced oxidative stress.

Several chemicals and drugs have been found to induce oxidative stress under specific conditions, where EA provided protection. For example, carbon tetrachloride (Girish and Pradhan 2012) and 4-hydroxy-17-β-estradiol and $CuCl_2$ (Aiyer et al. 2008) were used to induced oxidative stress in rat liver; isoproterenol (Kannan et al. 2012), and doxorubicin (Lin and Yin 2013) in rat heart; galactosamine (Ayhanci et al. 2016), selenite (Sakthivel et al. 2008), nickel (Ahmed et al. 1999) in kidneys; Adriamycin (Ceribasi et al. 2012) in testes and sperm of rats; dextran sulfate sodium

(Ogawa et al. 2002) and 1,2-dimethylhydrazine (Umesalma and Sudhandiran 2010) in the colons of rats; Beta-amyloid (Kiasalari et al. 2017) and 6-hydroxydopamine (Baluchnejadmojarad et al. 2017) in rat brain; ammonia (Iino et al. 2001) in rat stomach; and carrageenan and indomethacin in rat serum (El-Shitany et al. 2014). In these studies, EA used were in the range of 500 µg to 1 g per kg, while the duration of treatment varied from 1 h to 15 weeks. These studies suggested that EA has a protective role in oxidative stress induced by specific chemicals or drugs. The mechanism of the EA action had been similar to the one reported earlier. The extraction of celery leaf (ET) containing several phenolic compounds decreased lipid peroxidation (MDA) and reactive oxygen species (ROS) level, and elevated the antioxidant activities of the liver, spleen, and thymus in dexamethasone treated mice. Furthermore, ET increased the protein transcription of NF-E2-related factor 2 (Nrf2), hemeoxygenase-1 (HO-1) and glutathione S-transferase (GST) to against oxidation. These results suggested that ET can protect animals through the Nrf2/HO-1 signaling pathway from oxidative damage included by dexamethasone (Han et al. 2019).

Certain health conditions also caused oxidative stress, where EA can play a strong protective role. For example, in Dalton lymphoma and cancer-induced rats, the EA (40–80 µM) for 15 and 19 days were protective against oxidative stress (Mishra and Vinayak 2011, 2013, 2014, 2015). These studies reported that the mechanism of EA action was based on the expression of protein kinase C, protein carbonylation, measurement of carboxylated proteins, and lipid peroxidation. Similarly, the *in-vivo* lipid oxidation also caused oxidative stress (Zeb and Ullah 2015), which might result in significant health issues. Majid et al. (1991) reported the first study on the ellagic acid protective role against lipid peroxidation. Ellagic acid was protective against the toxic effects of the dietary oxidized lipids in rabbits (Zeb and Akbar 2018). The authors showed that the addition of EA to liver microsomes of mice resulted in a significant increase in the inhibition of NADPH dependent liver and lung lipid peroxidation. The oxidative stress was attenuated by EA in apolipoprotein E deficient mice (Ding et al. 2014); in hyperlipidemic mice (Yu et al. 2005); in pancreatic inflammation and fibrosis (Suzuki et al. 2009) and mice adipocytes (Makino-Wakagi et al. 2012). Dietary supplementation of EA-containing foods, such as sesame oil (Zeb et al. 2017) or berries (Aranaz et al. 2017) can also help to improve lipogenesis in animal models. These studies showed that EA or its food sources possess a significant role in preventing oxidative stress by activating the natural defense system.

14.5   Phenolic Compounds and Oxidative Stress in Human

In human volunteers, upon analyzing reported studies in which phenolic compounds were evaluated for *in-vivo* antioxidant potential, other significant constraints could be observed. Firstly, from the *in-vivo* experiments in animal models, only three of the studied phenolic compounds were further studied in human volunteers: gallic acid, ferulic acid, and rutin. The relative antioxidant efficiency (RAE) of those compounds varied between 17.2 and 27.0%, while for the others, RAE was lower than that. Only ellagic acid (RAE = 109.9%), caffeic acid (RAE = 40.2%), epigallocatechin gallate (RAE = 43.1%), eriodictyol (RAE = 40.8%), taxifolin (30.7%), apigenin (RAE = 35.3%), kaempferol (RAE = 58.8%), quercetin (RAE = 59.1%), sesamin (RAE = 109.9%) and sesaminol (RAE = 56.4%) evidenced a higher potential. However, as previously mentioned, it is important to highlight that those compounds are widely recognized for their antioxidant properties, being even used as food preservatives. Additionally, some researchers also use them as positive controls, to assess the bioactive properties of other compounds/substances. Interestingly, Wang and Goodman (1999) only used commercial compounds in their experiment, remaining unclear the real antioxidant effect of these phenolic compounds, isolated from natural sources, as well as their bioavailability and bio-efficacy. Several reports have shown that the recovery of phenolic compounds in plasma, urine, and feces, after ingestion of enriched-natural matrices in phenolic compounds, is poor (Zhao et al. 2009).

In relation to the *in-vivo* antioxidant potential of anthocyanins from *Vaccinium angustifolium* Ait., tested in human volunteers, only four of them were previously studied in animal models but isolated from a different natural source (*A. koreana* E.H. Wilson). Cyanidin-3-*O*-glucoside, delphinidin-3-*O*-glucoside, malvidin-3-*O*-glucoside, and petunidin-3-*O*-glucoside were the isolated anthocyanins from natural sources in which *in vivo* experiments, both using animal models and human healthy volunteers, were carried out. Interestedly, *A. koreana* E.H. Wilson presented a higher relative abundance of anthocyanins than *Vaccinium angustifolium* Ait.: 1.3625 g total anthocyanins/100 g (Ramirez-Tortosa et al. 2001) and 1.20 g total anthocyanins/100 g (Mazza et al. 2002), respectively. Considering these results, it is important to highlight that numerous matrices with *in vitro* and *in vivo* significant antioxidant potential, could be highly effective and bioavailable, in human individuals. But, as no clinical trials are available for the majority of them, their effect remains questionable.

There is limited literature available on the efficacy of ellagic acid in human oxidative stress. Most of the studies are based on specific human cell lines as mentioned in Table 1. The role of ellagic acid as an anticancer drug has been reviewed by Zhang et al. (2014), which showed EA as a strong anticancer agent for future research. Similarly, the pharmacological activities and molecular mechanism of EA involved in liver protection were recently reviewed by Garcia-Nino and Zazueta (2015). This review also stated the lack of knowledge on human trials involving EA as a drug. Limited studies are available regarding the EA containing substances fed to humans and its ability towards specific targets. For example, Stoner et al. (2005) reported a clinical trial of 11 human subjects to determine the safety/tolerability of freeze-dried black raspberries (BRB) and to measure specific polyphenolic compounds in plasma and urine. They suggested that intake of freeze-dried BRB daily is well tolerated and resulted in quantifiable anthocyanins and ellagic acid in plasma and urine and may be helpful against oxidative stress.

Morillas-Ruiz et al. (2005) tested the effects of antioxidant supplemented beverages containing EA on oxidative stress induced by exercise. There was no significant difference in the plasma TBARS in basal to post-exercise, however, at post-exercise, the amount of carbonyls decreased by 29% in the group receiving antioxidant containing beverage. The 8-OHdG increased in placebo, while significantly decreased in the antioxidant beverage. The study suggested that antioxidant containing EA, supplementation can counter the oxidative stress induced by exercise at 70% and thus is considered healthy. Similar results were also reported by the same group with EA antioxidant-containing beverage on exercise-induced oxidative stress (Morillas-Ruiz et al. 2006). These studies, however, lacked information about the mechanism of EA action in the studied subjects. Hayeshi et al. (2007) studied the effects of ellagic acid and curcumin on human serum GST levels. Ellagic acid and curcumin showed the time as well as concentration-dependent inactivation of GSTs M1–1, M2–2, and P1–1. These results enabled our understanding of the interaction of human GSTs with the selected polyphenolic compounds in terms of their role as chemo-modulators in cases of GST-overexpression in malignancies. Similarly, Chen et al. (2012) reported the protective role of pomegranate juice containing EA on oxidative stress during human placental development. The authors, however, studied the *in vitro* oxidative stress in placental trophoblast, suggesting that pomegranate juice may prevent placental injury during fetus development.

Polyphenols possess the anticancer activity and potential to interfere with the initiation, development, and progression of cancer by modulating various cell signaling pathways to induce apoptosis, cell cycle arrest, and inhibit angiogenesis (Abbaszadeh et al. 2019). Phenolic compounds interact with a wide range of molecules involved in cell proliferation pathways by affecting their expression or activity, including decreased expression of antiapoptotic proteins Bcl-2, Bcl-xL, myeloid cell leukemia-1, and survivin, increased expression of proapoptotic proteins Bax and Bcl-2 homologous antagonist/killer, release of cytochrome c with subsequent activation of caspase 9 and caspase 3, inactivation of NF-κB, activation of apoptosis signal-regulating kinase/JNK pathway, increase in the tumor suppressor gene p53, decrease in the cell cycle regulatory proteins such as cyclin D1, upregulation of CDK inhibitory proteins such as p21 and p27, release of apoptosis-inducing factor from mitochondria into cytosol, reduction in the levels of p-Akt, p-70S6K, and HIF-1α, inhibition of VEGF receptor 2 phosphorylation, PI3K/Akt/mTOR pathway, ERK, hyaluronidase, protein tyrosine kinase, topoisomerase, 20S proteasomal, proliferation-specific forkhead box M1, ribosomal protein S6 kinase and p38 MAPK signaling pathway, increase in the expression of TSP1, tissue inhibitor of metalloproteinases and Fas-associated death domain, and downregulation of intercellular adhesion molecule 1, MMPs, VEGF, iNOS, nitric oxide synthase 3, cyclooxygenase-1, and COX-2. Abbaszadeh et al. (2019) showed that the induction of apoptosis, cell cycle arrest, and inhibition of angiogenesis via modulation of various signal transduction pathways and transcription factors are the major mechanisms by which polyphenols cause cancer cell death and inhibit tumor growth and metastasis. Because anticancer and antiangiogenic mechanisms of polyphenols are still not fully known, therefore, further studies are needed to characterize the detailed mechanisms involved in anticancer and antiangiogenic activities of these natural compounds at the molecular and cellular levels. In addition, despite the promising results of in vitro and in vivo studies, only a few clinical studies about the anticancer efficacy of polyphenols have been conducted. Currently, most of the available clinical trials have focused on the polyphenols' pharmacokinetics (bioavailability, absorption, and metabolism) and only a few clinical studies show that polyphenols can be promising agents for the treatment of cancer. Therefore, in the future, more clinical trials are required to be carried out to provide more reliable evidence. Studies should also be continued to evaluate and improve the bioavailability of polyphenolic compounds.

14.6 Study Questions

- Write a short note on the basic biological mechanism of antioxidant action.
- What is the mechanism of phenolic antioxidants in in-vitro studies?
- Discuss the molecular mechanism involved in the in-vivo studies.
- How human can benefit from phenolic antioxidants?

References

Abbaszadeh H, Keikhaei B, Mottaghi S (2019) A review of molecular mechanisms involved in anticancer and antiangiogenic effects of natural polyphenolic compounds. Phytother Res 33(8):2002–2014

Ahad A, Ganai AA, Mujeeb M, Siddiqui WA (2014) Ellagic acid, an NF-kappaB inhibitor, ameliorates renal function in experimental diabetic nephropathy. Chem Biol Interact 219(1):64–75

Ahmed S, Rahman A, Saleem M, Athar M, Sultana S (1999) Ellagic acid ameliorates nickel induced biochemical alterations: diminution of oxidative stress. Hum Exp Toxicol 18(11):691–698

Aiyer HS, Srinivasan C, Gupta RC (2008) Dietary berries and ellagic acid diminish estrogen-mediated mammary tumorigenesis in ACI rats. Nutr Cancer 60(2):227–234

Anderson KC, Teuber SS (2010) Ellagic acid and polyphenolics present in walnut kernels inhibit in vitro human peripheral blood mononuclear cell proliferation and alter cytokine production. Ann N Y Acad Sci 1190:86–96

Anitha P, Priyadarsini RV, Kavitha K, Thiyagarajan P, Nagini S (2013) Ellagic acid coordinately attenuates Wnt/beta-catenin and NF-kappaB signaling pathways to induce intrinsic apoptosis in an animal model of oral oncogenesis. Eur J Nutr 52(1):75–84

Aranaz P, Romo-Hualde A, Zabala M, Navarro-Herrera D, Ruiz de Galarreta M, Gil AG, Martinez JA, Milagro FI, Gonzalez-Navarro CJ (2017) Freeze-dried strawberry and blueberry attenuates diet-induced obesity and insulin resistance in rats by inhibiting adipogenesis and lipogenesis. Food Functand 8(11):3999–4013

Atessahin A, Ceribasi AO, Yuce A, Bulmus O, Cikim G (2006) Role of ellagic acid against cisplatin-induced nephrotoxicity and oxidative stress in rats. Basic Clin Pharmacol Toxicol 100(2):121–126

Avello MA, Pastene ER, Bustos ED, Bittner ML, Becerra JA (2013) Variation in phenolic compounds of Ugni molinae populations and their potential use as antioxidant supplement. Rev Bras 23(1):44–50

Ayhanci A, Cengiz M, Mehtap Kutlu H, Vejselova D (2016) Protective effects of ellagic acid in D-galactosamine-induced kidney damage in rats. Cytotechnology 68(5):1763–1770

Bae JY, Choi JS, Kang SW, Lee YJ, Park J, Kang YH (2010) Dietary compound ellagic acid alleviates skin wrinkle and inflammation induced by UV-B irradiation. Exp Dermatol 19(8):e182–e190

Baluchnejadmojarad T, Rabiee N, Zabihnejad S, Roghani M (2017) Ellagic acid exerts protective effect in intrastriatal 6-hydroxydopamine rat model of Parkinson's disease: possible involvement of ERbeta/Nrf2/HO-1 signaling. Brain Res 1662:23–30

BenSaad LA, Kim KH, Quah CC, Kim WR, Shahimi M (2017) Anti-inflammatory potential of ellagic acid, gallic acid and punicalagin and B isolated from *Punica granatum*. BMC Complement Altern Med 17(1):47

Bhosle SM, Huilgol NG, Mishra KP (2005) Enhancement of radiation-induced oxidative stress and cytotoxicity in tumor cells by ellagic acid. Clin Chim Acta 359(1–2):89–100

Bhosle SM, Ahire VR, Henry MS, Thakur VS, Huilgol NG, Mishra KP (2010) Augmentation of radiation-induced apoptosis by ellagic acid. Cancer Investig 28(3):323–330

Cano M, Guerrero-Castilla A, Nabavi SM, Ayala A, Argüelles S (2019) Targeting pro-senescence mitogen activated protein kinase (Mapk) enzymes with bioactive natural compounds. Food Chem Toxicol 131:110544

Ceribasi AO, Sakin F, Turk G, Sonmez M, Atessahin A (2012) Impact of ellagic acid on adriamycin-induced testicular histopathological lesions, apoptosis, lipid peroxidation and sperm damages. Exp Toxicol Pathol 64(7–8):717–724

Chang WC, Yu YM, Chiang SY, Tseng CY (2008) Ellagic acid suppresses oxidised low-density lipoprotein-induced aortic smooth muscle cell proliferation: studies on the activation of extracellular signal-regulated kinase 1/2 and proliferating cell nuclear antigen expression. Br J Nutr 99(4):709–714

Chao PC, Hsu CC, Yin MC (2009) Anti-inflammatory and anti-coagulatory activities of caffeic acid and ellagic acid in cardiac tissue of diabetic mice. Nutr Metab 6:33

Chatterjee A, Chatterjee S, Das S, Saha A, Chattopadhyay S, Bandyopadhyay SK (2012) Ellagic acid facilitates indomethacin-induced gastric ulcer healing via COX-2 up-regulation. Acta Biochim Biophys Sin Shanghai 44(7):565–576

Chen CH, Liu TZ, Chen CH, Wong CH, Chen CH, Lu FJ, Chen SC (2007) The efficacy of protective effects of tannic acid, gallic acid, ellagic acid, and propyl gallate against hydrogen peroxide-induced oxidative stress and DNA damages in IMR-90 cells. Mol Nutr Food Res 51(8):962–968

Chen B, Tuuli MG, Longtine MS, Shin JS, Lawrence R, Inder T, Michael Nelson D (2012) Pomegranate juice and punicalagin attenuate oxidative stress and apoptosis in human placenta and in human placental trophoblasts. Am J Physiol Endocrinol Metab 302(9):E1142–E1152

Ding Y, Zhang B, Zhou K, Chen M, Wang M, Jia Y, Song Y, Li Y, Wen A (2014) Dietary ellagic acid improves oxidant-induced endothelial dysfunction and atherosclerosis: role of Nrf2 activation. Int J Cardiol 175(3):508–514

El-Shitany NA, El-Bastawissy EA, El-desoky K (2014) Ellagic acid protects against carrageenan-induced acute inflammation through inhibition of nuclear factor kappa B, inducible cyclooxygenase and proinflammatory cytokines and enhancement of interleukin-10 via an antioxidant mechanism. Int Immunopharmacol 19(2):290–299

Fedeli D, Berrettini M, Gabryelak T, Falcioni G (2004) The effect of some tannins on trout erythrocytes exposed to oxidative stress. Mutat Res Genet Toxicol Environ Mutagen 563(2):89–96

Garcia-Nino WR, Zazueta C (2015) Ellagic acid: pharmacological activities and molecular mechanisms involved in liver protection. Pharmacol Res 97:84–103

Gay NH, Phopin K, Suwanjang W, Songtawee N, Ruankham W, Wongchitrat P, Prachayasittikul S, Prachayasittikul V (2018) Neuroprotective effects of phenolic and carboxylic acids on oxidative stress-induced toxicity in human neuroblastoma SH-SY5Y cells. Neurochem Res 43(3):619–636

Girish C, Pradhan SC (2012) Hepatoprotective activities of picroliv, curcumin, and ellagic acid compared to silymarin on carbon-tetrachloride-induced liver toxicity in mice. J Pharmacol Pharmacother 3(2):149–155

Han DH, Lee MJ, Kim JH (2006) Antioxidant and apoptosis-inducing activities of ellagic acid. Anticancer Res 26(5A):3601–3606

Han L, Gao X, Xia T, Zhang X, Li X, Gao W (2019) Effect of digestion on the phenolic content and antioxidant activity of celery leaf and the antioxidant mechanism via Nrf2/HO-1 signaling pathways against dexamethasone. J Food Biochem 43(7):e12875

Hassoun EA, Walter AC, Alsharif NZ, Stohs SJ (1997) Modulation of TCDD-induced fetotoxicity and oxidative stress in embryonic and placental tissues of C57BL/6J mice by vitamin E succinate and ellagic acid. Toxicology 124(1):27–37

Hassoun EA, Vodhanel J, Abushaban A (2004) The modulatory effects of ellagic acid and vitamin E succinate on TCDD-induced oxidative stress in different brain regions of rats after subchronic exposure. J Biochem Mol Toxicol 18(4):196–203

Hayes JE, Stepanyan V, Allen P, O'Grady MN, O'Brien NM, Kerry JP (2009) The effect of lutein, sesamol, ellagic acid and olive leaf extract on lipid oxidation and oxymyoglobin oxidation in bovine and porcine muscle model systems. Meat Sci 83(2):201–208

Hayes JE, Stepanyan V, Allen P, O'Grady MN, Kerry JP (2010) Effect of lutein, sesamol, ellagic acid and olive leaf extract on the quality and shelf-life stability of packaged raw minced beef patties. Meat Sci 84(4):613–620

Hayeshi R, Mutingwende I, Mavengere W, Masiyanise V, Mukanganyama S (2007) The inhibition of human glutathione S-transferases activity by plant polyphenolic compounds ellagic acid and curcumin. Food Chem Toxicol 45(2):286–295

Ho CC, Huang AC, Yu CS, Lien JC, Wu SH, Huang YP, Huang HY, Kuo JH, Liao WY, Yang JS, Chen PY, Chung JG (2014) Ellagic acid induces apoptosis in TSGH8301 human bladder cancer cells through the endoplasmic reticulum stress- and mitochondria-dependent signaling pathways. Environ Toxicol 29(11):1262–1274

Hseu YC, Chou CW, Senthil Kumar KJ, Fu KT, Wang HM, Hsu LS, Kuo YH, Wu CR, Chen SC, Yang HL (2012) Ellagic acid protects human keratinocyte (HaCaT) cells against UVA-induced oxidative stress and apoptosis through the upregulation of the HO-1 and Nrf-2 antioxidant genes. Food Chem Toxicol 50(5):1245–1255

Hwang JM, Cho JS, Kim TH, Lee YI (2010) Ellagic acid protects hepatocytes from damage by inhibiting mitochondrial production of reactive oxygen species. Biomed Pharmacother 64(4):264–270

Iino T, Nakahara K, Miki W, Kiso Y, Ogawa Y, Kato S, Takeuchi K (2001) Less damaging effect of whisky in rat stomachs in comparison with pure ethanol. Role of ellagic acid, the nonalcoholic component. Digestion 64(4):214–221

Kannan MM, Quine SD, Sangeetha T (2012) Protective efficacy of ellagic acid on glycoproteins, hematological parameters, biochemical changes, and electrolytes in myocardial infarcted rats. J Biochem Mol Toxicol 26(7):270–275

Khanduja KL, Avti PK, Kumar S, Mittal N, Sohi KK, Pathak CM (2006) Anti-apoptotic activity of caffeic acid, ellagic acid and ferulic acid in normal human peripheral blood mononuclear cells: a Bcl-2 independent mechanism. Biochim Biophys Acta 1760(2):283–289

Kiasalari Z, Heydarifard R, Khalili M, Afshin-Majd S, Baluchnejadmojarad T, Zahedi E, Sanaierad A, Roghani M (2017) Ellagic acid ameliorates learning and memory deficits in a rat model of Alzheimer's disease: an exploration of underlying mechanisms. Psychopharmacology 234(12):1841–1852

Kim YS, Zerin T, Song HY (2013) Antioxidant action of ellagic acid ameliorates paraquat-induced A549 cytotoxicity. Biol Pharm Bull 36(4):609–615

Lee WJ, Ou HC, Hsu WC, Chou MM, Tseng JJ, Hsu SL, Tsai KL, Sheu WH (2010) Ellagic acid inhibits oxidized LDL-mediated LOX-1 expression, ROS generation, and inflammation in human endothelial cells. J Vasc Surg 52(5):1290–1300

Lin MC, Yin MC (2013) Preventive effects of ellagic acid against doxorubicin-induced cardio-toxicity in mice. Cardiovasc Toxicol 13(3):185–193

Majid S, Khanduja KL, Gandhi RK, Kapur S, Sharma RR (1991) Influence of ellagic acid on anti-oxidant defense system and lipid peroxidation in mice. Biochem Pharmacol 42(7):1441–1445

Makino-Wakagi Y, Yoshimura Y, Uzawa Y, Zaima N, Moriyama T, Kawamura Y (2012) Ellagic acid in pomegranate suppresses resistin secretion by a novel regulatory mechanism involving the degradation of intracellular resistin protein in adipocytes. Biochem Biophys Res Commun 417(2):880–885

Ma-On C, Sanpavat A, Whongsiri P, Suwannasin S, Hirankarn N, Tangkijvanich P, Boonla C (2017) Oxidative stress indicated by elevated expression of Nrf2 and 8-OHdG promotes hepa-tocellular carcinoma progression. Med Oncol 34(4):57

Martinez-Useros J, Li W, Cabeza-Morales M, Garcia-Foncillas J (2017) Oxidative stress: a new target for pancreatic Cancer prognosis and treatment. J Clin Med 6(3)

Martins N, Barros L, Ferreira ICFR (2016) In vivo antioxidant activity of phenolic compounds: facts and gaps. Trends Food Sci Technoland 48:1–12

Mazza G, Kay CD, Cottrell T, Holub BJ (2002) Absorption of anthocyanins from blueberries and serum antioxidant status in human subjects. J Agric Food Chem 50(26):7731–7737

Mertens-Talcott SU, Bomser JA, Romero C, Talcott ST, Percival SS (2005) Ellagic acid potentiates the effect of quercetin on p21waf1/cip1, p53, and MAP-kinases without affecting intracellular generation of reactive oxygen species in vitro. J Nutr 135(3):609–614

Mishra S, Vinayak M (2011) Anti-carcinogenic action of ellagic acid mediated via modulation of oxidative stress regulated genes in Dalton lymphoma bearing mice. Leuk Lymphoma 52(11):2155–2161

Mishra S, Vinayak M (2013) Ellagic acid checks lymphoma promotion via regulation of PKC signaling pathway. Mol Biol Rep 40(2):1417–1428

Mishra S, Vinayak M (2014) Ellagic acid induces novel and atypical PKC isoforms and promotes caspase-3 dependent apoptosis by blocking energy metabolism. Nutr Cancer 66(4):675–681

Mishra S, Vinayak M (2015) Role of ellagic acid in regulation of apoptosis by modulating novel and atypical PKC in lymphoma bearing mice. BMC Complement Altern Med 15:281

Morillas-Ruiz J, Zafrilla P, Almar M, Cuevas MJ, López FJ, Abellán P, Villegas JA, González-Gallego J (2005) The effects of an antioxidant-supplemented beverage on exercise-induced oxidative stress: results from a placebo-controlled double-blind study in cyclists. Eur J Appl Physiol 95(5):543–549

Morillas-Ruiz JM, Villegas García JA, López FJ, Vidal-Guevara ML, Zafrilla P (2006) Effects of polyphenolic antioxidants on exercise-induced oxidative stress. Clin Nutr 25(3):444–453

Ogawa Y, Kanatsu K, Iino T, Kato S, Jeong YI, Shibata N, Takada K, Takeuchi K (2002) Protection against dextran sulfate sodium-induced colitis by microspheres of ellagic acid in rats. Life Sci 71(7):827–839

Ou HC, Lee WJ, Lee SD, Huang CY, Chiu TH, Tsai KL, Hsu WC, Sheu WH (2010) Ellagic acid protects endothelial cells from oxidized low-density lipoprotein-induced apoptosis by modulat-ing the PI3K/Akt/eNOS pathway. Toxicol Appl Pharmacol 248(2):134–143

Pari L, Sivasankari R (2008) Effect of ellagic acid on cyclosporine A-induced oxidative damage in the liver of rats. Fundam Clin Pharmacol 22(4):395–401

Pavlica S, Gebhardt R (2005) Protective effects of ellagic and chlorogenic acids against oxidative stress in PC12 cells. Free Radic Res 39(12):1377–1390

Qiu Z, Zhou B, Jin L, Yu H, Liu L, Liu Y, Qin C, Xie S, Zhu F (2013) In vitro antioxidant and antiproliferative effects of ellagic acid and its colonic metabolite, urolithins, on human bladder cancer T24 cells. Food Chem Toxicol 59:428–437

Ramirez-Tortosa C, Andersen ØM, Gardner PT, Morrice PC, Wood SG, Duthie SJ, Collins AR, Duthie GG (2001) Anthocyanin-rich extract decreases indices of lipid peroxidation and DNA damage in vitamin E-depleted rats. Free Radic Biol Med 31(9):1033–1037

Rehman MU, Tahir M, Ali F, Qamar W, Lateef A, Khan R, Quaiyoom A, Oday OH, Sultana S (2012) Cyclophosphamide-induced nephrotoxicity, genotoxicity, and damage in kidney genomic DNA of Swiss albino mice: the protective effect of Ellagic acid. Mol Cell Biochem 365(1–2):119–127

Reuter S, Gupta SC, Chaturvedi MM, Aggarwal BB (2010) Oxidative stress, inflammation, and cancer: how are they linked? Free Radic Biol Med 49(11):1603–1616

Rong S, Hu X, Zhao S, Zhao Y, Xiao X, Bao W, Liu L (2017) Procyanidins extracted from the litchi pericarp ameliorate atherosclerosis in ApoE knockout mice: their effects on nitric oxide bioavailability and oxidative stress. Food Funct 8(11):4210–4216

Saji N, Francis N, Blanchard CL, Schwarz LJ, Santhakumar AB (2019) Rice bran phenolic compounds regulate genes associated with antioxidant and anti-inflammatory activity in human umbilical vein endothelial cells with induced oxidative stress. Int J Mol Sci 20(19):4715

Sakthivel M, Elanchezhian R, Ramesh E, Isai M, Jesudasan CN, Thomas PA, Geraldine P (2008) Prevention of selenite-induced cataractogenesis in Wistar rats by the polyphenol, ellagic acid. Exp Eye Res 86(2):251–259

Singh K, Khanna AK, Visen PK, Chander R (1999) Protective effect of ellagic acid on t-butyl hydroperoxide induced lipid peroxidation in isolated rat hepatocytes. Indian J Exp Biol 37(9):939–940

Soares ER, Monteiro EB, de Bem GF, Inada KOP, Torres AG, Perrone D, Soulage CO, Monteiro MC, Resende AC, Moura-Nunes N, Costa CA, Daleprane JB (2017) Up-regulation of Nrf2-antioxidant signaling by Açaí (*Euterpe oleracea* Mart.) extract prevents oxidative stress in human endothelial cells. J Funct Foods 37:107–115

Sonaje K, Italia JL, Sharma G, Bhardwaj V, Tikoo K, Kumar MN (2007) Development of bio-degradable nanoparticles for oral delivery of ellagic acid and evaluation of their antioxidant efficacy against cyclosporine A-induced nephrotoxicity in rats. Pharm Res 24(5):899–908

Soobrattee MA, Neergheen VS, Luximon-Ramma A, Aruoma OI, Bahorun T (2005) Phenolics as potential antioxidant therapeutic agents: mechanism and actions. Mutat Res/Fund Mol Mech Mutagen 579(1):200–213

Stoner GD, Sardo C, Apseloff G, Mullet D, Wargo W, Pound V, Singh A, Sanders J, Aziz R, Casto B, Sun X (2005) Pharmacokinetics of anthocyanins and ellagic acid in healthy volunteers fed freeze-dried black raspberries daily for 7 days. J Clin Pharmacol 45(10):1153–1164

Sudheer AR, Muthukumaran S, Devipriya N, Menon VP (2007) Ellagic acid, a natural polyphenol protects rat peripheral blood lymphocytes against nicotine-induced cellular and DNA damage in vitro: with the comparison of N-acetylcysteine. Toxicology 230(1):11–21

Sulaiman SF, Sajak AAB, Ooi KL, Supriatno SEM (2011) Effect of solvents in extracting polyphe-nols and antioxidants of selected raw vegetables. J Food Compos Anal 24(4):506–515

Suzuki N, Masamune A, Kikuta K, Watanabe T, Satoh K, Shimosegawa T (2009) Ellagic acid inhibits pancreatic fibrosis in male Wistar Bonn/Kobori rats. Dig Dis Sci 54(4):802–810

Turk G, Sonmez M, Ceribasi AO, Yuce A, Atessahin A (2010) Attenuation of cyclosporine A-induced testicular and spermatozoal damages associated with oxidative stress by ellagic acid. Int Immunopharmacol 10(2):177–182

Turk G, Ceribasi AO, Sahna E, Atessahin A (2011) Lycopene and ellagic acid prevent testicular apoptosis induced by cisplatin. Phytomedicine 18(5):356–361

Umesalma S, Sudhandiran G (2010) Differential inhibitory effects of the polyphenol ellagic acid on inflammatory mediators NF-kappaB, iNOS, COX-2, TNF-alpha, and IL-6 in 1,2-dimethylhydrazine-induced rat colon carcinogenesis. Basic Clin Pharmacol Toxicol 107(2):650–655

Umesalma S, Nagendraprabhu P, Sudhandiran G (2015) Ellagic acid inhibits proliferation and induced apoptosis via the Akt signaling pathway in HCT-15 colon adenocarcinoma cells. Mol Cell Biochem 399(1–2):303–313

Uzar E, Alp H, Cevik MU, Firat U, Evliyaoglu O, Tufek A, Altun Y (2012) Ellagic acid attenu-ates oxidative stress on brain and sciatic nerve and improves histopathology of brain in streptozotocin-induced diabetic rats. Neurol Sci 33(3):567–574

Vijayapadma V, Kalai Selvi P, Sravani S (2014) Protective effect of ellagic acid against TCDD-induced renal oxidative stress: modulation of CYP1A1 activity and antioxidant defense mecha-nisms. Mol Biol Rep 41(7):4223–4232

Wang W, Goodman MT (1999) Antioxidant property of dietary phenolic agents in a human LDL-oxidation ex vivo model: interaction of protein binding activity. Nutr Res 19(2):191–202

Wang Y, Ji S, Zang W, Wang N, Cao J, Li X, Sun C (2019) Identification of phenolic compounds from a unique citrus species, finger lime (*Citrus australasica*) and their inhibition of LPS-induced NO-releasing in BV-2 cell line. Food Chem Toxicol 129:54–63

Xueyan R, Jia Y, Xuefeng Y, Lidan T, Qingjun K (2018) Isolation and purification of five phenolic compounds from the Xinjiang wine grape (*Vitis Vinifera*) and determination of their antioxidant mechanism at cellular level. Eur Food Res Technol 244(9):1569–1579

Yang C-S, Tzou B-C, Liu Y-P, Tsai M-J, Shyue S-K, Tzeng S-F (2008) Inhibition of cadmium-induced oxidative injury in rat primary astrocytes by the addition of antioxidants and the reduction of intracellular calcium. J Cell Biochem 103(3):825–834

Yu YM, Chang WC, Wu CH, Chiang SY (2005) Reduction of oxidative stress and apoptosis in hyperlipidemic rabbits by ellagic acid. J Nutr Biochem 16(11):675–681

Yuce A, Atessahin A, Ceribasi AO, Aksakal M (2007) Ellagic acid prevents cisplatin-induced oxidative stress in liver and heart tissue of rats. Basic Clin Pharmacol Toxicol 101(5):345–349

Zeb A (2018) Ellagic acid in suppressing in vivo and in vitro oxidative stresses. Mol Cell Biochem 448(1–2):27–41

Zeb A (2020) Concept, mechanism, and applications of phenolic antioxidants in foods. J Food Biochem 44(9):e13394

Zeb A, Akbar A (2018) Ellagic acid suppresses the oxidative stress induced by dietary-oxidized tallow. Oxidative Med Cell Longev 2018:7408370

Zeb A, Ullah S (2015) Sea buckthorn seed oil protects against the oxidative stress produced by thermally oxidized lipids. Food Chem 186(1):6–12

Zeb A, Muhammad B, Ullah F (2017) Characterization of sesame (*Sesamum indicum* L.) seed oil from Pakistan for phenolic composition, quality characteristics and potential beneficial properties. Journal of Food Measurement and Characterization 11(3):1362–1369

Zhang HM, Zhao L, Li H, Xu H, Chen WW, Tao L (2014) Research progress on the anticarcinogenic actions and mechanisms of ellagic acid. Cancer Biol Med 11(2):92–100

Zhao Z, Xu Z, Le K, Azordegan N, Riediger ND, Moghadasian MH (2009) Lack of evidence for Antiatherogenic effects of wheat bran or corn bran in apolipoprotein E-knockout mice. J Agric Food Chem 57(14):6455–6460

Part III
Analysis of Phenolic Antioxidants

Chapter 15
Basics in Analysis of Phenolic Antioxidants

> **Learning Objectives**
> **In this chapter, the reader will be able to:**
> - Learn different extraction techniques used for the phenolic compounds in foods.
> - Know how phenolic compounds are analyzed using spectrophotometric methods.
> - Describe potentiometric methods for the analysis of phenolic compounds in foods.
> - Understand the applications of electrochemical methods for the analysis of phenolic antioxidants.

15.1 Introduction

Phenolic compounds are present in a diverse group of plant foods. The diverse occurrence and chemical structure make it difficult to use a single analytical method of extraction, separation, and identification. Therefore, several solvents, stationary phases, and methods have been used and were reviewed recently (Ciulu et al. 2018; Delgado et al. 2019; Khoddami et al. 2013). Significant developments in research focused on the extraction, identification, and quantification of phenolic compounds as dietary molecules have occurred over the last 25 years. Organic solvent extraction is the main method used to extract phenolic compounds. Chemical procedures are used to detect the presence of phenolic compounds, while spectrophotometric and chromatographic techniques are applied to identify and quantify individual phenolic compounds.

© Springer Nature Switzerland AG 2021
A. Zeb, *Phenolic Antioxidants in Foods: Chemistry, Biochemistry and Analysis*,
https://doi.org/10.1007/978-3-030-74768-8_15

15.2   Extraction of Phenolic Compounds

The first step of the analysis of phenolic compounds is their removal from the sample matrix. This step is far more critical for quantification than for identification, providing a detectable fraction of each compound is extracted. For a profiling method, as envisioned here, extraction conditions should be as mild as possible to maintain the integrity of the components, i.e., to accurately characterize the endogenous phenolic compounds (Harnly et al. 2007). Phenolic compounds are weak organic acids (pKa = 8–12), can range from hydrophilic to hydrophobic, and are usually readily extracted into aqueous alcohol. Glycosylation tends to render flavonoids less reactive and more water-soluble. Aqueous MeOH, EtOH, acetone, DMSO, and their mixtures are the most commonly reported solvents for extraction. The range of ratios reflects the variation in the polarity of the phenolic compounds, the specific compounds targeted by the research, and the range of endpoints used to characterize the extraction efficiency, e.g., chromatographic peak area of a single compound, the sum of peak areas for multiple compounds, or total phenolic assays, such as the Folin–Ciocalteu assay (Harnly 2017).

Aqueous MeOH is probably the most frequently used solvent, although extraction of anthocyanins is enhanced using acidified MeOH or acetone (Revilla et al. 1998) and some flavanones are known to be more soluble in DMSO. Acetonitrile-water and methanol-water in different ratios are commonly employed solvents for extractions (Zeb 2015). Different extractions techniques are used for the extraction of phenolic compounds. These include solid-liquid extraction, ultrasound-assisted extractions, microwave-assisted extractions, supercritical fluid extraction, and other methods.

15.2.1   Ultrasound-Assisted Extraction (UAE)

Ultrasonic radiation has frequencies higher than 20 kHz. These radiations have been found to facilitate the extraction of organic and inorganic compounds from solid matrices using liquid solvents (Khoddami et al. 2013). Sonication is the production of sound waves creating cavitation bubbles near the sample tissue, followed by a break down to disrupt cell walls, thus releasing cell contents (Toma et al. 2001). A suitable solvent is mixed with a food sample and sonicated under controlled temperature for a specified time. Extract recovery is influenced not only by sonication time, temperature, and solvent selection but also by wave frequency and ultrasonic wave distribution (Khoddami et al. 2013). Ultrasound (US) has been used in both static and dynamic modes to extract phenolic compounds from plant materials (Ha and Kim 2016). A static system is a closed-vessel extraction for which no continuous transfer of solvent occurs (Fig. 15.1). In dynamic extraction, a fresh solvent is supplied continuously, which allows efficient adsorption of phenolic compounds and their effective transfer from the extraction vessel. Continuous transfer of

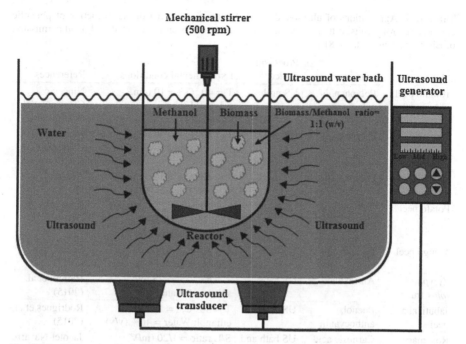

Fig. 15.1 Schematic diagram of the ultrasound-assisted extraction (UAE) process (reproduced with kind permission of Elsevier (Ha and Kim 2016))

extracted phenolic compounds prevents degradation of any thermo-labile compounds by the heat associated with sonication. Probe and bath systems are the two most common ways of applying ultrasound waves to the sample. Probe sonicators are constantly in contact with the sample and make reproducibility and repeatability difficult. Besides, the risk of sample contamination and foam production is higher. Bath sonicators can act on a range of samples simultaneously and allow for higher reproducibility (Corbin et al. 2015).

Compared to conventional methods, UAE is one of the most simple, inexpensive extraction systems and can be operated rapidly in a broad range of solvents for large-scale preparations suited for industrial purposes. Kwaw et al. (2018) observed that the US significantly ($p < 0.05$) improved the phytochemical, volatile profile, odor activity values, and sensory attributes of the fermented juice. Wen et al. (2018) reviewed the US-assisted extractions techniques used for the extraction of bioactive compounds from cash crops. The applications of UAE had been studied in pomegranate (Pan et al. 2012), *Zyzyphus lotus* fruit (Hammi et al. 2015), Jabuticaba peel (Rodrigues et al. 2015), orange peel (Khan et al. 2010), red raspberries (Chen et al. 2007), rosemary (Jacotet-Navarro et al. 2015), marjoram (Hossain et al. 2012), mango peel (Safdar et al. 2017), potato peel (Kumari et al. 2017), orange peel (Montero-Calderon et al. 2019), sour orange (Ana et al. 2018), and tomato peel (Ninčević Grassino et al. 2020) as shown in Table 15.1.

Table 15.1 Applications of ultrasound assisted extraction (UAE) in the extraction of phenolic compounds from fruits or their products. Part of table has been reproduced with kind permission of Elsevier (Wen et al. 2018)

Matrix	Extract	Processing device	Experimental conditions	References
Tomato peel	Polyphenols	US probe	T = 40 °C, t = 10 min, solvent: Ethanol	Ninčević Grassino et al. (2020)
Orange peel	Polyphenols	US probe	T = 50 °C, t = 30 min, solvent: Ethanol-water	Montero-Calderon et al. (2019)
Sour orange	Polyphenols	US probe	T = 40 °C, t = 12.5 min, solvent: Ethanol 50%	Ana et al. (2018)
Potato peel	Polyphenols	US bath	T = 30 to 45 °C, t = 30, 60, 180, 360 and 900 min, solvent: Methanol 80%	Kumari et al. (2017)
Mango peel	Polyphenols	US bath	T = 45 °C, t = 50 min, solvent: Ethanol 81%	Safdar et al. (2017)
Zyzyphus lotus fruit	Antioxidant	US bath	T = 63 °C, t = 25 min, solvent: Ethanol	Hammi et al. (2015)
Jabuticaba peel	Phenol, anthocyanin	US bath	T = 30 °C, t = 10 min, ethanol: Water = 46% (v/v)	Rodrigues et al. (2015)
Rosemary	Carnosic acid	US bath and probe	S/L ratio = 1/20 (m/V), t = 30 min, T = 40 °C	Jacotet-Navarro et al. (2015)
Pomegranate peel	Antioxidant	US probe	P = 59.2 W/cm^2, T = 25 °C, t = 60 min	Pan et al. (2012)
Marjoram	Rosmarinic acid	US probe	t = 5–15 min, T = 15–35 °C, pulse = 5 s on, 5 s off	Hossain et al. (2012)
Orange peel	Flavanone glycosides	US probe	T = 40 °C, P = 150 W, ethanol: Water = 4:1 (v/v)	Khan et al. (2010)
Red raspberries	Anthocyanins	US probe	Solvents: Materials = 4:1, t = 200 s, P = 400 W	Chen et al. (2007)

T temperature, *t* time

Ultrasound-assisted solid-liquid extraction (USLE) of phenols from olive fruit was more efficient than the US bath and agitation (Jerman et al. 2010). However, Das and Eun (2018) showed that as compared to the control, both ultra-sonication (UE) and agitation (AE) techniques significantly increased the yields of polyphenols (three-fold), catechins (two-fold), and flavonoids (two-fold), and resulted in higher antioxidant activity.

Klen and Vodopivec (2012) reported an extraction method based on ultrasound probe assisted liquid-liquid extraction (US-LLE) combined with a freeze-based fat precipitation clean-up for polyphenols in olive oil. The US probe super-agitation abilities have been once again proved to efficiently assist the phenols extraction. The ultra-high pressure (UHP) process is likely superior to thermo-sonication (TS) in bioactive compounds and antioxidant activity preservation (Dars et al. 2019).

Ultrasound-assisted water extraction of antioxidants from Korean black soybeans was found to be an efficient extraction method for anthocyanins (Ryu and Koh 2018, 2019). Sonication also resulted in the increased bioaccessibility of phenolic compounds in lettuce and green pepper, while no effect was observed for tomato, red pepper, and zucchini samples suggesting a matrix-dependent effect (Lafarga et al. 2019). Ahmed et al. (2020) used the US and AE for the extraction of phenolic compounds from amaranth. The author found that agitation at 70 °C could be used as an alternative for ultra-sonication to improve the bioactive compounds and antioxidant activities of amaranth.

The review of Dzah et al. (2020) showed that higher extraction temperatures above 50 °C in UAE are degradative to polyphenols in extracts; lower frequencies within the power ultrasound range below 40 kHz are most effective. The polyphenol yield generally increases with increasing power, but with a threshold, beyond which no significant increase can be observed; and higher ultrasound power produces free hydroxyl radicals which degrade polyphenols, especially in the presence of high water content. A comparative assessment of literature showed that UAE does not only contribute to increased extraction yield of polyphenols but also better preserves and increases the biological activity of polyphenol extracts compared to traditional maceration and Soxhlet extraction.

15.2.2 Microwave-Assisted Extraction (MAE)

Microwave is electromagnetic radiation with a wavelength from 0.001 m to 1 m (i.e. with a frequency from 3×10^{11} Hz to 3×10^{8} Hz), which can be transmitted as a wave. When microwave passes through the medium, its energy may be absorbed and converted into thermal energy. In general, the heating using microwave energy is based on two principles, ionic conduction and dipole rotation (Zhang et al. 2011a). Ionic conduction refers to the electrophoretic migration of the charge carriers (e.g., ions and electrons) under the influence of the electric field produced by the microwave. Microwaves induce molecular motion in materials or solvents with dipoles, resulting in sample heating. The heating causes plant cells to lose moisture through evaporation; the steam generated swells and eventually ruptures the cells, releasing their active components (Wang and Weller 2006). Apart from dipole materials of the plant cell, such as water molecules, the dipole rotation of the solvent molecules under the rapid change of the electric field plays an important role in MAE. During radiation, the wave electronic module changes 4.9×10^{4} times/s and the solvent molecules are induced to align themselves in the normal phase with the electric field. At such a great change in the speed of the electric phase the solvent molecules fail to realign and begin to vibrate, heating the sample due to frictional forces (Khoddami et al. 2013). The advantages of MAE techniques compared to conventional methods such as maceration and heat reflux, include reduced use of organic solvents, reduced extraction time (generally less than 30 min), and increased extraction yields. Compared with traditional reflux extraction, MAE reduced extraction time, decreased solvent consumption, and increased extraction yield of total phenolic compounds (Proestos and Komaitis 2008).

The hot solvents generated in MAE penetrate easily into the matrix and extract compounds from the lysed plant cells. For thermos-labile samples, transparent solvents such as hexane, chloroform, and toluene, or mixtures with non-transparent solvents, prevent degradation. It is important to select suitable solvents based on their boiling points, dissipation, and dielectric properties. The most commonly applied solvents in MAE are acetonitrile, water, ethanol, acetone, methanol, and 2-propanol (Khoddami et al. 2013). Polar solvents have a higher dielectric constant than non-polar solvents and can absorb more microwave energy, which can result in a higher yield of phenolic compounds.

The dissipation factor is also important to illustrate the solvent's power to release absorbed energy as heat to the sample material. Polyphenols are dipoles that can absorb microwave energy due to their hydroxyl groups; therefore MAE is a technique that can be used for the extraction of these compounds (Ajila et al. 2011; Venkatesh and Raghavan 2004). Aqueous acetone, ethanol, or their mixtures are employed to extract phenolic compounds using MAE. As MAE is influenced by many factors, several statistical optimizations have been performed to determine the best operating conditions to extract different phenolic compounds.

Microwave-assisted extraction is one of the important techniques for extracting valuable compounds from vegetal materials. Microwave-assisted extraction was also applied to the sample preparation for the determination of chlorogenic acid in *Lonicera japonica*. The nano-liquid chromatography-electrospray ionization mass spectrometry following microwave-assisted extraction was proven to be a fast and reliable method for quantitative analysis of chlorogenic acid in *L. japonica* (Zhang et al. 2011b).

Pinela et al. (2016) reported an optimized microwave-assisted extraction (MAE) process employing response surface methodology (RSM), to maximize the recovery of phenolic acids and flavonoids and obtain antioxidant ingredients from tomato. The optimum processing conditions were processing time of 20 min; at a temperature of 180 °C; and solid/liquid ratio of 45 g/L; provided tomato extracts with high potential as nutraceuticals. Similarly, the ethanol concentration of 36.9%, solvent/material ratio of 29.6 mL/g, extraction time of 71.0 min, the temperature of 40 °C, and microwave power of 400 W were found to be the optimal MAE conditions for the fruit of *Gordonia axillaris* (Li et al. 2017).

A microwave-assisted extraction (MAE) technology optimized by response surface methodology (RSM) was established to extract phenolic compounds from the fruit of *Melastoma sanguineum* (Zhao et al. 2018). Under optimal extraction conditions (31.33% ethanol, solvent/material ratio of 32.2 mL/g, 52.2 °C, 45 min, and 500 W), the total phenolic content was 39.0 mg of gallic acid equivalent (GAE)/g dry weight (DW). Similarly, the optimized MAE conditions for extraction of phenolic compounds from *Morus nigra* leaves were 20 mL of ethanol: water (1:1; v/v), 120 °C, 28 min, 0.414 g, and medium stirring speed (Radojkovic et al. 2018).

Microwave-assisted extraction of bioactive phenolic compounds from onion peel wastes employing ChCl:Urea: H_2O deep eutectic solvent was reported recently (Pal and Jadeja 2019). Under the MAE optimized conditions, the recovery of TPC and FRAP were 80.5 (mg GAE g^{-1} DW) and 636.1 (μmol AAE g^{-1} DW), respectively. Similarly, the impact of various parameters such as pH of solvent, sample to solvent ratio, irradiation time with/without cooling periods, and irradiation power were investigated individually and were studied in MAE extraction of phenolic compounds from the banana peel (Vu et al. 2019). The optimal conditions were pH of 1, the solvent ratio of 2:100 g/mL, 6 min irradiation, and microwave power of 960 W. Under these optimal conditions, approximately 50.6 mg of phenolic compounds were recovered from 1 g dried peel.

Using a response surface methodology, DiNardo et al. (2019) reported the optimized extraction parameters for MAE were 60% ethanol at a solid-to-solvent ratio of 0.1 g/mL. These conditions were able to yield maximum phenolic content and antioxidant activity in yellow European plums. Aliaño-González et al. (2020) reported two methods based on microwave-assisted extraction techniques for the extraction of both anthocyanins and total phenolic compounds from açai. The optimum conditions for the extraction of anthocyanins were 38% MeOH in water, 99.6 °C, pH 3.0, at 0.5 g: 10 mL of ratio, while for the extraction of total phenolic compounds they were 74.1% MeOH in water, 99.1 °C, pH 5.4, at 0.5 g: 20 mL of ratio. In another study, Box–Behnken design (BBD) with four-factor and three-level was used to estimate the effects of extraction time, microwave power, ethanol concentration, and liquid-solid ratio in strawberry leaf extracts on total phenolic content (TPC) and antioxidant capacity of strawberry leaves (Lin et al. 2020). The optimized conditions were extraction time of 40 s, ethanol concentration of 51.1%, microwave power of 300 W, and the liquid-solid ratio of 61.6 mL/g. These studies showed that microwave-assisted extraction can be used efficiently for the extraction of phenolic compounds in various foods.

15.2.3 Ultrasound/Microwave-Assisted Extraction (UMAE)

The coupling of two powerful radiation techniques i.e., ultrasonic and microwave is a new efficient approach to extract bioactive compounds. As mentioned earlier, MAE is a simple and rapid technique using dielectric mechanisms to heat samples and extract the plant bioactive compounds, whereas UAE forms cavitations, which increase mass transfer and improve penetration of the solvent into the sample. Thus, ultrasound/microwave-assisted extraction (UMAE) is a powerful technique that can reduce extraction time, consume lower volumes of solvents and result in higher extraction yields than conventional extraction, MAE, and UAE (Khoddami et al. 2013).

The yields of flavonoids from the *Spatholobus suberectus* obtained by UMAE were compared with MAE, UAE, Soxhlet, and heated reflux extraction methods under optimized conditions. The highest yield obtained for UAME was after 7.5 min using 20 mL/g solvent-sample ratios, while for other extractions, optimum yields depended on a higher solvent-sample ratio (40–120 mL/g) and longer time

(30–3600 min) (Cheng et al. 2011). Tomato paste lycopene has been extracted using UMAE and UAE. The optimized time needed to give the highest yield of extract (97.4% lycopene) with UMAE was 367 s, whereas the corresponding time for UAE was 1746 s and gave a lower yield (89.4% lycopene) (Lianfu and Zelong 2008). These results above imply that UMAE is a more efficient extraction method than the other extraction techniques tested (Table 15.2).

A schematic diagram of an apparatus for UMAE is presented in Fig. 15.2 (Cheng et al. 2011). Compared with commonly used extraction methods, UMAE showed higher efficiency and shorter extraction time for sample preparation. Both the ultrasound and microwave irradiations can be implemented continuously or intermittently. Intermittently, the ultrasound or microwave irradiation stops for an interval during the extraction process, and this on/off cycle would be repeated many times (Wen et al. 2020). The on/off cycles for the ultrasound and microwave irradiations can be synchronized or independent. The intermittent way is usually implemented when too high energy input into the extraction system, especially when no refluxing system is installed in the extraction system where too much evaporation of the solvent can occur, in particular for low boiling point solvents such as methanol, ethanol, etc. The intermittent way could help to reduce the evaporation or control the temperature of the solvent (Marić et al. 2018).

The optimum ultrasound-assisted extraction (UAE) parameters i.e., power, time, and solid/liquid ratio were evaluated for the extraction of antioxidant compounds from germinated chickpea using response surface methodology (Hayta and İşçimen 2017). The optimum conditions for UAE of germinated chickpea samples were solid/liquid ratio of 40%, amplitude power of 36.1%, and 20.1 min and the experimental values of TPC, DPPH radical scavenging activity, and TEAC were, 14.8 mg GAE/mL extract, 43.3% and 73.6%, respectively. Similarly, in another study reported by Kaderides et al. (2019) on the effects of solvent type, solvent/solid ratio, and microwave power on the yield of phenolics extraction from pomegranate peel. The optimum operating conditions were found to be: solvent type, 50% aqueous ethanol; solvent/solid ratio, 60/1 mL/g; power, 600 W.

Table 15.2 Comparison of UMAE and conventional extraction methods under the optimal conditions ($n = 3$) (reproduced with kind permission of Elsevier (Cheng et al. 2011))

Method	Extraction time	Extraction volume (mL/g)	Total flavonoids	
			Yield (mg/g)	RSD %
UMAE	450 s	20	18.79	1.2
MAE	0.5 h	40	13.66	2.4
UAE	1 h	120	9.85	2.2
SE	6 h	80	15.21	4.0
HRE	4 h	120	18.69	3.0

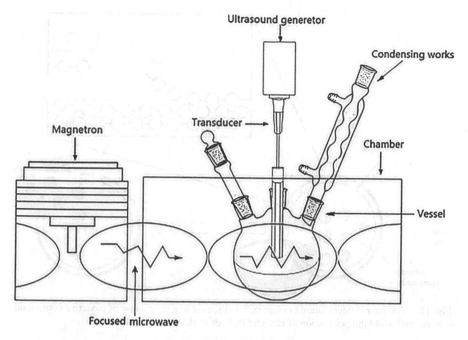

Fig. 15.2 Schematic diagram of simultaneous ultrasonic and microwave extracting apparatus (reproduced with kind permission of Elsevier (Cheng et al. 2011))

15.2.4 Ultrasound-Assisted Enzymatic Extraction (UAEE)

In recent years, some studies had reported the combination of ultrasound-assisted and enzyme-assisted extraction of bioactive compounds which is usually called ultrasound-assisted enzymatic extraction (UAEE) (Wen et al. 2020; Luo et al. 2019). UAEE is a combination of two complementary extraction methods so that it brings more advantages. An enzyme assisted extraction (EAE), enzymes promote recovery by degrading and disrupting the cell walls and membranes. However, enzymes cannot completely hydrolyze the matrix like the cell walls. UAE has a complementary effect for EAE, as the cavitation caused by power ultrasound in UAE can physically disrupt and break down the matrix to facility the enzymatic reaction and resultant release of target compounds (Fig. 15.3). In addition, enzymes alone cannot improve the mass transfer of solvent, target compounds, and enzymes themselves within or even outside the matrix (Gligor et al. 2019). Therefore, other physical measures like agitation, shaking are often employed in EAE to enhance mass transfer, among which UAE is an ideal option as it can enhance mass transfer not only outside but also within the matrix (Chemat 2011; Wen et al. 2020). Furthermore, the power of ultrasound in UAE increases the surface area of the matrix and contact the area between phases, which would expose more substrates in the EAE system to the enzymes and lead to the release of more target compounds. Studies also showed that the rates of enzymatic reaction are enhanced caused by ultrasonication in EAE can

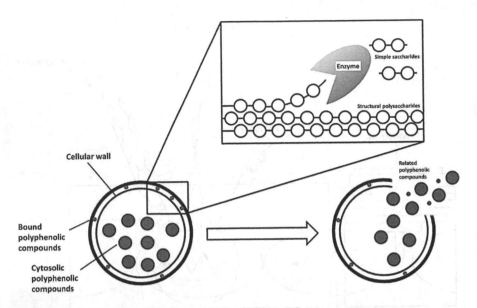

Fig. 15.3 Enzymatic degradation of plant cell walls, enabling the release of bioactive compounds (reproduced with kind permission of Elsevier (Gligor et al. 2019))

increase collisions between enzyme and substrate and imply a higher releasing rate. Also, low-frequency ultrasound under certain conditions was capable of improving cellulase activity. The viscosity of the extraction system (solvent + samples) would significantly affect the effect of ultrasound. The viscosity of the extraction system could be increased during extraction, especially when water is used as a solvent, due to the release of some un-targeted compounds like some hydrocolloids. Many of these un-targeted compounds could be degraded by some specific enzymes, reducing the viscosity and the resultant negative effect of viscosity on the ultrasound enhancement (Wen et al. 2020).

Enzyme assisted extraction offer several advantages as opposed to conventional methods. These include mild reaction conditions i.e., extraction occurring at low-temperature values and for short periods, the possibility of using the whole plant material, processes requiring fewer numbers of steps, a substrate specificity which in turn leads to extracting a large number of bioactive compounds (by bioaccessing even anchored molecules within cellular organelles such as vacuoles and plant cell walls, otherwise inaccessible) with a high bioavailability and quality (low residue levels), while also potentially lowering production costs by replacing multiple installations used for classical extraction processes (Gligor et al. 2019).

An analytical method of ultrasound-assisted enzymatic extraction (UAEE) was applied to mulberry must treated with processing variables of sonication frequency (FS), enzyme concentration (CE) and maceration time (TM) at constant temperature 20 °C and pulse durations of 10 s on and 5 s off (Tchabo et al. 2015). The optimum operating conditions were found to be $FS = 33.8$ kHz, $CE = 0.010\%$ (v/w) and TM = 11.6 min, under which it predicted TPC = 298.0 mg/100 ml; TFC = 379.2 mg/100 ml; TAC = 55.14 mg/100 ml.

Heidari and Dinani (2018) evaluated the suitability of ultrasound pretreatment in n-hexane solvent as well as enzymatic treatment with cellulase enzyme to extract oil from peanut seed powders. The optimum condition of ultrasound-assisted enzymatic extraction using n-hexane solvent (UAEE) was ultrasonic pretreatment time of 33.2 min, cellulase concentration of 1.47%, and pH of 4.6 before incubation process at a temperature of 56 °C for 120 min. Similarly, Nishad et al. (2019) utilized response surface methodology for the optimization of ultrasound-assisted extraction (UAE) and enzyme-assisted extraction (EAE) of bioactive from *Citrus sinensis* cv. Malta peel. The EAE enhanced a twofold higher yield of phenolic compounds than the UAE.

In finger millet treated with xylanase and US, the total flavonoids increased 1.4 fold in the US and 1.3 fold in xylanase, similarly, tannins also showed a significant increase (Balasubramaniam et al. 2019). Catechins, luteolin, and cyanidin were identified in the US and xylanase treated samples, whereas, shikimic acid derivatives, caffeoyl and di-caffeoyl were present only in xylanase extracts.

The application of cellulase and pectinase for the extraction of polyphenols from grape pomace was reported by Drevelegka and Goula (2020). The optimum extraction yield (48.7 mg GAE/g dry pomace) was achieved using ultrasound extraction at 56 °C, a solvent/solid ratio of 8 mL/g, an amplitude of 34%, and a time of 20 min with 53% v/v ethanol, after pretreatment with cellulase, at a concentration of 4% w/w, time of 240 min, and water/pomace ratio of 2 mL/g. A three-level, four-factor Box–Behnken design was used to optimize the extraction conditions of anthocyanins in mulberry wine residue (Zhang et al. 2020). The mathematical model reported in this study suggested a high coefficient of determination (R^2 = 0.9475) for the optimum conditions, namely 52 °C, 315 W, 0.22% enzyme, and 94 min incubation. The yield (5.98 mg/g) was close to the predicted value (5.87 mg/g). The two anthocyanins were cyanidin-3-glucoside and cyanidin-3-rutinoside consistent with those present in mulberry.

15.2.5   Microwave-Assisted Enzymatic Extraction (MAEE)

In recent times, EAE and MAE have been recognized as favorable techniques in the extraction of phenolic compounds due to the benefits of high extraction efficiency, easy handling, low solvent, and energy consumption (Wen et al. 2020). MAEE involves the combination of microwave irradiation and enzymolysis, which can disrupt cell wall structure, and increase the permeability of cell membranes or walls. Hence, the target compounds within the matrix cell can be transferred more easily into the solvent. The procedures of MAEE showed that enzymatic treatment can be conducted before and/or after microwave irradiation. In the case of adding enzymes after microwave irradiation, the samples may need a cooling process to maintain at a certain temperature to avoid inactivation of the subsequently added enzyme, or even directly to the enzyme optimal working.

An MAEE method was evaluated for the simultaneous extraction of corilagin (CG) and geraniin (GE) with deionized water (Yang et al. 2010). The optimal extraction conditions were irradiation power 500 W, the ratio of solvent to material 40 ml/g, irradiation temperature 33 °C, pH 5.2, amount of cellulase 3600 U/g, and irradiation time 9 min. Under these conditions, the extraction yields of CG and GE were 6.79 and 19.8 mg/g, which increased by 64.0% and 72.9%, respectively, as compared with the control ones.

In another study, the extraction efficiency of phlorotannins and antioxidant compounds of a South Australian brown seaweed *Ecklonia radiata* was carried out by enzymatic and microwave-assisted enzymatic extraction (Charoensiddhi et al. 2015). It was found that microwave-assisted Viscozyme extraction for 5–30 min was the most effective process with an extraction yield achieved of 52%. Chanioti et al. (2016) concluded that the MAEE method being environment-friendly was efficient for the polyphenols recovery in olive oil byproducts., while the polyphenols having significant antioxidant activity can be used as.

Görgüç et al. (2019) reported microwave-assisted enzymatic extraction of steviol glycosides and phenolic compounds Stevia leaf. The maximized yields of stevioside, rebaudioside A and total phenolic compounds were determined as 62.5, and 20.7 mg g^{-1}, respectively at optimum processing conditions which were estimated as 10.9 FBG unit g^{-1}, 53 °C, and 16 min. In conclusion, enzymatic extraction alleviates challenges posed by conventional extraction processes, reduce extraction time and give a better product, operational cost can be curbed by enzyme usage in combination with green technologies (Marathe et al. 2019). The coupled technique suits diverse food processing, pharma, and flavor industries.

15.2.6   Supercritical Fluid Extraction (SFE)

Once the fluid is forced to a pressure (Pc) and temperature (Tc) above its critical point, it becomes a supercritical fluid as shown in Fig. 15.4. Under these conditions, various properties of the fluid are placed between those of a gas and those of a liquid. Although the density of a supercritical fluid is similar to a liquid and its viscosity is similar to a gas, its diffusivity is intermediate between the two states (Herrero et al. 2006). The supercritical state of fluid has been defined as a state in which liquid and gas are indistinguishable from each other, or as a state in which the fluid is compressible (i.e. similar behavior to a gas) even though possessing a density similar to a liquid and, therefore, similar solvating power.

There is a wide range of compounds that can be used as supercritical fluids as shown in Table 15.3. Carbon dioxide is the most commonly used because of its moderate critical temperature (31.3 °C) and pressure (72.9 atm). Carbon dioxide is a gas at room temperature, so once the extraction is completed, and the system decompressed, substantial elimination of CO_2 is achieved without residues, yielding

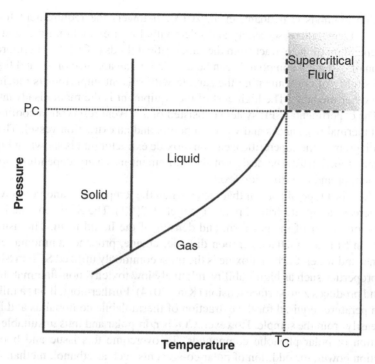

Fig. 15.4 Typical phase diagram for a pure compound (reproduced with kind permission of Elsevier (Herrero et al. 2006))

Table 15.3 Critical properties of several solvents used in supercritical fluid extraction (SFE) (reproduced with kind permission of Elsevier (Herrero et al. 2006))

Solvent	Critical property			
	Temperature (°C)	Pressure (atm)	Density (g/ml)	Solubility parameter δ_{SFC} (cal$^{-1/2}$ cm$^{-3/2}$)
Ethene	10.1	50.5	0.200	5.8
Water	101.1	217.6	0.322	13.5
Methanol	−34.4	79.9	0.272	8.9
Carbon dioxide	31.2	72.9	0.470	7.5
Ethane	32.4	48.2	0.200	5.8
Nitrous oxide	36.7	71.7	0.460	7.2
Sulfur hexafluoride	45.8	37.7	0.730	5.5
n-Butene	−139.9	36.0	0.221	5.2
n-Pentane	−76.5	33.3	0.237	5.1

a solvent-free extract (Herrero et al. 2006). On an industrial scale, when carbon dioxide consumption is high, the operation can be controlled to recycle it.

Supercritical fluid extraction (SFE) is another environmentally friendly extraction technique, which is a good alternative to conventional organic solvent extraction methods (Khoddami et al. 2013). It lowers the requirement for toxic organic solvents, increases safety and selectivity, lower extraction time, and facilitates separation of the extract from the supercritical fluids (SCF). Furthermore, degradation of extracted compounds can be avoided in the absence of air and light and the possibility of contaminating the sample with solvent impurities is much lower than in other methods. The high cost of the equipment is the main disadvantage of SFE. The experimental SFE system consisted of a solvent reservoir, a cooling bath, several thermal resistances and valves, a pump, and an extraction vessel. The schematic diagram of the supercritical carbon dioxide extractor unit is shown in Fig. 15.5 (Alvarez et al. 2019). The design of the equipment may vary depending upon the manufacturer and its utilization level.

An SCF is a type of solvent that forms when the temperature and pressure of the fluid increase above its critical point (Ignat et al. 2011). The SCF generated has the penetration power of the gas form and density of the liquid form. The usual SCF applied in SFE are methane, carbon dioxide, ethane, propane, ammonia, ethanol, benzene, and water. Carbon dioxide is the most commonly utilized SCF in SFE, due to its properties such as high stability, relatively low toxicity, non-flammability, low cost and produces zero surface tension (King 2014). Furthermore, it has a mild critical temperature required for the extraction of thermolabile compounds and is separated easily from the sample. However, CO_2 is non-polar and thus unsuitable for the extraction of polar phenolic compounds. To overcome this issue and boost CO_2 extraction power, the addition of polar co-solvents such as ethanol, methanol, ethyl acetate, and acetone is recommended (Bleve et al. 2008).

In the last decade, research has been conducted to optimize the extraction of phenolic compounds by SFE by varying pressure, temperature, extraction time, modifier, and the solvent/ modifier mixture ratio. For most phenolic materials, the highest yield was attained when the pressure was 50–600 bar, temperature 35–20 °C,

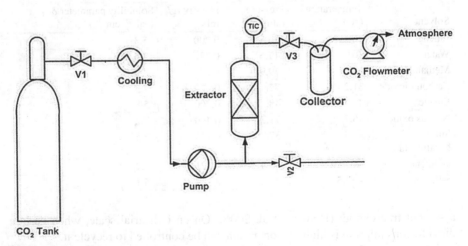

Fig. 15.5 Schematic diagram of the supercritical carbon dioxide extraction system (reproduced with kind permission of Elsevier (Alvarez et al. 2019))

and time 5–180 min (Junior et al. 2010). The application of supercritical fluid extraction of phenolic compounds from natural plant materials has been reviewed by (Tyskiewicz et al. 2018). The authors showed that co-solvent modified supercritical CO_2 is more efficient and mostly used in phenolic compounds extraction. Similarly, Gallego et al. (2019) also presented the updated perspective on the use of compressed fluids, mainly under sub- and supercritical conditions, for the extraction of bioactive components from natural matrices covering the period from 2015 to the present. The authors showed that these extraction technologies might have an important role in the development of sustainable and efficient extraction processes to cope with the high demand for natural bioactive compounds.

Different extraction methods including Soxhlet, MAE, and UAE, as well as SFE, have been applied to determine the total phenolic content of pomegranate seed oil. The different organic solvent extraction methods used in this study did not generate any significant differences in the total phenolics extracted, whereas the extracted oils from modified SFE gave a significantly higher yield of phenolic compounds (Abbasi et al. 2008). The SFE technique was applied and optimized for temperature, CO_2 pressure, and ethanol (modifier) concentration using orthogonal array design and response surface methodology for the extract yield, total phenols, and antioxidants from grape (Ghafoor et al. 2012). Optimum SFE conditions were the temperature of 44 ~ 46 °C temperature and 153 ~ 161 bar CO_2 pressure along with ethanol (<7%) as a modifier, for the maximum predicted values of extract yield (12.0%), total phenols (2.41 mg GAE/mL) and antioxidants (7.08 mg AAE/mL).

The uses of supercritical fluid extraction (SFE) of phenolic compounds from the pomace generated in the industrial processing of orange juice in Brazil were reported by Espinosa-Pardo et al. (2017). The authors showed that SFE required 78% less time and 10 times less ethanol than Soxhlet, so it is more economical in terms of time and energy consumption. In another study, a rapid, simple, and environmentally friendly SFE and supercritical fluid chromatography coupled to mass spectrometry (SFE-SFC-MS/MS) method have been developed for the analysis of nine phenolic compounds in garlic (Liu et al. 2018). In this method, the optimization of the SFE parameters using response surface methodology, fifteen phenolic compounds were successfully extracted at 50 °C in 9 min with the addition of 30% methanol. The method was automatic, efficient, green, and prevented oxidation.

The application of supercritical CO_2 extraction to obtain sage extracts rich in carnosic acid and carnosol was recently reported (Pavić et al. 2019). The optimal conditions according to RSM were a pressure of 29.5 MPa, a temperature of 49.1 °C, and a CO_2 flow rate of 3 kg h^{-1}, and the sage extract yield was calculated to be 6.54%, carnosic acid content 105 µg mg^{-1}, and carnosol content 56.3 µg mg^{-1}.

SFE at 300 bar proved to be the best option for obtaining a high-activity antioxidant extract from olive leaves, exhaust pomace, and pruning biomass, with hydroxytyrosol the prominent compound detected (Caballero et al. 2020). The optimal conditions for supercritical fluid extraction (SFE) for the extraction of trans-resveratrol from peanut kernels were studied (Jitrangsri et al. 2020). The optimal extraction parameters were a pressure of 7000 psi, a temperature of 70 °C, and a time of 50 min. The quantity of trans-resveratrol was predictable by a full quadratic regression equation with R2 (predict) = 95.56%. The predicted trans-resveratrol concentration in peanut samples was 0.7998 µg/g while the experimental concentration was 0.79 µg/g. The SFE was also found beneficial for the extraction of polyphenols from the winemaking byproducts (Aresta et al. 2020). Oils from apple seeds were extracted with the application of supercritical fluid (SFE) and Soxhlet techniques (Ferrentino et al. 2020). The optimum processing conditions for apple seed oil extraction were 24 MPa, 40 °C, 1 L/h of CO_2, 140 min. These studies showed that SFE is a valuable technique especially with co-extraction for the extraction of phenolic compounds in foods.

15.2.7 Subcritical Water Extraction (SCWE)

Another environmentally friendly extraction technique that has been utilized to efficiently isolate phenolic compounds is subcritical water extraction (SCWE) (Herrero et al. 2006), also known as superheated water, pressurized water, or hot liquid water extraction. The main advantages of SCWE over conventional methods are its simplicity, high extract quality, low extraction time, and environmental friendliness due to water being used as the solvent. With SFE, only non-polar compounds can be extracted from plant material using organic solvents as modifiers, and plant processing is likely to be more expensive than with SCWE (Khoddami et al. 2013). Water becomes subcritical when the temperature is 100–347 °C applied under sufficient pressure (normally 10–60 bar) to preserve its liquid form (below 220 Bar). The dielectric constant of water reduces under subcritical conditions due to the breakdown of intermolecular hydrogen bonds. By adjusting parameters like pressure and temperature, subcritical water displays different dielectric constant values and polarity (i.e., ethanol-water and methanol-water). Water at room temperature has high polarity and a dielectric constant close to 80. By applying suitable pressure to keep water in liquid form at 250 °C, the dielectric constant decreases to 27, which is similar to that of ethanol (Herrero et al. 2006).

The experimental device required for SWE is relatively simple (see Fig. 15.6). Mostly, the instrumentation consists of a water reservoir coupled to a high-pressure pump to introduce the solvent into the system, an oven, where the extraction cell is placed and extraction takes place, and a restrictor or valve to maintain the pressure. Plant extracts are collected in a vial placed at the end of the extraction system (Herrero et al. 2006). Besides, the system can be equipped with a coolant device for the rapid cooling of the resultant plant extract.

Treatment with SCWE is sufficiently powerful to extract a wide range of polar to low-polar compounds such as phenolic acids from grape skin (Luque-Rodríguez et al. 2007). For extraction of anthraquinones from *Morinda citrifolia*, the effectiveness of SCWE compared to that of other extraction methods, such as ethanol extraction in a stirred vessel, Soxhlet extraction, and ultrasound-assisted extraction has been studied (Pongnaravane et al. 2006). The results showed that SCWE extracts

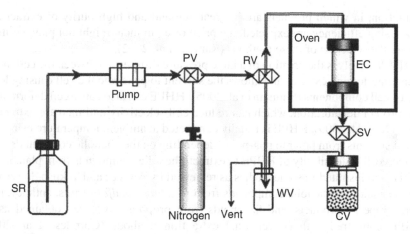

Fig. 15.6 Diagram of a subcritical water extraction system. *SR* solvent reservoir, *PV* purge valve, *RV* pressure relief valve, *EC* extraction cell; *SV* static valve, *CV* collector vial, *WV* waste vial (reproduced with kind permission of Elsevier (Herrero et al. 2006))

gave almost the same antioxidant activity as Soxhlet extracts, but SCWE extracts contained higher antioxidant activity than ethanol extracts and ultrasound-assisted extracts.

Stevia rebaudiana leaves were subjected to subcritical water extraction (Yildiz-Ozturk et al. 2014) Optimum extraction conditions were elicited as 125 °C, 45 min, 4 ml/min flow rate which yielded 38.7 mg/g stevioside and 35.6 mg/g rebaudioside A. In a study, SCWE was used to recover bioactive phenolic compounds from onion skins as reported by Munir et al. (2018). The authors found that the efficiency of extraction using the SCWE process was affected by particle size and pH. Extraction temperature was also found as a key factor affecting total phenolic content (TPC) and maximum TPC values were obtained at a lower temperature. In conclusion, SCWE could be a good alternative industrial method to use for the extraction of large amounts of phenolic compounds without toxic organic solvent residues. The products are ready to use as antioxidants for food products.

15.2.8 High Hydrostatic Pressure Extraction (HHPE)

Another novel technique that is in use recently to extract phenolic compounds from plants is called high hydrostatic pressure extraction (HHPE). This method utilizes non-thermal super-high hydraulic pressure (1000–8000 bar) and works based on mass transport phenomena (Shouqin et al. 2005). The pressure applied increases plant cell permeability, leading to cell component diffusivity according to mass transfer and phase behavior theories. The main disadvantage of methods such as HHPE, SCWE, and SFE is that expensive equipment is required; i.e., a solvent transporting pump, a pressure vessel and system controller, and a collection device for the extract (Khoddami et al. 2013). However, in the case of antioxidant extraction, in which products are in great demand and high purity of extract and processing efficiency are expected, the price of equipment might not play a critical role in the selection of these methods (Ramos et al. 2002).

HHPE involves the creation of a huge pressure difference between the cell membrane interior and exterior and allows the solvent to penetrate the cell causing leakage of cell components (Shouqin et al. 2005). HHPE can also cause cell deformation and protein denaturation, which can reduce cell selectivity and increase extraction yield (Xi et al. 2009). HHPE is usually conducted at ambient temperature using different solvents from polar to non-polar, depending on the phenolic compounds to be extracted. The feasibility of HHPE to extract phenolic compounds from plant material is demonstrated in some studies as reviewed by Altuner and Tokuşoğlu (2013). Higher yields of phenolic compounds from *Maclura pomifera* fruits, anthocyanins from grape by-products, and flavonoids from propolis have been obtained using HHPE compared with conventional extraction methods (Corrales et al. 2008; Shouqin et al. 2005). HHPE is also reported to be suitable to extract polyphenols from green tea leaves (Xi et al. 2009). A higher yield of soluble polyphenols in the juice of cashew apples has been obtained using HHPE compared to other methods (Queiroz et al. 2010).

Briones-Labarca et al. (2019) optimized the individual and interactive effect of operating high pressure and solvent polarity on yield extraction, flavonoid, and lycopene content from tomato pulp by using response surface methodology. Extraction at 450 MPa and 60% of hexane in the solvent mixture was considered the optimal HHPE conditions because it provided the maximum extraction yield (8.71%) of flavonoid (21.5 mg QE/g FW). Similarly, HHPE has been used for the extraction of polyphenols from tomato peel waste generated by the canning industry (Grassino et al. 2020). The impact of time (5 and 10 min), temperature (25, 35, 45, and 55 °C) and solvents (water, 1% HCl, 50 and 70% methanol with and without the addition of HCl, and 50 and 70% ethanol), at a constant pressure of 600 MPa, has been evaluated with respect to polyphenols' yields. The results of the study showed a significant variation in the contents of a large number of phenolic compounds in respect to the applied temperatures and solvents. Methanol (50 and 70%) at temperatures of 45 and 55 °C enhanced the recovery of polyphenols in comparison to other utilized solvents.

The effect of pressure level, extraction time, and solvent concentration was evaluated, as also the impact of HHPE on total phenolics (TPC), flavonoids, pigments, and antioxidant activity of stinging nettle leaves (Moreira et al. 2020b). The optimal conditions for the overall maximization of extraction yield, TPC, and antioxidant activity were 200 MPa, 10.2–15.6 min. The high hydrostatic pressure (HHP) extraction of noni fruits was reported by Jamaludin et al. (2020). The optimal HHP extraction conditions at the highest desirability for the simultaneous extraction of scopoletin, alizarin, and rutin were as follows: pressure 544 MPa, time 15 min, and ethanol concentration 65%, which gave the maximum yields of 82.4%, 77.2%, and 82.2% for scopoletin, alizarin, and rutin, respectively. In conclusion, Moreira et al. (2020a) showed that HHPE allows the use of room temperature, low volumes of organic solvents, reduction of extraction time, and energetic consumption, with higher yields and high-quality final extracts.

15.2.9 Cyclodextrin-Assisted Extraction

Cyclodextrins (CDs) are cyclic oligosaccharides produced by the enzymatic modification of starch, typically containing six, seven, or eight D-(+)-glucopyranose units (α-, β- and γ-CD, respectively) linked by α-1,4 glycosidic bonds. They contain a hydrophilic outer surface and a relatively hydrophobic central cavity. This peculiar structure characteristic enables CDs to form host-guest inclusion complexes with a wide range of compounds via non-covalent forces, such as van der Waals forces, hydrophobic interactions, and hydrogen bonds. Recently, CDs are being explored to extract PC such as phenolic acids, flavonoids, and stilbenes from various types of natural sources. The use of CDs in aqueous solution as extraction media can be considered as green extraction since water is the main solvent and the existence of CD hydrophobic cavity boosts the extraction of PC due to inclusion complex formation (Korompokis et al. 2017). It has been reported that CDs have the potential to be used as an alternative to organic solvents for the extraction of PC (López-Miranda et al. 2016). The green extraction of PC using CDs provides the possibility for more effective exploitation of natural plant resources and more widespread use of PC in the food and nutraceutical industry.

The theoretical basis for the use of CDs to extract PC and the schematic of CD-assisted extraction is shown in Fig. 15.7. After the natural materials are added to the aqueous solution of CD, the liquid water would be able to penetrate the cells, resulting in disassociation and diffusion of PC from the matrix to the extracting solution (Hu et al. 2016). The released PC molecules would interact with CD present in the extraction medium and form PC-CD inclusion complexes according to their corresponding affinity (Fumic et al. 2017). As the extraction system reaches equilibrium, the PC that forms more stable complexes with CD would remain in the solution predominately, thus being extracted efficiently.

Flavonols, flavan-3-ols, stilbenes, and phenolic acids were extracted from grape pomace using β-CD by Ratnasooriya and Rupasinghe (2012). The highest extraction of total quantified phenolics was obtained for β-CD, followed by γ-CD and α-CD. Specifically, the yield of flavan-3-ols recovered by β-CD was higher than those for γ-CD and α-CD; similar recoveries of flavonols and stilbenes were

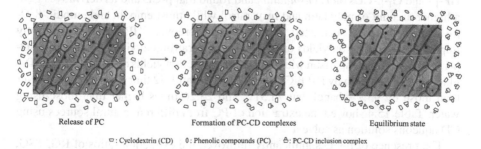

Release of PC   Formation of PC-CD complexes   Equilibrium state

▱ : Cyclodextrin (CD)   0 : Phenolic compounds (PC)   ᗡ : PC-CD inclusion complex

Fig. 15.7 The schematic of Cyclodextrins-assisted extraction of PC (reproduced with kind permission of Elsevier (Cai et al. 2018))

obtained with β-CD and γ-CD, which were higher than that for α-CD; in the case of phenolic acids, there was no significant difference observed between α-CD and β-CD. López-Miranda et al. (2016) used β-CD and HP-β-CD to recover catechin and epicatechin from red grape pomace. With respect to catechin extraction, no statistically significant differences between these two types of CDs were observed; while the HP-β-CD solution had extracted more epicatechin than β-CD. When used to extract polyphenols from pomegranate fruit, HP-β-CD resulted in a higher total phenolic content (TPC) (71.70 mg gallic acid equivalent (GAE)/g dry matter (DM)) than β-CD (58.70 mg GAE/g DM) (Diamanti et al. 2017). Similarly, the amount of 1.515 mg total flavonols/g DM was obtained under optimized conditions from apple pomace using β-CD at a concentration of 28 g/L (Mourtzinos et al. 2016). Gallic acid, protocatechuic acid, catechin, chlorogenic acid, caffeic acid, sinapic acid, epicatechin, and rutin were extracted from jujube fruit using six different CDs. The extraction efficiency of β-CD was relatively better than those of α-CD and γ-CD. Furthermore, in comparison to β-, HP-β- and NH_2-β-CDs, the extraction yields by M-β-CD were markedly higher, especially for the three flavonoids (catechin, epicatechin, and rutin) (Zhu et al. 2016).

Under optimized conditions, the total polyphenol ((luteolin glucosides, luteolin derivatives, apigenin glucosides, apigenin, rutin, and oleuropein) yield and the antiradical activity from olive leaf were 54.3 mg GAE/g DW and 352.7 μmol TE/g DW, respectively (Mourtzinos et al. 2016). The maximum extraction yields of EGCG and ECG from tea leaves were 118.7 mg/g and 54.6 mg/g, respectively, which were higher than those obtained using water and 50% ethanol.

The optimal polyphenol content (5.8 mg GAE/g DM) and antioxidant capacity (62.9 μmol TE/g DW) were obtained under optimized conditions. A polyphenol (gallic acid, 4-hydroxybenzoic acid, ferulic acid, *trans*-cinnamic acid) content of 2.6 mg GAE/g DM was reached after 5 h of extraction with β-CD, while it took 10 h with pure water to attain the same phenolic yield (Rajha et al. 2015). For the same polyphenol yield, a higher antioxidant capacity was found in the β-CD extracts compared to those obtained with 50% ethanol and water.

The β-CD extract contains resveratrol and showed the same radical scavenging activity and antioxidant capacity as the methanol extract (Mantegna et al. 2012). Gao et al. (2016) compared the extraction properties of α-, β-, γ-, M-β-, HE-β-, HP-β- and G_1-β-CDs on *P. cuspidatum* and found that β-CD and its derivatives gave a higher amount of polydatin and resveratrol while most emodin and emodin-8-*O*-β-D-glucoside were extracted with γ-CD. Moreover, HP-β-CD significantly improved the extraction yields of polydatin, resveratrol, emodin, and emodin-8-*O*-β-D-glucoside in comparison with β-CD. The extraction yields of emodin-8-*O*-β-D--glucoside and emodin by HP-β-CD were 25.40% and 1.42 times, respectively, lower than those by 95% ethanol, but 74.51% and 23.26 times higher than those using water. Table 15.4 showed the extraction of PC from different natural sources using CD aqueous solution as solvent.

The presence of β-CD significantly improved the extraction ratios of RG, TSG, emodin, nuciferine, and quercetin from Xue-Zhi-Ning, compared with aqueous extraction, especially those of TSG and emodin (Zhang et al. 2015). Similarly, the content of mangiferin in the water extract and β-CD extract were 66.4 mg/g and 72.5 mg/g DM, respectively (Mura et al. 2015). In comparison with water and ethanol, HP-β-CD resulted in an extract with both the highest total flavonoid content (daidzein, formononetin, genistein, and glycitein) and the lowest radical scavenging activity IC_{50} (Fumic et al. 2017).

In conclusion, the addition of CD to the water improves the extraction efficiency for PC and shortens the extraction time, and the obtained PC extract presents better antioxidant activity (Cai et al. 2018). Not only is the removal of CD unnecessary, but also the presence of CD favors the preparation of the PC extract in a solid-state, which are very beneficial since they will be expected to be used either to fortify foods or as a food supplement on an industrial scale. Cai et al. (2018) suggested that further studies are required to evaluate the possibility of combining CDs with other extraction methods, such as high hydrostatic pressure, pulsed electric fields, negative pressure cavitation, and pressurized liquid extraction, to extract PC from natural materials.

15.3 Spectrophotometric Analysis

The amount of phenolic compounds are measured in a variety of spectrophotometric methods. These methods are mostly limited to either a specific class of phenolic compounds or based on a single reagent. The basic idea behind all these methods is the formation of specific colored compounds from the phenolic analytes and their quantity is measured using a spectrophotometer. Aleixandre-Tudo et al. (2017) reviewed the spectrophotometric methods for the analysis of phenolic compounds in grape and wine. The selection of the author's review was based on spectrophotometry due to its ease of use as a routine analytical technique. These authors discussed the advantages and disadvantages of the reported methods. Details of some of these methods are given below.

15.3.1 Folin–Ciocalteu Index (FCI)

The colorimetric assay based on the protocol developed by Folin and Ciocalteu (1927) can be used to determine the concentration of soluble phenolics, such as anthocyanins in the example above, as well as complex phenolics such as hydrolyzable and condensed tannins. The Folin–Ciocalteu index (FCI) relies on the redox reaction of phenolic compounds with a mixture of phosphotungstic ($H_3PW_{12}O_{40}$) and phosphomolybdic ($H_3PMO_{12}O_{40}$) acids in an alkaline medium to create a blue-colored complex that can be quantified at 750 nm (Singleton and Rossi 1965). The Folin–Ciocalteu reagent (FC) reacts strongly with compounds that easily donate electrons such as mono- and dihydroxylated phenolics (Singleton et al. 1999).

Table 15.4 Extraction of PC from different natural sources using CD aqueous solution as solvent. Reproduced with kind permission of Elsevier (Cai et al. 2018)

Source	Extracted compounds	Extraction conditions	Main effects	References
Grape pomace	Flavonols, flavan-3-ols, stilbenes and phenolic acids	CD type: β-CD, CD concentration: 25 g/L, extraction technique: Shaking in a water bath (70 rpm), solid to liquid ratio: 1/20; for dry grape pomace powder, the best extraction temperature and time combination was at 72 °C for 12 h, while for grape pomace slurry, they were at 45 °C for 12 h, or at 60 °C for 6 h	Compared to the water, total quantified phenolic compounds recovered from slurry of grape pomace with aqueous solution of β-CD was 292-fold higher.	Ratnasooriya and Rupasinghe (2012)
Red grape pomace	Catechin and epicatechin	CD type: HP-β-CD, CD concentration: 50 mM (ca. 69.8 g/L), extraction temperature: 40 °C, extraction time: 90 min, solid to liquid ratio: 1/9, extraction technique: Ultrasound	The aqueous solutions of CDs could selectively extract catechin and epicatechin from grape pomace. The yield of catechin and epicatechin by using CDs was similar to that obtained with ethanol.	López-Miranda et al. (2016)
Apple pomace	Flavonols (quercetin and quercetin glycosides)	CD type: β-CD, CD concentration: 28 g/L, extraction temperature: 45 °C, extraction time: 25.6 h, solid to liquid ratio: 1/20, extraction technique: Ultrasound	The amount of 1.515 mg total flavonols/g DM was obtained under optimized conditions.	Mourtzinos et al. (2016)

(continued)

Table 15.4 (continued)

Source	Extracted compounds	Extraction conditions	Main effects	References
Pomegranate fruit	Polyphenols	CD type: HP-β-CD, CD concentration: 72 g/L, extraction temperature: 25 °C, extraction time: 363 min, solid to liquid ratio: 1/10, extraction technique: Using a Timatic semiautomatic extractor (solvent capacity: 1 L, compression time: 5 min, decompression time: 6 min, number of circles: 33, minimum pressure: 6 bar)	The addition of HP-β-CD to the water improved the total phenol content and the radical scavenging activity of pomegranate extracts.	Diamanti et al. (2017)
Jujube fruit	Gallic acid, protocatechuic acid, catechin, chlorogenic acid, caffeic acid, sinapic acid, epicatechin and rutin	CD type: M-β-CD, CD concentration: 40 g/L, extraction temperature: 60 °C, extraction time: 60 min, solid to liquid ratio: 1/5, extraction technique: Ultrasound	A better extraction efficiency of analytes was achieved by M-β-CD aqueous solution when compared with that by 70% (v/v) methanol.	Zhu et al. (2016)
Olive leaf	Polyphenols (luteolin glucosides, luteolin derivatives, apigenin glucosides, apigenin, rutin and oleuropein)	CD type: HP-β-CD, CD concentration: 70 g/L, glycerol content: 60% (w/v), extraction temperature: 60 °C, extraction time: 3 h, solid to liquid ratio: 1/50, extraction technique: Stirring at 600 rpm	Under optimized conditions, the total polyphenol yield and the antiradical activity were 54.33 mg GAE/g DW and 352.72 μmol TE/g DW, respectively.	Mourtzinos et al. (2016)
Tea leaves	EGCG and ECG	CD type: β-CD, CD concentration: 25 g/L, extraction temperature: 60 °C, extraction time: 60 min, solid to liquid ratio: 1/20, extraction technique: Ultrasound (300 W)	The maximum extraction yields of EGCG and ECG were 118.7 and 54.6 mg/g, respectively, which were higher than those obtained using water and 50% ethanol	Mourtzinos et al. (2016)

(continued)

Table 15.4 (continued)

Source	Extracted compounds	Extraction conditions	Main effects	References
Vine shoot cultivars	Polyphenols (gallic acid, 4-hydroxybenzoic acid, ferulic acid, *trans*-cinnamic acid)	CD type: β-CD, CD concentration: 37.7 g/L, extraction time: 48 h, extraction temperature: 66.6 °C, solid to liquid ratio: 1/20	The optimal polyphenol content (5.8 mg GAE/g DM) and antioxidant capacity (62.92 µmol TE/g DW) were obtained under optimized conditions. A polyphenol content of 2.6 mg GAE/g DM was reached after 5 h of extraction with β-CD, while it took 10 h with pure water to attain the same phenolic yield. For the same polyphenol yield, a higher antioxidant capacity was found in the β-CD extracts compared to those obtained with 50% ethanol and water.	Rajha et al. (2015)
Polygonum cuspidatum	Resveratrol	CD type: β-CD, CD concentration: 15 g/L, extraction temperature: 20 °C, extraction time: 1 h, ratio of solid to liquid: 1/50, extraction technique: Ultrasound (100 W in the first 5 min, then 50 W for 55 min)	The β-CD extract showed the same radical scavenging activity and antioxidant capacity as the methanol extract.	Mantegna et al. (2012)

(continued)

Table 15.4 (continued)

Source	Extracted compounds	Extraction conditions	Main effects	References
P. cuspidatum	Polydatin, resveratrol, emodin and emodin-8-*O*-β-D--glucoside	CD type: HP-β-CD, CD concentration: 100 g/L, extraction temperature: 80 °C, extraction time: 1 min, solid to liquid ratio: 1/50, extraction technique: Ultrasound-microwave-assisted extraction	The amount of polydatin and resveratrol in HP-β-CD extracts were 22.82% and 12.90%, respectively, higher than those in 95% ethanol, and 1.17 times and 1.55 times higher than those extracted by water, respectively. The extraction yields of emodin-8-*O*-β-D--glucoside and emodin by HP-β-CD were 25.40% and 1.42 times, respectively, lower than those by 95% ethanol, but 74.51% and 23.26 times higher than those using water.	Gao et al. (2016)
Xue-Zhi-Ning	RG, TSG, emodin, nuciferine and quercetin	CD type: β-CD, CD concentration: 2 g/L, extraction time: 2 h, number of extractions: 3, solid to liquid ratio: 1/25, extraction technique: Reflux extraction	The presence of β-CD significantly improved the extraction ratios of RG, TSG, emodin, nuciferine and quercetin, compared with aqueous extraction, especially those of TSG and emodin.	Zhang et al. (2015)

(continued)

Table 15.4 (continued)

Source	Extracted compounds	Extraction conditions	Main effects	References
Mangifera indica stem bark	Polyphenols (mainly mangiferin)	CD type: β-CD, CD concentration: 15 g/L, extraction temperature: 25 °C, extraction time: 1 h, extraction technique: Ultrasound	The content of mangiferin in the water extract and β-CD extract were 66.4 and 72.5 mg/g DM, respectively.	Mura et al. (2015)
Medicago sativa	Flavonoids (daidzein, formononetin, genistein, and glycitein)	CD type: HP-β-CD, CD concentration: 20 mM (*ca.* 27.92 g/L), extraction temperature: 20 °C, extraction time: 45 min, solid to liquid ratio: 2/25, extraction technique: Ultrasound (432 W)	In comparison with water and ethanol, HP-β-CD resulted in an extract with both the highest total flavonoid content and the lowest radical scavenging activity IC_{50}.	Fumic et al. (2017)
Sideritis scardica	Polyphenols (mainly EGCG, chlorogenic acid and myricetin)	CD type: HP-β-CD, CD concentration: 27 mM (*ca.* 37.692 g/L), solid to liquid ratio: 1/10, extraction technique: Boiled for 3 min and remained in warm water for extra 5 min	The presence of HP-β-CD enhanced the total phenol content and the radical scavenging activity of *Sideritis scardica* extracts.	Korompokis et al. (2017)

CD cyclodextrin, *DM* dry matter, *GAE* gallic acid equivalent, *TE* trolox equivalent, *EGCG* epigallocatechin gallate, *ECG* epicatechin gallate, *RG* rubrofusarin gentiobioside, *TSG* 2,3,5,4′-tetrahydroxy-stilbene-2-O-β-D-glucoside

Briefly, 0.5 mL of FC reagent is added to 1 mL of a diluted extract (1/5) mixed with 5 mL of distilled water. Then 2 mL of a 20% (m/v) Na_2CO_3 solution is added into the mixture. Finally, 2.4 mL of distilled water is incorporated into the test tube. Samples need to be incubated for 30 min before the absorbance is measured. The order of the additions needs to be strictly followed due to the instability of the FC reagent under alkaline conditions (Singleton and Rossi 1965). A sample blank with water instead of a sample is used to account for background interference. Moreover, if the absorbance is not placed around 0.9 AU, the dilution factor needs to be modified (Aleixandre-Tudo et al. 2017).

The main drawback of the method is due to the ability of some of the food components to donate electrons, causing overestimation of the reducing properties and making the direct comparison of samples with different phenolic profiles difficult (De Beer et al. 2004). The results of this method are commonly presented as milligrams per liter of gallic acid, although the comparison with the total phenolic index (TPI) by using a multiplying factor of 20 is also possible (i.e., A750 nm multiplied by the dilution factor times 20 will provide values comparable to the TPI). Besides, it is highly recommended to obtain the FC reagent from laboratory suppliers as it can be difficult to prepare. The validation of the FC method has been reported by Singleton et al. (1999). A coefficient of variation of 3.1% (N = 12) and accurate recovery tests were reported. Additionally, a positive correlation between the FCI and the total area under the HPLC chromatogram at 280 nm has been reported (De Beer et al. 2004).

Sánchez-Rangel et al. (2013) reviewed the Folin–Ciocalteu (FC) assay for total phenolic content (TPC) determinations and describes different approaches to improve its specificity. The author showed that when determining TPC in plant food extracts, the presence of reducing interferants such as ascorbic acid produces inaccurate estimations of TPC values. Different methodologies have been proposed to improve the specificity of the FC assay. These methodologies include: (i) the use of solid-phase extraction (SPE) cartridges to separate interferants from phenolics; (ii) the calculation of a corrected TPC value based on the ascorbic acid-reducing activity present in the extract; and (iii) the pre-treatment of extracts with oxidative agents before TPC quantification.

15.3.2 Total Phenolic Index (TPI)

The measurement of phenolic compounds using UV spectrophotometry was first proposed in the late 1950s by Ribéreau-Gayon. The absorbance at 280 nm was proposed as the best predictor for the phenolic content in wines because of the characteristic sharp absorbance peak at this wavelength (Aleixandre-Tudo et al. 2017). The ability of the aromatic rings of the majority of the phenolic substances to absorb UV light is used for the estimation of the TPI. A sample extract is diluted (50 times, in the case of wine) and measured at 280 nm. The TPI is calculated as the A280 nm times the dilution factor (DF). The total phenolic values are commonly reported as milligrams per liter of gallic acid by using a calibration curve. Although this is a simple, fast, and reproducible method, its main drawback relates to the UV light absorption ability of other wine constituents (nucleotides, proteins, peptides, and amino acids), which can lead to overestimation. However, Somers and Evans (1974) observed that the interference caused by these compounds was statistically constant, accounting for 4 index units, a value that can be subtracted from the index. The effect of other molecules, such as cinnamic acids and chalcones (with no maximum absorption at 280 nm), can be considered negligible as they are found in wine at very low concentrations.

15.3.3   Determination of Flavonoid Contents

15.3.3.1   Hydrochloric Acid Method

The estimation of total anthocyanin content in sample solutions containing other phenolic material is possible because of the characteristic absorption band exhibited by anthocyanins in the 490–550 nm region of the visible spectra (Giusti and Wrolstad 2001). The simpler and shorter methodology for the estimation of the total anthocyanin content makes use of hydrochloric acid (HCl); that is, the wine is diluted in a 1 M hydrochloric acid solution, and after a waiting period, the absorbance at 520 nm is recorded (Cliff et al. 2007). The values can be presented as absorbance units (AU) or as milligrams per liter of malvidin-3-glucoside equivalents by using the molecular weight and the molar absorptivity values. Reproducibility tests showed CV <2% with an estimation of uncertainty of 5% based on possible variables such as operator or daily operation.

15.3.3.2   pH Differential Method

A reliable method that exploits the structural transformations of the anthocyanin chromophore as a function of pH was developed by Giusti and Wrolstad (2001). The red-colored flavylium form predominates at a very low pH (pH 1), whereas the colorless hemiketal form is found as the predominant anthocyanin at pH 4.5 (Fig. 15.8). The accuracy of the method relies first on the use of the absorbance at the λmax, instead of quantifying at the fixed wavelength of 520 nm. Besides, the absorbance at 700 nm was included to avoid any possible interference caused by light scattering due to haze formation.

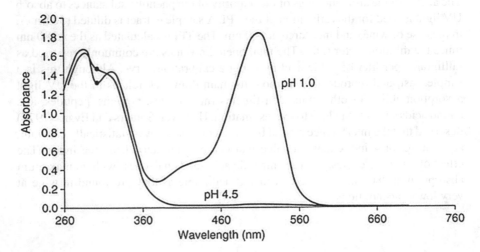

Fig. 15.8 Spectral characteristics of purified radish anthocyanins in pH 1.0 and pH 4.5 buffers

The method also indicates that the total anthocyanin content has to be calculated using the molecular weight (MW) and the molar absorption coefficient (ε) that correspond to the predominant anthocyanin in the sample, that is, malvidin-3-glucoside. Briefly, the samples need to be diluted in the 0.025 M KCl solution adjusted to pH 1 until the λmax is within the linear range of the instrument. A separate sample needs to be prepared with the same dilution factor in the 0.4 M $C_2H_3NaO_2$ solution adjusted to pH 4.5. After 15 min, the Aλmax and A700 nm are measured against a water blank. The concentration is calculated as follows:

$$\text{Monomeric anthocyanins}\left(\frac{mg}{L}\right) = \left(A \times MW \times DF \times 1000\right)/\left(\mu \times 1\right)$$

Where $A = (\lambda_{max} - A_{700\ nm})_{pH\ 1} - (\lambda_{max} - A_{700\ nm})_{pH\ 4.5}$, MW = molecular weight (493.2 g/mol), DF = dilution factor, and ε = molar extinction coefficient (28,000 L/cm·mol).

15.3.3.3 Bisulfite Degradation Method

This method allows for the estimation of additional indices by exploiting the bisulfite bleaching properties of the anthocyanins. These indices include pigment degradation, polymer color, and browning measurements, which makes this methodology useful to monitor color composition in grape extracts and wines (Giusti and Wrolstad 2001). Briefly, 100 μL of a 20% (w/v) $K_2S_2O_5$ solution is added to 1.4 mL of an appropriately diluted sample. A control sample is also prepared, but with the addition of water instead of $K_2S_2O_5$. After 15 min of equilibration, the A420 nm, Aλmax, and A700 nm are measured. The color density and polymeric color are determined as a function of anthocyanins.

15.3.3.4 Co-Pigmentation Assay

Copigmentation is based on weak associations among red-colored compounds (anthocyanins) and other mostly colorless substances including a wide range of phenolic compounds. The effect of these associations is observed by an increase in the absorbance at the λmax (hyperchromic effect) and a displacement of the λmax (bathochromic shift) toward higher wavelengths (bluish coloration). The extent of the co-pigmentation effect depends on many factors including the nature of the anthocyanin and the cofactor, the presence of metals, pH, and ethanol (Dangles et al. 1993). The co-pigmentation assay developed by Boulton (2001) has also received intense attention, as it is the only anthocyanin method that includes the measurement of the color that corresponds to co-pigmented anthocyanins. The method relies on the dissociation of the co-pigmented forms after dilution at constant pH to ensure the accurate direct comparison of the wine color. However, the adjustment of the pH individually for each sample increases the time of analysis and makes it very tedious if a large number of samples need to be analyzed.

15.3.4 Determination of Proanthocyanins or Condensed Tannins

15.3.4.1 Acid Hydrolysis Assay

A number of different chemical properties have been exploited to quantify the total content of proanthocyanidins or tannins in grape extracts and wines. The quantification of this specific group of phenolics has been challenging for researchers over the past years, as these compounds are of a very diverse nature, a fact that has a strong impact on the ability of the analytical principles to estimate their concentrations. The first proposed spectrophotometric method, based on acid hydrolysis, relies on the ability of the proanthocyanidins to be transformed into carbocations, which are later partially converted into anthocyanidins when heated in an acid medium (Bate–Smith reaction) (Aleixandre-Tudo et al. 2017). Briefly, 1 mL of wine diluted 50 times is added to a test tube with 0.5 mL of distilled water and 3 mL of 12 N HCl (treatment sample). The test tube is then heated to 100 °C in a water bath for 30 min. After the sample has been cooled, 0.5 mL of pure EtOH is added. Additionally, the control sample is also prepared but with no heating. The anthocyanidins are then quantified at 550 nm using the ε of the cyanidin-3-glucoside corrected to five grams per liter.

$$\text{Tannin}\left(\frac{g}{L}\right) = \left(\text{Treated samples} - \text{control samples}\right) \times 19.33$$

15.3.4.2 BSA Tannin Assay

The ability of tannins to precipitate through interactions with proteins has also been exploited as a quantification principle. The method was initially reported by Hagerman and Butler (1978). The method relies on the interactions of tannins with bovine serum albumin, precipitation from the wine matrix, and later re-dissolution in a buffer solution. The absorbance at 510 nm is recorded after a color reaction with ferric chloride (Aleixandre-Tudo et al. 2017).

One milliliter of BSA solution (1 mg BSA/mL dissolved in a 0.2 M acetic acid and 0.17 M NaCl buffer adjusted to pH 4.9) is added to 0.5 mL of the properly diluted sample. The dilution is made with a 12% v/v EtOH and 5 g/L tartaric acids adjusted to pH 3. After 15 min, the supernatant obtained after centrifugation is discarded. The pellet is then washed twice with 1 mL of the solution adjusted at pH 4.9. After the addition of 0.25 mL of the same solution, the sample is centrifuged for

1 min. The supernatant is again discarded, and the pellet is re-dissolved by adding 0.875 mL of a TEA buffer at pH 7 or 8. After the measurement of the A510 nm background, 0.125 mL of 10 mM ferric chloride in 10 mM aqueous HCl was added to the sample. The A510 nm FeCl was measured after 10 min. The total tannin content is calculated as follows and expressed as milligrams per liter of catechin equivalents:

$$\text{Tannin}\left(\frac{g}{L}\right) = A510^{FeCl} - 0.875 \times A510^{Background}$$

The method was modified for monomeric anthocyanins, small polymeric anthocyanins, and large polymeric pigments. Due to the complexity of the tannin mixtures, it is not possible to obtain true analytical standards. Classical validation methods are thus difficult to implement, this being why the available validation results focus mainly on repeatability tests.

15.3.4.3 Cellulose Precipitable Tannin (MCP) Assay

The use of other polymers to precipitate tannins has also been utilized in the methylcellulose precipitable tannin assay (MCP). The assay was developed and validated by Sarneckis et al. (2006). This method relies on polymer–tannin interactions in the presence of ammonium sulfate, leading to an insoluble complex that is separated by centrifugation.

Briefly, a treatment sample is prepared by adding 300 μL of an MCP solution (0.04% w/v; 1500 cP viscosity at 2%) to 25 μL of wine. After 2–3 min, 200 μL of a saturated solution of $(NH_4)_2SO_4$ and 475 μL of distilled water are added. A control sample is also prepared, but with distilled water (775 μL) instead of MCP solution. After 10 min, the samples are centrifuged for 5 min, and the tannin content is obtained by comparing the $A_{280\ nm}$ control and $A_{280\ nm}$ treatment referenced to epicatechin equivalents. This precipitation method appears to be simpler than the BSA tannin assay as only two simple solutions that do not require pH adjustment are needed (MCP and the $(NH_4)_2SO_4$ solutions).

15.4 Potentiometric Methods

The ionization constant is an important physicochemical parameter of a substance, and the knowledge of this parameter is of fundamental importance in a wide range of applications and research areas. Potentiometry is one of the important methods for the determination of the dissociation constant of an acid. Potentiometry is a highly accurate, and reproducible method of analysis. However, the analytical methods of Potentiometry for the determination of total phenolic compounds are limited. The potentiometric titration was used successfully for the determination of phenolic groups in isolated lignins (Sarkanen and Schuerch 1955).

A potentiometric technique had been reported which is suitable for the estimation of carboxylic acids in a 'neutral' solvent, such as a mixture of benzene and methanol, and the estimation of weaker acids, such as phenols, in basic solvents such as n-butylamine or a mixture of benzene and ethylene-diamine (Mathews and Welch 1958). Huma et al. (1999) reported a potentiometric titration method for the determination of phenol in an aqueous medium. The method incorporates titrants in a concentration range of several orders of magnitude greater than that of phenol in the aqueous phase. Figure 15.9 shows the predominant structural forms of anthocyanins present at different pH levels.

To overcome the lack of information related to the acid-base equilibria of this kind of compound in water-methanol medium, the pKa values of some hydroxybenzoic and hydroxycinnamic acids in water and methanol-water binary mixtures have been determined using the potentiometric method (Erdemgil et al. 2007). The phenolic compounds were seven hydroxybenzoic acids i.e., gallic, protocatechuic, gentisic, p-hydroxybenzoic, vanillic, isovanillic, and syringic acids, and four hydroxycinnamic acids i.e., caffeic, p-coumaric, ferulic, and sinapinic acids. The results showed a similar trend for the pKa values of all the studied compounds, as they increased with increasing concentration of organic modifier, which allows a linear relationship between pKa values and mole fraction of methanol to be obtained. The novel sensor membranes were prepared based on the use

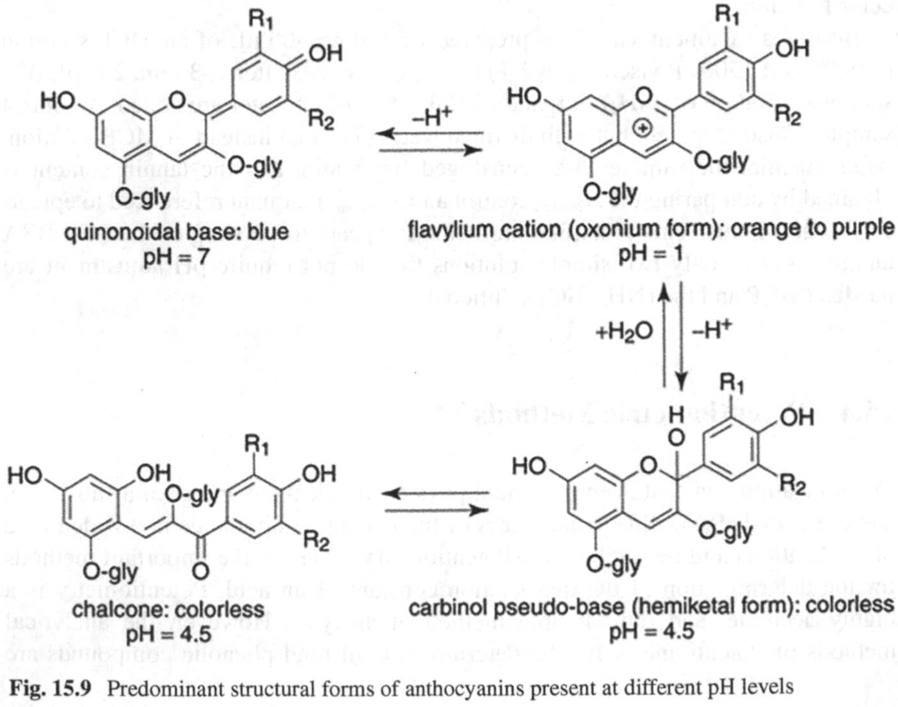

Fig. 15.9 Predominant structural forms of anthocyanins present at different pH levels

of either Cu-neocuproin/2,6-dichlorophenolindo-phenolate, or methylene blue/2,6-dichlorophenolindophenolate ion association complexes in a plasticized PVC matrix (Elbehery et al. 2019). The membrane sensors were highly selective for the determination of total phenolic contents.

The antioxidant capacity of chicoric acid or an ethanol/water-extract of Echinacea flowers was determined by potentiometric and UV-Vis absorption spectrophotometric titrations with ABTS˙ + radical cations as the oxidizing probe (Yi et al. 2020). The electrochemistry methods coupled with spectrophotometry can conveniently and reliably provide important quantitative and qualitative information on redox chemistry and are expected to find wider applications in accurately evaluating the redox activities of many other natural antioxidants.

15.5 Electrochemical Methods

Electrochemical techniques emerge as an alternative strategy for a quick, precise, and economic measurement of the AC of foods and beverages when we consider some of the characteristics of conventional spectrophotometric methods for determining their AC, like long analyses and sample preparation time, expensive equipment, use of expensive reagents unfriendly to the environment, and undefined reaction times (Hoyos-Arbeláez et al. 2017).

The phenolic antioxidants' electrochemical reactions have been investigated, and a relationship between their electron transfer properties and radical-scavenging activity has been found in many polyphenol groups, which include flavonoids, xanthonoids, phenolic acids, phenolic acid esters, and anthraquinones. The inherent redox-active phenol moiety, common to almost all these compounds, is what confers their antioxidant properties (Chiorcea-Paquim et al. 2020). All polyphenols present a common redox behavior, electrochemical oxidation occurring at the OH groups and influenced by the chemical substituents linked to the aromatic rings (e.g., -OCH_3, sugar, etc.). Among other factors, the pH of the environment is the most important, directly affecting the polyphenols' antioxidant capacity, redox behavior, and oxidation product formation (René et al. 2010). The main electrochemical methods used to elucidate the natural polyphenols' redox behavior are cyclic voltammetry (CV), differential pulse voltammetry (DPV), and square wave voltammetry (SWV), in stationary solution, and HPLC separation with electrochemical detection (HPLC–EC), in flow analysis.

15.5.1 Cyclic Voltammetry

Cyclic voltammetry (CV) is the most widely used technique for acquiring qualitative information about the properties and characteristics of the electrochemical processes. This technique provides considerable information on the thermodynamics of redox processes and kinetics of heterogeneous electron-transfer reactions, as well as on coupled chemical reactions or adsorption processes. Cyclic voltammetry consists of scanning linearly the potential of a stationary working electrode (WE) (in an unstirred solution) using a triangular potential waveform (Hoyos-Arbeláez et al. 2017). The potentiostat/galvanostat measures the current generated by the redox process resulting in a plot of current vs applied potential. Based on phenol oxidation, some basic principles that might be applied to all phenolic compounds can be highlighted (Enache and Oliveira-Brett 2011; Chiorcea-Paquim et al. 2020):

- Phenol is oxidized, in one-electron, one-proton irreversible step, to a phenoxy radical.
- As in homogeneous reactions, the activation energy of heterogeneous electrochemical reactions is driven by the stability of the intermediates. Therefore, in the case of phenol anodic oxidation, it depends on the stability of the electrogenerated phenoxy radical.
- The unstable phenoxy radical coexists in three resonant forms. The ortho- and para-phenoxy radicals have a larger spin density, their stability being far higher than the meta-radical that suffers secondary chemical reactions, that is, hydroxylation.
- The presence of an additional electroactive −OH group at the ortho- and para-positions leads to a two-electron and two-proton pH-dependent reversible process that, due to higher stability, appears at less positive potentials, than that of the meta-di-phenol or mono-phenol.
- The presence of non-electroactive substituents produces a small oxidation peak potential shift (generally not higher than 0.20 V); a greater shift occurs when the substituents are linked at the ortho- and para-positions.
- The electron donor substituents make the oxidation process easier, whereas electron-withdrawing substituents lead to a shift to higher anodic peak potential.
- As is usual in the electrochemical oxidation of organic species, the redox process often involves the participation of protons, thus, the higher the pH, the easier the electron loss.

The flavonoids' electroactivity and in vitro antioxidant capacity are mainly determined by the -OH groups' number and position on the aromatic rings. The -OH groups in para- and ortho-positions are oxidized at lower potentials than the -OH groups in the meta-position. Also, the mono-substituted -OH group on the A-ring requires a higher potential to be oxidized than the -OH group on the B-ring. Therefore, the electron donor ability of flavonoids is higher, in the flavonoids with a catechol moiety on B-ring, such as quercetin, quercitrin, rutin, luteolin, morin, and so on. Thus, since the deprotonation is easier, the resorcinol moiety on the B-ring oxidizes at a close, but slightly lower potential, than the same group on the A-ring. This relevant effect is illustrated in the electrooxidation of morin, a tetrahydroxy flavonoid, which occurs with the transfer of one-electron and one-proton (Chiorcea-Paquim et al. 2020).

An amperometric sensor for apigenin was developed, using nickel nanoparticles and activated carbon SPEs (Wang et al. 2017). Electrochemical sensors for the detection and quantification of flavonoids, using multi-walled CNT, and the electrochemical behavior of kaempferol and quercetin, were also investigated (Liang et al. 2017).

15.5.2 Differential Pulse Voltammetry

Differential pulse voltammetry (DPV) is a technique where fixed-magnitude pulses are superimposed on a linear potential ramp. The typical potential-time excitation signal for DPV, where the response current is sampled twice, just before the pulse application and again late in the pulse life. The first current is subtracted from the second and the current difference is plotted versus the applied potential. The pulse length usually takes values between 40 and 60 ms, and the interval between pulses varies from 0.5 to 5 s. By sampling the current just before the potential is changed, the amount of capacitive current (from the charging of the electrochemical double layer) is minimized in the current measurement (Hoyos-Arbeláez et al. 2017). DPV has been applied to antioxidant quantification due to the good detection limits achieved through it. Gao et al. (2015) reported a detection limit by using DPV of 1.5×10^{-7} M for gallic Acid by using a modified electrode. Glassy carbon, carbon paste, and modified glassy carbon are some of the most used electrodes. The first two kinds are reported in the study of wine samples for catechin quantitation with detection limits of 1.77 mg L^{-1} (Šeruga et al. 2011), as well as for the analysis of anthocyanins in the same drink, directly over a simple dilution with no need of buffer addition (Aguirre et al. 2010).

15.5.3 Square-Wave Voltammetry

Square Wave Voltammetry (SWV) is a large-amplitude differential technique in which a wave form composed of a symmetrical square wave, superimposed on a base staircase potential, is applied to the working electrode (Hoyos-Arbeláez et al. 2017). SWV has been used to identify and quantify antioxidants within a different food and beverage matrix, obtaining thereby very good detection limits as reported by Novak et al. (2010). They achieved a detection limit of 40 nM for epigallocatechin gallate in green and black teas.

15.5.4 Amperometric Method

The amperometric method or Chronoamperometry (CA) involves stepping the potential of the working electrode from a value at which no faradaic reaction occurs to a potential at which the surface concentration of the electroactive species is effectively zero. The response current is recorded as a function of time (Hoyos-Arbeláez et al. 2017). Glassy Carbon Electrode (GCE), Boron-doped Diamond (BDD), Multi-walled carbon nanotubes (MWCNT), and the mixture of these electrodes have been utilized for the determination of phenolic antioxidants.

15.6   Study Questions

• What are the different extraction techniques used for the phenolic compounds in foods?
• How phenolic compounds are analyzed using spectrophotometric methods?
• Describe potentiometric methods for the analysis of phenolic compounds in foods.
• Discuss the applications of electrochemical methods for the analysis of phenolic antioxidants.

References

Abbasi H, Rezaei K, Emamdjomeh Z, Mousavi SME (2008) Effect of various extraction conditions on the phenolic contents of pomegranate seed oil. Eur J Lipid Sci Technol 110(5):435–440

Aguirre MJ, Chen YY, Isaacs M, Matsuhiro B, Mendoza L, Torres S (2010) Electrochemical behaviour and antioxidant capacity of anthocyanins from Chilean red wine, grape and raspberry. Food Chem 121(1):44–48

Ahmed M, Ramachandraiah K, Jiang G-H, Eun JB (2020) Effects of ultra-sonication and agitation on bioactive compounds and structure of amaranth extract. Foods 9(8):1116

Ajila CM, Brar SK, Verma M, Tyagi RD, Godbout S, Valéro JR (2011) Extraction and analysis of polyphenols: recent trends. Crit Rev Biotechnol 31(3):227–249

Aleixandre-Tudo JL, Buica A, Nieuwoudt H, Aleixandre JL, du Toit W (2017) Spectrophotometric analysis of phenolic compounds in grapes and wines. J Agric Food Chem 65(20):4009–4026

Aliaño-González MJ, Ferreiro-González M, Espada-Bellido E, Carrera C, Palma M, Ayuso J, Barbero GF, Álvarez JÁ (2020) Extraction of anthocyanins and total phenolic compounds from açai (Euterpe oleracea mart.) using an experimental design methodology. Part 3: microwave-assisted extraction. Agronomy 10(2):179

Altuner EM, Tokuşoğlu Ö (2013) The effect of high hydrostatic pressure processing on the extraction, retention and stability of anthocyanins and flavonols contents of berry fruits and berry juices. Int J Food Sci Technol 48(10):1991–1997

Alvarez MV, Cabred S, Ramirez CL, Fanovich MA (2019) Valorization of an agroindustrial soybean residue by supercritical fluid extraction of phytochemical compounds. J Supercrit Fluids 143:90–96

Ana C-C, Jesús P-V, Hugo E-A, Teresa A-T, Ulises G-C, Neith P (2018) Antioxidant capacity and UPLC–PDA ESI–MS polyphenolic profile of Citrus aurantium extracts obtained by ultrasound assisted extraction. J Food Sci Technol 55(12):5106–5114

Aresta A, Cotugno P, De Vietro N, Massari F, Zambonin C (2020) Determination of polyphenols and vitamins in wine-making by-products by supercritical fluid extraction (SFE). Anal Lett 53(16):2585–2595

Balasubramaniam VG, Ayyappan P, Sathvika S, Antony U (2019) Effect of enzyme pretreatment in the ultrasound assisted extraction of finger millet polyphenols. J Food Sci Technol 56(3):1583–1594

Bleve M, Ciurlia L, Erroi E, Lionetto G, Longo L, Rescio L, Schettino T, Vasapollo G (2008) An innovative method for the purification of anthocyanins from grape skin extracts by using liquid and sub-critical carbon dioxide. Sep Purif Technol 64(2):192–197

Boulton R (2001) The copigmentation of anthocyanins and its role in the color of red wine: a critical review. Am J Enol Viticult 52(2):67–87

Briones-Labarca V, Giovagnoli-Vicuña C, Cañas-Sarazúa R (2019) Optimization of extraction yield, flavonoids and lycopene from tomato pulp by high hydrostatic pressure-assisted extraction. Food Chem 278:751–759

Caballero AS, Romero-García JM, Castro E, Cardona CA (2020) Supercritical fluid extraction for enhancing polyphenolic compounds production from olive waste extracts. J Chem Technol Biot 95(2):356–362

Cai R, Yuan Y, Cui L, Wang Z, Yue T (2018) Cyclodextrin-assisted extraction of phenolic compounds: current research and future prospects. Trends Food Sci Technol 79:19–27

Chanioti S, Siamandoura P, Tzia C (2016) Evaluation of extracts prepared from olive oil by-products using microwave-assisted enzymatic extraction: effect of encapsulation on the stability of final products. Waste Biomass Valori 7(4):831–842

Charoensiddhi S, Franco C, Su P, Zhang W (2015) Improved antioxidant activities of brown seaweed Ecklonia radiata extracts prepared by microwave-assisted enzymatic extraction. J Appl Phycol 27(5):2049–2058

Chemat F, Zill e H, Khan MK (2011) Applications of ultrasound in food technology: processing, preservation and extraction. Ultrason Sonochem 18 (4):813–835

Chen F, Sun Y, Zhao G, Liao X, Hu X, Wu J, Wang Z (2007) Optimization of ultrasound-assisted extraction of anthocyanins in red raspberries and identification of anthocyanins in extract using high-performance liquid chromatography–mass spectrometry. Ultrason Sonochem 14(6):767–778

Cheng X-L, Wan J-Y, Li P, Qi L-W (2011) Ultrasonic/microwave assisted extraction and diagnostic ion filtering strategy by liquid chromatography–quadrupole time-of-flight mass spectrometry for rapid characterization of flavonoids in *Spatholobus suberectus*. J Chromatogr A 1218(34):5774–5786

Chiorcea-Paquim A-M, Enache TA, De Souza GE, Oliveira-Brett AM (2020) Natural phenolic antioxidants electrochemistry: towards a new food science methodology. Compr Rev Food Sci Food Saf 19(4):1680–1726

Ciulu M, Cadiz-Gurrea ML, Segura-Carretero A (2018) Extraction and analysis of phenolic compounds in Rice: a review. Molecules 23(11):2890

Cliff MA, King MC, Schlosser J (2007) Anthocyanin, phenolic composition, colour measurement and sensory analysis of BC commercial red wines. Food Res Int 40(1):92–100

Corbin C, Fidel T, Leclerc EA, Barakzoy E, Sagot N, Falguiéres A, Renouard S, Blondeau J-P, Ferroud C, Doussot J, Lainé E, Hano C (2015) Development and validation of an efficient ultrasound assisted extraction of phenolic compounds from flax (*Linum usitatissimum* L.) seeds. Ultrason Sonochem 26:176–185

Corrales M, Toepfl S, Butz P, Knorr D, Tauscher B (2008) Extraction of anthocyanins from grape by-products assisted by ultrasonics, high hydrostatic pressure or pulsed electric fields: a comparison. Innovative Food Sci Emerg Technol 9(1):85–91

Dangles O, Saito N, Brouillard R (1993) Anthocyanin intramolecular copigment effect. Phytochemistry 34(1):119–124

Dars AG, Hu K, Liu Q, Abbas A, Xie B, Sun Z (2019) Effect of thermo-sonication and ultra-high pressure on the quality and phenolic profile of mango juice. Foods 8(8):298

Das PR, Eun J-B (2018) A comparative study of ultra-sonication and agitation extraction techniques on bioactive metabolites of green tea extract. Food Chem 253:22–29

De Beer D, Harbertson JF, Kilmartin PA, Roginsky V, Barsukova T, Adams DO, Waterhouse AL (2004) Phenolics: a comparison of diverse analytical methods. Am J Enol Viticult 55(4):389–400

Delgado AM, Issaoui M, Chammem N (2019) Analysis of Main and healthy phenolic compounds in foods. J AOAC Int 102(5):1356–1364

Diamanti AC, Igoumenidis PE, Mourtzinos I, Yannakopoulou K, Karathanos VT (2017) Green extraction of polyphenols from whole pomegranate fruit using cyclodextrins. Food Chem 214:61–66

DiNardo A, Brar HS, Subramanian J, Singh A (2019) Optimization of microwave-assisted extraction parameters and characterization of phenolic compounds in yellow European plums. Can J Chem Eng 97(1):256–267

Drevelegka I, Goula AM (2020) Recovery of grape pomace phenolic compounds through optimized extraction and adsorption processes. Chem Eng Process 149:107845

Dzah CS, Duan Y, Zhang H, Wen C, Zhang J, Chen G, Ma H (2020) The effects of ultrasound assisted extraction on yield, antioxidant, anticancer and antimicrobial activity of polyphenol extracts: a review. Food Biosci 35:100547

Elbehery NHA, Amr AE-GE, Kamel AH, Elsayed EA, Hassan SSM (2019) Novel potentiometric 2,6-Dichlorophenolindo-phenolate (DCPIP) membrane-based sensors: assessment of their input in the determination of total phenolics and ascorbic acid in beverages. Sensors (Basel) 19(9):2058

Enache TA, Oliveira-Brett AM (2011) Phenol and Para-substituted phenols electrochemical oxidation pathways. J Electroanal Chem 655(1):9–16

Erdemgil FZ, Şanli S, Şanli N, Özkan G, Barbosa J, Guiteras J, Beltrán JL (2007) Determination of pKa values of some hydroxylated benzoic acids in methanol–water binary mixtures by LC methodology and potentiometry. Talanta 72(2):489–496

Espinosa-Pardo FA, Nakajima VM, Macedo GA, Macedo JA, Martínez J (2017) Extraction of phenolic compounds from dry and fermented orange pomace using supercritical CO_2 and cosolvents. Food Bioprod Process 101:1–10

Ferrentino G, Giampiccolo S, Morozova K, Haman N, Spilimbergo S, Scampicchio M (2020) Supercritical fluid extraction of oils from apple seeds: process optimization, chemical characterization and comparison with a conventional solvent extraction. Innovative Food Sci Emerg Technol 64:102428

Folin O, Ciocalteu V (1927) On tyrosine and tryptophane determinations in proteins. J Biol Chem 73(2):627–650

Fumic B, Koncic MZ, Jug M (2017) Therapeutic potential of Hydroxypropyl-beta-Cyclodextrin-based extract of *Medicago sativa* in the treatment of Mucopolysaccharidoses. Planta Med 83(01–02):40–50

Gallego R, Bueno M, Herrero M (2019) Sub- and supercritical fluid extraction of bioactive compounds from plants, food-by-products, seaweeds and microalgae – an update. TrAC Trends Anal Chem 116:198–213

Gao F, Zheng D, Tanaka H, Zhan F, Yuan X, Gao F, Wang Q (2015) An electrochemical sensor for gallic acid based on Fe_2O_3/electro-reduced graphene oxide composite: estimation for the antioxidant capacity index of wines. Mater Sci Eng C 57:279–287

Gao F, Zhou T, Hu Y, Lan L, Heyden YV, Crommen J, Lu G, Fan G (2016) Cyclodextrin-based ultrasonic-assisted microwave extraction and HPLC-PDA-ESI-ITMSn separation and identification of hydrophilic and hydrophobic components of *Polygonum cuspidatum*: a green, rapid and effective process. Ind Crop Prod 80:59–69

Ghafoor K, Al-Juhaimi FY, Choi YH (2012) Supercritical fluid extraction of phenolic compounds and antioxidants from grape (*Vitis labrusca* B.) seeds. Plant Foods Hum Nutr 67(4):407–414

Giusti MM, Wrolstad RE (2001) Characterization and measurement of anthocyanins by UV-visible spectroscopy. Curr Protocol Food Anal Chem 2001:F1.2.1–F1.2.13

Gligor O, Mocan A, Moldovan C, Locatelli M, Crişan G, Ferreira ICFR (2019) Enzyme-assisted extractions of polyphenols – a comprehensive review. Trends Food Sci Technol 88:302–315

Görgüç A, Gençdağ E, Yılmaz FM (2019) Optimization of microwave assisted enzymatic extraction of steviol glycosides and phenolic compounds from Stevia leaf. Acta Periodica Technologica 50:69–76

Grassino AN, Pedisić S, Dragović-Uzelac V, Karlović S, Ježek D, Bosiljkov T (2020) Insight into high-hydrostatic pressure extraction of polyphenols from tomato Peel waste. Plant Foods Hum Nutr 75(3):427–433

Ha G-S, Kim J-H (2016) Kinetic and thermodynamic characteristics of ultrasound-assisted extraction for recovery of paclitaxel from biomass. Process Biochem 51(10):1664–1673

Hagerman AE, Butler LG (1978) Protein precipitation method for the quantitative determination of tannins. J Agric Food Chem 26(4):809–812

Haji Heidari S, Taghian Dinani S (2018) The study of ultrasound-assisted enzymatic extraction of oil from peanut seeds using response surface methodology. Eur J Lipid Sci Technol 120(3):1700252

Hammi KM, Jdey A, Abdelly C, Majdoub H, Ksouri R (2015) Optimization of ultrasound-assisted extraction of antioxidant compounds from Tunisian *Zizyphus lotus* fruits using response surface methodology. Food Chem 184:80–89

Harnly J (2017) Antioxidant methods. J Food Compos Anal 64:145–146

Harnly JM, Bhagwat S, Lin L-Z (2007) Profiling methods for the determination of phenolic compounds in foods and dietary supplements. Anal Bioanal Chem 389(1):47–61

Hayta M, İşçimen EM (2017) Optimization of ultrasound-assisted antioxidant compounds extraction from germinated chickpea using response surface methodology. LWT Food Sci Technol 77:208–216

Herrero M, Cifuentes A, Ibañez E (2006) Sub- and supercritical fluid extraction of functional ingredients from different natural sources: plants, food-by-products, algae and microalgae: a review. Food Chem 98(1):136–148

Hossain MB, Brunton NP, Patras A, Tiwari B, O'Donnell CP, Martin-Diana AB, Barry-Ryan C (2012) Optimization of ultrasound assisted extraction of antioxidant compounds from marjoram (*Origanum majorana* L.) using response surface methodology. Ultrason Sonochem 19(3):582–590

Hoyos-Arbeláez J, Vázquez M, Contreras-Calderón J (2017) Electrochemical methods as a tool for determining the antioxidant capacity of food and beverages: a review. Food Chem 221:1371–1381

Hu C-J, Gao Y, Liu Y, Zheng X-Q, Ye J-H, Liang Y-R, Lu J-L (2016) Studies on the mechanism of efficient extraction of tea components by aqueous ethanol. Food Chem 194:312–318

Huma F, Jaffar M, Masud K (1999) A modified potentiometric method for the estimation of phenol in aqueous systems. Turk J Chem 23(4):415–422

Ignat I, Volf I, Popa VI (2011) A critical review of methods for characterisation of polyphenolic compounds in fruits and vegetables. Food Chem 126(4):1821–1835

Jacotet-Navarro M, Rombaut N, Fabiano-Tixier AS, Danguien M, Bily A, Chemat F (2015) Ultrasound versus microwave as green processes for extraction of rosmarinic, carnosic and ursolic acids from rosemary. Ultrason Sonochem 27:102–109

Jamaludin R, Kim D-S, Md Salleh L, Lim S-B (2020) Optimization of high hydrostatic pressure extraction of bioactive compounds from noni fruits. J Food Measur Charact 14(5):2810–2818

Jerman Klen T, Mozetič Vodopivec B (2012) Optimisation of olive oil phenol extraction conditions using a high-power probe ultrasonication. Food Chem 134(4):2481–2488

Jerman T, Trebše P, Mozetič Vodopivec B (2010) Ultrasound-assisted solid liquid extraction (USLE) of olive fruit (*Olea europaea*) phenolic compounds. Food Chem 123(1):175–182

Jitrangsri K, Chaidedgumjorn A, Satiraphan M (2020) Supercritical fluid extraction (SFE) optimization of trans-resveratrol from peanut kernels (*Arachis hypogaea*) by experimental design. J Food Sci Technol 57(4):1486–1494

Junior MRM, Leite AV, Dragano NRV (2010) Supercritical fluid extraction and stabilization of phenolic compounds from natural sources–review (supercritical extraction and stabilization of phenolic compounds). Open Chem Eng J 4(1):51–60

Kaderides K, Papaoikonomou L, Serafim M, Goula AM (2019) Microwave-assisted extraction of phenolics from pomegranate peels: optimization, kinetics, and comparison with ultrasounds extraction. Chem Eng Process Intensif 137:1–11

Khan MK, Abert-Vian M, Fabiano-Tixier A-S, Dangles O, Chemat F (2010) Ultrasound-assisted extraction of polyphenols (flavanone glycosides) from orange (*Citrus sinensis* L.) peel. Food Chem 119(2):851–858

Khoddami A, Wilkes MA, Roberts TH (2013) Techniques for analysis of plant phenolic compounds. Molecules 18(2):2328–2375

King JW (2014) Modern supercritical fluid technology for food applications. Annu Rev Food Sci Technol 5(1):215–238

Korompokis K, Igoumenidis P, Mourtzinos I, Karathanos V (2017) Green extraction and simultaneous inclusion complex formation of *Sideritis scardica* polyphenols. Int Food Res J 24(3):1233–1238

Kumari B, Tiwari BK, Hossain MB, Rai DK, Brunton NP (2017) Ultrasound-assisted extraction of polyphenols from potato peels: profiling and kinetic modelling. Int J Food Sci Technol 52(6):1432–1439

Kwaw E, Ma Y, Tchabo W, Sackey AS, Apaliya MT, Xiao L, Wu M, Sarpong F (2018) Ultrasonication effects on the phytochemical, volatile and sensorial characteristics of lactic acid fermented mulberry juice. Food Biosci 24:17–25

Lafarga T, Rodriguez-Roque MJ, Bobo G, Villaro S, Aguilo-Aguayo I (2019) Effect of ultrasound processing on the bioaccessibility of phenolic compounds and antioxidant capacity of selected vegetables. Food Sci Biotechnol 28(6):1713–1721

Li Y, Li S, Lin S-J, Zhang J-J, Zhao C-N, Li H-B (2017) Microwave-assisted extraction of natural antioxidants from the exotic *gordonia axillaris* fruit: optimization and identification of phenolic compounds. Molecules 22(9):1481

Lianfu Z, Zelong L (2008) Optimization and comparison of ultrasound/microwave assisted extraction (UMAE) and ultrasonic assisted extraction (UAE) of lycopene from tomatoes. Ultrason Sonochem 15(5):731–737

Liang Z, Zhai H, Chen Z, Wang S, Wang H, Wang S (2017) A sensitive electrochemical sensor for flavonoids based on a multi-walled carbon paste electrode modified by cetyltrimethyl ammonium bromide-carboxylic multi-walled carbon nanotubes. Sensors Actuators B Chem 244:897–906

Lin D, Ma Q, Zhang Y, Peng Z (2020) Phenolic compounds with antioxidant activity from strawberry leaves: a study on microwave-assisted extraction optimization. Prep Biochem Biotech 50(9):874–882

Liu J, Ji F, Chen F, Guo W, Yang M, Huang S, Zhang F, Liu Y (2018) Determination of garlic phenolic compounds using supercritical fluid extraction coupled to supercritical fluid chromatography/tandem mass spectrometry. J Pharmaceut Biomed 159:513–523

López-Miranda S, Serrano-Martínez A, Hernández-Sánchez P, Guardiola L, Pérez-Sánchez H, Fortea I, Gabaldón JA, Núñez-Delicado E (2016) Use of cyclodextrins to recover catechin and epicatechin from red grape pomace. Food Chem 203:379–385

Luo X, Bai R, Zhen D, Yang Z, Huang D, Mao H, Li X, Zou H, Xiang Y, Liu K, Wen Z, Fu C (2019) Response surface optimization of the enzyme-based ultrasound-assisted extraction of acorn tannins and their corrosion inhibition properties. Ind Crop Prod 129:405–413

Luque-Rodríguez JM, Luque de Castro MD, Pérez-Juan P (2007) Dynamic superheated liquid extraction of anthocyanins and other phenolics from red grape skins of winemaking residues. Bioresour Technol 98(14):2705–2713

Mantegna S, Binello A, Boffa L, Giorgis M, Cena C, Cravotto G (2012) A one-pot ultrasound-assisted water extraction/cyclodextrin encapsulation of resveratrol from *Polygonum cuspidatum*. Food Chem 130(3):746–750

Marathe SJ, Jadhav SB, Bankar SB, Kumari Dubey K, Singhal RS (2019) Improvements in the extraction of bioactive compounds by enzymes. Curr Opin Food Sci 25:62–72

Marić M, Grassino AN, Zhu Z, Barba FJ, Brnčić M, Rimac Brnčić S (2018) An overview of the traditional and innovative approaches for pectin extraction from plant food wastes and by-products: ultrasound-, microwaves-, and enzyme-assisted extraction. Trends Food Sci Technol 76:28–37

Mathews DH, Welch TR (1958) Potentiometric titrations of weak acids in non-aqueous solvents. I. Benzoic acid, ε-cyclohexylcaproic acid, *p*-cresol and α-naphthol. J Appl Chem 8(11):701–710

Montero-Calderon A, Cortes C, Zulueta A, Frigola A, Esteve MJ (2019) Green solvents and ultrasound-assisted extraction of bioactive orange (*Citrus sinensis*) peel compounds. Sci Rep 9(1):16120

Moreira SA, Pintado M, Saraiva JA (2020a) High hydrostatic pressure-assisted extraction: a review on its effects on bioactive profile and biological activities of extracts. In: Barba FJ, Tonello-Samson C, Puértolas E, Lavilla M (eds) Present and future of high pressure processing. Elsevier, New York, pp 317–328

Moreira SA, Pintado ME, Saraiva JA (2020b) Optimization of high hydrostatic pressure assisted extraction of stinging nettle leaves using response surface methodology experimental design. J Food Measur Charact 14(5):2773–2780

Mourtzinos I, Anastasopoulou E, Petrou A, Grigorakis S, Makris D, Biliaderis CG (2016) Optimization of a green extraction method for the recovery of polyphenols from olive leaf using cyclodextrins and glycerin as co-solvents. J Food Sci Technol 53(11):3939–3947

Munir MT, Kheirkhah H, Baroutian S, Quek SY, Young BR (2018) Subcritical water extraction of bioactive compounds from waste onion skin. J Clean Prod 183:487–494

Mura M, Palmieri D, Garella D, Di Stilo A, Perego P, Cravotto G, Palombo D (2015) Simultaneous ultrasound-assisted water extraction and β-cyclodextrin encapsulation of polyphenols from *Mangifera indica* stem bark in counteracting TNFα-induced endothelial dysfunction. Nat Prod Res 29(17):1657–1663

Ninčević Grassino A, Ostojić J, Miletić V, Djaković S, Bosiljkov T, Zorić Z, Ježek D, Rimac Brnčić S, Brnčić M (2020) Application of high hydrostatic pressure and ultrasound-assisted extractions as a novel approach for pectin and polyphenols recovery from tomato peel waste. Innovative Food Sci Emerg Technol 64:102424

Nishad J, Saha S, Kaur C (2019) Enzyme- and ultrasound-assisted extractions of polyphenols from *Citrus sinensis* (cv. Malta) peel: a comparative study. J Food Process Preserv 43(8):e14046

Novak I, Šeruga M, Komorsky-Lovrić Š (2010) Characterisation of catechins in green and black teas using square-wave voltammetry and RP-HPLC-ECD. Food Chem 122(4):1283–1289

Pal CBT, Jadeja GC (2019) Microwave-assisted deep eutectic solvent extraction of phenolic antioxidants from onion (*Allium cepa* L.) peel: a box–Behnken design approach for optimization. J Food Sci Technol 56(9):4211–4223

Pan Z, Qu W, Ma H, Atungulu GG, McHugh TH (2012) Continuous and pulsed ultrasound-assisted extractions of antioxidants from pomegranate peel. Ultrason Sonochem 19(2):365–372

Pavić V, Jakovljević M, Molnar M, Jokić S (2019) Extraction of carnosic acid and carnosol from sage (*Salvia officinalis* L.) leaves by supercritical fluid extraction and their antioxidant and antibacterial activity. Plants (Basel) 8 (1):16

Pinela J, Prieto MA, Carvalho AM, Barreiro MF, Oliveira MBPP, Barros L, Ferreira ICFR (2016) Microwave-assisted extraction of phenolic acids and flavonoids and production of antioxidant ingredients from tomato: a nutraceutical-oriented optimization study. Sep Purif Technol 164:114–124

Pongnaravane B, Goto M, Sasaki M, Anekpankul T, Pavasant P, Shotipruk A (2006) Extraction of anthraquinones from roots of *Morinda citrifolia* by pressurized hot water: antioxidant activity of extracts. J Supercrit Fluids 37(3):390–396

Proestos C, Komaitis M (2008) Application of microwave-assisted extraction to the fast extraction of plant phenolic compounds. LWT Food Sci Technol 41(4):652–659

Queiroz C, Moreira CFF, Lavinas FC, Lopes MLM, Fialho E, Valente-Mesquita VL (2010) Effect of high hydrostatic pressure on phenolic compounds, ascorbic acid and antioxidant activity in cashew apple juice. High Pressure Res 30(4):507–513

Radojkovic M, Moreira MM, Soares C, Barroso MF, Cvetanovic A, Svarc-Gajic J, Morais S, Delerue-Matos C (2018) Microwave-assisted extraction of phenolic compounds from *Morus nigra* leaves: optimization and characterization of the antioxidant activity and phenolic composition. J Chem Technol Biot 93(6):1684–1693

Rajha HN, Chacar S, Afif C, Vorobiev E, Louka N, Maroun RG (2015) β-Cyclodextrin-assisted extraction of polyphenols from vine shoot cultivars. J Agric Food Chem 63(13):3387–3393

Ramos L, Kristenson EM, Brinkman UAT (2002) Current use of pressurised liquid extraction and subcritical water extraction in environmental analysis. J Chromatogr A 975(1):3–29

Ratnasooriya CC, Rupasinghe HP (2012) Extraction of phenolic compounds from grapes and their pomace using beta-cyclodextrin. Food Chem 134(2):625–631

René A, Abasq M-L, Hauchard D, Hapiot P (2010) How do phenolic compounds react toward superoxide ion? A simple electrochemical method for evaluating antioxidant capacity. Anal Chem 82(20):8703–8710

Revilla E, Ryan J-M, Martín-Ortega G (1998) Comparison of several procedures used for the extraction of anthocyanins from red grapes. J Agric Food Chem 46(11):4592–4597

Rodrigues S, Fernandes FAN, de Brito ES, Sousa AD, Narain N (2015) Ultrasound extraction of phenolics and anthocyanins from jabuticaba peel. Ind Crop Prod 69:400–407

Ryu D, Koh E (2018) Application of response surface methodology to acidified water extraction of black soybeans for improving anthocyanin content, total phenols content and antioxidant activity. Food Chem 261:260–266

Ryu D, Koh E (2019) Optimization of ultrasound-assisted extraction of anthocyanins and phenolic compounds from black soybeans (Glycine max L.). Food Anal Methods 12(6):1382–1389

Safdar MN, Kausar T, Nadeem M (2017) Comparison of ultrasound and maceration techniques for the extraction of polyphenols from the mango peel. J Food Process Eng 41(4):e13028

Sánchez-Rangel JC, Benavides J, Heredia JB, Cisneros-Zevallos L, Jacobo-Velázquez DA (2013) The Folin–Ciocalteu assay revisited: improvement of its specificity for total phenolic content determination. Anal Methods 5(21):5990–5999

Sarkanen K, Schuerch C (1955) Conductometric determination of phenolic groups in mixtures such as isolated lignins. Anal Chem 27(8):1245–1250

Sarneckis CJ, Dambergs R, Jones P, Mercurio M, Herderich MJ, Smith P (2006) Quantification of condensed tannins by precipitation with methyl cellulose: development and validation of an optimised tool for grape and wine analysis. Aust J Grape Wine Res 12(1):39–49

Šeruga M, Novak I, Jakobek L (2011) Determination of polyphenols content and antioxidant activity of some red wines by differential pulse voltammetry, HPLC and spectrophotometric methods. Food Chem 124(3):1208–1216

Shouqin Z, Jun X, Changzheng W (2005) High hydrostatic pressure extraction of flavonoids from propolis. J Chem Technol Biotechnol 80(1):50–54

Singleton VL, Rossi JA (1965) Colorimetry of total phenolics with phosphomolybdic-phosphotungstic acid reagents. Am J Enol Viticult 16(3):144–158

Singleton VL, Orthofer R, Lamuela-Raventós RM (1999) Analysis of total phenols and other oxidation substrates and antioxidants by means of folin-ciocalteu reagent. In: packer L (ed) methods in enzymology, vol 299. Academic press, pp 152-178

Somers TC, Evans ME (1974) Wine quality: correlations with colour density and anthocyanin equilibria in a group of young red wines. J Sci Food Agric 25(11):1369–1379

Tchabo W, Ma Y, Engmann FN, Zhang H (2015) Ultrasound-assisted enzymatic extraction (UAEE) of phytochemical compounds from mulberry (Morus nigra) must and optimization study using response surface methodology. Ind Crop Prod 63:214–225

Toma M, Vinatoru M, Paniwnyk L, Mason TJ (2001) Investigation of the effects of ultrasound on vegetal tissues during solvent extraction. Ultrason Sonochem 8(2):137–142

Tyskiewicz K, Konkol M, Roj E (2018) The application of supercritical fluid extraction in phenolic compounds isolation from natural plant materials. Molecules 23(10)

Venkatesh MS, Raghavan GSV (2004) An overview of microwave processing and dielectric properties of Agri-food materials. Biosyst Eng 88(1):1–18

Vu HT, Scarlett CJ, Vuong QV (2019) Maximising recovery of phenolic compounds and antioxidant properties from banana peel using microwave assisted extraction and water. J Food Sci Technol 56(3):1360–1370

Wang L, Weller CL (2006) Recent advances in extraction of nutraceuticals from plants. Trends Food Sci Technol 17(6):300–312

Wang Y, Wei Z, Zhang J, Wang X, Li X (2017) Electrochemical determination of apigenin as an anti-gastric cancer drug in *lobelia chinensis* using modified screen-printed electrode. Int J Electrochem Sc 12:2003–2012

Wen C, Zhang J, Zhang H, Dzah CS, Zandile M, Duan Y, Ma H, Luo X (2018) Advances in ultrasound assisted extraction of bioactive compounds from cash crops – a review. Ultrason Sonochem 48:538–549

Wen L, Zhang Z, Sun D-W, Sivagnanam SP, Tiwari BK (2020) Combination of emerging technologies for the extraction of bioactive compounds. Crit Rev Food Sci Nutr 60(11):1826–1841

Xi J, Shen D, Zhao S, Lu B, Li Y, Zhang R (2009) Characterization of polyphenols from green tea leaves using a high hydrostatic pressure extraction. Int J Pharm 382(1):139–143

Yang Y-C, Li J, Zu Y-G, Fu Y-J, Luo M, Wu N, Liu X-L (2010) Optimisation of microwave-assisted enzymatic extraction of corilagin and geraniin from *Geranium sibiricum* Linne and evaluation of antioxidant activity. Food Chem 122(1):373–380

Yi H, Cheng Y, Zhang Y, Xie Q, Yang X (2020) Potentiometric and UV-vis spectrophotometric titrations for evaluation of the antioxidant capacity of chicoric acid. RSC Adv 10(20):11876–11882

Yildiz-Ozturk E, Tag O, Yesil-Celiktas O (2014) Subcritical water extraction of steviol glycosides from *Stevia rebaudiana* leaves and characterization of the raffinate phase. J Supercrit Fluids 95:422–430

Zeb A (2015) A reversed phase HPLC-DAD method for the determination of phenolic compounds in plant leaves. Anal Methods 7(18):7753–7757

Zhang H-F, Yang X-H, Wang Y (2011a) Microwave assisted extraction of secondary metabolites from plants: current status and future directions. Trends Food Sci Technol 22(12):672–688

Zhang L, Liu J, Zhang P, Yan S, He X, Chen F (2011b) Ionic liquid-based ultrasound-assisted extraction of chlorogenic acid from *Lonicera japonica* Thunb. Chromatographia 73(1–2):129–133

Zhang H-J, Liu Y-N, Wang M, Wang Y-F, Deng Y-R, Cui M-L, Ren X-L, Qi A-D (2015) One-pot β-cyclodextrin-assisted extraction of active ingredients from Xue–Zhi–Ning basing its encapsulated ability. Carbohyd Polym 132:437–443

Zhang L, Fan G, Khan MA, Yan Z, Beta T (2020) Ultrasonic-assisted enzymatic extraction and identification of anthocyanin components from mulberry wine residues. Food Chem 323:126714

Zhao C-N, Zhang J-J, Li Y, Meng X, Li H-B (2018) Microwave-assisted extraction of phenolic compounds from *Melastoma sanguineum* fruit: optimization and identification. Molecules 23(10):2498

Zhu Q-Y, Zhang Q-Y, Cao J, Cao W, Xu J-J, Peng L-Q (2016) Cyclodextrin-assisted liquid-solid extraction for determination of the composition of jujube fruit using ultrahigh performance liquid chromatography with electrochemical detection and quadrupole time-of-flight tandem mass spectrometry. Food Chem 213:485–493

Chapter 16
Chromatography of Phenolic Antioxidants

Learning Objectives
In this chapter, the reader will be able to:
- Learn the basic concept of chromatography and its types.
- Understand thin layer chromatography for the analysis of phenolic compounds.
- Learn liquid chromatography, its types, and applications for the analysis of phenolic compounds in foods.
- Understand gas chromatography and its applications for the analysis of phenolic compounds in foods.

16.1 Introduction

Chromatography is an important technique that allows the separation, identification, and purification of the components of a mixture for qualitative and quantitative purposes. It is based on the principle that molecules in a mixture are applied onto the surface or into the solid and is separated from each other while moving with the help of a solvent. The factors effective in this separation process include molecular characteristics related to adsorption (liquid-solid), partition (liquid-solid), and affinity or differences among their molecular weights (Coskun 2016). Because of these differences, some components of the mixture retained longer on the solid phase, and they move slowly in the chromatography system, while others pass quickly with the solvent, and leave the system faster.

Chromatography is thus based on three basic components, which form the basis of the chromatography technique.

1. *Stationary phase*: This phase is composed of a "solid" phase or "a layer of a liquid adsorbed on the surface a solid support".

© Springer Nature Switzerland AG 2021
A. Zeb, *Phenolic Antioxidants in Foods: Chemistry, Biochemistry and Analysis*,
https://doi.org/10.1007/978-3-030-74768-8_16

2. *Mobile phase*: This phase is composed of "liquid" or a "gaseous component."
3. *Analytes*: The components of the mixture being separated.

The type of interaction between the stationary phase, mobile phase, and analytes contained in the mixture is the basic component effective in the separation of molecules from each other. Chromatographic methods based on partition are very effective for the separation, and identification of small molecules such as amino acids, carbohydrates, and fatty acids. However, affinity chromatography such as ion-exchange chromatography is more effective in the separation of macromolecules like nucleic acids, and proteins. Paper chromatography is used in the separation of proteins, and studies related to protein synthesis; gas-liquid chromatography is utilized in the separation of alcohol, ester, lipid, and amino groups, and observation of enzymatic interactions, while molecular-sieve chromatography is employed especially for the determination of molecular weights of proteins.

In chromatography, the stationary phase is a solid or a liquid coated on the surface of a solid phase. The mobile phase also known as eluent is flowing over the stationary phase. It may be a gas or liquid. If the mobile phase is liquid, it is called liquid chromatography (LC), and if it is gas then it is called gas chromatography (GC). Gas chromatography is applied for gases, and mixtures of volatile liquids, and solid material. Liquid chromatography is used especially for thermal unstable, and non-volatile samples such as phenolic compounds.

## 16.2	Types of Chromatography

Chromatography is classified based on its components and separation mode. There are two classes of chromatography, i.e., adsorption and partition chromatography. The adsorption chromatography is based on the competition between solid and liquid or gas. It is further classified into solid-liquid and solid-gas chromatography. Column chromatography and thin-layer chromatography are types of solid-liquid chromatography. The partition chromatography is based on the competition between liquid and liquid or gas. It is further classified into liquid-liquid and liquid gas chromatography. Paper chromatography and high performance liquid chromatography are examples of partition chromatography. The types of chromatography commonly used for the analysis of phenolic compounds in foods are thin-layer chromatography, liquid chromatography, and gas chromatography.

## 16.3	Thin Layer Chromatography

Thin layer chromatography (TLC) is a separation technique used to separate a mixture of analytes on a thin layer of the sheet. The sheet is usually of aluminum or glass, on which the solid phase is spread to separate analytes using liquid as a solvent phase. The stationary phase on the plate may be silica gel, modified silica gels, alumina, and cellulose. They may also be supplemented with characteristic compounds for visualization. TLC can be carried out usually in one dimension, for routine or preparative purposes (Zang et al. 2020). By adding a second development to perform two-dimensional TLC (2D TLC) allows even better resolution of complex samples (Rabel and Sherma 2016).

Analytes are extracted using an appropriate solvent with traditional techniques such as shake, homogenization, Soxhlet, or ultrasound-assisted extraction (UAE). Liquid samples such as wines can often be applied as packaged with no sample preparation. Solid-phase extraction (SPE) and supercritical fluid extraction (SFE) are more modern techniques applied for sample preparation (Sherma and Rabel 2018). Samples are applied to the plate as spots manually with a micropipette or syringe or automatically as bands or spots with a commercial instrument. Plate development is usually done at laboratory temperature in a large volume, mobile phase vapor saturated N-chamber in the one dimensional ascending or horizontal mode. Mobile phase compositions are in v/v proportions unless otherwise noted. If the analyte is not naturally colored or does not quench fluorescence on an F plate, detection of its separated zones is achieved by viewing plates in daylight or under 254 or 366 nm UV light after applying a derivatization reagent by spraying or dipping followed by heating if necessary. Chromatogram quantification is by slit scanning densitometry or image analysis (Sherma and Rabel 2018).

Preparative thin-layer chromatography (PTLC) is a method used to separate and isolate larger amounts of material (e.g., 10–1000 mg) on 0.5–4-mm layers than the nanogram to low milligram level that is separated in analytical TLC or HPTLC on 0.1–0.25-mm-thick layers (Rabel and Sherma 2017). The purpose of PTLC is to isolate pure compounds for further analysis. Most preparative planar chromatography continues to be done today using CPTLC, and most applications are in the isolation and characterization of natural products such as phenolic compounds. Typically, PTLC follows extraction and gradient open column chromatography, usually on silica gel but also columns such as diethylaminoethyl-cellulose, active charcoal, and CM-Sepharose, and RP-18 and Sephadex LH-20, to isolate, purify, and identify constituents using subsequent NMR spectrometry, IR spectrometry, UV–visible spectrometry, MS, analytical TLC/HPTLC, column liquid chromatography-MS, GC, GC-MS, or electrophoresis (Rabel and Sherma 2017).

TLC is a more powerful technique than paper chromatography to analyze phenolic compounds, especially in crude plant food extracts. TLC is widely used for purification and isolation of anthocyanins, flavonols, condensed tannins, and phenolic acids using different solvent systems (Naczk and Shahidi 2006). Phenolic compounds in crude plant extracts can be separated by several TLC techniques, which are cheap and supported by multiple detections on the same TLC plate in a short analysis time (Khoddami et al. 2013). Since the majority of the phenolic compounds are water-soluble, thus can be readily separated by TLC. A high-efficiency TLC with solid-phase similar to HPLC is now widely used for the analysis of phenolic compounds is called HPTLC.

16.3.1 Detection and Identification of Phenolic Compounds

Some phenolic compounds, for example, anthocyanins, aurones, chalcones can be visualized immediately on TLC plates by their colors. Many others appear in UV light either as dark, absorbing (e.g., flavones), or fluorescent (e.g., furanocoumarins) spots. Furthermore, most absorbing compounds give bright yellow, green, or brown colors in ultraviolet light when the plate is fumed with ammonia. Further information about the class of compound or structural substitution can be obtained by using one or more spray reagents. For flavones, flavonols, and their derivatives, is particularly useful, since it gives a range of fluorescent colors, observable in both daylights and in the ultraviolet, which varies according to the structure. Peck et al. (1986) showed that a TLC plate sprayed with p-nitroaniline produces different colors of the derivatives of hydroxybenzoic acids. The different di- and tri-hydroxy benzoic acids can be identified with characteristic colors as shown in Table 16.1.

Because direct identification and structural characterization of the analytes on the TLC plate through these methods are not possible, there has been long-held interest in the development of interfaces that allow TLC to be combined with mass spectrometry (MS)—one of the most efficient analytical tools for structural elucidation. So far, many different TLC–MS techniques have been reported in the literature; some are commercially available. According to differences in their operational processes, the existing TLC–MS systems can be classified into two categories: (i) indirect mass spectrometric analyses, performed by scraping, extracting, purifying, and concentrating the analyte from the TLC plate and then directing it into the mass spectrometer's ion source for further analysis; (ii) direct mass spectrometric analyses, where the analyte on the TLC plate is characterized directly through mass spectrometry without the need for scraping, extraction, or concentration processes (Cheng et al. 2011; Glavnik and Vovk 2019). Conventionally, direct TLC–MS analysis is performed under vacuum, but the development of ambient mass spectrometry has allowed analytes on TLC plates to be characterized under atmospheric pressure. Thus, TLC–MS techniques can also be classified into two other categories according to the working environment of the ion source: vacuum-based TLC–MS or

Table 16.1 Identification of derivatives of benzoic acid (BA) with diazotized p-nitroaniline (data are based on the data from Peck et al. (1986))

Derivative of Benzoic acid	Common name	Spot color
2,3-Dihydroxy-BA	2-Pyrocatechuic acid	Purple
2,4-Dihydroxy-BA	β-Resorcylic acid	Brown
2,5-Dihydroxy-BA	Gentisic acid	Yellow
2,6-Dihydroxy-BA	γ-Resorcylic acid	Orange
3,4-Dihydroxy-BA	Protocatechuic acid	Brown
3,5-Dihydroxy-BA	α-Resorcylic acid	Orange
2,3,4-Tri hydroxy-BA	Pyrogallolcarboxylic acid	Yellow
2,4,6-Trihydroxy-BA	Phloroglucinolcarboxylic acid	Orange
3,4,5-Trihydroxy-BA	Gallic acid	Brown

ambient TLC–MS. The review of Cheng et al. (2011) described the state of the art of TLC–MS techniques used for indirect and direct characterization of analytes on the surfaces of TLC plates.

Figure 16.1 showed a schematic representation of Laser desorption ionization (LDI), Matrix-assisted laser desorption ionization (MALDI), and Surface-assisted laser desorption ionization (SALDI) analyses of compounds on the surfaces of TLC plates (Cheng et al. 2011). In these techniques, the developed plate is attached to the sample probe using double-sided adhesive tape and then a pulsed laser beam is used to desorb and ionize the compounds present on the plate. LDI employs a pulsed laser beam to irradiate the spots of interest; in the absence of any matrix, the laser density required to produce analyte ions from the TLC plate is greater than that from a metal surface. In MALDI analysis, an analyte/matrix co-crystal is formed by applying an organic matrix solution to the spot surface; the matrix adsorbs energy from the pulsed laser beam to assist desorption/ionization of the analytes. In SALDI analysis, a suspension of carbon powder in a liquid is applied to the plate surface. This inorganic matrix adsorbs the UV laser energy to assist desorption/ionization of the separated compounds. TLC can also be coupled to HPLC. This hyphenation provides excellent results for phenolic compounds. For example, Móricz et al. (2020) screened lemon balm for multipotent bioactive compounds by HPTLC-EDA. The identification of most bioactive substances by HPTLC-HRMS. The HPLC-DAD-ESI-MS system has been expanded by installing a TLC-MS Interface with an oval elution head between the pump and the column. With this hyphenation, the compounds from the HPTLC zones can be eluted directly into the eluent for the

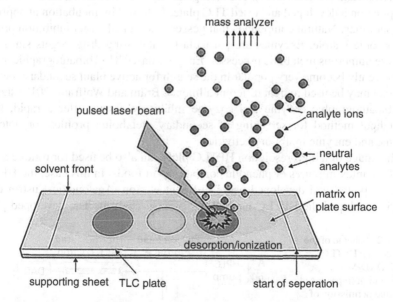

Fig. 16.1 Schematic representation of LDI, MALDI, and SALDI analyses of chemical compounds on the surface of a TLC plate (reproduced with kind permission of Elsevier (Cheng et al. 2011))

HPLC column as shown in Fig. 16.2. Zone elution from a dried HPTLC adsorbent inevitably injected a gas phase portion at the start of the elution into the eluent, and thus into the column.

Thin layer chromatography coupled with flame ionization detection (TLC–FID) is a unique system that combines the benefits of FID as a universal detector and the separation capability of TLC to make it a powerful analytical separation technique that can be used for a wide range of component analyses in foods. Anyakudo et al. (2020) reviewed the TLC–FID technique and found it to be a mature and reliable, state of the art technology that combines the separation power of TLC with FID as a universal detector, which can be applied to the analysis of a wide variety of organic compounds. In this case, the samples are prepared and spotted like classical TLC or may be automatic. The stationary phase is made up of quartz rods exclusively designed for Iatroscan. The rod holder comprises up to 10 chromarods (stationary phase), each made up of a quartz rod with a thin silica gel or alumina layer. After spotting followed by development and drying of the sample, the rods bearing the separated samples traverse the hydrogen/air flame resulting in the ionization of the sample by the flame and subsequent detection by FID or simultaneous FID/FPD, depending on the instrument used. Figure 16.3 showed a schematic diagram of TLC–FID Iatroscan equipped with FID and an FPD detector (Anyakudo et al. 2020).

Bioautographic assays represent an effect-directed analysis (EDA) tool and are defined as TLC-methods combined with biological detection. This procedure provides simultaneous chromatographic separation of a complex multi-component matrix and the localization of active constituents directly on a TLC plate in a short time (Bräm and Wolfram 2017). The biological target dissolved in a suitable medium is applied on a developed and dried TLC plate, followed by incubation at appropriate conditions. Natural compounds that possess biological target inhibition properties become visible. Enzymes are particularly interesting drug targets since they catalyze numerous metabolic processes. Enzyme-based TLC bioautographic assays have recently become very popular in the search for active plant secondary metabolites that may be used to treat different ailments. Bräm and Wolfram (2017) showed that bioautographic enzyme and enzyme inhibitory assays offer a rapid, high-throughput method for screening of secondary metabolite profiles for potential enzyme and enzyme inhibitory activities.

The image of a high-resolution HPTLC plate can also be used for routine analysis of complex mixtures of phenolic compounds in foods. In this case, the HPTLC plate is spotted and developed using different elution. Agatonovic-Kustrin et al. (2015) developed the HPTLC method with image analysis. The developed plates

Fig. 16.2 Scheme of the heart-cutting HPTLC-HPLC-DAD-MS configuration (reproduced with kind permission of Elsevier (Móricz et al. 2020))

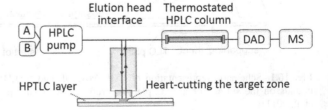

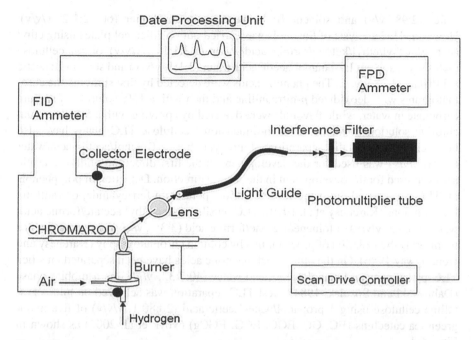

Fig. 16.3 Schematic diagram of TLC–FID Iatroscan equipped with FID and a FPD detector (reproduced with kind permission of Elsevier (Vovk et al. 2005))

were sprayed with the 2-aminoethyl diphenylborinate before and after. The images of plates were captured using a TLC-Visualiser (Camag, Muttenz, Switzerland) with a 12-bit camera under UV light at 366 nm. VideoScan Digital Image Evaluation software (Camag, Switzerland 2003) was used for the quantitative evaluation of plates and to transform images into chromatograms. The authors found the average concentration of caffeic acid, gallic acid, resveratrol, and rutin in the red wines were 2.15 mg/L, 30.17, 0.59, and 2.47 mg/L respectively with their concentration below the limit of quantification in the white wine samples. The highest concentration of resveratrol and rutin is found in the Cabernet and Shiraz wine samples. The combination with visual evaluation HPTLC could provide extremely rapid and cost-effective screening of wine samples.

16.3.2 Applications of TLC

There is a significant amount of literature on the applications of phenolic compounds in medicinal plants. The TLC is now widely used for the analysis of phenolic compounds in foods. Azar et al. (1987) have identified phenolic compounds of bilberry juice *Vaccinium myrtillus* using a two-dimensional TLC. Phenolic acids were chromatographed on a 0.1 mm cellulose layer with solvent A: acetic acid/

water (2:98, v/v) and solvent B: benzene/acetic acid/water (60:22:1.2, v/v/v). However, TLC analysis of flavonols was carried out on silica gel plates using ethyl acetate/methyl/ethyl/ketone/formic acid/water (5:3:1:1, v/v/v/v) or on cellulose plates using solvent I: *t*-butanol/acetic acid/water (3:1:1, v/v/v) and solvent II: acetic acid/water (15:85, v/v). The phenolic acids were detected by first spraying the chromatograms with deoxidized *p*-nitroaniline and then with a 15% solution of sodium carbonate in water, while flavonols were detected by spraying with a 5% aluminum chloride solution in methanol. Two-dimensional cellulose TLC plates have also been employed for the separation of procyanidins. *t*-Butanol/acetic acid/water (3:1:1, v/v/v) was used for the development in the first direction while 6% acetic acid was used for the development in the second direction. Detection of polyphenols on TLC was carried out using ferric chloride, potassium ferricyanide, or vanillin–HCl solutions (Karchesy et al. 1989). TLC on silica using ethyl acetate/formic acid/water (90:5:5, v/v/v) or toluene/acetone/formic acid (3:3:1, v/v/v) has been used for monitoring the isolation of procyanidins by column chromatography (Karchesy and Hemingway 1986). On the other hand, phenolic acids have been separated on silica TLC plates using *n*-butanol/acetic acid/water (40:7:32, v/v/v) as a mobile phase (Dabrowski and Sosulski 1984). Best TLC separation was achieved on microcrystalline cellulose using 1-propanol/water/acetic acid (20:80:1, v/v/v) of five major green tea catechins (EC, GC, EGC, ECG, EGCg) (Vovk et al. 2005) as shown in Fig. 16.4.

Sajewicz et al. (2012) reported that a silica gel TLC-based video imaging method is a valuable complementary fingerprint technique to identify phenolic acids and flavonoids fractions from different sage species.

Caffeoylquinic acid (CQA) derivatives were quantified in 295 varieties and breeding lines of sweet potatoes using HPTLC with digital images of the plates documented by the TLC Visualizer documentation system equipped with a high-resolution 12 bit CCD digital camera (Lebot et al. 2016). The plates were developed at room temperature with ethyl acetate, methanol, acetic acid, formic acid, and water (27:2:2:2:2, v/v/v/v/v) as the mobile phase (10 mL) for a maximum migration distance of 70 mm. The results presented in our study indicate that HPTLC is a simple, high-throughput, cost-efficient technique for the simultaneous quantitative estimation of CQA.

Densitometric HPTLC analysis was performed for the analysis of phenolic acids and flavonoids thermally processed sponge gourd (Yadav et al. 2017). The plates were developed in solvent system 1 (6.4 chloroform: 3.9 hexane: 2.0 methanol: 0.5 formic acid) for detection of gallic acid, caffeic acid, quercetin, apigenin, kaempferol, and chlorogenic acid; and solvent system 2 (4 chloroform: 1 hexane: 1 methanol: 1 formic acid) for detection of *p*-coumaric acid, luteolin, myricetin, catechin, and ellagic acid and solvent system 3 (4.5 acetonitrile: 1.0 methanol: 0.5 water) for detection of ferulic acid, benzoic acid, cinnamic acid, and vanillic acid) respectively rose to 80% of the plate height. HPTLC densitometry was found to be a useful method for the analysis of phenolic compounds.

A simple and suitable approach to the post-chromatographic derivatization in the analysis of alcohol and phenols mixtures by a combination of thin-layer

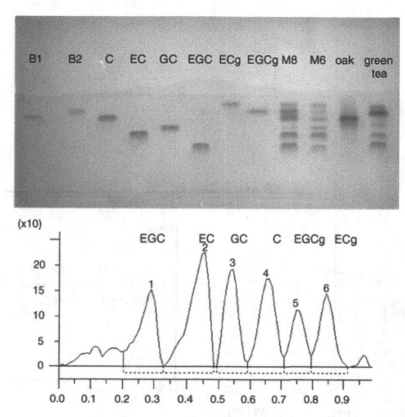

Fig. 16.4 The separation of flavan-3-ols test mixtures (M6 and M8), oak, and tea extracts on pre-washed HPTLC plates developed using 1-propanol–water–acetic acid (20:80:1, v/v) in the horizontal developing chamber (sandwich configuration). Lane M6 is presented also as a densitogram (reproduced with kind permission of Elsevier (Vovk et al. 2005))

chromatography and MALDI mass spectrometry was proposed by Esparza et al. (2018). The plate was developed in a solvent mixture of hexane: ethyl acetate in a ratio of 3: 0.5 for alcohols and terpenols and hexane: ethyl ether in a ratio of 3: 0.7 for phenols. The analytes were modified chemically by the treatment of eluted zones on thin-layer chromatograms using 3-bromopropionyl chloride in the presence of an excess of a base (pyridine or triethylamine). The resulting derivatives contained a residue with an ammonium fragment that ensures the efficient desorption of the derivatized phenolics from the sorbent layer during MALDI. The MS spectra of the derivatization products of menthol, eicosanol, and farnesol are shown in Fig. 16.5. It was shown that the proposed approach allows the recording of mass spectra with a high signal/noise ratio and the detection of alcohols and phenols of different structures.

An HPTLC method for the quantitative determination of phenolic compounds in honey was developed by Stanek et al. (2019). The authors detected and determined quantitatively seven phenolic compounds i.e., chlorogenic acid, myricetin, caffeic

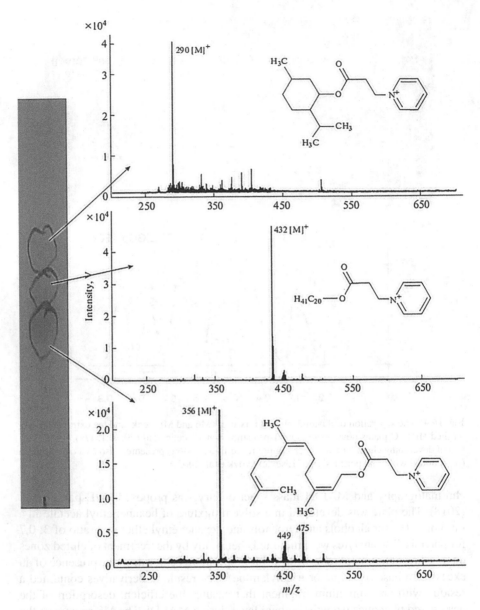

Fig. 16.5 A thin layer chromatogram of a mixture (left) and MALDI mass spectra of the derivatization products of menthol, eicosanol, and farnesol (derivatization was carried out after separation by TLC). Mass spectra were recorded using a combined matrix with the composition AT-graphite–glycerol (reproduced with kind permission of Springer Nature (Esparza et al. 2018))

acid, quercetin, ferulic acid, naringenin, and chrysin in honey samples. The proposed method was rapid, simple, and precise. Additionally, the combination of HPTLC fingerprints with statistical data analysis (PCA) led to the successful separation of acacia and lime honey samples.

The amount of chlorogenic acid was measured in four methanol extracts of various green coffees and one extract of black coffee using TLC with UV detection and TLC with effect-directed detection (Choma et al. 2019). TLC plates were developed to 8 cm distance with mobile phase consists of ethyl acetate/methanol/water (77:13:10, v/v). The chromatograms were documented using TLC Visualiser 2 and/ or TLC Scanner 4 controlled via WinCATS software. The evaluation of images obtained from TLC Visualiser 2 was performed by VideoScan Software. Densitometric measurements were done using WinCATS software. It was proved that thin-layer chromatography can be used as a quantitative (using densitometry) or semi-quantitative method (using other detection methods including effect directed detection) as well as for estimating total antioxidants or polyphenols content.

16.4 Liquid Chromatography

Liquid chromatography is a separation technique in which the mobile phase is liquid, while the stationary phase is solid. The separation is based on the affinity of analytes with respect to the mobile and stationary phases. The process is carried out either on a column or a plane. The sample mixture passes through the stationary phase and is separated depending on several factors, which classify them into different classes.

16.4.1 Types of Liquid Chromatography

Based on the choice of the stationary and mobile phase, several separation modes are used to separate desirous compounds in a mixture.

16.4.1.1 Reversed-Phase Chromatography

Reversed-phase chromatography utilizes a non-polar stationary phase and a relatively polar mobile phase. Therefore, non-polar analytes in the polar mobile phase tend to adsorb on the stationary phase, and the polar molecules in the polar mobile phase are transported through the column and elute earlier. Mixtures of water or aqueous buffers and organic solvents are used to elute components from a reversed-phase column. Gradient elution, in which the water-solvent composition changes as a function of time, is often used to separate a sample containing a wide range of components. Today, reversed-phase chromatography is the most widely used type of liquid chromatography.

16.4.1.2 Normal Phase Chromatography

In normal phase chromatography, the stationary phase is polar and the mobile phase is non-polar. The stationary phase is generally silica or organic moieties with cyano and amino functional groups and the mobile is hexane or heptane mixed with a slightly more polar solvent such as isopropanol, ethyl acetate, or chloroform. In normal phase chromatography, the least polar compounds elute first and the most polar compounds elute last. Normal phase chromatography is very useful to separate water-soluble compounds, geometric isomer, cis-trans isomers, and chiral compounds. It is also used for the separation of phenolic compounds in foods.

16.4.1.3 Ion Exchange Chromatography

In ion-exchange chromatography, the stationary phase contains ionic groups, for example, sulfonic or tetraalkylammonium, and the mobile phase is an aqueous buffer. The component molecule is retained by the stationary phase by coulombic attraction. This type of chromatography can be further divided into two categories: cation exchange chromatography retains positively charged cations and anion exchange chromatography retains negatively charged anions. Ion exchange chromatography is most commonly used to separate inorganic and organic anions and cations in an aqueous solution.

16.4.1.4 Size Exclusion Chromatography

In size exclusion chromatography, there is no chemical interaction between the sample molecules and the stationary phase. Instead, molecules are separated depending on their size relative to the pore size of the stationary phase. Largest molecules elute first and smallest molecules (which can permeate into the pores) elute last. Size exclusion chromatography is also known as gel-filtration chromatography and is widely used to separate polymer and proteins.

16.4.2 Stationary and Mobile Phases

Silica is one of the most popular adsorbent materials for liquid chromatography with particle sizes ranging from 3 to 50 microns. The porous silica particles may have a chemically bonded phase i.e., hydrophobic alkyl chains on their surface that interact with the components through chemical bonding. Figure 16.6 shows the

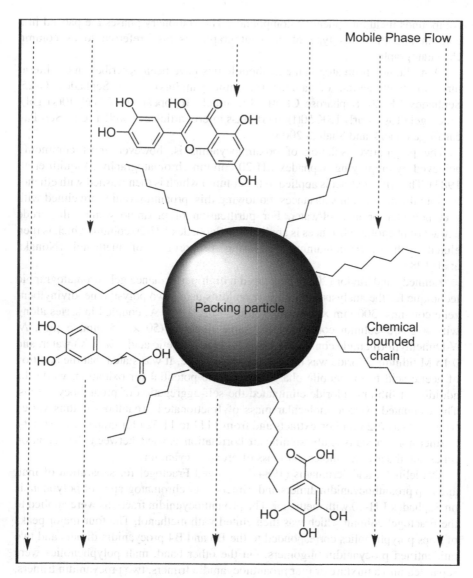

Fig. 16.6 Schematic representation of the separation of phenolic compounds using chemically bonded silica. Mobile phase flow is indicated by dashed arrows. Phenolic compounds with high polarity and low molecular size are separated earlier than high molecular weight compounds with relatively less polarity

schematic representation of the separation of phenolic compounds on chemically bonded silica. The three common chain lengths are C_4, C_8, and C_{18}. C_4 bonded stationary phase is mostly used for proteins, whereas C_8 and C_{18} are used for peptides or small molecules (Snyder et al. 2011; Meyer 2013). Recently the uses of C18 bonded stationary phases are more common for the separation of a wide range of compounds including phenolic compounds. The stationary phases are packed in a column and thus this type of chromatography is also referred to as column chromatography.

Many liquid chromatographic methodologies have been described in the literature for fractionation of tannins (proanthocyanidins) using Sephadex G-25, Sephadex LH-20, Sepharose CL-4B, Fractogel (Toyopearl TSK-HW 40(s) gel), Fractogel (Toyopearl) TSK 50(f), inert glass microparticles as well as C18 Sep-Pak cartridge (Naczk and Shahidi 2006).

The preparative isolation of proanthocyanins is, however, most commonly achieved by employing Sephadex LH-20 column chromatography (Asquith et al. 1983). The crude extract is applied to the column which is then washed with ethanol to elute the non-tannin substances. Following this, proanthocyanins are eluted with acetone/water or alcohol/water. For purification of proanthocyanins, the crude extract of phenolic substances is applied to a Sephadex LH-20 column which is then eluted with water containing increasing proportions of methanol (Nonaka et al. 1983).

Kennedy and Taylor (2003) developed a high-performance gel chromatographic technique for the analysis of proanthocyanidins using two polystyrene–divinylbenzene columns (300 mm × 7.5 mm, 5 μm, 500 Å × 100 Å) coupled in series along with a guard column containing the same material (50 × 7.5 mm, 5 μm). N, N-Dimethylformamide containing 1% (v/v) glacial acetic acid, 5% (v/v) water, and 0.15 M lithium chloride was used in a mobile phase. It was stated that the addition of acetic acid to the mobile phase reduced the potential for oxidation, while the addition of lithium chloride eliminated the self-aggregation of proanthocyanidins. The estimated average molecular mass of fractionated proanthocyanidins varied from 1193 to 7023 for hop extracts and from 1111 to 11,524 for grape skin extracts. Furthermore, a statistically significant correlation existed between the retention times and the average molecular mass of proanthocyanidins.

Derdelinckx and Jerumanis (1984) employed Fractogel for separation of malt and hop proanthocyanidin dimers and trimers after chromatography of polyphenols on Sephadex LH-20 with methanol. The proanthocyanidin fractions were applied to the Fractogel column which was then eluted with methanol. The four major peaks of hops polyphenolics corresponded to the B3 and B4 procyanidin dimers and two unidentified procyanidin oligomers. On the other hand, malt polyphenolics were separated into a mixture of four proanthocyanidin trimers, two procyanidin trimers, and an unknown procyanidin oligomer. According to Derdelinckx and Jerumanis (1984), Fractogel (Toyopearl TSK HW-40(s) gel) allows one to obtain the proanthocyanidins in an advanced state of purity.

Salagoïty-Auguste and Bertrand (1984) as well as Jaworski and Lee (1987) demonstrated that a C18 Sep-Pak cartridge can be used to separate grape phenolics into acidic and neutral fractions. Later, Sun et al. (1998) successfully used a C18 Sep-Pak cartridge for fractionation of grape proanthocyanidins according to their degree of polymerization. The procedure involved passing the extract of grape phenolics through two preconditioned neutral C18 Sep-Pak cartridges connected in series. The phenolic acids were then washed out with water; catechins and oligomeric proanthocyanidins were subsequently eluted with ethyl acetate and anthocyanidins and polymeric proanthocyanidins with methanol. The ethyl acetate fraction was redeposited on the same C18 Sep-Pak cartridges and catechins were first eluted with diethyl ether and then oligomeric proanthocyanidins were eluted with methanol.

Fulcrand et al. (1999) fractionated wine phenolic compounds into simple (phenolic acids, anthocyanins, flavonols, and flavanols) and polymeric components using a Fractogel (Toyopearl) HW-50(f) column (bed 12 × 120 mm). Simple phenolic were eluted from the column with ethanol/water/trifluoroacetic acid (55:45:0.005, v/v/v) while polymeric phenolics were then recovered with 60% (v/v) acetone. Later, Mateus et al. (2001) employed a Fractogel (Toyopearl) HW-40(s) column for fractionation of anthocyanin-derived pigments in red wines. Two liters of wine were directly applied onto the Toyopearl gel column (200 mm × 16 mm i.d.) at a flow rate of 0.8 mL/min. Anthocyanins were subsequently eluted from the column with water/ethanol (20:80, v/v). The elution of wine phenolics from the Toyopearl column yielded malvidin 3-glucoside and three derived pigments, namely malvidin 3-glucoside pyruvic adduct, malvidin -3-acetylglucoside pyruvic adduct and malvidin-3-coumarylglucoside pyruvic adduct. These glucosides accounted for 60% of the total monoglucosides content.

16.4.3 High Performance Liquid Chromatography

High performance liquid chromatography (HPLC) is an ideal technique for both separation and quantification of phenolic compounds. HPLC system consists of a solvent reservoir, a degasser to remove dissolved gases in the solvent phase, an injection system (manual or automatic), a column system, a detector, and a recording computer with a controlling software (Fig. 16.7). Zhou et al. (2020) reviewed that liquid chromatography has been a general separation method, however in most cases, the limited peak capacity and resolution achieved by 1-DLC could not meet the demand of complex phenolic separation. Thus, two-dimensional liquid chromatography (2-DLC), based on two independent separation mechanisms, has been proposed as a feasible solution. From a practical point, 2-DLC can be broadly divided into the off-line and on-line mode, according to the existence of an interface between D1 and D2. The interface includes sample loops, trapping columns, stop-flow, and vacuum evaporation interface.

In off-line mode, the sample is injected into a single 1-DLC system, and D1 fractions are manually collected and concentrated, then, injected into another 1-DLC system with a different separation mechanism. It does not require very rapid separations in the D2 compared to online 2-DLC, thus higher peak capacities can be achieved, although at the high cost of analysis time (Brandão et al. 2019). Nevertheless, off-line 2-DLC is more prone to sample contamination or losses than the on-line mode, which becomes more serious when dealing with complex samples. However, due to the unrestricted column size of the two dimensions and the freedom from solvent-incompatibility issues, preparative 2-DLC (pre-2-DLC) has

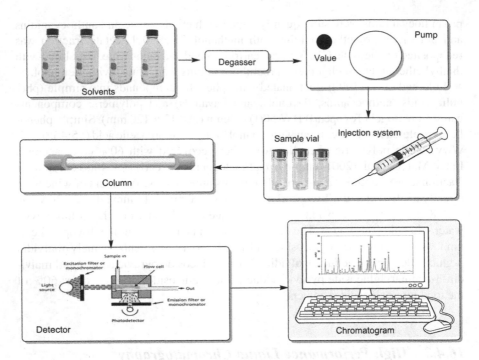

Fig. 16.7 Schematic representation of different parts of HPLC

been widely applied in the separation and purification of chemical monomers from plants, which is more advantageous than the on-line mode. The combination of different sizes of columns can meet the preparation demand of different sample loading quantity. The choosing of the stationary phase is the key point of constructing a 2-DLC system.

The on-line 2-DLC is set up by combining two different separation processes via an interface, which could continuously collect the effluents from the D1 column and subsequently re-inject D1 fractions into the D2 column. Most of the efforts and progress achieved in on-line 2-DLC have been devoted to the development of interfaces (Brandão et al. 2019). And due to the direct injecting of D1 effluents into the D2 column, the solvent compatibility between the two dimensions must be seriously considered, which is the key point for constructing a successful on-line 2-DLC (Li et al. 2013). The main advantage of on-line 2-DLC is the shorter treatment of unstable compounds and minimizing sample loss or contamination. It is also undeniable that the automation provided by on-line 2-DLC makes the separation more accurate, reproductive, and less labor-intensive than the off-line mode.

Various supports and mobile phases are available for the analysis of anthocyanins, procyanidins, flavonones and flavonols, flavan-3-ols, procyanidins, flavones, phenolic acids, and secoiridoids (Naczk and Shahidi 2004). Several factors affect HPLC analysis of phenolic compounds, including sample purification, mobile phase, column types, and detectors (Stalikas 2007). In general, purified phenolic compounds are subjected to an HPLC instrument utilizing a reversed-phase C18 column (RP-C18), diode array detector (DAD), and polar acidified organic solvents (Zeb 2015). Several reviews have reported the application of HPLC and the quantification of phenolic compounds in foods (Ignat et al. 2011; Ciulu et al. 2018; Alu'datt et al. 2017; Wang et al. 2020; Kalili and de Villiers 2011). Generally, the sensitivity and detection in HPLC are based on the purification of phenolic compounds and pre-concentration from complex matrices of crude extracts.

The purification stage includes removing the interfering compounds from the crude extract with partitionable solvents and using open column chromatography or an adsorption-desorption process (Khoddami et al. 2013). Sephadex LH-20, polyamide, Amberlite, solid-phase extraction (SPE) cartridges and styrene-divinylbenzene (XAD 4, XAD16, EXA-90, EXA 118, SP70), acrylic resins (XAD-7, EXA-31) are examples of regularly applied materials to purify phenolic compounds from crude sample extracts (Kalili and de Villiers 2011). However, in most studies, SPE is used for purification and partial concentration before separation using HPLC (Michalkiewicz et al. 2008). The use of SPE minimizes the noise to signal ratio and thus very helpful for accurate quantitation. However, due to the high cost of the SPE and limited applications, sample analysis using untargeted mass spectrometry provides more details of the composition and quantity of phenolic compounds present in foods.

16.4.3.1 HPLC Mobile Phases

Acetonitrile and methanol, or their aqueous forms, are the main mobile phases utilized in HPLC quantification of phenolic compounds. However, due to the high cost and toxicity of acetonitrile, acidified aqueous methanol is more preferred. Ethanol, tetrahydrofuran (THF), and 2-propanol have also been used. Attention is required to maintain the pH of the mobile phase in the range pH 2–4 to avoid the ionization of phenolic compounds during identification (Kalili and de Villiers 2011). Therefore, aqueous acidified mobile phases predominantly contain acetic acid but formic and phosphoric acids or phosphate, citrate, and ammonium acetate buffers at low pH are also reported (Zeb 2015). A gradient elution system is more commonly applied than an isocratic elution system (Zarena and Sankar 2012; Khoddami et al. 2013). Barwick (1997) reviewed the solvent selection for HPLC analysis. The authors showed that the polarity of the solvent can be a useful property while designing the HPLC method.

16.4.3.2 HPLC Stationary Phases

In HPLC, a large set of stationary phases are used in the form of columns. Proper column selection is a critical factor in identifying phenolic compounds in foods. Generally, based on the polarity, different classes of polyphenols can be detected using a normal phase C18 or reversed-phase (RP-C18) column having 10–30 cm in length, 3.9–4.6 mm ID, and 3–10 μm particle size. However, new types of columns (monolithic and superficially porous particles columns) from 3–25 cm length, 1–4.6 mm ID, and 1.7–10 μm particle size are employed in phenolic detection by advanced HPLC techniques like UHPLC (ultra-high pressure chromatography) and HTLC (high-temperature liquid chromatography) and two-dimensional liquid chromatography (LC × LC) (Klejdus et al. 2007). Most HPLC analyses of phenolic compounds are carried out at ambient column temperature. Recently, however, higher temperatures have also been recommended due to new columns and instrumentation. HPLC running time is the other factor that influences the detection of phenolic compounds and can range from 10 to 150 min. Roggero et al. (1997) highlighted that high reproducibility of results can be obtained when long analysis times are employed requires constant temperature. However, recently the short columns are more common for the analysis of phenolic compounds in a short time.

16.4.3.3   HPLC Detectors

There are several detectors coupled with HPLC reported in peer-reviewed publications. These are classified as *elemental detectors* (atomic absorption/emission, inductively coupled plasma–mass spectrometry and microwave-induced plasma); *optical detectors* (UV/visible, IR/Raman, optical activity, evaporative light scattering, and refractive index); *luminescent detectors* (fluorescence/phosphorescence, chemiluminescence/bioluminescence); *electrochemical detectors* (potentiometry, novel material/modified electrodes, array electrodes and pulsed and oscillometric techniques); *mass spectrometric detectors* (time-of-flight/MALDI, Fourier transform ion cyclotron resonance mass spectrometry, electrospray/thermospray, atmospheric pressure ionization, and particle beam); and *other detection systems* (nuclear magnetic resonance, radioactivity detectors, surface plasmon resonance) (Zhang et al. 2008). Swartz (2010) described the principles and attributes of many of the common HPLC detectors that in use today, and comparès and contrasts the advantages and disadvantages of the various detectors.

Phenolic compounds are often detected using UV-VIS and photodiode array (PDA) detectors at wavelengths 190–700 nm, fluorimetric (FLD), colorimetric arrays, PDA coupled with fluorescence, and chemical reaction detection techniques are other methods used (Khoddami et al. 2013). Diode array detector (DAD) has great sensitivity when the standard phenolic compounds are available. For example, 22 phenolic compounds were identified and quantified in thermally oxidized sunflower oil (Fig. 16.8). Mass spectrometric (MS) detectors attached to high performance liquid chromatography (HPLC–MS), electrospray ionization mass spectrometry (ESI-MS), matrix-assisted laser desorption/ionization mass spectrometry (MALDI–MS), fast atom bombardment mass spectrometry (FAB-MS), and electron impact mass spectrometry have also been utilized for structural characterization and confirmation of different phenolic classes (see Chap. 17). HPLC

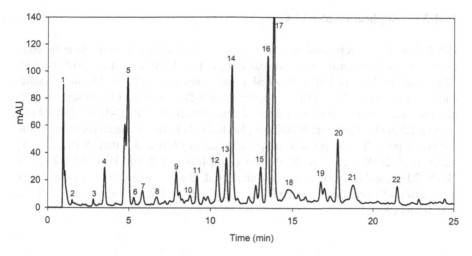

Fig. 16.8 Representative HPLC-DAD chromatogram of the phenolic profile of thermally heated sunflower oil samples containing spinach extract at 320 nm. Peak numbers representing phenolic compounds are 1 (*p*-hydroxybenzoic acid), 2 (gallic acid), 3 (vanillic acid hexoside), 4 (catechin), 5 (*p*-coumaric acid), 6 (syringic acid), 7 (tri-*O*-galloylquinic acid), 8 (caffeic acid), 9 (ellagic acid dihexoside), 10 (5-caffeoylquinic acid), 11 (spinacetine glucuronide), 12 (isorhamnetin-3-(hydroxyferuloylglucoside)-7-glucoside), 13 (isorhamnetin-3-(caffeoyl-diglucoside)-7-rhamnoside), 14 (apigenin), 15 (kaempferol), 16 (kaempferol-3-coumaroyl sinapoyldiglucoside), 17 (apigenin-2-pentoxide-8-hexoside), 18 (quercetin-3-(sinapoyldiglucoside)-7-glucoside), 19 (diometin-7-rutionside), 20 (quercetin-3-(*p*-coumaroyl-diglucoside)-7-glucoside), 21 (kaemferol-3-(*p*-coumaroyl-diglucoside)-7-glucoside) and 22 (quercetin-3-sinapolysophoroside-7-glucoside) (redrawn with the kind permission of John Wiley and Sons (Zeb and Ullah 2019))

coupled with MS detectors is highly sensitive and has the power to achieve high specificity due to the mass selectivity of detection.

HPLC–NMR and UHPLC are the other novel techniques to identify phenolic compounds in a variety of food sources. The new trends in the analysis of phenolic compounds are hydrophilic interaction liquid chromatography (HILIC) as well as two-dimensional liquid chromatography (2-D LC). HILIC may become more popular due to higher compatibility of applied mobile phase when linked to MS and enhanced accuracy to analyze polar components in complex matrices (Jandera et al. 2008; Jandera 2008). 2-D LC is a recent advance in chromatography that can afford separation and identification of structurally similar and minor compounds from complex matrices, enhancing peak capacity and selectivity (Zeng et al. 2012). A successful combination of 2-D LC × HILIC and 2-D LC × RP-LC has been used to detect polar and semi-polar fractions in traditional Chinese medicine. This combination of 2D-LC systems showed great potential to separate different components of a wide range of polarity from complex samples, which is not possible when using 1-D RPLC (Liang et al. 2012).

16.4.3.4 Application of HPLC

HPLC has been widely used for the separation and identification of phenolic compounds in foods. Several reviews revealed the detailed application of HPLC in foods (Kalili and de Villiers 2011; Pyrzynska and Sentkowska 2015). All those studies cannot be accommodated in this chapter. Arráez-Román et al. (2010) separated 23 phenolic acids and flavonoids in almond skin extracts by HPLC-TOF-MS in 9 min using a 2.5 μm C18 column. HPLC coupled to DAD have been used for the determination of phenolic compounds in spinach leaves (Zeb and Ullah 2019), chicory leaves (Zeb 2019; Zeb et al. 2019), processed foods (Zeb and Haq 2018; Zeb and Habib 2018), and leafy vegetables (Ahmad et al. 2020). Details of the studies on the application of phenolic compounds by HPLC can be found in the work of Nollet and Toldrá (2012).

16.4.4 Ultra High-Pressure Liquid Chromatography

Ultra high-pressure liquid chromatography (UHPLC or ultrahigh performance liquid chromatography) is an advanced form of liquid chromatography in which narrow-bore columns packed with very small particles (<2 μm) and mobile phase delivery systems operating at high back pressures are used (Pyrzynska and Sentkowska 2015). The major advantages of UHPLC over conventional HPLC are improved resolution, shorter retention times, and higher sensitivity. The theory behind the development of UPLC is the van Deemter equation. When the particle size decreases to less than 2.5 μm, there is a significant gain in efficiency, even when flow rates are increased or when linear velocities are increased. For the analysis of phenolic compounds in food samples, the most widely used analytical column is the BEH C18, based on the reversed-phase mode. The particles in this column are characterized by being hybrid with ethylene bridged and completely end-capped. The phenolic compounds are eluted accordingly to their polarity and molecule size. Besides the classical stationary phase with functional octadecyl groups, other more polar stationary phases with cyanopropyl and phenyl groups can be used for the separation and determination of phenolic compounds. Klejdus et al. (2008) compared different stationary phases for the separation of isoflavones. Chromatograms were acquired at the different chromatographic columns and linear-gradient elutions of 0.3% acetic acid/methanol. In the first experiment, Waters C18 BEH was used, in the second, Zorbax CN SB, and Waters Phenyl BEH columns were used in the third experiment. Other experimental conditions were UV–vis detection at 270 nm, injection volume of 0.5 μL, the concentration of analyzed compounds was 1 ng/mL. The linear gradient elution profiles are shown in Fig. 16.9.

Nováková et al. (2010) also employed a 1.7 μm C18 column and acidified water-methanol mobile phases for the separation of 29 catechins, flavonoids, and phenolic acids in 20 min. The separation of standard compounds revealed a good analytical approach as shown in Fig. 16.10. In a subsequent study, these authors used the same

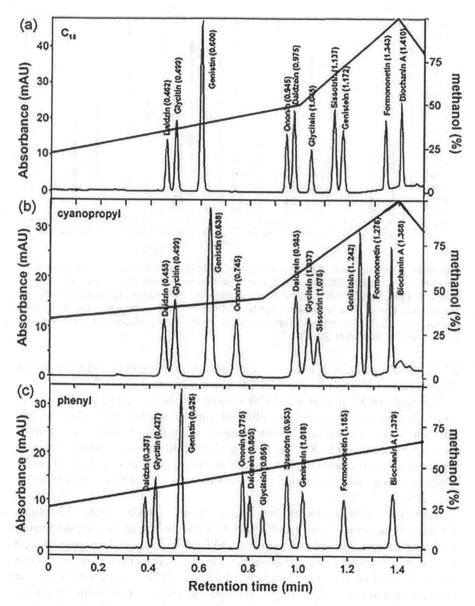

Fig. 16.9 Chromatogram of a standard mixture of 29 phenolic compounds and caffeine. (chromatographic conditions are UV 280 nm, 0.1% formic acid–methanol from 88.5:11.5 to 50:50 (v/v), 0.45 mL min⁻¹) (reproduced with kind permission of Elsevier (Nováková et al. 2010))

approach to separate 12 phenolic acids and flavonols in chamomile flower and tea extracts (Novakova et al. 2010). Gómez-Romero et al. (2010) tentatively identified a total of 135 phenolic compounds of which 21 were new, in different tomato extracts using a 46-min gradient analysis on a 150 mm, 1.8 μm column operated at

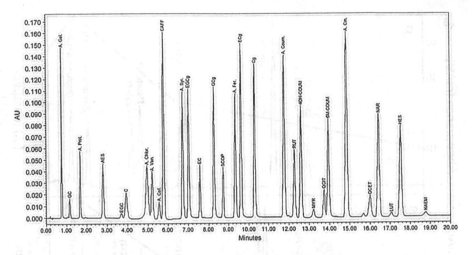

Fig. 16.10 Chromatograms of isoflavones acquired at the different chromatographic column and linear-gradient elutions of 0.3% acetic acid/methanol. In experiment (A) Waters C18 BEH, (B) Zorbax CN SB, and (C) Waters Phenyl BEH columns were used. Other experimental conditions: UV–vis detection at 270 nm was used; injection volume, 0.5 µl; concentration of analyzed compounds, 1 ng/ mL. The linear gradient elution profiles are shown (reproduced with kind permission of Elsevier (Nováková et al. 2010))

37 °C. Spáčil et al. (2010) achieved a rapid separation of eight tea catechins within 2.5 min using a 100 mm 1.7 µm BEH C18 column operated at 35 °C.

The study of Klejdus et al. (2008) showed a combination of small particles and high temperatures to achieve efficient, rapid resolution of isoflavones and phenolic acids on different stationary phases. In an earlier study, these authors used a 1.8 µm Zorbax C18 column along with acidified water-methanol mobile phases and a temperature of 80 °C to achieve the separations of up to 13 isoflavones in various soy preparations and plant extracts in less than 1 min (Klejdus et al. 2007).

Serra et al. (2009) used a 100 mm 1.8 µm high-strength silica (HSS) T3 column for the determination of procyanidins and their metabolites in rat plasma samples, following off-line solid-phase extraction (SPE). The method demonstrated high sensitivity and allowed accurate quantification of procyanidins and metabolites in plasma samples within 5 min. Similarly, Martí et al. (2010) used a similar column for the simultaneous analysis of procyanidins and anthocyanins in the rat plasma samples following µSPE. Recently, a simple and reproducible method for the qualitative and quantitative analysis of betalains in plasma samples, based on Solid Phase Extraction (SPE) and micro-high performance liquid chromatography coupled with mass spectrometry (micro-HPLC-MS/MS) was reported (Sawicki et al. 2017). The eight betalain compounds were detected and quantified and characterized in the fortified rat blood plasma samples. The developed method showed a good coefficient of determination ($R^2 = 0.999$), good recovery, precision, and appropriate limits of detection, and quantification for the identified compounds.

16.4.5 Hydrophilic Interaction Liquid Chromatography

Hydrophilic interaction liquid chromatography (HILIC) is another high-performance liquid chromatography (HPLC) technique for the separation of polar compounds. HILIC is a type of normal phase liquid chromatography, but the mechanism of separation in HILIC is more intricate than that in NP-LC. The mechanism of separation is either based on the partition between the mobile and stationary phase, and the adsorption of the analyte onto the surface of the adsorbent. The retention of the analyte may be based on the partition, and intermolecular interactions between solute and solvent, solute and stationary phase. The retention behavior has been reviewed by Jandera and Janás (2017). The sample structural effects on the retention are treated in terms of the Linear Solvation Energy Relationship model adopted for partially ionized polar compounds has been discussed in detail. The authors also reported the impact of the dual HILIC reversed-phase (RP) mechanism on a possible increase of the applications of some polar columns.

Generally, HILIC stationary phases consist of classical bare silica or silica gels modified with many polar functional groups such as amino or cyano, whereas the mobile phases are of similar nature of polarity as used in reversed-phase liquid chromatography. Qiao et al. (2016) reviewed the stationary phases used in HILIC. The authors showed that many new stationary phases have been developed recently by introducing a variety of functional groups like zwitterionic groups, hydrophilic macromolecules, and ionic liquids, which can provide a wide range of selectivity and applications. The special separation materials for HILIC showed a good selectivity and reproducibility for the separation of polar compounds such as phenolic compounds. HILIC also allows the analysis of charged substances, as in ion chromatography (IC) (Buszewski and Noga 2012). It is highly suitable for the analysis of specific compounds in complex systems that elute near the void in RP-HPLC. Polar samples always demonstrated good solubility in the aqueous mobile phase used in HILIC, which overcomes the drawbacks of the poor solubility often encountered in NP-LC. This lists solvents according to increasing elution strength. Relative solvent strengths in HILIC can be approximately summarized as follows:

Acetone<isopropanol~propanol<acetonitrile<ethanol<dioxane<DMF ~ metha nol<wateracetone<isopropanol~propanol<acetonitrile<ethanol<dioxane< DMF ~ methanol<water.

The mobile phase used in HILIC can be either as an isocratic or gradient elution. Certain additives can also be used such as ammonium acetate, ammonium formate, sodium acetate, and sodium perchlorate.

16.4.5.1 Application of HILIC

HILIC has been found to an effective technique for the separation of phenolic compounds and has significant potential in food analysis (Bernal et al. 2011; Sentkowska and Pyrzynska 2019). The Merck ZIC-pHILIC column was found to give better chromatographic peak shapes and generally improved resolution for a wider range of tea phenolic compounds than the silica-based HILIC column (Fraser et al. 2012). Marrubini et al. (2018) reviewed the applications of HILIC for the analysis of several bioactive metabolites including phenolic compounds.

The effect of several operating experimental parameters on the separation of 15 flavonoids was investigated using zwitterionic sulfobetaine functionalized methacrylate stationary phase (Sentkowska et al. 2013). Acetonitrile and methanol were compared as organic modifiers in the mobile phase. Both solvents provided the same elution order of the target flavonoids. However, increasing the content of organic solvent, methanol gave higher retention factors and allowed the stationary phase to retain more strongly the analytes in comparison with acetonitrile. Methanol and/or other low molecular weight alcohols-based mobile phases can be used under isocratic conditions but in this case, methanol could not help to resolve the critical couples of peaks represented by kaempferol and luteolin, and by hesperidin and quercetin. No significant changes in the retention were registered when pH was increased from 2.8 to 9.0. This was attributed to the low net negative surface charge at the boundary between the flowing and the static streams of mobile phase flowing over the SP, which were only a little affected by pH. Considering the effect of salt in the mobile phase, it was concluded that the ion-exchange mechanism did not contribute significantly to the overall retention of the more hydrophobic aglycons (*viz*, rhamnetin, kaempferol, and quercetin). For quercetin glycosides, only high salt concentrations resulted in a decrease in retention. The column temperature was also investigated for the flavonoids exhibiting sufficient retention under isocratic conditions (methanol or acetonitrile-10 mM ammonium formate, pH 7, 95:5, v/v), and only a slight decrease in retention was registered with the increasing of temperature. The selected temperature for the analysis of hesperidin and naringin in orange juices was 30 °C.

Willemse et al. (2013) used HILIC for the analysis of anthocyanins. After an evaluation of several HILIC stationary and mobile phases, the authors concluded that an amide functionalized stationary phase provided the best performance when a mobile phase of water-acetonitrile containing trifluoroacetamide (TFA, 0.4% v/v) was used. An acidic mobile phase was essential for ensuring satisfactory peak shapes and the prevalence of the flavilium cationic species of these molecules. The interesting mechanism underlying this observation is that the flavilium ion at neutral to basic pH through hydration could easily lead to the formation of the corresponding carbinol pseudo-base, which in turn undergoes further ring opening to yield the chalcone. In this case, further improvement in chromatographic performances may be obtained by increasing the column temperature to 50 °C. The retention behavior of anthocyanins resulted qualitatively predictable; in fact, higher is the degree of glycosylation, and thus higher is the hydrophilic character of the molecule, stronger is the retention, as already observed by many other groups for other classes of compounds (e.g., carbohydrates or nucleosides/nucleotides). HILIC was successfully used for the analysis of anthocyanins as demonstrated in the assessment of anthocyanin profiles of five different food matrices, namely blueberries, red cabbage, red radish, grape skins, and black beans.

Proanthocyanidins are oligomeric or polymeric compounds composed of flavan-3-ol (the most represented being catechin) and flavan-3,4-diol units. Their separation and determination are generally problematic because these molecules exist as complex mixtures of monomers, oligomers, and polymers which are present in vegetables and fruits to a degree of polymerization still not well defined. In recent years, HILIC has been successfully applied for the separation of condensed tannins in wild vegetables such as sea buckthorn berries (Yang et al. 2017). After extraction of the soluble fraction from dried material in acetone–water-acetic acid (90:19.5:0.5, v/v/v), proanthocyanidins were isolated from flavonols and other impurities using a column with hydroxypropylated dextran beads crosslinked to yield a polysaccharide network. Differently from RPLC and NPLC, HILIC mode provided the separation of proanthocyanidins based on the degree of polymerization and molecular size, with early eluting compounds having a lower degree of polymerization (starting from monomers, and dimers) up to a degree of polymerization as high as 11. All the condensed tannins observed were B-type with epigallocatechin being the monomeric unit more frequently observed. The quantification of dimers, trimers, and tetramers was carried out with ESI-MS-SIR (Kallio et al. 2014). Using the same purification steps, column, chromatographic and ESI-MS conditions, the proanthocyanidins profile and content were investigated in three different subspecies of wild and cultivated sea buckthorn from China, Finland, and Canada (Yang et al. 2016). The results indicated that the berries grown in northern Finland were the richest in B-type proanthocyanidins, evidencing also a strong influence of genetic background and interaction with growth location on proanthocyanidins composition and content.

16.5 Gas Chromatography

Gas chromatography (GC) is another important separation technique. It is used to separate both organic and inorganic compounds. However, it is widely used for volatile compounds. In gas chromatography, the mobile phase is a gas, while the stationary phase is usually a packed solid material. A simple mixture is vaporized and separated by a carrier gas. The components of the sample (called solutes or analytes) separate from one another based on their relative vapor pressures and affinities for the stationary phase (McNair et al. 2019). GC is performed using a long and very thin column coated on the inside with a silica-based solid phase applied as a thin film. The column is placed in an oven in which the temperature can be changed quickly and with great accuracy.

The samples are generally dissolved in a volatile organic solvent having a little affinity for the solid phase. A sample of volume (1–5 μL) is loaded onto the column through a septum using a syringe with a thin needle (Vermerris and Nicholson 2007). The sample is then moved through the column with an inert carrier gas such as helium (Fig. 16.11). The interaction of the compound with the solid phase is temperature-dependent, and elution of the compounds of interest is achieved by increasing the temperature. Detection of the compounds at the end of the column is nowadays mostly based on mass spectrometry, though flame ionization detectors (FID) are used for routine analyses of samples in which there are only a limited number of compounds that can be easily identified based on retention time. To increase the volatility of phenolic compounds, chemical modifications such as methylation, acetylation, or silylation can be employed.

16.5.1 Derivatization of Phenolic Compounds

Phenolic compounds are derivatized to form volatile products with high sensitivity for GC analysis. Farajzadeh et al. (2014) reviewed the derivatization of a variety of compounds for the analysis using GC. GC can be successfully applied for the separation, identification, and quantification of phenolic compounds such as phenolic acids, condensed tannins, and flavonoids (Khoddami et al. 2013). The major concerns of GC analysis, that do not apply to HPLC techniques, are the derivatization and volatility of phenolic compounds. With GC, quantification of phenolics from food matrices may involve clean-up steps such as lipid removal from the extract, release of phenolic compounds from the glycoside, and ester bonds in enzymatic, alkaline, and acidic media, and chemical modification steps, such as transformation to more volatile derivatives.

The derivatization is either before or after column separation. Pre-column derivatization reactions are performed "off-line" in a reaction vessel that is

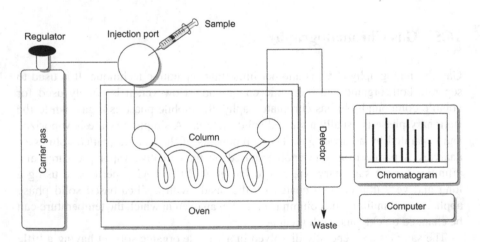

Fig. 16.11 Schematic representation of gas chromatogram (GC)

separated from the GC analysis hardware. As an alternative to off-line derivatization, "on-line" derivatizations are found. Compared with off-line derivatization, the on-line method eliminates time-consuming sample-processing steps, decreases the amount of reagent and organic solvents required for derivatization, and increases the speed and the efficiency of the analysis. Automation is another potential benefit of on-line derivatization. Injection-port derivatization or on-column derivatization is an example of an online mode. Alcohols and phenols possess several properties that make their GC analysis difficult.

Most derivatization agents used in the analysis of alcohols are also applicable for phenols. There are many derivatization methods, including acylation, silylation, and alkylation, for the derivatization of phenols and their derivatives. The acidity of phenols (pKa = 9–11) is higher than alcohols (pKa = 16–19), so permitting the use of additional derivatization reactions.

Acetylation is one of the procedures widely employed to convert alcohols and phenols into less polar compounds to increase their extraction efficiencies. Acyl derivatives of both alcohols and phenols are prepared with acid anhydrides and acid chlorides. Also, amide reagents have been used for the acylation of alcohols. This procedure can be catalyzed by bases or acids (Farajzadeh et al. 2014).

The hydroxyl group of alcohols and phenols can be readily silylated. The moisture-free conditions, high temperatures (60–80 °C), and long reaction times (30–50 min) are the limiting factors for silylation. Dimethylsilyl derivatives of alcohols are less volatile than TMS compounds, but they can be used as alternatives for alcohols whose TMS derivatives are not separable. A broad range of phenolic compounds including nitrophenols can be silylated quantitatively using silylating reagents. Silylation can be accomplished by using BTSA and BTMSTFA. It should be noted that trialkylsilyl groups increase the total ion current, and hence the sensitivity of positive-ion MS. Derivatization of hindered phenols requires disruptive conditions. Trifluoroacetic acid promotes the silylation of these compounds. Halogenated silyl derivatives of alcohols and phenols are useful for GC-ECD analysis. The iodo derivatives are more sensitive to ECD than the bromo derivatives (Farajzadeh et al. 2014). The silylation reaction is simple, free of unwanted side products, and produces tremendously volatile products with no interference with the analysis. Silyl derivatization is thus a very good option to identify phenolic compounds but more research is needed on the identification of silyl derivatives. Phenolic compounds from cranberry were derivatized using N, O-bis(trimethylsilyl) trifluoroacetamide (BSTFA) +1% trimethylchlorosilane (TMCS) reagents and 20 phenolics were thus identified which were mainly in conjugated forms (Wang and Zuo 2011).

Alkylation of the hydroxyl group yields a less polar derivative by nucleophilic displacement of active hydrogen with the alkyl group. Diazomethane and alkyl halides have been used as the alkylation reagents for alcohols and phenols. Phenols react with diazomethane without a catalyst.

p-Toluenesulfonyl chloride (TSC) is a chemical reagent for the tosylation of -OH and -NH groups, but this is scarcely used for analytical purposes, due to its insolubility in water. The derivatization reaction is performed under mild conditions and in the presence of organic solvents to modify TSC solubility in water. There are several types of reagents used to modify and create volatile derivatives. Ethyl and methyl chloroformate, diazomethane, and dimethyl sulfoxide in combination with methyl iodate are used to make methyl or ethyl esters of phenolics. However, in some studies, substantial confusion may occur due to the presence of methyl esters in a natural form (Hušek 1992). Another generation of reagents, which have advantages in the creation of volatile compounds, are the trimethylsilyl family of compounds, such as trifluoroacetymide, *N*-(tert-butyldimethylsilyl)-*N*-methyltrifluoroacetamide, and trimethylsilyl derivatives.

16.5.2   GC Stationary Phases

Fused silica capillaries of 30 m lengths, with internal diameters of 25–32 μm and stationary phase particle size of 0.25 μm are the most common columns used for phenolic quantification in GC techniques. There are exceptions, however, such as the column used with 15 m length and 10 μm film thickness (Shadkami et al. 2009).

16.5.3   GC Detectors

The use of a flame ionization detector (FID) is the most common method to detect phenolics but mass spectroscopy (MS) has become widespread recently (Robbins 2003). GC provides more sensitivity and selectivity when combined with mass spectrometry (Canini et al. 2007). For instance, the difficulties of flavonoid glycoside evaluation by conventional GC were solved when high-temperature–high-resolution GC–MS was applied (dos Santos Pereira et al. 2004). Another study indicated that GC-MS analysis of phenolic and flavonoid standards was more efficient than that of HPLC, providing a fast analysis with better resolution and baseline separation of all standards with minimum co-elution (Khoddami et al. 2013). However, it is not the case in all foods and other phenolic compounds. For example, phenolic compounds in olive oils were accurately separated and determined with LC-MS than GC-MS (Bajoub et al. 2016).

16.5.4   Applications of GC

Synthetic phenolic antioxidant was extracted using SPE and determined by gas chromatography-mass spectrometry (GC–MS) after derivatization with N-methyl-N-(*tert*-butyldimethylsilyl)-trifluoroacetamide (MTBSTFA) in a single run (Rodil et al. 2010). A new analytical approach based on gas chromatography coupled to atmospheric pressure chemical ionization-time of flight mass spectrometry was evaluated for its applicability for the analysis of phenolic compounds from extra-virgin olive oil (Garcia-Villalba et al. 2011). Figure 16.12 shows the Base Peak Chromatogram (BPC) of the Diol-SPE extract of a mixture of Arbequina and Picual oils achieved using the optimum GC–APCI-MaXis MS procedure. Garcia-Villalba et al. (2011) studied the effects of several parameters such as the concentration of derivatization reagent, reaction time, and temperature were studied. We have achieved the best performance adding 50 µl of BSTFA +1% TMCS to the dried sample at room temperature and incubation time 1 h. The effect of including an intermediate step of methoxyamination was adequately evaluated and no change in the peak area or stability was observed.

Twenty-one phenolic compounds were identified and quantified in olive oil. 20 phenolic compounds were identified in cranberry samples including benzoic acid, quercetin, and myricetin as the most abundant phenolics (Wang and Zuo 2011). A new GC-MS method for analyzing 16 microbial phenolic acid metabolites derived from grape flavanols was reported based on targeted analysis using QqQ/MS

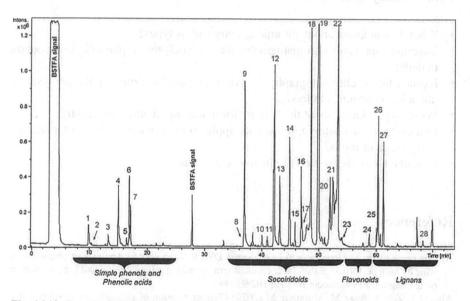

Fig. 16.12 Base Peak Chromatogram (BPC) of the Diol-SPE extract of a mixture of Arbequina and Picual olive oils. *Peak identification*: **1**, Ty-2H + 2TMS; **2**, isolated 4; **3**, *m/z* 281.0966/192.9388; **4**, Hyty-3H + 3TMS; **5**, Protocatechuic acid-3H + 3TMS + H; **6**, Dopac-3H + 3TMS + H; **7**, EA-1H + 1TMS + H; **8**, EA-1H + 1TMS + H/isolated 2; **9**, isolated 6 (D-Lig Agl); **10**, isolated 6; **11**, isolated 6; **12**, DOA-2H + 2TMS + H; **13**, *m/z* 501.3843/411.3312; **14**, Lig Agl-1H + 1TMS + H; **15**, methyl Ol Agl-2H + 2TMS + H; **16**, H-D-Ol Agl-3H + 3TMS + H; **17**, isolated 6 (Lig Agl); **18**, isolated 6 (Lig Agl); **19**, Ol Agl-2H + 2TMS + H; **20**, isolated 8; **21**, 10 H-Ol Agl-3H + 3TMS + H; **22**, Ol Agl-3H + 3TMS + H; **23**, Apigenin-3H + 3TMS + H; **24**, Luteolin-4H + 4TMS + H; **25**, Pinoresinol-2H + 2TMS + H; **26**, *m/z* 397.3825; **27**, acetoxy-pinoresinol-2H + 2TMS + H; **28**, Syringaresinol-2H + 2TMS + H. (For interpretation of the references to color in this figure legend, the reader is referred to the web version) (reproduced with kind permission of Elsevier (Garcia-Villalba et al. 2011))

under MRM mode showed high sensitivity, precision, and recovery (Carry et al. 2018).

A comprehensive characterization of *P. peruviana* calyx extracts, obtained by an optimized pressurized liquid extraction (PLE) procedure, was developed by Ballesteros-Vivas et al. (2019) applying first a UPLC-ESI-q-TOF-MS/MS method in positive and negative electrospray ionization (ESI) mode along with gas chromatography quadrupole time-of-flight mass spectrometry (GC-q-TOF-MS). The authors revealed the presence of several phenolic compounds. The results demonstrated the great potential of applying integrated identification strategies to high-resolution mass spectrometry (HRMS) data obtained from complementary LC- and GC-q-TOF-MS(/MS) platforms, as powerful identification tools for improving our understanding of the phytochemical composition of natural extracts intended to be used in functional foods.

16.6 Study Questions

- What do you know about chromatography and its types?
- Describe thin layer chromatography for the analysis of phenolic compounds in detail.
- Explain liquid chromatography, its types, and applications for the analysis of phenolic compounds in foods.
- What do you know about the high performance liquid chromatography?
- Discuss gas chromatography and its applications for the analysis of phenolic compounds in foods.
- Describe the derivitization of phenolic compounds.

References

Agatonovic-Kustrin S, Hettiarachchi CG, Morton DW, Razic S (2015) Analysis of phenolics in wine by high performance thin-layer chromatography with gradient elution and high resolution plate imaging. J Pharmaceut Biomed 102:93–99

Ahmad S, Zeb A, Ayaz M, Murkovic M (2020) Characterization of phenolic compounds using UPLC-HRM and HPLC-DAD and anticholinesterase and antioxidant activities of *Trifolium repens* L. leaves. Eur Food Res Technol 246:485–496

Alu'datt MH, Rababah T, Alhamad MN, Al-Mahasneh MA, Almajwal A, Gammoh S, Ereifej K, Johargy A, Alli I (2017) A review of phenolic compounds in oil-bearing plants: distribution, identification and occurrence of phenolic compounds. Food Chem 218:99–106

Anyakudo F, Adams E, Van Schepdael A (2020) Thin-layer chromatography–flame ionization detection. Chromatographia 83(2):149–157

Arráez-Román D, Fu S, Sawalha SMS, Segura-Carretero A, Fernández-Gutiérrez A (2010) HPLC/CE-ESI-TOF-MS methods for the characterization of polyphenols in almond-skin extracts. Electrophoresis 31(13):2289–2296

Asquith TN, Izuno CC, Butler LG (1983) Characterization of the condensed tannin (proanthocyanidin) from a group II sorghum. J Agric Food Chem 31(6):1299–1303

Azar M, Verette E, Brun S (1987) Identification of some phenolic compounds in bilberry juice *Vaccinium myrtillus*. J Food Sci 52(5):1255–1257

Bajoub A, Pacchiarotta T, Hurtado-Fernandez E, Olmo-Garcia L, Garcia-Villalba R, Fernandez-Gutierrez A, Mayboroda OA, Carrasco-Pancorbo A (2016) Comparing two metabolic profiling approaches (liquid chromatography and gas chromatography coupled to mass spectrometry) for extra-virgin olive oil phenolic compounds analysis: a botanical classification perspective. J Chromatogr A 1428:267–279

Ballesteros-Vivas D, Álvarez-Rivera G, Ibáñez E, Parada-Alfonso F, Cifuentes A (2019) A multi-analytical platform based on pressurized-liquid extraction, in vitro assays and liquid chromatography/gas chromatography coupled to high resolution mass spectrometry for food by-products valorisation. Part 2: characterization of bioactive compounds from goldenberry (*Physalis peruviana* L.) calyx extracts using hyphenated techniques. J Chromatogr A 1584:144–154

Barwick VJ (1997) Strategies for solvent selection — a literature review. TrAC Trends Anal Chem 16(6):293–309

Bernal J, Ares AM, Pól J, Wiedmer SK (2011) Hydrophilic interaction liquid chromatography in food analysis. J Chromatogr A 1218(42):7438–7452

Bräm S, Wolfram E (2017) Recent advances in effect-directed enzyme assays based on thin-layer chromatography. Phytochem Analysis 28(2):74–86

Brandão PF, Duarte AC, Duarte RMBO (2019) Comprehensive multidimensional liquid chromatography for advancing environmental and natural products research. TrAC Trends Anal Chem 116:186–197

Buszewski B, Noga S (2012) Hydrophilic interaction liquid chromatography (HILIC)—a powerful separation technique. Anal Bioanal Chem 402(1):231–247

Canini A, Alesiani D, D'Arcangelo G, Tagliatesta P (2007) Gas chromatography–mass spectrometry analysis of phenolic compounds from *Carica papaya* L. leaf. J Food Compos Anal 20(7):584–590

Carry E, Zhao D, Mogno I, Faith J, Ho L, Villani T, Patel H, Pasinetti GM, Simon JE, Wu Q (2018) Targeted analysis of microbial-generated phenolic acid metabolites derived from grape flavanols by gas chromatography-triple quadrupole mass spectrometry. J Pharmaceut Biomed 159:374–383

Cheng S-C, Huang M-Z, Shiea J (2011) Thin layer chromatography/mass spectrometry. J Chromatogr A 1218(19):2700–2711

Choma IM, Olszowy M, Studziński M, Gnat S (2019) Determination of chlorogenic acid, polyphenols and antioxidants in green coffee by thin-layer chromatography, effect-directed analysis and dot blot – comparison to HPLC and spectrophotometry methods. J Sep Sci 42(8):1542–1549

Ciulu M, Cadiz-Gurrea ML, Segura-Carretero A (2018) Extraction and analysis of phenolic compounds in Rice: a review. Molecules 23(11):2890

Coskun O (2016) Separation techniques: Chromatography. North Clin Istanb 3(2):156–160

Dabrowski KJ, Sosulski FW (1984) Quantitation of free and hydrolyzable phenolic acids in seeds by capillary gas-liquid chromatography. J Agric Food Chem 32(1):123–127

Derdelinckx G, Jerumanis J (1984) Separation of malt and hop proanthocyanidins on Fractogel TSK HW-40 (S). J Chromatogr A 285:231–234

dos Santos Pereira A, Costa Padilha M, Radler de Aquino Neto F (2004) Two decades of high temperature gas chromatography (1983–2003): what's next? Microchem J 77(2):141–149

Esparza CA, Polovkov NY, Borisov RS, Varlamov AV, Zaikin VG (2018) A new post-chromatographic derivatization approach to the identification of alcohols and phenols in complex mixtures by a combination of planar chromatography and matrix-assisted laser desorption/ionization mass spectrometry. J Anal Chem 73(13):1242–1247

Farajzadeh MA, Nouri N, Khorram P (2014) Derivatization and microextraction methods for determination of organic compounds by gas chromatography. TrAC Trends Anal Chem 55:14–23

Fraser K, Harrison SJ, Lane GA, Otter DE, Hemar Y, Quek S-Y, Rasmussen S (2012) Non-targeted analysis of tea by hydrophilic interaction liquid chromatography and high resolution mass spectrometry. Food Chem 134(3):1616–1623

Fulcrand H, Remy S, Souquet J-M, Cheynier V, Moutounet M (1999) Study of wine tannin oligomers by on-line liquid chromatography electrospray ionization mass spectrometry. J Agric Food Chem 47(3):1023–1028

Garcia-Villalba R, Pacchiarotta T, Carrasco-Pancorbo A, Segura-Carretero A, Fernandez-Gutierrez A, Deelder AM, Mayboroda OA (2011) Gas chromatography-atmospheric pressure chemical ionization-time of flight mass spectrometry for profiling of phenolic compounds in extra virgin olive oil. J Chromatogr A 1218(7):959–971

Glavnik V, Vovk I (2019) High performance thin-layer chromatography–mass spectrometry methods on diol stationary phase for the analyses of flavan-3-ols and proanthocyanidins in invasive Japanese knotweed. J Chromatogr A 1598:196–208

Gómez-Romero M, Segura-Carretero A, Fernández-Gutiérrez A (2010) Metabolite profiling and quantification of phenolic compounds in methanol extracts of tomato fruit. Phytochemistry 71(16):1848–1864

Hušek P (1992) Fast derivatization and GC analysis of phenolic acids. Chromatographia 34(11):621–626

Ignat I, Volf I, Popa VI (2011) A critical review of methods for characterisation of polyphenolic compounds in fruits and vegetables. Food Chem 126(4):1821–1835

Jandera P (2008) Stationary phases for hydrophilic interaction chromatography, their characterization and implementation into multidimensional chromatography concepts. J Sep Sci 31(9):1421–1437

Jandera P, Janás P (2017) Recent advances in stationary phases and understanding of retention in hydrophilic interaction chromatography. A review. Anal Chim Acta 967:12–32

Jandera P, Vynhuchalova K, Hajek T, Cesla P, Vohralik G (2008) Characterization of HPLC columns for two-dimensional LC x LC separations of phenolic acids and flavonoids. J Chemom 22(3–4):203–217

Jaworski AW, Lee CY (1987) Fractionation and HPLC determination of grape phenolics. J Agric Food Chem 35(2):257–259

Kalili KM, de Villiers A (2011) Recent developments in the HPLC separation of phenolic compounds. J Sep Sci 34(8):854–876

Kallio H, Yang W, Liu PZ, Yang BR (2014) Proanthocyanidins in Wild Sea buckthorn (*Hippophae rhamnoides*) berries analyzed by reversed-phase, Normal-phase, and hydrophilic interaction liquid chromatography with UV and MS detection. J Agric Food Chem 62(31):7721–7729

Karchesy JJ, Hemingway RW (1986) Condensed tannins: (4β-8;2β-O-7)-linked procyanidins in *Arachis hypogea* L. J Agric Food Chem 34(6):966–970

Karchesy J, Bae Y, Chalker-Scott L, Helm R, Foo L (1989) Chromatography of Proanthocyanidins. In: Hemingway R, Karchesy J, Branham S (eds) Chemistry and significance of condensed tannins. Plenum Press, New York, pp 139–151

Kennedy JA, Taylor AW (2003) Analysis of proanthocyanidins by high-performance gel permeation chromatography. J Chromatogr A 995(1):99–107

Khoddami A, Wilkes MA, Roberts TH (2013) Techniques for analysis of plant phenolic compounds. Molecules 18(2):2328–2375

Klejdus B, Vacek J, Benešová L, Kopecký J, Lapčík O, Kubáň V (2007) Rapid-resolution HPLC with spectrometric detection for the determination and identification of isoflavones in soy preparations and plant extracts. Anal Bioanal Chem 389(7):2277–2285

Klejdus B, Vacek J, Lojková L, Benešová L, Kubáň V (2008) Ultrahigh-pressure liquid chromatography of isoflavones and phenolic acids on different stationary phases. J Chromatogr A 1195(1):52–59

Lebot V, Michalet S, Legendre L (2016) Identification and quantification of phenolic compounds responsible for the antioxidant activity of sweet potatoes with different flesh colours using high performance thin layer chromatography (HPTLC). J Food Compos Anal 49:94–101

Li K, Zhu W, Fu Q, Ke Y, Jin Y, Liang X (2013) Purification of amide alkaloids from *Piper longum* L. using preparative two-dimensional normal-phase liquid chromatography × reversed-phase liquid chromatography. Analyst 138(11):3313–3320

Liang Z, Li K, Wang X, Ke Y, Jin Y, Liang X (2012) Combination of off-line two-dimensional hydrophilic interaction liquid chromatography for polar fraction and two-dimensional hydrophilic interaction liquid chromatography×reversed-phase liquid chromatography for medium-polar fraction in a traditional Chinese medicine. J Chromatogr A 1224:61–69

Marrubini G, Appelblad P, Maietta M, Papetti A (2018) Hydrophilic interaction chromatography in food matrices analysis: an updated review. Food Chem 257:53–66

Martí M-P, Pantaleón A, Rozek A, Soler A, Valls J, Macià A, Romero M-P, Motilva M-J (2010) Rapid analysis of procyanidins and anthocyanins in plasma by microelution SPE and ultra-HPLC. J Sep Sci 33(17–18):2841–2853

Mateus N, Silva AMS, Vercauteren J, de Freitas V (2001) Occurrence of anthocyanin-derived pigments in red wines. J Agric Food Chem 49(10):4836–4840

McNair HM, Miller JM, Snow NH (2019) Basic gas chromatography. Wiley, Boca Raton

Meyer VR (2013) Practical high-performance liquid chromatography. Wiley, Boca Raton

Michalkiewicz A, Biesaga M, Pyrzynska K (2008) Solid-phase extraction procedure for determination of phenolic acids and some flavonols in honey. J Chromatogr A 1187(1–2):18–24

Móricz ÁM, Lapat V, Morlock GE, Ott PG (2020) High-performance thin-layer chromatography hyphenated to high-performance liquid chromatography-diode array detection-mass spectrometry for characterization of coeluting isomers. Talanta 219:121306

Naczk M, Shahidi F (2004) Extraction and analysis of phenolics in food. J Chromatogr A 1054(1):95–111

Naczk M, Shahidi F (2006) Phenolics in cereals, fruits and vegetables: occurrence, extraction and analysis. J Pharmaceut Biomed 41(5):1523–1542

Nollet LM, Toldrá F (2012) Food analysis by HPLC. CRC Press

Nonaka G-I, Morimoto S, Nishioka I (1983) Tannins and related compounds. Part 13. Isolation and structures of trimeric, tetrameric, and pentameric proanthicyanidins from cinnamon. J Chem Soc Perkin Trans 1 (0):2139–2145

Nováková L, Spáčil Z, Seifrtová M, Opletal L, Solich P (2010) Rapid qualitative and quantitative ultra high performance liquid chromatography method for simultaneous analysis of twenty nine common phenolic compounds of various structures. Talanta 80(5):1970–1979

Novakova L, Vildova A, Mateus JP, Goncalves T, Solich P (2010) Development and application of UHPLC-MS/MS method for the determination of phenolic compounds in chamomile flowers and chamomile tea extracts. Talanta 82(4):1271–1280

Peck H, Stott A, Turner J (1986) Separation of dihydroxybenzoic acids and trihydroxybenzoic acids by two-dimensional thin-layer chromatógraphy. J Chromatogr A 367(1):289–292

Pyrzynska K, Sentkowska A (2015) Recent developments in the HPLC separation of phenolic food compounds. Crit Rev Anal Chem 45(1):41–51

Qiao L, Shi X, Xu G (2016) Recent advances in development and characterization of stationary phases for hydrophilic interaction chromatography. TrAC Trends Anal Chem 81:23–33

Rabel F, Sherma J (2016) A review of advances in two-dimensional thin-layer chromatography. J Liq Chromatogr Rel Technol 39(14):627–639

Rabel F, Sherma J (2017) Review of the state of the art of preparative thin-layer chromatography. J Liq Chromatogr Rel Technol 40(4):165–176

Robbins RJ (2003) Phenolic acids in foods: an overview of analytical methodology. J Agric Food Chem 51(10):2866–2887

Rodil R, Quintana JB, Basaglia G, Pietrogrande MC, Cela R (2010) Determination of synthetic phenolic antioxidants and their metabolites in water samples by downscaled solid-phase extraction, silylation and gas chromatography-mass spectrometry. J Chromatogr A 1217(41):6428–6435

Roggero J-P, Archier P, Coen S (1997) Chromatography of Phenolics in wine. In: wine, vol 661. ACS symposium series, vol 661. American Chemical Society, pp 6-11

Sajewicz M, Staszek D, Waksmundzka-Hajnos M, Kowalska T (2012) Comparison of Tlc and Hplc fingerprints of phenolic acids and flavonoids fractions derived from selected sage (salvia) species. J Liq Chromatogr Relat Technol 35(10):1388–1403

Salagoïty-Auguste M-H, Bertrand A (1984) Wine phenolics—analysis of low molecular weight components by high performance liquid chromatography. J Sci Food Agric 35(11):1241–1247

Sawicki T, Juśkiewicz J, Wiczkowski W (2017) Using the SPE and micro-HPLC-MS/MS method for the analysis of betalains in rat plasma after red beet administration. Molecules 22(12):2137

Sentkowska A, Pyrzynska K (2019) HILIC chromatography: powerful technique in the analysis of polyphenols. In: Watson RR (ed) polyphenols in plants. 2nd edn. Academic, New York, pp 341-351

Sentkowska A, Biesaga M, Pyrzynska K (2013) Effects of the operation parameters on HILIC separation of flavonoids on zwitterionic column. Talanta 115:284–290

Serra A, Macià A, Romero M-P, Salvadó M-J, Bustos M, Fernández-Larrea J, Motilva M-J (2009) Determination of procyanidins and their metabolites in plasma samples by improved liquid chromatography–tandem mass spectrometry. J Chromatogr B 877(11):1169–1176

Shadkami F, Estevez S, Helleur R (2009) Analysis of catechins and condensed tannins by thermally assisted hydrolysis/methylation-GC/MS and by a novel two step methylation. J Anal Appl Pyrolysis 85(1):54–65

Sherma J, Rabel F (2018) A review of thin layer chromatography methods for determination of authenticity of foods and dietary supplements. J Liq Chromatogr Relat Technol 41(10):645–657

Snyder LR, Kirkland JJ, Dolan JW (2011) Introduction to modern liquid chromatography. Wiley, Boca Raton

Spáčil Z, Nováková L, Solich P (2010) Comparison of positive and negative ion detection of tea catechins using tandem mass spectrometry and ultra high performance liquid chromatography. Food Chem 123(2):535–541

Stalikas CD (2007) Extraction, separation, and detection methods for phenolic acids and flavonoids. J Sep Sci 30(18):3268–3295

Stanek N, Kafarski P, Jasicka-Misiak I (2019) Development of a high performance thin layer chromatography method for the rapid qualification and quantification of phenolic compounds and abscisic acid in honeys. J Chromatogr A 1598:209–215

Sun B, Leandro C, Ricardo da Silva JM, Spranger I (1998) Separation of grape and wine Proanthocyanidins according to their degree of polymerization. J Agric Food Chem 46(4):1390–1396

Swartz M (2010) HPLC DETECTORS: A BRIEF REVIEW. J Liq Chromatogr Relat Technol 33(9–12):1130–1150

Vermerris W, Nicholson R (2007) Phenolic compound biochemistry. Springer Science & Business Media, New York

Vovk I, Simonovska B, Vuorela H (2005) Separation of eight selected flavan-3-ols on cellulose thin-layer chromatographic plates. J Chromatogr A 1077(2):188–194

Wang C, Zuo Y (2011) Ultrasound-assisted hydrolysis and gas chromatography–mass spectrometric determination of phenolic compounds in cranberry products. Food Chem 128(2):562–568

Wang Z, Li S, Ge S, Lin S (2020) Review of distribution, extraction methods, and health benefits of bound Phenolics in food plants. J Agric Food Chem 68(11):3330–3343

Willemse CM, Stander MA, de Villiers A (2013) Hydrophilic interaction chromatographic analysis of anthocyanins. J Chromatogr A 1319:127–140

Yadav R, Yadav BS, Yadav RB (2017) Phenolic profile and antioxidant activity of thermally processed sponge gourd (Luffa cylindrica) as studied by using high performance thin layer chromatography (HPTLC). Int J Food Prop 20(9):2096–2112

Yang W, Laaksonen O, Kallio H, Yang BR (2016) Proanthocyanidins in sea buckthorn (Hippophae rhamnoides L.) berries of different origins with special reference to the influence of genetic background and growth location. J Agric Food Chem 64(6):1274–1282

Yang W, Laaksonen O, Kallio H, Yang BR (2017) Effects of latitude and weather conditions on proanthocyanidins in berries of Finnish wild and cultivated sea buckthorn (*Hippophae rhamnoides* L. ssp rhamnoides). Food Chem 216:87–96

Zang Y, Cheng Z, Wu T (2020) TLC bioautography on screening of bioactive natural products: an update review. Curr Anal Chem 16(5):545–556

Zarena AS, Sankar KU (2012) Phenolic acids, flavonoid profile and antioxidant activity in mangosteen (*Garcinia mangostana* L.) pericarp. J Food Biochem 36(5):627–633

Zeb A (2015) A reversed phase HPLC-DAD method for the determination of phenolic compounds in plant leaves. Anal Methods 7(18):7753–7757

Zeb A (2019) Chemo-metric analysis of the polyphenolic profile of *Cichorium intybus* L. leaves grown on different water resources of Pakistan. J Food Measur Charact 13(1):728–734

Zeb A, Habib A (2018) Lipid oxidation and changes in the phenolic profile of watercress (*Nasturtium officinale* L.) leaves during frying. J Food Measur Charact 12:677–2684

Zeb A, Haq I (2018) Polyphenolic composition, lipid peroxidation and antioxidant properties of chapli kebab during repeated frying process. Food Measur Charact 12(1):555–563

Zeb A, Ullah F (2019) Effects of spinach leaf extracts on quality characteristics and phenolic profile of sunflower oil. Eur J Lipid Sci Technol 121(1):1800325

Zeb A, Haq A, Murkovic M (2019) Effects of microwave cooking on carotenoids, phenolic compounds and antioxidant activity of *Cichorium intybus* L. (chicory) leaves. Eur Food Res Technol 245(2):365–374

Zeng J, Zhang X, Guo Z, Feng J, Zeng J, Xue X, Liang X (2012) Separation and identification of flavonoids from complex samples using off-line two-dimensional liquid chromatography tandem mass spectrometry. J Chromatogr A 1220:50–56

Zhang B, Li X, Yan B (2008) Advances in HPLC detection—towards universal detection. Anal Bioanal Chem 390(1):299–301

Zhou W, Liu Y, Wang J, Guo Z, Shen A, Liu Y, Liang X (2020) Application of two-dimensional liquid chromatography in the separation of traditional Chinese medicine. J Sep Sci 43(1):87–104

Chapter 17
Spectroscopy of Phenolic Antioxidants

Learning Objectives
After reading this chapter, the reader will be able to:

- Learn the basic concept of spectroscopy and its classes.
- Conceptualize the basics in mass spectrometry, its instrumentation, applications, and advantages.
- Discuss the concept of Nuclear magnetic resonance (NMR) spectroscopy.
- Describe instrumentation and applications of NMR for the analysis of phenolic compounds in foods.
- Learn the basic concept of near-infrared spectroscopy, its fundamental concept, instrumentation, and applications for the analysis of phenolic compounds in foods.

17.1 Introduction

The interaction of matter and electromagnetic radiation is known as spectroscopy. The radiation has properties like wavelength and frequency. The spectroscopy was believed to be originated from the study of white light transmitted into colored lights by the prism. The transmitted light is represented by a spectrum. The spectrum may either continuous, emission, or absorption spectrum. The spectrum can be used for the characterization of chemical substances. Thus, spectroscopy is widely used in chemistry, biochemistry, physics, and several other related fields (Ball 2001; Penner 2010). Spectroscopy is classified into the following important types, which are commonly used in food science, and biochemistry:

- *Absorption spectroscopy:* In this case, the light absorbed is measured based on the interaction of light and matter. It is further classified based on radiations, for example, X-ray absorption spectroscopy, Ultraviolet-visible spectroscopy, infra-

© Springer Nature Switzerland AG 2021
A. Zeb, *Phenolic Antioxidants in Foods: Chemistry, Biochemistry and Analysis*,
https://doi.org/10.1007/978-3-030-74768-8_17

red spectroscopy, microwave absorption spectroscopy, and radio wave spectroscopy (nuclear magnetic resonance spectroscopy).

- **_Emission spectroscopy:_** When an atom or group of atoms reached an excited state, the electrons are pushed to higher energy orbitals and after reversion back to the original orbitals, they emit radiations. The energy released is in the form of photons. Different transitions resulted in different radiation wavelengths, forming an emission spectrum. The most common example of this type of spectroscopy is atomic emission spectroscopy.
- **_Resonance Spectroscopy:_** In this case, the radiation energy couples two quantum states of the material in a coherent interaction that is sustained by the radiating field. The examples of this type include nuclear magnetic resonance spectroscopy and laser spectroscopy.

Another term *"Spectrometry"* represents the measurement of the interactions between light and matter. It can be used to measure reactions, radiation intensity, and wavelength. In other words, spectrometry is a method of studying and measuring a specific spectrum, and it's widely used for the spectroscopic analysis of sample materials. Mass spectrometry is one of the important examples of spectrometry widely used in food analysis.

17.2 Mass Spectrometry

Mass spectrometry is an important analytical technique widely used for the analysis of phenolic compounds in foods. It is used for the determination of the structure of phenolic compounds together with absorption spectroscopy such as UV absorption. The structure of unknown compounds is determined either using searching on computer software in a set of a large library of the digitized mass spectrum or with help of other spectroscopic techniques. Mass spectrometry usually needs a very small amount of sample as compared to other techniques. The accuracy, high resolution, and reproducibility of the results of mass spectrometry have received a great deal of attention in the last two decades. Several good books can be consulted for details of the mass spectrometry (Thompson 2018). However, due to the high cost, limited universities, or research centers around the globe are equipped with mass spectrometer.

17.2.1 Instrumentation

Mass spectrometry (MS) works by placing a charge on a molecule, thus converting it to an ion in a process called ionization. The ions thus produced are then resolved according to their mass-to-charge ratio (m/z) by subjecting them to electrostatic fields (mass analyzer) and finally detected. An additional stage of ion fragmentation may be included before detection to elicit structural information in a technique known as tandem MS (Smith and Thakur 2010). The result of ion production, separation, fragmentation, and detection is displayed as a mass spectrum that can be interpreted to yield molecular weight or structural information.

Generally, a mass spectrometer performs three functions. The first function is the ionization of the analyte molecule to form positive or negative ions. This process is carried out using a variety of techniques, for example, electron impact, matrix-assisted-laser-desorption, or atmospheric pressure ionization. The second function is the separation of charge ions according to their m/z ratio. This process is carried out by the equipment part called mass analyzer which may be either based on quadrupoles, ion-traps, and Fourier transforms or time of flight. The last function of the instrument is the detection of the separated ions, which is carried out using a detector as shown in Fig. 17.1. Design and configuration may be different based on the different manufacturers, but the overall working mechanism is the same. Sample introduction may be either direct or through the insertion probe method. In phenolic compounds analysis, samples are usually separated earlier using liquid chromatography and directed to mass spectrometry.

Ionization in mass spectrometer is carried out using an ion source that consists of filament composed of rhenium or tungsten metal. When a direct current is applied to the filament, it is heated and emits electrons that move across the ion chamber towards the positive electrode. The compound of interest is exposed to an ion source during this process and become ionized. These ionized molecules are of high energies and maybe fragmented further. This process is called electron impact ionization. For phenolic compounds, both negatively charged and positively charged molecules can be detected based on the requirement. In LC-MS, the ionization may be carried out using electrospray ionization (ESI) or atmospheric pressure chemical ionization (APCI). The ESI source consists of the nozzle that contains a fused-silica capillary sample tube to transfer the LC effluent, which is coaxially positioned

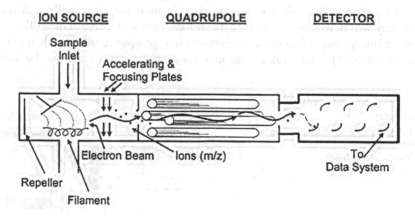

Fig. 17.1 Schematic of a typical mass spectrometer. The sample inlets (interfaces) at the top and bottom can be used for direct injection or interfacing to a GC (reproduced with kind permission of Springer Nature (Smith and Thakur 2010))

within a metal capillary tube to which a variable electrical potential can be applied against a counter-electrode. High velocity compressed nitrogen gas is introduced to help in the nebulization of the LC effluent as it exits the tip of the metal capillary tube (Smith and Thakur 2010). The relative velocity difference between the streams of nitrogen gas and LC effluent at the ESI tip results in the production of a fine spray of highly charged droplets (Fig. 17.2). The ESI interface is widely used for the determination of phenolic compounds in foods.

The APCI interface is a gas phase ionization technique used for compounds with relatively low polarity. The LC effluent carrying capillary tube projects about half-way inside a silicon-carbide vaporizer tube as shown in Fig. 17.3. The vaporizer tube is usually maintained at 400–500 °C and serves to vaporize the LC effluent. A high voltage is applied to a corona needle positioned near the exit of the vaporizer tube. The high voltage creates a corona discharge that forms reagent ions from the mobile phase and nitrogen nebulizing gas (Smith and Thakur 2010). These ions react with the sample molecules and convert them to ions.

The mass analyzer is separating the charged ions based on their m/z ratio. Several types of analyzers are now available in the market. These include quadrupoles (Q), ion traps (IT), time-of-flight (TOF), magnetic sectors, isotope ratio MS, Fourier-transform-based ion cyclotrons (FT-ICR), and Orbitraps (OT), and accelerator mass spectrometers (AMS) (Smith and Thakur 2010). The Q-analyzer consists of four poles/rods with positive and negative poles each placed opposite to each other. A positively charged ion upon reaching quadrupoles is attracted by the negatively charged poles and thus deflected. A stable ion upon entering the Q will flow through a sine wave pattern and will reach the detector. The potentials on the poles can be adjusted to detect a single ion of interest. Detector accurately detects the mass of the ions.

In a time-of-flight mass spectrometer, the masses of the ions are determined by how long it takes them to reach the collectors. All fragment ions receive the same kinetic energy and are accelerated by a high-voltage accelerator plate. Although the ions will arrive at the collector in nanoseconds, their velocities will be different since they will have different masses. Therefore, they will arrive at the collector at different times. This results in their separation into groups according to their masses (Thompson 2018). Each peak or fragment ion in a mass spectrum may be looked

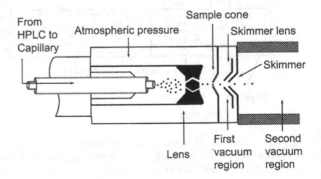

Fig. 17.2 Schematic of an electrospray LC-MS interface (reproduced with kind permission of Springer Nature (Smith and Thakur 2010))

Fig. 17.3 Schematic of an atmospheric pressure chemical ionization LC-MS interface (reproduced with kind permission of Springer Nature (Smith and Thakur 2010)

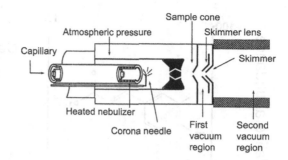

upon as a "piece" of the parent molecule, and the height of the peak is directly proportional to the abundance of the fragment ion (Fig. 17.4). Naturally, the more abundant fragments result from preferred modes of fragmentations. In the MS spectrum, the molecular or parent ion (M+) is produced by the loss of one electron from the parent molecule and is, therefore, a radical ion. Similarly, (M+1)+ ion usually represents the contribution of the 13C isotope, which has a natural abundance of about 1.1%. Adducts of the molecular ion or its fragments can also be observed in the spectrum.

The fragmentation pattern is an important factor in the elucidation of the structure of unknown phenolic compounds in foods. For this purpose, fragmentation of the standard phenolic compounds is taken into account. Recently, Cerrato et al. (2020) showed the fragmentation of flavonoids and B-type proanthocyanidins as shown in Fig. 17.5. The authors showed that O-glycosylated flavonoids, which produce Y fragments deriving from the loss of the sugar moiety, can be easily distinguished from C-glycosylated flavonoids, which generate ions from the internal cleavage of the carbohydrate, named X ions. The fragments derived from the cleavage of sugar moiety have been used to speculate the linkage of those on the aglycones. If more than one sugar was present, fragments derived from the loss of one of them have been used to distinguish the case of two sugars linked together, which produce only one Y1 ion, from those which were individually linked to the flavonoid, which produces two. For example, quercetin O-hexoside O-pentoside, produced fragments at m/z 463.0882 (quercetin -O-hexoside) and m/z 433.0776 (quercetin O-pentoside), while other compounds were tentatively identified as quercetin -O-deoxyhexosyl hexoside since they only generated one Y1 fragment due to the loss of the deoxyhexose. The fragmentation pathways of B-type proanthocyanidins generally consisted of quinone methide (QM) fissions, retro Diels-Alder (RDA) ring openings, and heterocyclic ring fissions. The QM fissions produced [MT-H]− and [ME-3H]− ions in negative polarity and [MT+H]+ and [ME-H]+ in positive polarity, which can help distinguish the flavanol monomers and their order in the polymeric structure. The RDA C-ring opening produces 1,3B losses (m/z 136.0524, 152.0473, and 168.0423 for afzelechin, catechin, and gallocatechin respectively), which also indicate the nature of the monomers, while HRF 1,4A losses produce [M-126.0317] confirming peaks (Cerrato et al. 2020).

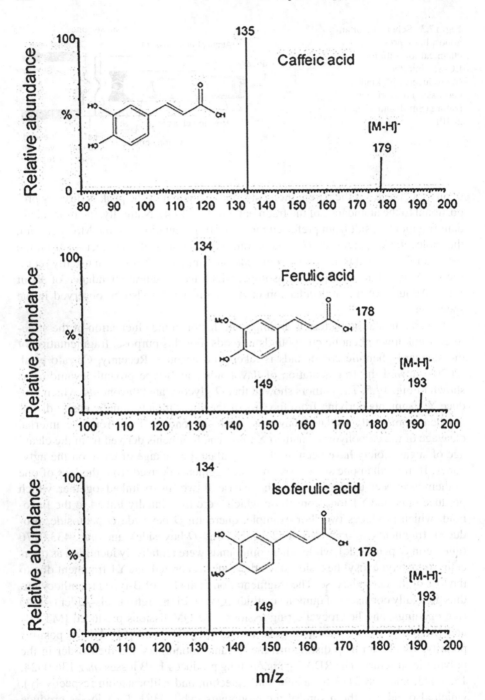

Fig. 17.4 Mass spectra of the deprotonated molecules of caffeic acid, ferulic acid, and isoferulic acid (redrawn with kind permission of John Wiley and Sons (Li et al. 2003))

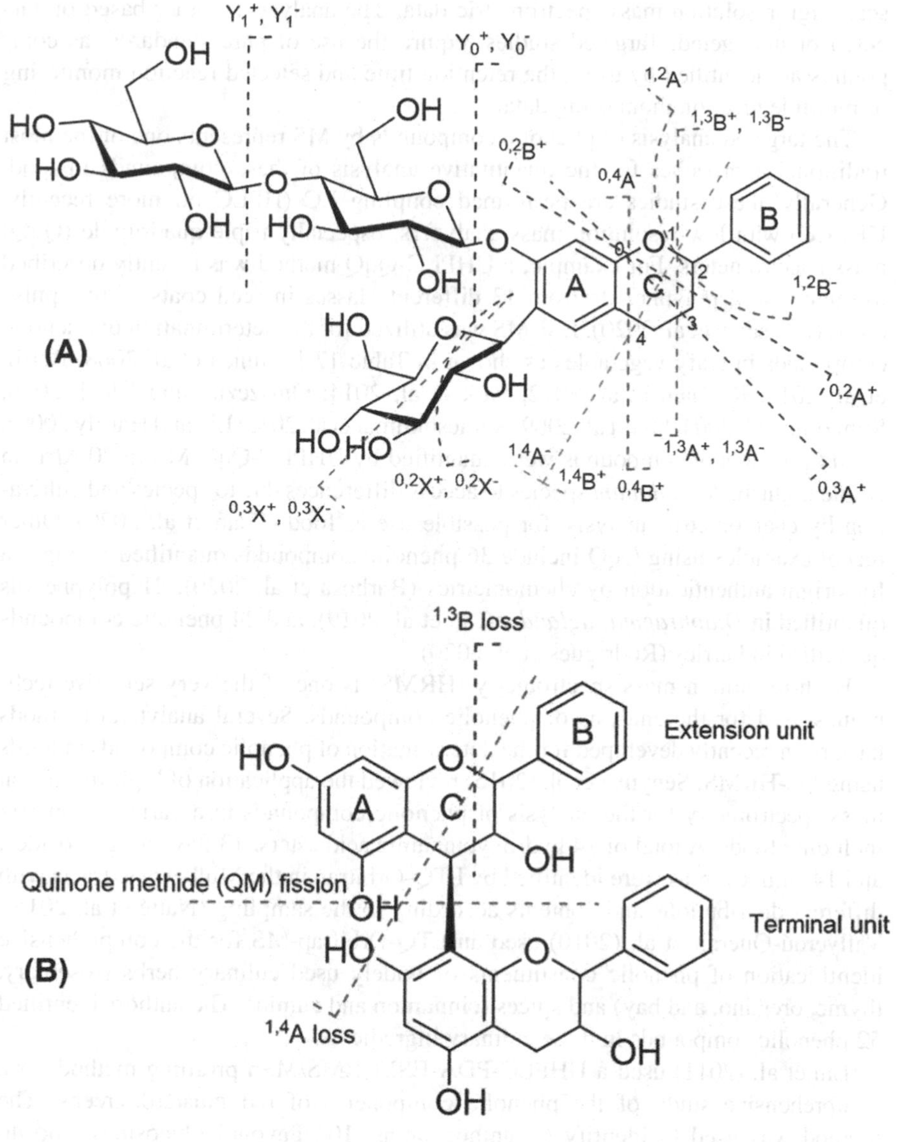

Fig. 17.5 Fragmentation pathways of phenolic compounds. (**A**) Flavonoid glycoside, and (**B**) B-type proanthocyanidin (reproduced with kind permission of Elsevier (Cerrato et al. 2020))

17.2.2 Applications of Mass Spectrometry

Mass spectrometry in combination with liquid chromatography is widely used for the determination of phenolic compounds in foods. Piovesana et al. (2020) reviewed the applications of LC coupled to MS for food analysis. The authors showed that most of the works in the field exploit the profiling approach by annotation of full scan high resolution mass spectrometric data. The analysis is either based on targeted or untargeted. Targeted studies require the use of pure standards, as compounds are identified by using the retention time and selected reaction monitoring or multiple reaction monitoring data.

The targeted analysis of phenolic compounds by MS represents one of the most traditional approaches for the quantitative analysis of these compounds in food. Generally, these studies are performed coupling LC (HPLC or, more recently, UHPLC) with low-resolution mass analyzers, especially triple quadrupole (QqQ) mass spectrometers. For example, a UHPLC-QqQ method was recently described to quantitate 8 polyphenols from 13 different classes in seed coats of five pulse crops (Elessawy et al. 2020). ESI-MS was utilized for the determination of phenolic compounds in leafy vegetables as shown in Table 17.1 (Bunea et al. 2008; Bertin et al. 2014; Khanam et al. 2012; Park et al. 2014; Oniszczuk and Olech 2016; Stintzing et al. 2004; Ola et al. 2009; Rameshkumar et al. 2013; Lin and Harnly 2009).

Six polyphenol compounds were quantified by UHPLC-QqQ MS in 20 African edible nightshades *Solanum* species to access differences due to species and cultivation by chemometric analysis, for possible use as food (Yuan et al. 2020). Other recent examples using QqQ include 36 phenolic compounds quantified in paprika for origin authentication by chemometrics (Barbosa et al. 2020), 21 polyphenols quantified in *Diaphragma juglandis* (Liu et al. 2019), and 24 phenolic compounds quantified in berries (Rodrigues et al. 2020).

High-resolution mass spectrometry (HRMS) is one of the very sensitive techniques used for the analysis of phenolic compounds. Several analytical methods have been recently developed for the determination of phenolic compounds in foods using LC-HRMS. Senyuva et al. (2015) reviewed the application of high-resolution mass spectrometry for the analysis of phenolic compounds in a variety of matrix including foods. A total of 14 hydroxycinnamic acid esters, 13 flavonol glycosides, and 14 anthocyanins were identified by LTQ-Orbitrap in the Mulberry extracts with different distributions and contents according to the sampling (Natić et al. 2015). Vallverdú-Queralt et al. (2010) used an LTQ-Orbitrap-MS for the comprehensive identification of phenolic constituents of widely used culinary herbs (rosemary, thyme, oregano, and bay) and spices (cinnamon and cumin). The authors identified 52 phenolic compounds in these culinary ingredients.

Lin et al. (2011) used a UHPLC-PDA-ESI/HRMS/MSn profiling method for a comprehensive study of the phenolic components of red mustard greens. The method was used to identify 67 anthocyanins, 102 flavonol glycosides, and 40 hydroxycinnamic acid derivatives. Interestingly, 27 acylated cyanidin-3-sophoroside-5-diglucosides, 24 acylated cyaniding-3-sophoroside-5-glucosides, three acylated cyanidin triglucoside-5-glucosides, 37 flavonol glycosides, and 10 hydroxycinnamic acid derivatives were detected for the first time in brassica vegetables. Barbosa et al. (2020) showed that UHPLC coupled with ESI-MS was very efficient for the separation and identification of 36 phenolic compounds in paprika as shown in Fig. 17.6. These compounds were (1) d-(−)-quinic acid, (2) arbutin, (3) gallic acid, (4) homogentisic acid, (5) protocatechuic aldehyde, (6) 4-hydroxybenzoic acid, (7) gentisic acid, (8) chlorogenic acid, (9) (+)-catechin, (10) caffeic acid, (11)

Table 17.1 Some important studies on the identification of phenolic acids in leafy vegetables using ESI-MS

Leafy Vegetable	Compounds	Observed (M + H)$^+$	Observed (M-H)$^-$	λ_{max}	References
Spinacea oleracea	p-Coumaric acid	–	163.2	225,310	Bunea et al. (2008)
	Ferulic acid		193	238, 295	
	o-Coumaric acid		163.2	215, 277, 325	
Sarcocornia ambigua	Cinnamic acid	149	–	–	Bertin et al. (2014)
	p-Coumaric acid	165			
	Vanillic acid	169			
	Caffeic acid	181			
	Ferulic acid	195			
	Syringic acid	199			
	Sinapic acid	225			
	Chlorogenic acid	355			
Komatsuna (Brassica rapa)	Vanillic acid	–	167	–	Khanam et al. (2012)
	Syringic acid		197		
Mizuna (Brassica rapa)	Chlorogenic acid		353		
	Caffeic acid		179		
Pok choi (Brassica rapa)	p-Coumaric acid		163		
	Ferulic acid		193		
Mitsuba (Cryptotaenia japonica)	m-Coumaric acid		163		
Horseno (Spinacea oleracea)	Sinapic acid		223		
	Ellagic acid		301		
Lettuce (Lactuca sativa)					
Red amaranth (Amaranthus tricolor)					
Green Amaranth (Amaranthus tricolor)					
Cabbages (Brassica oleracea var. capitata)	Caffeic acid	–	179.1	–	Park et al. (2014)
	p-Coumaric acid		163.1		
	Ferulic acid		193.1		
	Sinapic acid		223.1		

(continued)

Table 17.1 (continued)

Leafy Vegetable	Compounds	Observed (M + H)$^+$	Observed (M-H)$^-$	λ_{max}	References
Kale (*Brassica oleracea L.var sabellica*)	Protocatechuic acid	–	152.9	–	Oniszczuk and Olech (2016)
	4-Hydroxy benzoic acid		136.8		
	Vanillic acid		166.8		
	Trans-Caffeic acid		178.7		
	Cis-Caffeic acid		178.7		
	Trans-p-Coumaric acid		162.7		
	Cis-p-Coumaric acid		162.7		
	Trans-Ferulic acid		192.8		
	Salicyclic acid		136.8		
	3-Hydroxy cinnamic acid		162.8		
	Cis-Ferulic acid		192.8		
	Trans-Sinapic acid		222.8		
	Cis-Sinapic acid		222.8		
Amranthus spinosus	Caffeoyl-quinic acid	–	353	243, 302,327	Stintzing et al. (2004)
	Caffeoyl-quinic acid		353	234,314	
	Coumaroyl-quinic acid		337	233,301,314	
	Coumaroyl-quinic acid		337	232, 310	
	Feruloyl-quinic acid		367	239,302,328	
	Feruloyl-quinic acid		367	234,322	
Vernonia amygdalina	Caffeoyl quinic acid	–	353	330	Ola et al. (2009)
	Chlorogenic acid		353	330	
Manihot utilissima	Ferulic acid		193	330	
Corchorus olitorius	Caffeoyl quinic derivative		729	330	
	Chlorogenic acid		353	326	
	1,5-Dicaffeoylquinic acid		515	328	
	Dicaffeoyl quinic acid		515	326	
	Dicaffeoyl derivative		515	328	
Ocimum gratissimum	Caffeic acid		179	330	
	Rosmarinic acid		359	328	
	Cichoric acid		473	–	

(continued)

Table 17.1 (continued)

Leafy Vegetable	Compounds	Observed (M + H)⁺	Observed (M-H)⁻	λ_{max}	References
Merremia emarginata	Vanillic acid	168.31	–	259	Rameshkumar et al. (2013)
	Quinic acid	–	191.12	–	
	Trihydroxy(s) benzenpropanoic acid	–	197.04	346	
	Caffeic acid hexose	–	341.12	244, 324	
	3-*O*-Caffeoylquinic acid	–	353.16	326	
	Protocatechuic acid	–	153.19	218,260, 295	
	Caffeoyl glucose	–	341.81	–	
	4-*O*-Caffeoyl quinic acid	–	353.19	326	
	Rosmeric acid	–	359	328	
	5-*O*-Feruloylquinic acid	369.40	367.19	325	
	Caffeic acid	–	179.13	328	
	Ferulic acid	–	193	290,310	
	1,3-Dicaffeoylquinic acid	–	515	333	
	3,5-Dicaffeoylquinic acid	–	515	334	
	4,5-Dicaffeoylquinic acid	–	515.36	334	
	4-Feruloyl-5-caffeoylquinic acid	–	529	325	
	5-*O*-Coumaroylquinic acid	339	–	325	
Collard greens, kale and Chinese broccoli (*Brassica sps*)	3-Caffeoyl quinic acid	–	353	240,298,328	Lin and Harnly (2009)
	3-*p*-Coumarolyuinic acid		337	310	
	5-Caffeoylquinic acid		353	240,298,328	
	4-Caffeoylquinic acid		353	–	
	3-Feruloylquinic acid		367	–	
	5-*p*-Coumaroylquinic acid		337	310	
	5-Feruloylquinic acid		367	–	
	Caffeic acid		179	–	
	Hydroxy-ferulic acid		209	–	
	p-Coumaric acid		163	310	
	Sinapic acid		223	240,298,328	
	Ferulic acid		193	240,298,328	

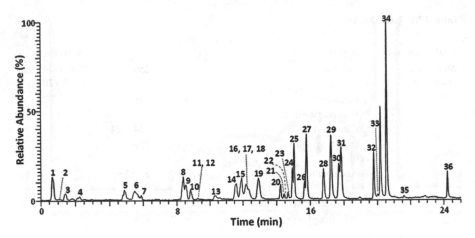

Fig. 17.6 UHPLC-ESI-MS chromatographic separation of the 36 phenolic compounds in Paprika (reproduced with kind permission of the American Chemical Society (Barbosa et al. 2020))

homovanillic acid, (12) syringic acid, (13) vanillin, (14) (−)-epicatechin, (15) ethyl gallate, (16) *p*-coumaric acid, (17) (−)-epigallocatechin gallate, (18) syringaldehyde, (19) umbelliferone, (20) procyanidin C1, (21) veratric acid, (22) ferulic acid, (23) sinapic acid, (24) polydatin, (25) rutin, (26) procyanidin A2, (27) nepetin-7-glucoside, (28) hesperidin, (29) homoplantaginin, (30) fisetin, (31) rosmarinic acid, (32) morin, (33) quercetin, (34) kaempferol, (35) asiatic acid, and (36) betulinic acid. Similarly, an accurate quantitative analysis of craft beers by LC-MS/MS identified phenolic acids and flavonoids mostly coming from malt, and bitter acids and prenylflavonoids coming from hop (Cortese et al. 2020). Gai et al. (2021) reported a UPLC-QqQ-MS/MS method was first set up to quantify phenolic compounds in pigeon pea.

De Paepe et al. (2013) used an LC-Orbitrap-HRMS for the identification and quantification of phenolic compounds in apples using ESI in negative-ion mode. A total of 39 compounds were analyzed with a mass accuracy of less than 1.5 ppm. Vallverdú-Queralt et al. (2010) used a combination of MS techniques with negative-ion detection i.e., LC/ESI-LTQ-Orbitrap-MS and LC/ESI-MS/MS for improved characterization of tomato polyphenols. The authors identified 38 phenolic compounds with very high mass accuracy. Castro-López et al. (2014) analyzed the natural antioxidants catechins and quercetin used in active packaging and functional foods using LC coupled to photodiode array and fluorescence detectors and compared the results with LTQ-Orbitrap-MS. Recent results also showed that the LC-HRMS and HPLC-DAD method revealed 29 phenolic compounds in *Trifolium repens* leaves (Ahmad et al. 2020). The highest amount present was of tyrosol, quercetin-3-glucuronide, formononetin-7-glucoside, quercetin-3-glucoside, 3,4-di-*O*-caffeoylquinic acid, formononetin-7-glucoside-acetate, quercetin-3-glucoside, and formononetin as shown in Table 17.2.

Table 17.2 Characteristics identification and quantification of phenolic antioxidants in *Trifolium repens* using HPLC-DAD and UPLC-HRMS (reproduced with kind permission of Springer Nature (Ahmad et al. 2020))

Peak	Rt	M+H⁺	MS2	Formula	Identity	Amount (mg/g)
1	1.1	168	137	$C_9H_{13}NO_2$	Tyrosol	11.1 ± 0.2
2	1.4	169	124, 138, 152	$C_8H_8O_4$	Vanillic acid	0.01 ± 0.001
3	1.8	155	137	$C_7H_7O_4$	Protocatechuic acid	0.01 ± 0.000
4	2.1	333	288, 290	$C_{20}H_{29}O_4$	Carnosic acid	0.04 ± 0.002
5	2.6	149	131, 164	$C_9H_9O_2$	Cinnamic acid	0.03 ± 0.002
6	3.4	355	131, 179, 191, 309	$C_{16}H_{19}O_9$	3-*O*-Caffeoylquinic acid	0.21 ± 0.003
7	4.3	195	123, 177, 179	$C_{10}H_{11}O_4$	Ferulic acid	0.02 ± 0.000
8	5.7	355	137, 177, 179, 309	$C_{16}H_{19}O_9$	5-*O*-Caffeoylquinic acid	1.47 ± 0.001
9	7	181	135, 163	$C_9H_9O_4$	Caffeic acid	0.01 ± 0.001
10	7.2	517	179, 337, 407, 471	$C_{25}H_{25}O_{12}$	3,4-di-*O*-Caffeoylquinic acid	4.69 ± 0.01
11	8.6	313	179, 203, 295	$C_{13}H_{13}O_9$	Caftaric acid	0.04 ± 0.00
12	10.5	339	147, 191, 293	$C_{16}H_{19}O_8$	3-*O*-Coumaroylquinic acid	0.74 ± 0.02
13	10.8	361	109, 123, 237, 251	$C_{16}H_{19}O_9$	Rosmarinic acid	1.76 ± 0.04
14	11.7	519	163, 179, 193, 341	$C_{22}H_{31}O_{14}$	Ferulic acid dihexoside	1.28 ± 0.04
15	12.4	463	123, 147, 163, 339	$C_{22}H_{23}O_{11}$	Isorhamnetin-7-*O*-rhamnoside	1.96 ± 0.04
16	13.2	611	147, 177, 285, 325, 433	$C_{27}H_{31}O_{16}$	Quercetin-3-*O*-rutinoside	3.62 ± 0.05
17	14.0	465	163, 179, 285, 301	$C_{21}H_{21}O_{12}$	Quercetin-3-*O*-glucoside	4.71 ± 0.02
18	14.4	479	177, 193, 285, 301	$C_{21}H_{19}O_{13}$	Quercetin-3-*O*-glucoronide	5.36 ± 0.03
19	14.8	449	163, 269, 285	$C_{21}H_{21}O_{11}$	Luteolin-7-*O*-glucoside	2.81 ± 0.04
20	15.3	551	109, 249, 301	$C_{24}H_{23}O_{15}$	Quercetin-3-*O*-malonylglucoside	4.26 ± 0.09
21	15.8	595	147, 177, 269, 325, 417	$C_{27}H_{31}O_{15}$	Kaempferol-3-*O*-rutinoside	3.93 ± 0.06
22	16.7	611	163, 179, 269, 285, 325	$C_{27}H_{31}O_{16}$	Kaempferol-3-*O*-sophoroside	3.24 ± 0.04
23	17.8	449	163, 169, 269, 285, 355	$C_{21}H_{21}O_{11}$	Kaempferol-3-*O*-glucoside	2.74 ± 0.02
24	18.4	625	163, 309, 315, 501	$C_{31}H_{29}O_{14}$	Isorhamnetin-3-caffeoyl-7-rhamnoside	2.14 ± 0.02
25	20.4	995	179, 223, 463, 531, 831	$C_{44}H_{51}O_{26}$	Quercetin-3-sinapolysophoroside-7-glucoside	2.63 ± 0.04

(continued)

Table 17.2 (continued)

Peak	Rt	M+H⁺	MS2	Formula	Identity	Amount (mg/g)
26	23.2	417	163, 179, 237, 253	$C_{21}H_{21}O_9$	Formononetin-7-glucoside	5.03 ± 0.03
27	24.3	459	135, 163, 179, 295, 323, 399	$C_{23}H_{23}O_{10}$	Formononetin-7-glucoside-acetate	4.29 ± 0.06
28	24.6	255	161, 237	$C_{15}H_{11}O_4$	Daidzein	2.15 ± 0.05
29	29.1	255	161, 237	$C_{15}H_{11}O_4$	Formononetin	3.64 ± 0.06

17.2.3 Advantages of Mass Spectrometry

Phenolic compound characterization in food is benefitting from the development of improved mass spectrometers, with increased resolution, and the development of untargeted metabolomics strategies for small molecule characterization (Piovesana et al. 2020). Due to the complexity of the food matrix and the increased information, which can be gathered by modern UHPLC-HRMS and multistage MS, the use of software for data interpretation is fundamental to increase the confidence and move from simple compound annotation to putative compound identification by matching productions spectra. Even if this approach still does not completely substitute the use of pure standards for compound validation, it is a very valuable strategy to provide the characterization of phenolic compound content with increased confidence. A more detailed structure characterization on the one hand reduces the final list of compounds compared to simple profiling on precursor exact mass, but the probability of a correct assignment is increased. Still, it is essential to understand that false positive and false negatives can arise in this process when a correct match is wrongfully discarded or when a compound is positively matched against the wrong candidate. False negatives typically originate from too-large differences between the experimental spectrum and the library against which it is matched, due to instrumental and experimental factors, especially in ESI, which is extensively used in the phenolic compound analysis. False positives happen when the available spectral data are not able to distinguish among several structural features. Even in cases where an exact identification of a compound is impossible, multiple candidates often belong to the same molecular class (De Vijlder et al. 2018). This issue is being considered in recent literature for metabolite annotation but is not still considered in the specific case of phenolic compound analysis. In any case, these issues are expected to be similar. For instance, a decoy strategy was recently suggested to estimate the false discovery rate of metabolite annotation, indicating that not only this approach is feasible for small molecules, but also that the number of confident annotations is reduced in large scale metabolite studies with strict false discovery rate thresholds, but this is necessary to improve quality of the obtained data (Piovesana et al. 2020). More confident compound identification is useful not only for a detailed qualitative description of the analyzed samples but it also helps in better elucidating which markers are significant in differentiating samples or which compounds are involved in a specific metabolism.

There are several advantages of using a mass spectrometer coupled to liquid chromatography. Kumar (2017) presented several advantages of the LC-MS method for the determination of phenolic compounds. The method offers various advantages over other chromatographic methods, which are described below:

- *Selectivity*: Co-eluting peaks can be isolated by mass selectivity and are not constrained by chromatographic resolution.
- *Peak assignment*: A molecular fingerprint for the compound under study is generated, which ensures correct peak assignment in the presence of complex matrices.
- *Molecular weight information*: Confirmation and identification of both known and unknown compounds even with low concentrations of the sample.
- *Structural information*: Controlled fragmentation pattern enables the structural elucidation of a chemical compound.
- *Rapid method development*: Providing easy identification of eluted analytes without retention time validation.
- *Sample matrix adaptability*: Decreasing sample preparation time and hence is less time-consuming.
- *Quantitation*: Quantitative and qualitative data can be obtained simultaneously with limited instrument optimization.
- *Cost-effective*: Another advantage of the LC-MS method is the capacity to multiplex several analytes within a single analytical run with a minimal incremental cost. This method has the potential to simplify laboratory set-up (e.g. creation of test panels) and provides additional useful information (e.g. metabolite profiles)

17.3 Nuclear Magnetic Resonance Spectroscopy

Nuclear magnetic resonance (NMR) spectroscopy is a powerful analytical technique having widespread applications. It is a non-destructive technique used for the structure determination of complex molecules. The applications of NMR to food analysis has increased over the last three decades. Magnetic resonance imaging (MRI) is a nondestructive technique that can be used to image product quality and changes during processing and storage. NMR deals with atomic nuclei of interest. The sample is positioned in a magnetic field and the NMR signal is produced by excitation of the nuclei sample with radio waves into nuclear magnetic resonance, which is detected with sensitive radio receivers. The intramolecular magnetic field around an atom in a molecule changes the resonance frequency, thus giving access to details of the electronic structure of a molecule and its individual functional groups (Atta-Ur-Rahman 2012). The principle of NMR usually contains three sequential steps: The alignment (polarization) of the magnetic nuclear spins in an applied, constant magnetic field. The agitation of this alignment of the nuclear spins by a weak oscillating magnetic field called a radio-frequency (RF) pulse.

When a spin nucleus is put in a magnetic field, an energy transfer arises between different energy levels corresponding to radiofrequency radiation at the appropriate frequency, as a result, the NMR phenomenon happens (Cao et al. 2021). Through different ways such as Fourier transform, phase adjustment, chemical shift calibration, integration, and time-domain experiment, the NMR signal created can be measured and processed to yield a certain NMR spectrum for the nucleus concerned as shown in Fig. 17.7. The most commonly used nuclei are 1H and 13C by high-field NMR spectroscopy.

NMR is used to identify proteins and other complex molecules including phenolic compounds. It provides detailed information about the structure, dynamics, reaction state, and chemical environment of molecules. The most common types of NMR are the proton and carbon-13 NMR spectroscopy, but it applies to any kind of sample that contains nuclei possessing spin. Despite its increasing popularity among food scientists, NMR is still an underutilized methodology in food science (Hatzakis 2019), mainly due to its high cost, relatively low sensitivity, and the lack of NMR expertise by many food scientists.

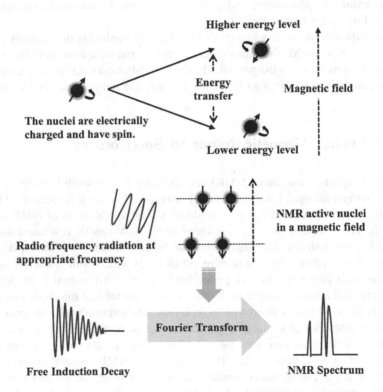

Fig. 17.7 Schematic diagram of Nuclear magnetic resonance (NMR) principle (reproduced with kind permission of Elsevier (Cao et al. 2021))

17.3.1 Instrumentation

A detailed description of NMR hardware can be found in previous literature (Atta-Ur-Rahman 2012; Sundekilde et al. 2019; Hodgkinson 2020), a brief introduction is presented here for understanding the applications of NMR in food science. According to the magnitude of magnetic field strength, NMR can be divided into three categories: high field NMR (magnetic field strength >1.0 T), midfield NMR (0.5 T < magnetic field strength <1.0 T), and low field (magnetic field strength <0.5 T). NMR systems generally consist of a magnet that generates the static magnetic field; an NMR probe that hosts the sample and is used to deliver the RF pulses, a console that contains all the electronics required for RF pulse generation, signal detection, and sample temperature control, and a computer that is used to control the console, the probe, and the magnet, as well as for the processing of the NMR data (Hatzakis 2019). A simplified schematic diagram of an NMR system is shown in Fig. 17.8.

The magnet is a cylindrical container that consists of various concentric tanks. In the inner one, there is a superconducting coil that generates the main magnetic field, B0, along the z-axis. Superconductivity is essential to generate strong magnetic fields appropriate for studies requiring high spectral resolution and sensitivity. To achieve superconductivity, the solenoid coil, made of niobium-titanium, is kept in a liquid helium bath in the inner tank. At this temperature (4 K) the coil resistance is practically close to zero and thus the coil is in the superconductive state. To reduce

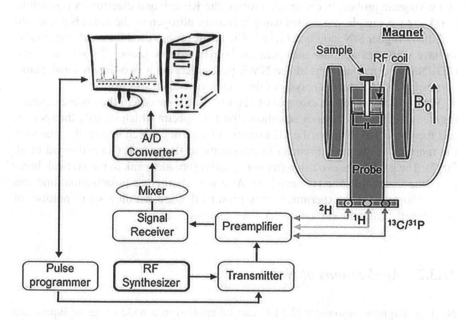

Fig. 17.8 Schematic diagram of Nuclear magnetic resonance (NMR) (reproduced with kind permission of John Wiley and Sons (Hatzakis 2019))

the helium boil-off, the inner container is surrounded by another tank, which is just a vacuum and is surrounded by the external Dewar which contains liquid nitrogen. The magnetic field is not constant and it drifts a few Hz per day. The size of the drift is comparable to the linewidths of the NMR signals and thus a correction must be applied. The so-called "lock" system is used to correct this drift. The lock system constantly measures the resonance of deuterium in the NMR solvent that appears in the sample and when the field drifts, an electric current is added or subtracted to bring it back to resonance. Field homogeneity is achieved by using another set of coils, called shim coils, which surround the sample. This is essential for acquiring spectra with sharp peaks and high resolution.

The RF coil used to excite the sample (transmitter) and receive the signal (receiver) is located inside the NMR probe. The probe is positioned in the middle of the magnet through a bore. There are several types of NMR probes and one of the most common ways to categorize them is as "inverse" and "observe" probes. In inverse probes, the coil which is closer to the sample (inner coil) is the one that is used to excite proton frequencies, whereas the outer coil is used for hetero-nuclei. Observe probes have the opposite structure. When proton nuclei are detected, as in the case of 1D 1H NMR or the most common 2D NMR experiments, an inverse probe is recommended. Observe probes are optimized for the detection of hetero-nuclei such as 13C and 31P in the inner coil while the much more abundant protons can be easily detected in the less-sensitive outer coil. Observe probes may be an attractive option for food analysis because hetero-nuclei are characterized by high spectral resolution. Another way to classify probes is as room temperature probes and cryogenic probes. In cryogenic probes, the RF coil and electronics (preamplifiers), not the sample, are cooled using helium or nitrogen so the noise is decreased and thus a higher S/N ratio is achieved. The NMR probes used for solid-state analysis have a different structure and operation from liquid state probes. Two-dimensional (2D) NMR techniques spread the NMR parameters on a two-dimensional plane, very helpful for in-depth analysis of the one-dimensional spectroscopy.

Figure 17.9 showed an example of 1H-13C WISE spectra of the cheese containing free catechin. The bottom axis shows the 13C spectra which identify the species, and the vertical axis shows the 1H evolution frequency which is directly related to the mobility of the proton nuclei in proximity to the carbon (Rashidinejad et al. 2017). The greater the evolution frequency ω_{HH} (broader peak in the vertical direction), the less mobile the 1H nuclei are. As a very short cross-polarization time was used (100 µs) in this experiment, only protons that are within a small number of bond lengths from the carbon nuclei were sampled.

17.3.2 Applications of NMR

Nuclear magnetic resonance (NMR) can be applied to a wide range of liquid and solid matrices without altering the sample or producing hazardous wastes (Marcone et al. 2013). Although the sensitivity and detection limits of NMR still need to be

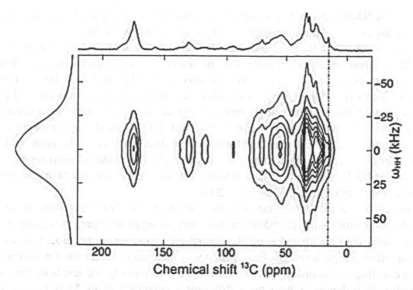

Fig. 17.9 1H-13C wide-line separation nuclear magnetic resonance spectra of cheese containing catechin. The dashed line indicates the location of the profile at 16 ppm (reproduced with kind permission of Elsevier (Rashidinejad et al. 2017))

improved, NMR still has several advantages relative to other common analytical tools such as high-pressure liquid chromatography (HPLC), gas chromatography (GC), and mass spectrometry (MS). NMR technology was initially used in the late 1940s to elucidate the structure of molecules in organic chemistry. However the diverse applications of NMR spectroscopy in food science were delayed until the 1980s, primarily due to lack of scientific expertise, high cost of equipment, and the absence of NMR parts designed specifically for food purposes, although pulsed NMR had been applied to foods and related materials earlier than this. With the development of NMR instrumentation and improved programs to collect and ana-lyze the data, NMR's applicability has recently rapidly expanded in the field of food science and technology. A wide range of NMR food-related research has covered various fields of food science, including food microbiology, food chemistry, food engineering, and food packaging (Kirtil and Oztop 2016). The progress in the research of NMR use on foods has been addressed in several recent reviews with a limited scope focusing on NMR applications either in particular foods, such as wine (Ogrinc et al. 2003) or specific applications, such as identification of food authenticity.

The use of NMR relaxation time not only permits compositional analysis of foods but also insights into food structure. In MRI, the NMR relaxation time allows for quantification of the image contrast, therefore enabling visualized monitoring of structural changes in foods during processing and storage. Information obtained from this data often leads to the determination of the desired structures, which could impact the final physical and mechanical properties of food. Two major subjects studied by NMR/MRI include (1) structural elucidation of specific food compounds; (2) quantification of microstructural changes occurring in foods.

The use of NMR for food authentication also extends to beverages as well. 1H NMR spectroscopy showed potential in the discrimination of green tea according to the country of origin or with respect to quality (Le Gall et al. 2004). 1H NMR was able to detect simultaneously the catechins, the amino, organic, phenolic, and fatty acids, and the sugars from a single green tea extract. It was also capable of detecting catechins, caffeine, 5-galloyl quinic acid, and 2-O-(â-l-arabinopyranosyl)-myo-inositol, all of which are associated with tea quality. Another application field for NMR spectroscopy is to examine the source of the raw material used for making juices. 1H NMR has also been shown to be very accurate in the determination of the origin or quality of juices (Cuny et al. 2008).

Tarachiwin et al. (2007) introduce new methods involving metabolic profiling and fingerprinting using 1H NMR. In this study, dried ground green tea leaves were mixed with D_2O, incubated at 60 °C, centrifuged to separate the supernatant containing the hydrophilic metabolites, which was then filtered and mixed with a buffer before testing. Subsequently, 1H NMR was able to identify the levels of chemical constituents in Japanese green tea in different frequency regions. More specifically, theanine (δ 1.10, 2.12, 2.39, 3.19 ppm) and quinic acid (δ 1.88, 1.97, 2.05 ppm) at δ 0.5–3.0 ppm, whereas caffeine (δ 3.27, 3.43, 3.89 ppm), arginine (δ 3.22, 3.47 ppm), myo-inositol (δ 3.30 ppm), chlorogenic acid (δ 3.89, 4.22 ppm), and quinic acid (δ 3.56 and 4.01 ppm) were observed at δ 3.0–4.5 ppm, after excluding sucrose and fructose signals, 2-O-β-l-arabinopyranosyl-myo-inositol (δ 5.19 ppm), p-coumaryl quinic acid and/or cinnamic acid (δ 7.51, 7.75 ppm), EGCG (δ 6.61, 7.02, 7.14 ppm), and ECG (δ 6.91, 7.02 ppm) were also observed (Fig. 17.10). The x-axis of the spectrum is the delta scale (δ) with units of ppm and the y-axis is an intensity scale. These above-mentioned compounds were targeted in this study because they are all relevant regarding the quality of green tea. Again, this 1H NMR based study proved the possibility of the use of NMR for quality evaluation of beverages by determining the quality-related chemical constitutes in food with simple sample preparation and short analysis time.

17.4 Near-Infrared Spectroscopy

Near-infrared light is non-visible electromagnetic radiation discovered as early as in the year 1800 in the famous experiment of Herschel. It was found that NIR light is absorbed by matter, it may be considered an odd strategy twist of scientific history that the actual development of NIR spectroscopy has lagged behind the techniques resorting to other spectral regions such as UV, visible or MIR as reviewed by Beć and Huck (2019). The potential of the NIR region remained not recognized at that time, but recent advances showed significant applications. Its universality, wide applicability, simple instrumentation, low time-to-result, and low-cost factors are prominent advantages from the point of view of qualitative and quantitative

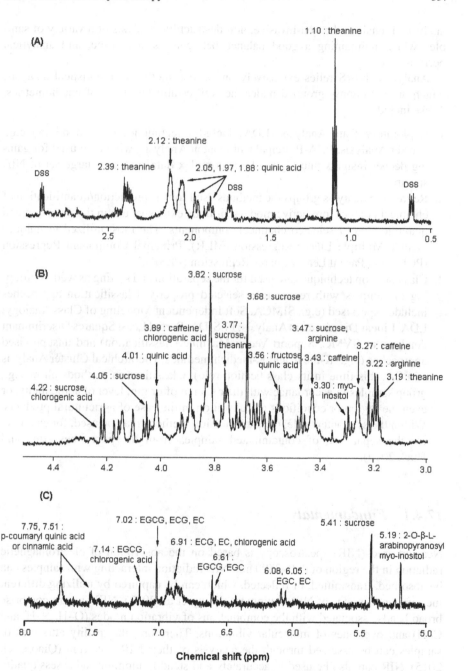

Fig. 17.10 1H NMR spectra (750 MHz, D_2O) of green tea extract from the highest quality sample ranking no. 1 in (**A**) high-, (**B**) middle-, and (**C**) low-frequency regions, measured at 25 °C (reproduce with kind permission of American Chemical Society (Tarachiwin et al. 2007))

analysis. It enables the non-invasive, non-destructible analysis of a variety of samples while maintaining a good balance between its cost, time, and analytical performance.

Analytical NIRS relies extensively on various methods of statistical analysis, which are commonly grouped under the well-established term of chemometrics. These include:

1. Exploratory Data Analysis (EDA) includes techniques of data mining e.g., Cluster Analysis, PCA-Principal Component Analysis, which are used for gaining deeper insights into high-volume complex data such as a large set of NIR spectra.
2. Regression analysis groups the methods used for the prediction/quantification of chemical content (predictive models); it finds extensive use in the detection and quantification of selected chemical components. The most utilized techniques include Multiple Linear Regression (MLR), Principal Component Regression (PCR), and Partial Least Squares Regression (PLSR).
3. Classification techniques are used for the separation and sorting as well as grouping of samples with regard to a selected property. Classification approaches include supervised (e.g., SIMCA, Soft Independent Modeling of Class Analogy; LDA, Linear Discriminant Analysis; PLS-DA, Partial Least Squares Discriminant Analysis or SVMC, Support Vector Machine Classification) and unsupervised approaches (e.g., K-mean and K-median methods, Hierarchical Cluster Analysis or PCA, this time in its classification role). Classification methods allow e.g., group samples in accordance with their source of origin, level of authenticity, or even the region or conditions of cultivation in the case of agricultural products. Within the set content rule, the classification methods may be used, for example, for the separation of contaminated samples from the pure ones (Beć and Huck 2019).

17.4.1   Fundamentals

Near-infrared (NIR) spectroscopy is based on the absorption of electromagnetic radiation in the region of 400–2500 nm. NIR radiation interacting with samples can be absorbed, transmitted, or reflected, which can be captured by utilizing different measurement modes of NIR equipment (Wang et al. 2017). NIR spectra comprise broad bands associated with the combinations of vibration modes (O–H, N–H, and C–H) and overtones of molecular vibrations. Therefore, the quality attributes of samples can be assessed through the analysis of their NIR spectrum (Qiao et al. 2015). NIR can also be used to accurately and steadily monitor and assess quality changes of raw materials and final products with simple sample preparation. Moreover, through the analysis of quality attributes, classification, and authentication of liquid foods can be analyzed. Overtone and combination vibrations of the first group dominate the NIR region (4000–$14,286$ cm^{-1}; 700–2500 nm), while those of the second group absorb in the mid-infrared region (MIR) (400–4000 cm^{-1}; 2500–$25{,}000$ nm) (Ozaki and Morisawa 2021). Electronic transitions absorb in the visible region ($14{,}286$–$25{,}000$ cm^{-1}; 400–700 nm) and in the ultraviolet region ($25{,}000$–$40{,}000$ cm^{-1}; 250–400 nm).

Reflectance mode determination is the easiest to acquire because it can be performing without contact with the food besides light intensities are fairly high. However, it is subjected to variations in surface characteristics. The advantage of the transmission cell is that it offers very accurate and reproducible spectroscopic determinations; however, the main drawbacks often require a destructive preparation or semi-preparation of the sample (Tahir et al. 2019). Generally, transmission mode determinations are better than reflectance mode for measurements of internal disorders of foods. However, the intensity of light penetrating the food is often very low, making it a challenge to acquire accurate transmission determination, mainly in environments of high ambient light levels. Transparent foods are usually analyzed in transmittance. Solid and semi-solid or turbid foods could be analyzed in diffuse transmittance, diffuse reflectance, or transflectance (D) based on their absorption and scattering properties. Also, pseudo absorbance (A) relative to standard reference material is determined (A = $\log(1/T)$ for transmittance and $\log(1/R)$ for reflectance spectra).

17.4.2 Instrumentation

Today, a varied selection of NIR spectroscopic instruments is available, and there are about 60 manufacturers of NIR spectrometer around the world. These instruments can be divided into three groups: (1) laboratory devices, (2) sorting and grading, and (3) portable devices. The main differences between these types of NIR devices and an overview of spectroscopy applications on fruits and vegetables based on the instrumental characteristics of the NIR devices employed for the studies can be found in Beghi et al. (2017). The literature showed that many applications of VIS-NIR spectroscopy involve the use of benchtop and portable full spectra devices, but recent studies have been conducted using simplified optical systems based on a small number of wavelengths as discussed recently (Beghi et al. 2017; Cortés et al. 2019). Irrespective of the type of instrument, the principal components are a sample holder, where the sample is placed, a light source, a detector to record the received light intensity, and a computer unit to register and process the spectral information obtained (Fig. 17.11). The light source is usually a tungsten light bulb or maybe light-emitting diodes (LEDs). The basic detector is a silicon detector with low sensitivity. Multiplied Si detector elements can be fitted with their filter each, with each such element tuned toward measuring its channel. Silicon-based detectors offer practical advantages, e.g., low power consumption. Other detectors include complementary metal-oxide-semiconductor (CMOS) and charge-coupled device (CCD), with CMOS requiring lower power consumption. Temperature sensors are also present to regulate heat produced during radiation. A wavelength selector such as a

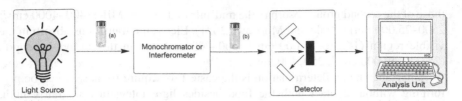

Fig. 17.11 Schematic representation of the near-infrared spectroscopic instrument. Sample and reference is represented by a and b.

monochromator is used to work under a specific wavelength of light. The use of fiber-optic probes is often desirable, as many current applications are based on their intensive use to simplify data acquisition procedures due to their multiplexing capacity, thus allowing them to monitor many points (Cortés et al. 2019; Huck 2021). According to the model used, light attenuation by the sample, relative to the reference, is known as reflectance (R) or transmittance (T). Commonly, R and T are transformed into absorbance (log 1/R or log 1/T) to perform chemometric analyses.

17.4.3 Chemometrics

The powerful NIR instruments currently available rapidly provide large amounts of information that need efficient pre-treatment and useful evaluation. Chemometrics is a discipline developed for this purpose. Generally, it involves three steps: (i) spectral data pre-treatment; (ii) construction of calibration models; and (iii) model transfer. The main objective of spectra pre-treatment is to transform the data into more useful information capable of facilitating its subsequent multivariate analysis. Some of the more frequent pre-treatments for NIR spectra include: (i) smoothing methods (for example, Gaussian filter, moving average, median filter, and Savitzky-Golay smoothing); (ii) derivation methods (usually first and second derivative); (iii) MSC; (iv) OSC; (v) SNV; (vi) wavelet transformation; (vii) normalization and/or scaling; and (viii) de-trending to eliminate the baseline drift in the spectrum. Moreover, different combinations of these methods applied simultaneously can also be used for signal processing. The first step of the data analysis is often principal component analysis (PCA), to detect patterns and outliers in the measured data. Another unsupervised pattern recognition technique that can be used is CA. Subsequently, a qualitative or quantitative approach to the data will be chosen according to the objectives of the particular study. Qualitative analysis involves classifying the samples according to their VIS-NIR spectra based on pattern recognition methods (Cortés et al. 2019). The classification model is created with a training set of samples with known categories, and subsequently, this model is evaluated by a test set of unknown samples. To do this, many qualitative methods are used, such as linear discriminant analysis (LDA), quadratic discriminant analysis (QDA), K-nearest neighbors (KNN), partial least squares discriminant analysis (PLS-DA), soft independent modeling of class analogy (SIMCA), artificial neural networks (ANN), and support vector machine (SVM). Of these techniques, partial least squares discriminant analysis (PLS-DA) is often commonly selected for optimal classification. For quantitative analyses, which focus on predicting some of the properties that, for example, can greatly influence fruit quality, methods such as multiple linear regression (MLR), principal component regression (PCR), partial least squares regression (PLS), or ANN are broadly used. Generally, a good regression model is characterized by higher coefficients in calibration (Rc), cross-validation (Rcv), and prediction (Rp) and lower error values in calibration (RMSEC), cross-validation (RMSECV), prediction (RMSEP), standard error of prediction (SEP), and standard error of calibration (SEC) (de Oliveira et al. 2018).

17.4.4 Application of NIR Spectroscopy

NIR spectroscopic techniques, combined with chemometric methods, have shown great potential due to their fast detection speed, and the possibility of simultaneously predicting multiple quality parameters or distinguishing between products according to the objectives (Cortés et al. 2019). Being able to automate processes is a great advantage compared to routine off-line analyses, mainly due to the savings achieved in time, material, and personnel. Recent demands, together with the advances being made in the technology and a reduction in the price of equipment, makes NIR technology an analytical alternative for continuous real-time food quality controls. NIR spectroscopy is effective in evaluating several agricultural products such as grains, fruits, vegetables, meat and fish products, and diverse food products (Ozaki et al. 2007; Wang et al. 2017). For example, the biochemical and nutritional composition of different wheat varieties were determined using NIRS (Khan et al. 2007). The result was found to be in accordance with other analytical techniques. Tian et al. (2021) developed the NIR model for rapid determination of wheat total phenolics. The authors found that the NIR method achieved comparable power as the Folin- Ciocalteu method. Phenolic contents in wine samples were measured with NIR spectroscopy (Cozzolino et al. 2004).

Mahesar et al. (2019) reviewed the applications of NIR spectroscopy for the determination of functional compounds in olive oil. Mora-Ruiz et al. (2017) applied NIR and MIR spectroscopy to study polar phenolic compounds of virgin olive oil (VOO) and their impact on the oil quality (Fig. 17.12). The authors obtained satisfactory multivariate test set validation algorithms for total polar phenolic (TPP) compounds (r-coefficient of determination = 0.91), hydroxytyrosol, and tyrosol secoiridoid derivatives (HtyrSec, TyrSec; r = 0.91 and 0.92, respectively) by NIR spectroscopy. Moreover, the authors pointed out that, in contrast to the NIR data, the chemometric analysis of the MIR spectra gave no satisfactory validation models (r = 0.43, 0.54, and 0.66 for HtyrSec, TyrSec, and TPP). This was unexpected as the calibration algorithms for MIR gave an even better correlation than NIR (r > 0.96 for all the polar phenolics studied). On the contrary, the optimization of phenolic

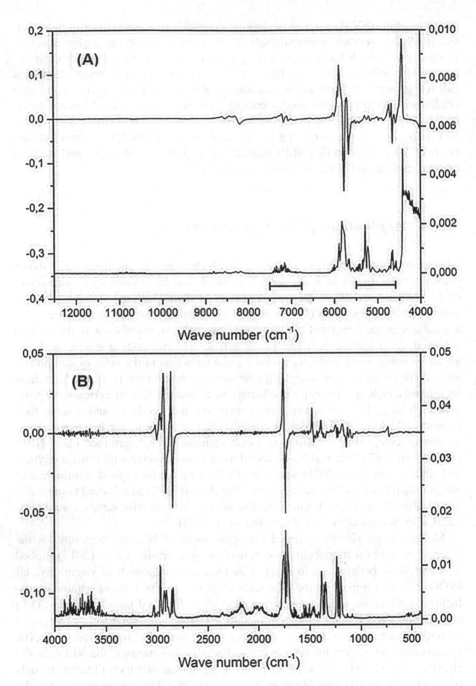

Fig. 17.12 NIR spectra of virgin olive oil (VOO). (**A**) First derivative (top) and standard deviation (bottom) of NIR spectra of the VOO samples studied ($n = 75$), (**B**) First derivative (top) and standard deviation (bottom) of mid-IR spectra of the VOO samples studied ($n = 75$) (reproduced with kind permission of John Wiley and Sons (Mora-Ruiz et al. 2017))

compounds extraction from EVOO was achieved by applying a response surface methodology as proposed recently (Fratoddi et al. 2018).

Similarly, very good MIR results for the quantification of virgin olive oil phenolic compounds were reported by Hirri et al. (2016). They obtained values for the correlation coefficient and the root means square errors of prediction of 0.99 and 0.11, respectively. This study underlined also that the spectral region in the range 3050–600 cm−1 was useful for predicting the total polyphenol content. Bellincontro et al. (2012) also focused on the total phenols. They developed a portable NIR-AOTF (Acousto Optically Tunable Filter) tool for the on-field and nondestructive measurement of specific and total phenols in olives for oil production. Models were developed for the main phenolic compounds (e.g., oleuropein, verbascoside, and 3,4-DHPEA-EDA) and total phenols by PLS. The results in terms of R^2 for the calibration, prediction, and cross-validation ranged between 0.930 and 0.998, 0.874 and 0.942, and 0.837 and 0.992, respectively. A recent preliminary study of Trapani et al. (2017) applied cost-saving NIR based on a discrete filter system for the rapid measurement of total phenolic content and oleuropein of olive fruits, in addition to the moisture, oil, and sugar. Although PLS models built for the latter ones were satisfactory, the instrument did not prove itself suitable for obtaining predictive models for phenolic compound contents. This is probably because the necessary wavelengths in the section of absorbance from 1100 to 1400 nm were not covered.

The ability of Fourier transform near-infrared (FT-NIR), attenuated total reflectance mid-infrared (ATR-MIR) and Fourier transform infrared (FT-IR) spectroscopies to predict compositional phenolic levels during red wine fermentation and aging was investigated by Aleixandre-Tudo et al. (2018). The authors showed that FT-NIR appeared as the most accurate technique to predict the phenolic content. The feasibility of reflectance near-infrared spectroscopy combined with partial least square (PLS) to non-destructively predict phenolic contents were evaluated (Quelal-Vásconez et al. 2020). All the analytes were generally well predicted with the prediction of flavanols (coefficient of determination for prediction 0.830–0.824; root mean square error of prediction 8.160–7.430% and bias −1.440 and −1.034 for catechin and epicatechin, respectively).

Total phenolic contents in red cabbage were analyzed using NIR (10000–4000 cm^{-1}) and ATR-FT-IR (3600–3200 cm^{-1}) by de Oliveira et al. (2018). The authors applied ordered predictors selection (OPS) and genetic algorithm (GA) for feature extraction before developing the PLS model. The results indicated that PLS combined with the OPS algorithm had better performance (RMSECV = 44.41 mg gallic acid equivalents (GAE) 100 g-1) (RMSEP = 42.46 mg GAE L^{-1}) with higher prediction accuracy (Rp = 0.99). In another study, NIR and FT-IR were investigated and compared based on the quantification of TPC in grape juice. Results showed that both techniques had comparable performance to determine TPC with low prediction errors. However, FT-IR showed an RMSEP (0.21 mg GAE100 mL^{-1}) which was moderately better than that of NIR (0.37 mg GAE100 mL^{-1}) (Caramês et al. 2017). In the case of the PLS model, both NIR and FT-IR spectroscopies had a similar satisfactory performance to predict TAC presenting low RMSEP = 4.22 mg 100 g^{-1} for FT-MIR and RMSEP = 4.44 mg 100 g^{-1} for NIR.

NIR spectroscopy combined with chemometric was applied to analyze nine individual catechins, total catechin, and gallic acid in green tea leaves. MPLS models presented high reproducibility and sensitivity for detection of catechins in the leaves of green tea, simultaneously improving its rapid and cost-effective features (Lee et al. 2014). Cross-validation models showed good correlations ($R2p = 0.58–0.97$) between reference measurements and NIR estimates. NIR spectroscopy with other coupled techniques can provide better quantitative and structural information of phenolic compounds in foods.

17.4.5 Advantages of NIR Spectroscopy

The main advantages of NIR spectroscopy over other spectroscopies are that they can provide spectra with high intensity, high resolution, precise spectral frequency measurement, being fluorescence-free, and ease of sample presentation. Furthermore, this technique is appropriate for online and in-line process monitoring and quality control of food products. Conversely, this system is characterized by poorly resolved spectra, the lack of information from nonpolar groups, an absence of structural selectivity and sensitivity, and spectra affected by temperature changes (Tahir et al. 2019). NIR coupled with hyperspectral imaging (HIS) is widely used for measurements of bioactive compounds in food. The advantage of NIR-HSI is the larger surface area being analyzed. Besides, it can acquire multi-constituent information and sensitive to minor chemical components. The main drawback of NIR-HSI is expensive and requires training.

17.5 Study Questions

- What is spectroscopy? How they are classified?
- What is mass spectrometry? Discuss instrumentation and applications of mass spectrometry.
- How mass spectrometry is advantageous as compared to other spectroscopic techniques.
- What is Nuclear magnetic resonance (NMR) spectroscopy?
- Discuss instrumentation and applications of NMR for the analysis of phenolic compounds in foods.
- What do you know about near-infrared spectroscopy?
- Discuss fundamental concepts, instrumentation, and applications for the analysis of phenolic compounds in foods.

References

Ahmad S, Zeb A, Ayaz M, Murkovic M (2020) Characterization of phenolic compounds using UPLC-HRM and HPLC-DAD and anticholinesterase and antioxidant activities of *Trifolium repens* L. leaves. Eur Food Res Technol 246:485–496

Aleixandre-Tudo JL, Nieuwoudt H, Aleixandre JL, du Toit W (2018) Chemometric compositional analysis of phenolic compounds in fermenting samples and wines using different infrared spectroscopy techniques. Talanta 176:526–536

Atta-Ur-Rahman (2012) Nuclear magnetic resonance: basic principles. Springer Science and Business Media

Ball DW (2001) The basics of spectroscopy, vol 49. Spie press

Barbosa S, Campmajó G, Saurina J, Puignou L, Núñez O (2020) Determination of phenolic compounds in paprika by ultrahigh performance liquid chromatography–tandem mass spectrometry: application to product designation of origin authentication by Chemometrics. J Agric Food Chem 68(2):591–602

Beć KB, Huck CW (2019) Breakthrough potential in near-infrared spectroscopy: spectra simulation. A review of recent developments. Front Chem 7:48

Beghi R, Buratti S, Giovenzana V, Benedetti S, Guidetti R (2017) Electronic nose and visible-near infrared spectroscopy in fruit and vegetable monitoring. Rev Anal Chem 36(4):20160016

Bellincontro A, Taticchi A, Servili M, Esposto S, Farinelli D, Mencarelli F (2012) Feasible application of a portable NIR-AOTF tool for on-field prediction of phenolic compounds during the ripening of olives for oil production. J Agric Food Chem 60(10):2665–2673

Bertin RL, Gonzaga LV, Borges GDC, Azevedo MS, Maltez HF, Heller M, Micke GA, Tavares LBB, Fett R (2014) Nutrient composition and, identification/quantification of major phenolic compounds in *Sarcocornia ambigua* (Amaranthaceae) using HPLC-ESI-MS/MS. Food Res Int 55:404–411

Bunea A, Andjelkovic M, Socaciu C, Bobis O, Neacsu M, Verhe R, Van Camp J (2008) Total and individual carotenoids and phenolic acids content in fresh, refrigerated and processed spinach (*Spinacia oleracea* L.). Food Chem 108(2):649–656

Cao R, Liu X, Liu Y, Zhai X, Cao T, Wang A, Qiu J (2021) Applications of nuclear magnetic resonance spectroscopy to the evaluation of complex food constituents. Food Chem 342:128258

Caramês ETS, Alamar PD, Poppi RJ, Pallone JAL (2017) Rapid assessment of Total phenolic and anthocyanin contents in grape juice using infrared spectroscopy and multivariate calibration. Food Anal Methods 10(5):1609–1615

Castro-López MM, López-Vilariño JM, González-Rodríguez MV (2014) Analytical determination of flavonoids aimed to analysis of natural samples and active packaging applications. Food Chem 150:119–127

Cerrato A, Cannazza G, Capriotti AL, Citti C, La Barbera G, Laganà A, Montone CM, Piovesana S, Cavaliere C (2020) A new software-assisted analytical workflow based on high-resolution mass spectrometry for the systematic study of phenolic compounds in complex matrices. Talanta 209:120573

Cortés V, Blasco J, Aleixos N, Cubero S, Talens P (2019) Monitoring strategies for quality control of agricultural products using visible and near-infrared spectroscopy: a review. Trends Food Sci Technol 85:138–148

Cortese M, Gigliobianco MR, Peregrina DV, Sagratini G, Censi R, Di Martino P (2020) Quantification of phenolic compounds in different types of crafts beers, worts, starting and spent ingredients by liquid chromatography-tandem mass spectrometry. J Chromatogr A 1612:460622

Cozzolino D, Kwiatkowski MJ, Parker M, Cynkar WU, Dambergs RG, Gishen M, Herderich MJ (2004) Prediction of phenolic compounds in red wine fermentations by visible and near-infrared spectroscopy. Anal Chim Acta 513(1):73–80

Cuny M, Vigneau E, Le Gall G, Colquhoun I, Lees M, Rutledge DN (2008) Fruit juice authentication by 1H NMR spectroscopy in combination with different chemometrics tools. Anal Bioanal Chem 390(1):419–427

De Paepe D, Servaes K, Noten B, Diels L, De Loose M, Van Droogenbroeck B, Voorspoels S (2013) An improved mass spectrometric method for identification and quantification of phenolic compounds in apple fruits. Food Chem 136(2):368–375

De Vijlder T, Valkenborg D, Lemière F, Romijn EP, Laukens K, Cuyckens FJMsr (2018) A tutorial in small molecule identification via electrospray ionization-mass spectrometry: the practical art of structural elucidation. Mass Spectrom Rev 37 (5):607–629

Elessawy FM, Bazghaleh N, Vandenberg A, Purves RW (2020) Polyphenol profile comparisons of seed coats of five pulse crops using a semi-quantitative liquid chromatography-mass spectrometric method. Phytochem Anal 31(4):458–471

Fratoddi I, Rapa M, Testa G, Venditti I, Scaramuzzo FA, Vinci G (2018) Response surface methodology for the optimization of phenolic compounds extraction from extra virgin olive oil with functionalized gold nanoparticles. Microchem J 138:430–437

Gai Q-Y, Jiao J, Wang X, Fu Y-J, Lu Y, Liu J, Wang Z-Y, Xu X-J (2021) Simultaneous quantification of eleven bioactive phenolic compounds in pigeon pea natural resources and in vitro cultures by ultra-high performance liquid chromatography coupled with triple quadrupole mass spectrometry (UPLC-QqQ-MS/MS). Food Chem 335:127602

Hatzakis E (2019) Nuclear magnetic resonance (NMR) spectroscopy in food science: a comprehensive review. Compr Rev Food Sci Food Saf 18(1):189–220

Hirri A, Bassbasi M, Souhassou S, Kzaiber F, Oussama A (2016) Prediction of polyphenol fraction in virgin olive oil using mid-infrared attenuated Total reflectance attenuated Total reflectance accessory–mid-infrared coupled with partial least squares regression. Int J Food Prop 19(7):1504–1512

Hodgkinson P (2020) Nuclear magnetic resonance. vol 46. Royal Society of Chemistry, London

Huck CW (2021) New trend in instrumentation of NIR spectroscopy—miniaturization. In: Ozaki Y, Huck C, Tsuchikawa S, Engelsen SB (eds) Near-infrared spectroscopy. Springer, Singapore, pp 193–210

Khan I, Ali I, Zeb A (2007) Analysis of wheat varieties by near infrared reflectance spectroscopy. J Chem Soc Pakistan 29(3):236–238

Khanam UKS, Oba S, Yanase E, Murakami Y (2012) Phenolic acids, flavonoids and total antioxidant capacity of selected leafy vegetables. J Funct Foods 4(4):979–987

Kirtil E, Oztop MH (2016) H nuclear magnetic resonance relaxometry and magnetic resonance imaging and applications in food science and processing. Food Eng Rev 8:1–22

Kumar BR (2017) Application of HPLC and ESI-MS techniques in the analysis of phenolic acids and flavonoids from green leafy vegetables (GLVs). J Pharm Ana 7(6):349–364

Le Gall G, Colquhoun IJ, Defernez M (2004) Metabolite profiling using 1H NMR spectroscopy for quality assessment of green tea, Camellia sinensis (L.). J Agric Food Chem 52(4):692–700

Lee M-S, Hwang Y-S, Lee J, Choung M-G (2014) The characterization of caffeine and nine individual catechins in the leaves of green tea (Camellia sinensis L.) by near-infrared reflectance spectroscopy. Food Chem 158:351–357

Li W, Sun Y, Liang W, Fitzloff JF, van Breemen RB (2003) Identification of caffeic acid derivatives in Actea racemosa (Cimicifuga racemosa, black cohosh) by liquid chromatography/tandem mass spectrometry. Rapid Commun Mass Spectrom 17(9):978–982

Lin L-Z, Harnly JM (2009) Identification of the phenolic components of collard greens, kale, and Chinese broccoli. J Agric Food Chem 57(16):7401–7408

Lin L-Z, Sun J, Chen P, Harnly J (2011) UHPLC-PDA-ESI/HRMS/MSn analysis of anthocyanins, Flavonol glycosides, and Hydroxycinnamic acid derivatives in red mustard greens (Brassica juncea Coss variety). J Agric Food Chem 59(22):12059–12072

Liu R, Zhao Z, Dai S, Che X, Liu W (2019) Identification and quantification of bioactive compounds in Diaphragma juglandis Fructus by UHPLC-Q-Orbitrap HRMS and UHPLC-MS/MS. J Agric Food Chem 67(13):3811–3825

Mahesar SA, Lucarini M, Durazzo A, Santini A, Lampe AI, Kiefer J (2019) Application of infrared spectroscopy for functional compounds evaluation in olive oil: a current snapshot. J Spectrosc 2019:5319024

Marcone MF, Wang S, Albabish W, Nie S, Somnarain D, Hill A (2013) Diverse food-based applications of nuclear magnetic resonance (NMR) technology. Food Res Int 51(2):729–747

Mora-Ruiz ME, Reboredo-Rodríguez P, Salvador MD, González-Barreiro C, Cancho-Grande B, Simal-Gándara J, Fregapane G (2017) Assessment of polar phenolic compounds of virgin olive oil by NIR and mid-IR spectroscopy and their impact on quality. Eur J Lipid Sci Technol 119(1):1600099

Natić MM, Dabić DČ, Papetti A, Fotirić Akšić MM, Ognjanov V, Ljubojević M, Tešić ŽL (2015) Analysis and characterisation of phytochemicals in mulberry (Morus alba L.) fruits grown in Vojvodina, North Serbia. Food Chem 171:128–136

Ogrinc N, Košir IJ, Spangenberg JE, Kidrič J (2003) The application of NMR and MS methods for detection of adulteration of wine, fruit juices, and olive oil. A review. Anal Bioanal Chem 376(4):424–430

Ola SS, Catia G, Marzia I, Francesco VF, Afolabi AA, Nadia M (2009) HPLC/DAD/MS characterisation and analysis of flavonoids and cynnamoil derivatives in four Nigerian green-leafy vegetables. Food Chem 115(4):1568–1574

de Oliveira IRN, Roque JV, Maia MP, Stringheta PC, Teofilo RF (2018) New strategy for determination of anthocyanins, polyphenols and antioxidant capacity of Brassica oleracea liquid extract using infrared spectroscopies and multivariate regression. Spectrochim Acta A 194:172–180

Oniszczuk A, Olech M (2016) Optimization of ultrasound-assisted extraction and LC-ESI-MS/MS analysis of phenolic acids from Brassica oleracea L. var. sabellica. Ind Crop Prod 83:359–363

Ozaki Y, Morisawa Y (2021) Principles and characteristics of NIR spectroscopy. In: Ozaki Y, Huck C, Tsuchikawa S, Engelsen SB (eds) Near-infrared spectroscopy. Springer, Singapore, pp 11–35

Ozaki Y, McClure WF, Christy AA (2007) Near-infrared spectroscopy in food science and technology. Wiley, New York

Park S, Arasu MV, Jiang N, Choi SH, Lim YP, Park JT, Al-Dhabi NA, Kim SJ (2014) Metabolite profiling of phenolics, anthocyanins and flavonols in cabbage (Brassica oleracea var. capitata). Ind Crop Prod 60:8–14

Penner MH (2010) Basic principles of spectroscopy. In: Nielsen SS (ed) Food analysis. Springer, Boston, MA

Piovesana S, Cavaliere C, Cerrato A, Montone CM, Laganà A, Capriotti AL (2020) Developments and pitfalls in the characterization of phenolic compounds in food: from targeted analysis to metabolomics-based approaches. TrAC Trends Anal Chem 133:116083

Qiao T, Ren J, Craigie C, Zabalza J, Maltin C, Marshall S (2015) Quantitative prediction of beef quality using visible and NIR spectroscopy with large data samples under industry conditions. J Appl Spectrosc 82(1):137–144

Quelal-Vásconez MA, Lerma-García MJ, Pérez-Esteve É, Arnau-Bonachera A, Barat JM, Talens P (2020) Changes in methylxanthines and flavanols during cocoa powder processing and their quantification by near-infrared spectroscopy. LWT Food Sci Technol 117:108598

Rameshkumar A, Sivasudha T, Jeyadevi R, Sangeetha B, Smilin Bell Aseervatham G, Maheshwari M (2013) Profiling of phenolic compounds using UPLC–Q-TOF-MS/MS and nephroprotective activity of Indian green leafy vegetable Merremia emarginata (Burm. F.). Food Res Int 50(1):94–101

Rashidinejad A, Birch EJ, Hindmarsh J, Everett DW (2017) Molecular interactions between green tea catechins and cheese fat studied by solid-state nuclear magnetic resonance spectroscopy. Food Chem 215:228–234

Rodrigues CA, Nicácio AE, Boeing JS, Garcia FP, Nakamura CV, Visentainer JV, Maldaner L (2020) Rapid extraction method followed by a d-SPE clean-up step for determination of phenolic composition and antioxidant and antiproliferative activities from berry fruits. Food Chem 309:125694

Senyuva HZ, Gökmen V, Sarikaya EA (2015) Future perspectives in Orbitrap™-high-resolution mass spectrometry in food analysis: a review. Food Add Contam: Part A 32(10):1568–1606

Smith JS, Thakur RA (2010) Mass spectrometry. In: Nielsen SS (ed) Food analysis, 4th edn. Springer, London

Stintzing FC, Kammerer D, Schieber A, Adama H, Nacoulma OG, Carle R (2004) Betacyanins and phenolic compounds from *Amaranthus spinosus* L. and *Boerhavia erecta* L. Zeitschrift für Naturforschung C 59(1–2):1–8

Sundekilde UK, Eggers N, Bertram HC (2019) NMR-based metabolomics of food. In: Gowda G, Raftery D (eds) NMR-based metabolomics, Methods in molecular biology, vol 2037. Humana, New York, pp 335–344

Tahir HE, Xiaobo Z, Jianbo X, Mahunu GK, Jiyong S, Xu J-L, Sun D-W (2019) Recent Progress in rapid analyses of vitamins, phenolic, and volatile compounds in foods using vibrational spectroscopy combined with Chemometrics: a review. Food Anal Methods 12(10):2361–2382

Tarachiwin L, Ute K, Kobayashi A, Fukusaki E (2007) 1H NMR based metabolic profiling in the evaluation of Japanese green tea quality. J Agric Food Chem 55(23):9330–9336

Thompson JM (2018) Mass spectrometry. Pan Stanford Publishing Pte. Ltd., Singapore

Tian W, Chen G, Zhang G, Wang D, Tilley M, Li Y (2021) Rapid determination of total phenolic content of whole wheat flour using near-infrared spectroscopy and chemometrics. Food Chem 344:128633

Trapani S, Migliorini M, Cecchi L, Giovenzana V, Beghi R, Canuti V, Fia G, Zanoni B (2017) Feasibility of filter-based NIR spectroscopy for the routine measurement of olive oil fruit ripening indices. Eur J Lipid Sci Technol 119(6):1600239

Vallverdú-Queralt A, Jáuregui O, Medina-Remón A, Andrés-Lacueva C, Lamuela-Raventós RM (2010) Improved characterization of tomato polyphenols using liquid chromatography/electrospray ionization linear ion trap quadrupole Orbitrap mass spectrometry and liquid chromatography/electrospray ionization tandem mass spectrometry. Rapid Commun Mass Spectrom 24(20):2986–2992

Wang L, Sun D-W, Pu H, Cheng J-H (2017) Quality analysis, classification, and authentication of liquid foods by near-infrared spectroscopy: a review of recent research developments. Crit Rev Food Sci Nutr 57(7):1524–1538

Yuan B, Dinssa FF, Simon JE, Wu Q (2020) Simultaneous quantification of polyphenols, glycoalkaloids and saponins in African nightshade leaves using ultra-high performance liquid chromatography tandem mass spectrometry with acid assisted hydrolysis and multivariate analysis. Food Chem 312:126030

Index

© Springer Nature Switzerland AG 2021
A. Zeb, *Phenolic Antioxidants in Foods: Chemistry, Biochemistry and Analysis*,
https://doi.org/10.1007/978-3-030-74768-8

Printed in the United States
by Baker & Taylor Publisher Services

Printed in the United States
by Baker & Taylor Publisher Services